Hermann von Helmholtz and the Foundations of Nineteenth-Century Science

CALIFORNIA STUDIES IN THE HISTORY OF SCIENCE

J.L. HEILBRON, Editor

The Galileo Affair: A Documentary History, edited and translated by Maurice A. Finocchiaro

The New World, 1939-1946 (A History of the United States Atomic Energy Commission, volume 1), by Richard G. Hewlett and Oscar E. Anderson, Jr.

Atomic Shield, 1947-1952 (A History of the United States Atomic Energy Commission, volume 2), by Richard G. Hewlett and Francis Duncan

Atoms for Peace and War, 1953-1961: Eisenhower and the Atomic Energy Commission (A History of the United States Atomic Energy Commission, volume 3), by Richard G. Hewlett and Jack M. Holl

Lawrence and His Laboratory: A History of the Lawrence Berkeley Laboratory, Volume 1, by J.L. Heilbron and Robert W. Seidel

Scientific Growth: Essays on the Social Organization and Ethos of Science, by Joseph Ben-David, edited by Gad Freudenthal.

Physics and Politics in Revolutionary Russia, by Paul R. Josephson

From c-Numbers to q-Numbers: The Classical Analogy in the History of Quantum Theory, by Olivier Darrigol

The Quiet Revolution: Hermann Kolbe and the Science of Organic Chemistry, by Alan J. Rocke

Hermann von Helmholtz and the Foundations of Nineteenth-Century Science, edited by David Cahan

Hermann von Helmholtz and the Foundations of Nineteenth-Century Science

Edited by
David Cahan

University of California Press
Berkeley / Los Angeles / London

University of California Press
Berkeley and Los Angeles, California

University of California Press
London, England

Copyright © 1993 by The Regents of the
University of California

Library of Congress Cataloging-in-Publication Data

Hermann von Helmholtz and the foundations of nineteenth-century science / edited by David Cahan.
      p. cm. — (California studies in the history of science ; 12) Includes bibliographical references and index.
      ISBN 0-520-08334-2 (alk. paper)
  1. Helmholtz, Hermann von, 1821–1894.
  2. Science—History—19th century.
  3. Scientists—Germany—Biography.
  4. Physiologists—Germany—Biography.
  I. Cahan, David. II. Series.
Q143.H5H47   1994
509.2—dc20
  [B]   92-16285
                                        CIP

Printed in the United States of America

2 3 4 5 6 7 8 9

The paper used in this publication meets the minimum requirements of American National Standard for Information Sciences—Permanence of Paper for Printed Library Materials, ANSI Z39.48-1984 ∞

"Ich bewundere den originellen,
freien Kopf Helmh[oltz].
immer mehr."

Albert Einstein, August 1899

Contents

Illustrations xi
Acknowledgments xiii
Contributors xv
Abbreviations xix
Chronological Listing of the Principal
Events and Publications of
Helmholtz's Life and Career xxi

Introduction: Helmholtz at the Borders of Science, *David Cahan* 1

Part One   **Physiologist**

1 **Helmholtz and the German Medical Community,** *Arleen Tuchman* 17

2 **Experiment, Quantification, and Discovery: Helmholtz's Early Physiological Researches, 1843–50,** *Kathryn M. Olesko and Frederic L. Holmes* 50

3 **The Eye as Mathematician: Clinical Practice, Instrumentation, and Helmholtz's Construction of an Empiricist Theory of Vision,** *Timothy Lenoir* 109

4 Consensus and Controversy: Helmholtz on the Visual Perception of Space, *R. Steven Turner* 154

5 Innovation through Synthesis: Helmholtz and Color Research, *Richard L. Kremer* 205

6 Sensation of Tone, Perception of Sound, and Empiricism: Helmholtz's Physiological Acoustics, *Stephan Vogel* 259

Part Two   Physicist

7 Helmholtz's *Ueber die Erhaltung der Kraft*: The Emergence of a Theoretical Physicist, *Fabio Bevilacqua* 291

8 Electrodynamics in Context: Object States, Laboratory Practice, and Anti-Romanticism, *Jed Z. Buchwald* 334

9 Helmholtz's Instrumental Role in the Formation of Classical Electrodynamics, *Walter Kaiser* 374

10 Between Physics and Chemistry: Helmholtz's Route to a Theory of Chemical Thermodynamics, *Helge Kragh* 403

11 Helmholtz's Mechanical Foundation of Thermodynamics, *Günter Bierhalter* 432

Part Three   Philosopher

12 Force, Law, and Experiment: The Evolution of Helmholtz's Philosophy of Science, *Michael Heidelberger* 461

13 Helmholtz's Empiricist Philosophy of Mathematics: Between Laws of Perception and Laws of Nature, *Robert DiSalle* 498

14 Helmholtz and Classicism: The Science of Aesthetics and the Aesthetics of Science,
Gary Hatfield 522

15 Helmholtz and the Civilizing Power of Science,
David Cahan 559

Bibliography 603
Index 637

Illustrations

Figure 1.1. Schematic drawing of the path taken by light as it travels through the ophthalmoscope.   31

Figure 2.1. Helmholtz's data for alcohol, water, and spiritous extracts from a muscle.   58

Figure 2.2. Weber's apparatus for studying the contraction of a muscle.   75

Figure 2.3. Helmholtz's muscle curve.   80

Figure 2.4. Helmholtz's apparatus for measuring the time course of muscle contraction and the propagation velocity of the nerve impulse.   86

Figure 2.5. Helmholtz's nerve propagation velocity data from 6 January 1850 with the results of least-squares calculations.   99

Figure 2.6. Helmholtz's graphical representation of data on the behavior of a stimulated muscle when no overload was applied.   102

Figure 2.7. Helmholtz's graphical representation of data on the effect of three degrees of muscle fatigue on the muscle's ability to lift a weight.   102

Figure 3.1. Wundt's ophthalmotrope.   135

Figure 3.2. Primary axes of rotation and Listing's Plane.   141

Figure 3.3. Motion around axes in Listing's Plane.   142

Figure 3.4. Projection of direction lines and objective verticals in the visual field.   144

Figure 3.5. Eye motion according to the Principle of Easiest Orientation. 145

Figure 4.1. Stereogram by Charles Wheatstone, purporting to refute the theory of identity. 161

Figure 4.2. Pattern of corresponding retinal meridians, eyes in the primary position. 164

Figure 4.3. Demonstration by Helmholtz that the horopter is the region in which the perception of relief is most acute (1863). 167

Figure 4.4. Afterimage experiment by Helmholtz confirming Listing's law of eye movements (1863). 170

Figure 4.5. Experimental test by Helmholtz of Hering's theory of retinal depth values. 192

Figure 4.6. The compensating rotation of the core plane, induced by the closing of one eye, according to Hering. 194

Figure 4.7. Experiment by Helmholtz, purporting to refute Hering's core-plane rotation (1867). 196

Figure 5.1. Brewster's absorption experiments (1823). 210

Figure 5.2. Brewster's triple spectrum (1831). 212

Figure 5.3. Helmholtz's absorption experiments (1852). 218

Figure 5.4. Helmholtz's "V-slit" mixed spectra (1852). 230

Figure 5.5. Helmholtz's revised mixing experiments (1855). 234

Figure 5.6. Helmholtz's barycentric curve for color mixing (1855). 235

Figure 5.7. Helmholtz's hypothetical response curves for Young's receptors (1860). 250

Figure 8.1. Herwig's apparatus for use in refuting Helmholtz's potential. 348

Figure 8.2a. Hertz's original diagram of a Wheatstone bridge with double-wound spiral inductors. 354

Figure 8.2b. A modern rendering of Hertz's original diagram. 355

Figure 8.3. A modern rendering of Hertz's apparatus with a floor circuit in place of a spiral inductor. 356

Acknowledgments

In the course of preparing this volume the University of Chicago generously agreed to sponsor a two-day conference (October 1990) on "Hermann von Helmholtz: Scientist and Philosopher." Most of the contributors to this volume were able to attend the conference, which gave us an excellent opportunity to discuss one another's papers and views on Helmholtz, and to explore some of the broader themes of his work. I want in particular to thank three individuals at Chicago for making the conference possible: George Stocking, Director of the Morris Fishbein Center for the Study of the History of Science and Medicine; Robert J. Richards, Chairman of the Committee on the Conceptual Foundations of Science; and Patricia Swanson of The John Crerar Library, which provided us with an ideal setting for stimulating discussion. It is a great pleasure to acknowledge Chicago's gracious support.

It is an equally great pleasure to acknowledge the help of three colleagues who acted as outside commentators on the final versions of the three Parts of this volume: John E. Lesch (Berkeley), who commented on the essays constituting Part One (Helmholtz as physiologist); Ole Knudsen (Aarhus), who commented on Part Two (Helmholtz as physicist); and Mitchell G. Ash (Iowa), who commented on Part Three (Helmholtz as philosopher). To all three I owe thanks for their extensive comments and judicious criticisms.

Finally, I owe general thanks to Richard L. Kremer and R. Steven Turner, two of my fellow contributors to this volume, for their counsel on numerous points concerning Helmholtz; and to Elizabeth Knoll, my sponsoring editor at the University of California Press, for her faith in this project and her patience in awaiting its completion.

—David Cahan

Contributors

Fabio Bevilacqua is Associate Professor of the History of Physics in the Department of Physics "A. Volta" at the University of Pavia. He has published articles on the history of nineteenth-century physics and on the use of the history of science in science education, as well as a book entitled *The Principle of Conservation of Energy and the History of Classical Electromagnetic Theory* (Pavia: LaGoliardica Pavese, 1983).

Günter Bierhalter, an independent scholar living in Pforzheim, in the Federal Republic of Germany, is the author of articles on nineteenth-century thermodynamics, including attempts to establish the mechanical foundations of thermodynamics and the problem of irreversibility. His principal scholarly interests are the history of the foundations of mechanics, extremum principles in physics, thermodynamics, and quantum physics. He is currently working on the history of nineteenth-century electrodynamics.

Jed Z. Buchwald is Bern Dibner Professor of the History of Science at the Massachusetts Institute of Technology and Director of the Dibner Institute for the History of Science and Technology. In addition to numerous articles on the history of nineteenth-century electromagnetism and optics, he is also the author of two book-length studies: *From Maxwell to Microphysics. Aspects of Electromagnetic Theory in the Last Quarter of the Nineteenth Century* (Chicago and London: The University of Chicago Press, 1985), and *The Rise of the Wave Theory of Light. Optical Theory and Experiment in the Early Nineteenth Century* (Chicago and London: The University of Chicago Press, 1989). He has recently completed a book entitled *The Creation of Scientific Effects: Heinrich Hertz and Electric Waves* (forthcoming).

David Cahan is Associate Professor of History at the University of Nebraska-Lincoln. His research concerns the history of physics and the social history of science from the early Enlightenment to the present. He has published articles on physics in nineteenth- and early twentieth-century Germany; has written a book entitled *An Institute for an Empire: The Physikalisch-Technische Reichsanstalt, 1871–1918* (Cambridge, New York, New Rochelle: Cambridge University Press, 1989); and has edited *Letters of Hermann von Helmholtz to His Parents: The Medical Education of a German Scientist, 1837–1846* (Stuttgart: Franz Steiner Verlag, 1993).

Robert DiSalle is Assistant Professor of Philosophy at the University of Western Ontario. His research concerns the history and philosophy of science, especially historical and philosophical issues connected with the foundations of mechanics and the development of general relativity. He is currently working on a critical account of space-time principles in the history of physics.

Gary Hatfield is Professor of Philosophy at the University of Pennsylvania, where he is also a member of the Graduate Group in History and Sociology of Science. He is the author of numerous articles in the history of philosophy and science and in the philosophy of psychology, as well as a book entitled *The Natural and the Normative: Theories of Spatial Perception from Kant to Helmholtz* (Cambridge, Mass.: MIT Press, 1990). He is currently researching the development of the distinction between *Naturwissenschaften* and *Geisteswissenschaften* by Helmholtz and his neo-Kantian contemporaries.

Michael Heidelberger teaches philosophy at the University of Freiburg, in the Federal Republic of Germany. He is the co-author of *Natur und Erfahrung*, 2nd ed. (Reinbek bei Hamburg: Rowohlt, 1985), a study of the Scientific Revolution; and the co-editor of and a contributor to *The Probabilistic Revolution*, vol. 1: *Ideas in History* (Cambridge, Mass.: MIT Press, 1987). He has published numerous articles in the history and philosophy of science, and has recently published a book entitled *Die innere Seite der Natur: Gustav Theodor Fechners wissenschaftlich-philosophische Weltauffassung* (Frankfurt am Main: Klostermann, 1993).

Frederic L. Holmes is Avalon Professor and Chairman of the Section of the History of Medicine in the Yale University School of Medicine. He is the author of several books and numerous articles in the history of chemistry and the life sciences in the eighteenth and nineteenth centuries. Most recently he has written *Hans Krebs: The Formation of a Scientific Life* (Oxford: Oxford University Press, 1991) and *Between Biology and Medicine: The Formation of Intermediary Metabolism* (Office for History of Science and Technology [Berkeley], 1991). He is presently working on Justus von Liebig and early nineteenth-century German chemistry; on topics in the early years of molecular biology; and, with Kathryn M. Olesko, on Hermann von Helmholtz's early scientific career, from which their essay in this volume is drawn.

Walter Kaiser holds the Lehrstuhl for the History of Technology at the Rheinisch-Westfälische Technische Hochschule Aachen. He has written articles on the history of electrodynamics and solid state physics, and, more generally, in the history and philosophy of science and the history of technology. He is the author of *Theorien der Elektrodynamik im 19. Jahrhundert* (Hildesheim: Gerstenberg Verlag, 1981); has edited Ludwig Boltzmann's *Vorlesungen über Maxwells Theorie der Elektricität und des Lichtes* (Graz: Akademische Druck- und Verlagsanstalt, 1982); and has co-authored the *Propyläen Technikgeschichte*, vol. 5 *1914–1990* (Berlin: Propyläen Verlag, 1992).

Helge Kragh is a researcher at the Roskilde University Centre in Denmark, where he works with a project on the history of technology and culture in Denmark since 1750. He was associate professor of physics and history at Cornell University from 1987 to 1989, and has written articles and books on the history of modern physical science. His most recent publications include *An Introduction to the Historiography of Science* (Cambridge, London, and New York: Cambridge University Press, 1987), and *Dirac: A Scientific Biography* (Cambridge, New York, and Port Chester: Cambridge University Press, 1990).

Richard L. Kremer is Assistant Professor of History at Dartmouth College. He has edited *Letters of Hermann von Helmholtz to His Wife, 1847–1859* (Stuttgart: Franz Steiner, 1990), and is completing a book-length monograph on the "culture of experiment" in physiology at German universities between 1800 and 1865.

Timothy Lenoir is Associate Professor of History and of History of Science in the Program in History of Science at Stanford University. He has published numerous articles on the history of the life sciences in the nineteenth century as well as a book: *The Strategy of Life: Teleology and Mechanics in Nineteenth-Century German Biology* (Dordrecht and Boston: D. Reidel, 1982; paperback reprint: Chicago and London: The University of Chicago Press, 1989).

Kathryn M. Olesko is Associate Professor of History at Georgetown University. She is the author of *Physics as a Calling: Discipline and Practice in the Königsberg Seminar for Physics* (Ithaca and London: Cornell University Press, 1991) and the editor of *Science in Germany:*

Problems at the Intersection of Intellectual and Institutional History (*Osiris 5* [1989]). She is currently writing a book entitled *The Meaning of Precision*, which examines the scientific, political, economic, and cultural meaning of precision measurement in Germany from 1780 to 1870. With Frederic L. Holmes she is preparing a monograph on Hermann von Helmholtz's early scientific career, from which their essay in this volume is drawn.

Arleen Tuchman is Assistant Professor of History at Vanderbilt University, where she teaches the history of science and medicine. She has published several articles on the rise of scientific medicine in nineteenth-century Germany, as well as a book entitled *Science, Medicine, and the State in Germany: The Case of Baden, 1815–1871* (Oxford and New York: Oxford University Press, 1993). Her current project is tentatively entitled *Gender and Scientific Medicine at a Crossroad: The Life and Times of Marie Zakrzewska, 1829–1902*, a biographical study of a German midwife who became one of the most prominent female physicians in nineteenth-century America.

R. Steven Turner teaches the history of science and technology at the University of New Brunswick in Fredericton, N.B., Canada. He has published widely on the development of the German university system in the eighteenth and nineteenth centuries; the history of sensory physiology and psychology in the nineteenth century; and the career and work of Hermann von Helmholtz.

Stephan Vogel is a research fellow in the Department of Linguistic Information Processing at the University of Cologne, in the Federal Republic of Germany. He is currently completing his Ph.D. degree in History and Philosophy of Science at Cambridge University with a thesis on the history of nineteenth-century acoustics.

Abbreviations

Note: Unless otherwise noted, all journal articles and books cited in this volume are by Hermann (von) Helmholtz. Most of Helmholtz's scientific articles were reprinted in his *Wissenschaftliche Abhandlungen* and are cited in this volume simply as "in *WA*." The following abbreviations are used in both the Notes and the Bibliography:

AO	*Archiv für Opthalmologie.*
AP	*Annalen der Physik [und Chemie].*
AW	Akademie der Wissenschaften, Berlin, Archiv, *Signatur*: Helmholtz Nachlaß.
BJHS	*British Journal for the History of Science*
DSB	*Dictionary of Scientific Biography.* Ed. Charles Coulston Gillispie. 16 vols. New York: Charles Scribner's Sons, 1970–1980.
Handbuch	Hermann von Helmholtz. *Handbuch der physiologischen Optik.* Leipzig: Leopold Voss, 1856–67.
Handbuch²	Hermann von Helmholtz. *Handbuch der physiologischen Optik.* 2nd rev. ed. Ed. Arthur König. Hamburg and Leipzig: Leopold Voss, 1896.
Handbuch³	Hermann von Helmholtz. *Handbuch der physiologischen Optik.* 3rd. ed., based on the text of the first, with supplementary material. A. Gullstrand, J. von Kries, and W. Nagel, eds. 3 vols. Hamburg and Leipzig: Leopold Voss, 1909–11.
HSPS	*Historical Studies in the Physical and Biological Sciences*

JfruaM	*Journal für reine und angewandte Mathematik*
JHBS	*Journal of the History of the Behavioral Sciences*
JHMAS	*Journal of the History of Medicine and Allied Sciences*
Kirsten	*Dokumente einer Freundschaft. Briefwechsel zwischen Hermann von Helmholtz und Emil du Bois-Reymond 1846–1894.* Eds. Christa Kirsten, et al. Berlin: Akademie-Verlag, 1986.
Koenigsberger	Leo Koenigsberger. *Hermann von Helmholtz*. 3 vols. Braunschweig: Friedrich Vieweg und Sohn, 1902–3.
MA	*Müller's Archiv für Anatomie, Physiologie, und wissenschaftliche Medizin*
MB	*Monatsberichte der Königlich Preussischen Akademie der Wissenschaften zu Berlin.*
NTM	*NTM—Zeitschrift für Geschichte der Naturwissenschaft, Technik und Medizin.*
PM	*Philosophical Magazine.*
SBB	*Sitzungsberichte der Königlich Preussischen Akademie der Wissenschaften zu Berlin.*
SBW	*Sitzungsberichte der Kaiserlichen Akademie der Wissenschaften. [Wien]. Mathematisch-Naturwissenschaftliche Classe.*
SHPS	*Studies in History and Philosophy of Science*
VR³	Hermann von Helmholtz. *Vorträge und Reden*. 3rd ed. 2 vols. Braunschweig: Friedrich Vieweg und Sohn, 1884.
VR⁴	Hermann von Helmholtz. *Vorträge und Reden*. 4th ed. 2 vols. Braunschweig: Friedrich Vieweg und Sohn, 1896.
VR⁵	Hermann von Helmholtz. *Vorträge und Reden*. 5th ed. 2 vols. Braunschweig: Friedrich Vieweg und Sohn, 1903.
WA	*Wissenschaftliche Abhandlungen von Hermann von Helmholtz.* 3 vols. Leipzig: J.A. Barth, 1882–95.
WZHUB	Issue devoted to "Hermann von Helmholtz' philosophische und naturwissenschaftliche Leistungen aus der Sicht des dialektischen Materialismus und der modernen Naturwissenschaften." *Wissenschaftliche Zeitschrift der Humboldt-Universität zu Berlin, Mathematisch-Naturwissenschaftliche Reihe*
ZPC	*Zeitschrift für Physikalische Chemie.*

Chronological Listing of the Principal Events and Publications of Helmholtz's Life and Career

1821 Birth (31 August) of Hermann Ludwig Ferdinand Helmholtz in Potsdam
1830 Enters Potsdam Gymnasium (Eastertime)
1838 Graduates Potsdam Gymnasium (September)
Enters the Königliches medicinisch-chirurgisches Friedrich-Wilhelms-Institut, Berlin, to study medicine and become a Prussian military medical doctor
1842 Defends dissertation—"*De fabrica systematis nervosi evertebratorum*"—and graduates medical school (October-November)
Begins year-long internship at the Charité hospital, Berlin
1843 Begins service as army staff surgeon with the Königliches Garde-Husaren-Regiment, Potsdam (October)
"Ueber das Wesen der Fäulniss und Gährung," *MA*
1845 On leave in Berlin from army unit to prepare for state medical examinations (October - February); simultaneously conducts research in Gustav Magnus's private laboratory
"Ueber den Stoffverbrauch bei der Muskelaction," *MA*
1846 Passes final state medical examination in February; returns to Potsdam to resume army duty
"Wärme, physiologisch," *Encyclopädisches Handwörterbuch der medicinischen Wissenschaften*
1847 Transfers to Königliches Regiment der Gardes du Corps, Potsdam (June)

"Bericht über 'die Theorie der physiologischen Wärmeerscheinungen' betreffende Arbeit aus dem Jahre 1845," *Fortschritte der Physik*
Ueber die Erhaltung der Kraft. Eine physikalische Abhandlung

1848 Leaves military service (September) to teach anatomy at the Kunstakademie and serve as assistant to Johannes Müller at the Anatomisches Museum, Berlin
"Ueber die Wärmeentwickelung bei der Muskelaction," *MA*
"Bericht über 'die Theorie der physiologischen Wärmeerscheinungen' betreffende Arbeiten aus dem Jahre 1846," *Fortschritte der Physik*

1849 Appointed extraordinary professor of physiology at the University of Königsberg
Marries Olga von Velten (1826–59)

1850 "Ueber die Fortpflanzungsgeschwindigkeit der Nervenreizung," *MB*
Birth of first child, Katharina ("Käthe") Caroline Julie Betty (1850–77)
"Messungen über den zeitlichen Verlauf der Zuckung animalischer Muskeln und die Fortpflanzungsgeschwindigkeit der Reizung in den Nerven," *MA*
Invents ophthalmoscope (October - December)
"Bericht über 'die Theorie der physiologischen Wärmeerscheinungen' betreffende Arbeiten aus dem Jahre 1847," *Fortschritte der Physik*

1851 Appointed ordinary professor of physiology at the University of Königsberg
"Ueber die Methoden, kleinste Zeittheile zu messen, und ihre Anwendung für physiologische Zwecke," *Königsberger naturwissenschaftliche Unterhaltungen*
Beschreibung eines Augenspiegels zur Untersuchung der Netzhaut in lebenden Auge
"Ueber die Dauer und den Verlauf der durch Stromesschwankungen inducirten elektrischen Ströme," *AP*

1852 Birth of second child, Richard Wilhelm Ferdinand (1852–1934)
"Messungen über Fortpflanzungsgeschwindigkeit der Reizung in den Nerven. Zweite Reihe," *MA*
"Ueber eine neue einfachste Form des Augenspiegels," *Vierordt's Archiv für Physiologische Heilkunde*
"Ueber die Theorie der zusammengesetzten Farben," *AP*
Delivers habilitation lecture at Königsberg University: "Ueber die Natur der menschlichen Sinnesempfindungen," *Königsberger naturwissenschaftlicher Unterhaltungen*

"Ueber Herrn D. Brewster's neue Analyse des Sonnenlichts," *AP*

"Ein Theorem über die Vertheilung elektrischer Ströme in körperlichen Leitern," *MB*

1853 Attends meeting of the British Association for the Advancement of Science; first trip abroad

"Ueber einige Gesetze der Vertheilung elektrischer Ströme in körperlichen Leitern mit Anwendung auf die thierisch-elektrischen Versuche," *AP*

Delivers first lecture on Goethe, before the Deutsche Gesellschaft, Königsberg: "Ueber Goethe's naturwissenschaftliche Arbeiten," *VR^5:1*

"Ueber eine bisher unbekannte Veränderung am menschlichen Auge bei veränderter Accomodation," *MB*

1854 Death of mother, Caroline (1797–1854)

"Erwiderung auf die Bemerkungen von Herrn Clausius," *AP*

Delivers lecture in Königsberg: "Ueber die Wechselwirkung der Naturkräfte und die darauf bezüglichen neuesten Ermittelungen der Physik," *VR^5:1*

"Ueber die Geschwindigkeit einiger Vorgänge in Muskeln und Nerven," *MB*

1855 Delivers Kant-Denkmal lecture in Königsberg: "Ueber das Sehen des Menschen," *VR^5:1*

Appointed professor of anatomy and physiology at the University of Bonn

"Ueber die Zusammensetzung von Spectralfarben," *AP*

"Ueber die Empfindlichkeit der menschlichen Netzhaut für die brechbarsten Strahlen des Sonnenlichts," *AP*

"Ueber die Accomodation des Auges," *AO*

1856 "Ueber Combinationstöne," *AP*

Handbuch der physiologischen Optik (Part 1)

1857 Appointed Corresponding Member of the Berlin Akademie der Wissenschaften

"Das Telestereoskop," *AP*

Delivers lecture in Bonn: "Ueber die physiologischen Ursachen der musikalischen Harmonie," *VR^5:1*

1858 Appointed professor of physiology at the University of Heidelberg

"Ueber Integrale der hydrodynamischen Gleichungen, welche den Wirbelbewegungen entsprechen," *JfruaM*

"Ueber die subjectiven Nachbilder im Auge," *Sitzungsberichte des naturhistorischen Vereins der preussischen Rheinlande und Westphalens*

"Ueber Nachbilder," *Amtlicher Bericht über die Versammlung deutscher Naturforscher und Aerzte*

1859 Death of father, August Julius Ferdinand (1792–1859)
Death of wife, Olga
"Ueber die Klangfarbe der Vocale," *AP*
"Ueber Luftschwingungen in Röhren mit offenen Enden," *JfruaM*
"Theorie der Luftschwingung in Röhren mit offen Enden," *JfruaM*

1860 "Ueber Reibung tropfbarer Flüssigkeiten," *SBW* (with G. v. Piotrowski)
Handbuch der physiologischen Optik (Part 2)

1861 Marries Anna von Mohl (1834–99)

1862 Birth of third child, Robert Julius (1862–89)
Appointed prorector of the University of Heidelberg for 1862–63; delivers prorectoral address before the University: "Ueber das Verhältniss der Naturwissenschaften zur Gesammtheit der Wissenschaften," $VR^5:1$
Delivers lecture series in Karlsruhe (1862–63): "Ueber die Erhaltung der Kraft," $VR^5:1$
"Ueber die Form des Horopters, mathematisch bestimmt," *Verhandlungen des naturhistorisch-medicinischen Vereins zu Heidelberg*

1863 *Die Lehre von den Tonempfindungen als physiologische Grundlage für die Theorie der Musik*
"Ueber die normalen Bewegungen des menschlichen Auges," *AO*

1864 Birth of fourth child, Ellen Ida Elisabeth (1864–1941)
Delivers lectures on the conservation of energy at The Royal Institution, London
"Ueber den Horopter," *AO*

1865 "Ueber stereoskopisches Sehen," *Verhandlungen des naturhistorisch-medicinischen Vereins zu Heidelberg*
Delivers lectures in Heidelberg and Frankfurt am Main: "Eis und Gletscher," $VR^5:1$
Populäre wissenschaftliche Vorträge (Heft 1)
Die Lehre von den Tonempfindungen (2nd ed.)
"Über die tatsächlichen Grundlagen der Geometrie," *Verhandlungen des naturhistorisch-medicinischen Vereins zu Heidelberg*

1867 *Handbuch der physiologischen Optik* (Part 3)

1868 Birth of fifth child, Friedrich ("Fritz") Julius (1868–1901)
Delivers lectures in Frankfurt am Main and Heidelberg: "Die

neueren Fortschritte in der Theorie des Sehens," *VR⁵:1*
"Ueber die Thatsachen, die der Geometrie zu Grunde liegen," *Nachrichten der königlichen Gesellschaft der Wissenschaften zu Göttingen*

1869 Delivers lecture in Innsbruck: "Ueber das Ziel und die Fortschritte der Naturwissenschaften," *VR⁵:1*

1870 Named Foreign Member, Berlin Akademie der Wissenschaften
Delivers lecture in Heidelberg: "Ueber den Ursprung und die Bedeutung der geometrischen Axiome," *VR⁵:2*
"Ueber die Gesetze der inconstanten elektrischen Ströme in körperlich ausgedehnten Leitern," *Verhandlungen des naturhistorisch-medicinischen Vereins zu Heidelberg*
"Ueber die Theorie der Elektrodynamik. Erste Abhandlung. Ueber die Bewegungsgleichungen der Elektricität für ruhende leitende Körper," *JfruaM*
"Ueber die Bewegungsgleichungen der Elektricität für ruhende leitende Körper," *JfruaM*
Die Lehre von den Tonempfindungen (3rd ed.)
Populäre wissenschaftliche Vorträge (Heft 2)

1871 Appointed professor of physics at the University of Berlin and named Ordinary Member of the Berlin Akademie der Wissenschaften
"Ueber die Fortpflanzungsgeschwindigkeit der elektrodynamischen Wirkungen," *MB*
"Ueber die Zeit, welche nötig ist, damit ein Gesichtseindruck zum Bewusstsein kommt, Resultate einer von Herrn N. Baxt in Heidelberger Laboratorium ausgeführten Untersuchung," *MB*
Delivers lecture in Berlin: "Zum Gedächtniß an Gustav Magnus," *VR⁵:2*
Delivers lectures in Heidelberg and Cologne: "Ueber die Entstehung des Planetensystems," *VR⁵:2*
Delivers lectures in Berlin, Düsseldorf, and Cologne: "Optisches über Malerei," *VR⁵:2*

1872 "Ueber die Theorie der Elektrodynamik," *MB*

1873 Receives Orden pour le mérite für Kunst und Wissenschaft
"Ueber die Theorie der Elektrodynamik. Zweite Abtheilung: Kritisches," *JfruaM*
"Vergleich des Ampère'schen und Neumann'schen Gesetzes für die elektrodynamischen Kräfte," *MB*
"Ueber galvanische Polarisation in gasfreien Flüssigkeiten," *AP*
"Ueber die Grenzen der Leistungsfähigkeit der Mikroskope," *MB*

1874 Receives Copley Medal from the Royal Society of London; and the Becquerel Medal, France

"Ueber die Theorie der Elektrodynamik. Dritte Abhandlung: Die elektrodynamischen Kräfte in bewegten Leitern," *JfruaM*

"Die theoretische Grenze für die Leistungsfähigkeit der Mikroskope," *AP*

"Zur Theorie der anomalen Disperion," *AP*

"Kritisches zur Elektrodynamik," *AP*

"Ueber das Streben nach Popularisirung der Wissenschaft. Vorrede zur Uebersetzung von Tyndall's *Fragments of Science*," VR^5:2

"Induction und Deduction. Vorrede zum zweiten Theile des ersten Bandes der Uebersetzung von William Thomson's und Tait's *Treatise on Natural Philosophy*, VR^5:2

1875 Delivers lecture in Hamburg: "Wirbelstürme und Gewitter," VR^5:2

"Versuche über die im ungeschlossenen Kreise durch Bewegung inducirten elektromotorischen Kräfte," *AP*

Populäre wissenschaftliche Vorträge (Hefte 1 and 2, 2nd ed.; Heft 3)

1876 "Bericht betreffend Versuche über die elektromagnetische Wirkung elektrischer Convection, ausgeführt von Hrn. Henry A. Rowland," *AP*

1877 Appointed advisory editor on mathematical and theoretical physics papers for the *AP*

Death of first daughter, Käthe

Appointed professor of physics at his alma mater; delivers lecture before same: "Das Denken in der Medicin," VR^5:2

Elected rector of the University of Berlin for the academic year 1877–78; delivers rectoral address before the University: "Ueber die akademische Freiheit der deutschen Universitäten," VR^5:2

"Ueber galvanische Ströme, verursacht durch Concentrationsunterschiede; Folgerungen aus der mechanischen Wärmetheorie," *AP*

Die Lehre von den Tonempfindungen (4th ed.)

1878 "Ueber die Bedeutung der Convergenzstellung der Augen für die Beurtheilung des Abstandes binocular gesehener Objekte," *AP*

"Telephon und Klangfarbe," *AP*

Delivers lecture to mark the *Stiftungsfeier* of the University of Berlin: "Die Thatsachen in der Wahrnehmung," VR^5:2

1879	"Ueber elektrische Grenzschichten," *MB*
	"Studien über elektrischen Grenzschichten," *AP*
1880	"Ueber die Bewegungsströme am polarisirten Platina," *AP*
1881	Receives honorary Doctor-of-Law degree, Cambridge University
	Delivers lecture (in English) before the Chemical Society, London: "Die neuere Entwickelung von Faraday's Ideen über Elektricität," *VR^5:2*
	Attends International Electrical Congress, Paris
	Delivers lecture before the Elektrotechnischer Verein, Berlin: "Ueber die Berathungen des Pariser Congresses, betreffend die elektrischen Maasseinheiten," *Elektrotechnische Zeitschrift*
	"Ueber die auf das Innere magnetische oder dielektrisch polarisirter Körper wirkenden Kräfte," *AP*
	"Ueber galvanische Polarisation des Quecksilbers und darauf bezügliche neue Versuche des Herrn Arthur König," *MB*
1882	"Die Thermodynamik chemischer Vorgänge," *SBB*
	"Zur Thermodynamik chemischer Vorgänge," *SBB*
	Delivers lecture before the Physikalische Gesellschaft, Berlin: "Bericht über die Thätigkeit der internationalen elektrischen Commission," *Verhandlungen der physikalischen Gesellschaft zu Berlin*
	"Ueber absolute Maasssysteme für elektrische und magnetische Grössen," *AP*
	Wissenschaftliche Abhandlungen (Volume 1)
1883	Ennobled by William I
	Appointed Honorary Member, National Academy of Sciences, Washington
	Participates in conference of the International Electrical Congress, Paris
	Participates in conference of the International Geodetic Congress, Rome
	Wissenschaftliche Abhandlungen (Volume 2)
	"Zur Thermodynamik chemischer Vorgänge. Folgerungen, die galvanische Polarisation betreffend," *SBB*
	"Bestimmung magnetischer Momente durch die Waage," *SBB*
1884	Represents the Berlin Akademie der Wissenschaften at the 300th anniversary of the founding of the University of Edinburgh
	"Studien zur Statik monocyclischer Systeme," *SBB*
	"Prinzipien der Statik monocyklischer Systeme," *JfruaM*
	Vorträge und Reden (3rd ed., originally entitled *Populäre wissenschaftliche Vorträge*)

1885 *Handbuch der physiologischen Optik* (2nd ed., appears in parts between 1885 and 1895)
1886 Appointed Vice-Chancellor of the Friedensclasse of the Orden pour le mérite für Kunst und Wissenschaft
"Ueber die physikalische Bedeutung des Princips der kleinsten Wirkung," *JfruaM*
Receives the first Graefe Medal, Ophthalmologische Gesellschaft, and delivers lecture in Heidelberg: "Antwortrede gehalten beim Empfang der Graefe-Medallie zu Heidelberg," *VR^5:2*
1887 "Zur Geschichte des Princips der kleinsten Action," *SBB*
Delivers lecture in Berlin: "Josef Fraunhofer. Ansprache gehalten bei der Gedenkfeier zur hundertjährigen Wiederkehr seines Geburtstages," *VR^5:2*
"Weiter Untersuchungen, die Elektrolyse des Wassers betreffend," *AP*
"Zählen und Messen, erkenntnisstheoretisch betrachtet," *Philosophische Aufsätze, Eduard Zeller zu seinem fünfzigjährigen Doctorjubiläum gewidmet* (Leipzig)
1888 Appointed first President of the Physikalisch-Technische Reichsanstalt; relinquishes full-time university teaching duties
Appointed Honorary Member of the Imperial Russian Academy of Medicine
"Ueber atmosphärische Bewegungen," *SBB*
1889 Death of Robert von Helmholtz
"Ueber atmosphärische Bewegungen. Zweite Mitteilungen," *SBB*
1890 Represents the University of Berlin at the 600th anniversary of the founding of the University of Montpellier
"Die Störung der Wahrnehmung kleinster Helligkeitsunterschiede durch das Eigenlicht der Netzhaut," *Zeitschrift für Psychologie und Physiologie der Sinnesorgane*
"Die Energie der Wogen und des Windes," *AP*
1891 Named "Wirklicher Geheimer Rat" with the predicate "Excellenz" by William II
Public celebration of Helmholtz's 70th birthday; numerous honors received
Serves as member on Prussian Commission on Issues of Higher Education
"Erinnerungen. Tischrede gehalten bei der Feier des 70. Geburtstages," *VR^5:1*
"Versuch einer erweiterten Anwendung des Fechnerschen Ge-

setzes im Farbensystem," *Zeitschrift für Psychologie und Physiologie der Sinnesorgane*

"Versuch, das psychophysische Gesetz auf die Farbenunterschiede trichromatischer Augen anzuwenden," *Zeitschrift für Psychologie und Physiologie der Sinnesorgane*

1892 Celebrates fiftieth anniversary of receiving his medical degree

Delivers second lecture on Goethe, before the Goethe Gesellschaft, Weimar: "Goethe's Vorahnungen kommender naturwissenschaftlicher Ideen," *VR^5:2*

"Das Princip der kleinsten Wirkung in der Elektrodynamik," *AP*

"Elektromagnetische Theorie der Farbenzerstreuung," *AP*

1893 Travels to America to serve as Germany's official representative at the International Electrical Congress in Chicago and tours America (August - October); sustains serious head injury on board ship while returning to Germany

"Zusätze und Berichtigungen zu dem Aufsatze: Elektromagnetische Theorie der Farbenzerstreuung," *AP*

"Folgerungen aus Maxwell's Theorie über die Bewegungen des reinen Aethers," *AP*

1894 Death (8 September); public memorial service held (12 December) in the Singakademie, Berlin

"Über den Ursprung der richtigen Deutung unserer Sinneseindrücke," *Zeitschrift für Psychologie und Physiologie der Sinnesorgane*

"Heinrich Hertz. Vorwort zu dessen Prinzipien der Mechanik," *VR^5:2*

1895 *Wissenschaftliche Abhandlungen* (Volume 3)

1897 *Vorlesungen über Theoretische Physik*, eds. Arthur König, et al. 6 vols. (Leipzig: J.A. Barth, 1897–1907).

Introduction: Helmholtz at the Borders of Science

David Cahan

Historians of modern science, philosophy, and cultural history have long recognized that Hermann von Helmholtz played a leading role in European cultural life during the second half of the nineteenth century. Helmholtz's genius profoundly altered his principal scientific disciplines of physiology and physics, and influenced the related disciplines of medicine, mathematics, physical chemistry, psychology, and meteorology. Philosophy and the fine arts of painting and music were also affected by his work. His views on science and society were listened to and solicited by ministers of state, and late in his career he participated in the interactions of science and industry.

Perhaps the most striking feature of Helmholtz's thinking was the ability to link findings within a given cognitive domain or between two or more domains. Although he helped shape several disciplines, his abiding intellectual interests were transdisciplinary in nature: what most engaged him were general problems of energy transformation, human perception, understanding nature as a mechanical system, and the foundations and limits of science itself. Moreover, his scientific work occasioned and, to some extent was also contoured by, his articulation of epistemological views and a philosophy of science and mathematics; even the nature of aesthetic experience and the place and role of science in society came within his cognitive purview. In May 1871, shortly after assuming his new position as professor of

Acknowledgments: I thank Jed Buchwald, Richard L. Kremer, R. Steven Turner, and (especially) Jean Axelrad Cahan for their comments on an earlier version of this Introduction.

experimental physics at the University of Berlin, he wrote to the philosopher Benno Erdmann:

> I am very glad that a better relationship between philosophers and natural scientists is again gradually developing, and hope that both [groups] will again become as close to one another as was once the case. I have always felt the need for such an alliance, since I have worked around the borders of science, in part on the most general geometrical and mechanical axioms on account of the conservation of force, in part on the theory of perceptions.[1]

Helmholtz consistently pursued scientific problems that stood at the common boundary of two or more sciences, using the methods or techniques of one science to work on problems in another, and sought to integrate the sciences and philosophy by articulating the nature and consequences of science for philosophy.

Although Helmholtz's work, like Darwin's and Einstein's, was characterized by the creation of overarching principles aimed at unifying large parts of the sciences, it differed from theirs in that it ranged across both the biological and the physical sciences. As a young man, Helmholtz (1821–94) had wanted to become a physicist. However, as he later reported his family's limited financial circumstances meant that preparing for a career in medicine was the closest that he could come to a life in science.[2] Nevertheless, his predilection for physics and his training in medicine, combined with a voracious intellect and ambition, and the propitious historical moment of post-1840 German science, ultimately allowed a scientific self-expression far broader than either Darwin or Einstein ever contemplated. Perhaps this was part of the meaning of Einstein's statement, quoted as the frontispiece to this collection of essays on Helmholtz, to Mileva Marić in August 1899: "I admire ever more the original, free thinker Helm[holtz]."[3]

One consequence of the broad range of Helmholtz's scientific research was, as several of the essays in this volume indicate, that his leadership in a given discipline was often short-lived. During the first

---

1. Helmholtz to Erdmann, Berlin, 6 May 1871, Sondersammlungen, Martin-Luther-Universität Halle-Wittenberg, Universitäts- und Landesbibliothek Sachsen-Anhalt, Sign. Yi 4 I 147. Cf. Gary Hatfield's essay "Helmholtz and Classicism: The Science of Aesthetics and the Aesthetics of Science," in this volume on 541.
2. "Erinnerungen," VR^5 2:1–21, on 7–9.
3. Albert Einstein to Mileva Marić, early August 1899, in *The Collected Papers of Albert Einstein*, eds., John Stachel, et al., 2 vols. to date (Princeton: Princeton University Press, 1987), vol. 1: *The Early Years, 1879–1902*:220–21, quote on 220.

half of his professional career, from the early 1840s to the late 1860s, Helmholtz forged a reputation as one of Germany's leading physiologists, medical scientists, and physicists. His pathbreaking discovery of the velocity of the nerve impulse in the late 1840s; his invention of the ophthalmoscope in 1850–51; his fundamental work on color theory and space perception in the 1850s and 1860s; and the completion of his historic, three-volume *Handbuch der physiologischen Optik* (1856–67) and epochal *Die Lehre von den Tonempfindungen als physiologische Grundlage für die Theorie der Musik* (1863) placed him among Europe's leaders in physiology and medicine. Yet after 1867 he largely abandoned work in physiology and medicine; it is particularly noteworthy that physiological optics and acoustics developed without his active participation.

With his move to Berlin in 1871, where he assumed one of the most important scientific posts in the new Reich, he was finally in a position to devote his full attention to physics. He had, of course, made his principal contribution to physics in 1847, with his essay on the conservation of force (energy), and again later in 1858, with a study on hydrodynamics. Yet during the 1850s and 1860s, as the centrality of energy physics became increasingly recognized, Helmholtz largely left development of that field to others. During the 1870s and 1880s he successively concentrated on studies in electrodynamics, chemical thermodynamics, the mechanical foundation of thermodynamics, and mechanics, making creative, critical contributions to each of these fields. But again his interest in each field waned relatively soon; specialists abandoned or transcended his ideas, and his leadership was assumed by others. Despite the critical nature of his contributions to electrodynamic theory, Helmholtz's electrodynamic ideas and viewpoint never spread very widely, and became passé after the late 1880s. Similarly, while his analysis of chemical thermodynamics helped paved the way for the new physical chemistry of the 1880s, and that on monocyclic systems helped clarify Ludwig Boltzmann's ideas on such systems, Helmholtz forsook further studies in these fields, both of which moved ahead without him. Much the same can be said for his more limited work in mathematics, psychology, and meteorology.

Circa 1871, Helmholtz's reputation among his fellow scientists probably stood at its height; few if any could deny his scientific status and authority within and beyond the Reich. Yet though he did much important and creative scientific work during the second half of his career, the effects of his restless scientific self began to take their toll. Although his reputation continued to rise among the educated public after 1871, and although he became the very embodiment of German *Wissenschaft*, this reputation was increasingly dependent on his pre-1871 work and on his institutional, political, and cultural standing in German science. Einstein's statement of 1899 notwithstanding, by the end of the century Helmholtz's stature was diminishing.

The broad range of Helmholtz's scientific work has also had the somewhat unusual consequence that few if any scholars to date have been able or willing to analyze the entire range of Helmholtz's scientific work and to place it within its appropriate scientific and philosophical contexts. Helmholtz's principal biographer, Leo Koenigsberger, who was a good friend of the Helmholtz family and gained privileged access to private correspondence, attempted in his three-volume biography *Hermann von Helmholtz* (1902–3) to portray all of Helmholtz's scientific accomplishments and to narrate his life. But the work remains largely an uncritical assemblage of extracts from Helmholtz's published writings and unpublished letters.[4] Of course there have been, as the Bibliography to the present volume attests, many valuable individual studies of selected aspects of Helmholtz's *oeuvre*. Yet there still does not exist a modern, critical treatment of the entire body of Helmholtz's work. The present volume of fifteen essays on Helmholtz seeks to meet precisely this historiographic goal. The essays describe, analyze, and interpret virtually all areas of his work: in medicine, heat and nerve physiology, color and vision theory, physiological optics and acoustics, energy conservation, electrodynamics, chemical thermodynamics, the mechanical foundation of thermodynamics, epistemology, philosophy of science and mathematics, and the relations of science and art and of science and society. Only studies of his relatively minor work in hydrodynamics and meteorology have been omitted.[5]

---

4. In addition to Koenigsberger, the other book-length biographies of Helmholtz are: John Gray McKendrick, *Hermann Ludwig Ferdinand von Helmholtz* (New York: Longmans, Green & Co., 1899); Julius Reiner, *Hermann von Helmholtz* (Leipzig: Theod. Thomas, 1905); Hermann Ebert, *Hermann von Helmholtz* (Stuttgart: Wissenschaftliche Verlagsgesellschaft, 1949); Petr Petrovich Lazarev, *Helmholtz* (in Russian) (Moscow: Akademia Nauk USSR, 1959); and A.V. Lebedinskii, U.I. Frankfurt, and A.M. Frank, *Helmholtz (1821–1894)* (in Russian) (Moscow: Akademia Nauk USSR, 1966). A special issue of *Die Naturwissenschaften*, "Dem Andenken an Helmholtz. Zur Jahrhundertfeier seines Geburtstages," *Die Naturwissenschaften 9:35* (1921):673–711; and a volume edited by Heinrich Scheel, *Gedanken von Helmholtz über schöpferische Impulse und über das Zusammenwirken verschiedener Wissenschaftszweige (= Sitzungsberichte des Plenums und der Klassen der Akademie der Wissenschaften der DDR Nr. 1, 1972)* (Berlin: Akademie-Verlag, 1972), contain collections of articles about various aspects of Helmholtz's work.

5. See, however, W.M. Hicks, "Report on Recent Progress in Hydrodynamics. Part I. General Theory," *Report of the Fifty-First Meeting of the British Association for the Advancement of Science; Held at York in August and September 1881* (London: John Murray, 1882), 57–88; Karl-Heinz Bernhardt, "Der Beitrag Hermann von Helmholtz'

This Introduction makes no pretense at either enumerating Helmholtz's many empirical scientific results and theoretical positions or at fully describing the interpretations of them by the contributors to this volume. It seeks instead to highlight two highly complex and, at points, interrelated themes that, as the volume's essays indicate, dominated various parts of Helmholtz's scientific work at various stages of his career: first, the combined use of instrumentation, measurement, and mathematical analysis to establish theoretical understanding; and second, a quest for intellectual synthesis.

Particularly during the first half of his career, Helmholtz creatively and interactively used scientific instrumentation, measurement, and mathematics to advance understanding of physiological phenomena; he brought an unprecedented degree of analytical rigor to his subject. As Kathryn M. Olesko and Frederic L. Holmes demonstrate in their essay on Helmholtz's early physiological research, Helmholtz employed a variety of sophisticated instruments and apparatus—multiplicators, muscle-contraction apparatus and measuring devices, apparatus for measuring the propagation velocity of nerve impulses and a myograph—to conduct that research. Central to his construction of an empiricist theory of vision, as Timothy Lenoir argues, was his use of an ophthalmotrope for studying eye movement and an ophthalmometer for measuring the curvature of parts of the eye. His views on spatial perception, in particular on understanding the horopter problem and eye movements were, as R. Steven Turner shows, buttressed by his use of a series of simple instrumental demonstrations. And in his work in physiological acoustics, as Stephan Vogel shows, Helmholtz again employed a series of instruments—tuning-fork apparatus, spherical resonators, polyphonic sirens, and sound synthesizers and analyzers—that proved crucial for his study of the physiology of sound. Although Helmholtz made any number of improvements on or adaptations of these instruments, virtually all were invented by others. There was of course one important instrument which he invented himself and for which he first achieved renown, and with which his name has been associated ever since: the ophthalmoscope. In her essay on Helmholtz and the German medical community, Arleen

---

zur Physik der Atmosphäre," *WZHUB 23:3* (1973):331–40; R. Wenger, "Helmholtz als Meteorologe," *Die Naturwissenschaften 10* (1922):198–202; and Elizabeth Garber, "Thermodynamics and Meteorology (1850–1900)," *Annals of Science 33* (1976):51–65, esp. 60–3. See also Heinz Stiller, "Zur Bedeutung der Arbeiten von H. v. Helmholtz für die geophysikalische Hydrodynamik und für die Physik des Erdinnern," in Heinrich Scheel, ed., *Gedanken von Helmholtz*, 45–8.

Tuchman discusses the medical context in which Helmholtz worked and how the ophthalmoscope significantly changed medical diagnostics. She argues that Helmholtz used his invention of the ophthalmoscope (and, to a lesser extent, the ophthalmometer) to show how physicians and physiologists could help one another establish a scientific basis for medicine and physiology as an autonomous discipline. Although Helmholtz left the development of the ophthalmoscope to others, it continually redounded to his benefit, not least, as Lenoir shows, when practicing ophthalmologists subsequently used it to produce evidence in favor of his empiricist theory of vision.

Helmholtz never lost sight of the larger purposes of scientific instruments: he built or had his instruments built in order to conduct measurements aimed ultimately at establishing laws and theories. At the same time, he consciously scrutinized the limitations of his instruments and measuring results, especially through mathematical data and error analysis. Olesko and Holmes show, for example, how Helmholtz used various quantitative techniques of precision as well as nonprecision measurement to determine the propagation velocity of the nerve impulse and to help convince his readers of his results. Through his rigorous, quantitative experiments on muscles; his precision measurements of heat formation in muscles; his precise, quantitative analysis of experimental error (especially the method of least squares); and his use of non-precision graphical analysis Helmholtz made a series of discoveries concerning muscular heat and nerve propagation velocity that simultaneously helped transform the very nature of experiment during the nineteenth century. Lenoir, again, shows that Helmholtz treated the eye itself as a measuring device which the brain uses to construct visual representations of the world with which it constantly interacts, and shows how Helmholtz used the principle of easiest orientation to analyze eye movement mathematically and to argue that the eye uses probability calculus for computing errors of eye motion. As the essays by Olesko and Holmes and by Lenoir show, Helmholtz brought the astronomer's and physicist's use of error analysis into physiology. Again, although markedly different from one another, Lenoir's and Turner's approaches to Helmholtz's deduction of Listings's and Donders's laws from the principle of easiest orientation respectively both call attention to Helmholtz's use of mathematics in physiology. Moreover, Vogel emphasizes Helmholtz's mathematical analysis of vowels, of the tone quality of musical instruments, and of sound perception. Finally, Robert DiSalle, in his essay on Helmholtz's empiricist philosophy of mathematics, analyzes the fundamental connection between Helmholtz's views on space perception and formal geometry, and how the latter developed out of the former: geometry,

Helmholtz maintained, derived from physical measurement. It was no accident, as DiSalle points out, that Helmholtz began his work in geometry in 1866, just as he was bringing the final volume of the *Handbuch* to a close.

Helmholtz's novel and exemplary employment of quantification and mathematical analysis in physiology was matched by its use in physics and chemistry, where it had a far more traditional and widespread use. In his analysis of the foundations of Helmholtz's electrodynamics, Jed Buchwald stresses the essential role of instruments and measurements for detecting the possible states of charged and current-carrying objects along with their interaction or system energies. This search after new states and energies—what Buchwald calls "Helmholtzianism"—allowed Helmholtz and his students, above all Heinrich Hertz, to look for new effects or to study known effects in new ways. The Helmholtzian physical laboratory became, Buchwald argues, a site for using instruments to manipulate systems in order to see what new effects they might yield. Moreover, as both Helge Kragh's essay on Helmholtz's route to a theory of chemical thermodynamics and Günter Bierhalter's on Helmholtz's mechanical foundation of thermodynamics indicate, during the 1880s Helmholtz brought an unprecedented degree of mathematical analysis to the study of chemical and thermodynamical phenomena. Indeed, as DiSalle argues, for Helmholtz there existed a deep connection between mathematics and the world of experience.

Empiricism was an essential theoretical and philosophical viewpoint for Helmholtz. Turner argues that Helmholtz's conflict with Ewald Hering over the sources of human spatial perception essentially created the famous nativist-empiricist controversy. That conflict, and its contemporary scientific context, Turner maintains, did as much to shape Helmholtz's approach to spatial perception as did commitment to any empiricist philosophy. The third part of Helmholtz's *Handbuch*, Turner argues, is in no small measure an exercise in scientific rhetoric, aimed at once to defend empiricism, refute Hering, and draw new theoretical outlines for the field. In a similar vein, Vogel argues that Helmholtz's empiricist epistemology developed as an organizing principle for and during his research in physiological acoustics (as well as optics). The central feature of his empiricist epistemology, Vogel holds, was the division of the sensory process into three parts: the physical, the physiological, and the psychological. Both Turner and Vogel regard Helmholtz's empiricist epistemology as much more a product of his daily scientific research than of specific philosophical precommitments on his part. Helmholtz's empiricist philosophy of mathematics, DiSalle finds, shows a strong interaction between mathematical and empirical truths yet did not simply reduce mathematical truths to empirical ones. Instead, DiSalle argues that Helmholtz's empiricist philosophy of mathematics was based on a strong interaction between the psychological processes underlying mathematics and the laws governing the objective natural world. Finally, Michael Heidelberger argues that empiricism—more precisely, experimental interactionism—was one of the two central features of Helmholtz's philosophy of science (the other being metaphysical realism). As an experimental interactionist, Helmholtz believed that knowledge could only be gained by intervening in nature: scientists and others must manipulate parts of the world to gain knowledge about it. The essays by Lenoir, Buchwald, and DiSalle well illustrate how this "experimental interactionism" worked in practice.

Helmholtz's quest for intellectual synthesis is the second overarching theme of the essays in this volume. His work attempted syntheses both within given disciplines and between parts of different disciplines with one another.

In Fabio Bevilacqua's essay we see that Helmholtz's epochal study *Ueber die Erhaltung der Kraft* (1847) represents an initial intellectual synthesis. Helmholtz here first emerged as a theoretical physicist, Bevilacqua argues, by combining a series of mechanical concepts and principles that had developed since the seventeenth century so as to deduce from them a version of the principle of conservation of force (energy), and by displaying a sophisticated methodology that distinguished clearly between theoretical and experimental physics. Moreover, Bevilacqua demonstrates precisely how Helmholtz applied his principle to a series of otherwise disparate topics in mechanics, heat, electricity, and magnetism, thereby helping to unite problems dispersed throughout physics. The essay of 1847, Bevilacqua argues, presented neither experimental nor mathematical novelties; it was, instead, intended as a study in and about theoretical physics. Though the precise relationship between his concurrent, heavily experimental, work in sensory physiology and that on the conservation of force remains unclear, there can be little doubt, as Bevilacqua points out, that Helmholtz hoped to apply his principle of force conservation to organic as well as inorganic nature.

Like Bevilacqua, Richard L. Kremer argues that the distinctive feature of Helmholtz's work in color research was his inclination and ability to theorize and synthesize. While acknowledging Helmholtz's skill as an observer and experimenter, Kremer shows that in his color research Helmholtz neither presented new fundamental observations nor employed new instruments. Rather, during the 1850s he exploited his deep understanding of physical optics to clarify old problems like absorption and color mixing while developing a theory of color vision that synthesized ideas first put forth by a series of nineteenth-century physicists and physiologists. In so doing, he again, as in his physical study of the conservation of force, united a set of disparate physiological phenomena. As a consequence, theories of color ultimately became theories of color vision. At the same time, Kremer shows that Helmholtz's theory of color was intellectually bifurcated: to explain simultaneous contrast, he ultimately resorted to psychological as well as physiological language. Kremer argues that while Helmholtz's synthesizing efforts indeed united a range of previously unrelated physical and physiological phenomena into a coherent whole, his combined use of psychological and physiological explanations also produced a measure of controversy and incoherence that haunted color theorists for the remainder of the century.

Helmholtz's work in space perception and physiological acoustics during the 1850s and 1860s further demonstrate his synthesizing efforts. In their different ways, both Lenoir and Turner argue that Helmholtz brought theoretical order to the study of eye movement and to the visual perception of space. Helmholtz's principle of easiest orientation linked physiology, physics, and psychology. And Vogel stresses the synthetic nature of Helmholtz's work in physiological acoustics: he shows that Helmholtz consciously set out to reform the entire field, developing theories of combination tones, consonance and dissonance, tone quality, and hearing which together were meant to constitute part of a larger theory of perception.

From the late 1860s onwards, when he essentially abandoned physiological research, Helmholtz devoted himself largely to theoretical work in physics, although he did still direct important experimental work in his laboratory. He had earlier shown himself to be a good experimental physiologist and placed considerable importance on the role of empirical work; that emphasis now faded. Indeed, even within physiology the *Handbuch der physiologischen Optik* and *Die Lehre von den Tonempfindungen als physiologische Grundlage für die Theorie der Musik* already marked his theoretical, synthesizing tendency; they are in effect theoretical treatises that brought order to their fields and helped establish future research problems for specialists.

It is thus not altogether surprising to see that when Helmholtz turned to physics full time, he did so largely as a theorist. In electrodynamics, as in his earlier work on the conservation of force, he aimed above all to bring order to a confusing field and to do so through the prosecution of theory rather than experiment. In his essay, Walter Kaiser argues that Helmholtz's work was crucial to the formation of classical electrodynamics. Although Helmholtz failed to convince many others of the validity of his own electrodynamic theory, he nonetheless did much to shape the field during the 1870s by calling attention to the shortcomings of Wilhelm Weber's theory and to the strengths of James Clerk Maxwell's, by proffering his own electrodynamic theory, and by directing his star student Heinrich Hertz to the field. As Kaiser shows, Helmholtz helped clarify and unify classical electrodynamics. At the same time, Helmholtz sought to unite electrodynamics and energy conservation, and, late in his career, to use the principle of least action to derive Maxwell's equations, here again seeking to bring disparate fields together.

Jed Z. Buchwald approaches Helmholtz's electrodynamics rather differently. He argues that Helmholtz's work in electrodynamics led him to reconceive the entire field of physics. Helmholtz came to view physics as the science of interacting objects whose physical states at any moment in time determine their energy systems. Helmholtzianism, according to Buchwald, was a way of doing physics, *not* a theory: it allowed physicists to operate on the micro-level without making any particular assumptions about the nature of micro-level objects; it eliminated both Weberian forces and Maxwellian fields and replaced them with energy systems; it stressed measurement (i.e., interaction energies) but deemphasized precision; and, because it required the perturbation of systems, it heightened experimentalists' awareness of potential new effects and encouraged their search after them. For Helmholtz, Buchwald argues, the aim of physics became the determination of object states and their interaction energies, or what Buchwald calls a taxonomy of interactions. Helmholtzianism stressed the relations between objects; found its intellectual source largely in British energy physics (especially that of William Thomson and Peter Guthrie Tait); and designated as its avowed enemy the romantic, anti-British approach to physics practiced by Weber's student, Friedrich Zöllner.

The increasingly abstract character of Helmholtz's work became evident through the 1880s. In the early 1880s, as Helge Kragh shows, Helmholtz pioneered a theory of chemical thermodynamics. Building on his much earlier work in physiological chemistry, energy conservation, electrochemistry, and electrolytic conduction, Helmholtz introduced the concept of "free energy" to develop a thermodynamic theory of chemical change. In chemistry, as in much of his physics, Helmholtz conducted little experimental work. Instead, he pursued theoretical lines of inquiry, helping to transform chemistry from a static into a dynamic science. In so doing, as Kragh further argues, Helmholtz helped legitimize theoretical chemistry as a subdiscipline and pave the way for the new physical chemistry, towards which ironically he came to show much skepticism.

In 1884 Helmholtz turned to the foundation of thermodynamics itself. As Bierhalter shows, Helmholtz sought to give a mechanical interpretation to thermodynamics. Using the highly abstract concept of monocyclic systems, he investigated the properties of these systems and their relation to heat theory. If nothing else, his study of monocyclic systems stimulated similar work by Boltzmann, which led to a critique of Helmholtz's own analysis. Ultimately, Bierhalter argues, Helmholtz aimed to refound the mechanical view of nature, to provide a mechanical understanding of all natural processes. He spent much of the last decade of his life (vainly) trying to use the principle of least action as the means to provide a unified view of all nature's forces, a new mechanical world picture.

The late 1860s marked a sea change in Helmholtz's philosophical thinking as well as in his areas of interest. The second half of his career found him almost exclusively theoretical in his scientific work but also more explicitly philosophical. In addition to his rethinking the goals of physics (Buchwald) and the nature of mathematics and its relation to the empirical world (DiSalle), Helmholtz now began to articulate his philosophy of science as well. Michael Heidelberger argues that, along with the well-known influence of Kantian epistemology on Helmholtz's philosophy of science, there were other discernible influences. While Kant's metaphysics of nature, with its abstract, a priori notions of force and matter as necessary general concepts of science, led the early Helmholtz to advocate metaphysical realism (i.e., that science deals with a reality hidden from human senses), in the late 1860s, mainly under the influence of Michael Faraday's empiricist views of force and matter, Helmholtz began to consider forces not as necessary abstractions hidden behind the appearances but rather as hypothetical lawful relations among the appearances. At the same time, Heidelberger further maintains, the influence of Johann Gottlieb Fichte's philosophy of action provided Helmholtz with a methodology that Heidelberger calls "experimental interactionism": our knowledge of the external world can only be gained through intervening in nature. Hence the central features of Helmholtz's mature philosophy of science became metaphysical realism and experimental interactionism; it was a philosophy of science, Heidelberger concludes, which had much to do with idealism and relatively little with materialism or positivism.

Helmholtz also endeavored to relate science and art and to demonstrate the utility of science for modern society. His attempt to explore the common ground of science and art is, for example, evident in the title of his book *Die Lehre von den Tonempfindungen als physiologische Grundlage für die Theorie der Musik*. Gary Hatfield's essay shows just how Helmholtz sought to apply sensory physiology and psychology to music and painting and how far Helmholtz thought science could go towards explaining aesthetic phenomena. Hatfield argues that Helmholtz's perception of the boundaries between science and art led him to a general distinction between the *Naturwissenschaften* and the *Geisteswissenschaften* and resulted in an (at best) prototheoretical aesthetics, one which placed severe limits on the explanatory powers of natural science for art but which stressed the importance of historical analysis for elucidating aesthetic principles. Hatfield further explores Helmholtz's comparison of the artist's and the scientist's cognitive activity by analyzing Helmholtz's psychological theories of "unconscious inference" and "artistic intuition." He maintains that the former theory resulted in Helmholtz's "classicist" aesthetics of scientific explanation and the latter in a naturalistic solution to one of the nineteenth century's primary aesthetic problems, that of the relation between thought and feeling or understanding and imagination. Hatfield concludes that for the mature Helmholtz both science and art are united in their aim of finding universal truths (the lawful or the ideal) amongst changing phenomena and that both do so above all through their use of the imagination.

Helmholtz's scientific work also provided the basis for his reflections on science and society. In this volume's final essay, the editor analyzes Helmholtz's numerous popular addresses on science and on science's relations with society. He argues that despite their diversity these addresses reveal an underlying, recurring theme of the civilizing power of science. More particularly, he argues that for Helmholtz a complex of four intimately related categories constitute the civilizing power of science: First, Helmholtz believed that science provides humankind with the capacity to understand the natural world and its place in it; second, that it enables humankind to control the world; third, that it forms the foundations for aesthetic life; and fourth, that it could help unite individuals socially and bind them to the larger polity of the nation-state. Though Helmholtz's views on science and society in good measure represented those of many nineteenth-century German liberals, he also spoke in part for the conservative German academic and cultural elite. The editor argues that Helmholtz sought to modernize the elite's understanding of the relations of scientific, socioeconomic, and political life by explaining to them the nature and institutional conditions for scientific advance; in turn, he hoped that they might understand his vision of the importance and centrality of natural science for building a modern industrial economy and for strengthening community bonds by stressing the importance of a rational (law-governed) world order and by forming an enlightened German citizenry. The complex theme of the civilizing power of science, then, was a grand, unifying vision of science and society, one fully worthy of the polymath who for over fifty years worked on the borders of science using instruments, measurement, and mathematics to establish new scientific theories, to synthesize the sciences, and to relate science and art.

Part One

Physiologist

1

Helmholtz and the German Medical Community

Arleen Tuchman

1. Introduction

"Medicine," Helmholtz wrote in 1877, "was once my intellectual home, the one in which I grew up, and the wanderer best understands and is best understood in his native land."[1] Yet only rarely does one think of Hermann von Helmholtz in a medical context. The traditional image of Helmholtz is rather of a scientist whose broad-ranging intellectual interests brought him into contact with physicists, physiologists, mathematicians, philosophers, and other learned individuals. Yet, as the above quotation indicates, Helmholtz began his intellectual wanderings within the nineteenth-century German medical community; he even practiced hospital medicine for five years before receiving his first academic appointment at the Berlin Akademie der Künste in 1848. From that point on Helmholtz never again practiced medicine, yet, as this essay seeks to demonstrate, his ties to the German medical community remained strong for most of his career. Not only did several of his discoveries—foremost among them his invention of the ophthalmoscope in 1850–51—have a direct impact on medical practice; his methodological approach to the study of nature influenced the medical community as well. Medical reformers in particular praised

---

Acknowledgments: I would like to thank David Cahan for his advice and criticism. The research for this paper stemmed in part from two larger projects funded by the Fulbright Commission and the National Science Foundation.

1. "Das Denken in der Medizin," in *VR³* 2:165–93, on 170.

17

the "new spirit of physiology," convinced that Helmholtz (and other experimental scientists) would help transform medicine into an exact science by teaching students new critical methods of analysis that would later be applicable to the study of disease.

The German medical community clearly benefited from Helmholtz's contributions, but, as this essay also aims to show, Helmholtz profited as well. Section 2 focuses on his early years in Berlin and Potsdam (1838–48), demonstrating that through his studies at the Königliches medizinisch-chirurgische Friedrich-Wilhelms-Institut and at the University of Berlin, and through his tenure as a staff surgeon to the Royal Hussars in Potsdam, he received perhaps the best medical education and training then available in Germany. According to Helmholtz's own account, this medical training influenced his early work and career. Section 3 examines this claim, turning specifically to the events surrounding his invention of the ophthalmoscope and its reception by the medical community. The ophthalmoscope significantly altered medical diagnostics, having its greatest impact on a growing number of physicians with specialist interest in ophthalmology. Keenly aware of the ophthalmoscope's diagnostic importance, Helmholtz actively publicized his instrument. Indeed, he frequently announced to the medical community the potential practical applications of his scientific work. His reasons for doing so, as Section 3 further shows, derived from his appreciation of the ways in which physicians and physiologists could mutually benefit one another. New instruments and the experimental method would help establish medicine on a scientific basis, at the same time that physicians would help physiologists in their fight to establish an autonomous discipline of physiological science. Seeking institutional support in the university medical faculties, Helmholtz and other experimental physiologists realized that their success rested in part on their ability to convince university and state administrators that their work was of potential significance to the medical community.

The importance of this strategy becomes evident in Section 4, which focuses on the events surrounding the institutionalization of experimental physiology at the University of Heidelberg and Helmholtz's appointment to a newly created chair for this subject in 1858. This appointment, it is argued, occurred squarely within a medical context, with Helmholtz remaining the favored candidate during a protracted period of negotiation largely because of his perceived link to the medical community, a perception that Helmholtz deserved—and cultivated.

This essay argues neither that Helmholtz is best understood within the nineteenth-century German medical community nor that his primary motivation in his early career was to improve medical practice.

Both claims are false. Throughout his years at Königsberg, Bonn, and Heidelberg, Helmholtz's intellectual interests centered on discovering the chemical and physical laws underlying physiological processes, and on determining the physical and philosophical foundations of sensory perception. Nonetheless, Helmholtz seldom lost sight of the potential practical applications of the knowledge he produced. He may only occasionally have pursued such practical applications himself, yet he appreciated and publicized them when they occurred, both recognizing the support he could win from the medical community and appreciating, like most individuals involved in "pure" research, that nothing better demonstrated the legitimacy of a scientific theory than the ability to derive practical consequences from it. Thus Helmholtz's motives for maintaining strong ties to the German medical community were both professional and intellectual, and they were ties he actively encouraged during the twenty-two years he spent as a member of German university medical faculties.

2. Helmholtz's Medical Education in Berlin and Practice in Potsdam, 1838–48

During his early teenage years, Helmholtz developed an interest in the natural sciences, particularly in physics. In the 1830s, however, pursuit of a career in science was difficult. Although some hoped to forge a link between scientific research and industrial advance, the German economy gave but scant support to scientific careers, and only a few individuals received academic appointments in the natural sciences. Consequently, more than a few who wanted to study the natural sciences chose instead to pursue medicine. This, at least, was the path Helmholtz selected, and in 1838, at the age of seventeen, he moved from Potsdam to Berlin to begin his medical studies at the Königliches medizinisch-chirurgische Friedrich-Wilhelms-Institut (formerly called the Pepinière), a medical school designed specifically for the training of military physicians.[2]

The Institut gave promising young men of insufficient financial means the opportunity to pursue a medical career. This arrangement greatly benefited Helmholtz, whose father, a Prussian Gymnasium teacher, reportedly did not earn enough to finance a university education for his son. Taking advantage of family connections to Christian

---

2. Biographical information on Helmholtz is taken from Koenigsberger *1*; and R. Steven Turner, "Hermann von Helmholtz," *DSB* 6:241–53. On the Institut, see Dr. Schickert, *Die militärärztlichen Bildungsanstalten von ihrer Gründung bis zur Gegenwart* (Berlin: Ernst Siegfried Mittler und Sohn, 1895).

Ludwig Mursinna, General Surgeon of the Prussian Army between 1789 and 1809 and professor of surgery at the Charité (the Berlin city hospital) and the Pepinière until 1820, in 1837 Helmholtz's father sought admission for his son to the Institut. Although Helmholtz's first love was physics, he was not, by his own admission, "at all opposed to the study of living nature."[3] There, as in the study of inorganic nature, he saw the possibility of pursuing the study of the causal relations among phenomena. In the fall of 1838, having received a stipend from the Institut, in exchange for which he promised to serve eight years in the Prussian military, Helmholtz left for Berlin, hoping that in the course of his medical education he might satisfy his scientific curiosity as well.

His hopes were not unfounded. German medical education during the 1830s was very much in transition. Whereas professors had previously focused their lectures on the recitation of a text (often their own), several now began to embellish their lectures with demonstrations and experiments. Among the younger lecturers and professors it was also not uncommon to provide instruction in microscopical, chemical, and even diagnostic skills. In the early 1830s, for example, Johann Lukas Schönlein, one of Germany's most prominent clinicians, encouraged his young assistants to provide instruction in the use of the stethoscope to medical students at the University of Würzburg.[4] (In 1840, Schönlein moved to the University of Berlin, where Helmholtz attended his classes.) Later in the decade, Jacob Henle, *Privatdozent* in general pathology and general anatomy at the University of Berlin, offered a class in microscopical techniques for which sixty students enrolled. Shortly thereafter, *Privatdozenten* at the universities of Bonn and Tübingen offered courses "in the use of the microscope" as well. By the early 1840s universities throughout Germany provided basic instruction in the chemical and microscopical investigations of normal and pathological specimens as well as in the new techniques of auscultation and percussion.[5] Yet despite these

---

3. "Erinnerungen," in *VR⁵* 1:1–21, on 9.
4. On Schönlein, see Johanna Bleker, *Die Naturhistorische Schule 1825–1845. Ein Beitrag zur Geschichte der klinischen Medizin in Deutschland* (Stuttgart: Gustav Fischer, 1981). On Helmholtz on Schönlein, see his letters of 12 May 1840 and 16 January 1843 to his parents in David Cahan, ed. *Letters of Hermann von Helmholtz to His Parents: The Medical Education of a German Scientist, 1837–1846* (Stuttgart: Franz Steiner Verlag, 1993), 76–7, 98–9, resp.
5. Josef Kerschensteiner, *Das Leben und Wirken des Dr. Carl von Pfeufer* (Augsburg: Lampart & Co., 1871), 9; Friedrich Merkel, *Jacob Henle. Ein deutsches Gelehrtenleben* (Braunschweig: Vieweg, 1891), 154–55, and W. Waldeyer, "J. Henle. Nachruf," *Archiv für mikroscopische Anatomie 26* (1886):i–xxxii, esp. iii. The two *Privatdozenten* at Bonn and Tübingen were August Meyer and Wilhelm Grube, respectively. (See Konrad Kläß,

important first steps toward introducing "practically" oriented courses into the medical curriculum, most universities still lacked adequate facilities for students interested in going beyond basic instruction so as to perfect their skills or conduct their own research. Small laboratories, minimal equipment and instruments (Henle had only a few microscopes for instructing a class of sixty students), and small university clinics made it difficult for most students to work directly with patients. Consequently, many students intent on increasing their clinical experience spent some time abroad in one of the large clinics in Vienna or Paris.

Still, in comparison to other German cities Berlin had much to offer. First, because it was a larger city it had more sick individuals to fill its clinical teaching wards. More to the point, it had, in addition to its university, several medical institutions and hospitals in its midst, including the Theatrum anatomicum (anatomical theater), the Collegium medico-chirurgicum, the Botanischer Garten, the Charité hospital, and the Friedrich-Wilhelms-Institut.[6] Designed primarily for the training of military physicians, all but the last of these institutions dated back to the reign of Frederick William I (1713–40), the so-called Soldier-King. Concerned about the heavy medical losses suffered annually by his beloved army, Frederick William instituted several measures to improve the quality of medical care available to his soldiers. He converted his father's Orangerie into the Botanischer Garten; founded the Theatrum anatomicum for public dissections (1713) and the Collegium medico-chirurgicum for theoretical instruction (1723); and established the Charité for clinical training (1726). To some extent these institutions served civilian students of medicine and surgery, yet their primary purpose was to train an advanced corp of surgeons who could supervise medical affairs in each of the army's regiments. To this end, each year eight young men received three-year fellowships to study at the Collegium. This educational system remained largely unchanged until the end of the century when, with the increasing militarism of the Prussian state, the need for more trained army surgeons grew. To meet these needs Frederick William II founded another medical school in 1795, the Pepinière, renamed the Königliches medizinisch-chirurgische Friedrich-Wilhelms-Institut in 1818. This new state

---

Die Einführung besonderer Kurse für Mikroskopie und physikalische Diagnostik (Perkussion und Auskultation) in der medizinischen Unterricht an deutschen Universitäten im 19. Jahrhundert [Med.diss., Göttingen, 1971].)

6. For a discussion of these institutes see Paul Diepgen and Edith Heischkel, *Die Medizin an der Berliner Charité bis zur Gründung der Universität* (Berlin: J. Springer, 1935), and Schickert, *Die militärärztlichen Bildungsanstalten.*

medical institute accommodated and financed the education of between eighty and ninety pupils per year; in return, its graduates obligated themselves to serve as military surgeons in the Prussian army for eight years.[7]

Although a student of the Institut, Helmholtz attended many classes at the University of Berlin. The two institutions had not always had close connections. In the years immediately following the founding of the university in 1810, the latter and the Institut had, due to their different goals, a tense relationship. The military medical institute had a practical goal—the training of medical surgeons for the Prussian army; by contrast, the university, founded on neo-humanist principles, sought to provide an environment that would awaken a spirit of inquiry in students.[8] Devotion to *Wissenschaft*, not to utilitarian or professional goals, defined the educational process. This process was dynamic, and had little to do with absorbing factual information. The teacher sought to arouse student curiosity while providing them with the proper critical tools for pursuing their new-found interests. The goal was to cultivate the individual's character; specific concrete achievements mattered little.

Despite their different goals, efforts to keep the university and the Institut separate soon failed. A few years after the university opened its doors, several members of the medical faculty, including the professor of surgery, Carl Ferdinand von Graefe, and the professors of anatomy, Christoph Knape and Karl Asmund Rudolphi, accepted adjunct positions on the professorial staff of the Institut—for which they received handsome honoraria.[9] The gradual breakdown of the barriers dividing the two institutions occurred not least because the new university was not simply modelled after the neo-humanists' vision. Its absolute dependence upon the state, cemented by the financial bond, as well as the state's urgent need for well-educated civil servants, physicians, and teachers, guaranteed that utilitarian concerns were not totally absent from university education.[10] By 1825 students of the two

---

7. Schickert, *Die militärärztlichen Bildungsanstalten*, 5, 49.
8. On neo-humanism see Helmut Schelsky, *Einsamkeit und Freiheit. Idee und Gestalt der deutschen Universität und ihrer Reformer*, 2nd ed. (Düsseldorf: Bertelsmann Universitätsverlag, 1971). On tensions between the university and the Institut see Arleen Tuchman, "Spannungen und Zusammenarbeit zwischen der Charité und der Berliner Universität in der ersten Hälfte des 19. Jahrhunderts," unpublished paper, read at the Deutsche Gesellschaft für Geschichte der Medizin, Naturwissenschaft und Technik, Bayreuth, September 1987.
9. Memorandum from Johann Wilhelm Wiebel, Chef des Militärmedicinalwesens, to Karl Freiherr Stein zum Altenstein, head of the Ministerium des Innern, 10 April 1827, in Geheimes Staatsarchiv Preußischer Kulturbesitz, Merseburg, Ministerium des Innern, Rep. 76 VIIIB, Nr. 4419, Bl.78.
10. The best treatment of the various interests surrounding the founding of the University of Berlin is Charles E. McClelland, *State, Society, and University in Germany*,

institutions shared many of the same medical and scientific institutes, and attended many classes together, listening to the same lectures from members of the university faculty.

Not all differences between the two institutions had disappeared, however. Whereas *Lernfreiheit* and *Lehrfreiheit* characterized university education, the course of study at the Institut was highly regimented. During the first of four years of instruction, students attended lectures on osteology, splanchnology, physics, botany, chemistry, physiology, general anatomy, and natural history. In the second year they continued to study these subjects, to which were added pathology and pharmacy, along with courses in anatomical dissection aimed at complementing their theoretical lectures. In the third year they advanced to general and special therapeutics, semiotics, surgery, and obstetrics. And in their fourth and final year they attended the polyclinic and the surgical, medical, ophthalmological, and obstetrical clinics in the Charité. Following this rigid four-year course of medical instruction, they spent a further year in the Charité doing rotation in the various wards. Although prohibited from treating their own patients, their duties were extensive. Under the supervision of the graduates of the Institut, they prescribed and administered medicines according to their superiors' orders; wrote case histories; and supervised the nursing staff in all hygienic matters.[11] Thus, at a time when university students had to go to Paris or Vienna to acquire extensive clinical experience, Institut students received extensive surgical and clinical training in the wards of a large city hospital.

Helmholtz valued the extensive medical training he received—indeed, he later attributed part of his success in inventing the ophthalmoscope to this training—but his real interests remained in the natural sciences, particularly in scientific research, which he pursued whenever opportunity availed.[12] Especially important here was his contact with Eilhard Mitscherlich, professor of chemistry at the university, and, above all, with Johannes Müller, professor of physiology. Berlin hired Müller in 1833 to help reform its medical faculty and to lend more support to a new direction in scientific research, one

---

1700–1914 (Cambridge and New York: Cambridge University Press, 1980), chap. 4.

11. Schickert, *Die militärärztlichen Bildungsanstalten*, 33–6, 48. See, also, Helmholtz's letter to his parents of 5 May 1839, in which he describes the forty-two hours of classes required of him during the second semester of his first year of study, in Cahan, ed., *Letters*, 56–60.

12. Helmholtz's appreciation of his medical education is evident in his letters of 7 and 16 October, 16 November, 8 and 19 December 1842, and 16 January 1843 to his parents, in Cahan, ed., *Letters*, 92–3, 93–4, 95, 95–6, 96–7, 98–9, resp.; and below, Section 3.

that moved beyond simple observation to the "creation of new phenomena through experiment."[13] This new direction appealed to Helmholtz, too; in 1841 he decided to write a dissertation under Müller's supervision. For this purpose he scraped together his savings and bought his first instrument—a microscope.[14] Displaying no sign of his later aversion toward microscopical research, Helmholtz investigated the structure of the nervous system in invertebrates, discovering that the nerve fibers originate in the ganglionic cells.[15] These cells had first been discovered in 1833 by the German physiologist Christian G. von Ehrenberg, who managed only to speculate that they give rise to nerve fibers; Helmholtz demonstrated the connection and suggested thereby that ganglia are more important developmentally than nerve fibers. Helmholtz's first contribution to the scientific community was thus a significant one, and it was made in microscopic anatomy.

Although this initial finding impressed Müller, he asked Helmholtz to investigate several more animals before submitting his dissertation so as to demonstrate his thesis conclusively. To this end, Müller offered him access to workspace in his anatomical museum. Every year Müller made such offers to a few talented students whose independent research he wished to encourage. Yet he could not provide them with much assistance. Compared to the research laboratories built in Germany after mid-century, the museums and laboratories of the early nineteenth century were extremely modest, usually consisting of only a few small rooms equipped with the simplest of equipment and instruments. Müller's museum, with its cramped quarters, few instruments (beyond microscopes), and small budget was typical.[16] Still, his museum provided a virtually unmatched intellectual atmosphere, allowing talented students to work and to discuss their ideas with one another. In the 1830s the list of medical students who frequented Müller's museum included Jacob Henle, Theodor Schwann, and Robert Remak, all important contributors to the cell theory; in the 1840s they were followed by Rudolph Virchow, Emil du Bois-Reymond, Ernst Brücke, and Hermann Helmholtz.

Although all these individuals made outstanding contributions to physiological research, their emphases differed. Several, particularly

---

13. Manfred Stürzbecher, "Zur Berufung Johannes Müllers an die Berliner Universität," *Jahrbuch für die Geschichte Mittel- und Ostdeutschlands 21* (1972):184–226, quote on 193.
14. Koenigsberger *1*:52.
15. *De Fabrica systematis Nervosi Evertebratorum* (Med. Diss., Berlin, 1842), in *WA* 2:663–79.
16. Axel Genz, *Zur Emanzipation der naturwissenschaftlichen Physiologie in Berlin* (Med. diss. Magdeburg, 1976).

Henle and Virchow, became leaders of the medical reform movement that flourished in Germany in the 1840s.[17] They sought to give medicine a scientific basis by demonstrating that disease was nothing more than a deviation from normal physiological processes brought about by abnormal conditions. Convinced that physiological processes followed deterministic laws of nature, they argued that scientific medicine should aim at ascertaining how bodies, subject to these laws, behave under altered (i.e., abnormal) conditions. In their vision of scientific medicine, experiments provided a means of artificially creating a diseased condition. Thus, the programmatic statements of Henle, Virchow, and others stressed the use of physiological experiments, pathological anatomy, microscopy, chemistry, and clinical observation as the tools for analyzing bodily functions and how they become diseased.

While sympathetic to the goals of Henle and Virchow, Helmholtz, du Bois-Reymond, and Brücke focused not on the reform of medicine but rather on the means for establishing an autonomous physiological science. In particular, they wanted to transform physiology into an "organic physics." In 1847 the three friends, along with Karl Ludwig, took an oath to reduce physiology to its chemical and physical foundations, creating an "organic physics" based exclusively upon mathematical, physical, and chemical laws.[18] In developing this new approach "the 1847 group" (as Paul Cranefield has labelled them) relied heavily upon sophisticated instruments and instrumental techniques. Du Bois-Reymond's astatic galvanometer and induction apparatus, Helmholtz's myograph, ophthalmoscope, and ophthalmometer, and Ludwig's kymograph and vacuum pump illustrate a few of the many instruments conceived of and in part constructed by this new generation of physiologists.[19] These instruments simplified the conditions for experimenting on isolated organs, allowed better control of phenomena, aided in the establishment of causal connections, and permitted exact measurements and graphical representation of organic functions. They required, moreover, a level of mathematical

---

17. Claudia Huerkamp, *Der Aufstieg der Ärzte im 19. Jahrhundert* (Göttingen: Vandenhoeck und Ruprecht, 1985); E. Ackerknecht, "Beiträge zur Geschichte der Medizinalreform von 1848," *Sudhoffs Archiv 25* (1932):61–109, 113–83; and Arleen Tuchman, *Science, Medicine, and the State in Germany: The Case of Baden, 1815–1871* (Oxford and New York: Oxford University Press, 1993).
18. Paul Cranefield, "The Organic Physics of 1847 and the Biophysics of Today," *JHMAS 12* (1957):407–23.
19. K.E. Rothschuh, "Emil Heinrich du Bois-Reymond," *DSB 4* (1971):200–5; Heinz Schröer, *Carl Ludwig. Begründer der messenden Experimentalphysiologie* (Stuttgart: Wissenschaftliche Verlagsgesellschaft, 1967), 104–14, 170–80; Stanley Joel Reiser, *Medicine and the Reign of Technology* (Cambridge and London: Cambridge University Press, 1978), 100–1; and below, Section 3.

skill and technical dexterity that clearly distinguished the methodological approach of the "organic physicists" from the microscopical, chemical, and vivisectional skills of individuals like Henle and Virchow.

Despite Helmholtz's commitment to the transformation of physiology into an exact science, he could not devote his energy solely to this pursuit. In late September 1842, having completed his course work at the Institut, he began his year-long medical "internship" in the Charité, an experience he found both difficult and challenging.[20] His first assignment in the ward for internal diseases proved to be his most demanding. He began each day at seven o'clock, examining his own patients before making the rounds with his supervisor and other interns, at which time he reported on his findings and suggested appropriate therapeutic measures. The entire procedure lasted over four hours and occurred twice daily. During the remaining hours, he performed autopsies and recorded the daily events in the hospital journal. His day did not end until eight o'clock in the evening.[21]

Helmholtz's most difficult months were the first two, when his immediate supervisor proved to be a tyrant and most of his patients suffered from chronic illnesses for which he could do little more than prescribe opium. Although he welcomed the opportunity to study these illnesses firsthand, he was relieved when rotations toward the end of the year placed him in a ward with more varied cases and a supervisor who permitted the interns greater responsibility for their patients. He found this work more rewarding—not only did he come into contact with "a selection of the most interesting patients"; he also found time to further his medical knowledge by visiting Schönlein's clinic, "in which one receives multiple stimulations to a deeper understanding of the disease process."[22] He clearly enjoyed his work in the Charité, his only disappointment being that he had so little time to conduct his own research. During the fall of 1842 he managed only to put the finishing touches on his dissertation, which he defended on 2 November 1842. Entitled *De fabrica systematis nervosi evertebratorum* ("The Structure of the Nervous System in Invertebrates"), the thesis went beyond Helmholtz's initial study by demonstrating the link between

---

20. Helmholtz's attitude toward his internship is revealed in several letters to his parents: 7 October, 8 December, and 19 December 1842, and 16 January and 8 February 1843, in Cahan, ed., *Letters*, 92–3, 95–6, 96–7, 98–9, 99–100, resp.
21. Helmholtz to his parents, 7 October 1842, in ibid., 92–3. Rudolf Virchow's experience as an intern in the Charité was similar to Helmholtz's. See Virchow's letter to his father of 14 May 1843, in Marie Rabl, ed., *Rudolf Virchow. Briefe an seine Eltern 1839 bis 1864* (Leipzig: W. Engelmann, 1907), 64.
22. Helmholtz to his parents, 16 January 1843, in Cahan, ed., *Letters*, 98–9.

ganglionic cells and nerve fibers in a host of organisms, both invertebrate and vertebrate.[23]

Beginning in February 1843, Helmholtz was able to allot more time to his scientific studies. Having completed his tour of duty in internal medicine, he moved on to the children's ward, followed by obstetrics, where there was little to do.[24] He took advantage of this situation by spending more time in Müller's laboratory, where he conducted experiments on putrefaction and fermentation, a topic he viewed as fertile ground for furthering the goal of the "organic physicists." At the time, scientists disagreed as to whether these processes were essentially chemical or dependent upon the presence of living organisms. Helmholtz's experimental results, probably to his dismay, were equivocal.[25] Although they seemed to demonstrate that putrefaction involved a process of decay that occurred independently of any life forms, they also indicated that fermentation, although a form of putrefaction, resulted only when a living organism was introduced to the putrifying medium. Thus Helmholtz had not succeeded in reducing all organic processes to their chemical and physical components.

Helmholtz's clinical internship ended on 1 October 1843, at which time he was promoted to staff surgeon to the Royal Hussars at Potsdam and transferred to an army hospital there.[26] He remained in his native city for the next five years, dividing his time between his medical responsibilities and scientific research. The former were not overly demanding. Helmholtz managed to establish a small laboratory in the army barracks so as to conduct experiments. During his Potsdam years (1843–48) he shifted his focus of research from putrefaction and fermentation to the source of heat production during muscle contraction, conducting experiments that led ultimately to his articulation of the principle of the conservation of force. He now recognized that the ability to explain animal heat as a function of the chemical transformations occurring within the muscles was better suited for his plans to create an "organic physics." Between 1843 and 1847 he showed experimentally that chemical changes occur in the muscle tissues during activity; that heat emission accompanies this process; and that this heat is not brought to the muscles by the nerves or blood but rather is produced in the tissues themselves. Having demonstrated these

---

23. *De Fabrica.*
24. Helmholtz to his parents, 8 February 1843, in Cahan, ed., *Letters*, 99–100.
25. "Ueber das Wesen der Fäulniss und Gährung," *MA* (1843):453–62, in *WA* 2:726–34. For discussion of this work see the essay by Kathryn M. Olesko and Frederic L. Holmes, "Experiment, Quantification, and Discovery: Helmholtz's Early Physiological Researchs, 1843–50," in this volume.
26. Koenigsberger *1*:54.

chemical physiological facts, he then analyzed and quantified this physiological process, deriving a mechanical equivalent for the amount of heat produced. This work formed an important part of his monograph, *Ueber die Erhaltung der Kraft*, which he published in 1847.[27] By establishing the principle of the conservation of force, Helmholtz had effectively demonstrated the superfluousness of evoking any special life forces as explanatory principles.

Although Helmholtz's work on the conservation of force received a mixed review, there could be no doubt that he was a man of much promise. Apart from Müller's support, he now won that of the influential Alexander von Humboldt, who always had a keen eye for talented young scientists. When Brücke, who had been teaching anatomy at the Akademie der Künste in Berlin, received a call to Königsberg as extraordinary (*außerordentlicher*) professor of physiology and pathology, von Humboldt arranged to have Helmholtz released from his military duties so that he could succeed Brücke.[28] Thus, three years before the projected completion of his official military obligations, Helmholtz entered into a civilian, academic career.

Although Helmholtz never again practiced medicine after he left the army in 1848, his ties to the medical community remained unbroken. This connection manifested itself first and foremost institutionally: until he moved to Berlin in 1871 as professor of experimental physics, he taught in medical faculties, even lecturing on pathology during the six years he spent in Königsberg (1849–55). This medical connection, however, extended well beyond an institutional affiliation. In his work, particularly in physiological optics, Helmholtz made several significant contributions to the medical world. The most important of these was his invention of the ophthalmoscope, an instrument which, more than any of his other discoveries, made his reputation in the medical community and beyond. It was, as he later said, "the most popular of my scientific achievements."[29] Indeed, it was the invention that launched him on his stupendously successful career. Its invention and the medical community's response to it are the focus of Section 3.

---

27. *Ueber die Erhaltung der Kraft* (Berlin: G. Reimer, 1847). Helmholtz's experiments between 1843 and 1847 are published in "Ueber den Stoffverbrauch bei der Muskelaktion," *MA* (1845):72–83; and in "Ueber die Wärmeentwicklung bei der Muskelaction," ibid. (1848):144–64, both in *WA* 2:735–44 and 745–63, respectively. On his principle of the conservation of force see Fabio Bevilacqua's essay, "Helmholtz's *Ueber die Erhaltung der Kraft*: The Emergence of a Theoretical Physicist," in this volume.
28. Koenigsberger *1*:93–110.
29. "Erinnerungen," in *VR*[5] *1*:12.

3. Helmholtz's Ophthalmoscope: Its Invention and the German Medical Community's Response

Helmholtz remained at the Akademie der Künste for only one year. In the summer of 1849 Brücke transferred to Vienna as professor of anatomy and physiology, and his position in Königsberg went to Helmholtz. In his new capacity as extraordinary professor in the medical faculty, Helmholtz had responsibility for teaching not only physiology but pathology as well. He held this position for the next six years (1849–55).

During his Königsberg years, Helmholtz continued his studies of nerve physiology (measuring the velocity of the nerve impulse), and also moved on to physiological optics and acoustics—two areas that occupied his attention for nearly twenty years.[30] His interests continued to center on the physical foundations of sensory stimulation and the epistemological foundations of sense perception; yet while preparing a lecture on optics for his medical students, Helmholtz realized that certain rather simple laws of geometrical optics permitted him to construct an instrument of potentially great diagnostic importance to the medical community—the ophthalmoscope.[31]

---

30. "Messungen über den zeitlichen Verlauf der Zuckung animalischer Muskeln und die Fortpflanzungsgeschwindigkeit der Reizung in den Nerven," *MA* (1850):276–364; "Messungen über Fortpflanzungsgeschwindigkeit der Reizung in den Nerven. Zweite Reihe," ibid. (1852):199–216, both in *WA* 2:764–843 and 844–61, respectively; *Handbuch*; and *Die Lehre von den Tonempfindungen als physiologische Grundlage für die Theorie der Musik* (Braunschweig: F. Vieweg and Sohn, 1863). For analyses of Helmholtz's work in these areas see the respective essays in this volume by Olesko and Holmes, "Experiment, Quantification, and Discovery"; R. Steven Turner, "Consensus and Controversy: Helmholtz on the Visual Perception of Space"; Richard L. Kremer, "Innovation through Synthesis: Helmholtz and Color Research"; and Stephan Vogel, "Sensation of Tone, Perception of Sound, and Empiricism: Helmholtz's Physiological Acoustics."

31. *Beschreibung eines Augen-Spiegels zur Untersuchung der Netzhaut im lebenden Auge* (Berlin: A. Förstner, 1851). On Helmholtz's invention of the ophthalmoscope see, for example, Richard Greef, "Historisches zur Erfindung des Augenspiegels," *Berliner klinische Wochenschrift 38:48* (1901):1201–2; Koenigsberger *1*:133–43; Ernest Engelking, ed., *Dokumente zur Erfindung des Augenspiegels durch Hermann von Helmholtz im Jahre 1850* (Munich: J.F. Bergmann, 1950); Albert Esser, "Zur Geschichte der Erfindung des Augenspiegels," *Klinische Monatsblätter für Augenheilkunde 116* (1950):1–14; Wolfgang Jaeger, ed., *Die Erfindung der Ophthalmoskopie dargestellt in den Originalbeschreibungen des Augenspiegel von Helmholtz, Ruete und Giraud-Teulon* (Heidelberg: Brausdruck, n.d. [1977]); Reiser, *Medicine and the Reign of Technology*, 46–8; Klaus Klauß, "Ein neuentdecktes frühes Dokument zur Geschichte der Erfindung des Augenspiegels durch Hermann v. Helmholtz," *NTM 18:1* (1981):58–61; George Gorin, *History of Ophthalmology* (Wilmington, Del.: Publish or Perish, 1982), 129; and Frank W. Law, "The Origin of the Ophthalmoscope," *Ophthalmology 93:1* (1986):140–41.

Helmholtz wanted to describe the phenomenon, observed by both William Cummings, an English physician, and Brücke, wherein the human eye glowed in a dark room when light was directed at the eye and an observer stood near the light source. Neither Cummings nor Brücke had managed to see the eye's inner structure; whenever they approached the eye closely enough to peer inside it, the glare from the light source diffused over the entire pupil. While preparing his lecture, Helmholtz asked himself how the light rays reflected back from the illuminated eye produce an optical image; he was thus led to analyze the rays' paths. He discovered that the rays followed an identical path when entering and leaving the eye; this allowed him to explain Brücke's inability to see the internal structure: to do so Brücke would have had to stand directly in the path of the light rays, thus blocking the light source.

It took Helmholtz merely eight days to circumvent the problem and invent an instrument that permitted him to see the retina and vessels within the living eye. He used a plane-polished glass surface which both reflects and transmits light, and thus acts as a partial mirror. By looking through this glass, which he placed at an angle, and using one surface to reflect light into the observed eye, he was able to eliminate the glare while allowing enough light to return so as to permit a clear view of the retina. Figure 1.1 presents Helmholtz's own schematic arrangement of the ophthalmoscope and ray paths. Here **C** is the plane-polished glass. A candle **A** illuminates the glass plate, whereby most of the light is reflected into the eye **D** being observed. The dorsal area of the eye (i.e., the retina) in turn reflects this light back along the same path by which the rays had entered. Some light returns to **A** while some continues in a straight line (the reverse path of entry) through the glass plate and on to the eye of the observer **G**. Since the latter must stand very close to the person being observed in order to see through the pupil's small opening, the light rays entering the observer's eye converge, resulting in a blurred picture. To circumvent this problem, Helmholtz placed a concave lens **F** between the observer and the glass plate. His ophthalmoscope, at least in its initial, highly primitive form, was complete.

In December 1850 Helmholtz wrote his father expressing his surprise that no one before him had figured out how to construct such an instrument.[32] He reported that he had needed only the most elementary knowledge of optics. Indeed, he had actually worked out some of the optical laws while still a student at the Potsdam Gymnasium.

---

32. Koenigsberger *1*:133–34.

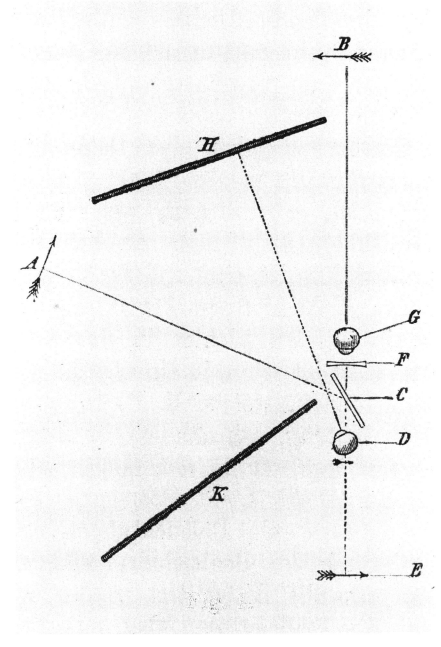

Figure 1.1. Schematic drawing of the path taken by light as it travels through the ophthalmoscope. Source: Beschreibung eines Augen-Spiegels zur Untersuchung der Netzhaut im lebenden Auge *(Berlin: A. Förstner, 1851), 47; reprinted in* Klassiker der Medizin *(Leipzig: J.A. Barth, 1910), 12.*

Yet he underestimated the mathematical knowledge needed to comprehend the geometrical optics upon which the ophthalmoscope was based. Some of the physicians who voiced opposition toward the instrument (about which more presently) may have done so because of their inability to understand the physical and mathematical principles involved in the instrument's construction.

Yet Helmholtz's geometrical knowledge did not alone lead to the idea of the ophthalmoscope. As he himself later noted, his success resulted from his hybrid education: he knew more physics than physicians, and more physiology and medicine than physicists and mathematicians.[33] His medical studies had familiarized him with the practical problems confronting ophthalmologists, leading him to recognize the diagnostic potential of an instrument that would permit physicians to investigate changes in the retina. He was aware, moreover, of recent developments in medical diagnostics which had begun to free physicians from their dependence upon gross physical symptoms and patients' own accounts of their ailments. Of particular importance here were new types of analyses and techniques, such as physical diagnosis, pathological anatomy, and microscopy, which seemed to provide physicians with more objective information about the internal organic changes accompanying the diseased state (although the interpretation of this information often remained disputed). No instrument better exemplified this transition than the stethoscope, which was invented in the early nineteenth century and which made audible what was invisible to the physician's gaze.[34]

Helmholtz realized immediately that an ophthalmoscope would similarly reveal what had previously remained hidden—the anatomical and physiological changes of the eye's interior. He wrote to his father:

> Until now a series of the most important eye diseases, included under the name "black cataract," have been *terra incognita* because one could learn nothing about the changes in the eye either in the living [state] or even after death. My invention will make possible the finest investigation of the internal structures of the eye.... Where possible, I shall examine patients with the chief ophthalmologist here and then publish the material.[35]

Helmholtz published the results of his work in 1851 in a small pamphlet entitled *Beschreibung eines Augen-Spiegels zur Untersuchung der*

---

33. "Erinnerungen," *VR*[5] *1*:13.
34. Reiser, *Medicine and the Reign of Technology*, chap. 2; and Michel Foucault, *The Birth of the Clinic* (New York: Vintage Books, 1973).
35. Koenigsberger *3*:142f.

Netzhaut im lebenden Auge. The medical community responded immediately. Among general practitioners, particularly older ones, some voiced opposition to the instrument. Helmholtz reported that one colleague condemned the ophthalmoscope as a dangerous instrument for letting too much light into the eye. Indeed, one of the loudest objections was that prolonged exposure of the retina to light could lead to blindness; some critics thus charged that the ophthalmoscope actually created diseases. Other complaints stemmed from difficulties in understanding and using the instrument. One of Helmholtz's colleagues reportedly went so far as to claim that only those with poor eyesight needed the assistance of such an instrument.[36]

These hostile responses aside, the ophthalmoscope received extensive positive publicity in scientific and medical journals soon after its invention. Although general practitioners probably did not use the ophthalmoscope in their routine patient examinations until the end of the century, it appealed immediately to a sizable group of physicians with specialist interests in optometry and ophthalmology. By early December 1851 Helmholtz had received eighteen orders for his instrument. Between 1851 and 1856 at least sixteen books and eighteen articles were written on the ophthalmoscope. Several of these books aimed specifically at teaching physicians how to use the instrument: for example, *Der Augenspiegel und das Optometer für praktische Aerzte* (1852), by Christian Georg Theodor Ruete, extraordinary professor of ophthalmology at the University of Göttingen; and *Ueber die Anwendung des Augenspiegels, nebst Angabe eines neuen Instruments* (1853), by Ernst Adolf Coccuis, *Privatdozent* in ophthalmology at the University of Leipzig.[37] These specialists felt deeply indebted to Helmholtz and his instrument for their own recent successes. In 1854, for example, in a review of Helmholtz's work in the *Medicinische Centralzeitung*, one author adopted an ecstatic tone, characterizing Helmholtz as the "emancipator and liberator" of the field of ophthalmology.[38] And in the same year, Albrecht von Graefe, then a practicing ophthalmologist in Berlin and lecturer at the university, founded a specialist journal, *Archiv für Ophthalmologie*, in which he attributed the field's great upswing to Helmholtz's invention.[39]

---

36. "Das Denken in der Medizin," *VR³* 2:179. For other discussions of opposition to the ophthalmoscope, see Reiser, *Medicine and the Reign of Technology*, 50; and Gorin, *History of Ophthalmology*, 129.

37. These and other works are listed in F. Heymann, "Die Augenspiegel, ihre Construktion und Verwendung," *Schmidt's Jahrbücher der In- und Ausländischen gesammten Medicin 89* (1856):105–22.

38. *Medicinische Centralzeitung*, 1 November 1854, noted in *Mittheilungen des badischen ärztlichen Vereins 9* (1855):last page (no pagination).

39. Albrecht von Graefe, "Vorwort," *AO 1* (1854):v–x.

The ophthalmoscope revolutionized the study of eye disease. Von Graefe, hailed as one of the founders of modern ophthalmology, praised the instrument for "revealing to us a new world," and for allowing the investigation of what had previously remained opaque— the various anatomical, physiological, and pathological changes of the inner eye.[40] The *terra incognita*, for example, to which Helmholtz had referred—black cataract—lost much of its mystery. The condition represents an advanced cataract state in which sclerosis causes the lens to change color, eventually turning a very dark brown, if not black. It was a condition, one ophthalmologist claimed, in which "the patient sees nothing, but neither does the physician."[41] Ophthalmoscopic investigations permitted early diagnosis of the disease and made it easier to distinguish between cataracts and other eye disorders, such as amblyopia, which are also marked by the gradual loss of sight. Most importantly, increased knowledge of cataracts led to therapeutic advances. In 1865 von Graefe improved upon the standard procedure for removing cataracts, which had involved a corneal flap incision, by substituting a linear incision in the sclera. This technique significantly reduced the morbidity associated with cataract surgery.[42]

Cataract victims were neither the first nor the only individuals to benefit from the ophthalmoscope. As early as 1856 von Graefe performed the first successful iridectomy for acute glaucoma.[43] He removed a portion of the iris in order to reduce the ocular tension responsible for the discomfort and blindness associated with the disease. The ophthalmoscope helped von Graefe develop this therapeutic technique by allowing him to ascertain definitively that glaucoma arose not, as some had claimed, from changes in the optic nerve, but rather from increased intraocular pressure. Experiments with alternative procedures for reducing this pressure eventually led him to consider and develop the technique of iridectomy.

In addition to aiding the development of surgical techniques, the ophthalmoscope also advanced the study of optic pathology. Within a decade of the instrument's invention, ophthalmologists identified and described thrombosis and embolies of the retinal arteries, pigmentary retinitis, retinal detachment, and degenerative diseases of

---

40. Quoted in Eduard Michaelis, *Albrecht von Graefe. Sein Leben und Wirken* (Berlin: G. Reimer, 1877), 39.
41. Greeff, "Historisches zur Erfindung des Augenspiegels," 1201.
42. George E. Arrington, Jr., *A History of Ophthalmology* (New York: MD Publications, Inc., 1959), 93.
43. Albrecht von Graefe, "Ueber die Iredectomie bei Glaucom und über den glaucomatömsen Process," *AO 3:2* (1857):456–560.

the retina.[44] Furthermore, they explored pathological conditions of the eye associated with general systemic diseases. Richard Greeff, professor of ophthalmology at Berlin around the turn of the century, viewed the inner eye as an ideal window to other bodily diseases because alterations in the fine structures of the retina often reflected pathological changes elsewhere in the body. To support this claim, he cited a host of discoveries made within a decade of the ophthalmoscope's invention, including retinal changes associated with diabetes mellitus, leukemia, syphilis, and diseases of the brain and kidneys.[45]

If the ophthalmoscope was Helmholtz's most important contribution to the medical community, it was certainly not the only one. He also did important work on accommodation—the process by which the eye adjusts its focus to maintain a clear picture of objects at different distances. Prior to his work in 1856 on accommodation, there existed several contradictory theories about the process of accommodation. Some claimed accommodation occurred through contraction of the pupil; others that it depended on changes in either the curvature of the cornea, the shape of the eyeball, the position of the lens, or the shape of the lens. To test these competing theories, Helmholtz invented an instrument—the ophthalmometer—that permitted him to measure changes in the curvature of the cornea and in the anterior and posterior surfaces of the lens. Most importantly, the ophthalmometer permitted these measurements in the living eye. Helmholtz found that changes in the shape of the lens alone account for accommodation, and he described how the anterior surface of the lens became more convex, the apex moved forward, and the axial portion grew thicker when the eye focused on objects nearby. When objects were far away, the reverse occurred. He postulated, moreover, that tension and relaxation of the ciliary muscles were primarily responsible for the changes in the shape of the lens.[46]

Helmholtz published his results in 1856 in von Graefe's *Archiv für Ophthalmologie*. Aware of the journal's medical readership, he closed

---

44. Wolfgang Münchow, *Geschichte der Augenheilkunde* (Stuttgart: Ferdinand Enke Verlag, 1984), 584; and Gorin, *History of Ophthalmology*, 132–39. The individuals most responsible for this work included Coccius, Frans Cornelis Donders, Albrecht von Graefe, Julius Jacobson, Eduard Jaeger, Hermann Jakob Knapp, Richard Liebreich, and Ruete.
45. Greeff, "Historisches zur Erfindung des Augenspiegels," 1201; and Münchow, *Geschichte der Augenheilkunde*, 584.
46. "Ueber die Accommodation des Auges," *AO 2* (1856):1–74, in *WA 2*:283–345. See also *Handbuch 1*:103–25; and "Die neueren Fortschritte in der Theorie des Sehens (1868)," in *VR³ 1*:233–331 (Helmholtz discusses the ophthalmometer on 243).

his article on accommodation with a discussion of the medical significance of his work. His theory, he noted, offered a reasonable explanation of the fact, well known to clinicians, that people with severe mydriasis (enlargement of the pupil) or iridectomies still retain some ability to accommodate. Helmholtz, who knew that greater knowledge of these diseased conditions would assist him in his work, appealed to ophthalmologists to carry out research "in the interests of the physiological theory of accommodation, and determine exactly the power of accommodation of these eyes."[47] Moreover, he encouraged several of his advanced students to pursue clinical studies in ophthalmometry. One in particular, Hermann Jakob Knapp, applied the ophthalmometer to studies of the curvature of the human cornea, confirming Helmholtz's claim that accommodation occurred through changes in the lens and not the cornea.[48] Other individuals, above all Franciscus Cornelis Donders, one of Europe's leading ophthalmologists and co-editor of Graefe's *Archiv*, soon used the ophthalmometer to study common eye disorders. Donders worked out exact determinations for nearsightedness, farsightedness, and for astigmatic conditions, in which irregularities in the curvature of the cornea create optical distortions.[49]

Helmholtz's work in physiological optics thus made an indisputable impact on medical practice. He appreciated this and cultivated his medical connections wherever possible. It was, for example, to the Königsberg Gesellschaft für wissenschaftliche Medizin, which had elected him president in 1850, that he first presented the ophthalmoscope. Shortly thereafter, while touring physiological institutes in Germany, he contacted members of the scientific and medical communities in order to demonstrate his instrument. The ophthalmologist Ruete, whom Helmholtz first met in Göttingen, was particularly pleased with what he saw. "For my trip," Helmholtz wrote his wife, "the ophthalmoscope has been splendid. I demonstrated it this morning and created a sensation here as well."[50]

Indeed, whenever Helmholtz's research led him to a discovery of potential medical significance, such as his work on accommodation,

---

47. "Ueber die Accommodation des Auges," in *WA* 2:345.
48. Hermann Jakob Knapp, "Über die Lage und Krümmung der Oberflächen der menschlichen Kristallinse und den Einfluss ihrer Veränderungen bei der Akkommodation auf die Dioptrik des Auges," *AO 6:2* (1860):1–52; 7:2 (1860):136–38. On Knapp, see *Biographisches Lexikon der hervorragenden Ärzte aller Zeiten und Völker*, 2nd ed., 5 vols. (Berlin and Vienna: Urban and Schwarzenberg, 1928–34), s.v. "Knapp, Hermann Jakob."
49. "Die neueren Fortschritte," *VR³* 1:243–44.
50. Helmholtz to his wife Olga, 6 August 1851, in Richard L. Kremer, ed., *Letters of Hermann von Helmholtz to His Wife 1847–1859* (Stuttgart: Franz Steiner Verlag, 1990), 51.

he advertised it to the medical community. In 1852, moreover, he simplified the ophthalmoscope's external fittings, and published his results in Karl Vierordt's *Archiv für physiologische Heilkunde*, a journal that reached a wide medical audience.[51] Yet nowhere was Helmholtz's interest in publicizing the medical significance of his work more obvious than in his *Handbuch der physiologischen Optik*, the first volume of which appeared in 1856. In 1854 he told his friend Adolph Fick, *Privatdozent* for physiology at the University of Zurich, that although he had decided against writing the *Handbuch* in a popular fashion for physicians, he did arrange it so that everything of medical significance was grouped together.[52] Thus, physicians, uninterested and usually unqualified to understand the mathematical parts of Helmholtz's text, could easily focus on the sections which included an encyclopedic presentation of the anatomy of the eye, its dioptric, its various imperfections resulting in nearsightedness, farsightedness, and astigmatism, as well as a description of the ophthalmoscope and ophthalmometer. Helmholtz's success in reaching the medical community led several ophthalmologists to hail the *Handbuch* as their bible, the foundation text for the newly emerging science of ophthalmology.[53]

Helmholtz's interest in encouraging close ties to the medical community derived from his appreciation of how physicians and physiologists could mutually benefit each other.[54] Concerning the contributions physiologists made to medicine, he could point to his own invention of the ophthalmoscope or to Ludwig's invention of the kymograph, which permitted the graphical representation of physiological processes, such as respiration and blood pressure changes.[55] But more important to physicians, experimental physiology exemplified the proper methodological approach to the study of organic processes—whether healthy or diseased. As one physician wrote in the *Mittheilungen des badenischen ärztlichen Vereins* in 1852, the student who learned the exact method in the physiological laboratory had "so cultivated his sense for the proper way of looking at things ... that the only task left would be to give instruction on how to direct his experience through the great labyrinth of pathology and therapy."[56]

51. "Ueber eine neue einfachste Form des Augenspiegels," *Vierordt's Archiv für Physiologische Heilkunde 11* (1852):827–52, in *WA* 2:261–79.
52. Koenigsberger *1*:265.
53. Noted in Greeff, "Historisches zur Erfindung des Augenspiegels," 1202.
54. Cf. Timothy Lenoir's essay, "The Eye as Mathematician: Clinical Practice, Instrumentation, and Helmholtz's Construction of an Empiricist Theory of Vision," in this volume.
55. On Ludwig's invention of the kymograph see Reiser, *Medicine and the Reign of Technology*, 100–1.
56. Anon., "Wie sollen die Aerzte gebildet werden?," *Mittheilungen des badischen ärztlichen Vereins 9* (12 May 1852):65–9, quote on 66.

Helmholtz shared this conviction. He believed that the method medical students learned in the laboratory provided them with the necessary mental outlook for making accurate prognoses. The physician, he explained,

> must strive to know in advance what the result of his intervention will be if he proceeds in one way or the other. In order to determine in advance what has not yet happened or what has not yet been observed to happen, there is no other method than to learn through observation the laws governing phenomena; and these can be learned through induction—through the careful search, production, and observation of those cases which fall under the law.[57]

The proof of these claims for Helmholtz rested in the improvements medicine had already accrued through the adoption of the experimental method, including advances in microscopy, pathological anatomy, and physiology. But he believed no branch of medicine demonstrated his point more clearly than ophthalmology. In a popular lecture delivered in 1869 entitled, "Ueber das Ziel und die Fortschritte der Naturwissenschaften" ("On the Aim and Progress of Natural Science"), he reminded his audience that correct knowledge of the structure and composition of the eye, acquired through the inductive method, had permitted the construction of corrective lenses, as well as the early diagnosis and treatment of diseases which had previously resulted in blindness. For Helmholtz, ophthalmology had become for the other branches of medicine "as brilliant an example of the capabilities of the true method as astronomy had long been for the other sciences."[58]

If the medical community benefited from physiology, the reverse was also true. For one, medical research occasionally provided valuable information on normal bodily functions, as in Charles Bell's experimental work of 1811 in which he demonstrated the functional specificity of the peripheral nerves of the brain. Bell had derived his idea from exact observations at the bedside, leading one clinician to predict that "it may easily happen that more light will come to the theory of the movement of the heart through pathological observations by means of the stethoscope than through all the physiological experiments done to date."[59] Moreover, physiologists occasionally received confirmation

---

57. "Das Denken in der Medizin," *VR³* 2:183.
58. "Ueber das Ziel und die Fortschritte der Naturwissenschaften," in *VR³* 1:333–63, on 362.
59. Carl Pfeufer, "Ueber den gegenwärtigen Zustand der Medizin. Rede gehalten bei dem Antritt des klinischen Lehramts in Zürich den 7 November 1840," *Annalen der*

of the validity of their theories from physicians engaged in clinical research, as in Knapp's validation of Helmholtz's theory of accommodation. Finally, physiologists turned to the medical community for assistance in their struggle for disciplinary autonomy. Several members of the 1847 group had appreciated the help physicians could offer them in their fight to win institutional support in the university medical faculties. Early in his career Ludwig, for example, initiated a steady correspondence with Henle, then professor of anatomy, physiology, and pathology at the University of Heidelberg. Henle, who directed his work toward the elimination of boundaries between physiology and pathology, had strong ties to the medical community. When Ludwig first contacted him in 1846, Henle was already co-editor, together with the clinician Karl Pfeufer, of the *Zeitschrift für rationelle Medizin*, a journal dedicated to convincing physicians of the necessity of founding medicine upon a physiological basis. Ludwig believed Henle was ideally positioned to help him persuade physicians of the relevance of experimental research for medicine. "It has become obvious to me," he wrote in his first letter to Henle, "that we calculators and experimentalists would not be able to live at all without a person like you."[60]

Like Ludwig, Helmholtz was also aware of the advantages that support from the medical community could offer the experimental physiologists. He admonished his friend du Bois-Reymond, whose research concentrated almost exclusively on nerve physiology, to keep this concern in mind. Indeed, he believed du Bois-Reymond's difficulties in the 1850s in landing an appointment as professor of physiology issued from his apparent indifference toward the medical community. "You can well imagine," he wrote his friend,

> that men of practice lash out here and there at your lecturing style in physiology. ... I will leave it to you to judge whether it is worth your while to further reduce the time you spend on animal electricity in physiology in order to acquire a better opinion from the [medical] faculty.[61]

Helmholtz's advice to du Bois-Reymond stemmed from his own positive experience. Six years had passed since he had invented the ophthalmoscope and earned a reputation in medical circles. The ophthalmoscope, he knew, had in many ways made his career. "For

---

städtischen allgemeinen Krankenhäuser in München 1 (1878):395–406, quote on 404–5.

60. Ludwig to Henle, 19 July 1846 and 27 March 1849, in Astrid Dreher, *Briefe von Carl Ludwig an Jacob Henle aus den Jahren 1846–1872* (Med. diss. Heidelberg, 1980), 43–5 and 67–72, respectively, quote on 44.

61. Helmholtz to du Bois-Reymond, 26 May 1857, in Kirsten, 171–73, on 172.

my place in the world," he later wrote, "the construction of the ophthalmoscope was very decisive. Among authorities and colleagues, I found, thereafter, such an appreciation of and benevolence toward my requests, that I was able to pursue the inner drive of my intellectual curiosity much more easily."[62]

Helmholtz's favorable reputation in medical circles permitted him to define his research projects as he wished, and even to move slowly away from teaching routine medical courses. This became evident in 1855 when he expressed interest in leaving Königsberg for a position as professor of anatomy and physiology in Bonn. He wanted to move for personal and professional reasons: he was concerned about the adverse effects of the cold northern Prussian winters on his wife's health and believed that Bonn would provide milder weather; he expected to have more influence on the scientific community from Bonn than from Königsberg; he wanted a slightly higher salary; and he wanted to redefine his teaching responsibilities so as to exclude pathology.[63] As he wrote in 1854 to Johannes Schulze, the Prussian minister of cultural affairs, he no longer felt fully capable of or interested in representing this subject:

> It has been my wish for a long time to be able to substitute general pathology, on which I lecture here, with anatomy, because the latter lies more within my interests than the former. I sense more and more by the questions and viewpoints that have recently arisen in pathology that the notions that I acquired during my earlier medical practice are no longer totally sufficient, and I must fear that this will grow worse every year.[64]

Helmholtz got his wish, and in 1855 he moved to Bonn. This move marked his first step away from the teaching of medical subjects proper. However, from the moment he arrived in Bonn he had problems with the research and teaching facilities. The anatomical institute, he wrote du Bois-Reymond, "is in a gruesome condition. Physiological instruments are few, and [Julius] Budge [the professor of physiology] has let these perish in dirt."[65] The situation was so bad that he had to set up

---

61. Helmholtz to du Bois-Reymond, 26 May 1857, in Kirsten, 171–73, on 172.
62. "Erinnerungen," VR^5 *1*:12–3.
63. See Helmholtz to du Bois-Reymond, 5 November 1854, and to Johannes Schulze, 3 and 19 December 1854, and Alexander von Humboldt to Johannes Schulze, 24 March 1855, and to Helmholtz, 24 March 1855, in Koenigsberger *1*:225–31, 248–51.
64. Helmholtz to Schulze, 19 December 1854, quoted in ibid., *1*:230–31.
65. Helmholtz to du Bois-Reymond, 14 October 1855, in Kirsten, 157. For a discussion of Helmholtz's problems at the University of Bonn see Tuchman, *Science, Medicine, and the State*, chap. 7.

a laboratory at home in order to conduct experiments. For three years he fought with the Bonn administration and the Prussian ministry to get funds to purchase instruments and to finance a new institute for anatomy and physiology. Several times he almost succeeded, yet each time something delayed funding appropriation. To make matters worse, he began to be blamed for the atrocious state of Bonn's institutional facilities.[66] Frustrated and disgusted, he decided to leave Prussia in 1858 for the University of Heidelberg in Baden. Heidelberg offered him not only a newly created chair in experimental physiology and a salary of 3,600 gulden (which was the highest salary paid at the university), but promised him a brand new physiological institute equipped with workspace for himself and students, and a substantial yearly endowment to pay for supplies and instruments for both his teaching and his research.[67] His appointment and the motives behind Heidelberg's generosity fell within the larger context of medical reform in Baden during the 1850s.

4. Medical Reform in Baden and the Search for an Experimental Physiologist

Baden's decision to invest heavily in experimental physiology was closely tied to broader interests in medical reform. In 1858, the very same year in which Helmholtz was hired, the government totally revamped its state medical examination and licensing requirements. The benefits which Helmholtz anticipated through involvement in the medical community continued to accrue. Yet experimental physiologists were not the only beneficiaries of this connection. Medical reformers perceived an advantage to themselves as well. In their attempt to turn medicine into a science, they stressed the importance of incorporating the exact method of the experimental sciences into medical education and research. They believed that by learning this method in the laboratory, students would acquire an analytical tool that they could later apply at the bedside. Thus the institutionalization of the experimental sciences at the University of Heidelberg (and at other

---

66. Helmholtz discusses the deplorable conditions in Bonn in a letter to Justus Olshausen, 16 April 1858, Darmstaedter Collection, File F 1 a 1847: Hermann von Helmholtz, Staatsbibliothek Preußischer Kulturbesitz, Berlin, Handschriftenabteilung.

67. The negotiations surrounding Helmholtz's appointment as professor of physiology in Heidelberg are discussed in Rep. 235/29872, Badisches Archiv. On his salary being the highest in the university see the report from the Ministerium des Innern to the Staatsministerium, 17 February 1858, ibid.

universities as well) reflected a broader-based conviction that training in the experimental method would ultimately translate into practical benefits.

Interest in medical reform in Baden dated back to the 1840s, when university-trained physicians throughout Germany united to fight for the standardization of stricter educational and licensing requirements. They sought to convince state governments to grant them a monopoly over health care; yet at the time they could hardly claim greater effectiveness than their competitors. By arguing for stricter educational requirements, physicians were promising to acquire those skills that would make them superior to their competitors; and they saw increased training in the natural sciences, particularly in the laboratory and clinic, as a means of acquiring that superiority.[68]

In 1848, during the height of the medical reform movement and one year after Helmholtz and his friends took their oath to transform physiology into an "organic physics," several members of the medical faculty at the University of Heidelberg attempted to convince the Baden government to create a new position for an experimental physiologist. The most ardent of these advocates was Henle, who, since his appointment in 1843, taught anatomy, physiology, general pathology, and pathological anatomy. He had agreed to teach these courses at a time when physiological research had involved little more than microscopical anatomy accompanied by occasional chemical tests and investigations.[69] But Helmholtz, du Bois-Reymond, Brücke, and Ludwig had begun to change physiology, and Henle, a man of the microscope, accepted and supported the growing divergence between their different methodological approaches to the study of organic processes. Between 1848 and 1852 Henle tried repeatedly to bring a young experimental physiologist to Heidelberg; Ludwig, Helmholtz, and du Bois-Reymond headed his list.

Henle's efforts received strong backing from the state health commission. Consisting primarily of physicians, the commission controlled the state medical examination and granted medical licenses. Moreover, through its advisory capacity it influenced the kind of education and training offered to medical students. In the early 1850s, it began simultaneously to push for reform of educational and licensing

---

68. See, for example, "Wie sollen die Aerzte gebildet werden?" For a general discussion of the medical reform movement of the 1840s see the references in note 17.

69. Karl Rothschuh, "Von der Histomorphologie zur Histophysiologie unter besonderer Berücksichtigung von Purkinjěs Arbeiten," in *J.E. Purkyně 1787–1869. Centenary Symposium, Prague, 8–10 September 1969*, ed. Vladislav Kruta (Brno: Universita Jana Evangelisty Purkyně, 1971), 197–212.

requirements and to speak out strongly in support of the new experimental sciences. For example, in response to the government's request that it recommend whether the university should hire a zoologist or an experimental physiologist, the commission declared:

> To stay in the spirit of the times it is necessary to focus on cultivating the exact method in true scientific research and to train students more in the method of examining and utilizing natural objects. This being so, he [a zoologist] would not teach the subject of physiology in an up-to-date fashion . . . , and another teacher, acquainted with the exact method and capable of heading a physiological institute with success, must be called in his place.[70]

Although the commission did not succeed in convincing the government in 1852 to hire a physiologist, the issue did resurface four years later. The director of the health commission again centered his argument on the different methodological approaches of anatomists and physiologists. "Physiology," he wrote

> can be taught only by someone who is familiar with the exact sciences. The anatomist, even if he is very highly educated in his subject, is still engaged in a descriptive science, and this does not qualify him as a teacher of physiology. In physiology things must be explained; description is not enough. And if the [anatomy] teacher would nevertheless insist upon [teaching physiology] he would do so in contradiction to the spirit of physiology, even hampering its development.[71]

As these quotations attest, physicians too believed that changes in physiology, in particular the emergence of an exact method of investigation, held great consequence for medicine. This conviction reflected diverse assumptions and hopes. Those interested in the nature of disease, such as Henle and Virchow, believed that disease represented little more than the body's normal response to abnormal stimuli; thus, knowledge gained about the chemical and physical laws underlying physiological processes brought with it a greater understanding of the disease process. For those practicing medicine, by contrast, an increase in theoretical knowledge meant less than the gains they hoped for and envisioned in diagnostics and therapeutics. Practitioners could already

---

70. Die Sanitäts-Commission to the Ministerium des Innern, 23 June 1852, Rep. 235/3133, Badisches Archiv.
71. Die Sanitäts-Commission to the Ministerium des Innern, 10 December 1856, Rep. 235/29872, Badisches Archiv.

point to such diagnostic instruments as the ophthalmoscope, kymograph, and spirometer as evidence of the advantages experimental physiology could bring to medicine.[72] Although similar progress had yet to occur in therapeutics, physicians were convinced that experimental physiology had much to offer. Direct benefits perhaps belonged to the future, but experimental physiology still introduced students to "the spirit of the times" and "the spirit of physiology." By conducting physiological experiments the student learned a critical method that, medical reformers believed, would later help in the analysis of disease. The laboratory offered an opportunity to learn, under simplified and controlled conditions, an exact method of investigation; armed with these skills, the young student could enter the clinic prepared to confront the more complex phenomena of disease.

In 1856, when Heidelberg began searching for an experimental physiologist who could teach physiology in "the proper way," it seriously considered all four members of the 1847 group. Yet there were clear favorites. Several members of the medical faculty, for example, expressed serious reservations concerning du Bois-Reymond, whom they criticized for his "excessive" specialization in nerve physiology. They were especially concerned that his teaching would be as specialized as his research—a serious criticism given their desire to attract medical students and to teach them the new methods of investigation. By contrast, the entire medical faculty favored Helmholtz, whom they thought the most capable of establishing a link between experimental physiology and medicine. Helmholtz, rather surprised at hearing this, wrote du Bois-Reymond that for some reason the Heidelberg medical faculty "does me the very questionable honor—I do not know why—of holding me to be less of a physicalist than the rest of our friends."[73]

Helmholtz's surprise is difficult to understand. Although his research had advanced the physicalist program as much as anyone else's, his invention of the ophthalmoscope and his work in physiological optics, as well as his five years of service as an army surgeon in Prussia, distinguished him from the other members of the 1847 group. No other member of the group would have been greeted as effusively as Helmholtz was when he finally accepted Heidelberg's offer. Shortly after he arrived in Heidelberg, a newly formed ophthalmological society honored him with a ceremony and a cup on which were inscribed (in

---

72. Although historians of medicine have argued convincingly that experimental physiology had little impact on therapeutics before the late nineteenth century, they have tended to overlook its impact on diagnostics.
73. Helmholtz to du Bois-Reymond, 5 March 1858, in Kirsten, 176–78, on 176.

German) the words: "To the Creator of Modern Science, the Benefactor of Mankind, in grateful remembrance of the Discovery of the Ophthalmoscope."[74]

That Helmholtz's appointment occurred squarely within a medical context is nowhere more evident than in the government's decision to institute new educational and licensing requirements in the very year that Helmholtz began teaching at Heidelberg. As already noted, local physicians' groups, backed by the state health commission, had been trying for almost a decade to convince the government to reform educational and licensing requirements so as to exert greater control over medical practice in Baden. The regulations in effect at mid-century had been issued in 1803 and had received only minor modifications since then. According to these regulations students had absolute freedom to study what they wanted (*Lernfreiheit*), meaning that a set curriculum could not be established. Moreover, to qualify for the state medical examination the student had only to fulfill the ambiguous requirement of having acquired "a thorough knowledge of the natural sciences and medicine."[75] Shortly after Helmholtz's appointment, the Baden government replaced the lax curricular structure and the vague state examination procedure with specific detail and an exact system of requirements. In many ways the new curriculum moved in the direction of the one Helmholtz had followed at the Institut. Where *Lernfreiheit* had formerly ruled, students were now required to attend the university for at least four years before qualifying for the state examination. Moreover, the curriculum received more structure. During the first two years students had to attend general courses in the natural sciences, take two courses in dissection, and spend one semester each in the physiology and chemistry laboratories. In the latter two years they advanced to "medical courses" proper, including one year of practical work in the medical, surgical, and obstetrical clinics.[76]

The institutionalization of experimental physiology and the hiring of Helmholtz at the University of Heidelberg were thus important elements in the reform of medical education in Baden, and cannot be understood divorced from that context. Helmholtz's position was in

---

74. Quoted in Koenigsberger *1*:314.
75. *Großherzoglich Badisches Staats- und Regierungsblatt*, 5 August 1828 (Karlsruhe: Macklot, 1828). The laws and statutes dictating medical educational and licensing requirements until 1859 are collected in C.A. Diez, *Zusammenstellung der gegenwärtig geltenden Gesetze, Verordnungen, Instructionen und Entscheidungen über das Medicinalwesen und die Stellung und die Verrichtungen der Medicinalbeamten und Sanitätsdiener im Großherzogthum Baden* (Karlsruhe: A. Bielefeld, 1859).
76. *Mittheilungen des badischen ärztlichen Vereins 3* (10 February 1848):17–20.

the medical faculty; his appointment coincided with a total restructuring of medical education; and one semester in the physiology laboratory became a requirement for medical students. In the "spirit" of scientific medicine, students were introduced during laboratory instruction to common microscopical and chemical techniques, such as the preparation of bone sections and injections, and the analysis of urine and creatine. Moreover, they conducted physiological experiments proper, inducing the contraction of frog legs, cutting superficial nerves, and simulating digestion, for example. Two to three hours of instruction were offered each week, and the laboratory, under the watchful eye of a supervisor, was open every morning from eight to twelve o'clock in order to provide students with the opportunity to perfect their techniques or to do their own experiments.[77]

The supervisor was not, however, Helmholtz, but one of his assistants (e.g., Wilhelm Wundt). Indeed, Helmholtz soon expressed skepticism about the wisdom of this laboratory requirement. Shortly after his arrival in Heidelberg he characterized (to du Bois-Reymond) the "legal regulation which turns the physiology course into a required course for Baden students" as an "exaggeration of enlightened principles." He feared especially that "it could become very burdensome" for him.[78] He soon solved this problem, however: he made sure he had little to do with this laboratory instruction. Unlike Ludwig, who involved himself directly in his students' work, and whose interest and excellence in teaching techniques and methods attracted individuals from all over the world, Helmholtz remained aloof from the routine drills conducted in his laboratory. These became the sole responsibility of his assistants. Moreover, he further distanced himself from his "medical" duties by requiring his assistant to teach his courses in microscopical anatomy, justifying this by his lack of histological knowledge and his tendency to get headaches from looking through a microscope.[79] In the years Helmholtz spent at Heidelberg, he lectured only on subjects he enjoyed, alternating between a general human physiology course one semester and a more specialized course on the physiology of the sense organs the following semester.[80] Much

---

77. Wolfgang G. Bringmann, Gottfried Bringmann, David Cottrell, "Helmholtz und Wundt an der Heidelberger Universität 1858–1871," *Heidelberger Jahrbücher 20* (1976):79–88, on 81–5; and Wilhelm Wundt, *Erlebtes und Erkanntes* (Stuttgart: A. Kröner, 1921), 154.
78. Helmholtz to du Bois-Reymond, 29 October 1858, in Kirsten, 193–94, on 193.
79. Bringmann, et al., "Helmholtz und Wundt," 79–88.
80. *Anzeige der Vorlesungen auf der Großherzoglich Badischen Ruprecht Karolinischen Universität zu Heidelberg* (Heidelberg: n.p., 1858–71).

had changed since his days in Königsberg, where he had been responsible for anatomy, pathological anatomy, and pathology, in addition to physiology. By the 1850s and 1860s, experimental physiology had gained disciplinary autonomy, permitting Helmholtz (and others) to focus their time and energy on their specialized areas of interest. A paradoxical situation thus arose: the institutionalization of experimental physiology had occurred because of its perceived medical significance, yet this same institutional support granted Helmholtz more license than ever to pursue his own interests, be they medical or otherwise. Baden had hired him and built him a brand new laboratory as part of its reform of the medical curriculum, yet for the first time in his career Helmholtz did not have to teach routine medical courses.

Helmholtz did not, however, cut all ties with medical students. On the contrary, he welcomed advanced students into his laboratory who were interested in clinical work in ophthalmology and ophthalmometry. Emanuel Mandelstamm, M. Woinow, and Knapp, for example, all came to Heidelberg after studying with von Graefe in Berlin in order to learn the theoretical underpinnings of their specialty. Knapp even stayed on to write his *Habilitation* on the curvature of the human cornea and to teach ophthalmology at the university.[81]

Nevertheless, in the thirteen years Helmholtz spent at Heidelberg, his work gradually took him out of a medical context. Despite his pursuit of physiological optics and acoustics until 1863, his research increasingly turned to questions of epistemology, aesthetics, and ultimately to mathematics and physics. In 1871 he finally left Heidelberg to accept a position as professor of experimental physics in the philosophical faculty at the University of Berlin.

5. Epilogue

During the last twenty-three years of his life, Helmholtz focused his teaching and research on physics, the subject that had intrigued him since his childhood. Much of his time in Berlin was spent, moreover, in planning and then presiding over the Physikalisch-Technische Reischsanstalt, which opened its doors in 1887.[82] These new activities may have weakened Helmholtz's links to the medical community, but

---

81. *Biographisches Lexikon*, s.v. "Mandelstamm, Emanuel"; "Woinow, M."; and "Knapp, Hermann Jakob."
82. David Cahan, *An Institute for an Empire: The Physikalisch-Technische Reichsanstalt 1871–1918* (Cambridge: Cambridge University Press, 1989), chaps. 1–3.

his old ties were never entirely broken. He continued to receive honors for the work he had done in ophthalmology until well into the 1880s;[83] and in 1877 he accepted the professorship in physics at the Institut. Asked to give the keynote address that year to the incoming class, he spoke about his own medical education, recent changes in medicine, and the direction he believed (and hoped) future medical education and research would take. Although the occasion and setting doubtless encouraged some rhetoric on Helmholtz's part, he nonetheless sincerely claimed that at a time when education had been based largely on book learning, his medical studies had "taught me more forcibly and more convincingly than any other training could have done the eternal principles of all scientific work."[84] These principles included independent observations, the construction of theories based on such observations, and the testing of these theories in practice. Physicians, he concluded, must for this reason always

> play a prominent role in the work of true enlightenment, for among those who must continually and actively verify their knowledge by testing it against nature, physicians begin with the best mental preparation.[85]

Helmholtz himself chose a career that did not require him constantly to test his theories "in practice," yet this may be one reason he appreciated the importance throughout his years as a physiologist of maintaining a link with the medical community. Medicine, Helmholtz once wrote, was little more than the practical side of physiology.[86] Here he echoed the views of medical reformers who believed medicine would become a science when practitioners began applying at the bedside both the knowledge and the methods they had learned in physiological laboratories. For these reformers, Helmholtz had been one of the outstanding symbols of the new scientific medicine. As we have seen, this image—and it was one Helmholtz had helped popularize—had assisted Helmholtz in achieving his career goals, being largely responsible for his appointment to the chair of physiology at the University of Heidelberg. Yet the advantages he accrued from his connection to the medical community extended well beyond these professional gains. Through the application of his theories and instruments in a clinical setting, Helmholtz also received evidence not only of the

---

83. In 1885, e.g., he received the Graefe-Medaille from the Ophthalmologische Gesellschaft. See "Antwortrede gehalten beim Empfang der Graefe-Medaille zu Heidelberg 1886," VR^5 2:311–20; and Koenigsberger 2:337–41.
84. "Das Denken in der Medicin," VR^3 2:169.
85. Ibid., 190.
86. "Ueber das Ziel und die Fortschritte der Naturwissenschaften," VR^3 1:361.

validity but also of the usefulness of the knowledge he produced. That this pleased him is nowhere more evident than in the popular speech he gave in Heidelberg in 1862 on the occasion of his election to the office of university rector. As he told his audience:

> Knowledge itself is not the object of people on earth. . . . Only action gives a person a dignified life; therefore his goal must be either the practical application of his knowledge or the increase in knowledge itself.[87]

Helmholtz may have dedicated his life to the latter, but his contributions to the former were significant as well. Indeed, without them it is impossible to understand his early career.

---

87. "Ueber das Verhältniss der Naturwissenschaften zur Gesammtheit der Wissenschaft," in VR^3 1:117–45, quote on 140.

2

Experiment, Quantification, and Discovery

Helmholtz's Early Physiological Researches, 1843–50

Kathryn M. Olesko
Frederic L. Holmes

1. Introduction

Before Hermann Helmholtz took up the work for which he became best known in sensory physiology, he published four investigations in general physiology between 1843 and 1850. The first, on putrefaction and fermentation, was well conceived but inconclusive. The next three—those on muscle and nerve physiology (1845, 1848, and 1850)—became classics. Each defined a problem of fundamental importance. Through an elegant experimental design often requiring precise quantitative results, each resolved issues posed simply yet decisively. Moreover, each germinated a thriving subfield of investigation within the burgeoning field of mid-nineteenth-century experimental physiology.

Historians have examined Helmholtz's experiments of 1843, 1845, and 1848, addressing them primarily within the context of broader theoretical issues, especially theories of animal heat, Helmholtz's formulation of the principle of conservation of force, and his opposition to conceptions of vital force.[1] Although he pondered these and other conceptual issues as he worked on these experiments, his investigative pathway was also shaped by an increasingly penetrating understanding of the limitations posed by his research methodology's techniques.

---

1. The most recent studies are Timothy Lenoir, *The Strategy of Life: Teleology and Mechanics in Nineteenth-Century German Biology* (Dordrecht: D. Reidel, 1982), 197–215, 231–35; and Richard Kremer, *The Thermodynamics of Life and Experimental Physiology, 1770–1880*, Harvard Dissertations in the History of Science, ed. Owen Gingerich (New York and London: Garland Press, 1990), 237–55, 275–307.

Even though in each of these cases he either pushed existing methods to their limits or surpassed current practice so as to redefine a problem at a higher level, he was also acutely aware that his results sometimes only established the boundaries within which exact solutions resided. This awareness was especially prominent in his adaptation to physiology of precision-measuring techniques, an area in which he quickly became a leading craftsman and innovator.

Historians have also examined Helmholtz's deployment of precision techniques, especially his modification of instruments; yet several issues remain unexamined.[2] These include his use of various forms of quantification associated with precision measurement and their interaction over time, and how his methods evolved from his earliest investigations up through his painstaking and delicate measurements of the propagation velocity of the nerve impulse in 1849 and 1850. The historically significant role of precision measurement in mid-nineteenth-century German science itself justifies a detailed study of Helmholtz's early investigative techniques; for Helmholtz, like many other investigators, recognized the power of precision. An examination of his techniques is also important for gaining a perspective on the crucial transformation in German science during the second half of the century to non-precision forms of experimentation; for Helmholtz, in the course of his nerve investigations, did not always achieve the precision he believed he needed, nor did he necessarily believe that quantitative exactitude was always essential or even appropriate for convincing others of his findings. His ability to extract from simpler techniques what he considered to be adequately certain results helped to diminish reliance on certain forms of precision measurement in the construction of scientific knowledge. As one of his rare and hitherto unstudied laboratory notebooks reveals, his passage to simpler forms of error and data analysis ironically began during the course of an investigation whose results were made possible only by the most complex techniques of contemporary precision measurement, that of measuring the velocity of the nerve impulse.[3]

---

2. Here, too, the most recent studies are by Timothy Lenoir, "Models and Instruments in the Development of Electrophysiology," *HSPS* 17 (1986):1–54; and Kremer, *Thermodynamics of Life*, esp. 275–307. Kremer's solid study, with its comprehensive view of Helmholtz's early physiological studies, especially of his use of instruments, greatly aided us in the preparation of the present essay. One way in which this essay differs from Kremer's is in its analysis of Helmholtz's treatment of data.

3. AW, No. 547: Untersuchungen über Muskeln und Nerven. We treat this laboratory notebook in detail in our book on Helmholtz's early scientific career (in preparation). There may be other extant laboratory notebooks written either by Helmholtz or his students; see, e.g., AW, No. 544: Versuche über Muskelton (Mai 1866-Februar 1867).

This essay examines Helmholtz's early physiological experiments and focuses especially on the techniques that dominated his early investigations: the instrumental and quantitative techniques of exact experiment. Precision measurement became an issue for Helmholtz as early as his third scientific publication, that on muscle physiology, when the physiological questions that he posed began to exceed the technical means at his disposal. As Section 2 of this essay shows, through his investigations from 1843 to 1845 Helmholtz came to see that he could conduct deeper, more rigorous, and more quantitative physiological investigations than others before him, thereby helping to transform the nature of experiment in physiology. Moreover, by criticizing previous studies in physiological heat, Helmholtz sharpened his own understanding of the conditions under which data is produced and the nature and quality of data, as Section 3 demonstrates through its analysis of Helmholtz's 1845 review article on these studies.

Section 4 of this essay then turns to an analysis of Helmholtz's route to his 1848 paper on the formation of heat in muscles. It shows how Helmholtz adopted and adapted a series of instruments and apparatus—including a multiplicator, thermoelectric circuits, and an induction coil—and how he learned to adjust his experimental set-up to meet a series of experimental goals, including the source of animal heat. Having just promulgated in 1847 his principle of force conservation and working within a broad physicalist program, Helmholtz further adapted, as Section 5 shows, instruments and measuring techniques, specifically Eduard Weber's muscle-contraction apparatus and Carl Ludwig's apparatus for graphically recording physiological effects, that in 1850 allowed him to measure muscle-contraction times. Section 6 concerns Helmholtz's preliminary results of 1849–50 on measuring the propagation velocity of the nerve impulse. It especially draws attention to the ways in which Helmholtz tried to convince others of the certainty and demonstrability of his findings. Section 7 then analyzes his use of quantitative error analysis (in particular, the method of least squares) in his discovery of the propagation of the nerve impulse. Finally, Section 8 briefly reflects on the larger meaning of that discovery, both for Helmholtz's own emergence as a master of instrument analysis and design and for the stimulation it brought to several scientific fields.

2. Helmholtz's First Quantitative Muscle Experiments

Before precision measurement became an issue for Helmholtz in his work on muscle physiology, he had already mastered the principles of anatomical observation and the techniques of chemical experimentation in the Berlin laboratories of Johannes Müller, Eilhard Mitscherlich, and Gustav Magnus. Under Müller's guidance in particular, Helmholtz had displayed considerable technical skill in the delicate dissection of small invertebrate tissues and in microscopic observation.[4] Although historians have recently made much of the philosophical differences between Müller and his students,[5] there is no evidence that any of them quarrelled with Müller or that he lessened his high regard for them. Müller and his students shared a common commitment to rigorous standards of investigation, employing the most effective methods available from anatomy, experimental physiology, physics, and chemistry. Müller's example reportedly inspired Helmholtz to decide, even before he completed his medical degree (November 1842), to pursue methodologically rigorous investigations whose results he could bring to bear on the broad questions about vital processes that stirred him.[6] After graduation, Helmholtz took up the problem of "the basis of the so-called spontaneous decomposition processes [fermentation and putrefaction] of organic substances deprived of life," a question over which he thought that "among chemists and physiologists highly contradictory views have prevailed."[7] The new problem also required him to acquire additional scientific skills. To the microscope he now added chemical methods of investigation, probably learned under Mitscherlich.

Helmholtz's definition of the problem of investigating fermentation and putrefaction was neither original in its definition nor its methods.[8] His aim was not to discover novelty but rather to exert an experimental critique capable of resolving a disputed issue. By adopting and refining

---

4. See, for example, Helmholtz's 1842 doctoral dissertation under Müller, *De Fabrica Systematis Nervosi Evertebratorum*, Inaugural-Dissertation (Berlin: Typis Nietackianis, 1842), in *WA* 2:663–79.
5. Müller's students have recently been described as in "rebellion" against him; see Lenoir, *Strategy of Life*, 195; cf. Kremer, *Thermodynamics of Life*, 305n.
6. Koenigsberger *1*:51–2.
7. "Ueber das Wesen der Fäulniss und Gährung," *MA* (1843):453–62, on 453, in *WA* 2:726–34.
8. Cf. the work of his predecessors, including: Theodor Schwann, "Vorläufige Mittheilung, betreffend Versuch über die Weingährung und Fäulniss," *AP 11* (1837):184–93; Joseph Gay-Lussac, "Extrait d'un mémoire sur la fermentation," *Annales de Chimie 76* (1810):245–59; Justus Liebig, "Ueber die Erscheinungen der Gährung, Fäulniss und Verwesung und ihre Ursachen," *Annalen der Pharmacie 30* (1839):250–87, 363–68. For a comprehensive treatment of the history of fermentation, see Joseph Fruton, *Molecules and Life: Historical Essays on the Interplay of Chemistry and Biology* (New York: Wiley-Interscience, 1972), 22–86. Helmholtz's later experiments on the matter are outlined in AW, No. 666: Versuch über Gährung bei Magnus, which describes experiments performed in Gustav Magnus's Berlin laboratory during the winter of 1845–46.

methods that others had already devised, he attained results similar to theirs but which closed loopholes and reduced uncertainties. In principle, Helmholtz's experiments appear to have provided sufficient evidence to show conclusively the dependence of both fermentation and putrefaction upon microorganisms. In fact, the issue remained disputed, and Justus Liebig's chemical decomposition theory retained an influence for nearly two more decades until overtaken by Louis Pasteur's dramatic refutations.[9]

Although the historical effect of Helmholtz's published work on fermentation and putrefaction was limited, it was, for the evolution of his own investigative powers, a major step. Undertaking for the first time to solve a central contemporary problem through experimental means rather than observation alone, Helmholtz improved methods that his able predecessors had introduced. He showed signs of a capacity to master experimental techniques from another field; to adapt and combine them to center in on the key points of a disputed question; to maintain controls; and to arrive at results more definitive than those of his predecessors. Nonetheless, his reported results did not sufficiently emphasize the power of his methods. Still, the investigation of putrefaction and fermentation provided an entrée into more controlled observation, in the form of experiment, than he had previously practiced.

Not until late 1843 did Helmholtz begin to develop more exacting and quantitative procedures of investigation. While working as an army physician in Potsdam, he embarked upon experiments aimed at establishing whether or not the chemical composition of a muscle changes when it is stimulated to prolonged activity.[10] At the opening of his 1845 paper on the subject, he presented his reasons for having taken up this investigation in 1843:

> One of the highest questions in physiology, touching directly upon the nature of the vital force—namely, whether the life of the organism is the effect of its own self-regenerating, purposefully acting force, or the result of forces that are active also in lifeless nature, only specially modified by the nature of their combined actions—has in recent times taken on, with special clarity in Liebig's attempt to derive physiological facts from known chemical and physical laws, a much more concrete form; that is, whether or not the mechanical force and the heat created in the organism can be derived completely from the *Stoffwechsel*.[11]

---

9. Fruton, *Molecules and Life*, 49–63.
10. Koenigsberger *1*:54–5.
11. "Ueber den Stoffverbrauch bei der Muskelaction," *MA* (1845):72–83, on 72, in *WA* 2:735–44. On the long-range implications of this investigation see Frederic L.

Helmholtz hoped to answer experimentally not the entire question that he had posed, but rather the more limited one of whether it could be shown that "in the production of mechanical effects . . . matter is consumed." Although physiologists had long assumed such a consumption from the common experience of fatigue after exertion and gradual recovery, "scarcely any ideas had been established about the nature of the matter consumed and the location of the transformations." Moreover, Helmholtz noted that "there is still lacking a knowledge of all of the beginning and intermediate links of the process and the place in which they are formed, and since inference based on the end products found in the excretions must always remain problematic, I decided to try an entirely direct way to investigate them."[12] He adopted a strategy of stimulating electrically one pair of isolated frog muscles to contract repeatedly until it was exhausted, and then comparing its chemical composition with that of the non-stimulated muscle. Since galvanic currents themselves produced chemical changes that might be mistaken for those he sought to detect, Helmholtz reverted to the eighteenth-century method of charging a Leyden jar from an electrostatic generator and stimulating the frog muscle by means of its rapid intermittent discharges.[13]

Far more difficult technically was the problem of identifying the chemical changes that might have occurred through the activity of the stimulated muscle. Jöns Berzelius, who pioneered quantitative work on this problem, was acutely aware of the inadequacies of such analyses. In January 1830, as he completed the work necessary to write the section on animal chemistry for his textbook, he wrote to Mitscherlich:

> I have been occupied this winter with little else than the chemistry of animals, which is not the most pleasant [of subjects], because in this area, no matter how much interest one may take in it, the uncertainty of the results and the impossibility of controlling them sufficiently is always unsatisfying; but to know something with certainty is the greatest satisfaction that an investigation can bring to one.[14]

The situation had changed little when Helmholtz began his own investigation thirteen years later. Relying on methods similar to those

---

Holmes, *Between Biology and Medicine: The Formation of Intermediary Metabolism* (Berkeley, Calif.: Office for History of Science and Technology, 1992).
12. "Stoffverbrauch," 73.
13. Ibid., 74.
14. J.J. Berzelius to E. Mitscherlich, 19 January 1830, Mitscherlich Nachlaß 165, Deutsches Museum, Sondersammlungen, Munich.

that Berzelius had used to extract and identify substances from animal matter, Helmholtz found that the discriminating quantitative results he needed were not easy to achieve. Even if he was already familiar with the methods of animal chemistry he must have spent a considerable part of the two years he devoted to his investigation of muscle contraction gaining sufficient proficiency with these methods to deploy them on a problem that made unprecedented demands on them for quantitative discrimination.

Helmholtz considered it "a priori" probable that "the muscle fiber [identifiable with Berzelius's "solid part" of the muscle] takes part in the decomposition," "because we generally find protein compounds as carriers of the highest vital energies," and because the increased quantities of phosphates and sulfates in the urine found after muscular exertion might well have derived from the decomposition of the protein.[15] Although Helmholtz placed his discussion of muscle fiber near the end of his published paper, his initial expectation suggests that he may well have tried early on in the investigation to find out if it is decomposed. "A direct decision through experimentation has not been possible until now," he reported,

> because the error, which arises from the uncontrollable greater or lesser filling [of the muscle] with blood and the greater or lesser absorption of moisture, makes the relative proportion of the solid part not accurately enough comparable. The observed weighings varied, so that sometimes one side, sometimes the other side was around one-fourth to one-half percent greater, and the possible decomposition of the muscle fiber might not exceed this amount.[16]

That he sought to detect a change expected to be a fraction of one percent of the quantity of muscle fiber present is itself a revealing indication of the extent to which Helmholtz was pressing the limits of the available analytical means; for normally, errors in similar experiments could be ten to one hundred times higher.

"Among the soluble constituents," Helmholtz wrote, "the albumin was investigated first." He found, however, that the uncontrollable variations, including the range in the amount of albumin, so exceeded the average percentage difference as to prohibit drawing decisive conclusions from his results. His dismissal of these results, which had been difficult to achieve, seems to have been based merely on an *intuitive* sense that they were indecisive. Yet from previous analyses of

---

15. "Stoffverbrauch," 82.
16. Ibid.

Berzelius and others, Helmholtz knew that there remained in the solution from which the albumin had been separated the substances known collectively as the "extractive matter" of the muscle. Following approximately Berzelius's procedures, he separated them into "aequous," "spiritous," and "alcoholic" extracts. When he dried and weighed these various extracts, he again encountered irregularities that interfered with an accurate determination of their absolute weights; but in this case the difficulty did not prevent him from reaching significant comparative results. He wrote:

> If one takes care that all of these operations are carried out in exactly the same manner and under exactly the same conditions with both portions of muscle, one obtains correct figures for the relative proportions, even with a less-careful determination of the absolute quantities. To determine the latter involves great difficulties, because one does not always succeed in washing the filtrate completely out of the organic matter, which is not easily filterable. I have, however, convinced myself through special experiments that the quantities remaining behind are too small to have an influence on the result.
>
> For these extracts the result turns out to be, in all experiments without exception, that the aequous extract is diminished in the electrified portion of muscle and, inversely, the spiritous and alcohol extract are increased in comparison to the non-electrified portion.[17]

Helmholtz summarized the results of nine "more accurate trials" (Figure 2.1). "The result stated above emerges clearly in these figures," he asserted,

> even though the ratios $a:b$ and $c:d$ in the second table still vary greatly, which may derive partly from the greater or lesser intensity of the contractions that may be induced in the unelectrified muscle through its preparation, air, or warm water. It should be noticed that the average difference in the aequous extracts of 0.3 corresponds rather well to the spiritous extract of 0.24.[18]

In his determination and analysis of the extracts Helmholtz thus realized that obtaining meaningful results depended on what was then known as the method of repetition: maintaining "exactly the same conditions" in one trial after another. Although his quantitative assessment of experimental errors was not as sharp as those routinely performed in physics, it seems to have been primarily the nature of

17. Ibid., 76–7.
18. Ibid., 78.

Nummer des Versuchs.	Alkoholextrakt auf 100 Theile des frischen Fleisches.		
	a) im elektrisirten Fleische.	b) im nicht elektrisirten Fleische.	Verhältniss a:b
I	0,752	0,606	1,24
II	0,569	0,427	1,33
III	0,664	0,481	1,38
IV	0,652	0,493	1,32
V	0,575	0,433	1,33
Auszug mit 95 procent. Alkohol.			
VI	1,020	0,748	1,36

Nummes des Versuchs.	Wasserextrakt			Spiritusextrakt		
	a) im elektrisirten Fleische.	b) im nichtelektrisirten Fleische.	Verhältniss a:b	c) im elektrisirten Fleische.	d) im nichtelektrisirten Fleische.	Verhältniss c:d
VII	1,21	1,63	0,79	1,69	1,50	1,13
VIII	0,93	1,23	0,76	1,65	1,35	1,22
IX	0,72	0,90	0,80	1,76	1,53	1,15
Mittel	0,95	1,25	0,78	1,70	1,46	1,16

Figure 2.1. Helmholtz's data for alcohol, water, and spiritous extracts from a muscle. Source: "Ueber den Stoffverbrauch bei der Muskelaktion," MA (1845):72–83, on 78. Courtesy of Yale University Photographic Services.

the problem confronting him rather than his developing skills that was most responsible for shaping his experimental style. Evaluating the extracts demanded craftsmanlike control over the experiment's material conditions. Hence he could assign to variations in his numerical results only probable causes that he could not then quantify or even estimate. Still, given the daunting complexity of what Helmholtz was attempting, his was a stunningly successful outcome. He himself acknowledged: "Scarcely hoping for any positive result at the beginning,

I was all the more surprised when even the first imperfectly executed trial experiments yielded striking results, which were completely confirmed by careful repetitions."[19] And he concluded his paper with the following claim:

> I believe that I have, through the facts set forth, delivered the promised proof that during the action of muscles a chemical transformation in the compounds contained in them takes place. The knowledge gained stands, to be sure, isolated and without inner connections, but I have limited myself here to its presentation because my wider investigations on this point, from which a deeper understanding of the process might emerge, appear to me still capable of a more accurate foundation and more particular execution, and that to that end a more accurate investigation of the extractive matters is necessary. For that reason I will postpone their publication.[20]

But what had Helmholtz really discovered? He had certainly not found that the mechanical force and the heat created in the organism derived from its *Stoffwechsel*. The importance of this investigation lay rather in a subtler issue. In the course of his experiments on muscle physiology he had come to realize that it was possible to investigate these deeply hidden phenomena in a manner more rigorous and quantifiable than he or others had heretofore utilized. Most important, in these investigations he began to discriminate results partly on the basis of raw estimations of error or even on the qualitative, but probable, identification of their sources. Hence, concurrent with the introduction of quantitative analyses, Helmholtz's conception of physiological experimentation, to the extent that it began to include a discussion of experimental errors, moved further away from the observational ideal of anatomy and toward the exact forms of experiment already more common in chemistry and, especially, physics. Yet although he had achieved the kind of control that Berzelius had earlier sought, he could not yet claim the certainty of results that he (and others engaged in similar investigations) desired.

3. Helmholtz as Critic of Studies on Physiological Heat

In the fall of 1845, before leaving Potsdam for Berlin, Helmholtz had completed writing a review article on "Physiological Heat" for an

---

19. Ibid., 73.
20. Ibid., 83.

encyclopedia of the medical sciences issued by members of the Berlin medical faculty. It was a critical, searching review of the subject of physiological heat, grounded in a penetrating discussion of the physical nature of heat itself. It included the most lucid treatment of the problem of "the source of animal heat" available in the current physiological literature.[21] Of all topics in contemporary physiology, animal heat was preeminently suited for—in fact, peremptorily demanded—investigation by precise quantitative measurements. The temperature of warm-blooded animals, Helmholtz noted, is "subject to only small deviations of about 1° C." Since thermometers accurate to 0.1 or 0.01° C were available to measure temperatures within animals, physiologists could, in principle, detect variations in internal body temperature due to such factors as time of day, rest and movement, and the effects of external heat or cold; moreover, they could establish temperature differences in different bodily organs and in different animal species. Helmholtz reviewed critically the measurements that had been made over the previous decades, and found most of them wanting. Part of their weakness derived from an inadequate consideration of the conditions under which they had been taken. "Concerning the temperature of the internal organs," he wrote, "we have a few series of investigations, but most of these are uncertain because of unavoidable observational errors and the neglect of temporal variations." Yet a more problematic weakness was the thinness of and variations within the data. "For the exact determination of the mean temperature of various animal species," Helmholtz wrote, "this variation can only be eliminated by a series of observations which must be taken with precautionary measures." By "precautionary measures" he meant generally maintaining control over the conditions under which data was taken. He also seemed to want to create data that would obviate the problems and uncertainties that arose when data sets taken by different observers were combined, for he noted that "until now we have . . . only several isolated observations taken in different ways and, so it appears, also with very different thermometers, so that the uncertainty is often as high as 2–3°."[22]

Such results were hardly surprising in investigations on animals and their parts which could not be rigorously controlled. Yet Helmholtz seemed somehow to want to surmount disturbing conditions afflicting the data. Thus he believed that the solution could be found, as he had

---

21. "Wärme, physiologisch," *Encyklopädisches Handwörterbuch der medicinischen Wissenschaften* (Berlin: Berlin Medicinische Fakultät, 1845), 523–67, in *WA* 2:680–725.
22. Ibid., 681–85, on 681.

shown in his work on muscle physiology, partly in controlling experimental conditions; data taken under conditions where "precautionary measures" were not observed were "to be considered unreliable."[23] Yet the review essay, in contrast to his earlier writings, makes evident his growing sense of what properties a good set of data possessed, including its density—he considered inadequate the isolated cases then available—and the conditions under which data taken by different individuals could be combined. These were difficult issues to consider for two reasons. First, the data could not be improved entirely by instrumental means, for thermometric calibration, especially the proper interpolation of scales, was a pressing concern in physics and would remain so for decades to come. Second, methods for combining observations had diffused only slowly from astronomy, where they were commonly used, to other experimental fields.

Although Helmholtz paid particular attention to the measurements that Antoine-César Becquerel and Gilbert Breschet had made by thermoelectric means rather than with thermometers, his concern was not instrumentation per se. He did note that one would expect "thermoelectrical measurements in living animals to yield greater accuracy, yet we still find in [their] results very significant variations"; still, what he identified as the probable cause of the variation, "unrest in the animals," shows that he was interested primarily in understanding the nature of the data produced. He thus examined three sets of measurements that Becquerel and Breschet had made in the thigh muscle, thorax, abdomen, and skull of a dog. He observed that the differences between the temperatures measured in these four locations were too small in comparison to the differences between individual measurements in the same location "to be able to prove anything at all." Reviewing the various comparisons that other investigators had made of the temperature in the left and right ventricles of the heart, Helmholtz concluded similarly that "it is not possible to judge with certainty if these values fall within the limits of error of such trials and in how far their results about the higher temperature of the arterial blood are correct."[24]

His review of the question of the source of animal heat also required him "to evaluate theoretical views on the nature of this heat" by coming to an understanding of the physical nature of heat[25] and of the role of heat in vital processes. "It is clear," he wrote,

23. Ibid., 692–93.
24. Ibid.
25. Ibid., 696.

how great the influence of the question of the origin of animal heat is for theoretical views of vital processes, since it is in this form that by far the greatest part of the *Kraftequivalent* appears that lies in the chemical *Kraft* of the ingested matter, and therefore how important the empirical resolution of this question is. . . . Unfortunately the question has not yet been fully clarified, because observers have for the most part hitherto conceived the question one-sidedly as whether the animal heat derives from the combustions in the respiratory organs.[26]

Helmholtz considered all the physical and chemical processes that might contribute to an increase in the temperature, eliminating such processes as the decrease in the specific heat of a gas when it is compressed because it "does not occur in the body."

The only "direct experimental" tests of "the question of the chemical origin of animal heat" had been carried out independently by Pierre Dulong and César Despretz in Paris during the 1820s. Helmholtz subjected their results and the assumptions underlying them, as well as the results of others, to a more deeply probing critique than had anyone before him. One of the central points of his analysis was to show that the heats of combustion of carbon and hydrogen as elements did not necessarily equal that which could be released from the same quantities of carbon and hydrogen incorporated into the organic molecules that constitute the foodstuffs. Given all the uncertainties that he pointed out, Helmholtz found the overall result, that the heat given off by animals was "on the average about 1/9 greater than that creatable by means of the combustion of carbon and hydrogen with the inspired oxygen," "conforms completely with the consequences of the chemical theory of organic heat as we have derived it above from the most probable assumptions." Experiments adequate to determine whether an "accurate" agreement could be reached between "the entire heat developed chemically" and the measured heat were not yet possible, so that "we must for the present be satisfied" to conclude that the chemical processes in the organism produce "very nearly" as much heat as the organism gives off.[27]

Helmholtz's analysis of these questions was far more probing than most contemporary discussions of quantitative reasoning in physiology. He rigorously assessed the possible sources of constant errors in the experiments of Dulong and Despretz. The most profound aspect

---

26. Ibid., 700.
27. Ibid., 707–11; see Frederic L. Holmes, "Introduction," in Justus Liebig, *Animal Chemistry* (New York: Johnson Reprint Co., 1964), xxxix–xl; and Richard L. Kremer, "Defending Lavoisier: The French Academy's Prize Competition of 1821," *History and Philosophy of the Life Sciences* 8 (1986):41–65.

of his analysis, however, concerned the simplifying physiological and chemical assumptions that had been made in equating the quantity of heat released in the combustion of carbon and hydrogen with the heat released by the overall chemical processes of the animal body. The actual processes were so much more complex than those included in the experimental calculation that one could not expect to arrive at an exact equality. Helmholtz did not claim that the measurements themselves proved quantitatively the chemical source of animal heat, but only that they were compatible with a belief in the chemical source of heat inferred from broader considerations. By carefully examining the quantitative results of experiments performed by others, he may well have drawn lessons pertinent to his own investigative direction: One had the choice either of applying so many simplifying assumptions to a very complex phenomenon that one had to be satisfied with "approximate" answers, with measuring the "greater part" of a process; or of arranging experimental conditions so as to isolate individual factors from a complex whole in order to achieve an accurate measurement of a simpler process. Helmholtz chose the latter.

In his paper of 1845 on the *Stoffverbrauch* in muscles, Helmholtz had moved the question from that of the overall process and the end products of the chemical transformations to that of the "intermediate links." Here, after completing his discussion of Dulong and Despretz's results, he made a closely parallel move:

> We have up until now occupied ourselves with the organic heat that derives from the sum of all of the chemical processes of the body, and that must therefore be calculable from the end products of the processes, and we have seen that the experiments and observations previously instituted do not contradict the chemical theory. Now it is much more difficult to discover the individual processes through which heat is created, because for most of them we have no idea what substances are transformed and what substances are produced, and in all cases we lack knowledge of the quantity of heat that can be released in that way; we must therefore limit ourselves to a few particulars.[28]

Further echoing his *Stoffverbrauch* paper, he emphasized that "heat is released during muscle action"; and, according to the experiments of several investigators, it is released "above all in muscle itself." Since, however, the acceleration of other processes, such as respiration and circulation, probably contribute as well to the rise in overall body temperature, "it must for the present remain undecided whether the

28. "Wärme, physiologisch," 714.

increase in heat in muscle contraction may not derive from an increased flow of arterial blood" into the muscle.[29]

From his subsequent activities it is clear that the preceding paragraphs concerning "the individual processes through which heat is created" in animals should be seen as the outline for a research program that Helmholtz adopted as his own experimental goal. It is more difficult to tell whether writing the review article itself led him to identify this as his investigative enterprise or whether these paragraphs were only a public statement of a plan that he had already conceived privately himself. However that may be, the experience of writing this essay review gave Helmholtz the opportunity to subject the data of others to the kind of independent third-party analysis his own data could have used and, in so doing, to sharpen his sense of discrimination regarding data. In addition to the conditions under which data was produced, Helmholtz now came to consider the quality of data itself.

4. Adjusting Instruments to Experimental Goals: The Formation of Heat in Muscles

Of the three processes that Helmholtz had described as "closely connected" in muscle action, he chose first to take up the relation between the chemical change that he had identified and the electric current that his friend Emil du Bois-Reymond had found in muscles.[30] Du Bois-Reymond lent Helmholtz, probably as soon as he obtained a copy, a massive article on "*Muskelbewegung*" that Eduard Weber, the leading German expert on the subject, had prepared for Rudolph Wagner's *Handwörterbuch der Physiologie*. Weber's treatment of muscle movement exerted a lasting influence on the course of Helmholtz's research on muscle action. Yet what initially interested him most, as he wrote du Bois-Reymond in late July 1846, was the detailed description of Weber's "rotation apparatus," capable of producing variable currents; Helmholtz referred to the apparatus as "very important for my own investigations."[31] Weber had designed this apparatus to deliver, from a pair of coils rotated beneath the poles of a horseshoe magnet, a current that oscillated rapidly in intensity between zero and a pre-set maximum. When he used this current to stimulate a muscle, the latter entered a steady, prolonged contraction, enabling him to

---

29. Ibid., 714–15.
30. Emil du Bois-Reymond, "Vorläufiger Abriß einer Untersuchung über den sogenannten Froschstrom und über die elektromotorischen Fische," *AP 58* (1843):1–30.
31. Helmholtz to du Bois-Reymond, 24 July 1846, in Kirsten, 73.

study what he termed the "active" state of a muscle more effectively than was possible during the momentary contractions obtained with an ordinary galvanic current.[32] The advantage of this apparatus for Helmholtz's study of muscles was potentially great, not only because of the possibility of sustaining contraction, but also because Helmholtz could vary the current delivered to the muscle better than he could with the charges from a Leyden jar that he had earlier used.

In early October Helmholtz wrote du Bois-Reymond that he was beginning to encounter difficulties not apparent in his first experiments:

> In the meantime I have continued the analyses of closed and unclosed frog muscles [that is, muscles placed or not placed in the multiplicator circuit], but they are not all in agreement, as are the first four whose results I recently communicated to you. In order to be able to continue these experiments with a prospect for more certain results, I consider it necessary to ascertain first with the multiplicator the conditions under which the *Froschstrom* is most rapidly and consistently weakened through its own action; but I doubt that my [multiplicator] will be adequate to detect smaller differences in the currents.[33]

Helmholtz here sought conditions analogous to those that had enabled him to detect a consistent difference in the proportions of the extracts from a muscle stimulated to contract and a resting muscle. Rapidly repeated stimuli had quickly exhausted the active muscle, so that he could then conduct the chemical analyses before postmortem changes in the muscle tissue had time to interfere with the results. The *Froschstrom*, however, did not exhaust itself in so regular a manner, a factor to which Helmholtz attributed the irregularity of his latest results. To attain conditions under which the *Froschstrom* did so, he needed to measure very small changes in a current already so small that its detection pressed against the limits of his multiplicator. Like du Bois-Reymond before him, Helmholtz was being drawn more deeply into the question of how his multiplicator functioned, and how it could be set up so as to achieve the great sensitivity and precision that his experimental objectives demanded from it.

During the next two and a half months Helmholtz fully immersed himself in the problems attending his use of the multiplicator as a precision instrument. In late December 1846 he reported his progress to du Bois-Reymond, hoping that what he had to say would be useful

---

32. Eduard Weber, "Muskelbewegung," in *Handwörterbuch der Physiologie mit Rücksicht auf physiologische Pathologie*, ed. Rudolph Wagner, 4 vols. (Braunschweig: Vieweg, 1842–53), *3.2*:10–2.
33. Helmholtz to du Bois-Reymond, 5 October 1846, in Kirsten, 74–5, on 74.

"in the construction of your own instrument or in its calibration" despite the fact that he "fell upon persistent new difficulties that have not allowed me to reach an adequate resolution. I am not yet by any means finished with the calibration," he added, "but from my trials to date I have at least learned how to go about it." Through his adjustments of the multiplicator, Helmholtz learned that the production of precision measurements was a delicate combination of eliminating or reducing constant errors produced by the instrument; accurately calibrating the instrument and finding the range within which it was most reliable; accounting for some constant errors by the judicious use of theory (in this case, of the effect of geomagnetism); and developing a skill that could only be acquired after repeated trials with the instrument such that the practitioner became one with it. "Exercise makes the master," he told du Bois-Reymond. Noteworthy in his calibration was his use of graphs: he plotted current intensity as a function of the degree of deflection of the multiplicator's needle over the first 15° of deflection. Although he found that his instrument was most reliable in the range from 2° to 8°, owing to the weak currents that frequently occurred in an investigation of this type, he still needed greater accuracy in the range below 4°. For that range he replaced observed angles with corrected ones, thus producing *corrected* measurements for the range below 4° in addition to *direct* precision measurements for the range above 4°. At the same time, Helmholtz also revealed to du Bois-Reymond that he had already started another experimental problem: "Through the adjustments of the casing and the scale [that is, as a consequence of his calibrations of the multiplicator] I have also taken up experiments on the temperature changes in contracting frog muscles."[34]

Helmholtz initially employed thermoelectric currents as a procedure for calibrating his multiplicator. Apparently in the course of conducting these experiments he realized that he could combine features of his earlier experiments on isolated frog muscles with measurements of the increase in temperature during muscle contractions by inserting thermocouples into the muscles of intact animals and humans. By so doing, he could now answer the question that he had posed fifteen months earlier in his review of "physiological heat": that is, whether the increase in heat in muscle contraction arises from the muscle itself or from an increased flow of arterial blood into the muscle. The passage in which he reported this experiment to du Bois-Reymond suggests that he carried it out by adapting instruments already on hand to a

---

34. Helmholtz to du Bois-Reymond, 21 December 1846, in ibid., 75–8, on 78.

new purpose. The significance of his preliminary results induced him to plan further experiments in which he would carefully design both the apparatus and the conditions under which the measurements would be taken in order to maximize the observed effect.[35]

By early February 1847, Helmholtz had already completed a draft of what became his famous essay on the conservation of force—*Ueber die Erhaltung der Kraft*—and sent it off to du Bois-Reymond for comments.[36] Then he began to design a special apparatus to use in his further "thermal muscle experiments." By mid-April he had "pretty well" finished putting it together.[37] The new apparatus, a "thermoelectric circuit," departed "completely from the previous form of these circuits." In order to multiply the thermoelectric effects he placed three bimetallic elements (each containing two junctions) in series. Physically, however, he placed them in parallel, close enough so that one set of junctions could be inserted through a single frog muscle. He connected the three elements electrically through loops of flexible copper wire. Instead of making the elements themselves from the usual combination of copper and iron, as he had earlier done, he now chose iron and silver, because that combination "possessed 2-½ times as much thermoelectric force" as iron and copper did. To achieve a minimum internal resistance in the circuit while internally coming as close as possible to the temperature of the muscle in which it was inserted, he made the elements out of strips of metals instead of the customary wires. He fastened the strips into a glass platform, with ivory insulators and thumbscrews arranged in such a manner that he could quickly insert the strips through the muscle and mount them on the apparatus.[38]

In constructing this apparatus, Helmholtz displayed for the first time a feeling for the material craft of experimental design. Effectively using the apparatus, however, was another matter altogether. While attempting to determine whether activity in isolated frog muscles and nerves resulted in the formation of measurable amounts of heat, he wrote his fiancée Olga von Velten in late July 1847 about his difficulties in producing the effect he wished to measure:

---

35. Ibid., 75–8.
36. Helmholtz to du Bois-Reymond, 12 February 1847, in ibid., 78; see Fabio Bevilacqua's essay "Helmholtz's *Ueber die Erhaltung der Kraft*: The Emergence of a Theoretical Physicist," in this volume. If Helmholtz did think of his investigation of muscle action as a means to verify his conservation principle, it could only have been as the starting point of a long research program, not as the immediate goal of his present experiments.
37. Helmholtz to du Bois-Reymond, 10 April 1847, in Kirsten, 79–81, on 81.
38. "Ueber die Wärmeentwickelung bei der Muskelaction," *MA* (1848):147–64, on 147–50, in *WA* 2:745–63.

I became doubtful about whether the success of my experiment on heat formation in the nerves was not caused by further accidental disturbances, and had devised various changes in the apparatus in order to diminish these as much as possible. I did not always succeed. Now, however, I have found an arrangement of the apparatus which appears to protect the results completely from accidental disturbances.

Yet Helmholtz's craftsmanlike skill in controlling his apparatus and the environment of his experiment did not lead to results he had expected, for he further reported to Olga that he "had the unexpected surprise that what I had considered heat formation in the nerve had now completely disappeared."[39]

Helmholtz did not build his own instrument for stimulating the muscle contractions; nor did he adopt Weber's rotation apparatus, in spite of his earlier interest in it. Instead, he ordered from the Berlin instrument maker Johann Georg Halske a so-called Neeffian apparatus, that is, a commercially available induction coil introduced in 1839 by Christian Ernst Neeff and commonly used by other physiologists for similar purposes.[40] Helmholtz modified the apparatus's circuit through which the secondary discharges were delivered to the muscle in such a way that their duration would be short enough to avoid interference with the multiplicator's operation. With this modified apparatus he was able to attain the same effect that Weber had designed the rotation apparatus to achieve. The series of short stimuli came so close together that the muscle entered a state of sustained contraction.[41]

By the time Helmholtz was ready to begin these experiments, he had abandoned his earlier effort to calibrate his multiplicator "accurately and completely." Not only did he find the calibration a tedious process, but the condition of his very sensitive instrument did not remain stationary long enough to make the task worthwhile. He therefore contented himself with an approximate calibration relying on the assumption that within the first 20° the deflections of the needle were "almost proportional" to the intensities of the current. "The greatest deviation from proportionality," he asserted, "was only 3/100 of the measured magnitude, and in the degree of exactitude sought here it can be neglected." He estimated that "in the vicinity of the null point temperature differences of less than 1/1000 degree can be recognized."[42] Helmholtz thus brought to this physiological investigation

---

39. Helmholtz to Olga von Velten, 30 July 1847, in Richard L. Kremer, ed., *Letters of Hermann von Helmholtz to His Wife, 1847–1859* (Stuttgart: Franz Steiner, 1990), 21–4, on 22. See also his letter of 4 August 1847, in ibid., 24–5.
40. C.E. Neeff, "Ueber einen neuen Magnetoelectrometer," *AP 46* (1839):104–9.
41. "Wärmeentwickelung," 154–56.
42. Ibid., 152–54.

the standard of thermometric precision present in physics, and had thus improved considerably on what Becquerel and Breschet had expected when they measured the increase in the temperature of muscles in warm-blooded animals: "In order to observe that effect well," they stated, "the apparatus must be able to indicate tenths of a degree of temperature."[43]

Spring frogs probably began arriving by mid-April, so that Helmholtz could begin his experiments as soon thereafter as his Neeffian apparatus was ready. The experiments he conducted with it were broadly analogous to his 1845 experiments on the chemical changes in muscle. Now as then, he prepared the two gastrocnemius muscles of a frog so that he could compare the state of one of the muscles put into a state of contraction with an otherwise nearly identical muscle remaining at rest, thereby identifying the effect sought. In this case he left one of the muscles attached to the spinal cord, through which he stimulated it. He then pushed the three thermoelectric elements through both muscles so that each muscle covered one of the two sets of three bimetallic junctions. He then clamped the elements in place, covered the apparatus to protect the nerve from drying and the muscles from external temperature changes, connected up the circuits, and waited for $1/2$ to $3/4$ of an hour until the multiplicator needle had settled at the null point. This showed that the temperatures in the two muscles were now equal. He then stimulated the innervated muscle to tetanic contractions for 2–3 minutes. The needle deflected by 7–8°, indicating a temperature difference between the two muscles of 0.14° to 0.18° C.[44]

To guard against the possibility that the stimulating current had itself caused these readings, Helmholtz performed a control experiment similar to the one he had used in his experiments on the chemical changes in muscle. Stimulating the nerve after the muscle had lost its irritability caused "not the slightest movement of the magnetic needle." Helmholtz later found that he could stimulate the muscle directly rather than through the nerve. In this case the stimulating current did cause a small deflection after the muscle was exhausted, but it was less than a tenth of what the fresh-contracting muscle caused. Direct stimulation of the muscle had the advantage that its irritability lasted longer, so that he could "observe the decrease in the thermal effect in proportion to that of the energy of the contraction."[45] Helmholtz did not say whether he had achieved these results as soon as he began using his carefully designed apparatus, or whether he had first

---

43. Antoine Becquerel and Gilbert Breschet, "Premier mémoire sur la chaleur animale," *Annales des sciences naturelles 32* (1835):257–73, on 272.
44. "Wärmeentwickelung," 156–57.
45. Ibid., 157–58.

to go through a period of adjusting his instruments and procedures. He did not fully realize the hope he had expressed in December that the improvements he introduced into the experiment would allow him to reach "much larger amplitudes" than his initial results did. The new range of temperature differences (0.14° to 0.18° C) was only slightly larger than his highest earlier readings (0.08° to 0.12° C); but, assuming that all his results really fell within the range reported, then his measurements appear to have constituted decisive evidence that contracting muscles produce heat.

Helmholtz concentrated his attention on the formation of heat in muscles not because he thought they were the only source of animal heat, but rather because he expected them to be the largest, and therefore most easily measurable, source. He had already indicated in his review of the subject in 1845 that heat was probably produced in other organs through other chemical processes. In this respect he had mentioned particularly the actions of the motor nerves that stimulate muscle action, and summarized a longstanding debate about whether "nervous action" itself was a source of heat.[46] By the late summer of 1847 he had a more immediate reason to think that the "proof" of the development of heat in muscles that he had just delivered "could also be carried out in nerves." Du Bois-Reymond had discovered in nerves a current similar to the *Froschstrom* in muscles, and he told Helmholtz that this current too was modified when nerves are stimulated. "Hence," Helmholtz thought, "one would expect chemical changes and the development of heat here too to be closely connected with these currents."[47] Since the quantities would, however, be expected to be far smaller than in muscles, the situation presented a strong challenge to the sensitivity of his experimental method.

Early in the summer of 1847 Helmholtz obtained deflections of "½– 1° in the sense of a warming of the nerve."[48] For a time he believed that he had "made visible the heating of nerves in frogs." By early August, however, he had to inform du Bois-Reymond that the observation "seems to have rested on a disturbing influence of another kind, for since I have taken care to fasten the nerves properly through the pieces of cork all the deflections have disappeared."[49] He later attributed his misleading readings mainly to the fact that the nerve preparations had been insufficiently isolated; "so that besides the nerve a second conducting path existed between the spinal cord and the muscle, through which . . . a part of the induction current passed and

46. "Wärme, physiologisch," 715–17.
47. "Wärmeentwickelung," 158–59.
48. Ibid., 161.
49. Helmholtz to du Bois-Reymond, 6 August 1847, in Kirsten, 84.

developed a quantity of heat detectable through such a sensitive method."[50] Helmholtz still wished, he wrote du Bois-Reymond, to test the matter further on "fresh, strong frogs; in any case it would be possible to carry the test to 1/1000 of a degree." He also wished to measure again the heat produced in muscles with fresh frogs, because "my frogs are already too weak from hunger." Nevertheless, he was confident enough about the results already obtained—he described to du Bois-Reymond the increase in temperature in frog muscles as "going easily up to as much as 0.2° C"—to say that "I shall probably enter the field with these investigations very soon, that is [send them to] *Müller's Archiv.*"[51]

Helmholtz's paper "Ueber die Wärmeentwickelung bei der Muskelaction" appeared in *Müller's Archiv* in 1848; but he first presented it on 12 November 1847, at a meeting of the Berlin Physikalische Gesellschaft. It began with a summary of Becquerel and Breschet's thermoelectric measurements in the muscles of the human body, along with a few thermometric measurements made by others in animals. In a passage closely echoing his discussion of 1845, he asserted that "from these experiments one cannot decide what the source of the increase in the temperature of the muscles might be: whether it derives from processes occurring in the muscle itself or is only a consequence of an increased general body heat through strengthening of the vegetative functions and a richer flow of arterial blood into the muscles." That the flow of arterial blood could explain the temperature rise was made plausible by the measurements of Becquerel and Breschet and others showing arterial blood to be slightly warmer than venous blood. "In order to attain a certain and decisive answer to the question, it seemed necessary that the formation of heat be investigated in muscles that are removed from circulation."[52]

In comparison with his extensive discussion of his apparatus, Helmholtz's treatment of the experiments themselves is surprisingly brief. He did not mention the circumstances under which he had begun the investigation, nor did he even outline its subsequent course. He did not indicate that he had performed the initial experiment before he constructed the apparatus employed to obtain the definitive results. Indeed, he did not even describe any specific experiments that he had performed or list individual results. Instead, he summarized his experimental experience in the form of a generic instruction on how one *ought* to carry out such an experiment. He continued in the same manner to describe the procedures through which such an experiment

50. "Wärmeentwickelung," 161–62.
51. Helmholtz to du Bois-Reymond, 6 August 1847.
52. "Wärmeentwickelung," 145.

is to be performed, leading up to the succinct statement that "during the time [of the tetanus contraction] the needle deviates 7–8° toward the side that corresponds to the warming of the stimulated muscle. Through this deviation a temperature difference of 0.14° to 0.18° is indicated."[53]

In his discussion of experiments on the formation of heat in nerves, the second section of his essay, Helmholtz likewise allowed only a few traces of his personal experience into his text. "All of the experiments that I instituted in this regard," he acknowledged, "have had a purely negative result." He would describe the procedures anyway, he wrote, partly because his result showed that the development of heat in nerves, if such takes place, "would be vanishingly small in comparison to that in the muscles" and, if it did occur, it would not be more than "a few thousandths of a degree," and partly "in order to draw attention to several sources of error, the neglect of which can easily lead to an apparently opposite result."[54] That he wanted to publish this negative result is an indication of the central importance he attached to the accurate determination of sources of error in such delicate experimental investigations where spurious results, often arising from faulty protocols and analyses, could be mistaken as decisive. Hence in both sections of the essay, on heat development in the muscles and in the nerves, he devoted much attention to the description of the apparatus and the procedures associated with the component instruments, and none to the individual experimental trials or results. The only table he reported concerned his calibration.[55]

The strategic importance of error analysis—in the sense of using protocol and instrument design as a means to avoid errors or, as Helmholtz put it, "to protect the determination of results as much as possible from foreign influences"[56]—to his investigation also accounts for his style of reporting: it is largely devoid of narrative elements normally present in experimental essays. By contrast, he had composed his 1845 paper on the chemical changes in muscle action in a form that included prominent narrative elements. There he had described how he had proceeded experimentally "at the beginning" and the changes he had made along the way. He had described what had worked and what had not worked, and he had listed in tabular form the results of nine numbered experiments (Figure 2.1). But in 1845 he had a positive answer, and to explain it Helmholtz could reorder his protocol and

53. Ibid., 156–57.
54. Ibid., 164, 159.
55. Ibid., 153.
56. Ibid., 156.

trials in a way that explained the appearance of his results. Decisive positive results were not the central issue in 1847; what mattered in the reporting was the meticulous design of the experiment or what might be called the "purification" of the experimental apparatus, design, and protocol (including the attempt to convince his readership of such).

Even so, "*Ueber die Wärmeentwickelung bei der Muskelaction*" became a classic paper in nineteenth-century physiology, not only because of the elegance and rigor with which Helmholtz had measured a very small effect, but because he had marshalled his measuring devices in order to answer a clearly conceived physiological question of fundamental importance. Unlike his earlier work on the chemical changes in muscle action, in which he produced the first concrete evidence for what physiologists generally believed already, his demonstration that a muscle produces heat through processes within itself did not merely confirm the expected. The prevailing view at the time was that animal heat was produced by the respiratory oxidation of foodstuffs in the blood as it circulated through the body. By demonstrating that the heat-producing processes took place in the same organ that produced mechanical work, Helmholtz laid the groundwork for the refutation of such separable functions and for the acceptance of a unitary view of chemical processes in the tissues that yield energy in the form of both work and heat. At almost the same time, his paper on the conservation of force provided a powerful theoretical foundation for such a view. Decisive though these experiments now appear to be, however, they did not immediately overthrow the view that respiratory oxidations release the animal heat in the blood. Like most "crucial experiments," they did not "compel assent"; yet they did provide the most conspicuous starting point for a gradual shift in opinion toward the point of view for which they provided such impeccable evidence.

As Helmholtz became more involved in instrument construction, his expectations that both experimental protocol and instrumentation were perfectible rose, and thereby grew his expectation that the "correct" numerical result would eventually emerge. Helmholtz experienced difficulties in instrument construction and experimental protocol, to be sure. But it was also becoming ever more apparent that he placed his faith in the craftsmanlike qualities of experimentation as the means to experimental success. Undoubtedly his theoretical convictions concerning the conservation of force partly shaped the kinds of questions he asked, but those convictions did not dominate the evolution of experimental design and procedures so as to compel automatic assent to them. Theory was also not so dominant in another respect. Absent from Helmholtz's discussion of his instruments and protocol were theoretical discussions of possible sources of error and the analytical determination of them. Although in the case of his multiplicator he compensated for the effect of geomagnetism, in these delicate thermal investigations he did not try to compensate theoretically for heat loss to the atmosphere, although he did try by material means to prevent heat loss from the surface of the nerve. This meant, too, that Helmholtz did not analyze his data, but seemed rather to accept it as sufficiently refined by instrumentation and experimental protocol so as to be regarded as nearly errorless. Even though his results were cast in the form of ranges, Helmholtz seems to have viewed even those ranges as a sharper view of reality than had been hitherto available because they gave a more refined location in which the "real" value might be found.

5. Graphs and the Temporal Recording of Muscle Contraction

By 1848, when he applied for Brücke's position at the Akademie der Künste in Berlin, Helmholtz was actively pursuing the problem of muscular activity. He continued to view that problem as a combination of highly connected "mechanical, ... electrical, thermal, and chemical" processes. Having already addressed the chemical, electrical, and thermal aspects of the phenomenon, he now turned to the mechanical. His primary point of departure was Weber's work on muscular motion as the latter had comprehensively treated it in Wagner's *Handwörterbuch*. "All muscle movements produced by outside causes [such as the stimulation of a nerve by a galvanic current] have, until now," Weber asserted, "differed from natural movements in that their duration has been only momentary, whereas through the will contractions can be effected that last for longer times." By means of his rotation apparatus, however, he claimed, that "I have succeeded in producing through outside stimuli muscle contractions of long duration, in fact as complete as those we produce naturally through the will."[57] The central thrust of his work on muscle motion was to study

---

57. Weber, "Muskelbewegung," 10–11. For a discussion of Weber's work on muscle action as a continuation of an investigative tradition begun by Theodor Schwann in 1835, see Kremer, *Thermodynamics of Life*, 315–33. Kremer notes the connection between Weber's article of 1846 on "Muskelbewegung" and Helmholtz's experiments beginning in 1848 on the mechanical aspects of muscle action in ibid., 326–33, and in idem, ed., *Letters*, 51n.

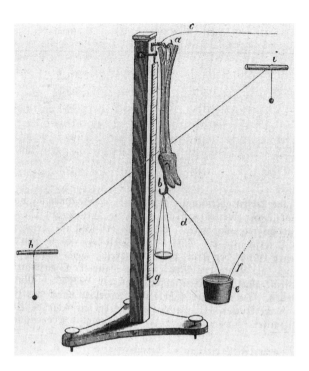

Figure 2.2. Weber's apparatus for studying the contraction of a muscle. Source: *Eduard Weber, "Muskelbewegung," in* Handwörterbuch der Physiologie mit Rücksicht auf physiologische Pathologie, *ed. Rudolph Wagner, 4 vols. (Braunschweig: Vieweg, 1842–53), 3.2:1–122, on 86. Courtesy of Yale University Photographic Services.*

through precise quantitative measurements the nature and causes of the sustained muscle contractions as produced by his rotation apparatus.

Devising an apparatus from which he could suspend a frog muscle—usually the thin hypoglossal muscle attached normally to the frog's tongue—from a hook **a** (see Figure 2.2), Weber hung weights from a hook attached to the other end of the muscle at **b**, so that he could examine the effects that occur when the muscle movements take place against resistance. He stimulated the muscle directly by attaching one end of the circuit **c** of his rotation apparatus to the upper end of the muscle and the other end through a wire **d** and a mercury cup **e**. Behind the muscle was a scale **g**. Weber passed a black thread **ih** through the muscle in such a manner that its movement up and down the scale measured the shortening or lengthening of the muscle. By observing the thread through a microscope he could determine these distances to tenths of a millimeter.[58] With this apparatus Weber examined various relationships between the shortening of the muscle, the duration of the contraction, and the amount of weight lifted.

The aim and design of his experiments—namely, to study muscle contractions of long duration—led Weber to concentrate on the contracted and relaxed *states* of muscles, and pay much less attention to the movement itself as a process occurring in time. With the exception of his measurements of the gradual extension of a muscle after it had reached its maximum contraction—a process that he could observe over intervals of 6 to 8 seconds at a time—Weber regarded the movements of contraction and relaxation as essentially instantaneous. "When one stimulates a muscle through a motor nerve," he wrote, its movement "occurs at the same moment," that is, "the movement begins and ends with the stimulus." When he produced a tetanic contraction with his rotation apparatus and then broke off the current, "the tetanus disappears at the same moment." The situation was different when he stimulated a muscle indirectly within an intact animal by means of a sensory nerve. Then "a sensible time passed" between the stimulus and the movement.[59] The strongest contrast he drew, however, was between "animal" (or voluntary, in modern terminology) muscles and "organic" (or involuntary) muscles.

> Animal muscles shorten . . . during the continuing stimulus of the rotation apparatus rapidly and to a considerable degree. . . . Their [individual] bundles all enter into contraction simultaneously and at the moment in which their nerves or their own substance is stimulated, and this contraction lasts as long as the stimulus. But if the stimulus stops, if one interrupts the current, these muscles return just as instantly to their previous condition of length and inactivity as they had shortened at the beginning. . . . Organic muscles behave in all these respects in exactly the opposite way. If a galvanic current acts on them, they contract slowly, and not at the moment at which the nerve or their substance is stimulated, but only after some time, so that the stimulus has often ceased before the contraction becomes visible.[60]

The motives that led Helmholtz to study muscle contraction and the prominence of Weber's work in the background of his study are probably accurately, if retrospectively, reflected in the opening paragraph of a paper he wrote two years later (1850), the first part of which

---

58. Weber, "Muskelbewegung," 28–9.
59. Ibid., 19.
60. Ibid., 22–3.

dealt with this topic. He had, he wrote, begun the work as part of a "broader plan." "Eduard Weber," he continued,

> has discovered the laws according to which muscles act in the resting state and in an enduring state of contraction, and has thereby laid the foundations for a knowledge of their mechanical actions. But one of the central questions in this field cannot be resolved through investigations of continuously excited muscles: that is, concerning the mechanical work that they are able to perform. A steadily excited muscle produces, even through the most exhaustive exertion, no work in the sense of mechanics, it only causes the parts of the body to remain at rest in a new position of equilibrium. In order to perform work, either motions of its own body or changes in the outside world, the muscle must alternate between rest and excitation, and the magnitude of its work will depend sensibly upon the rate of change. From this point of view I began to study the processes occurring in the simple contraction of a muscle; by such an action I mean one that results from a stimulus of vanishingly small duration.[61]

Helmholtz began this work sometime before the early summer of 1848, probably not long after he had completed his experiments on the formation of heat in muscles late in 1847. His plan thus seems to have taken shape in the aftermath of his formulation of the law of conservation of force, and it appears self-evident from the above passage that his interest in the problem was in part a desire eventually to apply the physical concepts he had derived theoretically to his longstanding physiological involvement with the action of muscles. Helmholtz set out to reverse the move that Weber had made: where Weber had found a means to study the lasting state of contraction that he believed to be the "normal" action of a muscle, Helmholtz returned to the momentary action resulting from a single (or a steady) outside stimulus. That posed for him the problem of how he could observe a process that occurred in "the blink of an eye." At just this time, however, another physiologist who had recently entered the circle of du Bois-Reymond, Brücke, and Helmholtz provided the means that Helmholtz required.

In 1847, Carl Ludwig had published in *Müller's Archiv* a paper entitled "Beiträge zur Kenntniss des Einflusses der Respirationsbewegungen auf den Blutlauf im Aortensysteme." The specific physiological effects that he reported, though interesting in themselves, have been overshadowed by the pathfinding apparatus that he employed in order

---

61. "Messungen über den zeitlichen Verlauf der Zuckung animalischer Muskeln und die Fortpflanzungsgeschwindigkeit der Reizung in den Nerven," *MA* (1850):276–364, on 276–77, in *WA* 2:764–843.

to record the rapidly changing pressures of the arterial blood and the thoracic cavity. Adopting a Poiseuille manometer, one end of which he connected to the pressure-sensing instruments, he mounted on the other end a vertical shaft that floated on top of the manometer fluid. To the upper end of this shaft he attached a horizontal arm, the tip of which held a goose quill. The quill contacted a drum that rotated at nearly constant speed, driven by a falling weight and regulated by a rotating pendulum. Ludwig could adjust the relation between the drum and the quill so that the latter pressed on the drum just enough to make a steady mark on its smoked surface. With this apparatus he was able to record "curves which fulfill all requirements."[62]

In the spring of 1847 Ludwig traveled to Berlin to meet Johannes Müller and three of his most-promising former students—du Bois-Reymond, Brücke, and Helmholtz. There the four younger men formed a strong friendship and an informal alliance to pursue a "reductionist" physiology aimed at advancing the physicalist viewpoint earlier espoused by Theodor Schwann.[63] A year later du Bois-Reymond wrote in the introduction to the first volume of his *Untersuchungen über die thierische Elektricität* what has often been regarded as the "manifesto" of this group. One of du Bois-Reymond's key arguments was that physiologists must not merely make measurements, but also establish the changing magnitudes of observed effects as functions of all the conditions that influence them. He wrote:

> The dependence of the effect on each condition presents itself in the form of a curve, whose exact law remains, to be sure, unknown, but whose general character one will be able in most cases to trace. It will almost always be possible to decide whether a function increases or decreases with the investigated variables. In other cases one may be able to discover distinctive points of the curves which constitute its bending with respect to the abscissa, whether it approaches asymptotically a constant value, etc.[64]

Du Bois-Reymond thus hoped that from a continuous sequence of data one could deduce the general nature of causal relationships and

---

62. C. Ludwig, "Beiträge zur Kenntniss des Einflusses der Respirationsbewegungen auf den Blutlauf im Aortensysteme," *MA* (1847): 242–301, esp. 257–67. See also Heinz Schröer, *Carl Ludwig: Begründer der messenden Experimentalphysiologie* (Stuttgart: Wissenschaftliche Verlagsgesellschaft, 1967), 104–14.

63. Schröer, *Ludwig*, 36. See also Paul Cranefield, "The Organic Physics of 1847 and the Biophysics of Today," *JHMAS* 12 (1957):407–23; and William Coleman, *Biology in the Nineteenth Century: Problems of Form, Function, and Transformation* (New York: John Wiley, 1971), 150–54.

64. Emil du Bois-Reymond, *Untersuchungen über die thierische Elektricität*, 2 vols. (Berlin: Reimer, 1848–49), *1*:xxvi–xxvii.

obtain a general overview of the progress of a phenomenon. Helmholtz probably saw something similar in Ludwig's graphical method: the means to obtain automatically the denser series of "observations" that he had earlier desired and believed he needed, and to do so with that experimental technique in which he had placed most faith: instrumentation. Sometime in late 1847 or early 1848, the confluence of a measuring device, a broad physicalist research program, Helmholtz's own prior interest in muscle physiology, his newly formulated conservation principle, his developed experimental style, and his familiarity with Weber's work on muscle contraction emerged.

From his brief description of the "simple" apparatus that he first constructed to record the movement of a simple muscle contraction, it appears that Helmholtz directly combined elements of Ludwig's apparatus with elements of the experimental setup that Weber had used to study muscle contraction. Helmholtz stated explicitly that he had, "in a manner completely similar to that in which Ludwig recorded the heights of blood pressure," recorded "the heights to which a weight hung from a muscle is raised in successive points of time by a contraction." Like Weber, he suspended an isolated frog muscle from a support and hung a weight on its other end. He added, however, "intermediate pieces" between the muscle and the weight that were linked to a "well-polished metal rod that moves vertically between two metal plates." This rod was the equivalent of the rod in Ludwig's apparatus that floated on the manometer fluid. The rest of Helmholtz's apparatus—a horizontal crossarm ending in a steel point that made a mark on a rotating cylinder driven by a falling weight—apparently imitated, in simplified form, Ludwig's design.[65] By the early summer of 1848 Helmholtz had progressed far enough in the construction of his apparatus to begin testing it. As he wrote his fiancée in mid-July:

> I have now finished building my frog-tracing machine and have already carried out a few tracing experiments on mica sheets. Instead of the frog muscles, I inserted a spring. The weight hung from it, oscillated up and down, and recorded its movements. The traces are much prettier than the earlier ones, very fine and regular.[66]

The curves that Helmholtz obtained were so small that he had to observe them with a microscope and make enlarged freehand copies of them (Figure 2.3). Nonetheless, his tiny curves were sufficient to establish, in contradiction to Weber's views, "the previously unknown

65. "Messungen," 280–81.
66. Helmholtz to Olga von Velten, in Kremer, ed., *Letters*, 42–4.

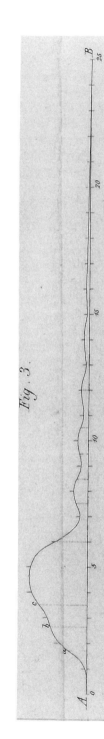

Figure 2.3. Helmholtz's muscle curve. Source: "Messungen über den zeitlichen Verlauf der Zuckung animalischer Muskeln und die Fortpflanzungsgeschwindigkeit der Reizung in den Nerven," MA (1850):276–364, Figure 3. Courtesy of Yale University Photographic Services.

fact that in animal muscles too, as is the case in much longer time intervals in organic muscles, the energy of the muscle does not develop completely at the moment of an instantaneous stimulus. Rather, in most cases after the stimulus has already ceased, it increases gradually, reaches a maximum, and again subsides."[67] The wording of this statement reveals the degree to which Helmholtz developed his point of view in response to that of Weber. By adopting a method that enabled him to follow a process lasting only "a small fraction of a second"—one that appeared "momentary" to Weber—Helmholtz could apply to animal muscles a close paraphrase of the description that Weber had given of the much slower process occurring in organic muscles.

By the time he reached this point in his investigation, Helmholtz stood on the threshold of a major change in his professional life. In the fall of 1848 he assumed Brücke's former position at the Akademie in Berlin. He had not been there long before an opportunity arose to obtain another position more favorable to his real scientific interests. Brücke was moving again, this time to Vienna; that left his professorship at Königsberg open. In early May 1849 Helmholtz applied for the position.[68] The situation was somewhat delicate because du Bois-Reymond and Ludwig were also interested in the position. Müller wrote to the Kultusministerium supporting all three of his former students. His reasons warrant full quotation:

> I hasten to report in obedience to the request of 16 April concerning the appointment of a professor of physiology at Königsberg University. The experimental method in physiology has in recent times entered into an intimate connection with the physical sciences, especially with physics and chemistry, and the same demands for exactness in the methods of investigation made in physics are now made in the field of physiology. And so those who can contribute most to progress in physiology are those who, equipped with all of the knowledge from organic fields accessible only to the physician and physiologist, have acquired an equally fundamental education [*Bildung*] in physics and are capable of applying the methods and means of physics with certainty in experiments.
>
> The most promising young talents in this direction in Germany are Brücke, du Bois-Reymond, Helmholtz, and Ludwig. Through his classic work on animal electricity *Privatdozent* du Bois has acquired the strongest claim on the professorship of physiology, but he seems not inclined to accept the position in question before the completion of his present work.

---

67. "Messungen," 282–83.
68. Helmholtz to Kultusministerium, 4 May 1849, Staatsbibliothek Preussischer Kulturbesitz, Handschriftenabteilung, Darmstaedter Sammlung, Signatur F 1 a (1847).

> Concerning Helmholtz, who is presently an assistant at the anatomical museum and teacher of anatomy at the Akademie der Künste, I have already had the honor to report in detail to the ministry. I regard him as one of the most significant talents in physiology. So far as I could ascertain, he would only be inclined to accept the position in question if his appointment as teacher at the Akademie der Künste can be extended for some time. Concerning his qualifications as a teacher I have not the slightest doubt. If he can obtain and accepts the professorship at Königsberg, I would recommend Dr. du Bois to you as assistant and also as teacher of anatomy at the Akademie der Künste.
>
> Professor Ludwig stands at the same level as Brücke, du Bois, and Helmholtz and would also be thoroughly suited to the position in question. He has distinguished himself through his physiological investigations in decisively recognized ways, is a very beloved teacher, and has already trained many students to do fundamental work. I believe with certainty that he would accept a call to the chair in Königsberg.[69]

Müller's letter of recommendation is of considerable significance, for he was doing much more than attempting to maneuver academic positions in favor of former students. In the strongest terms possible he was backing the achievements and program of the four young physiologists whom historians have often viewed as having just declared themselves in opposition to Müller's own view of physiology. Müller's letter clearly shows that any philosophical differences that may have existed between himself and his former students were far less important to him than the methods of experimental investigation that they were pursuing and of which he thoroughly approved. Decades earlier Müller himself had at times yearned for exacting results, but he knew that the technical capabilities to achieve them were not available. Now his own students were creating the resources that would lead towards goals that Müller himself favored. In the event, du Bois-Reymond withdrew from the competition, and Ludwig was ruled out because of his youthful political activities. In mid-May of 1849 Helmholtz was named extraordinary (*außerordentlicher*) professor of physiology and director of the physiological institute at Königsberg.[70] He was ordered immediately to begin his physiological lectures for the summer term. Helmholtz temporarily put his muscle contraction investigation aside.

---

69. Johannes Müller to Kultusministerium, 7 May 1849, ibid., Signatur 3K 1826(2).
70. Koenigsberger *1*:109–11.

6. Precision, Certainty, and the Velocity of the Nerve Impulse

Although involved in preparing his winter-semester lectures at Königsberg University in the fall of 1849, Helmholtz nonetheless found time in spare moments and on holidays to continue the muscle-contraction experiments. The new environment brought mixed blessings. Unimpressed by the medical faculty's lack of scholarly productivity, Helmholtz felt early on that he would "probably become closer to the mathematicians here." Although it at first appeared to him "difficult to get close to the most important of them," the physicist and mineralogist Franz Ernst Neumann, whose mathematico-physical seminar at Königsberg pioneered the integration of mathematical and exact experimental techniques in physics, Helmholtz eventually came to know Neumann professionally and benefited from Neumann's expertise in physics. By contrast, Helmholtz struck up a friendship more quickly with the astronomer and director of Königsberg's observatory, Alexander Busch, whom Helmholtz believed was "very highly regarded as an accurate observer." Although Helmholtz did not immediately gain funding for his projects, he did manage to acquire a room in the main university building, where he set up his instruments and experiments. The most highly valued addition to his research enterprise, however, was the assistance of his newly wed and beloved wife Olga. "She stands by me faithfully as director of protocol of the observed divisions of my scale in the trials," Helmholtz told du Bois-Reymond. Helmholtz found her assistance "very necessary" because he became "completely confused when I am supposed to pay attention to so many things at the same time" in the muscle-contraction experiments, which by the fall of 1849 had become more complicated technically.[71]

To a certain degree, Helmholtz's graphical method had been extremely useful, for it enabled him to interpret the gross features of the patterned mapping of the time course of the muscle's lifting of the weight in physical terms, thus extending his understanding of the relation between the muscle's mechanical work and energy. Significantly, he inferred from the curve that the "force of the muscle was not strongest directly after the stimulation, but rather increases for a time and then falls."[72] It was the latter that attracted his attention in October 1849, when he realized that the graphical method he had used until

---

71. Helmholtz to du Bois-Reymond, 14 October 1849, in Kirsten, 86–8, on 86, 88.
72. "Messungen," 283; and Helmholtz to du Bois-Reymond, 14 October 1849, in Kirsten, 86–8.

then could not be further exploited to resolve the fine details of the initial time interval of the muscle contraction, in particular the apparent delay between the electrical stimulus and the muscle's response. Helmholtz's graphs, which had the advantage of being fast and easy to use, had the disadvantage of a "smaller claim to accuracy," and accuracy, he believed, was required to probe the time-lag more closely.[73] Actually measuring the time-lag, however, proved problematic. "Our senses are not capable of directly perceiving an individual moment of time with such a small duration," he pointed out, so "we must apply artificial methods to its observation and measurement."[74] His customary "artificial method," the graphical method, had limitations in its present form that were not, in fact, to be overcome. The method's limitations, such as those stemming from the friction of the stylus moving along the drum, were either intractable or too large to obtain reliable measurements of the time interval in question. Unable to eliminate the errors completely, Helmholtz learned that even persistent manipulation of the apparatus was counterproductive because "even with the highest possible perfection of the construction of an apparatus a greater accuracy of the measurements is thwarted."[75]

While working in October 1849 on new trials on the motion of a muscle following a single stimulation, he reported to du Bois-Reymond that he was doing so "according to an entirely new method" that demonstrated the time-lag "in a very obvious way."[76] The time-lag in question, Helmholtz found, could be examined by adapting to his investigation C.S.M. Pouillet's method of using the deflection of a galvanometer needle to measure small time intervals. Pouillet's principle was that the length of the arc through which the magnetic needle of a multiplicator swings when a momentary current passes through its coil is proportional to the duration of the current. Helmholtz thus had to arrange his apparatus in such a way that the beginning and end of the time interval marking the current's duration coincided with the time interval that began with the application of the stimulus and ended with the onset of the muscle's mechanical action. He explained that he managed to make these two time intervals coincide by ensuring that the onset of mechanical action—in this case the lifting of a weight by the muscle—broke the electrical current. During this time interval, a magnet with a mirror was set in motion, and the deflection of the

73. "Messungen," 283.
74. Ibid., 278.
75. Ibid., 284.
76. Helmholtz to du Bois Reymond, 14 October 1849, in Kirsten, 87.

magnet was proportional to the time interval in question. Helmholtz varied the degree of mechanical work performed by the muscle by attaching a scale, on which could be put a small weight, to one end of the hanging muscle. If no weight were on the scale, the magnet moved very little; the deflection grew as the weight increased.[77] As this description of his new method makes clear, in October 1849 Helmholtz employed Pouillet's method within the context of a continuing investigation of the *mechanical* action of the muscle and he did so in order to examine at a more refined level the properties of the curves he initially had produced by his graphical method.

The task proved to be non-trivial. "The adjustment of the instrument gave me a great deal of trouble and cost me much time," he told du Bois-Reymond.[78] Still, even though he had to lay his investigation aside when the winter semester began in late October, he had by then perfected his apparatus and protocol so that he was able to produce more refined measurements. His apparatus (Figure 2.4), classical in its form, allowed a muscle to hang suspended from the screw **I** to which was attached a series of screws and contact surfaces that would break the flow of the current when the muscle raised the weight suspended on the scale-pan **K**. The apparatus was placed in a container with humidity-enriched air in order to prevent the muscle from drying out; this set-up remained in a usable state for three to four hours. At a certain point following stimulation of the nerve **w**—see the enlargement on the right-hand side of Figure 2.4, where **v** is the current-carrying wire to the nerve—the "energy" (*Spannung*) of the muscle would equal the load suspended from its lower end. After that point, any increase in the energy of the muscle would elevate the load "a little" and separate point **m** from point **n** on the apparatus. If, however, a weight were put on the scale-pan **K**, such that the muscle were acted upon by an additional overload, then the stimulated muscle could raise the combined weight only if its energy (elastic *Spannung*) equaled the sum of the weights of the load and the overload. Helmholtz arranged his apparatus so that the current, whose time interval was to be measured, would break when the elastic *Spannung* of the muscle increased by an amount sufficient to raise the weight of the overload.[79]

The experience of working with a more refined apparatus, whose protocol differed markedly from that of the graphical method, brought much more than the accurate measurements Helmholtz needed to

77. Ibid.
78. Ibid.
79. "Messungen," 284–89.

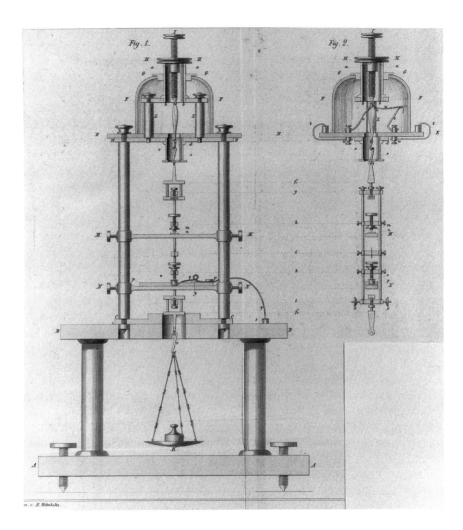

Figure 2.4. Helmholtz's apparatus for measuring the time course of muscle contraction and the propagation velocity of the nerve impulse. On the left, Figure 1 shows the entire apparatus; on the right, Figure 2 shows the arrangement when the nerve w is attached and more than one point on the nerve can be stimulated. Source: "Messungen über den zeitlichen Verlauf der Zuckung animalischer Muskeln und die Fortpflanzungsgeschwindigkeit der Reizung in den Nerven," MA (1850):276–364, Figures 1 and 2. Courtesy of Yale University Photographic Services.

characterize in greater detail the initial features of the curve depicting the mechanical action of a muscle as a function of time. Around Christmas 1849, when he again resumed his experiments, Helmholtz learned with his new apparatus that when he stimulated a frog muscle to contract by applying a current to the nerve attached to it, the time that elapsed between the stimulus and the moment when the muscle exerted sufficient force to be able to lift the overload varied according to *where* he stimulated the nerve. Stimulating the nerve a greater distance from the muscle resulted in a longer time interval than stimulating it closer to the muscle.[80] Helmholtz temporarily put aside further examination of the mechanical work performed by the muscle and concentrated instead on this time difference, made so evident by his new apparatus and protocol which had been designed initially for other purposes, that of investigating the mechanical action of the muscle. This time difference suggested to Helmholtz that he might be able to determine the propagation velocity of the nerve impulse, something Müller and others had hitherto believed almost instantaneous and not accessible through measurement.[81] Helmholtz's new apparatus thus gave him the opportunity to make a major scientific breakthrough, one made possible by more precise measurements.[82]

In late December 1849 and early January 1850, Helmholtz conducted trials in which he compared the deflection of the magnet when a nerve was stimulated in two different places. In addition to measuring, by Pouillet's method, the current's time interval, he also measured the height to which the muscle raised the overload in order to guarantee that both the nearer and the more-distant parts of the nerve attained an equal mechanical effect. If the nerve's irritability decreased or if the electrical stimulation became weak, the muscle would not elevate the overload as high; Helmholtz used this as a means to ensure

---

80. Ibid., 328–29.

81. "The time [the stimulus takes to travel to the brain and back] is infinitely small and unmeasurable." (Johannes Müller, *Handbuch der Physiologie des Menschen*, 4th ed., 2 vols. [Coblenz: J. Hölscher, 1840–44], *1*:583.) Although Leo Koenigsberger believed that Müller and others thought that the propagation velocity of the nerve impulse would never be measurable because it was close to if not in fact instantaneous (Koenigsberger *1*:118–19), Müller himself thought that knowledge of the velocity of the nerve impulse was held up for want of the technical means to examine it.

82. Other historians as well as some nineteenth-century scientists have maintained that Helmholtz *discovered* the nerve velocity through his muscle curves; see, e.g.,: Lenoir, "Models and Instruments," 20; Kremer, *Thermodynamics of Life*, 298, n. 57; and H. Kronecker, "Ueber graphische Methoden in der Physiologie," *Zeitschrift für Instrumentenkunde 1* (1881):26–33, on 27. In our view all the curves did was to *suggest* a time-lag in the mechanical action of the muscle, a time-lag that Helmholtz considered in further need of investigation if he were to understand the complete course of the mechanical action of the muscle.

the constancy of the stimulation. He measured the height by adding to his apparatus a wooden bar that could be raised; with the aid of a microscope, its elevation could be measured to within 1/20th of a millimeter. He observed the magnet's deflection with a telescope. The observations were complicated enough to require two people to conduct them, and so Olga Helmholtz, now pregnant with their first child due in June, continued to assist him. Some additional modifications also had to be made in the currents applied to each part of the nerve—they could not always be equal—in these preliminary trials because Helmholtz was working with frogs that had become weakened due to four months of captivity.[83]

The earliest set of trials that Helmholtz reported were conducted on 29 December 1849 and again on 4 and 6 January 1850. For the first (29 December), Helmholtz used a 40-millimeter nerve taken from a frog that had been in captivity for four months. For the right muscle with an overload of 20 grams, the time interval was 0.00130 seconds and the propagation velocity 30.8 meters per second. For the left muscle with an overload of 100 grams, the time interval was 0.00125 seconds and the velocity, 32.0 meters per second. He obtained similar results on 4 January when he used a nerve 43 millimeters long, kept the same muscle, but ran trials with two overloads (20 and 100 grams) obtaining a velocity of 31.4 and 38.4 meters per second, respectively. Two days later he again used a nerve 43 millimeters long, but applied the same current to both places on the nerve and used an overload of 180 grams in all of the 22 trials that he conducted. Here the time interval was 0.00175 seconds and the velocity 24.6 meters per second.[84]

These were Helmholtz's initial, early results as he reported them over a half-year later. With ranges in the velocity of the nerve impulse from 24.6 to 38.4 meters per second, they were scarcely decisive on the issue of what the nerve velocity was, or even if it had a single value. His measurements and even the curves produced by each stimulus of the nerve strongly suggested, however, that the propagation velocity of the nerve impulse was measurable and finite. This result had profound physiological implications. Helmholtz knew that "as long as physiologists insist on reducing the nerve effect to the propagation of an imponderable or psychic principle, it will appear unbelievable that the velocity of the current should be measurable." He interpreted the measurability of the nerve impulse as opening the way to an alternative explanation for the action of the nerve, even though

---

83. "Messungen," 331–34.
84. Ibid., 339–44.

the actual "processes in the nerve" that took place in the course of the stimulation were "mostly still unknown" to him. Invoking du Bois-Reymond's theory that a stimulated effect could be understood as a change in the arrangement of the material molecules, Helmholtz likened the propagation of effects in the nerve to physical processes, such as "the conduction of sound in the air and in elastic matter or the burning of a tube filled with an explosive mixture."[85] Yet in the long course of time over which he conducted these and similar experiments, Helmholtz did not develop these physical analogies into a hypothetical theory on the operation of the nerve impulse.

Instead, in the days following these trials he exploited his findings to advance his career as a young extraordinary professor by establishing priority for what he now considered to be a very important discovery, one that he undoubtedly considered sufficiently well established by the few trials he had conducted thus far. Following less than three weeks of trials, on 15 January 1850, Helmholtz dispatched the first of his preliminary reports. All told, five announcements of his findings appeared in scientific journals in Germany and France.[86] Reactions to these reports stimulated Helmholtz to rework and rewrite his argument and, more importantly, to rethink the demonstrability—and hence at some level also the reality—of the process he thought he had isolated and of whose existence he was now trying to convince others. Techniques associated with precision measurement played a central role in the evolution of his argument over the next half-year or so before its final presentation to the Berlin Physikalische Gesellschaft on 19 July 1850 and its publication in full in *Müller's Archiv* later that year.

In the first of these reports, presented by Helmholtz's esteemed teacher Müller to the Berlin Akademie der Wissenschaften on 21 January 1850, Helmholtz presented his results in terms of the specific conditions of his experiment. There was a "measurable time," he

---

85. Ibid., 330–31.
86. The first was sent to du Bois-Reymond, whom Helmholtz instructed to read the report to the Berlin Physikalische Gesellschaft in order to establish his priority. (Helmholtz to du Bois-Reymond, 15 January 1850, in Kirsten, 90–2, on 90.) Du Bois-Reymond read "Ueber die Fortpflanzungsgeschwindigkeit des Nervenprinzips" to the Gesellschaft on 1 February 1850. Müller read Helmholtz's report to the Berlin Akademie der Wissenschaften on 21 January 1850; this version appeared in the Akademie's *Bericht* and was republished in Müller's *Archiv* and J.C. Poggendorff's *Annalen der Physik und Chemie*: "Ueber die Fortpflanzungsgeschwindigkeit der Nervenreizung," *MB* (1850): 14–15, in *WA 3*:1–3; "Vorläufiger Bericht über die Fortpflanzungsgeschwindigkeit der Nervenreizung," *MA* (1850):71–3; and "Ueber die Fortpflanzungsgeschwindigkeit der Nervenreizung," *AP 79* (1850):329–30. The fifth announcement, which Alexander von Humboldt transmitted to the Paris Académie des sciences, was: "Note sur la vitesse de propagation de l'agent nerveux dans les nerfs rachidiens," *Comptes rendus 30* (1850):204–6.

found, in the transmission of the stimulus; specifically, "for the case of large frogs, whose nerves were 50–60 millimeters long, and which I had prepared at 2–6° C while the temperature of the observation room was between 11° and 15°, this interval of time was 0.0014 to 0.0020 seconds." That indeed "the duration of the current [was] exactly equal to the time between the stimulation of the nerve and the first mechanical effect of the muscle," Helmholtz sought to convince his readers by providing a thumbnail sketch of his apparatus and its protocol (which the Akademie's physicists probably understood better than its physiologists). He carefully mentioned that an error afflicting the currents did not exceed one-tenth the time interval marking the delayed response of the frog's muscle. This was, however, the only experimental error that he mentioned. In addition, he was sensitive to the variability in the data produced in his trials. When the muscle lifted a 20-gram weight, the time intervals varied widely; Helmholtz reasoned that a large series of measurements was required to take averages that would smooth out the irregularities. By contrast, the time intervals measured when the muscle lifted weights between 100 and 180 grams showed remarkable consistency, so that Helmholtz believed that any single result in the trials could represent them all.[87]

At the time of this first preliminary report, Helmholtz focused primarily on the time required for the stimulus to traverse a given length of nerve and the variation of this interval with temperature.[88] Even his brief presentation made it clear that the isolation and measurement of this time interval depended upon a complicated array of coinciding events, especially an electrical (the breaking of the current) and a mechanical one (the lifting of the weight). Despite the brevity of his description of this complex experiment, Helmholtz's preliminary report generated remarkable praise. Müller, the first to respond, found that Helmholtz's results pleased him to the highest degree: "I am convinced," he explained, "that therewith a great stride has occurred, the first in the art of measuring the nerve effect." Leaping ahead of Helmholtz, Müller boldly suggested a research program: to extend Helmholtz's methods to the measurement of the time of reflex nerve action and of the propagation of stimuli in the sympathetic nervous system.[89] Alexander von Humboldt praised Helmholtz's "acumen and talent in

---

87. "Fortpflanzungsgeschwindigkeit der Nervenreizung," 14. See also AW, No. 541: Vorläufiger Bericht über die Fortpflanzungsgeschwindigkeit der Nerventhätigkeit, which differs insignificantly from the published version.
88. "Fortpflanzungsgeschwindigkeit der Nervenreizung," 15.
89. Müller to Helmholtz, 7 February 1850, AW, No. 318.

experimenting with the most delicate apparatus. . . . Such a noteworthy discovery speaks through the astonishment that it has stimulated."[90] Carl Ludwig, du Bois-Reymond's comrade-in-arms in the reform of physiology along exact lines, told Helmholtz that his discovery "has filled me with such pride," and then noted that whereas Helmholtz had climbed up to "the superior circle of physicists and mathematicians," Ludwig was satisfied that at least he had "sprung into the class of chemists."[91]

Yet du Bois-Reymond himself was initially not as enthusiastic, and he became the self-appointed translator and interpreter of Helmholtz's findings. In mid-February 1850, Humboldt, who was to transmit Helmholtz's preliminary report to the *Comptes rendus* of the Académie des sciences in Paris, reported that du Bois-Reymond, who was translating the report into French, was clarifying the report without changing its meaning.[92] Nor was that all. Although Müller had given Helmholtz no indication that the presentation of his report elicited anything out of the ordinary,[93] du Bois-Reymond told Helmholtz "with pride and sorrow" that Helmholtz's report was understood and valued by no one but himself because

> you represented the matter—do not take this in the wrong way—in so extravagantly obscure a manner, that your report could at most serve as a short introduction to the rediscovery of the method. The consequence was that Müller did not discover it again, and after his lecture members of the Akademie thought that you had not eliminated the time which elapses during the process in the muscle. I had to clarify the matter to one person after another: to [Peter] Riess, [Heinrich Wilhelm] Dove, Magnus, [Johann Christian] Poggendorff, Mitscherlich, and finally to Müller himself who wanted nothing to do with it. I read it to the [Berlin Physikalische] Gesellschaft and at least this difficulty did not surface. But Humboldt was entirely out of his element and refused to send your piece to Paris, whereupon I offered to rework it and make it understandable.[94]

It is difficult to determine if du Bois-Reymond's clarifications had in fact been responsible for the overwhelmingly positive responses that

---

90. Humboldt to Helmholtz, 12 February 1850, partially reprinted in Koenigsberger *1*:118.
91. Ludwig to Helmholtz, 10 [February?] 1850, AW, No. 293.
92. Humboldt to Helmholtz, 12 February 1850.
93. Müller to Helmholtz, 7 February 1850.
94. Du Bois-Reymond to Helmholtz, 19 March 1850, in Kirsten, 92–4, on 92.

Helmholtz had thus far received for his paper—all were dated after the presentations that du Bois-Reymond mentioned—or if he exaggerated the lack of understanding of Helmholtz's report so as to position himself as a crucial intermediary between Helmholtz and the scientific community at large. If the latter were the case—and it appears that at least for a while that it was so—then du Bois-Reymond did not always play the role consistently; for while he continued to assist Helmholtz during various stages in the publication of his results, he failed to report on Helmholtz's findings in the numerous reviews on the progress in electrophysiology in the Berlin Physikalische Gesellschaft's *Fortschritte der Physik*.[95]

The immediate effect of du Bois-Reymond's intervention came from his and Humboldt's rewriting of the report for publication in the *Comptes rendus*. Humboldt took the length of the nerve and divided it by the times elapsed, thus producing actual velocity measures for the nerve impulse, 25 and 43 meters per second. This was a small modification, certainly, but it was not one that Helmholtz had thought of doing himself. It served the purpose of casting Helmholtz's results even more in physical terms. Du Bois-Reymond, on the other hand, claimed not to have changed a single detail, but to have "developed [the essay] inductively." In du Bois-Reymond's version, Helmholtz's preliminary report began not with Helmholtz's quantitative findings (which were now noted last), but by announcing the existence of the time interval between the electrical stimulus and the muscle's action, a time interval that was "not too difficult to estimate." Du Bois-Reymond then went on to describe Helmholtz's experiment, the error Helmholtz had considered, and Helmholtz's qualitative conclusions.[96]

Du Bois-Reymond's editing prompted Helmholtz to reflect on the nature of scientific writing. Helmholtz revealed to du Bois-Reymond that he had found the wording of such a preliminary report to be "difficult" because such a "short notice for the experts," intended to

---

95. See *Fortschritte der Physik* 3 (1847; published in 1850); 4 (1848; published in 1852); 5 (1849; published in 1853); and 6/7 (1850/51; published in 1854).
96. "Note sur la vitesse de propagation"; and AW, No. 526: Emil du Bois-Reymond. Uebersetzung eines Briefkonzepts von Helmholtz in die französische Sprache. Du Bois-Reymond's rewriting of Helmholtz's report is of a piece with Brigitte Lohff's characterization of du Bois-Reymond's conception of experiment: he placed primary emphasis on the *Anschaulichkeit* of the experiment, the technical improvement of instruments, and the control of experimental conditions; and he assigned secondary importance to quantitative results which, for him, had to say something about a functional relation between variables (hence he assigned relatively small weight to tables). (Brigitte Lohff, "Emil du Bois-Reymond's Theorie des Experiments," in *Naturwissen und Erkenntnis im 19. Jahrhundert: Emil du Bois-Reymond*, ed. Gunther Mann [Hildesheim: Gerstenberg, 1981], 117–28, esp. 118, 120, 121.)

"lead them to discover the particulars" of the experiment, was not easy to compose. "It is not easy to judge," Helmholtz lamented, "at which level of knowledge one must aim for" in these brief notes, whereas when composing the final, completed essay, "one reckons on the lowest possible level."[97] What is real for the discoverer, he seemed thereby to be saying, may not be so for peers and still less so for other audiences. Yet it seems that du Bois-Reymond's reaction and rewriting had profound implications not only for how Helmholtz constructed his final argument concerning his results, but also for how he shaped the subsequent course of the trials that constituted the chief repository of evidence for that argument. The degree of precision and certainty that a reader would acknowledge in his results depended upon the final structure of this argument and the nature of these trials. Precision, certainty, and demonstrability were all responsible for conveying the reality of the effect Helmholtz had found.

Du Bois-Reymond's reaction, however, was not the only one that broached the issue of the certainty and the demonstrability of Helmholtz's findings. Helmholtz's father, August Ferdinand Julius Helmholtz, found the finiteness of the propagation velocity peculiar, having himself regarded mental thoughts and bodily actions as simultaneous.[98] Helmholtz tried to convince his father of the propagation velocity's reality in two ways. One was through analogy. Just as sound is carried by the elasticity of the air to the ear's nerves, so, Helmholtz thought, following du Bois-Reymond's theory of nerve propagation that the nerve stimulus is propagated by means of an alteration of molecules, some kind of ponderable matter in the nerve propagates a message along its path. Analogy, however, offered at best a means to visualize the effect; it scarcely offered convincing evidence of its reality. For this Helmholtz turned to the techniques of error analysis that had become so important in the analysis of precision measurements in astronomy and that, primarily through Magnus in Berlin, Wilhelm Weber in Göttingen, and Neumann in Königsberg, had become an integral part of instruction in exact experimental physics.[99] "The inaccuracy of our

---

97. Helmholtz to du Bois-Reymond, 22 April 1850, in Kirsten, 96–7. According to Koenigsberger, Helmholtz rewrote his essays several times before he achieved the "logical" form that suited him. (Koenigsberger *1*:145.) Using Helmholtz's laboratory notebook from these experiments, we treat the relation of scientific writing to the historical unfolding of the experimental investigation in our study of Helmholtz's early scientific career.
98. August Ferdinand Julius Helmholtz to Helmholtz, 3 April 1850, cited in Koenigsberger *1*:121–23, on 122.
99. On error analysis and physics instruction at Königsberg, see Kathryn M. Olesko, *Physics as a Calling: Discipline and Practice in the Königsberg Seminar for Physics* (Ithaca, N.Y.: Cornell University Press, 1991).

time-perceptions by two different sense-organs has recently been demonstrated in the most striking manner," Helmholtz told his father. "Astronomers vary in their estimation of the moment at which a star crosses a web of their telescopes by more than a whole second, while the estimates of any individual taken by himself agree within one-tenth of a second if frequently repeated."[100] Helmholtz was here referring to the so-called personal equation, which had been identified in 1823 by the Königsberg astronomer Friedrich Wilhelm Bessel as an unavoidable error contributed by the observer in the act of measuring. Helmholtz thus invoked one of the principal techniques by which the epistemological *un*certainties in measurement were determined in order to certify his evidence for the reality of the effect he had isolated experimentally.[101]

Helmholtz could not afford to dismiss du Bois-Reymond's letter and criticisms. Du Bois-Reymond was well known as a pioneer and master in the field that Helmholtz had just entered, electrophysiology, and for that reason Helmholtz needed his respect, professional support, recognition, and especially his endorsement of the discovery that had emerged from Helmholtz's study of the mechanical action of muscles. Helmholtz could thus not afford to ignore the academicians' question concerning whether he had taken into account the time course of the effect through the muscle, if only because he knew from his previous experience with heat production in the nerves that neglecting experimental errors could lead to spurious results dramatically like real effects. Even though he may have thought that in this set of experiments he had meticulously calibrated and assessed disturbing factors, apparently more had to be done to make his argument more convincing.

Over the next six months, before the near-final version of his report was presented to the Berlin Physikalische Gesellschaft on 19 July 1850, Helmholtz developed his investigation along two lines. One was to use the quantitative techniques associated with precision measurement to test the reliability of his data and enhance the rigor of his argument. The other was to vary the style of his experimental trials in ways that diminished the importance of certain precision techniques and to return to the kind of graphical analysis with which he had begun his

---

100. Helmholtz to his father, cited in Koenigsberger *1*:123–25, on 124.
101. Ironically, Müller had also known of Bessel's personal equation and even mentioned it prior to his discussion of the probable unmeasurability of the propagation velocity of the nerve impulse. He did not, however, view the equation's existence as possible evidence for the finiteness and measurability of the propagation velocity of the nerve impulse. (Müller, *Handbuch 1*:582.)

study in order to examine the physiological significance of his discovery and, more importantly, to demonstrate it. In short, even as Helmholtz found it necessary to use the most complicated forms of reasoning associated with precision measurement, he was also developing less rigorous forms of quantitative experimentation that could explore more fully the properties of the muscle curve and, inter alia, of the nerve impulse as well. Each line of investigation served a different purpose and addressed a different audience.

7. Precision, Demonstrability, and the Role of Error Analysis

Helmholtz's deployment of error analysis reveals especially well what precision measurement meant to him. Unlike his late eighteenth-century predecessors in quantification, Helmholtz did not identify the fineness of his measurements with the number of decimal places he was able to measure (or, more likely, merely generate). Practitioners of precision measurement had, by the 1840s, come to view the meaning of precision more in terms of the techniques that delimited the validity of data and that supplied assurances as to the epistemological soundness of the data they had obtained. Although the properties of data had occasionally concerned Helmholtz in previous investigations, he here sought with greater effort to remove or to define the influence of spurious effects on his data, especially those that adversely influenced its consistency, regularity, or pattern.

Helmholtz accounted for the errors in his measurements in three ways: by computing or in some other way compensating for his experiment's constant (or systematic) errors; by understanding in physical and, if possible, quantitative terms the operation of his protocol and by perfecting that protocol; and finally, by attempting to come to terms with irregularities in the data that could not be treated by either of these two methods. By far the greater part of his effort was devoted to the first two of these, especially to the modification of his apparatus so that he could guarantee that the stimulating process coincided "accurately" with the beginning of the time-measuring current.[102] He also

---

102. See his modifications of his apparatus and of Pouillet's method, especially his use of induced currents whose duration, he noted, "has hitherto evaded all measurements and has always been considered in all applications as vanishingly small," so that he had to "convince" himself "that the duration of the applied current does not attain as small an interval of time, as that which I am measuring." ("Messungen," 290–95, quotes on 296, 297.)

determined how certain environmental features of his experiment affected its outcome. Air currents affected his time measurements in the amount of not more than 1/10,000 second, and colder temperatures, he deduced, probably resulted in a greater time lapse and a lower propagation velocity.[103] He "delayed" his proof "that the measurements made could not be inflicted with considerable errors" until *after* a discussion of these external factors, indicating that no matter what the original historical order of his treatment of these factors in his experimental trials had been, they had a definite place in his argumentative strategy.[104] This "proof" consisted of a discussion of the errors stemming from the imperfect mechanical operation of his apparatus and the muscle, especially the muscle's elastic after-effect, and the mechanical errors that were due to the "weak disposition of the apparatus," including its organic components (i.e., the muscle), when their operation did not exactly fulfill the mechanical laws under which they were assumed to operate.[105]

Error analysis thus became for Helmholtz not a way to achieve an exact value of the time interval that could be used to calculate the nerve velocity, but rather an interpretive device, a way to determine his data's condition and reliability. He believed that these analyses proved that his measurements were "not essentially falsified" by the operation of the apparatus or the execution of the protocol.[106] Hence, when he summarized "the sources of error ... according to their type of influence," he considered error's "influence" not in terms of its physical effect on the apparatus or protocol, but in terms of its impact on his data's condition; specifically, according to whether or not the error changed the data's average value.[107] But what was the meaning of that average value if it was in some way afflicted with errors? It seems that Helmholtz had so constructed his argument up to this point that he *had* to answer that question, or one like it, in order to determine, at a deeper level, the epistemological certainty of his data. And to do that he had to turn to another technique associated with precision measurement, one derived from probability calculus: the method of least squares.

In 1850, least squares was used almost exclusively in the exact sciences of astronomy, where it became common in the 1810s, and physics and chemistry, where it was introduced in the 1830s and 1840s.

---

103. Ibid., 296 (for air currents), 344–45, 358–62. Helmholtz seems to have suspected the role of temperature in his trials as early as his preliminary report in January, when he provided the temperature at which his trials had been conducted.
104. Ibid., 309.
105. Ibid., 309–21, on 315.
106. Ibid., 317.
107. Ibid., 321–22.

Due to his acquaintance with exact scientists at both Berlin and Königsberg, Helmholtz could have learned of the method from any number of sources. That he did not use the method in earlier instances where he could have suggests that he may have first become aware of the method's power in precision measurement at Königsberg, where he had done his pathbreaking trials on the velocity of the nerve impulse. Here the method had been popular in several exact sciences since the late 1830s when Bessel, Neumann, and the mathematician Carl Gustav Jacobi were each exploring the method and its applications. At the same time, in 1838, Neumann introduced it into his mathematico-physical seminar where, as one of the defining features of his teaching and of several of his students' investigations, the epistemological certitude of data in precision measurement, as determined by the method of least squares, became a matter of deep concern in the type of exact experimental and theoretical physics practiced by several of his students.[108] The method was, however, scarcely known in the life sciences, where data were often more irregular than in the physical sciences and hence not especially suited for the method, and where quantitative reasoning was generally less prevalent. Now speaking from the perspective of someone familiar with the method's usefulness, Helmholtz remarked in his nerve impulse study that "one is *accustomed* to measuring the irregularities of observations through the most probable error of the observations, that is to say, through that value by which individual observations deviate from the average of a sufficiently large number of [observations]," these observations exceeding the average value as often as they fail to attain it.[109]

Two issues of concern to Helmholtz in the delicate measurements of the time interval were his data's reliability and accuracy. Both are difficult to define; they often depend on the individual investigator's faith in the instruments at hand, sense of what the data are *supposed* to be (sometimes based on theoretical expectations), and judgments regarding the data's accuracy. The values guiding all three are community-based, often modified by individual experiences. Helmholtz appeared to gauge data's reliability by its density (a reasonably extended series of observations) and by the agreement of individual observations among one another.[110]

Accuracy was another matter. By the time of this investigation, Helmholtz had become sensitive to the subtleties involved in distinguishing those properties of data attributable to instrument operation and design from those stemming from other factors. It had not always

---

108. Olesko, *Physics as a Calling*, esp. 159–61, 167–68.
109. "Messungen," 322–23 (emphasis added).
110. Ibid., 337.

been clear, however, exactly how accuracy could be determined. Like other German investigators, Helmholtz used the word *"Genauigkeit"* to mean both accuracy (roughly, the degree of agreement between theoretical and experimental values) and precision (the close agreement of refined observations to one another). From time to time (and later with increased frequency) he used the word *"Präzision"* to describe data's fineness and degree of exactitude. In one early instance, for example, he praised Wilhelm Weber's "wonderful precision" in confirming Ampère's law by measuring trials. Yet he also recognized that precise measurements were not all alike, for he noted that Weber's electrodynamometer was "not calculated for great sensitivity, but for the arranging of absolute measurements."[111] Helmholtz did not use the word *"Präzision"* in describing the nerve-velocity data. Instead, in discussing the trials used to calculate the nerve-velocity impulse, he explained that, for the trials with one or two different overloads in which an extended series of numbers could be recorded, he calculated "the average value of the time intervals between the [electrical] stimulation and the [mechanical] action of the muscle in both places on the nerve *and besides that for the judgment of accuracy [I] have calculated the most probable error of all these values according to the rules of probability calculus.*"[112] He appended a lengthy footnote to his argument at this point, wherein he defined the meaning of the probable error in terms of the likelihood of winning a bet as to where the true value was found; this spoke to his audience's ignorance of this kind of quantitative reasoning.[113] Thus, rather than emphasizing the fineness of his measurements or their degree of exactitude, he chose to define

---

111. Helmholtz to du Bois-Reymond, 10 April 1847.
112. "Messungen," 337 (emphasis added). Helmholtz here uses the word *"Genauigkeit"* to mean "accuracy."
113. Helmholtz's footnote read: "For those of my readers who are not familiar with the rules of probability calculus, I note here the example, from the nine series of trials, that the value of the time difference with regard to the propagation would be 0.00175 second with the probable error ±0.00014. According to a popular expression it would be a 1 to 1 bet that the true value of this difference lies between 0.00189 and 0.00161 seconds. Furthermore, it is a 10 to 1 bet that the deviation is at most 2.5X, 100 to 1 that it is at most 3.8X, 1000 to 1 that it is 4.8X as great as the most probable error. Therefore, the value lays with the probability

1 to 1 between 0.00189 and 0.00161
10 to 1 between 0.00210 and 0.00140
100 to 1 between 0.00228 and 0.00122
1,000 to 1 between 0.00242 and 0.00108."

The trial in question was that of 6 January 1850, the third of the early reported trials but the first propagation velocity trial mentioned in his final essay (Figure 5). (Ibid., 337n-38n.)

Reihe IX.

Am Oten Januar mit dem Muskel eines vier Monate aufbewahrten Frosches angestellt. Durch beide Stellen des Nerven wird der gleiche Strom geleitet, die Entfernung derselben ist 43 mm. Einstellung des Muskels ungeändert; Ablenkung vorher 121,24, nachher 118,61, im Mittel 119,97.

No.	Uebertastung.	Erhebungshöhe.	Differenz der Ausschläge bei Reizung der entfernteren Nervenstelle.	Differenz der Ausschläge bei Reizung der näheren Nervenstelle.
1	180	0,88	186,83	
2	—	0,87	189,71	
3	—	0,83		180,66
4	→	0,82		181,83
5	—	0,80	190,79	
6	—	0,80	189,99	
7	—	0,80		186,62
8	—	0,80		182,09

No	Uebertastung.	Erhebungshöhe.	Differenz der Ausschläge bei Reizung der entfernteren Nervenstelle.	Differenz der Ausschläge bei Reizung der näheren Nervenstelle.
9	180	0,78	193,06	
10	—	0,76	193,94	
11	—	0,77	191,85	
12	—	0,72		180,38
13	—	0,70		182,43
14	—	0,70		184,20
15	—	0,68	192,80	
16	—	0,65	190,64	
17	—	0,65		186,27
18	—	0,65		181,87
19	—	0,65	190,89	
20	—	0,65	191,14	
21	—	0,65		181,58
22	—	0,65		183,96
Mittel			191,13	183,44
Wahrscheinlicher Fehler des Mittels			± 0,89	± 0,42
Derselbe der einzelnen Beobachtung			± 1,81	± 1,89
Zeitdauer zwischen Reizung und Erhebung des Gewichts			0,04394	0,04219
Wahrscheinl. Fehler derselben			0,00009	± 0,00010

Daraus bestimmt sich endlich der Zeitunterschied wegen der Fortpflanzung: 0,00175 ± 0,00014

die Fortpflanzungsgeschwindigkeit: 24,6 ± 2,0 Mt. in der Sekunde

Figure 2.5. Helmholtz's nerve propagation velocity data from 6 January 1850 with the results of least-squares calculations. The third of the preliminary trials that Helmholtz conducted, this one was the first to appear with nerve velocity calculations in his final essay. Note the consistency of the heights of elevation when the muscle raised an overload of 180 grams. Source: "Messungen über den zeitlichen Verlauf der Zuckung animalischer Muskeln und die Fortpflanzungsgeschwindigkeit der Reizung in den Nerven," MA (1850):276–364, on 339–40 (Reihe IX). Courtesy of Georgetown University Photographic Services.

accuracy in terms of his measurements' *uncertainties*. For the earliest trials reported in the final essay—those from 29 December 1849 and 4 and 6 January 1850 which were presumably the basis of his preliminary report—Helmholtz computed the probable error of the magnet's deflection, the time interval, and, finally, the velocity (Figure 2.5).[114]

In Helmholtz's argument, precision's power to convince the reader of the significance of his findings relied on a basic understanding of probabilistic reasoning and a shrewd calculation of the unknown, of

---

114. Ibid., 339–44. The uncertainties in the velocities he reported were, in some cases, actually quite large. For the trials on 29 December 1849 and 4 and 6 January 1850, the velocities (all in meters per second) and their uncertainties were: 24.6 ± 2.0; 30.0 ± 6.4; 32.0 ± 9.7; 31.4 ± 7.1; and 38.4 ± 10.6.

the uncertainties always present in the act of measuring. Error analysis was thus for Helmholtz an essential stage in the process of discovery, of shaping one's findings. At the most basic level, error analysis secured the regularity, reliability, and certainty of his data. Error analysis also served a far more important epistemological function. It established the reality of the effect he had measured, and hence served as a powerful technique of argumentative persuasion. A reader essentially had to understand and believe in the value of error analysis in order to appreciate how Helmholtz's data *meant* something. From his analysis of constant errors, Helmholtz believed that he had "certified" that "the energy of the muscle develops gradually at first."[115] His use of the method of least squares was a lesson in how fineness and precision of measurement were not in themselves guarantors of certainty. Even a reference to the personal equation, echoed in his letter some months earlier to his father, constituted de facto evidence for the effect he had discovered; for the well-known fact that "most practiced astronomers differ around an entire second in comparative visual and audio perceptions" was corroborative evidence for the impossibility of directly perceiving the time interval in question. That this time interval remained after all errors were accounted for meant that the delay was a real property of the nerve, not the apparatus. As Helmholtz wrote: "Evidently this [time] difference cannot be conditional on some one of the earlier mentioned sources of error, which have their basis in the mechanical and electrical processes of our method of measuring. . . . The reason must rather be found in the [physiological] processes within the nerve itself."[116]

Yet precision measurement, and the error analysis that was a part of it, did not and could not address deeper physical and especially physiological questions concerning the properties of the nerve-velocity impulse. The large number of trials needed to compute the nerve velocity and its error by least squares were, ironically, limited in their ability to yield information concerning the dependency of the result on such factors as the weight of the overload, the decline in irritability of the muscle, and the temperature of the nerve—all issues which had to be examined in order to explore, at a deeper level, processes within the nerve itself. To explore these properties more deeply Helmholtz had to employ other techniques and other investigative strategies designed to elicit primarily *qualitative* rather than quantitative results. Thus, between the end of March 1850, when he

---

115. Ibid., 317. Helmholtz uses "energy" here to mean the mechanical effect of the muscle contraction.
116. Ibid., 329.

began new trials at the beginning of a six-week vacation, and the end of June 1850, Helmholtz, with his wife's assistance, conducted several series of trials, the runs of which were less numerous but whose variables (e.g., the overload, the temperature of the nerve, the condition of the frogs) changed more often.[117] He notably also began, on 11 April 1850, to stimulate the muscle directly.[118] In addition, he initiated investigations on human nerves, but at first had only "irregular" success in comparison to his trials with frogs, and so "a beautifully exact proof [was] not to be had." By using warm-blooded animals, he hoped to find the lower limit of the propagation velocity.[119] In sum, this part of Helmholtz's investigative strategy was characterized by a willingness to alter features of the apparatus and the protocol of his experiment not in order to achieve more precise numbers, but rather to use these changes as a way to elicit the full scope of the properties of the effect that he had isolated.

In fact, only a small number of trials designed primarily to supplement the preliminary trials of the nerve-velocity impulse from December and January were actually completed; in none of these later trials did Helmholtz compute errors by the method of least squares.[120] Still, these later trials agreed remarkably with his original ones. Their average was 27 meters per second; the average from the original trials, computed by the method of least squares, was 26.4 meters per second. The agreement between the less quantitatively rigorous trials and his earlier ones undoubtedly impressed upon Helmholtz the limited returns that the time-consuming operations of precision measurement and the tedious calculations required by least squares actually yielded.

The argument from precision measurement and error analysis had another shortcoming, too: its power to demonstrate its import and meaning to others was not guaranteed, so much did it rely on a readership sophisticated in quantitative reasoning. The demonstrative power of graphical analysis, on the other hand, was greater for two reasons. For readers, its message was accessible and understandable. For Helmholtz, it provided a quick way to detect patterns, and so to ask more detailed questions about the course of the muscle's action. The visual information contained in his muscle curves, situated at the beginning of his essay, served the didactic function of familiarizing the reader with the gross outlines of his findings, of presenting them with

117. Helmholtz, AW, No. 547.
118. Ibid., Bl. 27–54.
119. Helmholtz to du Bois-Reymond, 5 April 1850, in Kirsten, 94–5, on 94.
120. These trials were from 20, 24, and 25 May 1850. (AW, No. 547, Bl. 63–7, 73–95; "Messungen," 345–50.)

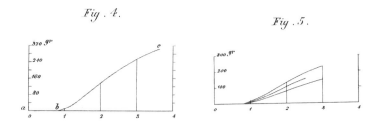

Figure 2.6. Helmholtz's graphical representation of data on the behavior of a stimulated muscle when no overload was applied. Abscissa is in hundredths of a second; ordinate, in grams. Source: *"Messungen über den zeitlichen Verlauf der Zuckung animalischer Muskeln und die Fortpflanzungsgeschwindigkeit der Reizung in den Nerven," MA (1850):276–364, Figure 4. Courtesy of Yale University Photographic Services.*

Figure 2.7. Helmholtz's graphical representation of data on the effect of three degrees of muscle fatigue on the muscle's ability to lift a weight. Abscissa is in hundredths of a second; ordinate, in grams. Source: *"Messungen über den zeitlichen Verlauf der Zuckung animalischer Muskeln und die Fortpflanzungsgeschwindigkeit der Reizung in den Nerven," MA(1850):276–364, Figure 5. Courtesy of Yale University Photographic Services.*

a puzzle whose solution had to be found. From time to time throughout his essay Helmholtz even found it advantageous to compare these directly recorded curves to others constructed from observations. One such instance concerned the behavior of the stimulated muscle when no overload was applied. From his numerical observations Helmholtz derived "a new significant result," that "a time [about 1/100 second] after the stimulation passes before the energy generally begins to climb." The general features of the curve representing these observations (Figure 2.6), Helmholtz concluded, had "great similarity" with the "curve of the height of equilibrium" obtained by his graphical method (Figure 2.3), although the ordinates were not quite proportional to one another.[121] Nevertheless, the fact that these two curves did not contradict one another undoubtedly indicated to Helmholtz

---

121. "Messungen," 308–9.

that graphical and quantitative evidence were comparable even though produced by two different apparatus.

At other times Helmholtz used curves constructed from data to understand or infer *visually* the physical process of the muscle's action, including its ability to raise an overload, its diminished response due to irritability, and so on. For example, in examining the effect of fatigue upon the muscle's ability to raise a weight, Helmholtz used his data to draw ascending curves corresponding to three levels of fatigue (Figure 2.7). The discrete data entries represented in tabular form could not tell at "a glance" what these curves could: how the ordinates behaved.[122] Hence patterns discerned in these constructed curves, and even in those produced directly by a stylus on a drum, were important foundations for inferences. Helmholtz expressed "the ascent and descent of the energy for two different nerve places through curves" that were superimposed so that both their congruencies and the longer abscissa of the curve belonging to the more distant stimulated nerve point became apparent. "Out of this type of temporal progression which offers us the effect of stimulation in the muscle," he then reasoned, "we could make an inference about the progression of the corresponding, mostly still unknown, processes in the nerve." He continued:

> The delay in the effect can only rest on the time that passes before it propagates from the distant position to the muscle. Through these experiments we are therefore in the position to ascertain the propagation velocity of the stimulation in the motor nerves, if one understands by stimulation those processes in the nerve which develop in it in consequence of a stimulating external effect.[123]

In possession of such curves prior to the production of exact quantitative data, Helmholtz found that he could use them to plan the course of his trials and ascertain where his next set of measurements should be taken.

Over the next few years, as Helmholtz extended this study and disseminated his results, in writing and in person throughout the universities of Germany, presentations and arguments based on graphical analysis edged out those based on the techniques and reasoning associated with precision measurement. He used his frog-drawing apparatus, which he named the myograph in 1854,[124] to illustrate his

122. Ibid., 325.
123. Ibid., 330.
124. Helmholtz to du Bois-Reymond, 13 June 1854, in Kirsten, 144–45.

findings and to demonstrate the reality of the propagation velocity to a larger public of both scientists and students. "I am demonstrating my frog curves everywhere," he wrote in 1851.[125] Significantly in the second half of his nerve propagation velocity study, published in *Müller's Archiv* in 1852, Helmholtz opened his essay with a criticism of the electromagnetic method of measuring time that had been so central to establishing the finiteness and measurability of the nerve's velocity. This method, he argued, was the "best guarantee" of the execution of "accurate measurements"; but, he continued, it had the disadvantage of requiring an extended, tiresome series of individual trials using especially good frog preparations which were difficult to achieve consistently. So, he continued, he turned to the graphical method of measuring time because it was "a great deal simpler and easier execution of the proof of the nerve's propagation time."[126] He devoted his essay almost entirely to a description of his more refined curve-producing apparatus, which was designed to produce two curves simultaneously, and to an explanation of the curves produced. About the absolute value of the propagation velocity thus derived, Helmholtz knew that it was "not measured with very great accuracy. Still, we find the value of the velocity [27.25 meters per second] approximately as large as in the earlier [electromagnetic] method [26.4 meters per second]."[127] The accuracy and certainty that he had earlier sought through precision measurement thus became too tedious and time-consuming to warrant spending extra time and effort just to get a result that, for Helmholtz's purposes, was only marginally more precise.

In the course of his continuing, multifaceted investigations on the nerve-velocity impulse, then, Helmholtz moved away from precision measurement and its complicated techniques. Precision was certainly powerful but it could not be the foundation of all arguments and reasoning in this intricately detailed investigation. For Helmholtz's purposes, precision was especially limited in its argumentative versatility. Although he initially used both precision and graphical analysis, each

---

125. Helmholtz to Olga Helmholtz, 6 August 1851, in Kremer, ed., *Letters*, 46–53, on 52.

126. "Messungen über Fortpflanzungsgeschwindigkeit der Reizung in den Nerven. Zweite Reihe," *MA* (1852):199–216, in *WA* 2:844–46.

127. Ibid., 215. It was principally the comparison of the two curves, not the time-lag itself, that suggested the finiteness of the propagation velocity. Helmholtz had first suggested this new, improved cylinder apparatus that could produce two curves near the end of his first essay on the nerve velocity. ("Messungen," 363, 358.) He described this apparatus's principles in a lecture delivered before the Königsberg Physikalisch-ökonomische Gesellschaft on 13 December 1850: "Ueber die Methoden, kleinste Zeittheile zu messen, und ihre Anwendung für physiologische Zwecke," *Königsberger naturwissenschaftliche Unterhaltungen, 2:2* (1851):169–89, on 185–86, in *WA* 2:862–80.

had served different purposes in his essay. Precision assumed a more professional and sophisticated readership or audience. It had confirmed the reality of the time interval through the quantification of the extraneous and the unknown. Graphs, on the other hand, could be presented to a more general, even popular, audience, and they were certainly well suited for teaching. They suggested the existence of the time interval through visual evidence, and hence their demonstrative power was inherently greater.[128] There were certainly points in Helmholtz's essay where curves were only considered approximations because the ultimate arbiter of what the curves "said" was more precise data.[129] Yet precision measurements were not epistemologically isolated; for their interpretation was not in any sense direct, automatic, or transparent. Reasoning from and interpreting precision depended on two strategies, neither of them dispensable. One was error analysis, especially (in the initial stages of the nerve-impulse study) the method of least squares. The other was graphs. Precision achieved by electromagnetic means was certainly, as Helmholtz himself admitted, a "more perfect method." But it was also one which "would offer perhaps greater difficulties without knowledge of the mentioned facts" that had been presented by the graphs of the muscle curves which offered a "simpler chain of reasoning to a comprehensive overview."[130] Successful use of different forms of quantitative experimental evidence thus depended on different preconditions. In the case of the nerve-impulse discovery, precision measurements would not have been so easily understood without first having had the "facts," and at least a part of their interpretation, on hand in the form of direct graphical recordings of the process Helmholtz so closely examined.

8. The Meaning of the Nerve-Impulse Discovery

Olga Helmholtz recorded Helmholtz's last trials on the nerve-impulse velocity on 19 June 1850, three days before she gave birth to their first child, a daughter.[131] Toward the end of June, Helmholtz finished composing his final arguments on the nerve-impulse discovery

---

128. See, e.g., Helmholtz to Olga Helmholtz, 10 August 1851, in Kremer, ed., *Letters*, 53–60, on 56, where he explains that the Heidelberg physiologist Hermann Nasse doubted Helmholtz's nerve velocity findings until Helmholtz showed him the frog curves, which Nasse thought made the findings at least "plausible."
129. See, for example, "Messungen," 352–53.
130. Ibid., 280.
131. Helmholtz, AW, No. 547, Bl. 104–9.

and gave his completed essay to the Königsberg chemist Bernhard Rathke, who was then traveling to Berlin.[132] There du Bois-Reymond was to see Helmholtz's results through publication. Before dispatching the final version of the essay, though, du Bois-Reymond communicated Helmholtz's findings to the Berlin Physikalische Gesellschaft so that they "did not come into the world without a baptismal certificate." "Will you continue to exploit your method?" he asked Helmholtz, adding that "an entire field of discoveries lays before you—you need only to choose in order to pluck the most beautiful."[133]

Helmholtz's continuing investigations in electrophysiology did in fact lead him to new discoveries concerning induction currents, sensory nerves, and other related problems. Their stories cannot be told here.[134] His discovery of the nerve-impulse velocity carried meaning, however, beyond his own research agenda. The delightful way in which he popularized his findings to the Königsberg Physikalisch-ökonomische Gesellschaft in December 1850 was an appropriate beginning to the widespread dissemination of his method and results.[135] His experiment proved to be a beautiful one: classic in its instrumental design; possessing a formulaic protocol that, over time, was duplicated and imitated; and decisive in both its precise and, after 1852, graphical, results. Practitioners from many different fields found something further to explore or exploit in it. Psychologists found the question that had been first raised by his father—a form of the mind-body problem—worth pondering in the context of his discovery, which also helped in understanding the relationship between stimuli and responses. For the physicist Friedrich Kohlrausch, duplicating Helmholtz's later experiments on sensory nerves became an exercise in understanding notions of error and precision.[136] For advanced students in Helmholtz's Heidelberg physiological laboratory, the experiment became an exercise that introduced them to the intricacies of quantitative reasoning in physiology.[137] Canonized early on in textbooks, the experiment remained, until early in the twentieth century, a classic that offered the

---

132. Helmholtz to du Bois-Reymond, 28 June 1850, in Kirsten, 92.
133. Du Bois-Reymond to Helmholtz, 25 August 1850, in ibid., 98–100, on 98. Helmholtz agreed with du Bois-Reymond's decision to read it to the society before publication; see Helmholtz to du Bois-Reymond, 28 August 1850, ibid., 100–2, on 100.
134. We treat these long-term implications of Helmholtz's discovery in our forthcoming book.
135. "Ueber die Methoden, kleinste Zeittheile zu messen."
136. Friedrich Kohlrausch, "Ueber die Fortpflanzungsgeschwindigkeit des Reizes in den menschlichen Nerven," *Zeitschrift für rationelle Medizin 28* (1866):190–204.
137. N. Baxt, "Mittheilung, betreffend Versuch über die Fortpflanzungsgeschwindigkeit der Reizung in den motorischen Nerven des Menschen," *MB* (1867):228–234; idem, "Neue Versuch über die Fortpflanzungsgeschwindigkeit der Reizung in den mo-

practitioner something to ponder and the student the opportunity to learn the skills of the electrophysiological craft.

It is clear from the interaction of precision measurement, achieved by Helmholtz's modification of Pouillet's method, and graphical analysis, achieved by what became belatedly known as the myograph, that the "discovery" of the nerve-impulse velocity was a protracted process. That process extended from the first indication of the time lapse in the muscle curves through the construction of his final argument with its intricate quantitative reasoning. Following Ludwik Fleck,[138] it might be ventured that the process of discovery also extended through the period during which Helmholtz disseminated his results to various audiences; for only through widespread community recognition could the "fact" of the finite propagation of the nerve velocity be firmly established. Techniques of persuasion and demonstrability were thus also a part of the discovery's process. If there was a crucial stage in this process, in the isolation of the "fact," it was not the early use of the myograph, when the time lapse became visible. Rather, the turning point in the establishment of velocity's "factness" came between February and July 1850 when Helmholtz applied the techniques of error analysis, especially the method of least squares, to his quantitative findings. Error analysis separated fact from artefact and assigned probable certainties to the results Helmholtz had achieved.

It is noteworthy that Helmholtz's sensitivity to the meaning of quantitative results had become more acute over the course of his early physiological investigations and had reached a climax in the nerve-impulse study where the interpretation of his findings depended upon a judicious use of the techniques of precision measurement. Although techniques of graphical analysis, with its less exact results and less rigorous form of reasoning, superseded those of precision in his study of frog nerves, Helmholtz did not abandon precision's techniques. He retained Pouillet's method for measuring the propagation velocity of human nerves, a study begun in the spring of 1850 while still analyzing the frog nerve velocity measurements. Here once again Helmholtz used the probable error, as computed by the method of least squares, as an indicator both of the reliability of his measurements and of consistency of his results.[139] His periodic return to precision techniques alternated

---

torischen Nerven der Menschen," ibid. (1870):184–91; idem, "Ueber die Zeit, welche nöthig ist, damit ein Gesichtseindruck zum Bewusstsein kommt," ibid. (1871):333–37.

138. Ludwik Fleck, *Genesis and Development of a Scientific Fact*, trans. Fred Bradley and Thaddeus J. Trenn (Chicago: University of Chicago Press, 1979).

139. AW, No. 540: Mitteilung für die physikalische Gesellschaft in Berlin betreffend Versuche über die Fortpflanzungsgeschwindigkeit der Reizung in den sensiblen Nerven

with a relaxation of rigorous quantitative methods; the choice depended upon the problem under investigation and the audience he addressed. The lessons of the nerve-impulse discovery were not, however, forgotten. In the end Helmholtz treated precision measurement, with its rigorous techniques of quantification, not as a primary or epistemologically superior method or form of experimental reasoning, but rather as one among several, assigning each their just measure.[140]

---

des Menschen [15 Dezember 1850]. Klaus Klauß has transcribed and commented upon this document in "Die erste Mitteilung von Helmholtz über die Fortpflanzungsgeschwindigkeit der Reizung in den sensiblen Nerven des Menschen," unpublished manuscript. (1991).

140. Helmholtz's association with and support for precision measurement continued throughout his lifetime. Late in the century, for example, he became the founding director of the Physikalisch-Technische Reichsanstalt; see David Cahan, *An Institute for an Empire: The Physikalisch-Technische Reichsanstalt, 1871–1918* (Cambridge, New York, New Rochelle: Cambridge University Press, 1989). For an analysis of the strategic role of a different type of Helmholtzian experiment in German physics after 1870 see the essay by Jed Z. Buchwald, "Electrodynamics in Context: Object States, Laboratory Practice, and Anti-Romanticism," in this volume.

3

The Eye as Mathematician

Clinical Practice, Instrumentation, and Helmholtz's Construction of an Empiricist Theory of Vision

Timothy Lenoir

1. Introduction

Helmholtz first sketched his theory of vision in a lecture in 1855 entitled "Über das Sehen des Menschen," four years after he had entered the field of sensory physiology. As Helmholtz himself pointed out, his lecture was ironic in that, given the recent divisions between philosophers and natural scientists, a lecture honoring the 100th anniversary of the inaugural dissertation of Immanuel Kant, a philosopher, was being delivered by a natural scientist. Indeed, the lecture may not have seemed at all appropriate for a Kant celebration. For Kant had never written on vision, and, on the surface at least, empirical consideration of the processes connected with vision scarcely raised any genuinely philosophical problems. Moreover, Helmholtz seriously discussed Kant's work only at the close of his lecture; instead, he devoted almost his entire lecture to describing the results of recent research on vision. Furthermore, the theory of vision Helmholtz outlined was deeply *anti*-Kantian; to call the empiricist theory of perception outlined in the lecture "Kantian" would have shocked many in the audience. Helmholtz did not attempt to avoid controversy in his lecture.[1]

Acknowledgments: I am grateful to Peter Galison, Simon Schaffer, and M. Norton Wise for many helpful comments on previous drafts of this paper. The final draft has benefited greatly from David Cahan's firm editorial hand.

1. "Über das Sehen des Menschen," VR^5 *I*:85–117.

As this essay seeks to show, Helmholtz's purposes in his 1855 lecture were twofold: First, he used it to announce a program statement for simultaneously extending and reforming the views on sensory physiology defended by his great teacher, Johannes Müller. Helmholtz's work in physiological optics was characterized by the relentless pursuit of a single, central question: what is *given* in the act of visual perception? Müller had addressed this issue before Helmholtz. While initially agreeing with a number of Müller's points, Helmholtz's subsequent research eventually forced him to oppose some of his mentor's views. Section 2 of this essay reviews Müller's position on visual perception—itself complex and sometimes misunderstood—in order to highlight the points of agreement and disagreement between Müller and Helmholtz.

Second, Helmholtz used the 1855 lecture to strengthen his developing empiricist theory of perception by evoking a radical new direction in Kant scholarship then emerging in the 1850s, one that contributed to the articulation of the sign theory (*Zeichentheorie*) central to Helmholtz's empiricist theory of vision. The mature version of the empiricist theory articulated by Helmholtz argued that visual spatial perception is completely learned, and that space is empirically constructed through a process of learning to coordinate eye movements, retinal stimulations, and various sorts of tactile sensations. The act of seeing itself, for Helmholtz, is a purely psychological phenomenon, no part of which is determined by innate or inborn anatomical and physiological mechanisms. Moreover, the visual image for Helmholtz is a sign, no passive copy of external things such as a photographic image, but rather a purely symbolic representation constructed by the mind to facilitate our physical interaction with things. This position contrasted sharply with the views of Müller, who argued for the interaction of innate physiological mechanisms with psychological factors and for the determination of certain crucial features of space by these anatomical and physiological mechanisms. Müller, too, employed a sign theory in his discussion of perception, but, as this essay argues, there were crucial differences between the approaches of Müller and Helmholtz to the sign theory. Section 3 of this essay explores the background to Helmholtz's empirical construction of space through a discussion of the intellectual influence of Hermann Lotze's theory of local signs and of Johann Friedrich Herbart's empirical psychology. Both Lotze's sign theory and Herbart's conception of the symbolic character of space served Helmholtz as the basis for going beyond Müller and for the cooperation between the natural sciences and philosophy which he called for in the 1855 lecture.

Eye movements were crucial to the theory of local signs. In breaking with Müller's position on the innate mechanisms controlling spatial perception Helmholtz sought to establish a psychological principle capable of guiding eye movements consistent with an empiricist approach to vision as learned behavior constantly in need of some self-corrective learning procedures to adapt it to the exigencies of visual practice. Section 4 discusses the challenges to Müller's position on visual perception during the 1850s as a result of new empirical studies of the retina and of eye movements, studies obtained in part by clinical practitioners operating with ophthalmoscopes, an instrument first invented by Helmholtz in 1850–51. Section 5 then goes on to show how Helmholtz, in good part building on diverse concepts and empirical results first put forth by others, gave theoretical unity to this diversity through his principle of easiest orientation, the psychophysical principle which served as the cornerstone of his mathematical analysis of eye movements. Helmholtz treated the eye as a measuring device, and in so doing helped advance, in opposition to both Kant and Müller, his empiricist theory of vision.

The essay's conclusion, Section 6, seeks to stress how Helmholtz connected his work on vision with other strands in his culture, including the role of practice, measurement, and instrumentation in shaping and defining an idea only broadly sketched as an investigative goal in his 1855 lecture. Through discussions with eye surgeons, clinicians, and practitioners of ophthalmology—discussions initiated by his invention of the ophthalmoscope and his improvement of the ophthalmometer—Helmholtz acquired important evidence for his theory while simultaneously constructing in concert with those practitioners measuring practices and concepts crucial to the articulation of his own views as well as to their own emerging ophthalmological science. In short, in Helmholtz's hands ophthalmological theory and practice became mutually constitutive of one another. The tools and practices in terms of which instruments were fashioned, concrete measurements were made, and evidence and standards of evaluation were negotiated became powerful resources for theory. Particularly crucial in this regard for Helmholtz were a series of models employed as demonstration devices for eye movements. Considerations of eye movements and of the ophthalmotrope, as this demonstration device was called, served as the resource for one of the central theoretical components of Helmholtz's mature empirical theory of vision. Helmholtz's consideration of this device provided the mathematical underpinnings for his principle of easiest orientation, a psychophysical principle which was the cornerstone of the empiricist theory of vision.

2. Johannes Müller's Nativism and Helmholtz's Empirical Theory of Vision

Müller, himself an embryologist, held a developmental view of spatial perception. The genesis of visual perception, he argued, depended upon the prior development of the perception of the body as extended. Implying that development of spatial perception begins in the womb with self-initiated bodily movement and the sensation of our own internal organs, Müller stated that the apperception of our own body as spatially extended and differentiated from other, external bodies in space is immediate and fully present at birth.[2] He wrote:

> The experience of our own corporality is the measure according to which we determine immediately in the sense of touch the extension of all resisting bodies. The question whether the idea of space is originally and independently present in the sensorium and acts upon all sensations or whether the idea of space only emerges successively through experience can be passed over for the moment. ... But here one thing is certain, even if the representation of space is not dimly present originally in the sensorium and is only awakened by sensation and applied, it already has to emerge empirically in the first processes of sensation connected with the sense of touch.
>
> The dim representation of a sensing body opposed to the outer world, which fills space itself, and of the spatiality of external objects is already present and developed to a certain degree of clarity and certainty before the sense of sight enters into activity at birth.[3]

Müller claimed that the sensory nerves penetrating the extremities and the body's other major organs responsible for the sense of touch are at the same time the source of the generalized representation of the body as spatially extended, "for every point in which a nerve ends is represented in the sensorium as a spatial particle [*Raumteilchen*]."[4] The senses of touch and sight are closely related in Müller's view. Moreover, he believed that "the eye's retina senses its own extension and position without any external affection, as darkness in front of the eyes."[5] Helmholtz, Wilhelm Wundt (Helmholtz's assistant at Heidelberg in later years), and others read these passages as providing an anatomical and physiological underpinning for Kant's view, expressed

---

2. Johannes Müller, *Handbuch der Physiologie des Menschen*, 2 vols. (Coblenz: J. Hölscher, 1838–40), 2:2:362.
3. Ibid., 2:2:269.
4. Ibid., 2:2:262.
5. Ibid., 2:2:264.

most forcefully in the "Transcendental Aesthetic" of the *Kritik der reinen Vernunft*, that space is given immediately as the a priori organizing feature of experience.[6]

Müller's grounding of the sense of space in the neuroanatomical wiring of the sensory organs encountered a potential difficulty in the fact that we have two retinas but perceive visual space as a single, unified space. Müller found an ingenious anatomical solution to the problem of explaining how single points are seen *as* single even though two point images are cast on the retinas: He argued, through detailed comparative anatomies of the visual apparatus of numerous types of animals, that the nerve endings in the nasal half of the left eye join with the nerve endings of the temporal half of the right eye after passing through the optic chiasma. Similarly, the nerves leading from the nasal half of the right eye unite with their counterparts from the temporal half of the left eye after passing through the chiasma. With the macula of each eye considered as center, each nerve ending in the temporal half of the left eye, according to Müller, has an identical corresponding nerve ending located an equal distance in the same direction from the center of the retina in the nasal half of the right eye. Single vision thus occurs because the corresponding points of each retina are simply branches of a common organic unity, either the same root nerve separated into two parts at the optic chiasma or some part of the brain. Vision, as Müller neatly put it, had "an organic ground."[7] He concluded:

> The congruence of identical positions of both retinas is therefore inborn, and it remains forever unchanged. Both eyes are, as it were, two branches of a simple root, and every particle of the simple root is as if divided into two branches for both eyes.[8]

Müller was here arguing explicitly against a position developed by Johann Georg Steinbuch, one which had many of the same elements as Helmholtz's empiricist theory. Following John Locke and the associationists, Steinbuch treated visual space as constructed. He proposed that light falling on the retina does not give rise to the immediate experience of a spatial continuum, but rather that the continuity of visual space is constructed from a learned association of the combined

---

6. *Handbuch*, 249, 594; and Wilhelm Wundt, *Beiträge zur Theorie der Sinneswahrnehmung* (Leipzig and Heidelberg: C.F. Winter'sche Verlagshandlung, 1862), 93–7, stressed this point.
7. Müller, *Handbuch* 2:2:380.
8. Ibid., 2:2:381.

muscle motions required to bring the visual axis of each eye successively to different points of external space. We learn to see, Steinbuch argued, by associating the sense of touch and other kinesthetic sensations with the degrees of muscular contraction required to bring the visual axes into these different positions. In Steinbuch's theory an "action idea" (*Bewegungsidee*) symbolizing the combined motions required to bring the visual axis to each point was assigned through the learning process. Visual space in his account was the system of "action ideas" associated with both retinas.[9]

Müller's Kantianism led him to reject Steinbuch's empiricist strategy. Müller constantly insisted that space could not be an empirically derived construct. In 1826, for example, he wrote:

> The concept of space cannot be learned; rather, the intuition of space and time are necessary presuppositions, even forms of intuition for all perceptions. Whenever something is perceived, it is perceived in those forms of intuition. . . . Thus the spatial intuition of our own corporality lies at the basis of the concept of motion rather than motion giving rise to the concept of space.[10]

In 1840, in his *Handbuch der Physiologie des Menschen*, Müller again stated this position with equal force. Arguing against Steinbuch and Caspar Theobald Tortual, he wrote:

> If one considers that a newborn animal has immediate intuitions of spatial juxtaposition through the visual sense and perceives images in that it seeks the mother's teats, then I believe it an indisputable fact that prior to all education the spatial is perceived as spatial in the retina.[11]

By emphasizing space and time as a priori forms of perception through which things are given in sensuous intuition, Müller unmistakably allied himself with Kant.

The identity of corresponding retinal points was one of the two aspects of Müller's theory of vision that constituted what Helmholtz called "nativism"; the other was the doctrine of specific nerve energies. That doctrine maintained that each nerve type, such as the optic nerve,

---

9. Johannes Georg Steinbuch, *Beitrag zur Physiologie der Sinne* (Nürnberg: Johann Leonhard Schragg, 1811), 156–62, 241–43. On the theories of Müller and Steinbuch, see the excellent study by Gary Hatfield, *The Natural and the Normative: Theories of Spatial Perception from Kant to Helmholtz* (Cambridge, Mass.: The MIT Press, 1990), 131–43, 152–58.

10. Johannes Müller, *Zur vergleichenden Physiologie des Gesichtsinnes des Menschen und der Thiere nebst einem Versuch über die Bewegungen der Augen und über den menschlichen Blick* (Leipzig: Cnobloch, 1826), 54–5.

11. Müller, *Handbuch* 2:3:558.

produced only one type of response when stimulated. Whether stimulated by light (from the visible spectrum), electricity, or pressure, the optic nerves respond in accord with their own particular organization, and this organization is responsible for the sensation of light, dark, and color. The doctrine served Müller and others who adopted the nativist thesis as the basis for arguing that visual experience is a purely subjective matter. Müller held that in the act of vision the brain or "sensorium" is not *seeing* the external world; rather, for the sensorium, the retina *is* the external world: "In the act of seeing only the condition of the retina is experienced and nothing else; moreover, the retina is identical with the visual field, dark in the condition of rest, bright in the condition of excitement."[12] The claim that in the act of vision we are seeing our retinas immediately raised a fundamental problem, namely, that of the relationship between the retina and the sensorium:

> This part of sensory physiology can simply be called metaphysical, because at the present time we are lacking sufficiently empirical tools to account for this interaction. Where is the condition of the retina experienced? In the retina itself or in the brain [*Gehirn*]?[13]

Although he proclaimed the issue undecidable with the tools then available, Müller believed that the *retina* was the site of several crucial structural elements of visual perception and that the visual sensorium was literally embedded in the optic nerve's apparatus, with the brain being its central part and the retinal nerve endings its periphery. The problem of the interaction between mind, brain, and retina would best be solved, Müller thought, if the sensory nerves were regarded as participating in the activity of the soul, "so that the soul continues to act into the nerve endings of the retina, in that the nerve endings of the sensory nerves are simply extensions of the sensorium."[14] He argued that the crucial feature of vision was a direct perception of order, one satisfied by the retina's ordered mosaic of nerve endings. Were the perception of spatial ordering to occur in the brain, then the retinal arrangement of nerve endings would have to be preserved intact through the optic canal into the lateral geniculate body and even into the visual cortex. He regarded this as unlikely and, in any case, as empirically undemonstrated.[15] "All order in the visual field," he declared, "depends upon the order of the affected parts of the retina."[16]

12. Ibid., 2:2:350. See also idem, *Zur vergleichenden Physiologie*, 52.
13. Müller, *Handbuch* 2:2:351.
14. Ibid.
15. Ibid., 2:2:352.
16. Ibid.

For Müller, spatial order was based in the retina's nerve apparatus.

Müller's theory held that the retina's organization conditioned at least two other spatial components of vision. Foremost among these was the sense of place or visual direction, which Müller declared to be a direct consequence of the identity of corresponding retinal points: the image of any object affecting points on the retina having the same relationship to the entire surface of the retina would be automatically seen in the same direction.[17] Two-dimensional shape was a further aspect of spatial organization given immediately in perception through the organization of the retina. This followed from the assumption that we have a direct perception of our own retinas. "Since the form of the visual image depends completely upon the extent of the affected retinal parts, mere perception suffices to distinguish simple, two-dimensional shapes, such as that of a square or a circle."[18]

For Müller, the perception of space in general—the idea of space as unified, connected, and ordered—and the perceptual immediacy of visual direction and two-dimensional shape were all given through the organization of the visual apparatus. Yet this did not prevent him from arguing that crucial conceptual and emotional[19] components also entered into vision. These were the brain's contributions to visual perception—described above as the "central part" of the visual apparatus. Moreover, Müller believed that learning also plays an important role in combining these conceptual elements with the elements given through the neuroanatomical organization of the visual apparatus. For instance, he argued that depth perception cannot be derived from the retinal image; rather, it results from a judgment learned through experience and training. Similarly, the apparent size of an object is given by the size of the image on the affected part of the retina, but perceptions of actual size are a matter of judgment based on experience and comparison with previous representations of near and distant objects.[20] And further, the judgment that the retinal image we see corresponds to an object outside and beyond the retina is a projection we learn to make.[21]

Müller's theory of vision also contained a sign theory of perception. Müller argued that a visual perception consists of two components, a

---

17. Ibid., *2:2*:359–61.
18. Ibid., *2:2*:362.
19. On this important aspect of Müller's views, but one irrelevant to the present discussion, see the sections on "Phänomene der Wechselwirkungen," in the *Handbuch*, *3:2*:559–88, and in Müller's earlier work, *Über die fantastische Gesichtserscheinungen* (Coblenz: Hölscher, 1826).
20. Müller, *Handbuch 2:2*:362.
21. Ibid., *2:2*:355.

representation (*Vorstellung*) and a sensation (*Empfindung*) produced by the sense energy of the affected retinal nerve endings. This distinction paralleled the distinction drawn by Kant between the form and content of a sensory perception. Müller wrote:

> The representation is related to the sensation rather like the sign for a thing, but like a sign which is only used for a particular thing, and the manner of which is therefore dependent on the sensation.

Furthermore:

> The representation of the sensory is, therefore, qualitatively distinct from the sensation; it [the representation] is something merely in consciousness, while the sensation is something both in the sense energy and in consciousness, the former being the sign of the latter. That representation can produce traces [*Zeichnungen*] in the sensory organs is indeed correct, but this is a composite appearance.[22]

Müller explained that this symbol theory of perception did much work: the sensations themselves need not be stored and remembered as long as the same sign is always associated with the same sensation. The process of perception can be treated analogously to reading, where the same symbol evokes a particular mental representation. Moreover, like reading, the complex of symbols need not in any way be identical or even similar to the things represented. Thus conceived as analogous to the symbolic practices of reading, the mental representation of a visual perception need not itself be extended in order to represent the sensation of extension in the retina:

> The representation of spatial objects need not be extended in space. Rather, the representation can be related to the sensed object like the expression of a figure in an algebraical equation to the figure itself or like the infinitesimally small differential to the integral in analysis. Regarding the uncertainty whether vision takes place in the visual apparatus or in the brain, we can still put forth the view that the representations of sensed objects always also occur in the sensory organs in which the impressions are produced, and therefore they recur in spatial relationships.[23]

One of Müller's strategic goals in treating space as given in terms of inborn anatomical structures was to avoid the derivation of the

22. Ibid., *2:3*:526–27.
23. Ibid., *2:3*:527.

spatial from something inherently non-spatial. Yet this strategy had a problem: at some point in the chain of relations between retina, brain, and sensorium the transition from neuroanatomy to the psychological phenomena of vision required a conversion from the physical to the mental, and hence required explaining the spatial in terms of the non-spatial. Müller's claim that when we see we are looking at our retinas, simply pushed the problem back one layer deeper into the brain. Müller attempted to sidestep the problem by having a non-extended mental token stand for things that are inherently spatial. Insisting that signs are uniquely linked to sensations and that the latter are somehow both in consciousness and materially embedded in sense energy, Müller had gambled that, while the representation is not itself extended, it is linked to structures in the retina that are.

In sum, Müller's theory of vision, characterized by Helmholtz as "nativism," had two cornerstones: the doctrines of corresponding retinal points and of specific sense energies. In addition, it featured an emphasis on the role of learning in some aspects of vision and a sign theory of perception. Helmholtz, as Section 3 shows, adhered faithfully to the doctrine of specific sense energies; but he radically expanded Müller's ideas on the role of learning in vision, and he used a different, more complex sign theory in his own, ultimately divergent account.

Helmholtz fully agreed with Müller that the specificity of sense energies based on the individual anatomical organization of each sense should be the fundamental doctrine of sensory physiology, and much of his own research during the next few years in the area of physiological color theory and in physiological acoustics helped establish Müller's theory as one of the central dogmas of the field.[24] Yet unless it was united with appropriate empirical measures, the doctrine of specific sense energies could lead Helmholtz to a feared physiological version of subjective idealism. Helmholtz believed that nativism led to a version of pre-established harmony between perceptions and the real world as the basis for the objective reference of knowledge claims: two worlds, an objective physical world and a subjective world of intuition, somehow causally related to one another but existing as independent, parallel worlds. Believing that it was necessary to escape the subjectivity and idealism of Müller's world of sense energies, Helmholtz asked in his 1855 Kant lecture: "But how is it that we escape

---

24. On these subjects see the respective essays by Richard L. Kremer, "Innovation through Synthesis: Helmholtz and Color Research," and Stephan Vogel, "Sensation of Tone, Perception of Sound, and Empiricism: Helmholtz's Physiological Acoustics," both in this volume.

from the world of the sensations of our nervous system into the world of real things?"[25] His answer—the empiricist theory of vision—was only sketched in the 1855 lecture; not until the completion of the final Part of his *Handbuch der physiologischen Optik* in 1867 would he present his fully developed epistemology.

While Müller attributed considerable importance to judgment, experience, and learning, he also vigorously defended the claim that certain features of spatial perception are determined by the anatomical and physiological organization of the visual apparatus. The most crucial aspects of this position are the identity of corresponding retinal points and the direct sense we have of our own retinas as spatially extended. Helmholtz, by contrast, wanted to eliminate all elements of physiological determinism from the perception of space. While he recognized that the brain makes use of the neuroanatomical and muscular systems connected with the visual apparatus in spatial perception, he considered spatial perception to be a purely psychological phenomenon. He did not deny the relevance of the anatomical hardware and its functioning; but he did deny the view that it determines anything about spatial perception. Indeed, conspicuously absent from his 1855 lecture was any reference to corresponding retinal points. He did not think these corresponding regions in the retina cause the identity of corresponding points. While he agreed with Müller that corresponding points play a crucial role in single vision, Helmholtz became convinced by evidence emerging from ophthalmological research in the years between 1855 and 1862 that the assignment of particular points in the two retinae as identical corresponding points is *learned* rather than given, and that it can be altered so that corresponding points are not the identical corresponding nerve endings of Müller's theory.[26]

Paradoxically, it was by eliminating the physiological givens in Müller's account that Helmholtz tried to escape into the world, to give up subjective idealism for scientific realism. He accomplished this by remaining ruthlessly consistent with Müller's sign theory which he linked to a rigorous empiricism based on hypothesis, experiment, and learning. Helmholtz argued that seeing is learned through experimentation and must be constantly reinforced through practice. The kernel of his answer of how we get outside the world of our retinas and into the real world of things was that we do it through practical action.

---

25. "Über das Sehen des Menschen," VR^5 *1*:116.
26. *Handbuch*, 698–701, esp. 701 (n1). In establishing his position, Helmholtz referred to works by A. W. Volkmann, Albrecht von Graefe, Alfred Graefe, Franz Cornelius Donders, and Nagel, all of which are discussed below.

Visual space for Helmholtz is in no way given or determined by physiological considerations but rather empirically constructed, and its construction depends on learning to associate various types of sensory stimuli, such as retinal and tactile stimuli, with subjectively initiated hand and eye movements. Successful practical repetition of these coordinated groups results in the establishment of a system of learned dispositions to act, which, eventually, become unconscious inferences. Ironically, it was through this argument that he claimed to heal the wounds between philosophy and natural science. A new generation of Kantians could regain a foothold in scientific discussions through empirical psychology.

Two lines of research led Helmholtz to abandon Müller's nativism in favor of the empiricist approach. The first was Helmholtz's own experimental research in 1850 on the speed of nerve transmission.[27] Müller himself had been impressed with Helmholtz's experiments, and he had personally read Helmholtz's paper on the speed of nerve transmission to the Berlin Akademie der Wissenschaften. In 1850 Müller wrote Helmholtz to congratulate him on the experiment; he observed that the experiment opened up the possibility of further investigating the "interaction" (*Wechselwirkung*) between sensorium and retina that he had called for in his *Handbuch*.[28] Yet Müller was apparently unaware of the difficulty this experiment created for the supposedly unmediated quality of certain perceptions; nor was he aware of the direction this line of reasoning would lead Helmholtz. In that same year Helmholtz wrote his incredulous father to report on the unexpectedly slow speed of nerve transmission, relating it to several well-known anomalies of perception, such as the personal equation in astronomy, the problem of determining whether two watches are ticking synchronously when held against different ears, and experiments that he had conducted on the time required to initiate a response to an unpleasant electrical shock on the finger. All these phenomena had led Helmholtz to conclude that a great deal of mental attention, thought, and judgment must go into the act of perception.[29] Perception, even tactile perception, could not be immediate as Müller had argued.

The second line of research was the experiments on binocular vision with Wheatstone's stereoscope.[30] With illusions such as stereoscopic

---

27. For an analysis of this work see the essay by Kathryn M. Olesko and Frederic L. Holmes, "Experiment, Quantification, and Discovery: Helmholtz's Early Physiological Researches, 1843–50," in this volume.
28. Müller to Helmholtz, 7 February 1850, AW, Nr. 318.
29. Koenigsberger *1*:123–25.
30. For analysis of this point see the essay by R. Steven Turner, "Consensus and Controversy: Helmholtz on the Visual Perception of Space," in this volume.

lustre, in which separate black and white retinal images, each presenting a different perspectival representation, are combined into a single perception of a shiny metallic object, Helmholtz supported his view that in addition to physiological processes, such as nerve action, muscle contraction, and accommodation, visual perception also requires an additional psychological process of judgment and practical reasoning. Unpacking these psychophysical processes was the declared program of Helmholtz's 1855 lecture:

> The determination of the nature of the psychic processes which transform the sensation of light into a perception of the external world is a difficult task. Unfortunately, we find no assistance from the psychologists, ... for we can conclude the existence [of the psychic processes that interest us] more readily from the physiological investigation of the sensory apparatus. The psychologists have attributed the mental acts which are my concern almost entirely to sensory perception and have not attempted to gather further insight into them.[31]

Helmholtz here announced the course of his research interests for the future.

3. The Roots of Helmholtz's Empirical Construction of Space: Lotze's Local Signs, Practical Utility, and Herbart's Empirical Psychology

Central to Helmholtz's empiricist approach was the sign theory of representation and its associated pragmatic brand of realism. Merely hinted at in the 1855 lecture,[32] by 1863 the sign theory would emerge as the centerpiece of the empiricist theory of vision. As we have seen (pp. 116–18), Müller had already introduced the sign theory into his sensory physiology. Yet Müller's was a far simpler theory than Helmholtz's. In the act of vision Müller distinguished between a sign, which was present in the sensorium and different from its signified (in that it was not itself spatial), and a sense energy, which was present in the nerves and retina. These were just different processes occurring at the peripheral and central ends of the visual apparatus. For Müller, the visual intuition of spatial order rested on the fact that the retinal nerve endings are situated and directly perceived by us as next to one another.

---

31. "Über das Sehen des Menschen," VR^5 *1*:111.
32. See especially ibid., *1*:112–14.

The order among nerve endings in the retina is simply duplicated by the signs of their sense energies in the sensorium.[33]

Helmholtz, by contrast, based his theory of visual representation on the definition of *local signs* first proposed by Hermann Lotze.[34] For Lotze, and for Helmholtz, each sensory impression of color on the retina—red, for example—produces the same sensation of redness on all parts of the retina, **a, b, c**. . . . But in addition to this impression at the parts **a, b, c,** . . . the light source also makes an accessory impression, . . . $\alpha, \beta, \gamma,$ independent of the nature of the color seen and dependent entirely on the place excited. The accessory impressions are different from and completely independent of the color impressions. These "local signs" are assigned by the succession of feelings associated with eye movements. When light—a candle flame, for instance—stimulates a peripheral spot on the retina **P**, the eye rotates to fix the light in the spot of sharpest vision **S**—ordinarily the macula in the visual axis. While the eye passes through the arc **PS**, the sensorium receives at each instant a feeling of its momentary situation, a feeling of the same sort that we have in locating our limbs in a dark room. Similarly, we feel the effort expended in contracting the eye muscles. Corresponding to the arc **PS** is a series of constantly changing feelings of position, the first of which is π and the last of which is σ. This series $\pi\sigma$, the local sign, is stored in memory and repeated along with **PS** whenever **P** is stimulated. In this way, a local sign, consisting of the practical actions necessary to relate the direction and position of each sensitive retinal point with respect to the visual axis, is associated with each point in visual space.[35]

Lotze's and Helmholtz's theory of local signs differed from Müller's theory of signs in two key respects. First, for Lotze and Helmholtz local signs are learned associations of several types of sensory input, including retinal stimulations, eye movements, and head and body position. Müller, by contrast, explicitly denied any connection between eye movements and the perception of visual space.[36] Second, the theory of local signs did not presuppose the identity of corresponding retinal points. Local signs were referred to the center of sharpest vision on the retina. While the macula was the normal origin for a system of local signs, implicit in this operational definition was the possibility

---

33. Müller, *Handbuch* 2:3:557.
34. Hermann Lotze, *Medicinische Psychologie oder Physiologie der Seele* (Leipzig: Weidmann, 1852).
35. Ibid., Book II, Chapters 1 and 4, esp. p. 331. For discussion of Lotze's theory of local signs see Hatfield, *The Natural and the Normative,* 158–64.
36. Müller, *Handbuch* 2:3:558.

that, under special circumstances (such as those associated with ametropias or pathological conditions involving muscle paralysis), some other region of the retina could also serve as the center of sharpest vision. An entirely different system of local signs could thus be constructed with the same neural wiring.

Vision for Helmholtz is a purely psychological phenomenon in which the brain uses the eye as a measuring device for the purpose of constructing a practically efficient map of the external world. For Helmholtz, the world of our visual experience is in no sense a passive copy of the external world, simply delivered to the mind as given by the optical properties of the eye and the physiological mechanisms involved in its functioning. Rather, perceptions are pictures (*Bilder*), purely symbolic representations constructed by the mind to handle the temporal sequences of sensory input it receives in the most efficient manner possible for gaining practical mastery over the world of external objects. As he wrote in 1867:

> I have described sense perceptions as symbols for the relations in the external world, and I have denied to them any form of similarity or equivalence with the things they signify. . . . To require that a perception reproduce unchanged the nature of the perceived object, and therefore that it be true in an absolute sense, would be to require an action that would be completely independent of the nature of the object upon which it is acting, which would be a patent contradiction. Thus human perceptions no less than the perceptions of any intelligent being we can imagine are pictures [*Bilder*] of objects, the nature of which is essentially co-dependent on the nature of the perceiving consciousness and co-determined by its characteristics.[37]

The sole truth of a visual symbol, according to Helmholtz, is its practical utility as a guide for active orientation in the world of things. He declared:

> I maintain, therefore, that it cannot possibly make sense to speak about any truth of our perceptions other than practical truth. Our perceptions of things cannot be anything other than symbols, naturally given signs for things, which we have learned to use in order to control our motions and actions. When we have learned how to read those signs in the proper manner, we are in a condition to use them to orient our actions such that they achieve their intended effect; that is to say, that new sensations arise in an expected manner.[38]

37. *Handbuch*, 442–43.
38. Ibid., 443 (quote) and 446.

Helmholtz was unabashed in his realism, but his was a realism of a strongly pragmatic sort. Our visual perceptions have real import, but rather than being properties abstracted or copied from things, the contents of visual perceptions are constructed by us through the practical actions we undertake in successfully orchestrating our interactions with the world.

The sensible images of visual experience are a symbolic shorthand for aggregates of sensory data, which Helmholtz called "*Empfindungscomplexe.*" Perceptions of objects in space, for instance, link together information on direction, size, and shape, with that of color, intensity, and contrast. None of these classes of information is simply given; rather, they are the result of measurements carried out by the components of the visual system. Moreover, these data aggregates are not linked with one another by an internal logic given in experience; rather, the connections are constructed by trial, error, and repetition. The more frequently the same linkages of sensory data are effected, the more rapidly the linkages are carried out by the brain; for the conscious mind in this process, they come to have the same force of necessity as logical inference. To facilitate the speed of this combinatory process, the brain assigns symbols to sensory complexes. We learn to associate these symbols with the operations necessary to produce the linkage of sensory complexes, and we become so familiar with this symbolic calculus that we simply operate with it instead of with the production of the sensory complex itself, much as a mathematician learns to operate with matrices or vector additions and products. The constructive labor involved in seeing is thus itself hidden from view; nature comes to stand as the immediately given.[39]

Helmholtz carefully distinguished these unconscious inferences, as he called them, from the logical inferences of the conscious mind. The primary difference, he noted, was that the linkages that play a part in perception cannot be articulated in words. Indeed, he maintained that "it is precisely the impossibility of describing sensations ... in words which makes it so difficult to discuss this area of psychology at all."[40] Though inexpressible in terms of concepts, sensory knowledge results all the same from mental operations. Helmholtz argued that the mental activities upon which the unconscious inferences giving rise to sensory knowledge are grounded chiefly involve the will and control over voluntary movements of exactly the same sort, to cite some examples

---

39. "Die neueren Fortschritte in der Theorie des Sehens," VR^5 1:265–365, esp. 365, trans. Russell Kahl in his edition of, *Selected Writings of Hermann von Helmholtz* (Middletown, Conn.: Wesleyan University Press, 1971), 144–222, esp. 213–22.

40. "Die neueren Fortschritte," VR^5 1:217.

given by Helmholtz, as in learning to walk on stilts, sing an aria, or play a complicated violin piece.[41]

The sign theory of representation and Helmholtz's pragmatic realism were the basis for the cooperation of philosophy and natural science that Helmholtz had promised in the introduction to his 1855 Kant lecture. But the philosophical orientation required to enable that cooperation not only demanded rejecting the speculative philosophy of German idealism; it also meant abandoning the transcendental philosophy of pure reason in Kant's first critique—the *Kritik der reinen Vernunft*—in favor of the practical philosophy of the third critique— the *Kritik der Urteilskraft*. Helmholtz was undoubtedly aware that his gesture toward Kant's practical philosophy was consonant with the turn taken by academic philosophy after the Revolution of 1848.[42] One of the leaders of this pragmatic turn was Immanuel Fichte, son of Johann Gottlieb Fichte and Ferdinand Helmholtz's close friend. Even before the Revolution, Immanuel Fichte had used his journal, *Die Zeitschrift für Philosophie und philosophischen Kritik*, as a forum for advocating the rejection of the old "monarchical" system of philosophy dominated by a single all-encompassing philosophy—he meant the speculative idealism of the Hegelians—and its replacement by a "republican form of government" which recognized a plurality of philosophical viewpoints.[43] After the collapse of the Frankfurt Parliament, Fichte and others advocated a re-examination of the foundations of knowledge. To reorient themselves, it was necessary to return to Kant, "the Atlas holding up the world of German speculative philosophy."[44] The individual who had contributed most to understanding Kant's practical philosophy was Johann Friedrich Herbart, whose work had been insufficiently appreciated in its own day. The publication of Herbart's collected works in 1850 and 1851 now fanned the realist interest in following Herbart's path back to Kant. More to the point, Helmholtz's constructivist theory of vision was deeply inspired by Herbart's conception of the *symbolic* character of space and of the deep connection between motion and the construction of visual space.

---

41. Ibid., *VR*[5] *1*:218.
42. On the practical turn see the excellent study by Klaus Christian Köhnke, *Entstehung und Aufstieg des Neukantianismus: Die deutsche Universitätsphilosophie zwischen Idealismus und Positivismus* (Frankfurt am Main: Suhrkamp, 1986).
43. See Hermann Ulrici, "Die wissenschaftlichen Tendenzen im Verhältniß zu den praktischen Interessen," *Zeitschrift für Philosophie und philosophischen Kritik 17* (1847):24–37, esp. 33. Ulrici was Fichte's co-editor of the *Zeitschrift*.
44. Johann Eduard Erdmann, "Schopenhauer und Herbart, eine Antithese," *Zeitschrift für Philosophie und philosophischen Kritik 21* (1852):206–29, esp. 225.

Herbart argued that each modality of sense is capable of a spatial representation. Color could be represented as a triangle in terms of three primary colors, tone as a continuous line, the sense of touch as a manifold (*mannigfaltikeit*) defined by muscle contractions, and still other spaces as associations of hand-eye movements. "To be exact," he wrote, "sensory space is not originally a single space. Rather, the eyes and the sense of touch *independently from one another* initiate the production of space; afterward both are melted together [*verschmolzen*] and further developed. We cannot warn often enough against the prejudice that there exists only one space, namely, phenomenal space."[45] Therefore, for Herbart, "space is the *symbol of the possible community of things standing in a causal relationship.*"[46] He insisted that for empirical psychology space is not something real, a single container in which things are placed. Rather, it is a tool for representing the various modes of interaction with the world through our senses. Such a spatial representation was shown to succeed or fail by an experiment establishing its practical utility in negotiating the world. Like Helmholtz, Herbart treated a successful visual representation as a projected expectation or "hypothesis" formed through an unconscious inference and a psychological compensatory mechanism for adjusting that initial hypothesis in terms of empirical feedback.[47] Again like Helmholtz, he treated vision as an experiment constantly performed by the brain with the aid of the eye as its measuring device.[48] Motion was thus crucial to both Helmholtz's and Herbart's empirical construction of space. In light of the current revival of Herbart's approach to Kant and the fact that Herbart had held Kant's chair in Königsberg from 1809 to 1833, Helmholtz's first public announcement—in Königsberg in 1855—of his theory of vision was invested with broad implications for the future relationship between philosophy and the natural sciences.

4. Challenges to Müller's Identity Theory: New Empirical Results in Ophthalmology and on Eye Movements

Helmholtz did not begin to elaborate his empiricist theory of vision until after he had completed the second Part of the *Handbuch der*

---

45. Johann Friedrich Herbart, *Psychologie als Wissenschaft*, vol. 5 of *Sämmtliche Werke*, ed., Gustav Hartenstein, 6 vols. (Leipzig: Leopold Voss, 1850–51), 5:349–50.
46. Ibid., 5:306.
47. Ibid., 5, esp. 118–32.
48. The relevant discussion is in *Handbuch*, 533. In his lecture "Die neueren Fortschritte," VR^5 *1*:321, Helmholtz linked his ideas directly to Herbart.

physiologischen Optik (1860), which treats visual sensation, including stimulation of the retina by light, and the phenomena of color, contrast, and adaptation.[49] Two lines of research in physiological optics from 1855 to 1860, when Helmholtz began to concentrate on the empirical theory, proved crucial to his own, post-1860 work: investigations of the validity of Müller's doctrine of the identity of corresponding retinal points, and more detailed studies of eye movements. According to Helmholtz's empiricist theory, the identity of corresponding retinal points could not be physiologically determined as Müller had asserted. Strong support for the empiricist position began to surface just as Helmholtz turned his attention to this area. Second, in place of the notion of inborn localization due to identical retinal points, the empiricist theory focused on the notion of local signs, assigned by eye movements required to rotate the center of sharpest vision to any stimulated retinal point (see above, pp. 122–25). It was for this reason that Helmholtz included eye movements in the third, psychological Part of the *Handbuch* rather than in earlier sections of the work.

Helmholtz's own invention of the ophthalmoscope opened up the line of work which eventually provided evidence that proved fatal for the identity theory. The principal contributor to this new field of ophthalmological studies was Albrecht von Graefe, a former student of Müller's who received his medical degree in 1847. Hoping eventually to obtain a position in the medical faculty of the University of Berlin, von Graefe became a *Privatdozent* while simultaneously establishing his own clinical practice in ophthalmology. In 1851, von Graefe visited Ernst Brücke in Vienna to discuss potential improvements to Brücke's failed attempt to construct an ophthalmoscope. Brücke told von Graefe that Helmholtz had recently solved the problem; that in fact he had constructed a cardboard model ophthalmoscope; and that a description of the instrument would soon appear. Von Graefe then contacted Helmholtz directly,[50] who supplied him with an instrument constructed by his instrument maker in Königsberg. In turn, von Graefe became, so to speak, the clearing house for this instrument's distribution to Franz Cornelius Donders in Utrecht, Louis-Auguste Desmarres in Paris, and William Bowman in London, while he himself modified the ophthalmoscope so as to simplify its use for the busy practitioner.

---

49. Helmholtz's general strategy of a psychological approach to vision is a theme beyond the scope of the present essay. It was partially worked out during the 1850s through his research into color theory and physiological acoustics, as the essays in this volume by Kremer, "Innovation through Synthesis," and Vogel, "Sensation of Tone," show.

50. Albrecht von Graefe to Helmholtz, 7 November 1851, AW, Signatur Nr. 172.

With this new instrument von Graefe now launched one of the most successful medical practices of the nineteenth century. Enabled to see into the living eye, von Graefe diagnosed numerous pathological conditions of the retina as well as various glaucomas; pioneered in the improvement of surgical techniques for treating cataracts; and studied pathological conditions of the eye muscles connected with various strabisms, along with developing surgical procedures for their correction. By October 1852, barely ten months after opening his small, private, ambulatory clinic, von Graefe reported to Donders that he had 1,900 patients upon whom he had performed 160 operations for cataract removals and strabism corrections. The following year he reported seeing over 4,000 patients. He increased the size of his clinic to thirty-six beds, so that he now regularly performed between ten and fifteen operations daily.[51] Moreover, the new instrument provided an unprecedented quantity of new clinical material and observations relevant to test the identity theory. Indeed, the explosion of new work in the field encouraged von Graefe to found a new journal, the *Archiv für Ophthalmologie*, the first volume of which appeared in 1854. In explicating the journal's mission, von Graefe pointed to Helmholtz's invention of the ophthalmoscope as the beginning of a new era of exact research in ophthalmology and scientific medicine.[52]

Von Graefe's contributions to the journal's first volume included one article on the pathology of the eye's oblique muscles, and a second, long article on the identity of corresponding retinal points.[53] The first article disputed Müller's claim that the eye does not undergo cyclorotation around the visual axis. The point of the critique was to support the validity of Donders's law, which stated that, with respect to the eye's vertical and horizontal meridians in its primary position of fixation, the meridians have a different but definite orientation in each position of the visual axis, that is, that the primary meridians make a wheel-like rotation around the visual axis. Donders's law, as further discussed below (pp. 139–41), was to play a central role in Helmholtz's treatment of eye movements.

In the second article, von Graefe defended Müller's doctrine of the identity of corresponding retinal points; indeed, he made it the cornerstone of his entire practice. The ophthalmoscope opened the possibility of investigating the validity of the identity theory in the living

---

51. See letters of Albrecht von Graefe to Donders from 1852 and 1853 in *Die Briefe Albrecht von Graefe's an F.C. Donders (1852–1870)*, eds., H.J.M. Weve and G. ten Doesschate (Stuttgart: Enke, 1935), 8–16.
52. Albrecht von Graefe, "Vorwort," *AO 1* (1854):vi.
53. Albrecht von Graefe, "Beiträge zur Physiologie und Pathologie der schiefen Augenmuskeln," *AO 1* (1854):1–81; and idem, "Über das Doppelsehen nach Schiel-Operationen und Incongruenz der Netzhäute," ibid., 82–120.

eye. With the new instrument, suitably modified, it was now possible to determine where the retinal area of fixation is located in normal eyes (that is, in the fovea) and in eyes with various problems, including strabisms, deformed retinas, and various stigmatisms. Deviations between the pupillary axis and the line-of-fixation now also became detectable, thus making it possible to evaluate the relationship between corresponding retinal points and single vision. Furthermore, it was well known that most patients suffering from congenital strabisms or even from longstanding aberrations in the conformity of direction of the visual axes do not experience double images. Müller and his followers argued that such findings did not contradict the identity theory because vision in the divergent eye is normally weak-sighted or the image from it is simply suppressed. Von Graefe, however, established that the affected eye continues to contribute to the visual image by providing a wider visual field and by increasing the total quantity of light admitted.[54] He used this finding to explain the well-documented but surprising fact that after corrective surgery double images appeared, particularly in eyes in which visual acuity had not degenerated. He thought that the double images were caused by the reduction in eccentricity of the retinal image in the affected eye.[55]

Moreover, he argued that, due to an adjustment in the power of accommodation, the double images following an operation gradually fuse. Still, he could not actually produce any concrete evidence supporting his claim that the shape of the retina is altered in strabisms. In an ingenious use of the (Coccius) ophthalmoscope, in which, while closing one eye, the patient fixated with the other, amblyopic eye on the hole in the ophthalmoscope's mirror, von Graefe established that the patient fixated with a retinal area on the fovea's nasal side. He regarded this as a case of "incongruence" of the two retinas, manifest in the displacement of the most sensitive spot on the retina to a region eccentric from the macula lutea, giving rise to an anomalous fixation area. He argued that an alternate fixation area is associated with a "vicarious" fixation axis, different from the normal visual axis. Thus, von Graefe argued, incongruency of the two retinas played a causal role in the generation of the strabism.[56]

The 1850s thus brought growing doubts and dissatisfaction with Müller's identity theory. Carl Ludwig's *Lehrbuch der Physiologie des Menschen* (first edition 1852; second edition 1858), which was intended as a programmatic statement of the views of Ludwig, Brücke, Emil du Bois-Reymond, and Helmholtz, serves as a useful barometer of the

54. Idem, "Über das Doppelsehen," esp. 90–1.
55. Ibid., 91.
56. Ibid., 105–14.

discussion just prior to Helmholtz's attack on the problem of eye movements. Ludwig's text shows that many of the elements of the empiricist theory were beginning to coalesce around serious challenges to Müller's theories. Ludwig noted, for example, that the work of von Graefe, Donders, Adolf Fick (Ludwig's student), and Georg Meissner[57] had demonstrated that, while some restrictions obtain on the possible muscle combinations in eye movements, the eye muscles are nonetheless controlled by the will. Deeper consideration of eye movements, Ludwig declared, would open a wide field of research.[58]

While Ludwig's *Lehrbuch* incorporated most of Müller's views, Ludwig nonetheless apportioned the relationship between the anatomical and psychological components rather differently: for Ludwig, vision was completely psychological, even though it was occasioned by the sensations of the retina. After describing the anatomical and physiological components of the visual apparatus, Ludwig opened the section on vision with two plausible hypotheses about the relation between retina and brain: the first considered the connection between retina, brain, and vision as built into the anatomical and physiological apparatus; the second, by contrast, treated retinal and brain processes as separate but closely correlated. By 1858 Ludwig favored this latter interpretation. He asserted that while vision contained a physiological component, it was in principle an act of judgment and, hence, properly a psychological affair. He wrote: "The physiologist investigates only those factors which take part in the combined sensations of vision and leaves to the psychologist questions and answers as to the specific manner in which these factors are linked to one another and to the soul."[59] Where Müller had argued that direction and localization in space are given directly through the stimulation of the retina, Ludwig maintained that vision is psychological, and so he introduced learned coordination between the muscle sense and stimulated retinal points as the basis for spatial location. He emphasized the role of feelings associated with eye and head movements as essential to spatial localization.[60] Moreover, he noted that the recent efforts to supersede Müller's theory of corresponding retinal points were consistent with this position. Yet he also acknowledged that the

---

57. Though not discussed here, Georg Meissner, "Zur Lehre von den Bewegungen des Auges," *AO 2* (1855):1–123, and idem, *Beiträge zur Physiologie des Sehorgans* (Leipzig: Engelmann, 1854), constituted important texts, especially for Helmholtz, for the debate on eye movements.

58. Carl Ludwig, *Lehrbuch der Physiologie des Menschen*, 2nd ed., 2 vols. (Leipzig and Heidelberg: C.F. Winter'sche Verlagshandlung, 1858) *1*:240–41.

59. Ibid., *1*:315–16.

60. Ibid., *1*:326.

facts were still too uncertain to decide between one of the alternative hypotheses.[61]

Ludwig was not alone in his doubts about Müller's identity theory. In 1859, a year after the publication of the second edition of Ludwig's *Lehrbuch*, A.W. Volkmann, one of the most respected voices in the field of physiological optics, published a lengthy study in which he outright rejected Müller's theory.[62] Volkmann's study constituted a complete recantation of the position he had defended in 1846 in his article on vision in Rudolf Wagner's *Handwörterbuch der Physiologie*, an article generally considered the most comprehensive work on vision published after Müller's *Handbuch*.[63] In his study of 1859, Volkmann presented no less than thirty-nine experiments supporting several important departures from Müller's views: first, that retinal points which are only slightly different in position and which normally behave like corresponding points in producing single visual images can, under exceptional conditions, be made to produce double images through increased attentiveness to the sensory process;[64] second, that different (non-corresponding) meridians of both retinas can fuse when the meridians are nearly perpendicular;[65] and third, that the differences in limits of the visual field in different directions is not inborn but rather acquired through experience.[66] Volkmann summed up the main conclusion of his experiments as follows:

> The perception of images falling on non-corresponding retinal points as a single image is of psychological origin.... The unification of images falling on non-corresponding retinal points is an act of the soul which presupposes experience of the real unity of the object seen, and which we can only acquire through the training of the senses.[67]

Were the identity theory to fall, Volkmann claimed, then the foundations of ophthalmological practice would be endangered and, in particular, it would be necessary to construct an entirely new theory of the visual horopter.

---

61. Ibid., *1*:327.
62. A.W. Volkmann, "Die stereoskopischen Erscheinungen in ihrer Beziehung zu der Lehre von den identischen Netzhautpunkten," *AO 5:2* (1859):1–100.
63. A.W. Volkmann, "Sehen," in *Handwörterbuch der Physiologie mit Rücksicht auf physiologische Pathologie*, ed., Rudolf Wagner, 4 vols. (with vol. 3 in 2 parts) (Braunschweig: Vieweg, 1846), *3:1*:265–351, esp. 316–18, where Volkmann argues that a pure perception of space is given as inborn and that single vision is anatomically organized in terms of the corresponding retinal points.
64. Volkmann, "Die stereoskopischen Erscheinungen," 20.
65. Ibid., 53.
66. Ibid., 59.
67. Ibid., 86–7.

Volkmann's sense of crisis about the identity theory was fully and thoroughly documented by Albrecht Nagel in his long, book-length review entitled *Das Sehen mit Zwei Augen und die Lehre von den identischen Netzhautsstellen*.[68] Nagel had studied with von Graefe in Berlin and with Helmholtz in Bonn, and he dedicated his book to his mentors. Drawing primarily on the work of Charles Wheatstone with the stereoscope, Nagel carefully reviewed the experiments of Müller, von Graefe, Volkmann, Alfred Graefe (Albrecht von Graefe's nephew), and many others, in order to demonstrate that these physiologists and pathologists had implicitly assumed the validity of the identity theory when testing it; that they had overlooked crucial disconfirming evidence, particularly from strabisms; and that they had completely neglected the role of muscle feeling in assigning local position. Nagel assembled massive support for what Helmholtz would soon call the empirical theory of vision.

No less important to ophthalmological practice during the 1850s than investigations of the retina were studies on eye movements. They, too, challenged Müller's theory. Studies by von Graefe, Donders, Volkmann, Meissner, and others filled the pages of the *Archiv für die holländische Beiträge zur Natur- und Heilkunde*, the *Archiv für rationelle Medicin*, and the *Archiv für Ophthalmologie*. As early as 1845 attempts had been made to provide an exact quantitative treatment of eye movements so as to improve clinical procedures for correcting strabisms. Ludwig Böhm, a close friend of Helmholtz's and du Bois-Reymond's, and an early member of the Berlin Physikalische Gesellschaft, published a voluminous, 450-page, clinical study of the surgical treatment of strabisms.[69] Theodor Ruete, Göttingen clinical professor, working closely with Johann Benedict Listing, his physics colleague at Göttingen, produced the most important work of this early period. Ruete devised an instrument, the ophthalmotrope, for studying how the six eye muscles work together to bring about the eye's motion. Knowledge of the combinations of muscles active in any ocular rotation, the strength of the muscles acting, and the axes of the eye's rotation in the normal condition was considered essential for treatment of pathological strabisms. As Ruete noted: "A relatively exact determination of the contribution of the individual muscles to the eye's different positions is not only necessary for the exposition of normal function, but also extremely important for the determination of the degree of

---

68. Leipzig: Winter'sche Buchverhandlung, 1861.
69. Ludwig Böhm, *Das Schielen und der Sehnenschnitt in seinen Wirkungen auf Stellung und Sehkraft der Augen* (Berlin: Duncker and Humblot, 1845).

shortening of the individual muscles in abnormal positions of the visual axes in strabisms. For in these cases it does not suffice for the purposes of a fundamental cure through an operation to know which muscles are exercising a disproportionate force; rather, the degree of the shortening of the individual muscles must be examined."[70] The eye's mechanical replica was to serve students as a demonstration device; but it was also intended to facilitate research into the theory of eye movements.

Adolf Fick was the first to attempt the sort of exact physical treatment of eye movements Ruete had called for. Fick began his medical studies in Cassel, before working in Zurich with Ludwig and in Berlin with du Bois-Reymond. After Ludwig was appointed to the Josephs-Akademie in Vienna in 1855, Fick became his successor at Zurich. Du Bois-Reymond described Fick to Helmholtz as "an outstanding student of Ludwig, a mathematician and physician, who has never imagined the world other than as organized by the principle of the conservation of force, and who feels himself called upon someday to continue the firm of organic physics."[71] Fick proposed to treat eye movements as a problem in statics, wherein the eye's moment of rotation at rest is the sum of the moments of rotation produced by the six muscles holding it in equilibrium. For each muscle he constructed an axis of rotation perpendicular to the plane determined by the muscle's points of attachment to the eye and to the eye's center. The resultant moment was then constructed along the axis perpendicular to the line-of-sight in terms of the parallelogram rule for forces as the sum of these six individual moments. Fick then proceeded to establish a set of equations representing the eye's equilibrium in the initial position (when it is looking directly forward). He repeated this same exercise for each subsequent position of the visual axis. However, he could not provide a general solution linking primary and secondary positions in terms of the moments of rotation, and he was forced to proceed on a case-by-case basis. One source of his difficulty was the infinite number of combined muscle movements possible in arriving at the result. Moreover, Fick could specify neither the individual forces of contraction nor the rotational axis used by the eye in any specific eye movement. In order to carry through a specific calculation, Fick

---

70. C.G. Theodor Ruete, "Das Ophthalmotrop, dessen Bau und Gebrauch," *Göttinger Studien 1* (1845):111–50, esp. 128–29. Ruete presented a much improved ophthalmotrope in 1857; see idem, *Ein neues Ophthalmotrop zur Erläuterung der Functionen der Muskeln und Brechenden Medien des menschlichen Auges* (Leipzig: Teubner, 1857).
71. Du Bois-Reymond to Helmholtz, 30 December 1849, in Kirsten, 89.

assumed that whenever a particular motion is made the same muscles are always employed in that particular motion and, furthermore, that the motion is made with the least total expenditure of force.[72] Ludwig, for one, regarded Fick's attempt as extremely promising, even if inconclusive, and he devoted five pages to discussing it in the second edition of his *Lehrbuch.* "It remains disappointing," Ludwig noted, "that it is still impossible to arrive at the solution to this problem by means of a general expression, every examination of the thing resulting in a sea of special cases."[73]

In 1862, Wilhelm Wundt attempted to improve on Fick's model. Wundt sought to develop his theory directly on the ophthalmotrope as a model of the eye, to construct the instrument in such a manner that springs completely replicate the activity of the eye muscles (see Figure 3.1). Wundt believed that

> the artificial, experimental production of phenomena observed in nature is the final goal of all natural science. The final goal of physiology is the *homunculus.* And even if the construction of an entire homunculus will most probably always belong to the realm of impossible dreams, still, it has to a certain degree already been accomplished piecewise, and physiologists will only be satisfied when they have all the pieces in hand.[74]

Wundt took seriously Ludwig's injunction to construct a general expression for the action of the eye muscles. He stressed the need to develop an equation incorporating a measure of the forces actually exerted by contracting muscles. Fick's notion of basing eye movements on the least expenditure of force could only be formulated as a general rule capable of being fulfilled in an infinite number of ways. Moreover, as a subjective guiding principle, there was no measure by which to quantify it. Attempting to remain close to sound experimental practice in measuring forces, Wundt proposed deducing eye movements from a general principle of statics which incorporated the resistance the muscles encounter in moving the eye. "In every voluntarily determined position of the visual axis," he pointed out, "the eye always adopts that position in which the opposition to its muscles is least."[75] In formulating what he referred to as the principle of least resistance,

---

72. Adolf Fick, "Die Bewegungen des menschlichen Augapfels," *Zeitschrift für rationelle Medicin 4* (1854):101–28, esp. 120.
73. Ludwig, *Lehrbuch 1*:237.
74. Wilhelm Wundt, "Beschreibung eines künstlichen Augenmuskelsystems zur Untersuchung der Bewegungsgesetze des menschlichen Auges im gesunden und kranken Zustanden," *AO 8:1* (1862):88–114, esp. 91.
75. Ibid., 92.

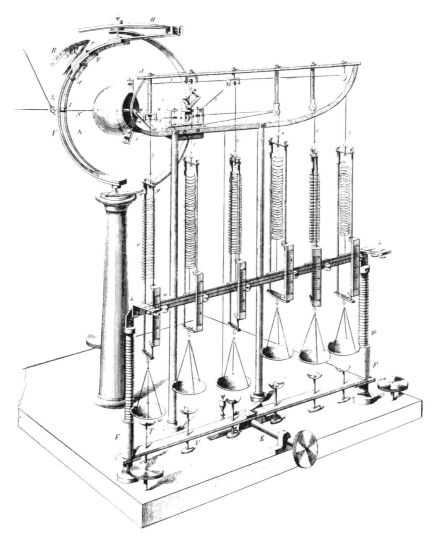

*Figure 3.1. Wundt's ophthalmotrope. (*Source: AO 8 *[1862]:no page number.)*

he had to include the contractile and elastic forces of muscles, the opposition between agonist and antagonist muscles in a given movement, and the resistance to motion offered by external obstacles, such as the optic nerve, fat, and so on.

To derive a general quantitative expression, Wundt concentrated on the muscle's contractile forces as measured in terms of the resistances they encounter. Both studies on muscles done by the brothers Ernst Heinrich and Wilhelm Weber and on muscle contraction by

Wundt (when working previously under the direction of du Bois-Reymond), showed that when a muscle shortens or lengthens by a magnitude e, the resistance dw it experiences in each infinitesimally small section of the path de is proportional to the product ede. The form of the equation describing the muscle contraction was similar to that used in describing the motion of springs, thus justifying the use of the ophthalmotrope as an appropriate physical model of the eye. The earlier work had also established that the resistance is directly proportional to the diameter q of the muscle and inversely proportional to the length l of the muscle. Based on these considerations Wundt wrote the following equation describing the momentary resistance dw experienced by the muscle contracting through an infinitesimally small distance de:

$$dw = \mu \frac{q}{l} ede,$$

where μ is the coefficient of muscle elasticity. Integration of this equation gives the total resistance w experienced by the muscle contracting through distance e:

$$w = \mu \frac{q}{l} \int_0^e ede.$$

If the resistances, lengths, cross-sectional diameters, and displacements are similarly represented for each of the six muscles, we then have

$$w_1 = \mu \frac{q_1}{l_1} \frac{e_1^2}{2}, \quad w_2 = \mu \frac{q_2}{l_2} \frac{e_2^2}{2}, \ldots,$$

where w_1 represents the total resistance for one of the eye muscles through distance e_1, w_2 represents the total resistance for a second eye muscle through distance e_2, and so on. Wundt assumed the eye to be a system in which all of the muscles act like elastic springs and bring the visual axis to a particular position according to the condition that the sum of the resistances is a minimum, that is, that $W_1 + W_2 + \ldots + W_6$ is a minimum. Neglecting the constant $\mu/2$, Wundt expressed this condition as:

$$\frac{q_1}{l_1} e_1^2 + \frac{q_2}{l_2} e_2^2 + \ldots + \frac{q_6}{l_6} e_6^2 = \text{minimum} \quad (1)$$

or

$$\sum_{i=1}^{6} \frac{q_i}{l_i} e_i^2 = \text{minimum},$$

an equation which Wundt wrote (*sans* index) to symbolize the condition of equilibrium for total resistance of all six eye muscles acting simultaneously.

In his discussion of an alternative interpretation of the minimum equation and its significance as a regulative principle for guiding the visual axis, Wundt noted—and this was perhaps the most interesting aspect of his analysis for Helmholtz—that equation (1) expressing the condition of equilibrium for the eye agrees in form with the general equation expressing the least squares of the observational error for six measurements, each weighted differently. He wrote:

> If we were to suppose $e_1, e_2, \ldots, e_6$ to be the observational errors for any phenomenon $q_1/l_1, q_2/l_2, \ldots, q_6/l_6$, those magnitudes which are assigned as weights of observation, our formula would be nothing other than the fundamental equation of the method of least squares, which asserts that the sum of the products of the squares of the observational errors multiplied by the weighted observations must be a minimum. Therefore, whenever we move the visual axis into a new position the eye proceeds just like a mathematician when he compensates for errors according to the rules of the probability calculus. The individual muscles behave like the individual observations, the lengthenings and shortenings which they experience in the transition to the new position behave like the unavoidable errors in observation and the coefficients of resistance of the muscles behave exactly like the observational weightings.
>
> Indeed, it is easy to see that it would have been possible for us to have grounded the principle in a completely different way than we have followed. We could have deduced it directly from the fundamental postulates of the probability calculus in the same way that the method of least squares is grounded.[76]

This was doubtless a conclusion that most readers of the *Archiv für Ophthalmologie* found unfamiliar. Physicists, however, would have immediately recognized Wundt's source to be the work of Carl Friedrich Gauss. In 1829 Gauss had published a paper entitled "Über ein neues allgemeines Grundgesetz der Mechanik." The new general law referred to in the title was the principle of least constraint. The principle stated that each mass in any system of masses connected to one another through any sort of external constraints moves at every instant in the closest agreement possible with free movement or under the movement each would follow under the least possible constraint, with the measure of constraint which the entire system experiences at each instant being considered as the sum of the products of the masses and

---

76. Wilhelm Wundt, "Über die Bewegung der Augen," *AO 8:2* (1862):1–87, esp. 58–9.

the squared deviation from the free path. Gauss had concluded his brief but powerful paper as follows:

> It is very remarkable that free motions, if they cannot exist with the necessary conditions, are modified by nature in exactly the same manner in which the mathematician, calculating according to the method of least squares, compensates experiences which have a necessary quantitative dependence on one another. This analogy could be followed further in still other directions, but this does not accord with my present purpose.[77]

Wundt's conclusion relating the probability calculus to his own principle of least resistance in eye movements was extraordinarily similar to Gauss's conclusion to his paper on the principle of least constraint. The comparison was completely appropriate: for the eye and the six attached muscles operate as a system of constrained motion closely analogous to the situation described by Gauss in deriving his law.

By calling the eye a mathematician operating in terms of the probability calculus Wundt signaled his intent to incorporate a psychological component to vision. In the forward to his *Beiträge zur Theorie der Sinneswahrnehmung* Wundt asserted that his studies on eye movements discussed above were to be considered as part of an investigative program to treat perception as a psychological problem. The goal, he explained, was to disassemble the phenomena of perception into elemental psychic processes, which were to be explored with the tools of physiological experiment.[78] Gustav Theodor Fechner's recently published, two-volume *Elemente der Psychophysik* (Leipzig: Engelmann, 1860), exploring applications of the psychophysical law proposed by Fechner and his colleague, Ernst Heinrich Weber, was exemplary of the most elementary level of work in physiological psychology. The Weber-Fechner law, according to Wundt, opened the domain of psychophysics by correlating an external stimulus with an internal, mental sensation. But Wundt believed it possible to go further in discovering other analogous psychophysical laws in purely internal, psychological domains, such as the link between sensations and perceptions.[79] Wundt did not attempt to demonstrate experimentally that Fechner's law holds for the relation between retinal sensations and space perception as called for in the above passage.[80]

---

77. Carl Friedrich Gauss, "Über ein neues allgemeines Grundgesetz der Mechanik," in *Carl Friedrich Gauss: Werke*, 12 vols. (Göttingen: Königliche Gesellschaft der Wissenschaften, 1863–1929), 5 (1877):5–28, quote on 28.
78. Wundt, *Beiträge*, iv–v.
79. Ibid., xxx.
80. On which see Ernst Mach, "Über das Sehen von Lagen und Winkeln durch die Bewegung des Auges. Ein Beitrag zur Psychophysik," *SBW 34* (1861):215–24.

If Wundt admired Fechner's approach, he nonetheless wanted to carry psychophysics to deeper levels of mental activity. Wundt's papers on eye movements and visual perception were conceived in this spirit, with the principle of least resistance functioning as a psychophysical law analogous to the Weber-Fechner law. In terms of Wundt's research goal, the analogs to Fechner's stimulus and sensation were the resistance encountered by the eye as it moves and the sensations generated by our eye muscles as they contract in moving to a particular position. Wundt implied that the principle of least resistance had both a physical and an equivalent psychological manifestation. From the point of view of the mechanics of eye movement Wundt's principle of least resistance was assimilable to Gauss's principle of least constraint. On the other hand, by calling the eye a mathematician operating in terms of the probability calculus and by claiming that the principle of least resistance could have been derived from the principles of error distribution in the probability calculus—that is, from a *mental* procedure of calculation—Wundt was in effect making the powerful suggestion that these processes were simply different sides (the mental and the physical) of the same physiological process. He was implying that while the brain guides the voluntary movements of the eye in terms of the principle of least squares, the mechanics of eye movement are carried out in an equivalent fashion in terms of the principle of least constraint.

Wundt's treatment of the problem was, as Section 5 shows, extremely suggestive; but it had its shortcomings. By emphasizing least resistance as the regulative principle guiding the eye, he had reduced the problem of vision to one of mechanics. Although Wundt seemed, at least metaphorically, to invoke a psychological process, the principle of least resistance could not directly illuminate the process of vision by providing, for instance, an explanation for how single vision occurs. In the passage quoted above (p. 137), he claimed that it would be possible to start from the probability calculus and arrive at constraints on eye movements; but he did not provide this derivation. A year later, in a paper for von Graefe's *Archiv* for 1863, Helmholtz made the derivation of the conditions of eye movement from a psychophysical principle his goal.

5. Helmholtz's Principle of Easiest Orientation and Analysis of Eye Movements

Helmholtz introduced his analysis of eye movements by noting what he found to be problematic in the work of Fick and Wundt. "I am not at all averse to accepting the principle of least exertion of the muscles as highly probable," he wrote, "... still, even if the principle ... should prove to be completely applicable, it would not follow that an optical principle is not actually decisive."[81] By "optical" Helmholtz meant a principle directly and primarily concerned with visual perception but which would produce the same effect in coordinating muscle movements as Wundt's principle of least resistance. The primary goal in vision, Helmholtz noted, was to avoid double images by fusing the images in both eyes into a single visual object. To do so, the eyes must be properly oriented. Given that all sorts of variations in the strengths of the eye muscles would be compatible with Wundt's principle of least resistance, it seemed more reasonable to consider the purposes of achieving singular vision as primary in the selection among the myriad muscle actions compatible with Wundt's principle in bringing about visual orientation. Indeed, Helmholtz could cite numerous experimental examples where voluntary exertion produces the eye position best suited to vision. Helmholtz called the principle actively guiding this process the principle of easiest orientation. His principle was neither merely a semantic variation of Fick's and Wundt's proposals nor a metaphor like Wundt's idea that the eye is a mathematician.

Crucial to Helmholtz's development of the principle of easiest orientation were the laws of Donders and, especially, of Listing. Previous investigators had paid insufficient attention to these laws—Meissner was the one exception here—while Fick and Wundt had not considered them at all. Donders's law, it will be recalled, states that there is one and only one orientation for each position of the line-of-fixation: the visual axis. Helmholtz pointed out that this fact alone greatly facilitated orientation, but it did not ensure optimal conditions for a stable visual field during eye movements. For it would not guarantee a smooth transition from one configuration of points to an adjacent or remote configuration. Listing's law states that every eye movement from the primary to any other position occurs by rotation around an axis perpendicular to the meridional plane that passes through the final position of the line-of-fixation. Each of these axes of rotation lies in an equatorial plane (known as Listing's Plane) perpendicular to the visual axis, as indicated in Figure 3.2. The eye has three primary degrees of freedom, Helmholtz noted. It performs abductions (horizontal motions in the temporal direction) and adductions (horizontal movements in the nasal direction) around a central vertical axis through the eye; it performs elevations and depressions around a central horizontal axis; and, finally, it performs torsions and cyclorotations (or

---

81. "Ueber die normalen Bewegungen des menschlichen Auges," *AO 9:2* (1863):153–214, esp. 160, in *WA 2:*360–419.

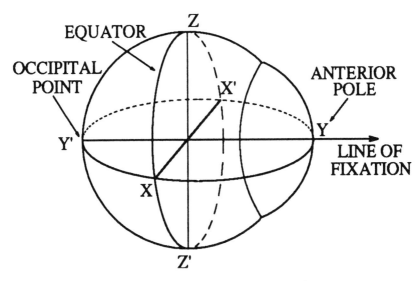

*Figure 3.2. Primary axes of rotation and Listing's Plane. **XX'**, transverse horizontal giving elevation and depression; **YY'**, antero-posterior, giving rolling or torsion; **ZZ'**, vertical giving abduction or adduction. (Source: Adapted from Handbuch: 468, Fig. 156.)*

wheel-like rotations) around the visual axis itself, the eye's anterior-posterior axis of the eye (see Figure 3.3). This latter point, as noted above (p. 128), had been demonstrated by von Graefe. Helmholtz considered that the work of achieving single vision would be greatly simplified if, in order to maintain the eye's orientation as it moves from one position to another, the mind did not have to attend to elevations, abductions, and cyclorotations, but rather that cyclorotations would be excluded, thus leaving only two variables to be attended to. This could be accomplished if the eye's cyclorotations were completely dependent upon and determined by elevations and abductions. Were Donders's and Listing's laws to be confirmed, they would guarantee this dependence of eye movements on two variables alone and at the same time exclude cyclorotations.[82]

Helmholtz's principle of easiest orientation is a rule that restricts eye movements as closely as possible to axes in Listing's Plane. In order to discuss motions limited to rotational axes in Listing's Plane, Helmholtz introduced what he termed the *atropic* or non-rotated line, a line perpendicular to the plane of the rotational axes of the eye (see

---

82. Ibid., 156, 196.

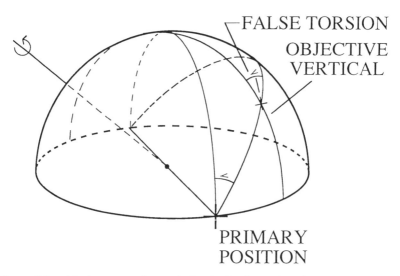

Figure 3.3. Motion around axes in Listing's Plane. Simple rotations around axes in Listing's Plane induce no cyclorotation with respect to the primary position, although oblique rotations induce "false torsion," that is, cyclorotations with respect to objective verticals.

the **XZ** plane in Figure 3.2). In the primary position the atropic line coincides with the visual axis. Helmholtz later showed that if the visual field coincides with a sphere, the atropic line and the visual axis coincide. In treating the eye's motion from a primary to a secondary position Helmholtz showed that it was possible to resolve the motion into two rotations by the parallelogram of forces. Thus, a secondary position could be reached by first executing an elevation α around the horizontal axis **XX'** in the **XY** plane, and then an abduction ω around the vertical axis **ZZ'**. Alternatively, the motion could be represented as a single rotation around an axis in the equatorial **XZ** plane (Listing's Plane when the eye is in the primary position) intermediate between the horizontal and vertical axes and perpendicular to the plane containing the primary and secondary points. The direction and magnitude of that rotation are determined by the parallelogram of forces according to familiar rules of projection in terms of direction cosines— the modern vector product. In terms of the atropic line, this meant that, in general, for motions occurring in Listing's Plane, any motion of the visual axis could be decomposed into rotations around axes perpendicular to the atropic line. If, however, the motion of the visual axis was generated around an axis not in Listing's Plane, Helmholtz showed that it would always be possible to decompose that displacement into a rotation around an axis perpendicular to the atropic line (that is, one axis in Listing's Plane) and a rotation ρ around the atropic line itself. The principle of easiest orientation demanded that the components of rotational motion having the atropic line as axis must be zero.[83]

To establish whether eye movements do indeed obey Listing's law, Helmholtz placed a gray cloth with a centric horizontal and vertical line on a wall in front of an observer.[84] With eyes looking straight ahead perpendicular to the wall at the center of the horizontal and vertical lines, the observer fixated on a thin black strip of paper aligned horizontally, thereby creating an afterimage. According to Listing's law, as the eye moves away horizontally (to the left or to the right) from the primary position the afterimage should be seen to move along the horizontal line directly opposite on the wall without deviation. To check eye movement to secondary positions not on the horizontal, Helmholtz tilted the observer's head to the horizontal by having him bite into the board upon which the fixation strip was attached! The board itself was fixed to a stand capable of being rotated so that angular measurements could be taken. In this way the head and its coordinate frame would remain fixed. The observer, with head tilted, then performed the same inward and outward eye movements as before with the afterimage. For the different head positions—Helmholtz selected angles of 10°—represented in Figure 3.4, a radial, wheel-like grid was obtained, each line representing the path the eye should follow were it to move in accordance with Listing's law. Helmholtz showed that those lines consistent with motion around axes in Listing's Plane are the right projections onto the wall of great circles passing through the point Y' in Figure 3.2, which he called the occipital point. For reasons which will become apparent in their connection with visual orientation, Helmholtz called these *directional circles*. These are, in a sense, the eye's own internal coordinate system. Also represented in Figure 3.4 are the objective verticals—that is, lines perpendicular to the "horizon line" on the wall-grid—as projected from the occipital point. Except for the primary vertical, which is also a meridian passing through the occipital point, the projections of these lines onto the wall are not straight lines but rather arcs (as indicated in the diagram), which have been added to Helmholtz's diagram for the sake of completeness in representing orientation with respect to objective horizontal and vertical lines in different parts of the visual field. In practical terms, the implication of the grid construction was that if the eye in the primary position fixates on a cross—indicated by the cross at the

---

83. Ibid., 166, 212–14.
84. Helmholtz describes these procedures in ibid., 172–80.

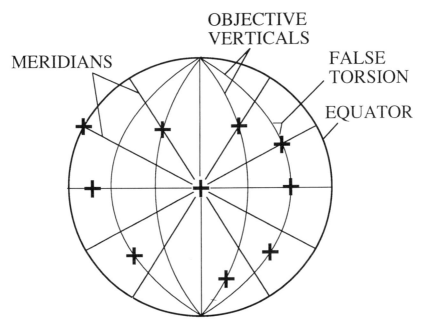

Figure 3.4. Projection of direction lines and objective verticals in the visual field. Listing's law demands that the cross be upright in the map for all positions of the line of fixation. False torsion is the angle that the vertical arm of the cross makes with the circles that represent the objective verticals in the map. (Source: Adapted from **Handbuch**, Figs. 154 and 155, pp. 465–66.)

center of Figure 3.4—the cross's afterimage will retain the same orientation with respect to the primary position.[85]

Helmholtz set up this grid in order to investigate the fundamental question of whether in fact the visual axis can move freely from one point to any other point in the visual field—and not simply from a primary to a secondary position—without producing a cyclorotation. Citing Meissner's work, Helmholtz argued that, for the most general case, if eye movements are restricted to a very small section of the visual field, the principle can be satisfied. For infinitesimally small movements, in which the sine of the angle of rotation is nearly identical with the rotational arc, it could be shown that the axes of rotation all lie in the same plane. For larger displacements, however, this was not in general the case. The result was that, for a continuous movement of the visual axis over a large angle of the visual field, the succession of axes for each of the infinitesimally small rotations constituting it

---

85. Ibid., 180–82.

The Eye as Mathematician   145

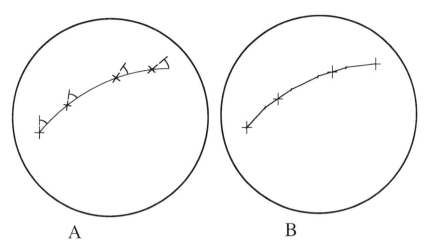

Figure 3.5. Eye motion according to the Principle of Easiest Orientation. A: a simple rotation along a great circle (not a meridian) induces cyclorotation; B: cyclorotation is avoided along line segments approximating the arc of the great circle (or any other curved path).

could not all lie in the same plane. Hence, cyclorotations are unavoidable. Helmholtz did show, however, that there are some paths which are torsion free, namely, those along the directional circles. A cross moving along one of these lines always remains upright. Motion along the directional circles can be generated around axes in Listing's Plane, so that cyclorotation will not occur. The directional circles are crucial to visual orientation. In any given part of the field, the path between two points not connected by a meridian can be approximated as a succession of small paths along neighboring directional circles (see Figure 3.5). Any deviation from a directional circle would be regarded as an error.[86] The task in visual orientation as the visual axis moves from point to point is to keep the sum of these errors to a minimum, a task that can be accomplished by the process of successive approximations along directional circles. Helmholtz wrote:

> If we consider, therefore, every rotation of the eye around the atropic line as an error, we can reduce the principle of easiest orientation to the following demand: *The law of eye movements must be so determined that the sum of the squares of the errors for the total of all the possible infinitesimally small movements of the eye taken together shall be a minimum.*[87]

86. Ibid., 195.
87. Ibid., 169.

Helmholtz then went on to reclaim Wundt's metaphor:

> The squares of the errors have to be understood here in the same sense as the well-known *principle of least squares* in the probability calculus.[88]

In an appendix to his paper, Helmholtz derived this proposition mathematically. He developed an expression for a small rotation around the atropic line as a function of the angles of abduction α, elevation θ, and cyclorotation ω in the primary coordinate axes of the eye, postulating as a condition for motion that ω be a single-valued function of α and θ. He showed that when α, θ, and ω all undergo a small displacement, the resultant rotation ρ around the atropic line constructed in terms of the composition of small rotations can be represented as:

$$\rho = d\omega + \cos\alpha\theta = \frac{d\omega}{d\alpha}d\alpha + (\frac{d\omega}{d\theta} + \cos d\alpha)d\theta.$$

The sums of the squares of these magnitudes for all infinitely small displacements ds of the atropic line over the entire visual field were supposed to be minimized. Helmholtz carried out this complex calculation over the next five pages of the appendix, thereby confirming the result that a minimum value for this equation would indeed produce a rotation obeying Listing's law.[89]

In a sweeping, magisterial move, Helmholtz had accomplished what Wundt had claimed as the goal of his studies on eye movement;

---

88. Ibid.
89. Ibid., 195–206. Ewald Hering subjected Helmholtz's derivation to a scathing critique; see his *Beiträge zur Physiologie* (Leipzig: Engelmann, 1861–64), esp. 274–83. This critique apparently shook Helmholtz's confidence in the rigor of his mathematical treatment: the derivation of Listing's law from the principle of easiest orientation provided in the *Handbuch* (497–510) differed markedly from the original derivation in the 1863 paper ("Über die normalen Bewegung"); and while preparing this paper in 1883 for inclusion in his *Wissenschaftliche Abhandlungen*, Helmholtz replaced the old derivation with the new one from the *Handbuch* (see *WA:* 2: 397–412), though in the second edition of the *Handbuch*, begun in 1885 and completed posthumously by Arthur König in 1896, he dropped the derivation altogether. König stated that Helmholtz devoted his major efforts in the second edition to developing completely new derivations for the mathematical parts, but that in July 1894, while in the midst of reworking the mathematical material for the law of eye rotations he became ill and never recovered. (See *Handbuch*², ix–x.) A generation later, the mathematician Horace Lamb provided a simple and elegant proof of the correctness of Helmholtz's theorem using the method of quaternions in place of Helmholtz's cumbersome geometrical techniques; see Horace Lamb, "The Kinematics of the Eye," *The London, Edinburgh, and Dublin Philosophical Magazine and Journal of Science* 38 (1919):685–95.

namely, Helmholtz had demonstrated that the mechanics of eye movement obeyed laws which were consistent with a psychological principle, his principle of easiest orientation. Wundt, as this essay has argued, sought—but was unable—to ground the relation between elementary sensations and visual perceptions in a psychophysical law. Helmholtz had superseded Wundt's attempt by showing that the principles the mind employs in making judgments in pursuit of the fundamental aim of visual orientation, single vision, have as their physical consequence motion according to Listing's and Donders' laws, both of which are the optimal conditions for eye movements in a mechanical system operating under least constraint. The psychological and physical aspects were intimately linked in Helmholtz's solution.[90]

In Helmholtz's empiricist theory visual sensations involve psychological acts of judgment. The collection of points we judge to be a stable object is generated by stimulation of the retina. But that judgment is based on confirmation of a sensory hypothesis: To see a collection of points in a visual image as a stationary object, the points in the group must retain their identical configuration with respect to one another when the eye makes successive passes over the supposed object, or, in the case of a moving object, the collection of points must retain its character as a connected group as it moves across the visual field. In effect, the mind is performing a series of experiments with the eye, testing the hypothesis of an object; and the outcome is judged a success if the squares of the errors after several passes or in different parts of the field are an acceptable minimum. The comparison base with respect to which we make this judgment of "fit" is the distribution of departures from Listing's law in different parts of the visual field as indicated by the cross experiments discussed above. Listing's law is crucial, for in a mechanical system such as the eye, it is the simplest method for approximating a distortion-free mapping of connected points on a sphere as the sphere adopts different successive orientations. As infants we may not have made such controlled experiments with horizontal and vertical lines, but we undoubtedly did some sort of eye exercises with the similar outcome of our becoming intimately familiar with the direction lines of our visual system, those distortion-free lines in the visual field which obey Listing's law perfectly. This measuring device embedded in the eye is in fact sensory knowledge, part of the system of local signs we carry around with us

---

90. Hatfield, *The Natural and the Normative*, 326, n. 19, suggests that to call Helmholtz's work in this context pyschophysics is anachronistic. Although Helmholtz did not call his work psychophysics, he did share the same definition of problems that Fechner and Wundt described as psychophysical.

at all times, active but below the level of consciousness, like the violin player's skill, ready to be activated whenever we view external objects.

6. Conclusion: The Practices of Vision

Helmholtz's principle of easiest orientation was one of the key components of his empiricist theory of vision. The principle illustrated his concern to treat the eye as a measuring device. Particularly crucial for his theory was the notion that vision consists (in part) in the formation, testing, and correction (on the basis of empirical feedback) of hypotheses. He showed that in carrying out its function the eye is controlled by principles of the probability calculus used by mathematicians in computing error. The principle of easiest orientation was not merely a rule for managing empirical practice, however. It derived, as this essay has argued, from tools of measurement familiar to practicing physiologists and physicists, such as the method of least squares, and models of eye movement, such as the ophthalmotrope. In Helmholtz's hands, networks of practice served as resources for theory.

First and most obvious were the ways in which physiologists drew on well-established practices of measurement and instrumentation from optics and astronomy. The use of least squares was a well-known method of data reduction introduced by Gauss into astronomy. Investigators such as Volkmann and Wundt utilized the method to correct for errors of measurement; and, while Helmholtz does not seem to have discussed statistical methods in his data reduction, he did indeed, as Kathryn M. Olesko and Frederic L. Holmes have shown, at times practice the method.[91] Furthermore, Helmholtz explained that the ophthalmometer developed as a modification of the heliometer used by astronomers to determine the minute dimensions of planets in motion.[92] The heliometer accomplishes this determination by simply doubling the images; the same is true of Helmholtz's ophthalmometer used to measure the curvature of the lens and other dimensions of the eye. That the instrument derived from astronomical instrumentation was well appreciated by Donders, who noted that Helmholtz had developed the ophthalmometer directly on the model of the heliometer.[93] Indeed, Helmholtz himself described the ophthal-

---

91. See their "Experiment, Quantification, and Discovery."
92. "Über die Accommodation des Auges," *AO 1:2* (1855):1–74, see esp. 4; in *WA* 2:283–345.
93. F.C. Donders, *On the Anomalies of Accommodation and Refraction of the Eye* (London: New Sydenham Society, 1864), 18.

mometer as essentially a telescope modified for observations at short range.[94] The system of axes used by Helmholtz to treat eye movements was explicitly modeled on the rotational axes used for orienting telescopes. He drew upon and modified for the purposes of physiological optics the practices and instrumentation of precision measurement in astronomy. For those unable to follow the details of Helmholtz's mathematical and mechanical arguments, calling the eye a probability calculator embedded theory in the new domain more persuasively and securely within the familiar network of practices of astronomy and optics. Helmholtz played a major role in constructing these networks of instrumentation and practice crucial to both the development of his theory and its acceptance.

No less significant than the invention of the ophthalmoscope and other instruments was Helmholtz's participation in their refinement and dissemination through close work with colleagues and students. Helmholtz's interaction with von Graefe, Volkmann, and Nagel are typical of the expansion of these networks of practice. The development of the ophthalmometer, for example, was central to work on accommodation, not only for Helmholtz's work but especially for that of Donders, who made this field his primary area of research. Upon learning in 1852 that Helmholtz had observed that the reflected image on the front surface of the eye's lens changes during accommodation, an observation previously reported by Donders and his student Antonie Cramer, Donders entered immediately into correspondence with Helmholtz[95] and arranged a visit to Bonn to work with him in order to learn how to make measurements with the instrument and to improve its design for practical work. Helmholtz explained the theory of his instrument in his essay on the accommodation of the eye in the first volume of von Graefe's *Archiv für Ophthalmologie*.[96] Donders confessed himself barely able to follow the details of Helmholtz's physical and mathematical arguments and, believing that the average clinical practitioner of ophthalmology was in an even worse position, he undertook the task of generating rule-of-thumb methods which were nonetheless related to high theory. Donders explained:

> For the oculist it is perhaps an additional advantage that I am no mathematician. I freely admit that I am not competent to follow the investigations of Gauss and of [Friedrich Wilhelm] Bessel in this department,

---

94. *Handbuch*, 8.
95. See Donders to Helmholtz, 6 June 1853 and 20 September 1855, in AW, Nr. 116.
96. "Über die Accommodation des Auges."

and even the study of the physiological dioptrics of Helmholtz required an effort on my part. I have, therefore, sought a way of my own, and, as I believe, I have found it. The whole theory of the cardinal points of compound dioptric systems, as it is here put forward, is quite explicit and elementary, depending almost exclusively upon the mutual comparison of similar triangles. If the road has thus become somewhat longer, it presents the advantage that it lies open to all.

... Practice, in connexion with science, here enjoys the rare, but splendid satisfaction, of not only being able to give infallible precepts based upon fixed rules, but also of being guided by a clear insight into her actions.[97]

Donders's volume laid the foundations of clinical practice for prescribing corrective lenses for myopias, astigmatisms, and various strabismic conditions. To help assure his readers—the clinical ophthalmologists at whom he directed his study—of the validity of his rules, he opened his volume with an acknowledgment pertinent to this essay's analysis: "To guard against the possibility of its leading on any point to error," he noted, "I have requested my friend [Martinus] Hoek, our Professor of Astronomy, to look over it."[98]

Donders here signaled a further domain of practice crucial to the development and acceptance of theory: the labor involved in forging links between high theory, measurement practices, and clinical routine. Perhaps the best record of Helmholtz's engagement in this sort of activity within physiological optics is provided by his extensive correspondence with Volkmann. During an eight-day visit to Halle over the Easter holiday in 1851, du Bois-Reymond laid the groundwork for the interaction between Helmholtz and Volkmann. Du Bois-Reymond had attempted to school Volkmann in the measurement of forces in muscle physiology, the central notion being "that the measure of force is the change in velocity."[99] Volkmann initiated contact with Helmholtz while the former was working on the set of thirty-nine experiments discussed above (see pp. 131–32) in connection with the identity theory and the theory of eye movements. Volkmann's revealing letter deserves a lengthy quotation:

For years I have concerned myself with research on muscles and have obtained an enormous quantity of material without being in a position to command it theoretically. At the end of last summer, for the first time I finally made decisive progress in that, in light of certain theoretical

97. Donders, *On the Anomalies*, viii–ix.
98. Ibid., viii.
99. Du Bois-Reymond to Helmholtz, 16 May 1851, in Kirsten, 113.

assumptions, I measured the force of muscle contraction in a particular manner and thereby came to values in which a law is expressed. The strange thing now is that the theoretical assumptions from which I had found the law engender doubts, and it is accordingly unclear what my law actually says.

I have spoken with various individuals about this without much essential progress. One individual is too little a physicist, while the other is too little a physiologist. You are both in an exceptional degree and perhaps the only person from whom I can get advice.[100]

Volkmann concluded his letter by arranging to visit Helmholtz in Bonn to discuss his work.

The importance of such personal contact in forging links between Helmholtz's theoretical work and the work of experimentalists becomes evident in their common analysis of eye movements according to Listing's law. Between 1863 and 1867 the two men conducted an extensive correspondence,[101] which itself was largely of a preliminary nature to their personal discussions at meetings in Heidelberg. Among the topics they discussed are Volkmann's criticism of the instruments used by Wundt in his measurement of eye motions.[102] Even more important for the acceptance of Helmholtz's views were the crucial discussions on cyclorotation in which the two men sought to agree on operational definitions for terms like "visual plane," "horizontal meridian," "vertical meridian," "cyclorotation," and so on.[103] Central to establishing Helmholtz's claims was agreement about whether a measurable cyclorotation occurs, and this in turn depended on establishing a measurement protocol. In measuring eye position, for instance, both Fick and Wundt had adopted procedures which did not treat the coordinate frame of the primary position as fundamental. Rather, they related cyclorotations to the visual plane—which is variable—instead of to the horizontal in the primary position. Since cyclorotations or deviations from Listing's law are measured in Helmholtz's system with respect to the orientation of the primary position, corrections had to be introduced into Fick's and Wundt's measurements in order to obtain agreement with Helmholtz's results. Helmholtz apparently struggled to enforce consistency in these measurement practices when determining cyclorotations.

    100. Volkmann to Helmholtz, 6 March 1856, AW, Nr. 490.
    101. Between 1856 and 1871 Volkmann sent Helmholtz twenty-two letters; see ibid.
    102. Volkmann to Helmholtz, 21 March 1864, ibid.
    103. Volkmann to Helmholtz, 26 July and 4 November 1864, and 17 February 1865, ibid.

Although Volkmann initially disagreed with Helmholtz on the question of cyclorotations and on Listing's law, their dialogue eventually led Volkmann to accept Helmholtz's view. After a long series of exchanges on what counts as a cyclorotation and how to measure it, in August 1864 Volkmann wrote Helmholtz: "Your last letter has convinced me that our experiments do not contradict one another and that I have drawn false conclusions from my observations. That's the way it goes unfortunately when one is so little schooled in geometrical intuition as I am!"[104] Perhaps the most salient feature of the closure of such debates was the assurance that Volkmann's experimental results could be solidly reinterpreted in the light of Helmholtz's quantitative theory. In March 1864, Volkmann expressed the features of this exchange of experimental and theoretical results in explicit terms:

> In the next volume [that is, Volkmann's *Physiologische Untersuchungen im Gebiete der Optik*] I will treat the eye's cyclorotations. I would have liked to include my experiments in the present volume, but I must admit that I have not dared to do so because, in the course of writing, I ran into too many mathematical difficulties for which I am not equipped. Since I only observe and do not calculate, it could easily occur that I pay attention to something which is already generally grounded in your theoretical investigations. Unfortunately, it is impossible to come to agreement about these matters through letters; however, I do want to offer you a special case where my conception and yours contradict one another and where by all back-and-forth consideration I cannot determine by myself what in my way of looking at the problem is erroneous.[105]

The enormous effort invested in linking the practices of other skilled experimenters into his own theoretical framework was, of course, crucial to establishing the authority of Helmholtz's claims beyond the walls of his own laboratory. Volkmann—and in this he was not alone—wanted to see his hard-won experimental results preserved under the legitimating aegis of high theory, even if he did not fully understand the full internal workings of that theory. The advantages to Helmholtz, the theorist, in gaining allies skilled in the practices needed to appreciate his work became evident in the next few months, as the controversy between Helmholtz and Ewald Hering over nativism and empiricism erupted.[106] Hering had dismissed the principle of easiest orientation as meaningless, and indeed had claimed that Helmholtz was simply trying to gain allies for his empiricist theory by disguising

---

104. Volkmann to Helmholtz, 7 August 1864, ibid.
105. Volkmann to Helmholtz, 21 March 1864.
106. For analysis of this controversy see Turner's "Consensus and Controversy."

his vacuous claims with the dazzle of fancy mathematics. In August 1864, Volkmann, wavering slightly in his allegiances, wrote Helmholtz:

> Many of us are hoping for your response to Hering's objections; for there are only a few physiologists who trust their own judgments in such difficult mathematical questions. The vehemence with which he has opposed you is obviously most unfortunate. By the way, I hear from [Gustav] Fechner that he [Hering] had the calculations done for him by the younger [Hermann] Hankel, who is thought to be a good mathematician in Leipzig.[107]

Ultimately, however, Volkmann did not desert the empiricist cause. Indeed, through the activities of Helmholtz, Brücke, Ludwig, Donders, Fick, Wundt, Nagel, and Volkmann the empiricist theory had become embedded in a network of theoretical technologies and empirical practices linking it to broader and deeper questions about scientific materialism and progress in German society, a point obliquely indicated by Volkmann to Helmholtz in October 1864:

> Naturally, it would be extremely interesting to me to hear what you have to say to the Hering Opposition. Only lately has it become clear to me what grave issues are at stake in these differences.[108]

Helmholtz would deliver that response in the introduction and in the closing of the third part of his *Handbuch der physiologischen Optik* (1867), wherein he criticized other theories and recorded his final break from both Kant and Müller.

---

107. Volkmann to Helmholtz, 7 August 1864.
108. Volkmann to Helmholtz, 6 October 1864, AW, Nr. 490.

4

Consensus and Controversy

Helmholtz on the Visual Perception of Space

R. Steven Turner

1. Introduction

The scientific study of vision assumed its modern outlines between 1838 and 1868.[1] The chief architect of this new science was Hermann Helmholtz, and its paradigmatic statement was Helmholtz's epic *Handbuch der physiologischen Optik*.[2] The successive volumes of the *Handbuch* offered readers a brilliant theoretical synthesis, unified by a single philosophical perspective and buttressed by Helmholtz's dazzling analytical skills. At the same time the *Handbuch* brought war as well as peace to physiological optics. While it generated consensus and unity on many points, it also gave rise to fundamental and historically chronic disputes, above all concerning Helmholtz's interpretation of the visual perception of space.

This essay describes Helmholtz's work on spatial perception and its component problems. It focuses upon research done and published between 1860 and 1865, for these earlier papers reveal more sharply than the later, synthetic presentation in the *Handbuch* how heavily

---

Acknowledgments: I thank David Cahan for his encouragement and invaluable editorial advice.

1. R. Steven Turner, "Paradigms and Productivity: The Case of Physiological Optics, 1840–94," *Social Studies of Science* 17 (1987):35–68.

2. For publishing details on the *Handbuch*, see the Abbreviations for this volume. In addition, an English edition of the *Handbuch* appeared as *Helmholtz's Treatise on Physiological Optics, Translated from the Third German Edition*, ed. James P.C. Southall, 3 vols. in 2 (New York: Optical Society of America, 1924–25; reprinted New York: Dover, 1962). The English edition is here cited as *Optics*.

Helmholtz drew upon other contemporary work, and how his own positions changed and matured in response to the polemical exchanges into which he was forced.

The essay also explores the role of controversy in the forging of Helmholtz's theoretical views. His work on the visual perception of space led Helmholtz into an immediate confrontation with his lifelong nemesis Ewald Hering. From the positions adopted by these two combatants emerged the infamous nativist-empiricist controversy, which was to dominate scientific and philosophical discussions of perception for the next century. At one level, Helmholtz's position in this controversy constituted a series of particular approaches to and explanations of various phenomena connected with vision and space perception. On another, deeper level, it expressed a general epistemological stance which he had begun to develop in his popular lectures during the 1850s and was to defend in semi-popular presentations for the rest of his career.[3] Helmholtz himself drew the connection between these two levels in the third Part of the *Handbuch*. There he portrayed the science of physiological optics as polarized, both historically and logically, between the two methodological stances he labelled nativism and empiricism. With consummate rhetorical skill he portrayed his empirical results and his particular positions on the major issues of the field as logical consequences, natural outgrowths, of the empiricist methodology and philosophical stance.

Influenced by Helmholtz's own representation of his work and its context, much subsequent historiography has dealt with Helmholtz's role as a philosopher and epistemological theorist and found its major challenge in defining Helmholtz's place in a continuing philosophical tradition. This essay seeks the other Helmholtz: the experimentalist, polemicist, and working scientist. It contends that the course of Helmholtz's early work on perception was dictated less by deep methodological or philosophical commitments—though these were certainly present—than by the context of contemporary empirical debates, by the practical necessity of imposing order on a chaotic new field of investigation, and, most of all, by the polemical exchanges with Hering. During the three crucial years (1864–66) leading up to the publication of the third Part of the *Handbuch*, the development of Helmholtz's

---

3. "Ueber die Natur der menschlichen Sinnesempfindungen," *Königsberger naturwissenschaftliche Unterhaltungen 2* (1852):1–20, in *WA 2*:591–609; and *Ueber das Sehen des Menschen* (Leipzig: L. Voss, 1855). Cf. the essays by Timothy Lenoir, "The Eye as Mathematician: Clinical Practice, Instrumentation, and Helmholtz's Construction of an Empiricist Theory of Vision"; Michael Heidelberger, "Force, Law, and Experiment: The Evolution of Helmholtz's Philosophy of Science"; and Gary Hatfield, "Helmholtz and Classicism: The Science of Aesthetics and the Aesthetics of Science," all in this volume.

views reflected what Robert Westman has called "the local rationality of the battlefield": the pressing need to consolidate and fortify positions against the attacks of a skilled and implacable opponent.[4] Accordingly, the third Part of the *Handbuch* must be understood strategically in this light, as a rhetorical vehicle, intended simultaneously to forge a controversy and to finish it. Helmholtz there defined and refuted a methodological position he called nativism; his real target, however, was the flesh-and-blood figure of Ewald Hering.

2. The Problem of Visual Space before 1862

The scientific problem of visual spatiality largely reduces to the problem of localization: When an external object makes a visual impression on our retinas, what physiological and psychological processes mediate our determination of the *direction* of the object with respect to the central planes of our bodies? What processes mediate our determination of how far the object lies from us (depth perception) or its distance relative to other objects in the visual field (the perception of relief)?

These problems have had a venerable history within philosophy and speculative psychology.[5] Because the visual image itself is extended over the retinal surface in two dimensions, the problem of two-dimensional localization was traditionally considered less problematic than that of how, from this two-dimensional evidence, the eye perceives depth. Most writers tacitly accepted the correlation between the visual and the retinal image (and hence our perception of two-dimensional space) as innate and inborn. They sometimes added the proviso that we must *learn*, often with the help of the sense of touch, how to orient this innate perception of visual direction to the central planes of the body.[6]

---

4. Robert S. Westman, et al., "The Rational Explanation of Historical Discoveries (Panel Discussion)," in *Scientific Discoveries: Case Studies*, ed. Thomas Nickles (Dordrecht and Boston: D. Reidel, 1978), 21–49, quote on 44.

5. See especially Gary Hatfield, "Mind and Space from Kant to Helmholtz," (Ph.D. diss., University of Wisconsin, 1979); idem, *The Natural and the Normative: Theories of Spatial Perception from Kant to Helmholtz* (Cambridge, Mass.: MIT Press, 1990); David E. Leary, "Immanuel Kant and the Development of Modern Psychology," in William R. Woodward and Mitchell G. Ash, eds., *The Problematic Science: Psychology in Nineteenth-Century Thought* (New York: Praeger, 1982), 17–42; and Edwin G. Boring, *Sensation and Perception in the History of Experimental Psychology* (New York: Appleton-Century-Crofts, 1942), 221–311.

6. For example, Johann Samuel Traugott Gehler, *Physikalisches Wörterbuch oder Versuch einer Erklärung der vornehmsten Begriffe und Kunstwörter der Naturlehre mit kurzen Nachrichten von der Geschichte...*, 6 vols. (Leipzig: Schwickertschen Verlage, 1787–96), "Sehen," 4:10–29; and A.W. Volkmann, "Sehen," in *Handwörterbuch der*

Because the retina is not extended in the third dimension, our capacity for depth perception had always seemed more paradoxical. At least since the time of Bishop George Berkeley (1685–1753), it had been taken for granted that this ability rests on learned, monocular cues such as the superposition of objects in the visual field, the visual angle subtended by familiar objects, atmospheric perspective, the perceived sharpness of a seen object, or the change of accommodation necessary to fixate successively on different objects in the visual field. When both eyes are used, our awareness of the degree of convergence of the optic axes was believed to provide an additional, binocular cue. Theorists speculated that we have learned what these visual cues mean about the distance of objects by associating them with perceptions of distance built up from the more primitive spatial sense which resides in touch; visual space was often regarded as derived from tactile space. Theorists disagreed as to whether the processes that intervene between raw sensation and our conscious perceptions of objects in depth are "judgmental" in nature, and so proceed from the faculty of the understanding, or "associational" in nature, and thus more closely related to basic sensory processes. Our awareness of depth appears to us as an immediate and primitive sensation only because these intervening processes, whatever their exact nature, proceed so rapidly and habitually as to be inaccessible to introspection.[7]

At the end of the eighteenth century, Immanuel Kant attempted to endow spatiality with a profoundly different epistemological and ontological status. Kant insisted that space was a "concept" or a "pure intuition"—something imposed upon sensory data by the understanding rather than discovered in the world by either the visual or the tactile sense.[8] But Kant did not regard our ability to *localize* objects visually within this intuitional space (at least in the third dimension) as innate or a priori; his views on the empirical psychology of vision scarcely differed from those of his contemporaries. Kantian doctrine, therefore, did not drastically affect the proximate physiological and psychological issues involved in visual spatiality.

What did transform these proximate issues was Charles Wheatstone's epochal invention in 1838 of the stereoscope or opticon, a device which took Victorian parlors as well as physiological laboratories by storm.[9] The stereoscope is an instrument for presenting

---

Physiologie mit Rücksicht auf physiologische Pathologie, ed. Rudolph Wagner, 4 vols. (Braunschweig: Friedrich Vieweg und Sohn, 1842–54), *3.1* (1846):264–351, esp. 316–17.
   7. Hatfield, "Mind and Space," 46–54.
   8. Ibid., 70–118.
   9. Charles Wheatstone, "Contributions to the Physiology of Vision—Part the First: On some Remarkable, and hitherto Unobserved, Phenomena of Binocular Vision,"

separately to the eyes two different photographs or line drawings of a single object or scene which have been made from separated points in space. The slight disparity between the two flat fields presented to each eye is resolved in the fused binocular field into a vivid, sometimes stunning perception of the object or scene in three-dimensional relief.

The impact of the stereoscope on the theory of vision can scarcely be exaggerated. Before 1838 the significance of binocularity and the slight disparity of the two monocular fields for the perception of relief had scarcely been appreciated. Wheatstone's instrument not only showed that binocularity was crucial in many cases to depth perception; it also suggested to some observers, including Wheatstone, that we may perceive depth visually in two quite different ways. One way obtains when we view objects with one eye only, or at distances so great that the binocular disparity between the two eyes is negligible. The other way obtains under circumstances of binocular disparity, and can give rise, apparently even in the absence of obvious empirical cues, to perceptions of relief or depth which are far more immediate and compelling.

Wheatstone's discovery raised the question of how the mechanism of perception translates the two flat, monocular fields into a combined, binocular field which shows objects in relief. The first response to this theoretical challenge was to generalize to the binocular situation the old "theory of projection" which had long been applied to monocular vision. Imagine an external point-object imaging at points on the two retinas. According to the projectionist formulation, the mind is able to imagine straight lines from the two stimulated retinal points passing through the visual centers of the eyes. It can further imagine these lines projected out to their point of intersection at the (true) position of the external body, and to know the orientation of those lines with respect to the head. How the dioptrics of the eye determines these lines of projection had long been a staple item of scientific discussion. A knowledge of the relevant angles and the interocular distance provides the trigonometric data necessary to determine the direction of the object and to calculate its distance from the body. The two visual images of the object could then be thought of as "projected" outward to the point of intersection of the direction-lines and then "seen" as one at the proper distance and in the proper direction from the body. In this way projection theorists like David

---

Philosophical Transactions of the Royal Society 28 (1838):371–94; reprinted in Nicholas J. Wade, ed., *Brewster and Wheatstone on Vision* (London: Academic Press, 1983), 65–93.

Brewster claimed to explain both single vision with two eyes and the binocular perception of depth or relief.[10]

The very language of projection theory suggested that the binocular visual field is an intellectual construct which the mind has somehow inferred from evidence present in the two monocular fields. Nevertheless, projection theorists rarely claimed to be describing the *actual* physiological or psychological processes which mediate depth perception; they originally offered only a quasi-mathematical understanding of how *in principle* depth perception was possible. By 1860, however, critics had begun to evaluate projection by the stronger criteria of physiological and psychological explanation. They found the chief problem to be that projection explained too much. Projection provides the geometrical information necessary for all objects to be seen at their true distance and to be seen single. In fact, however, we see most external objects in double images. An object which appears at one place in the monocular field of the left eye, and at another place in the monocular field of the right eye, will often be seen double, at two different places, in the combined binocular field. These double images are very indistinct and so we are usually unaware of them; nevertheless, on a strict projectionist formulation, it is hard to see why any objects should be seen in double images.

To meet this objection, subsequent defenders of projection sometimes turned to the hypothesis of a "projection surface." Objects are seen in double images in the binocular field because the two monocular images are not projected fully to the point of intersection, but instead are projected only as far as an imaginary surface in space. That surface was taken to be either a frontal plane through the point of fixation or, in the ingenious theory of Albrecht Nagel in 1861, the surfaces of two "projection spheres" centered on the two eyes and passing through the point of fixation.[11]

Wheatstone's invention affected more directly a second venerable theory of binocular vision, a theory widely associated in the 1840s with the German physiologist Johannes Müller, under whom Helmholtz had studied at the University of Berlin between 1838 and 1842. According to Müller, the retinas possess pairs of "corresponding" or "identical" points, each pair feeding the visual sensation of *one* point

---

10. Sir David Brewster, "On the Law of Visible Position in Single and Binocular Vision, and on the Representation of Solid Figures by the Union of Dissimilar Plane Pictures on the Retina," *Transactions of the Royal Society of Edinburgh 15* (1844):349–68; reprinted in Wade, ed., *Brewster and Wheatstone*, 93–114. See the other articles by Brewster reprinted in Wade, along with Wade's "Assessment," 301–28.

11. Albrecht Nagel, *Das Sehen mit zwei Augen und die Lehre der identischen Netzhautstellen* (Leipzig: C.F. Winter, 1861), esp. 12–9.

in the unified visual field. An external point-object located in space so that it images on a pair of corresponding points of the two retinas will always be seen single in the combined binocular field. Another small object, located at an external point in space from which its images fall on non-corresponding points, will be seen indistinctly in double images. Müller suspected that the retinal correspondence was organic, and he pointed to the partial crossover of nerves in the optic chiasma as confirmation.[12]

Müller's theory faced two problems. First, it did not explain enough. Although it explained why some external objects are seen single and some double, it could not explain why those seen single were seen in visual relief. Second, even as an explanation of single- and double-vision it seemed inconsistent with the evidence of the stereoscope, as Wheatstone charged in 1838. Wheatstone's instrument demonstrated undeniably that points and lines which did not image on corresponding points (and hence showed binocular disparity) could nevertheless be seen fused and single if the contour relationships were appropriate and the images neither too different nor too disparate. Indeed, it was precisely the fusing of slightly disparate and hence non-corresponding images which gave rise to the perception of binocular relief. In ordinary visual experience, Wheatstone pointed out, we routinely see small, real objects in the binocular field single; yet, according to the rigorous theory of identity, at most only a series of points—never a real macroscopic object—can be seen single at one time. Thus Wheatstone believed that he had proved that objects not imaging on corresponding retinal points can be seen single in the binocular field.[13]

Worse still for Müller's theory, Wheatstone also claimed to have proved that point-objects which do image on corresponding retinal points need not invariably be seen single. He presented the readers of his 1838 treatise with the notorious stereogram shown in Figure 4.1. When this stereogram is placed in a stereoscope, Wheatstone claimed that most viewers will be able to fuse the two strong lines S_l and S_r, so that they see a single strong line in three-dimensional relief, the top of the line inclined backward. The faint line F_l presented to the left eye only will be seen vertical in the binocular field and passing through the midpoint of the strong line S_l-S_r. Wheatstone reasoned as follows: Because F_l and S_r are positioned symmetrically on the stereogram, they must fall onto corresponding rows of points of the left and right retinas, respectively. Hence they ought to be seen fused in the binocular

---

12. Johannes Müller, *Handbuch der Physiologie des Menschen für Vorlesungen*, 2 vols. (Coblenz: J. Hülscher, 1837–40), 2:351–85.

13. Wheatstone, "Contributions," in Wade, ed., *Brewster and Wheatstone*, 88–90.

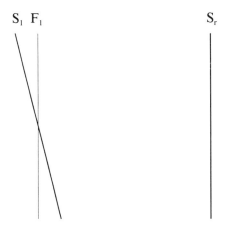

Figure 4.1. Stereogram by Charles Wheatstone, purporting to refute the theory of identity.

field. Yet they are not; S_r fuses with S_l, not F_l, and the two monocular images are seen at different positions in the binocular field. Wheatstone concluded that "this experiment affords another proof that there is no necessary physiological connection between the corresponding points of the two retinae,—a doctrine which has been maintained by so many authors."[14]

Wheatstone's paper elicited relatively little reaction during the 1840s. In England, Brewster responded in 1844 to Wheatstone's invention by setting forth the projectionist theory of binocular vision. In Germany, Müller's student Ernst Brücke and his Berlin colleague Heinrich Dove, a physicist, exchanged an important series of papers on the significance of eye movements in binocular fusion; but, in general, through the 1840s Dove continued to work almost alone on the problems of stereoscopy.

By the early 1850s, however, interest in physiological optics in general and in binocular vision in particular suddenly boomed. The scientific literature on vision began to grow exponentially, and continued to do so at least until the 1880s. In Germany, the 1850s saw major figures like Friedrich von Recklinghausen, Alexander Rollet, Wilhelm Wundt, A.W. Volkmann, Peter Ludwig Panum, Albrecht von Graefe, Georg Meissner, and Nagel move aggressively into the field. From 1845 to 1849 papers on stereoscopy, binocular vision, depth perception, and eye movements had constituted up to 19.3 percent of all scientific

14. Ibid., 83.

publications on vision and the eye; from 1850 to 1854 the comparable percentage increased to 24.5 percent; and from 1855 to 1859 it reached 29.2 percent. Equally impressive was the growing preponderance of German researchers in the field; by the mid-1860s German contributions may have reached 70 percent of the total literature.[15]

Institutional changes partly explain this explosion of interest in physiological optics. The boom coincided with the "take-off" of the German natural sciences in the 1840s and the rapid growth of the German university system with its strong research orientation. In their well-known model, Joseph Ben-David and Awraham Zloczower describe the German university system as supporting successive waves of disciplinary creation, as one new medical specialization after another was accorded separate professorships and institutes for teaching and research. As each wave brought the promise of new career opportunities, the institutional dynamic drew young researchers into the currently emerging specialty and challenged them to open up the field with new and revolutionary work which would both build the discipline and place them at the crest of the institutional wave.[16] The study of vision benefited from two such waves, the establishment of autonomous chairs of physiology—previously physiology had been combined with anatomy—which occurred between 1850 and 1868, and the institutionalization of ophthalmology, which occurred mainly between 1870 and 1880.

Helmholtz's own meteoric career illustrates how effectively some young German scientists could exploit this institutional dynamic. Helmholtz had completed his medical study under Müller at Berlin in 1842, and following a stint as army surgeon at Potsdam, received his first university appointment, to the University of Königsberg, in 1849 at the age of twenty-eight. His epochal invention of the ophthalmoscope followed almost immediately, and over the following decade his interests began to shift away from nerve and muscle physiology toward sensory physiology and psychology. The *Handbuch* itself originated with Helmholtz's acceptance of an invitation to contribute to the multi-volumed *Allgemeine Encyclopädie der Physik*, edited by the physicist Gustav Karsten, whom Helmholtz had known from Berlin in the 1840s. The three Parts of Helmholtz's *Handbuch* collectively

---

15. Turner, "Paradigms and Productivity," 38–44.
16. Avraham Zloczower, "Konjunktur in der Forschung," in Frank Pfetsch and Avraham Zloczower, *Innovation und Widerstände in der Wissenschaft* (Düsseldorf: Bertelsmann, 1972); Joseph Ben-David and Avraham Zloczower, "Universities and Academic Systems in Modern Societies," *European Journal of Sociology* 3 (1972):45–84; and see also Steven Turner, Edward Kerwin and David Woolwine, "Careers and Creativity in Nineteenth-Century Physiology: Zloczower *Redux*," *Isis* 75 (1984):523–29.

constituted volume nine of Karsten's *Encyclopädie*. Part one appeared in 1856, shortly after Helmholtz's transfer to the chair of anatomy and physiology at Bonn; it dealt with the dioptrics of the eye. Part two, delayed by Helmholtz's move to Heidelberg in 1858 and personal difficulties engendered by the death of his first wife, appeared in 1860; it dealt with the sensations of light: color vision, afterimages, adaptation, and contrast. The final Part did not appear until late 1866; it dealt with space perception.[17]

Helmholtz's research into the problems of physiological optics followed the sequence of topics treated in the *Handbuch*. He seems not to have begun intensive research on specific problems of binocularity and space perception until around 1860, immediately after he had completed the second Part of the *Handbuch* on color theory. By then, Helmholtz was the acknowledged master of physiological optics; yet he had written little on binocular perception and he seemed unfamiliar with the field's controversies. His popular lectures of 1852 and 1855 had discussed depth perception, binocularity, and the significance of Wheatstone's stereoscope either cursorily or not at all. In 1857 Helmholtz had announced the invention of his so-called telestereoscope, an instrument intended to produce an exaggerated perception of landscape relief.[18] In this paper, too, Helmholtz had treated stereoscopic vision as unproblematic. Yet once he turned his attention almost exclusively to these problems, important results emerged readily from his laboratory.

3. The Horopter Problem

Helmholtz never enjoyed the luxury of addressing particular problems of his own choice in physiological optics. Karsten's encyclopedia project compelled him to undertake a synthesis of the entire field and to dedicate himself to those problems already established by his contemporaries as the most salient and controversial. He began naturally enough with the horopter problem, a highly controversial issue around 1860, and one to which his mathematical talent naturally inclined him.

When Helmholtz first turned to this problem, the horopter was usually defined as the locus of all external points which can be seen

---

17. Koenigsberger *1* and *2*. See du Bois-Reymond's comments to Helmholtz (30 May 1853 and 25 March 1862) on the encyclopedia project in Kirsten, 142–44, on 143–44, and 200–3, on 200, resp. For Helmholtz's work on color vision and hearing, see the respective essays by Richard L. Kremer, "Innovation through Synthesis: Helmholtz and Color Research," and Stephan Vogel, "Sensation of Tone, Perception of Sound, and Empiricism: Helmholtz's Physiological Acoustics," both in this volume.

18. "Das Telestereoskop," *AP 102* (1857):167–75, in *WA 2*:484–92.

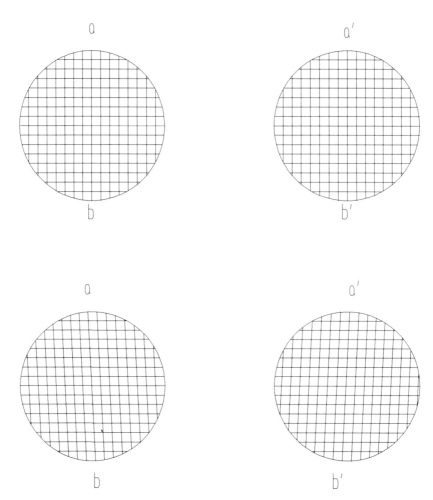

Figure 4.2. Pattern of corresponding retinal meridians, eyes in the primary position, (top) as traditionally assumed, and as accepted by Hering, (bottom) as deduced by Helmholtz from the retinal incongruity. (Schematic representation)

single (not in double images) at any instant. On the theory of retinal correspondence, therefore, it also constituted the locus of all external points which image on corresponding retinal points for a fixed position of the two eyes in the head. Any mathematical deduction of the form of the horopter therefore depends on the distribution of the corresponding points across the two retinas. Theorists typically assumed that when the eyes are in the "primary position"—head erect, visual plane horizontal, visual axes parallel to the median plane—infinitely distant points will image on corresponding retinal points. This makes the vertical and horizontal meridians of each retina as defined in the primary position corresponding meridians, and it ensures that corresponding points would be those equidistant from the respective axes in each eye formed by these corresponding meridians. Thus, if the retina were flat, the corresponding points would constitute the points of intersection of two orthogonal grids, as suggested heuristically in Figure 4.2 (top).

Using this assumption about the distribution of corresponding points, Alexandre Prevost, Gerhard U.A. Vieth, and Müller had mathematically deduced the form of the horopter—normally a circle and an intersecting line—for the simpler cases when the point of fixation lies in the horizontal or median planes. Meissner had investigated its form empirically.[19] Helmholtz also used that assumption in his first horopter paper of 1862. That paper limited itself to simpler, special configurations of the horopter, and like much of Helmholtz's other work during this period it reflected a significant debt to Wilhelm Wundt, who worked as his assistant in the physiology institute at Heidelberg between 1858 and 1864.[20]

By 1864, however, Helmholtz had learned of the so-called retinal incongruity, a phenomenon detected almost simultaneously by Recklinghausen, Meissner, and Volkmann. In eyes that display this phenomenon, the apparent vertical meridians—the meridians which correspond and which, when separately stimulated, give the visual impression of a vertical line—are not physically vertical when the eyes are in the primary position. Instead, they are rotated outward and diverge from each other and from the objectively vertical meridian at an angle which varies from individual to individual. The retinal incongruity implies, Helmholtz concluded, that the distribution of iden-

---

19. Georg Meissner, *Beiträge zur Physiologie des Sehorgans* (Leipzig: Engelmann, 1854); see also the historical discussion in Wilhelm Wundt, *Beiträge zur Theorie der Sinneswahrnehmung* (Leipzig and Heidelberg: C.F. Winter, 1862), reprinted in *The Origins of Psychology. A Collection of Early Writings*, ed. Wolfgang G. Bringmann, mul. vols. (New York: Alan R. Liss, 1976), 5:240-52.

20. "Ueber die Form des Horopters, mathematisch bestimmt," *Verhandlungen des naturhistorisch-medizinischen Vereins zu Heidelberg 3* (1862):51-5, in *WA* 2:420-26; and Wilhelm Wundt, "Ueber binoculares Sehen," *AP 192* (1862):617-26. On Helmholtz's relationship to Wundt see Wolfgang G. Bringmann, Gottfried Bringmann, and David Cottrell, "Helmholtz und Wundt an der Heidelberger Universität, 1858-1871," *Heidelberger Jahrbücher 20* (1976):79-88; and R. Steven Turner, "Helmholtz, Sensory Physiology, and the Disciplinary Development of German Psychology," in Woodward and Ash, eds., *The Problematic Science*, 147-66.

tical points is symmetric around the apparent vertical meridians, not around the true vertical meridians as had always been assumed. Helmholtz thought of the correspondence as similar to the grid of points suggested heuristically in Figure 4.2 (bottom), where lines **ab** and **a'b'** constitute the apparent vertical meridians.

These altered assumptions about the pattern of corresponding points necessitated an entirely new attack upon the horopter problem. In 1864 Helmholtz published a mathematical treatment which presupposed the retinal incongruity, introduced a new analytical system of coordinates, and deduced the general form of the horopter curve when the point of fixation lies in a "tertiary position" (i.e., not in the median or horizontal planes).[21]

Helmholtz then confronted a fundamental issue: Was the horopter problem simply "theory-spinning," as some had suggested, or did the horopter possess real sensory significance? He pointed out that, rigorously understood, the horopter cannot have its old definition based on the theory of identity; it is not strictly the locus of all external points seen single, since the stereoscope has proved that some noncorresponding images can be fused in the binocular field.

Instead, Helmholtz based the sensory significance of the horopter on an entirely new principle: the horopter is, he claimed, the locus of points in space around which the perception of relief is most acute. He supported his position by experiments made with three needles placed vertically into a small wooden block, not in a straight line but along a cylindrical surface of large curvature (see Figure 4.3). In Figure 4.3 the eyes are fixated on the center needle; the circle through the eyes and that needle will be part of the horopter. When the three needles are situated as shown, or anywhere on or near the horopter curve, Helmholtz announced, the eyes show an exquisite ability to judge the three-dimensional relief of the needles and in this way to determine whether the base is oriented such that the needles present a concave or a convex surface. But when the base is displaced even slightly from the horopter locus, the accuracy of judgment falls off. In describing this experiment, Helmholtz offered his readers a warning. If the outer needles are spaced too far from the inner one, then what he referred to as a "peculiar visual illusion" spoils the experimental outcome: if the three needles are arranged such that they lie precisely along the horopter curve, then they appear to lie in a straight line; if they are arranged in a straight line, the line appears convex, and if objectively convex, more convex than it really is.[22]

21. "Ueber den Horopter," *AO 10*, Part 1 (1864):1–60, in *WA* 2:427–77.
22. Ibid., *WA* 2:448–54.

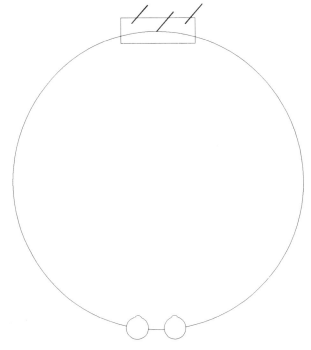

Figure 4.3. Demonstration by Helmholtz that the horopter is the region in which the perception of relief is most acute (1863). (Not to scale)

Helmholtz aimed not only to deduce the form of the horopter which followed from the retinal incongruity, but also to show its functional significance for vision. This preoccupation with function runs through all his early papers and at points borders upon the teleological. In Figure 4.2 (top), if we think of the apparent vertical meridians **ab** and **a'b'** as defining straight lines through the center of each eye and lying in a vertical plane, then these lines, when extended, will intersect somewhere below the level of the eyes. Helmholtz showed mathematically that if the eyes are in the primary position, the horizontal plane passed through this point of intersection will be part of the horopter. After examining a small number of individuals (including his children), he concluded that the angle of the retinal incongruity closely correlated with the individuals' heights, so that this plane coincided very nearly with the ground. Hence an individual walking on level ground gazing at the distant horizon would have the ground before him lying in the horopter. This, Helmholtz believed, endowed the retinal incongruity with great functional significance, for it would vastly aid the individual's ability to walk and to perceive small irregularities on the ground in indirect vision. Significantly, Helmholtz spent little time discussing how this fortunate adjustment might be acquired; for him, the most important thing was demonstrating the functional significance per se of the retinal incongruity.

The horopter problem also led Helmholtz into his first confrontation with his lifelong nemesis, Ewald Hering, then an unknown *Privatdozent* at the University of Leipzig. In 1864, independently of Helmholtz, Hering published a general solution to the horopter problem using the techniques of projective geometry developed by Jakob Steiner at Berlin.[23] Moreover, he began a running critique of Helmholtz's work on the horopter, which goaded the latter into a reply in 1864 and into an aloof review of the whole controversy in Part three of the *Handbuch*.[24]

The issues of the controversy were largely trivial and personal. They concerned murky questions of priority, minor mathematical errors, terminological disagreements, and several embarrassing mistakes made by Helmholtz in verbally describing his mathematical results. Hering's tone was alternately sycophantic and pugnacious; Helmholtz's grew increasingly exasperated as it became obvious that neither acknowledgment nor praise would placate Hering. One significant issue did divide the two men. Helmholtz criticized Hering's treatment as incomplete because it did not take the retinal incongruity into account. Hering, on his part, dismissed the retinal incongruity as one of the many potential eccentricities among individual eyes for which no general horopter derivation should try to account. Hering argued sardonically that the angle of incongruity varied widely among individuals, that there was little correlation between individual height and the angle of retinal incongruity, and, hence, that the functional significance which Helmholtz read into the phenomenon was mostly imaginary. Stung and disconcerted by this cogent criticism, Helmholtz nevertheless refused to retreat from his earlier claims; but his ambivalent discussion of the functional significance of the retinal incongruity in the *Handbuch* shows that his confidence in them had been shaken.[25]

This confusion over the size and prevalence of the retinal incongruity typified a methodological problem endemic to contemporary physiological optics. Resolution of the central controversies in the field

---

23. Ewald Hering, *Beiträge zur Physiologie*, 5 parts (Leipzig: Wilhelm Engelmann, 1861–64), esp. Part 3, "Vom Horopter," 171–224, reprinted in Hering, *Wissenschaftliche Abhandlungen von Ewald Hering*, ed. Sächsische Akademie der Wissenschaften zu Leipzig (Leipzig: Georg Thieme, 1931), item 27 (no consecutive pagination).

24. Hering, *Beiträge*, Part 4, pp. 225–86, esp. 241–47, 274–86, and Part 5, "Vom binocularen Tiefsehen. Kritik einer Abhandlung vom Helmholtz über den Horopter," 287–358; Helmholtz, "Bemerkungen über die Form des Horopters," *AP 123* (1864):158–61, in *WA* 2:478–81; and *Optics* 3:484–85.

25. *Optics* 3:421–25, 466–82.

hinged on difficult subjective experiments which required an exquisite ability to hold rigid fixation of the eyes, observe phenomena in indirect vision, and avoid muscle fatigue. These experiments could be performed only by skilled observers, usually the principal experimenter and a few assistants in his institute. It was notoriously the case that observations made in any one institute were mutually consistent and supported the theoretical position of the principal investigator there. Physiologists tended to dismiss anomalous observations by other scientists as resulting from inattention, inability to hold hard fixation, or individual eccentricities of vision. Experiment acted only as a loose restraint upon theory-building and the outcome of controversies.

4. Eye Movements

In 1860, eye movements ranked second only to the horopter problem as the most controversial topic in physiological optics. Helmholtz accordingly addressed the problem of eye movements almost simultaneously with his work on the horopter, publishing his first major paper on the theme in 1863.[26]

Normally three coordinates are necessary to describe fully the position of one eye in the head. Two of these specify the elevation and horizontal rotation of the line-of-sight from the true horizontal and vertical, or from a reference setting usually taken as the primary position of the eye. The third specifies the rotation of some reference plane (thought of as fixed to the eyeball) from its initial orientation when the eye is in the primary position. How this third coordinate varies with the movement of the eye is partly a convention, as its values depend upon how the initial position is defined and especially upon how the net movement is conceived as occurring. The most physiologically interesting change of this coordinate's value issues from a rotation of the eye around the line-of-sight, a movement called a cyclorotation, or *Rollung* in Hering's terminology. In 1863 Helmholtz defined one convention for establishing this third coordinate and called the coordinate value the *Raddrehungswinkel* or "angle of torsion." Neither his terminology nor his coordinate system distinguished clearly between general changes in the angle of torsion and those caused by cyclorotations; this ambiguity subsequently caused much confusion.[27]

---

26. "Ueber die normalen Bewegungen des menschlichen Auges," *AO* 9, Part 2 (1863):153–214, in *WA* 2:361–419. See also G. Westheimer, "Helmholtz on Eye Movements," *Human Neurobiology 3* (1984):149–52.

27. See Franz Bruno Hofmann, *Die Lehre vom Raumsinn des Auges*, 2 vols. (Berlin: Julius Springer, 1920–25), 2:273–76; and the comments of Johannes von Kries in *Optics* 3:136–39.

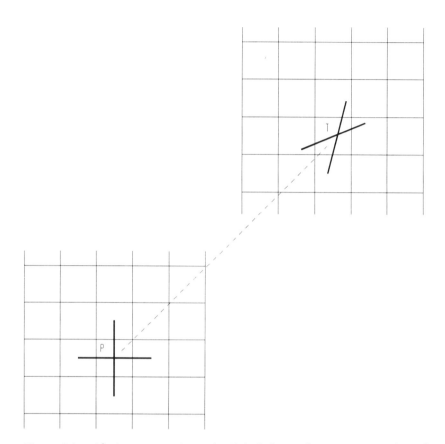

Figure 4.4. Afterimage experiment by Helmholtz confirming Listing's law of eye movements (1863).

Obscure as the whole issue of eye movements was in 1860, everyone accepted the empirical principle formulated by the great Dutch ophthalmologist Franz Donders and later known as Donders's law. This principle states that the angle of torsion, however measured, is uniquely determined by the angles of elevation and horizontal rotation; for every tertiary position of the eye, no matter how that position is reached, the eye assumes a unique degree of torsion.

Nevertheless, the question of why the eye assumes a particular angle of torsion at any tertiary position remained open. The only hypothesis purporting to solve this problem was that informally proposed in 1856 (but never published) by the Göttingen physicist Johann Benedikt Listing. He speculated that at any tertiary position the eye assumes just the angle of torsion it would have had, had it moved to that position out of the primary position, through a rotation around an axis through the center of the eye and perpendicular to the initial and final positions of the lines-of-sight. When the eye performs such a motion around such an axis (a "Listing axis") there is no component of rotation at all around the line-of-sight, i.e., no cyclorotation. If the eye moves to the tertiary position from some other initial position, that movement will normally involve a true cyclorotation, but the degree of torsion reached at the end will always be *as if* the eye had reached the position through a Listing rotation out of the primary position.

Listing's proposal had a checkered history. It had been rejected by Wundt, Adolf Fick, and Recklinghausen; Meissner had first rejected and then accepted it, but his experimental data purporting to confirm it was considered by some actually to refute it.[28] Nevertheless, in 1863 Helmholtz championed Listing's proposal as the one principle around which studies of eye movements could be ordered. Helmholtz knew that Listing's hypothesis could be experimentally tested most directly by deducing the theoretical angle of torsion to be expected at tertiary positions and comparing these to observations. Meissner and Wundt had already followed this strategy, however, and had obtained disconfirming results. Helmholtz therefore turned to a new experimental approach. Adopting the afterimage techniques introduced by Volkmann and Donders, he introduced experimental refinements to ensure that the head was held rigid during the experimental observations, and offered the following ingenious proof.

Helmholtz sat with his eyes in the primary position and stared at an orthogonal grid of vertical and horizontal lines mounted on a wall (see Figure 4.4). He stared at two crossed, colored ribbons, so as to imprint the afterimage of a right-angled cross on his retinas. When he shifted his gaze directly from **P** to a tertiary position in the upper-right-hand quadrant of the wall grid (say to point **T** in Figure 4.4), the arms of the afterimage cross were observed to have undergone the separate and opposed rotations illustrated in Figure 4.4. Three factors can be thought of as inducing these rotations: projective distortion of the grid, changes in the angle of torsion not involving a cyclorotation, and changes in the angle of torsion arising from a possible cyclorotation. Helmholtz then negated the contribution of the first two factors by rotating the ribbons and the grid pattern on the wall, so that their formerly horizontal components lay parallel to the direction of the line

---

28. Wundt, *Beiträge*, Part 3, 105–42; Friedrich von Recklinghausen, "Netzhautfunctionen," *AO 5*, Part 2 (1859):127–79; idem, "Zur Theorie des Sehens," *AP 110* (1860):65–92; and Georg Meissner, "Zur Lehre von den Bewegungen des Auges," *AO 2*, Part 2 (1855):1–123.

PT. Now when the afterimage was imprinted with the gaze at **P**, and the gaze then moved to **T**, the arms of the (oblique) afterimage cross remained right-angled and continued parallel and perpendicular to the lines of the grid. This implied necessarily that no cyclorotation occurred, which in turn meant that the eyes in both cases had moved in conformity with Listing's principle. Helmholtz promptly elevated the principle to the status of Listing's law.

In 1863 Helmholtz saw his main task as demonstrating the functional significance of Listing's law. He first pointed out that a functional principle obviously underlies Donders's law. Suppose we glance at an object and look away, and then glance back at it and find it has undergone an apparent rotation. If Donders's law were not in effect, then we would be unable to judge whether the rotation were a true rotation of the object or a result due merely to a different torsional setting of our eyes between the first and second glance. Helmholtz called this teleological principle the principle of easiest orientation.

It has an obvious extension. Suppose we move the gaze from point **P** to point **P′** and want to ensure that during the motion other objects seen indirectly in the visual field undergo no subjective rotation or distortion. The condition for this is that all such points in the visual field should seem to move parallel to the line of motion of the point of fixation as it travels from **P** to **P′**, and this parallel motion is possible only if the rotation is carried out about a Listing axis. Yet it can also be shown that all finite rotations of the eye cannot be performed around Listing axes if they are simultaneously to satisfy Donders's law. By recourse to a complex analytical technique of error-minimization, Helmholtz showed that these unavoidable violations of the principle of easiest orientation are minimized if and only if the eye has a primary position and obeys Listing's law with respect to it. The analysis also specified the axes around which non-Listing rotations must be carried out, subject to the condition that they obey Donders's law and minimize the amount of cyclorotation.[29]

Helmholtz's ostensible deduction of Listing's law and Donders's law from the principle of easiest orientation did not escape criticism. Probably few contemporaries understood the analytical treatment. Hering, who did understand it, criticized it as artificial, and dismissed it by showing that Listing's law could in fact be deduced more easily from teleological principles other than easiest orientation.

On the other hand, nearly all practitioners immediately accepted Helmholtz's experimental demonstration of Listing's law. Hering

---

29. "Ueber normalen Bewegungen," *WA* 2:396–419; and *Optics* 3:85–100.

called the demonstration elegant and decisive, and the law came to play a paradigmatic role in the continuing tradition of research into eye motions in the 1860s and 1870s. Much of that research—led mostly by Hering—consisted of studying the deviations from Listing's law which occur when the visual plane is raised or lowered, when the head is tilted, when the gaze is directed to the periphery of the field, and particularly when the eyes are converged. Helmholtz saw a deep teleological importance in Listing's law and was reluctant to admit the significance of these departures from it. His discussion in the *Handbuch* of 1867 largely recapitulated the results of 1863, except that in his new examination of convergent eye positions Helmholtz and his assistants consistently reported deviations from Listing's law more than nine times smaller than those detected by even so reliable an observer as Volkmann.[30]

5. Ewald Hering and the Theory of Identity

By 1864 the early sparring between Helmholtz and Hering over the form of the horopter and the functional principles governing eye movements had given way to a slugging match based upon much deeper and more fundamental questions. What alternative views did Hering offer to those of Helmholtz?

Karl Ewald Konstantin Hering (1834–1918) had studied medicine in Leipzig during the 1850s and had turned to research on vision in the early 1860s because, as he said, "it was cheap."[31] Although he went on to a distinguished career as professor of physiology at the Josephinum in Vienna, at Prague, and at Leipzig, Hering nurtured an image of himself as the irascible outsider struggling against the physicalist establishment entrenched in Berlin physiology. He passed on this sense of grievance and of mission to many of his fanatically loyal students, some of whom inherited his unsurpassed skills as a polemicist and introspective observer.

Throughout his life Hering opposed all theoretical accounts which portrayed the processes of sensory perception as analogous to higher, intellectualistic functions of the mind, rather than viewing them as

---

30. *Optics* 3:37–126, esp. 52.
31. On Hering see Leo M. Hurvich and Dorothea Jameson, "Introduction," in Ewald Hering, *Outlines of a Theory of the Light Sense*, ed. and trans. Hurvich and Jameson (Cambridge, Mass.: Harvard University Press, 1964), i–xxv; Hurvich, "Hering and the Scientific Establishment," *American Psychologist* 24 (1969):497–514; and Vladislav Kruta, "Hering, Karl Ewald Konstantin," *DSB* 6:299–301.

immediate expressions of the life forces in the sense organs themselves.[32] In accordance with this principle, much of his early work consisted of frontal assaults upon theories of projection and their defenders. We do not, Hering insisted, possess the exquisite and subliminal awareness of the tension in our eye muscles which is required to know the instantaneous orientation of the eyes in the head and hence of the hypothetical lines of visual projection. Only with the most elaborate of ad hoc hypotheses could projection theories explain the obvious fact of double images, and they cannot at all explain why half-images are usually seen at the distance of the object that produces them. Projection theories reify the geometrical abstractions of the physicist and substitute them for real subjective perceptions and their underlying physiological mechanisms.[33]

In keeping with his central explanatory principle, Hering also mounted a strong defense of the theory of identity. Corresponding pairs of retinal points wholly determine unique visual directions in the binocular field, as if all visual directions are referred to an imaginary "central eye" located equidistant between the two real eyes.[34] This "Law of Identical Visual Direction" gave the retinal correspondence a new and unsuspected significance for perception. The existence of a retinal correspondence, Hering insisted, is an experiential fact which must be accepted even in the absence of a clearly established organic basis for it. He subjected Wheatstone's classical arguments and demonstrations against the theory of correspondence to a penetrating analysis and refutation.[35]

Hering's most important contribution to physiological optics lay in his extension of the theory of identity to make it a comprehensive theory of depth perception. He drew a new and sharp distinction between absolute depth localization—by which we estimate the distance of the instantaneous point of fixation—and relative depth localization— by which we perceive objects to lie nearer or farther than the point of

---

32. Hering's publications on binocularity and space perception include the *Beiträge*; *Der Raumsinn und die Bewegungen des Auges (mit Nachtrag über binoculare Farbenmischung und binocularen Contrast, und über Irradiation)*, in *Handbuch der Physiologie des Menschen*, ed. Ludimar Hermann (Leipzig: Vogel, 1879–80), *3*, Part 1:343–601; *Die Lehre vom binocularen Sehen* (Leipzig: Wilhelm Engelmann, 1868), reprinted in Hering, *Wissenschaftliche Abhandlungen*, item 34, and the English translation of this work, *The Theory of Binocular Vision*, ed. and trans. Bruce Bridgeman, commentary by Lawrence Stark (New York and London: Plenum Press, 1977). For Hering's philosophy of biology see Hering, *Fünf Reden*, ed. H.E. Hering (Leipzig: Wilhelm Engelmann, 1921), esp. the "Antwortsrede" (1908), 133–40, and "Ueber das Gedächtnis als eine allgemeine Funktion der organisierten Materie," (1870), 5–32.
33. Hering, *Beiträge*, Part 2 (1862):132–58.
34. Ibid., 81–170; and Hering, "Das Gesetz der identischen Sehrichtungen," *MA* (1864):27–51, reprinted in his *Wissenschaftliche Abhandlungen*, item 29.
35. Hering, *Beiträge*, Part 2 (1862):81–96.

fixation. The former depends largely upon learned, empirical cues, Hering conceded, but the latter depends upon binocular disparity which may be supplemented by empirical cues. Point-objects not lying on the horopter will cast their images onto non-corresponding retinal points, and these pairs of retinal points will be increasingly disparate as the objects lie increasingly behind or in front of the point of fixation. The retina, Hering postulated, has innate physiological mechanisms to assess this degree of image-disparity and to convert it to a perception of objects in three-dimensional relief. If the disparity is not too great we may see the object single in the binocular field; otherwise we will see it in double images. In either case, the retinal mechanisms in question give us a direct awareness of the object's depth vis-à-vis the instantaneous point of fixation.[36]

These views merely expressed an emerging consensus among researchers on the relationship of binocularity and depth perception. In 1865, however, Hering elaborated this relationship through an ingenious extension of the old doctrine of "local signs" (see Section 6 below). He hypothesized that every point on each retina is endowed with three separate, spatial-sensory qualities called "space values" (*Raumwerthe*), so that when a given point is stimulated it awakens the perception of a particular direction (the "height value" [*Hohenwerth*] and the "breadth value" [*Breitenwerth*] associated with the point) and of the relative distance vis-à-vis the point of fixation (the "depth value" [*Tiefwerth*]). Hering hypothesized that these space values vary continuously across the retinas, and that identical point-pairs possess equal height values and breadth values. Pairs of identical points lying on the corresponding vertical meridians possess the same depth values but of opposite sign. The vertical meridians passing through the foveas divide each retina into areas on which the depth values are increasingly negative (temporal side) and increasingly positive (nasal side), with the depth value of the points lying on the vertical meridians passing through the foveas being zero. On Hering's model, when any pair of corresponding retinal points is stimulated, the resulting sensation of direction and of relative distance is the algebraic mean of the three sets of space values associated with these points.[37]

Hering's scheme was ingenious. First, it predicted that every pair of corresponding points defines a unique visual direction in the monocular or binocular field, in conformity to Hering's law. Second, any

---

36. Ibid., Part 5 (1865):287–358 ("Vom binocularen Tiefsehen"); and idem, "Die Gesetze der binocularen Tiefwahrnehmung," *MA* (1865), 79–97, 152–65, reprinted in his *Wissenschaftliche Abhandlungen*, item 32.
37. Hering, *Beiträge*, Part 5 (1864):287–315, 323–29; and "Tiefwahrnehmung," 79–97, 152–65.

external point-object lying on the horopter (including the point of fixation) will yield a combined depth value of zero for its apparent position in the binocular field. This means, Hering explained, that all such objects will normally be seen as lying at the same distance as the point of fixation and that the absolute distance of the point of fixation can be estimated only through learned, empirical cues and not by any innate physiological mechanism. He called the point of subjective, visual space at which we localize the point of fixation the "core point" (*Kernpunkt*). He boldly hypothesized that all external points imaging on identical retinal points (and so lying in the horopter) will seem to lie in a frontal plane passing through the core point; he called this plane the "core surface" (*Kernflache*). Ironically, the sole piece of empirical evidence which Hering introduced to support this hypothesis was Helmholtz's 1864 experiment, in which he reported that three vertical needles situated along the horopter-circle appeared to lie in a plane through the point of fixation.

Third, Hering pointed out that the model of retinal depth values ensures that a point-object lying nearer or farther than the point of fixation will image on non-corresponding retinal points, and that the algebraic mean of the depth values of those retinal points will be increasingly negative or positive as their distance from a frontal plane through the point of fixation increases. In short, Hering believed that his ingenious system of depth values explained the physiological cues which allow us to perceive objects in depth and that it expanded the theory of retinal correspondence into a comprehensive theory of binocular vision.[38] Ingenious as it was, however, the overtones of organic determinism implicit in Hering's scheme were calculated to offend some of Helmholtz's deepest convictions.

6. The Helmholtz-Hering Controversy before the *Handbuch*

In their studies of space perception, both Helmholtz and Hering faced the daunting task of establishing order in a field frequently described by participants as confused and chaotic. The large and mostly German literature of the late 1850s and early 1860s consisted principally of exchanges between proponents of the various projection and identity theories. It was replete with controversies: over the interpretation of stereoscopic lustre, the validity of Wheatstone's "crucial"

---

38. Hering, *Beiträge*, Part 5 (1864):287–315, 323–29; and idem, "Tiefwahrnehmung," 79–97, 152–65.

experiment refuting the identity theory, the significance of contours and psychological factors in stereoscopic fusion, the meaning and mathematical derivation of the horopter, our subliminal awareness of eye-muscle tension as a guide to localization, and convergence as a cue to depth perception.

Between 1862 and 1866 Hering tried to impose order by establishing the theory of retinal identity as the cornerstone of the field. His strategy involved attacking the defenders of projection and publicly cajoling and bullying the indecisive into accepting his extended theory of identity. The greatest prize was Helmholtz himself, the "coryphaeus of physiological optics," as Hering called him in 1864.[39] Hering's strategy met with some success. In 1863 Helmholtz repudiated projectionist interpretations of the horopter, citing Hering's work in support of this stand, and the *Handbuch* leaves no doubt that he and Hering were in general agreement on this issue.[40]

Indeed, seen against the confusion of the 1850s, Helmholtz and Hering held similar positions on many of the major issues in the field. Both abandoned projectionist explanations and accepted an approximate retinal correspondence, derived similar forms of the horopter curve, reached similar conclusions about the nature and purpose of eye movements, and agreed in dismissing muscle-kinesthesia as a significant guide to localization. Around the work of the two men, and especially around the third Part of the *Handbuch*, a theoretical consensus on key issues began to coalesce that brought an end to many of the controversies which had marked the literature of the 1850s and early 1860s.

But as Helmholtz and Hering moved toward consensus on these key questions, they simultaneously moved toward irreconcilable opposition over a set of issues which Helmholtz in 1867 dubbed the "nativist-empiricist controversy." By 1867 Helmholtz was insisting that the processes which mediate our visual perception of space are psychological or quasi–inferential in nature, occur high in the central nervous system rather than in the end-organs, and depend heavily on learning and experience acquired by the individual during his lifetime. Hering, too, acknowledged the role of learning and experience, but he insisted on a primitive residue of spatial perception which is mediated in the peripheral organs or low in the central nervous system, which is largely fixed by inherited organic structures, and which is prior to

---

39. Hering, *Beiträge*, Part 4 (1864), vi.
40. "Ueber den Horopter," *WA* 2:450–51; *Optics* 3:267–68; and see the comments of von Kries, in *Optics* 3:569–70.

all learning and experience.⁴¹ How did these differences between the protagonists emerge?

Key elements of this controversy predated Helmholtz and Hering. The controversy had been loosely prefigured in the old debate as to whether the process of depth perception was associative or judgmental in nature, and in a still looser sense in the debate over the Kantian claim that spatiality is an a priori form of the intuition. In the 1850s, however, the arena of controversy had been limited by the tacit (though not universal) consensus that two-dimensional spatiality is "given" in the retinal image, while, on the contrary, the localization of objects in depth (if not always the intuition of depth itself) is an empirical and acquired capacity.

Several developments around mid-century undermined this consensus. The stereoscope suggested that the perception of depth arising from binocular disparity had a different, more immediate quality than that arising from obviously empirical cues as superposition, apparent size, and so forth. How could stereoscopic depth perception be understood as an acquired facility? Conversely, stereoscopic evidence that non-corresponding retinal images could be seen single in the binocular field seemed to prove that binocular fusion was not an innate organic mechanism, as the old theory of retinal identity purported. Did even binocular fusion demand recourse to higher-order mental processes in order to explain why some non-corresponding images are fused and others are not?

The mid-nineteenth century also brought a flurry of theoretical speculation that the capacity for visual localization in two-dimensional space is not innate, as the textbook consensus assumed, but rather is acquired through experience from aspatial visual sensations.⁴² Theorists in this tradition usually had recourse to the hypothesis of "local signs": non-spatial and usually non-sensory qualities associated with particular retinal points which allowed the mind to distinguish between identical sensations arising from the stimulation of different retinal points. From the local signs, and from the different degrees of muscular exertion required to bring peripheral visual sensations onto the fovea,

---

41. On the nativist-empiricist controversy see Hatfield, "Mind and Space"; Boring, *Sensation and Perception*, 28–34, 233–38, and passim; Julian E. Hochberg, "Nativism and Empiricism in Perception," in *Psychology in the Making: Histories of Selected Research Problems*, ed. Leo Postman (New York: Alfred A. Knopf, 1962), 255–330; and William Woodward, "From Association to Gestalt: The Fate of Hermann Lotze's Theory of Spatial Perception, 1846–1920," *Isis* 69 (1978):572–82. The following discussion of the nativist-empiricist controversy is particularly indebted to Hatfield.

42. These efforts are carefully analyzed in Hatfield, "Mind and Space," 119–285.

we gradually acquire knowledge of the ordering of the various local signs of the fovea, and we represent this ordering to ourselves in the visual image as a two-dimensional, spatial distribution. Perhaps the most influential of such theorists was the philosopher-psychologist Hermann Lotze, who developed his theory of vision between 1846 and 1865, and who influenced Wundt to develop similar ideas between 1858 and 1862.

Theorists employing "local signs" in this way purported to explain how retinal images get localized as to direction, not how the spatial sense per se arose. One might regard the spatial sense as presupposing an a priori intuition of space, as did Lotze, or as a "psychic synthesis" emerging de novo in the ordering process, as did Wundt, or as the adaptation to vision of a sense of space already present in the sense of touch, as Helmholtz later occasionally suggested. The advantage of such theories, of course, was that they placed all three dimensions of spatial awareness on the same epistemological footing, as capacities acquired through experience.

While some issues of the later nativist-empiricist controversy were historically prefigured in this way, discussions of spatiality were not sharply polarized around issues of this kind prior to 1865. Gary Hatfield has cogently demonstrated that practically all contributors to the field combined "nativist" and "empiricist" points of view: Müller, for example, regarded depth perception as acquired, while Lotze accepted with Kant the a priori status of spatiality per se (although not the localization within that space). Hering initially regarded physiological optics as divided, not on the issue of nativism versus empiricism, but between defenders of projection and defenders of identity.[43] Nor did this alignment correspond to the later nativist-empiricist division: Panum, criticized by Helmholtz as a nativist, was also a projectionist; Wundt, the arch-empiricist of the early 1860s, freely used the notion of the retinal correspondence.

These important distinctions suggest that Helmholtz and Hering did not simply choose sides in a controversy that was already sharply joined and defined. Like Hering, Helmholtz did not initially regard the problem-field of binocular perception as polarized, either in principle or in practice, between nativists and empiricists. That interpretation of the field found full expression only in the third Part of the *Handbuch*. Nor can Helmholtz's evolving positions on the outstanding issues of the field be fully and consistently deduced from his empiricist epistemology or empiricist conception of mind. His early papers on

---

43. Ibid., esp. 228–30.

the horopter and on eye movements, for example, show a concern with the functional significance of the phenomena, but much less with how they are acquired or with the psychological mechanisms which mediate them. Helmholtz's differences with Hering, and the conviction that those differences collectively reflected two fundamentally antagonistic approaches to the field, surfaced abruptly around the mid-1860s. This conviction coalesced initially around a series of quite specific empirical problems and controversies, rather than around philosophical or methodological disputes per se. The first of these issues concerned the nature of binocular fusion.

When different external point-objects image on each of two corresponding points (as is always the case unless the object lies on the horopter), one of the monocular point-images will normally be suppressed in favor of the other in the binocular field. In particular, the fusion of non-corresponding images always requires the suppression of the conflicting half-images. Stereoscope experiments show that just what types of half-images will fuse and which will be suppressed depends heavily on factors such as the contours of the field, the similarities of the images to be fused, and sometimes the conviction that one is or is not looking at the outlines of a discrete object. Volkmann meticulously investigated this complex phenomenon and concluded that binocular fusion was psychological in nature and could not be explained wholly by organic factors such as the degree of disparity or the existence of the "circles of sensitivity" postulated by Panum.[44]

In 1864 Hering attacked Volkmann and Wundt for this line of psychological explanation, objecting that it made physiology the "stepdaughter" of psychology. Hering also criticized the appeal to "unconscious inferences" to explain perceptual processes, a line of psychological explanation he traced back to Kant and the philosopher J.F. Herbart, and which he then identified with Wundt.[45] This set the stage for the controversy with Helmholtz, who accepted Volkmann's psychological interpretation of binocular fusion and also became a champion of unconscious inference.

The nativist-empiricist confrontation also coalesced around several of Hering's claims concerning retinal space values. Helmholtz was

---

44. A.W. Volkmann, "Die stereoskopischen Erscheinungen in ihrer Beziehung zu der Lehre von den identischen Netzhautpunkten," *AO 5*, Part 2 (1859):1–100; and P.L. Panum, *Physiologische Untersuchungen über das Sehen mit zwei Augen* (Kiel: Schwerssche Buchhandlung, 1858).

45. Hering, *Beiträge*, Part 5 (1864):iii–v; and Wundt, *Beiträge*, in Bringmann, ed., *Origins of Psychology* 4:109–206, 5:1–375.

prepared to accept Hering's scheme insofar as it merely summarized the known facts of binocular perception, but he rejected the overtones of physiological determinism it seemed to imply. What rankled Helmholtz most was that Hering should base his hypothesis of the core-surface on Helmholtz's own experiments with the vertical needles. This forced Helmholtz to repeat his experiments and to recant publicly his original result. The new version of the experiment used suspended black threads aligned along the horopter circle. In a major reply to Hering of 1865 and again in the *Handbuch*, Helmholtz argued that when the set of hanging threads lies on the horopter circle it is usually *not* perceived as plane, as he had erroneously reported earlier and as is required by the hypothesis of the core-surface, but in other apparent curvilinear configurations. That the phenomenon occurs at all, Helmholtz attributed to a complex optical illusion. Under the conditions of the experiment the eyes must rely on convergence alone as a cue to distance; therefore they consistently overestimate the distance of the fixation point. The perceptual processes can reconcile this distance-cue with the retinal disparity of the threads only by assuming that the threads lie in curvilinear configurations other than the objectively correct one.[46] This tortured interpretation left Helmholtz publicly committed to an important role for psychological factors in depth perception and to an implausible explanation of a phenomenon central to Hering's model.

Not all the exchanges found Helmholtz on the defensive. Hering's scheme of depth values predicted that only horizontal retinal disparities, not vertical ones, provide cues to depth perception. Helmholtz disputed this. By attaching small beads to the suspended threads, and so creating a vertical disparity between retinal points, Helmholtz showed that the eye can then immediately and accurately judge the true configuration of the set of hanging threads. In the *Handbuch* he provided a series of stereograms designed to prove that vertical disparities could generate the perception of stereoscopic relief.[47]

Hering's Law of Identical Visual Directions also did not escape Helmholtz's critique. That we refer all visual directions to an imaginary central eye, even in monocular vision, merely reflects the fact that we normally look at objects with both eyes, Helmholtz argued. We have therefore learned to refer visual direction to a point midway

---

46. "Ueber stereoskopisches Sehen," *Verhandlungen des naturhistorisch-medizinischen Vereins zu Heidelberg 3* (1865):8–11, in *WA* 2:492–96; *Optics* 3:318–22, and cf. von Kries's comments on the subsequent history of the problem, *Optics* 3:488–90.
47. "Ueber stereoskopisches Sehen," in *WA* 2:492–96; and *Optics* 3:318–22.

between the two real eyes, and now persist in doing so even when one eye is closed.[48] Showing that Hering's law was merely an acquired response undermined the status which Hering had accorded it of being the central principle of space perception.

The most important issue around which the nativist-empiricist dispute coalesced concerned Helmholtz's conviction that the eyes are separate organs which in principle may be moved wholly independently of each other. In 1863 and again in London in his Croonian Lecture of 1865, Helmholtz argued that adherence to Donders's and Listing's laws is a habit acquired by the eyes of an individual during his lifetime in order to facilitate clear and easy visual orientation. Once acquired and ingrained, however, the facility cannot be overridden by acts of the will. Accordingly, some eye movements are impossible for adults to make voluntarily, notably cyclorotations, vertical divergences, and horizontal divergences beyond parallel lines-of-sight. Such movements are not anatomically impossible, Helmholtz insisted, for through the use of prism glasses, which produce a constant, abnormal angular separation of the lines-of-sight entering the eye, we can induce the eyes to perform vertical and absolute divergences in order to bring the distorted images onto corresponding retinal places in the interests of clear vision.

Hering did not immediately contest these experimental claims, but he heaped sarcasm upon Helmholtz's favorite supporting argument. In 1863 and frequently thereafter, Helmholtz reported how, while drowsing over a book late in the evening, he had observed the lines of print to dissolve into double images and then to incline at angles to one another. With this involuntary relaxation of the will, Helmholtz concluded, the two lines-of-sight had diverged vertically because his eyes had performed independent and opposite cyclorotations, movements we are incapable of performing voluntarily. Hering responded contemptuously that it was much more likely that the great Helmholtz, in his dozing state, had simply failed to notice that he had allowed his head to nod to one side, which would produce the same result! Deeply stung by this exchange, Helmholtz seems to have concluded that henceforth any concession on the theoretical principle involved would expose him not merely to charges of inconsistency but of ridicule as well. In the *Handbuch* he defiantly repeated and strengthened these claims, including his view "that the rotation of the eye round the visual axis cannot be effected by our will, because we have not learnt by which exertion of our will we are to effect it, and that the inability does not

---

48. "Ueber stereoskopisches Sehen," in *WA* 2:492–96; and *Optics* 3:318–22.

depend on any anatomical structure either of our nerves or of our muscles which limits the combination of movements."[49]

7. The Nativist-Empiricist Controversy in the *Handbuch*

Until 1866 the exchanges between Helmholtz and Hering hinged mostly on specific scientific issues, not on abstract methodological principles. The tone of these exchanges was pointed, yet mild in comparison to some of Hering's concurrent polemics with Wundt, Volkmann, and others. On many of the specific issues of the field, Helmholtz and Hering were still in close agreement. Yet the exchanges had left their mark on Helmholtz, placing him for the most part on the defensive. Hering repeatedly had picked out forced or implausible arguments, subjected them to experimental refutation or public ridicule, and thus made it increasingly difficult for Helmholtz to retreat. As Helmholtz labored during 1866 over the final Part of the *Handbuch*, he had no reason to expect that Hering would mitigate his future attacks. Perhaps that explains why he allowed the preparation of the *Handbuch* to drag on and on. Helmholtz's correspondence with du Bois-Reymond from this period abounds in melancholic, agonized passages in which he describes his work on binocular space perception as a "burden" or an "ordeal." He stated that, like no other aspect of his physiological optics, the work drew him into disparaging philosophical questions about which it was impossible to "persuade people" of any definitive answer.[50] To be sure, Helmholtz only rarely mentioned Hering's name in this correspondence with du Bois-Reymond—he was never entirely open with the latter on that sore subject; yet the specter of Ewald Hering hung over Helmholtz's program for the reformation of sensory physiology. In the final Part of the *Handbuch*, Helmholtz knew, that specter would have to be laid to rest.

When the final Part appeared in early 1867, it largely recapitulated and extended Helmholtz's earlier results. Yet he presented those results in a radically new framework. The third Part of the *Handbuch*

---

49. "On the Normal Motions of the Human Eye in Relation to Binocular Vision (the Croonian Lecture, April 14, 1864)," *Proceedings of the Royal Society of London 13* (1863–64):186–99, in *WA 3*:25–43, on 34–5; see also "Ueber die normalen Bewegungen des menschlichen Auges," *AO 9* (1863), Part 2:153–214, in *WA 2*:360–95; "Ueber die Augenbewegungen," *Heidelberger Jahrbücher 58* (1863):255–59, in *WA 3*:44–48; and *Optics 3*:54–62, and passim.

50. Helmholtz to du Bois-Reymond, 26 February 1864, 3 January and 13 February 1865, and 11 January 1866, in Kirsten, 207–8 (quote), 213–14, 215, and 219–21, resp.

grounded those results in a deep theory of epistemological and perceptual processes, presented and interpreted them as explicit consequences of empiricist methodology, and reinterpreted the whole field of visual perception as polarized around the competing methodologies of nativism and empiricism.

Helmholtz set the stage by opening Part three with one of his most famous epistemological discussions, "Von den Wahrnehmungen im Allgemeinen."[51] There he restated his conviction that our perceptions of objects localized in space reach consciousness already psychologically "processed" by factors of expectation, habit, and memory acquired through past visual and tactile experience. He described the psychological operations through which primitive sensation is processed to conscious spatial perception as "unconscious inferences." Both these views were quite unexceptional and merely reflected contemporary consensus. Wundt, to give an outstanding example, had published very similar ideas between 1858 and 1862.[52]

By contrast, Helmholtz now fully elaborated for the first time his more unusual conviction that these "unconscious inferences" are syllogistic in nature.[53] He argued that their major premise consists of an inductive generalization drawn from previous visual experience and built up through the process of association; the minor premise is provided by current sensory data; and the conclusion is an inference to the existence of an external object and its particular localization in space. These "unconscious inferences" differ from logical inferences, according to Helmholtz, only in that they are imagistic rather than linguistic in nature, and so they give our perceptions of objects and space the phenomenological quality of primitive experience.[54]

Helmholtz quickly moved from epistemology to methodology. He enunciated three strong methodological tenets for the study of perception which, in his view, followed immediately from the empiricist theory. The mind's perceptual task, he assumed, is wholly pragmatic and utilitarian: it allows us to localize and identify external objects sufficiently for the purposes of life. This implies, however, that the

---

51. *Optics* 3:1–36.
52. Hatfield, "Mind and Space," 318–45.
53. See especially the discussion in ibid., 286–345, to which this analysis is indebted.
54. In the very large literature on Helmholtz's epistemology see especially ibid., 286–351; Richard M. Warren and Roslyn P. Warren, "Introduction," in their *Helmholtz on Perception: Its Physiology and Development* (New York: John Wiley & Sons, 1968), 3–23; Carlos-Ulises Moulines, "Hermann von Helmholtz: A Physiological Approach to the Theory of Knowledge," in *Epistemological and Social Problems of the Sciences in the Early Nineteenth Century*, eds. H. Jahnke and M. Otte (Dordrecht: D. Reidel, 1981), 65–73; and Herbert Hörz and Siegfried Wollgast, "Hermann von Helmholtz und Emil du Bois-Reymond," in Kirsten, 11–66.

unconscious inferences which mediate perception are susceptible to a functional analysis; a perception is "explained" when we can plausibly show that perceptions of that kind normally yield correct or useful information about external objects. The functional emphasis of his early work now became an explanatory principle.

This first tenet also served direct polemical ends, for it distanced Helmholtz's approach from Hering's phenomenological orientation. One unexpected consequence of the tenet, Helmholtz noted, is that introspective analysis offers a poor guide to understanding our perceptual processes. In the radical pragmatism of perceptual experience, unconscious inference routinely ignores or suppresses sensations which do not give us useful information about external objects. Helmholtz insisted that we are interested in our complex perceptions only as they are "signs" of external objects, and so we normally cannot decompose perceptual compounds into their uninterpreted sensory elements.[55] The phenomenological immediacy of a perceptual experience therefore permits no inference to its simple, uncompounded nature.

Helmholtz's second methodological tenet proclaimed that what we call an "illusion of the senses" is really an "illusion in the judgement of the material presented to the senses, resulting in a false idea of it."[56] Helmholtz distinguished his discussion of this commonplace claim through a rigorous insistence on functional explanation. He insisted that every perceptual illusion arises out of the inappropriate application of some inferential procedure, which under ostensibly similar stimulus conditions is known to yield correct information about objects. A methodologically legitimate explanation of a visual experience we call an illusion must specify the normal inference involved and also show how unusual patterns of stimuli "tricked" the perceptual processes into an inappropriate application of it. This tenet, too, served polemical ends. It anticipated the objections of Hering and Panum, already directed against Volkmann and Wundt, that appeals to "psychological" explanations of perceptual regularities must be arbitrary and hypothetical.[57]

Helmholtz's third tenet was a directive for detecting the role of learning and experience in the formation of our perceptions. He postulated that whenever practice, concentration, or external aids cause a familiar pattern of external stimuli to be perceived differently, we may conclude that the original perception was itself an interpreted experience, somehow conditioned by expectation and prior learning.

55. *Optics* 3:6.
56. Ibid., 4.
57. Ibid.

Even this principle, he insisted, constituted a sufficient, not a necessary, condition for distinguishing perceptions from primitive or uninterpreted sensations. This principle had, as Helmholtz well knew, already been discussed, and rejected, by Hering.[58]

This discussion also set out for the first time Helmholtz's vision of a physiological optics starkly polarized between the methodologies of nativism and empiricism. The central problem of a theory of human visual perception, Helmholtz asserted, is to determine how much of perception is due directly to sensation and how much to experience and training. This dictum in itself merely echoed that of many previous writers; Helmholtz, however, went on to portray the problem as the central cleavage of the field. The methodological predisposition to "concede to the influence of experience as much scope as possible, and to derive from it especially all notions of space," is the "empirical theory" of perception. By contrast, researchers who "believe it is necessary to assume a system of innate apperceptions that are not based on experience, especially with respect to space-relations," may be said to adhere to the "intuition theory" (*nativistische Theorie*) of the sense perceptions.[59]

Nativists, Helmholtz went on, try to explain the relative constancy of an individual's perceptual experience (which really arises from "fixed and inevitable associations of ideas") by seeking "some mechanical mode of origin for this connection (between the sensation and the conception of the object) through the agency of imaginary organic structures." Hence, as in the doctrines of Kant, Müller, and Hering, he wrote, nativism does not *explain* our perceptual images of space, it merely postulates innate mechanisms, which in the presence of proper neural stimulation give rise to them. Thus nativism provides no true explanation of anything, violates the methodological canon of explanatory economy, and multiplies hypotheses extravagantly.[60]

The substantive chapters of the *Handbuch* laid out Helmholtz's empirical findings on the nature of perception in such a way as massively to confirm the empiricist theory and to portray those findings

---

58. Ibid., 13. Helmholtz often represented hue and brightness as primitive sensations, but his earlier insistence on the role of unconscious inference in color constancy, adaptation, and simultaneous contrast shows that he regarded these experiences also as frequently (if not always) mediated by unconscious inference. One recent writer attempts to resolve this apparent inconsistency by arguing that Helmholtz was groping toward the concept of "information processing" as the basis of perception; see Theodoor Meijering, "Naturalistic Epistemology: Helmholtz and the Rise of a Cognitive Theory of Perception," (Ph.D. thesis, University of California at Berkeley, 1981); and Theo C. Meyering, *Historical Roots of Cognitive Science* (Norwell, Mass.: Kluwer, 1989).
59. *Optics* 3:10.
60. Ibid., 17–8.

as direct consequences of his three methodological tenets. His long discussion of geometrical and other illusions, for example, attributed most of these effects to eye movements or to the principle of psychological contrast—the mind's tendency to perceive clearly demarcated divisions or differences to be larger than those more vaguely demarcated or not demarcated at all. This discussion included a blistering refutation of Hering's unfortunate early suggestion that certain geometrical illusions arise because the eye estimates distances on the retina by the chord of the arc subtended rather than by the arc itself. That suggestion epitomized for Helmholtz the dangerous nativist tendency to endow the retina with direct awareness of its spatial attributes.[61]

In his discussion of the monocular visual field, Helmholtz followed Lotze and Wundt in arguing that our knowledge of distance and direction on that field is gradually acquired from intrinsically non-spatial cues. His explanation differed from Wundt's mainly in attributing a minimal role to muscle kinesthesia and in denying that the local signs are known through qualitative variation in sensation. Here as elsewhere in his writings, Helmholtz tacitly ignored as overly speculative the potential distinction between spatial awareness per se and the visual assessment of the distance and direction of points in that space. He usually treated the former as if it were given in the sense of touch and merely assimilated to the visual sense.[62]

Like others before him, Helmholtz found support for empiricism in selected case-studies of blind persons who had their sight surgically restored in one eye. One of these was the celebrated Cheselden case (1728), in which the recovering patient was at first supposedly incapable of all visual perception of distance, and to have initially reported that all objects seen "touched his eyes (as he expressed it) as what he felt did his skin." In a second case the inability of the patient immediately to identify visually such familiar objects as a ring or a key—well-known through the sense of touch—spoke, according to Helmholtz, against the existence of "an innate power in the retina of recognizing the form of images there," since "the surface of the key, with its ring and tag, must have been represented on the retina in the same form as it feels to the touch."[63]

Ironically, this same evidence compelled Helmholtz to accept the existence of *some* innate mechanisms for spatial perception. The newly sighted acquire monocular spatial vision so quickly, Helmholtz ad-

61. Ibid., 192.
62. Ibid., 154–232; and cf. the discussion in *Handbuch*² (1896), 583–88.
63. *Optics* 3:220, 226.

mitted, that the local signs which make that learning possible cannot be "disconnected and unsystematic signs, whose connection with the adjacent retinal points can only be acquired by experience." The local signs must vary continuously across the retina, so that, prior to all experience, adjacent retinal points must be recognized in the sensation as adjacent. From a logical perspective, this startling concession to the nativists effectively rendered Helmholtz's position almost indistinguishable from Hering's on matters concerning directional localization. Yet it scarcely touched the real issues of power and program to which the nativist-empiricist controversy was addressed.[64]

In discussing the perception of visual direction, Helmholtz presented an ingenious new demonstration that our sensation of a change in visual direction is cued by the effort of the will in moving the eyes, not by the resulting tension in the muscles, change in the visual sensation, or kinesthetic awareness in the eye muscles that might result from the actual movement.[65] That demonstration supported empiricism no more than nativism; but in the course of the discussion, Helmholtz reasserted his belief that visual depth perception is constantly adjusted by, and so dependent on, the more reliable evidence of touch. Nativists, Helmholtz countered, insist that primitive visual perceptions of space are independent of tactile experience, and for this reason they must assume a "pre-established harmony" between visual and tactile localization in order to explain the coordination of the two. Such an approach, he cautioned, opens the way to the "wildest speculations."[66]

In his key discussion on binocular double vision, Helmholtz speculated on how the retinal correspondence itself must be acquired by the individual during its lifetime. The infant, Helmholtz implied, is simultaneously conscious of both monocular fields. In order to perceive the external world more clearly, it learns to focus both eyes on a single object so that the object images on the two foveas; through the sense of touch, it comes to identify these foveal images as signs of a single object at a particular distance before it in space. The infant then superimposes the two monocular fields onto one another so that the two foveal images correspond and are seen single. Through a similar process all individuals have learned to see images on the retinal horizons and apparent vertical meridians single, and so they build up an entire pattern of correspondence, including an experiential knowl-

---

64. Ibid., 227.
65. Ibid., 242–70, esp. 243.
66. Ibid., 252.

edge of what retinal disparities mean in terms of the relative distance of objects.[67]

As proof of the acquired nature of the retinal correspondence, Helmholtz introduced ophthalmological evidence drawn mostly from the young Berlin physician von Graefe, from his friend and colleague Donders, and from his student Nagel. It centered on patients who suffer from fixed-angle strabismus, or squint. Such individuals cannot fixate, that is, when they bring the image of an external object onto one fovea it will not image on the fovea of the other eye, so that external objects can never be imaged on corresponding points. Most such individuals simply suppress the image from one eye and see the world monocularly. A few, however, continue to see the world with both eyes simultaneously. They ought to see everything in double images, yet they do not. This suggests, Helmholtz argued, that in response to their handicap they acquire an anomalous retinal correspondence, different from the correspondence of normal individuals.[68]

In 1838 the first operation to correct for fixed-angle strabismus was performed in Berlin. This allowed the patient to converge his lines-of-sight and image an external object upon both foveas simultaneously. As the operation became more common through the 1850s with use of anesthesia, it was frequently reported that following such operations the patient saw double. Within weeks, however, single vision gradually returned, suggesting that the patient had not only developed an anomalous correspondence to correct his or her handicap, but that after the operation he or she had reacquired a normal correspondence in the interests of clear vision. Helmholtz cited these cases as being "of fundamental importance for the theory of binocular vision," and prime support for the empiricist theory.[69]

As further evidence that the retinal correspondence is acquired, Helmholtz argued that it may be overridden by empirical factors. Although he criticized Wheatstone's famous experiment as inconclusive, he conceded that Wheatstone's conclusion "cannot be well contradicted," and he put forward stereograms of his own purporting to show that points imaging on corresponding retinal points could be seen at different places in the binocular field. He agreed with Volkmann that binocular fusion was a psychological act arising from the unconscious conclusion that the eyes are viewing a familiar object or image in binocular perspective.[70]

67. Ibid., 400–88.
68. Ibid., 405–7.
69. Ibid., quote on 406.
70. Ibid., 222.

Helmholtz concluded his case for empiricism with the phenomenon of retinal rivalry, the condition which arises when the separate monocular fields "cannot plausibly be combined into the image of a single object."[71] After refuting the views of Panum and Hering that rivalry suggests an underlying organic mechanism, Helmholtz turned to the phenomenon which he regarded as the single strongest support of the empiricist interpretation: stereoscopic lustre. Look at a stereogram which shows the outlines of an object or area, with one half of the stereogram showing the object or area darkly colored and the other half the same scene lightly colored. The binocular perception, he argued, is not that of an intermediate gray; rather, the object or area will seem to shine with a metallic lustre. Now the circumstances under which a real object appears lustrous, Helmholtz explained, are those in which the object has the property of a mirror, reflecting light regularly, rather than diffusely, off its surface. Under such circumstances, it frequently occurs that more light from the object will fall onto one eye than onto the other, so that one of the two binocular fields appears much brighter than the other. When this occurs artificially, as in viewing a specially prepared stereogram, the judgment infers from past experience that it is looking at an object which is lustrous, and so the binocular apperception is of a lustrous object, not of an object of brightness intermediate between the two fields.[72]

Helmholtz regarded binocular lustre as a crucial experiment supporting the empiricist position. It and other phenomena associated with rivalry proved "that the content of each separate field comes to consciousness without being fused with that of the other field by means of organic mechanisms; and that therefore, the fusion of the two fields in one common image, when it does occur, is a psychic act."[73] Above all, stereoscopic lustre supposedly proved that binocular perception depends heavily upon our prior experience of viewing real objects under normal viewing conditions. This was the essence of the empiricist position.

8. The Refutation of Hering

Helmholtz's achievement in the third Part of the *Handbuch* consisted not only of his logical and empirical arguments in favor of individual points. For the *Handbuch* as a literary whole was, in two quite different senses, a masterpiece of scientific rhetoric.

---

71. Ibid., 493–528, quote on 494.
72. Ibid., 512–13.
73. Ibid., 499, 512–13.

First, the third Part functioned as vehicle of persuasion, aimed at the scientific community and intended to destroy Ewald Hering forever. In February 1865, Helmholtz wrote to du Bois-Reymond with uncharacteristic smugness that he had so far refrained from "putting Hering in his place," partly out of respect for the intelligent and consequent nature of Hering's writings, and partly because of rumors that he had earlier been mentally disturbed.[74] By 1866, however, Helmholtz had clearly decided to stay his hand no longer. The *Handbuch* set out not only to refute Hering or to anticipate his objections; indeed, prior to the final chapter, Hering's name is only occasionally mentioned. The rhetorical strategy of the work rather reduced Hering to a vessel, a temporary instrument, of an idea which Helmholtz branded as the enemy of science, reason, and human potential. Moreover, Helmholtz enhanced the rhetorical force of this impassioned argument by an aloof and olympian tone of scrupulous fairness.

Helmholtz's dramatic juxtaposition of empiricism and nativism, the leitmotif of the whole work, constituted the volume's second rhetorical strategy. This literary strategy imposed a thematic unity upon a chaotic mass of material, as required by the *Handbuch*-genre to which Helmholtz was committed. On the other hand, the strategy imposed on the work a narrative tension which led onward from chapter to chapter, tightening and intensifying at every turn, toward its final confrontation and resolution. The work achieved that culmination in the final chapter, Helmholtz's famous "Kritik der Theorien." Here nativism, previously portrayed mostly as a methodological abstraction, now appeared incarnated as Ewald Hering and was subjected to one final, massive critique.[75]

From the start of that critique Helmholtz judiciously positioned his own empiricist perspective on the inductivist high ground, from which he assailed his nativist opponents with charges of philosophical idealism, theoretical dogmatism, and disregard of Ockham's razor. He represented Hering as the latest and most distinguished member of a tradition reaching back to Kant, Müller, and Panum. He traced the opposing empiricist tradition to the British empiricists of the eighteenth century by way of Herbart and Lotze, also including in it his contemporaries Wundt, Nagel, and Classen. Although Helmholtz offered a series of more specific criticisms of Hering's position, only three need concern us.

---

74. Helmholtz wrote: "das hat mich bisher zurückgehalten, ihn abzutrumpfen, wie er es stellenweise verdient hätte"; see Helmholtz to du Bois-Reymond, 13 February 1865, in Kirsten, 215.

75. *Optics* 3:531–59.

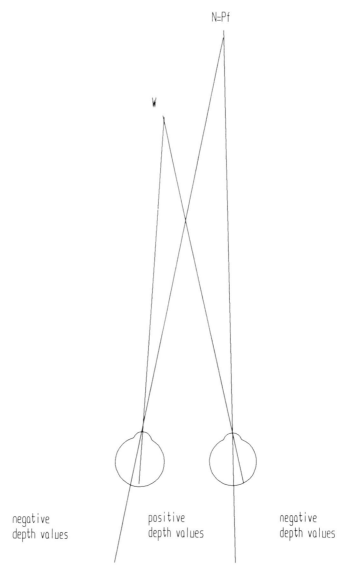

*Figure 4.5. Experimental test by Helmholtz of Hering's theory of retinal depth values. The half-images of the vertical wire **W** should seem to lie both nearer and farther than the needle **N** at the points of fixation. (Not to scale)*

Hering's one crucial experiment in defense of his system of retinal depth values consisted of a vertical wire **W** positioned slightly to the left of a pin or needle **N** and somewhat nearer the eyes (see Figure 4.5). When the needle is fixated, the wire will be seen in double images. Because the image which falls on the right retina will fall on the outer or temporal hemisphere, it will induce, according to Hering's scheme, negative depth values, and so should be seen to lie nearer than the frontal plane containing the needle (the core plane). The other half-image will induce positive depth values on the nasal side of the left retina; the theory predicts that it should be seen lying further than that frontal plane. In fact, it is extremely difficult to form any clear idea of the relative positions of the double images, because of the great difficulty in maintaining rigid fixation on the needle while the apparent distances of the two half-images are compared. Nevertheless, after long fixation and rigid attention to the phenomenon, Hering had claimed success in observing the double images at different distances as predicted by his theory.[76]

Helmholtz cited this passage from Hering and wrote of his attempts to confirm it:

> I have gazed at the pin so long and so fixedly that at last everything was extinguished by the negative afterimages. There is a stage when all that can be seen are the nebulous individual parts of the double images of the wire emerging now and then in the course of the conflict with the ground on which they lie and with the afterimages; and then I have noticed that these parts appear sometimes to be far and sometimes to be near, one just as often as the other and just as energetically. But I have never been able to persuade myself that this phenomenon occurred in the main as it ought to occur according to the Hering theory; and I never should have ventured to lay the foundation of a new theory of vision on an observation made with images that are half-extinguished in this fashion. However, I admit that I may have been unskillful. Only, Mr. Hering will have to forgive me for not being able to say that I have been convinced by this "overwhelming" proof of the correctness of the theory, as he himself puts it.[77]

By impugning Hering's competence and veracity as an expert, introspecting witness, Helmholtz delivered the field's ultimate personal attack. Helmholtz's rhetorical skills, combined with his own reputation as an observer, carried the day. Although Hering's defenders later cited

76. Hering, *Beiträge*, Part 5 (1864):340–41.
77. *Optics* 3:554.

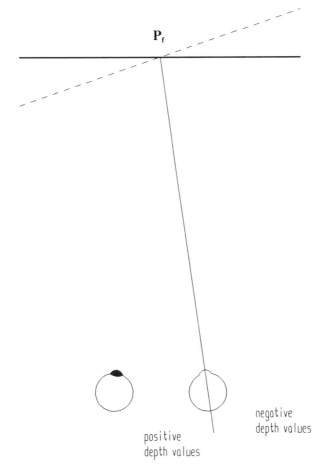

Figure 4.6. The compensating rotation of the core plane, induced by the closing of one eye, according to Hering. (Not to scale)

the experiment occasionally, no subsequent writer disputed Helmholtz's judgment of its outcome, and the issue of the depth-localization of double images largely vanished from the literature.

The nativist position, even as Helmholtz represented it, could not be equated simply with Hering's scheme of retinal depth values. Helmholtz seized upon this scheme, however, as Hering's most vulnerable quarter. When we look at a frontal plane (for example, the wall of a room) with one eye closed, then according to the theory of depth values the wall ought to appear rotated clockwise about a vertical line through the point of fixation P_f (see Figure 4.6). This is because our perception of the wall's relief is governed only by the depth values of one retina, half of which are positive and half negative. Hering himself had anticipated this objection and had argued that in this case the perception is overcome by experiential factors, which cause us to "reorient our visual space" and rotate the core surface so as to preserve its perpendicularity to the visual axis.[78]

Helmholtz responded with a clever experiment in which we look at the wall with two eyes, but with a strip of black paper cut the same width as the interpupillary distance held before the eyes (see Figure 4.7). Thus each eye has the inner half of its normal visual field cut off; points from the visible left-hand side of the wall will image only on the nasal side of the left retina; points from the visible right-hand side of the wall will image only on the nasal side of the right retina; and, according to Hering's theory, the visible portions of the wall should seem inclined to one another, as in **AA'**, **BB'**. Yet they do not. Experiential factors again seem to override the primitive space values, and this time in a more complicated way than by simply rotating the core surface.[79]

On the basis of these and similar observations Helmholtz drew his moral. Whenever the depth values conflict with experiential factors, they seem to cease entirely and are totally overridden by the latter. "Are we not then forced to conclude," he inquired rhetorically, "that those depth-feelings, if they exist at all, are at least so weak and so vague that their influence is negligible in comparison with the factors derived from experience? And therefore, that the apperception of depth might arise just as well *without* them as *with* them, as is supposed to happen on Hering's assumptions?"[80]

The volume culminated with Helmholtz's ultimate argument. His preference for the empiricist position, he wrote, followed from the highest principles of scientific method. "I think it is always advisable to explain natural processes on the *least* possible number of hypotheses and on those which are as *definitely formulated* as possible."[81] Having spent "a large part of my life" in subjective visual experimentation, Helmholtz noted that he had acquired ever greater control over the movements of his eyes and the focusing of his attention. For him, unexpected realms of perceptual experience had been brought under rational control and to scientific understanding by discipline, practice, and the reasoned exercise of the will. With this greater control came the conviction that "the essential phenomena in this field" could not

78. Hering, *Beiträge*, Part 5 (1864):346.
79. *Optics* 3:545–55.
80. Ibid., 556.
81. Ibid., 558.

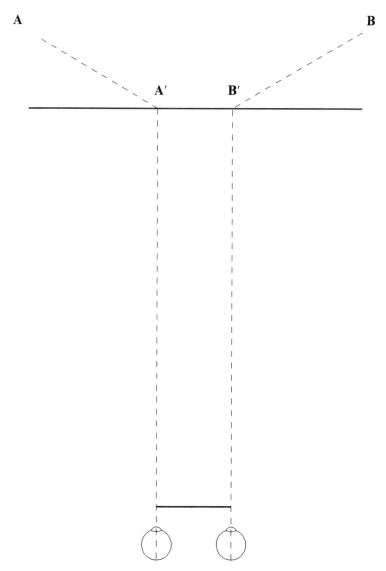

Figure 4.7. Experiment by Helmholtz, purporting to refute Hering's coreplane rotation (1867). The distant surface continues to be seen as a perpendicular frontal plane, even with a screen placed before the eyes. (Not to scale)

be explained by innate neural mechanisms. On this confident appeal to the rational human mastery of self and experience, Helmholtz rested his defense of the empiricist theory.[82]

To describe the arguments of the *Handbuch* as a brilliant exercise in scientific rhetoric implies no insincerity about them on Helmholtz's part. It does emphasize, however, the need to evaluate them from a larger perspective. Helmholtz represented the nativist-empiricist division as merely continuing the old dispute between British empiricist-associationism (with which he identified himself) and German idealism (with which he implicated Hering). That assertion was rhetorically effective but historically misleading for two reasons. First, Hering's intellectual roots lay with Müller and Gustav Theodor Fechner, not directly in any tradition of philosophical idealism or even in Kantianism.[83] Second, the polemical exchanges between Helmholtz and Hering as much created the later nativist-empiricist dispute as they carried on any existing controversy. Before Hering "extended" the theory of retinal identity in the 1860s, no major theorist had made the claim which became the central issue of the later dispute, namely, that the eye possesses innate physiological mechanisms which compel us to see objects in three-dimensional relief. Similarly, no theorist before Helmholtz had ever portrayed the field of physiological optics as so starkly polarized around the nativist-empiricist question. Even Hering later expressed surprise that Helmholtz had drawn the battlelines so sharply around this particular issue.[84]

9. The Nativist-Empiricist Controversy after the *Handbuch*

If Helmholtz created the nativist-empiricist controversy in 1867, he did so only that he might simultaneously finish it. That strategy ultimately failed, but for at least a decade he was widely seen to have been victorious in his debate with Hering. Even Hering sensed this; after 1867 he retreated from his hypothesis of retinal depth values, and by 1870 his research interests had shifted to color theory and

---

82. Ibid.
83. R. Steven Turner, "Fechner, Helmholtz, and Hering on the Interpretation of Simultaneous Contrast," in *G.T. Fechner and Psychology*, eds. Josef Brožek and Horst Gundlach (Passau: Passavia Universitätsverlag, 1987), 137–50.
84. See Hering, *Zur Lehre vom Lichtsinne. Sechs Mittheilungen an die kaiserlichen Akademie der Wissenschaften in Wien* (Vienna: Carl Gerold's Sohn, 1878), 1–5.

contrast, fields in which he found the Helmholtzian orthodoxy more vulnerable to attack.

Helmholtz's influence on physiological optics remained, if anything, even stronger outside Germany than within. The English translation of the *Handbuch*, done in New York in 1924–25, deeply entrenched Helmholtz's views in American and British science.[85] In Anglo-American psychology, empiricist approaches to the problem of spatial perception dominated elementary textbooks well into the twentieth century, the nativism of William James notwithstanding. These approaches, whether derived from Helmholtz or from Wundt, proved congenial to the British associationist tradition and to the temper of American pragmatism.

In Germany the subsequent debates took a more complex course. Despite Helmholtz's initial victory and the continuing dominance of his views in elementary textbooks, the scientific literature of the later 1880s slowly began to swing toward nativism. Eulogies written on the occasion of Helmholtz's death in 1894 noted this fact, and the writings of Helmholtz's greatest German disciple, Johannes von Kries, grew increasingly defensive in tone.[86]

A range of factors influenced the subsequent German debate, and one of the more important was the growing acceptance of evolutionary theory. Evolution convinced many observers that the nativist-empiricist debate had been unnecessarily polarized. From an evolutionary perspective, the empirical acquisition of a spatial sense with its supporting patterns of eye movements and retinal correspondence can be regarded as having occurred over the evolutionary life of the species rather than as occurring anew during the first months in the life of every individual. Thus, both positions in the controversy could be

---

85. See note 2 (above) and the Abbreviations to this volume on the subsequent editions of the *Handbuch*. For an English view of the controversy see James Sully, "The Question of Visual Perception in Germany," *Mind. A Quarterly Review of Psychology and Philosophy 3* (1878):1–23, 167–95.

86. Carl Stumpf, "Hermann von Helmholtz and the New Psychology," *Psychological Review 2* (1895):1–12; Emil du Bois-Reymond, *Gedächtnisrede auf Hermann von Helmholtz, aus den Abhandlungen der königlichen preussischen Akademie der Wissenschaften zu Berlin vom Jahre 1896* (Berlin: Georg Reimer, 1896), 39–40; and Johannes von Kries, "Helmholtz als Physiolog," *Die Naturwissenschaften 9* (1921):673–93, and esp. the supplementary material by Kries ("Empiricism und Nativism," "Ueber den Ursprung der Gesetze der Augenbewegung," and "Historisch-kritische Bemerkungen") in *Handbuch³* (1910), reprinted in the American edition, *Optics 3*:607–51. The leading elementary textbook of the period, Ludimar Hermann's *Grundriss der Physiologie des Menschen*, 4th ed. (Berlin: A. Hirschwald, 1872) and 13th ed. (Berlin: A. Hirschwald, 1905), gave Hering wider coverage in each of its successive editions, while remaining Helmholtzian in its overall tone.

considered partly correct. Helmholtz's friend and ally du Bois-Reymond advanced this opinion publicly in 1871, as did Donders in 1872.[87]

Nativists were especially ready to accept the phylogenetic compromise. Hering was an enthusiastic evolutionist, and his school regarded evolution as strongly supporting its position, since it made the mechanisms of perception innate from the standpoint of the individual born with them. Helmholtz, by contrast, accepted the evolution of our sensory mechanisms but regarded it as irrelevant for the nativist-empiricist debate, since evolutionists usually could not specify the experiential factors which had caused the mechanisms of perception to develop as they had. Furthermore, the evolution argument said nothing about the degree of residual plasticity left to the individual to be shaped by experience and learning, and this was what Helmholtz and his students took to be the main issue in dispute.[88]

The empirical question bearing most directly on the controversy concerned eye movements. Though thwarted on other fronts, in 1868 Hering published a powerful rejoinder to Helmholtz's claim that in the newborn the eyes are essentially separate organs that may be moved independently of one another by acts of the will. Hering's famous *Die Lehre vom binocularen Sehen* argued that the eyes must be regarded as a single organ, that they always move in tandem by a single innervation from the central nervous system. The structure of the musculature, not learning and experience, ensures that the eyes conform to Donders's and Listing's laws, whatever adaptive principles may underlie the phylogenetic origins of that conformity. Hering's persuasive case fed the general resurgence of nativist sentiment which began in the 1880s, and by the early twentieth century Helmholtz's defenders had beaten a substantial retreat from his original position on eye movements.[89]

Observations on animals and human infants played an ambiguous role in the subsequent debate. As evidence for an innate spatial sense, nativists frequently pointed to the high degree of spatial coordination evinced among the newly born or newly hatched of many species of

---

87. Emil du Bois-Reymond, "Leibnizische Gedanken in der neueren Naturwissenschaften (1870)," in his *Vorträge über Philosophie und Gesellschaft*, ed. Siegfried Wollgast (Berlin: Akademie Verlag, 1974), 25–44; and Franz Cornelius Donders, "Ueber angeborene und erworbene Association," *AO 18* (1872):153–64.

88. *Optics* 3:535.

89. Hering, *Die Lehre vom binocularen Sehen* (Leipzig: Wilhelm Engelmann, 1868), 60–5 and passim; and von Kries, "On the Origin of the Laws of the Ocular Movements," *Optics* 3:625–34.

animals and birds.[90] This kind of evidence, however, rarely went beyond the anecdotal, and recourse to direct animal experimentation was rare during the nineteenth century.[91] Observations on human infants, especially their eye movements, provoked much discussion but furnished ambiguous results. Jena physiologist William Preyer, the leading researcher in this field, originally sided entirely with Helmholtz. By 1884, however, Preyer was prepared to concede that the infant may be born with phylogenetic predispositions to certain visual behaviors, which are triggered with the gradual maturation of the neural and visual apparatus during the months after birth.[92]

Far more important for the debate, however, was clinical evidence from the blind who had their sight restored, usually through operation for cataract, and from patients recovering from corrective surgery for fixed-angle strabismus. Helmholtz had relied heavily on such evidence to support his view that the retinal correspondence was empirically acquired; but by 1890 the weight of the ophthalmological evidence was shifting. Although some squinters do acquire an anomalous correspondence and achieve single vision without wholly suppressing one monocular field, nativists could now produce evidence to show that this acquired correspondence is rarely a true correspondence. It does not permit binocular depth perception or fusional movements and only rarely does it wholly replace the normal pattern of correspondence. By the First World War the debate over strabismus was a front on which the empiricists were in retreat.[93]

Disciplinary fragmentation within the field of vision studies also influenced the nativist-empiricist debate. During the 1860s vision research had been pursued on a broad cooperative front by physicists, ophthalmologists, philosophers, psychologists, and, most important of

90. E.g., Hering, "Der Raumsinn und die Bewegungen des Auges," in *Handbuch der Physiologie*, ed. Ludimar Hermann (Leipzig: F.C.W. Vogel, 1879), *3*, Part 1:341–601, on 564; Armin Tschermak-Seysenegg, *Einführung in die physiologische Optik* (Berlin: Springer, 1942), 125; and see Nicholas Pastore, "Helmholtz's 'Popular Lectures on Vision'," *JHBS 9* (1973):190–202; and Hochberg, "Nativism and Empiricism," 318–30.
91. But see Donders, "Ueber Association," 153–64.
92. William Thiery Preyer, *Die Fünf Sinne des Menschen. Eine Populäre Vorlesung. Gehalten im akademischen Rosensaal in Jena am 9. Februar 1870* (Leipzig: Fues's Verlag, 1870), 33–53; and idem, *Die Seele des Kindes. Beobachtungen über die geistige Entwicklung des Menschen in den ersten Lebensjahren*, 2nd ed. (Leipzig: Grieben Verlag, 1884), esp. 1–51.
93. Franz Bruno Hofmann, "Die neueren Untersuchungen über das Sehen der Schielenden," in *Ergebnisse der Physiologie 1*, Part 2 (*Biophysik und Psychophysik*):801–46; von Kries, in *Optics 3*:578–80, 584–89; and Oscar Zoth, "Augenbewegungen und Gesichtswahrnehmungen," in *Handbuch der Physiologie des Menschen*, ed. Wilhelm Nagel (Braunschweig: F. Vieweg und Sohn, 1905), *3*:283–437, on 396.

all, physiologists. After 1880, however, a rapid compartmentalization of vision studies ensued, partly as a result of the growing isolation of the disciplines which contributed to it. At mid-century, physicists, for example, had been active contributors to vision studies and had been traditionally sympathetic to Helmholtz's approach; by 1900, they had largely abandoned the field.[94]

Experimental psychologists found the issues of the nativist-empiricist controversy most unavoidable to their disciplinary concerns. A marginal group numerically in 1860, by the 1890s German psychologists had begun to differentiate themselves sharply from philosophers, to secure specialized chairs and institutes, and to play an increasingly dominant role in vision studies.[95] The new psychologists found Helmholtz's empiricism, despite its congenial emphasis on learning and experience, to operate with a quasi-rationalistic concept of mind and its unconscious activities that gave psychology little purchase. Hering's nativism, on the other hand, threatened to reduce psychology to physiology. What appealed to the new psychologists in Hering's writings was, however, the primacy he gave to phenomenological experience: his insistence that the primitive sensory capacities of consciousness have to be studied as we find them and not "reduced" to rationalistic inferences or physico-chemical mechanisms. Although the influential school of psychology at Leipzig continued to represent Wundt's version of the empiricist orthodoxy, by 1900 other experimental psychologists like Carl Stumpf, Hermann Ebbinghaus, and Oswald Külpe were moving toward more nativistic positions.[96] Like Hering, they found it impossible to believe that a sensation so apparently primitive as that of space could really be "emergent," that is, an intellectual construct out of other sensory capacities. Psychologist Franz Hillebrand in particular

---

94. Turner, "Paradigms and Productivity," 45.

95. "Zur Einführung," *Zeitschrift für Psychologie und Physiologie der Sinnesorgane 1* (1890):1-4; Mitchell Graham Ash, "The Emergence of Gestalt Theory: Experimental Psychology in Germany, 1890-1920," (Ph.D. diss., Harvard University, 1982); idem, "Academic Politics in the History of Science: Experimental Psychology in Germany, 1879-1941," *Central European History 13* (1980):255-86; Joseph Ben-David and Randall Collins, "Social Factors in the Origin of a New Science: The Case of Psychology," *American Sociological Review 31* (1966):451-72 (with a critique by Dorothy Ross); and Turner, "Paradigms and Productivity," 45-50, 62-5.

96. Carl Stumpf, *Ueber den psychologischen Ursprung der Raumvorstellung* (Leipzig: S. Hirzel, 1873), 97-103, and passim; Hermann Ebbinghaus, *Grundzüge der Psychologie*, 4th ed. (Leipzig: Veit & Comp., 1919), 490-94; and Oswald Külpe, *Grundriss der Psychologie* (Leipzig: Wilhelm Engelmann, 1893), 385-87. See also Ash, "Emergence of Gestalt Theory," 85-135; and Turner, "Helmholtz, Sensory Physiology, and the Disciplinary Development of German Psychology," in *The Problematic Science*, eds. Woodward and Ash, 147-66.

revered Hering; in the 1890s he re-examined many points of the original dispute between Helmholtz and Hering and mounted an impressive defense of Hering's theory of retinal depth values and his concept of the core surface.[97]

Larger changes in the intellectual climate of central Europe also influenced the course of the nativist-empiricist dispute. The intellectual "mood" of Germany in the 1850s and 1860s had been one of buoyant materialism, one in which philosophy had rejected the speculative idealism and in which liberalism had turned from a program of idealistic political reform to the pragmatic goals of economic and social modernization. Helmholtz's empiricism offered a theory of mind and perception nicely congruent to this intellectual mood. It owed an obvious debt to British empiricism; it emphasized individualism, learning, the flexibility and plasticity of human capacities; and it insisted that perception itself proceeded on a rational basis through "unconscious inferences."

But the Germany of 1890 was a very different place. With its period of heroic industrialization and political unification behind it, Germany was now struggling with the problems of economic recession, urban poverty, a strong and restless socialist movement, and a national leadership prone to mask fundamental political antagonisms behind appeals to ever-more-fervent nationalism. At the end of the nineteenth century the rising mood of cultural pessimism, including its fascination with the irrational and the instinctual, was detectable all over Europe. In this intellectual climate Helmholtz's appeals to experience, utilitarianism, and quasi-rationalistic processes of unconscious inference rang increasingly hollow. Conversely, the nativist position that our perceptual experience flows from innate predispositions, neural structures, and instincts seemed increasingly acceptable or even obvious.

Two developments in academic philosophy also encouraged the resurgence of nativist sentiment. The strong neo-Kantian movement of the later nineteenth century reasserted the dictum that the categories of space and time are "pure intuitions" imposed by the mind upon experience. Superficially understood, Kantianism seemed to favor Hering's insistence upon an innate spatial sense rather than Helmholtz's position that our knowledge of space is learned. This threat sent Helmholtz's disciple von Kries, himself an avowed neo-Kantian, scrambling to reemphasize the alleged Kantian elements in Helmholtz's views and

---

97. See Franz Hillebrand, *Lehre von den Gesichtsempfindungen auf Grund hinterlassener Aufzeichnungen*, ed. Franziska Hillebrand (Vienna: Julius Springer, 1929), and the literature cited there; see also idem, *Ewald Hering. Ein Gedenkwort der Psychophysik* (Berlin: Julius Springer, 1918).

to relocate Hering into a phenomenological and organismic tradition far removed from Kant.[98] The positivism of Ernst Mach stood at the philosophical antipodes of neo-Kantianism. Nevertheless, Machism's phenomenological approach to knowledge echoed Hering's lifelong insistence that the spatiality of our perceptions is phenomenologically given, and that to try to go beyond that given to hypothetical inferences about its origins or to unconscious experiential deductions is to go beyond the realm of legitimate science. Mach supported all Hering's central positions and revered Hering as a formative influence on his own thought, while he was persistently critical of Helmholtz.[99]

Finally, among all the factors which influenced the subsequent course of the nativist-empiricist debate, the characters and careers of its two leading protagonists may have been the most important. Hering trained a large and loyal school of prominent physiologists and ophthalmologists and was personally elaborating and defending his views in an increasingly receptive scientific climate almost to his death in 1918.[100] Helmholtz's career after 1867 was quite different. The disputes with Hering and the ordeal of preparing the final volume of the *Handbuch* affected him deeply. Partly because of those frustrations, Helmholtz abandoned sensory physiology for physics after 1867 and accepted the prestigious chair of experimental physics in Berlin in 1871.

Down to his death in 1894, Helmholtz elaborated and defended his empiricist epistemology on a philosophical and popular level. Yet except in the field of color theory (to which he returned only late in his career) he did not keep up with subsequent research on the nature of perception or personally participate in technical debates.[101] Numerous students in several scientific fields worked under him on physiological optics; but as a teacher of physiology Helmholtz never inspired the

---

98. Thomas E. Willey, *Back to Kant: The Revival of Kantianism in German Social and Historical Thought, 1860-1914* (Detroit: Wayne State Press, 1978); and von Kries, in *Optics* 3:640-51. See also Nicholas Pastore, "Reevaluation of Boring on Kantian Influence, Nineteenth-Century Nativism, Gestalt Psychology and Helmholtz," *JHBS 10* (1974):375-90; and J.H. Hyslop, "Helmholtz's Theory of Space-Perception," *Mind 16* (1891):54-79.

99. Ernst Mach, *The Analysis of Sensations and the Relation of the Physical to the Psychical*, trans. C.M. Williams (New York: Open Court, 1897; reprint ed. New York: Dover, 1959), 69, 168, and passim; and John T. Blackmore, *Ernst Mach. His Work, Life, and Influence* (Berkeley: University of California Press, 1972), 55-63, 180-203.

100. R. Steven Turner, "Vision Studies in Germany: Helmholtz vs. Hering" in *Research Schools*, eds. Gerald Geison and Frederic L. Holmes, *Osiris 8* (forthcoming).

101. Helmholtz died while revising the *Handbuch* for the second edition. His assistant, Arthur König, stated that Helmholtz had intended to reproduce the material of the third Part unchanged from the first edition. (See *Handbuch²*, ix-xi.) He did, in fact, slightly revise the introductory chapter, "Von den Wahrnehmungen im Allgemeinen."

fierce loyalty or intellectual devotion that was so common among Hering's students. His influence derived more from his great scientific reputation and the immense synthetic and rhetorical power of the *Handbuch*. That influence sufficed brilliantly to inaugurate modern investigations into the physiology and psychology of space perception, but not to impose a lasting theoretical consensus upon it.

5

Innovation through Synthesis

Helmholtz and Color Research

Richard L. Kremer

1. Introduction

At the end of his 1672 paper in which he presented a new theory of light and colors, Isaac Newton admitted: "But, to determine more absolutely, what light is, . . . and by what modes or actions it produceth in our minds the phantasms of colours, is not so easie. And I shall not mingle conjectures with certainties."[1] As Newton realized, his theory of color was primarily a theory of light, a physical theory with light rays, prisms, angles of refraction, lenses, and barycentric diagrams as its chief "working objects." Although the *Opticks* did occasionally mingle conjectures about how the human eye "sees" colors, Newton never systematically examined color as a phantasm of the human mind. Not until the nineteenth century did color become a subjective phenomenon, an aspect of nature impossible to consider apart from human verbal reports of visual experience. The color-seeing eye became both the instrument of and the object under analysis. This shift

---

Acknowledgments: Research for this essay has been supported by grants from the German Academic Exchange Service (DAAD), the Alexander-von-Humboldt Foundation, and Dartmouth College. For their helpful criticisms and suggestions I thank O.-J. Grüsser, Michael Hagner, Renato G. Mazzolini, and the other authors in this volume, especially David Cahan, Gary Hatfield, and R. Steven Turner.

1. Isaac Newton, "New Theory about Light and Colours," *Philosophical Transactions* (1672):3075–87, reprinted in I. Bernard Cohen, ed., *Isaac Newton's Papers & Letters on Natural Philosophy*, 2nd ed. (Cambridge, Mass.: Harvard University Press, 1978), 47–59, at 57.

in the conceptual location of color, from the physical to the physiological and psychological, perhaps began with Goethe's *Zur Farbenlehre* (1810) but it became canonical only with the appearance of the second Part of Helmholtz's *Handbuch der physiologischen Optik* (1860).[2] As in so many other areas of physiological optics, Helmholtz's immediately influential *Handbuch* consolidated, organized, and directed research in color vision until early in the twentieth century.[3] Not until the 1920s, when Edgar Adrian developed instruments to measure electrical responses from single sensory nerves in vivo, would so-called objective sensory physiology begin to replace the subjective methods of Helmholtz's *Handbuch*.[4]

This essay examines Helmholtz's development as a color theorist from 1852, when he began work on color research, to 1860, when his comprehensive survey of the field appeared in the *Handbuch*. It argues that Helmholtz, despite his considerable skill in conducting observations and experiments, worked most effectively as a color theorist. Unlike other areas of sensory physiology where he developed new apparatus and made important discoveries in the laboratory,[5] Helmholtz (through 1860) offered few new observations of fundamental importance for color research. Further, this essay claims that if Helmholtz as color theorist saw further he did so by building on the work of others: in color theory, as in so many other fields in which he toiled, Helmholtz acted as a creative borrower and syntheziser who clarified others' ideas, removed conceptual problems from others' experiments, collected scattered researches of others into unified programs of research, and sought to define orthodoxy by formulating an authoritative discourse for the field. "Order and coherence," wrote Helmholtz at the end of his *Handbuch*, "even when founded on an untenable principle, are better than contradictions and incoherence."[6] The diverse group of scientists on whom Helmholtz depended included, besides Newton,

---

2. See Jeffrey Barnouw, "Goethe and Helmholtz: Science and Sensation," in *Goethe and the Sciences: A Reappraisal*, ed. Frederick Amrine, et al. (Dordrecht: Reidel, 1987), 45–82.
3. R. Steven Turner, "Paradigms and Productivity: The Case of Physiological Optics, 1840–94," *Social Studies of Science 17* (1987):35–68.
4. Richard Jung, "Sensory Research in Historical Perspective," in *Handbuch of Physiology*, Section 1: *The Nervous System*, Vol. 3: *Sensory Processes*, ed. John M. Brookhart and Vernon B. Mountcastle (Bethesda: American Physiological Society, 1984), 1–74, at 53.
5. See the essays by R. Steven Turner, "Consensus and Controversy: Helmholtz on the Visual Perception of Space," and Stephan Vogel, "Sensation of Tone, Perception of Sound, and Empiricism: Helmholtz's Physiological Acoustics," both in this volume.
6. *Handbuch*, vi.

most leading nineteenth-century color researchers to 1860: Jan Purkyně, Johannes Müller, Thomas Young, Heinrich Dove, Ernst Brücke, Gustav Fechner, and James Clerk Maxwell. Many earlier studies of Helmholtz's color research have, by following too closely Helmholtz's own accounts of the history of the field, overemphasized his originality.[7] The stories of victors, however, may conceal as much as they illuminate. This essay seeks to provide a contextual review of eighteenth- and nineteenth-century color research as the background against which Helmholtz's synthesizing efforts may readily be seen.

The first half of this essay examines Helmholtz's early experiments aimed at preserving Newtonian color theory. Sections 2 and 4 review, respectively, David Brewster's challenge to Newton's theory of color and post-Newtonian, often conflicting, traditions of color mixing, the two sets of problems Helmholtz chose to address from 1852 to 1855 in his early work on color. Ostensibly experimental, these early papers, considered in Sections 3 and 5, demonstrate Helmholtz's ability to reconceptualize long-standing problems, drawing especially on a deep knowledge of physical optics. They show him either uninformed about his predecessors' work or somewhat less than generous in citing that work. More importantly, they reveal an awareness of color as a subjective phenomenon, even as Helmholtz retained the Newtonian thematic of color as a physical problem and rejected Young's physiological hypothesis of tripartite color vision.

The second half of the essay analyzes Helmholtz's "conversion" to Young's hypothesis, and his elaboration of it into a comprehensive explanatory tool soon christened by his contemporaries as the "Young-Helmholtz" theory of color vision. In contrast to other recent studies, Section 6 suggests that the conversion occurred already in 1855, and that Helmholtz's crucial modification of Young's hypothesis, which made the latter truly workable, derived not from Maxwell or Fechner but rather from Müller's doctrine of specific sense energies. Section 7 considers Helmholtz's deployment of the modified theory to guide his broad review of color in the second Part of his *Handbuch*. This Section also argues that Helmholtz's commitment to Müller's doctrine, understood as positing one-fiber, one-sensation, forced him finally to disrupt

---

7. John G. McKendrick, *Hermann Ludwig Ferdinand von Helmholtz* (New York: Longmans, 1899); Koenigsberger; E.C. Millington, "History of the Young-Helmholtz Theory of Colour Vision," *Annals of Science 5* (1941):167–76; Edwin G. Boring, *Sensation and Perception in the History of Experimental Psychology* (New York: D. Appleton, 1942); and Paul D. Sherman, *Colour Vision in the Nineteenth Century: The Young-Helmholtz-Maxwell Theory* (Bristol: Adam Hilger Ltd., 1981).

the theoretical unity of the *Handbuch*'s treatment of color. To explain the subjective phenomena of color contrast, Helmholtz added mental processes of "unconscious judgments" to the physiological mechanisms of Young's receptors. Here, too, precedent existed for such a move. But it meant that the language of color theory in the *Handbuch* contained both physiological and psychological terms, a bifurcation which, as indicated in Section 8, would generate ever more controversy among color theorists during the remainder of the century.

2. Absorption and Brewster's Challenge to Newton

As Kathryn M. Olesko and Frederic L. Holmes argue elsewhere in this volume,[8] in all of his early researches (except for the essay on conservation of force) Helmholtz sought by experimental means to answer widely different yet always fundamental questions in contemporary physiology, questions posed by such leading figures as Justus Liebig or Johannes Müller. In considering absorption and the nature of the solar spectrum, Helmholtz addressed a problem defined thirty years earlier by another leading natural philosopher, David Brewster. During the first half of the nineteenth century Brewster had deservedly acquired a reputation, even among his critics, as the "Father of Modern Experimental Optics."[9] Influenced by Scottish common sense philosophy and deeply adverse to "hypotheses" such as a luminiferous aether, Brewster formulated what he called the "laws of the phenomena" for polarization, double refraction, spectroscopy, colors of thin plates, and phosphorescence. Yet with his claim, based on a series of absorption experiments, that sunlight consists of three differently colored rays of light, each of which is everywhere present in the visible spectrum, Brewster fundamentally challenged Newton's theory of color. Helmholtz began his color research by confronting the most serious objection to Newton's program since Goethe's *Farbenlehre*.

For Newton and his followers the passage of light through variously colored media was an unproblematic and insignificant optical phe-

---

8. "Experiment, Quantification, and Discovery: Helmholtz's Early Physiological Researches, 1843–50."

9. G.B. Airy, "On the Theoretical Explanation of an Apparent New Polarity in Light," *Philosophical Transactions* (1840):225–44, at 226. See Edgar William Morse, "Natural Philosophy, Hypotheses and Impiety: Sir David Brewster Confronts the Undulatory Theory of Light" (Ph.D. diss., Berkeley, 1972); David Hargreave, "Thomas Young's Theory of Color Vision: Its Roots, Development, and Acceptance by the British Scientific Community," (Ph.D. diss., Wisconsin, 1973); and A.D. Morrison-Low and J.R.R. Christie, eds., *'Martyr of Science': Sir David Brewster, 1781–1868* (Edinburgh: Royal Scottish Museum, 1984).

nomenon. From his famous experiments with prisms, Newton distinguished two types of light, compound and simple (or homogeneous). Simple rays, according to Newton's theory, each exhibit a unique color, possess a unique refrangibility, and cannot be modified by interaction with bodies, including absorptive interactions. Depending on their thickness and the characteristics of their constituent particles, colored transparent solids and liquids may reflect some homogeneous rays, transmit others, or, as Newton added in the *Opticks*, may "stop and stifle" rays. But absorption itself never alters a simple ray. For Newton, absorption yielded nothing of interest; his theory of colors rested rather on the features of light made visible by the prism—refrangibility and color.[10]

Practical problems rather than a critique of Newton's theory first turned Brewster's attention to absorption. To avoid chromatic aberration in microscopes, Brewster in the early 1820s tried to produce monochromatic light by means of colored glass. After systematically exploring the absorptive characteristics (effects on intensities of solar prismatic rays) of multitudes of colored glasses and liquids, he discovered, to his dismay, that glass filters seldom yielded pure monochromatic rays (see Figure 5.1). Almost as an aside, Brewster noted that absorption might answer the residual question of whether yellow light "has a separate and independent existence" in sunlight. In 1802, William Hyde Wollaston wrote that he had seen four primary divisions in the solar spectrum (red, "yellowish green," blue, and violet) bounded by the dark lines he was first to observe (later to be named "Fraunhofer lines"). In discussing Wollaston's observations, Thomas Young named the four colors as red, green, blue, and violet, and added that his friend Wollaston had also seen a narrow band of yellow between the red and the green. Young explained the yellow as a mixture (due to slight overlapping) of the red and the green. Challenging this explanation, Brewster found that the yellow band did not disappear when he absorbed first the red and then the green light in the solar spectrum. Brewster concluded that yellow could not be a mixture but must be an "independent" spectral color which the prism alone could not separate from the red and the green which overlapped it.[11]

---

10. See Alan E. Shapiro, "The Evolving Structure of Newton's Theory of White Light and Color," *Isis 71* (1980):211–35; John Hendry, "Newton's Theory of Colour," *Centaurus 23* (1980):230–51; and Michel Blay, *La conceptualisation Newtonienne des phénomènes de la couleur* (Paris: Vrin, 1983).

11. David Brewster, "On the Action of Transparent Bodies upon the Differently Coloured Rays of Light," *Transactions of the Royal Society of Edinburgh 8* (1818):1–23; idem, "Description of a Monochromatic Lamp for Microscopical Purposes," ibid. *9* (1823):433–44, at 438; idem, *A Treatise on Optics* (London: Longman & Co., 1831),

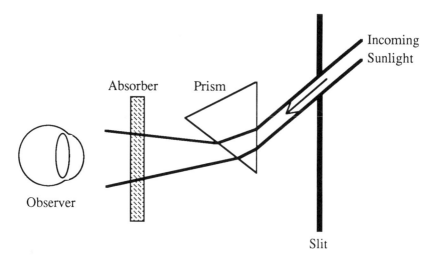

Figure 5.1. Brewster's absorption experiments (1823) as described in his "Description of a Monochromatic Lamp for Microscopical Purposes," Transactions of the Royal Society of Edinburgh 9 (1823):433-44, on 434.

By 1831, following some prodding from his friend John Herschel, Brewster developed this aside into an anti-Newtonian theory of the solar spectrum. In many experiments with various absorbing media, Brewster found he could "insulate" three colors (red, yellow, and blue) and white bands at several places along the dispersed solar spectrum. These observations convinced him that the solar spectrum consists of three coterminal, overlapping spectra of red, yellow, and blue light. Since the mixture of Brewster's three "primaries" produces white, this color too must be present throughout the solar spectrum, a white which prisms cannot decompose. The existence of such a non-prismatic white, Brewster boasted, "has never even been conjectured." He then proposed three hypothetical (not measured) curves of intensity as a function of refrangibility for his three primaries; the peak of each corresponded to the location of that color in Newton's

---

71-2; William Hyde Wollaston, "A Method of Examining Refractive and Dispersive Powers by Prismatic Reflection," *Philosophical Transactions* (1802):365-80, at 378; and Thomas Young, *A Course of Lectures on Natural Philosophy and the Mechanical Arts*, 2 vols. (London: William Savage, 1807), *1*:438.

spectrum. Summed, the curves yielded the intensities that Josef Fraunhofer had measured across the solar spectrum (see Figure 5.2). Brewster thus claimed to have reduced Newton's infinitude of homogeneous solar rays to just three, each of which can have any degree of refrangibility within the visible spectrum. Although Brewster presented his theory as a physical account of light rays, he did suggest a physiological consequence of the new view. Since red–color-blind individuals can see light in the red region of the spectrum, Brewster suggested that their visual sense still functions for the yellow and blue rays present in that part of the spectrum. Not expanding on this speculation and not mentioning Young's theory (see below, p. 225), Brewster tacitly assumed that the visual system contains three components, corresponding to his primaries, each of which may independently become dysfunctional.[12]

Although a few skeptics like the German physicist J.C. Poggendorff quietly suggested that someone repeat Brewster's observations,[13] most British optical researchers in the 1830s unquestioningly accepted the phenomena of the triple spectra. The first controversy provoked by Brewster's observations concerned competing theories of light. This debate—could undulationist or emissionist accounts better explain selective absorption by transparent media—played a crucial role in the eventual acceptance of the wave theory in Britain.[14] Yet no one challenged, at least publicly, the reliability of Brewster's observations. In 1834, the Royal Society of Edinburgh awarded its Keith Biennial Prize to Brewster for his work on the triple spectrum.

Late in the 1840s, however, Brewster's triple spectrum began to arouse public criticism. The British Astronomer Royal, George Airy; the Italian experimental physicist, Macedonio Melloni; the American chemist, John Draper; and a Bordeaux lycée teacher, Félix Bernard, all tried with varying degrees of success to repeat Brewster's observations. William Whewell, and the French publisher, François Moigno, compiled the arguments against Brewster. Even earlier, a Berlin Gym-

---

12. David Brewster, "On a New Analysis of Solar Light" [1831], *Transactions of the Royal Society of Edinburgh 12* (1834):123–36, at 129.
13. See Poggendorff's "Zusatz" to his translation of Brewster's 1831 essay in *AP 23* (1831):441–43.
14. Jack Morrell and Arnold Thackray, *Gentlemen of Science: The Early Years of the British Association for the Advancement of Science* (Oxford: Clarendon Press, 1981), 466–72; Frank A.J.L. James, "The Debate on the Nature of the Absorption of Light, 1830–1835: A Core-Set Analysis," *History of Science 21* (1983):335–68; and G.N. Cantor, "Brewster on the Nature of Light," in Morrison-Low and Christie, eds., *'Martyr of Science'*, 67–76.

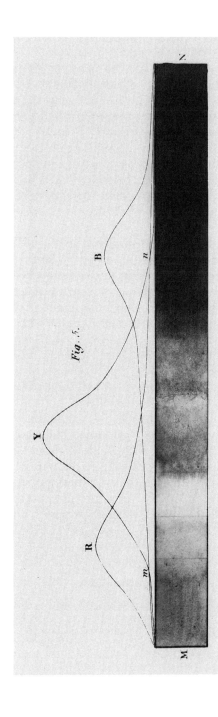

Figure 5.2. Brewster's triple spectrum (1831). Intensity curves for Brewster's three colors as a function of refrangibility. Reprinted from Brewster, "On a New Analysis of Solar Light," [1831], Transactions of the Royal Society of Edinburgh 12 (1834):123-36, Plate II, Figure 5.

nasium mathematician, Emil Wilde, had rejected Brewster's conclusions. These critics found logical, physical, and physiological flaws in Brewster's work. Their arguments would define the problem for the young Helmholtz.[15]

Brewster's critics noted that he had tacitly accepted the conventional view of red, yellow and blue as the three primary colors,[16] although his observations in no way limited the coterminal spectra to only three, or to those particular colors. They asked what Brewster could mean when he claimed that "simple inspection" reveals the "existence of red" in other zones of the solar spectrum. For Moigno, a spectral color by definition was primitive and uncompounded; to say that red was "contained" in primitive orange was to utter a "paralogism." Moigno also pointed out that Brewster assumed his absorbers passively filter something out of light and do not actively modify or alter the transmitted light. If absorbers were to act in the latter manner, then Brewster's experiments would reveal not coterminal spectra but alterations in homogeneous rays already completely separated by the prism. Hence, even if Brewster's observations were repeatable, they would not justify belief in the physical existence of coterminal triple spectra.

The critics also asked whether Brewster's spectra were "real" or only artifacts of flawed experimental design. If, for example, Brewster had not excluded all reflected or scattered white light from his apparatus, such light might well have mixed with the prismatic spectrum to create the "white" Brewster had postulated throughout the spectrum. The critics complained that Brewster never specified the physical dimensions of his apparatus. A non-collimated or relatively wide beam of sunlight falling upon the prism might create a series of overlapping spectra, each of which would contain separated homogeneous colors. When superimposed they would produce mixed colors at every point

---

15. William Whewell, *History of the Inductive Sciences*, 3 vols. (London: Parker, 1837), 2:361; idem, *History of the Inductive Sciences*, 2nd rev. ed., 3 vols. (London: Parker, 1847), 2:504–6; George Airy, "On Sir David Brewster's New Analysis of Solar Light," *PM*, 3rd ser. *30* (1847):73–6; [Macedonio] Melloni, "Essais d'une analyse calorifique du spectre solaire," *Bibliothèque universelle 49* (1844):141–68; idem, "Recherches sur les radiations des corps incandescents et sur les couleurs élémentaires du spectre solaire," *Supplément à la bibliothèque universelle 5* (1847):238–58; John Draper, "On the Production of Light by Heat," *PM*, 3rd ser. *30* (1847):345–60; Félix Bernard, "Thèse sur l'absorption de la lumière par les milieux non crystallisés," *Annales de chimie et de la physique* 3. sér. *35* (1852):385–438; [François] Moigno, *Répertoire d'optique moderne*, 4 parts cont. pag. (Paris: A. Frank, 1847–50), 460–69; and Emil Wilde, *Geschichte der Optik* [1838–43], 2 vols., fac. reprint ed. (Wiesbaden: Sändig, 1968), 2:170–71.

16. See Alan E. Shapiro, ed., *The Optical Papers of Isaac Newton*, vol. 1: *The Optical Lectures, 1670–1672* (Cambridge: Cambridge University Press, 1984), 506, for earlier concepts of "primary colors."

except the very ends of the combined spectra. Thus Brewster might well have seen white in every region of the spectrum, but sloppy experimental design rather than three overlapping physical rays could have produced this phenomenon.

Finally, the critics identified several physiological effects which might have created Brewster's phenomena. "The eye has no memory for colors," asserted Airy, who wondered how Brewster could accurately compare the unmodified (i.e., not passing through an absorber) and modified spectra sequentially. Both Airy and Bernard sought to avoid this problem by placing the absorber over only part of the slit in order that the unmodified and modified spectra would appear side by side and simultaneously on a screen or directly on the retina. Yet Bernard discovered a complication with such a setup, for contrast effects between the two strips of color could generate subjective colors or "illusions," as he called them. Airy also noted that variations in intensity of colored lights create "apparent effects" to the eye, a phenomenon first identified in 1825 by the Breslau physiologist Jan Purkyně (see below, p. 219). Brewster had not considered whether changes in intensity as light passed through his absorbing media might affect his observations of hue; nowhere had he even commented on the relative intensities in his modified spectra. Even though Brewster himself had published several papers on subjective color effects, he apparently thought such factors could be ignored in the study of absorption.[17]

Along with such general criticisms, the experimentalists also sought to reproduce Brewster's observations, which, since Brewster never described his observational arrangements in detail, was no easy task. Airy compared thirty different absorbing media, and found no change in the "qualities of the colours" of the modified spectra, thereby flatly contradicting Brewster. Melloni and Bernard could repeat some of Brewster's observations: certain colors did disappear when the spectrum passed through given absorbers. Yet these two experimenters attributed such changes to the various experimental effects discussed above, and refused to follow Brewster in challenging Newton's theory of the correspondence of color and refrangibility in the solar spectrum.

Never one to shy away from controversy, Brewster quickly responded to these critiques. He chastised the experimentalists for not repeating his experiments exactly, and accused Airy of being colorblind if he could not see changes in the modified spectrum. Against the charge that he had observed overlapping spectra, Brewster replied

---

17. Airy, "Brewster's New Analysis," 74; and Bernard, "Thèse," 402.

that he had not used large prisms and that, since he always had observed distinct Fraunhofer lines, his spectra must have been "pure" and "unmixed." Furthermore, Brewster asserted that the phenomena he had observed remained unchanged even as the "refracting face of the prism is reduced to the smallest possible dimension." "I feel assured," he concluded in 1848, "that my analysis of the spectrum will be confirmed by future observers who may repeat my experiments with the same care with which they were made, and without any prejudice in favour of their own speculations."[18] By 1852, none of the earlier critics had responded. Yet at least one observer, the German physicist Gustav Karsten, considered the issue of the triple spectrum completely unsettled as he surveyed the controversy for the *Fortschritte der Physik*. One hoped, he wrote, that "such an important point in optics soon would experience a fundamental investigation so that we can know whether color and refrangibility are different things.... If Herr Brewster's view is correct, then a great revolution in our views on the nature of color is a necessary consequence."[19] Karsten's close friend Hermann Helmholtz sought to provide such a fundamental investigation.

3. Facts and Artifacts: Helmholtz meets Brewster's Challenge

As Paul Sherman has noted, by carefully following the recent debate Helmholtz understood Brewster's experimental techniques better than any previous critic.[20] Helmholtz also possessed the advantages of being able to consult several knowledgeable Königsberg colleagues, of being well-grounded in the physical optics of the eye after developing the ophthalmoscope a year earlier, of studying concomitantly the mixing of colored lights (see Section 5), and of having a close friend, Ernst Brücke, who had studied subjective colors since 1850. By exploiting such resources, Helmholtz was able to control what had been an unstable experimental situation and to refute Brewster's notion of the

---

18. David Brewster, "Reply to the Astronomer Royal on the New Analysis of Solar Light," *PM*, 3rd ser. *30* (1847):153–58; idem, "Observations on the Analysis of the Spectrum by Absorption," ibid., 461–62; and idem, "Observation on the Elementary Colours of the Spectrum, in Reply to M. Melloni," ibid., *32* (1848):489–94.
19. Gustav Karsten, "Optische Phänomene," *Fortschritte der Physik 4* (1852):150–64, at 163; see also idem, "Optische Phänomene," ibid., *3* (1850):117–37, at 131–34. Karsten was not alone in worrying about the status of Newton's color theory: at least eight journals, issued in Berlin, Halle, Geneva, Paris, and London published full or abridged versions of Helmholtz's 1852 essay on absorption. See *Fortschritte der Physik 8* (1855):251.
20. Sherman, *Colour Vision*, 45.

triple spectrum so convincingly that discussion of the phenomenon essentially disappeared from optical discourse after 1852.

As in his 1847 essay on the conservation of force, Helmholtz explicitly directed his 1852 essay to physicists, whose attention, he wrote, had been attracted to the topic because of Brewster's fame and his anti-Newtonian conclusion.[21] Like Karsten, Helmholtz defined the issue as the integrity of Newton's theory of colors; nowhere did he mention theories of light or the initial controversy which Brewster's observations had provoked. In contrast to previous critics, Helmholtz began by defending Brewster, whom he lauded as a "proven observer." Brewster's choice of fundamental colors, Helmholtz argued, was not important for the question of whether absorption alters the color of homogeneous light. The "purity" of Brewster's spectrum was vouchsafed by the observed sharp Fraunhofer lines. And one's "color memory" is more than adequate for such experiments since the changes in color are prominent. More importantly, Helmholtz found (like Melloni and Bernard) that he could repeat Brewster's experiments and could observe in nearly every case exactly the phenomena Brewster had reported. In Helmholtz's hands, Brewster's phenomena appeared stable and repeatable.

Yet two "doubts" arose for Helmholtz as he repeated Brewster's experiments. Both had been considered by Brewster's earlier critics, a point Helmholtz ignored. First, Helmholtz asked whether the white Brewster had observed at every point along the spectrum might have been sunlight scattered by the apparatus or the observer's eye. Second, he wondered whether "physiological optical influences" might have generated some of the phenomena both he and Brewster had observed, so that one would not need to posit the existence of coterminal spectra to explain the experiments.

Helmholtz's first doubt derived from the thorough knowledge of physical optics that he had acquired while developing the ophthalmoscope and studying Newton's *Opticks* and *Optical Lectures*. Slight imperfections in the glass or surface of the prism could scatter unrefracted white light from the incoming solar beam directly into the retina where it could mix with the refracted light. Internal reflections from the non-polished sides of a prism could throw white light over the refracted spectrum. Imperfections in the absorbing media could also scatter light. In addition, reflections between the surface of the

---

21. "Ueber Herrn D. Brewster's neue Analyse des Sonnenlichts," *AP 86* (1852):501–23, in *WA* 2:22–44. Cf. Fabio Bevilacqua's essay, "Helmholtz's *Ueber die Erhaltung der Kraft*: The Emergence of a Theoretical Physicist," in this volume.

observer's cornea and the absorber might produce imperfectly overlapping spectra on the retina. Not noting that Newton too had warned against each of these problems,[22] Helmholtz suggested that scattering could be reduced by altering the experimental set-up. Yet one final source of physical scattering (which Newton had not discussed) could not be avoided: dispersion of light within the eye itself. Referring to his treatise on the ophthalmoscope, Helmholtz noted that whenever any bright light falls on part of the retina, light of the same color appears as a "weaker light fog" diffused over a larger portion of the visual field.[23] Such fog might arise from diffraction of light passing through the narrow pupil, from scattering off slight imperfections in the ocular medium (membranes, cells, fibers), or from reflections of light between nonparallel surfaces of the retina and the cornea. The eye itself, therefore, might mix incoming separated light and produce weak sensations of white everywhere across the visual field.

To support his contention that scattering had produced "Brewster's white," Helmholtz modified Brewster's apparatus by adding a second prism and a second slit, creating an arrangement very similar to Newton's *experimentum crucis*, an experiment widely discussed in contemporary textbooks (see Figure 5.3).[24] On Brewster's interpretation, light passing through the second slit would contain the selected refracted color mixed with "Brewster's white" not decomposed by the first prism. The introduction of absorbers would therefore modify the selected hue despite the second prism. If, however, Newton's theory were correct and if scattering existed as Helmholtz suspected, then the second prism would refract the scattered white light and prevent it from overlapping the color selected by the second slit. That is, in Helmholtz's modified set-up the introduction of absorbers should produce no change in the selected hue. Helmholtz had devised a crucial experiment for his hypothesis of scattering. Upon placing absorbers at various positions along the light's path, he observed in each case no change in the "purity and saturation" of the selected color, even as he

---

22. Isaac Newton, *Opticks*, 4th London ed. [1730] (New York: Dover, 1952), 71–7.
23. *Beschreibung eines Augen-Spiegels zur Untersuchung der Netzhaut im lebenden Auge* (Berlin: Förstner, 1851), in *WA* 2:229–58, at 250–51.
24. See Newton, "A Second Letter of P. Pardies . . . and Mr. Newton's Answer," *Philosophical Transactions* (1672):5012–18, reprinted in Cohen, ed., *Newton's Papers*, 97–103, at 101; Newton, *Opticks*, 47; Richard S. Westfall, *Never at Rest: A Biography of Isaac Newton* (Cambridge: Cambridge University Press, 1980), 214. For a contemporary illustration of a double-prism arrangement similar to Helmholtz's, see Johann Müller, *Grundriß der Physik und Meteorologie*, 2nd enl. ed. (Braunschweig: Vieweg, 1850), 216.

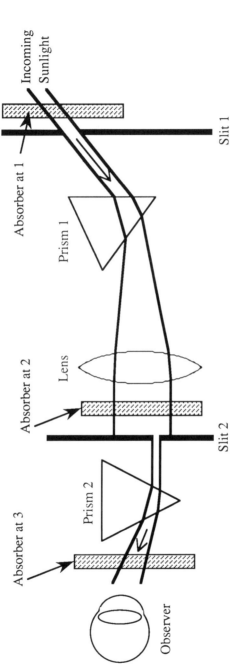

Figure 5.3. Helmholtz's absorption experiments (1852), as described in his "Ueber Herrn D. Brewster's neue Analyse des Sonnenlichts," AP 86 (1852):501–23, in WA 2:22–44, on 35–7.

increased the thickness of the absorber by fourfold. For Helmholtz this demonstrated that scattered sunlight was an artifact of Brewster's experimental arrangement, and removed the need to posit a white component of sunlight not decomposed by a prism.

In dealing with his first "doubt," Helmholtz had deployed the laws of physical optics to separate fact from artifact. A second "doubt," however, challenged more fundamentally the reliability of any facts derived from Brewster's experiment. A year earlier, Helmholtz's friend Brücke had completed a lengthy study of subjective colors, a topic which by 1850 had become a major problem for optical researchers and one which Helmholtz suspected might be relevant to Brewster's work.[25] Repeating Brücke's observations of simultaneous contrast (see below, p. 253), Helmholtz found that a light entering the eye induced sensations of the same or complementary color in adjacent dark areas of the visual field, depending on the intensity of the incoming light. He now asked, as had Bernard, whether such subjective inductions might have produced Brewster's (and his own) observations of color "extending" into a neighboring zone of the spectrum. To explore this suggestion, Helmholtz added a telescope to the experiment to isolate selected regions of the spectrum. When his eye could not see other inducing colors, the "extension" effects disappeared.[26]

Finally, Helmholtz suggested that other observations by Brewster might have derived from subjective alterations of hue arising from objective changes in the intensity of light entering the eye. Not mentioning Jan Purkyně, who in 1825 had first described such phenomena, or his former Berlin physics professor, Heinrich Dove, who had recently read a paper to the Berlin Akademie der Wissenschaften on this very topic, Helmholtz did note that his Königsberg colleague, the physicist Ludwig Moser, had published on these effects. These scientists had shown that with increasing intensity all colors appear white to the eye, with the violet end of the spectrum going white first; and that with decreasing intensity all colors become black, with the red end of the spectrum going black first. By attenuating sunlight through multiple reflections from uncoated, uncolored glass (which could not be suspected of modifying the color of the light), Helmholtz found that he could alter the width of the yellow visible in the spectrum.

Hence, scattered white light and subjective contrast and intensity effects might have produced the phenomena Brewster and Helmholtz observed. For Helmholtz, this possibility—he had not proved that such

25. Ernst Brücke, "Untersuchungen über subjective Farben," *AP 84* (1851):418–47.
26. "Ueber Brewster's Analyse," 38–9.

effects actually existed in Brewster's observations—eliminated the need to postulate coterminal triple solar spectra. Helmholtz concluded that he had removed "all the argumentative force" from Brewster's research.[27] The young Königsberg physiologist had separated fact from artifact in Brewster's experiments by borrowing three constellations of ideas from his predecessors: Newton's *experimentum crucis*; Brücke's analysis of subjective complementary colors; and Dove's and Moser's examination of the Purkyně effect. And Helmholtz must have drawn on the knowledge of the Königsberg physicist, Franz Neumann, an expert on theoretical optics, from whom he borrowed a prism.[28] Unlike his earlier experimental papers on fermentation and animal heat, Helmholtz's work on Brewster's spectrum appeared entirely convincing to contemporary readers.[29] At the 1855 meeting of the British Association for the Advancement of Science, the seventy-four-year-old Brewster tried to defend his work against Helmholtz's and Bernard's criticisms; he reaped only scorn in the ensuing uproar.[30] Only once more did the triple spectra briefly surface, in an 1856 disagreement between Cambridge professors James Challis and George G. Stokes concerning fluorescence. Neither protagonist, however, accepted Brewster's postulation of a triple spectrum.[31] Thus, in his first paper on color Helmholtz did not open a new research front but rather closed an earlier, controversial one.

4. Color Mixing from Newton to Helmholtz

Helmholtz's first 1852 essay on color ended a controversy by repeating and reinterpreting a well-known experiment. His second essay

---

27. Ibid., 44.
28. Ibid., 29.
29. For enthusiastic, if not impartial, announcements of Helmholtz's victory over Brewster, see Carl Ludwig, *Lehrbuch der Physiologie des Menschen*, 2 vols. (Heidelberg: Winter, 1852–56), *1*:225; and Heinrich Dove, *Darstellung der Farbenlehre und optische Studien* (Berlin: G.W.F. Müller, 1853), 25–6. F.C. Donders in Utrecht verified Helmholtz's results, and on 26 June 1853 declared that "no one still admits the theory of Brewster." (Donders to Helmholtz, AW, 116)
30. Brewster, "On the Triple Spectrum," *Report of the Twenty-Fifth Meeting of the British Association for the Advancement of Science; Held at Glasgow in September 1855* (London: John Murray, 1856) *2*:7–9; and idem, *Memoirs of the Life, Writings and Discoveries of Sir Isaac Newton*, 2 vols. (Edinburgh: Thomas Constable and Co., 1855), *2*:117–26. For the public uproar surrounding Brewster's 1855 speech, see *The Glasgow Herald* (19 September 1855), 3; and for Maxwell's account, see Lewis Campbell and William Garnett, *The Life of James Clerk Maxwell* (London: Macmillan, 1882), 215–16.
31. Challis, "A Theory of the Composition of Colours on the Hypothesis of Undulations," *PM*, 4th ser. *12* (1856):329–38; and Stokes, "Remarks on Professor Challis's Paper," ibid., 421–25.

of 1852 sought to order a confusing miscellany of observations and experiments, extending back to Newton, concerning color mixing. By distinguishing between what he called "additive" and "subtractive" mixing, Helmholtz brought a new level of order to the phenomena even though he displayed, at best, a truncated knowledge of his predecessors and overemphasized his own originality. Furthermore, his own experiments on color mixing were not immediately "successful," that is, they did not correspond to Newton's theory of color mixing. Only by 1855, after enduring sharp criticism of his 1852 work, did he achieve experimental results corroborating Newton's theory.

Although not as chaotic as Helmholtz portrayed it, the field of color mixing during the eighteenth century had developed at least five separate experimental and conceptual traditions, each of which derived from the cornucopia of experience presented in Newton's *Opticks* (1704) and *Optical Lectures* (published in 1729). Fundamental to Newton's theory of colors was his claim that spectral colors, once separated by refraction, could be recombined to form white light. Employing lenses, prisms, mirrors or total internal refraction to recombine the spectrum, Newton found he could always obtain white light, provided all the spectral colors were mixed. Pairs of colored lights never mixed to produce white, and Newton guessed that a minimum of four or five colors might be required for making white. Accepting the common seventeenth-century view that lights and pigments mix identically, Newton found similar results when combining colored powders. Mixing powders of the five "principal" colors (for Newton: red, yellow, green, blue, and purple), however, yielded gray rather than white. Only under intense, direct sunlight did this mixture appear white, because, Newton explained, colored bodies reflect light of their own color less copiously than do white bodies. Newton did not, however, follow the standard seventeenth-century practice of seeking the minimum number of "primary" pigments required to mix all other colors. Although he denominated only seven colors in the spectrum, his theory required an infinite number of "colors" in white light.[32]

The painterly analyses of pigment mixing became the best-known post-Newtonian tradition of color mixing.[33] The German engraver and artist, Jakob LeBlon (1730); the natural philosophers Tobias Mayer (1758) and Johannes Lambert (1772); the British entomologist, Moses Harris (1766); the Romantic painter, Philip Otto Runge (1810); and the French chemist and director of dyes at the national tapestry workshop, Michel E. Chevreul (1839), each produced an influential scheme

---

32. Newton, *Opticks*, 150–54; and Shapiro, ed., *Optical Papers*, 109–13, 473–75, 506–9.
33. For a concise summary, see Sherman, *Colour Vision*, 60–80.

for classifying and standardizing colored pigments. Based on mixing a limited number (usually three) of "primary colors" (usually red, yellow, and blue), and, after Mayer, on the addition of white or black to vary saturation, these colorists sought quantitative and graphical means to characterize all possible pigment colors. Although occasionally one of these individuals might note the conflict between Newton's theory of white light as the mixture of all spectral colored lights and the fact that mixed pigments never produce white, this painterly tradition was predominantly practical. Its adherents seldom considered other forms of color mixing.

For the mixing of colored lights a richer tradition emerged, exploiting a variety of experimental tools derived from Newton. A spinning top with sectors painted in different colors enabled Newton to present to the eye a "successive intermixture" of colored lights. Newton's top with sectors of all the spectral colors appeared white. The color top, more commonly designated "color wheel," became a standard means of mixing colors in the nineteenth century, especially after observers such as Joseph Plateau and Dove began using it to explore subjective phenomena.[34] Unlike the pigment mixers, many color-wheel users admitted they could not produce pure white, no matter how carefully they tried to match the spectral colors on their painted wheels. The standard explanation for this echoed Newton's for mixed powders: pigments "are too turbid and absorb too much light" to match a white disk which reflects all the incoming light.[35]

A second means of mixing colored lights in the eye, but simultaneously rather than successively, deployed Newton's arrangements of prisms and lenses to recombine selected portions of spectral light. The best known of these observers was Christian Ernst Wünsch, a professor of mathematics and technology at the university in Frankfurt an der Oder. In a major study published in 1792, Wünsch tried to copy Newton's procedures yet could not reproduce Newton's results. Unlike the Englishman, Wünsch argued that pigments and lights do not mix identically, since the former reflect in small amounts of all colors present in lights which illuminate them. He made white from numerous com-

---

34. Newton, *Opticks,* 141; Young, *Lectures* 1:440–41; Joseph Plateau, "Ueber einige Eigenschaften der vom Lichte auf das Gesichtsorgan hervorgebrachten Eindrücke," *AP* 20 (1830):304–32; Heinrich Dove, "Ueber Darstellung des Weiss aus Complementarfarben und über die optischen Erscheinungen, welche in rotirenden Polarisationsapparaten sich zeigen," *AP* 71 (1847):97–111; and idem, "Ueber Scheiben zur Darstellung subjectiver Farben," ibid. 75 (1848):526–28.

35. Johannes Müller, *Handbuch der Physiologie des Menschen*, 2 vols. (Coblenz: Hölscher, 1834–40), 2:297.

binations of four, or sometimes only three, spectral lights, and in several cases he needed only two spectral lights (violet + "greenish yellow" or red + blue) to produce white. Influenced by the painterly tradition, however, Wünsch argued from his experimental results that the solar spectrum consisted of only three colors (red, green, and violet) in three partially overlapping beams. Even as he defended Newton's theory of white light, he challenged Newton's claims that monochromatic rays have single refrangibilities and that no two colors can have the same refrangibility. Wünsch's work, made notorious by Goethe's ridicule, attracted considerable attention early in the nineteenth century as several other German and French scholars confirmed his observations. This tradition of mixing colored lights simultaneously, although it thoroughly exploited Newton's experimental methods, increasingly rejected the master's claim that lights and pigments mix identically.[36]

Two additional traditions emphasized color mixtures as physiological phenomena. In one, observers simply looked at different colored lights or pigments with each eye, so that any combination had to occur at a "higher" level within the sensory system than the retina. (Newton had never reported such experiments, although he had speculated about their results.) One might view a white ground through two differently colored plates of glass, as had been done since the mid-eighteenth century, or one might place differently colored glasses into the eyepieces of a stereoscope, the instrument developed in the 1830s by Charles Wheatstone and Dove to explore binocular vision. Many observers found that different colors entering each eye produced a "contest of visual images" in which first one and then the other color appeared as an unmixed sensation. Others, including Dove, reported circumstances in which colors in the stereoscope fused. Given such contradictory observations, by 1850 binocular mixing was increasingly rejected as a means to explore color mixing.[37]

36. Christian Ernst Wünsch, *Versuche und Beobachtungen über die Farben des Lichtes* (Leipzig: Breitkopf, 1792), viii–ix, 24, 32. In "Xenion," Goethe included a couplet on Wünsch's color theory: "Gelbroth und grün macht das Gelbe, grün und violblau das Blaue! / so wird aus Gurkensalat wirklich der Essig erzeugt." *Goethes Werke*, 133 vols. (Weimar: Hermann Böhlaus, 1887–1912), *5/1*:230; and for Goethe's comments in the *Farbenlehre*, see ibid., 2nd sec. 2:267, 4:366–75.

37. [François] du Tour, "Discussion d'une question d'optique," *Mémoires de mathématique et de physique, présentés à l'Académie Royale des Sciences* 3 (1760):514–30, *4* (1763):499–511; Karl Scherffer, *Abhandlung von den zufälligen Farben* (Vienna: Trattnern, 1765), 47–50; [Georg Wilhelm] Muncke, "Gesicht" [1828], in *Gehler's physikalische Wörterbuch, neu bearbeitet*, 12 vols. (Leipzig: Schwickert, 1825–45), *4/2*:1364–1485, at 1478–81; A[lfred] W. Volkmann, *Neue Beiträge zur Physiologie des Gesichtssinnes* (Leipzig: Breitkopf, 1836), 92–6; Müller, *Handbuch* 2:338; Charles Wheatstone, "On some Remarkable, and Hitherto Unobserved, Phenomena of Binocular Vision,"

The second physiological tradition, which would become vitally important for Helmholtz, was less experimental than theoretical and derived from Cartesian mechanical philosophy. For color vision, two types of theories emerged, each of which postulated processes of mechanical mixing in the sensory system. One approach, espoused by Descartes, Newton, Euler, and eighteenth-century associationists like David Hartley, considered retinal particles as passive transducers of any incoming vibration, with analysis or mixture occurring "higher" in the brain or the Cartesian soul. In his earliest manuscript on colors, Newton speculated that incoming light causes vibrations in the retina, which are carried as vibrations of aether in the optical nerve, like fluid in a pipe, to the brain where combinations occur. Newton thought that vibrations from blue in one eye and yellow in the other would combine in the brain to form a sensation of green. In the *Opticks*, Newton more explicitly suggested that the frequencies of retinal vibrations are proportional to the color and refrangibility of incoming rays. Hartley and Euler also postulated retinal particles capable of receiving any frequency of visible light via sympathetic vibrations from the luminous aether; Hartley even suggested how such vibrations could undergo three types of mixing in the brain.[38]

The other mechanical approach considered retinal particles as analyzers, each capable of responding only to a single frequency of incoming light. Charles Bonnet in 1760 postulated such a theory of specific excitability for the nerve fibers of all the senses. In 1765 Karl Scherffer, a Viennese Jesuit and associate of Roger Boscovich, proposed a theory of successive colored afterimages by assuming that each retinal element vibrates only with one characteristic frequency. Since

---

Philosophical Transactions (1838):371–94, at 386–87; Heinrich W. Dove, "Ueber die Combination der Eindrücke beider Ohren und beider Augen zu einem Eindruck," *Bericht über die Verhandlungen der kgl. preussischen Akademie der Wissenschaften zu Berlin* (1841):251–52; idem, "Ueber das Binocularsehen prismatischer Farben," ibid. (1850):152–54; August Seebeck, "Beiträge zur Physiologie des Gehör- und Gesichtssinnes," *AP 68* (1846):449–65; and Léon Foucault and J. Regnault, "Note sur quelques phénomènes de la vision au moyen des deux yeux," *Comptes rendus 28* (1849):78–80.

38. Michael Hagner, "Die Entfaltung der cartesischen 'Mechanik des Sehens' und ihre Grenzen," *Sudhoffs Archiv 74* (1990):148–71; J.E. McGuire and Martin Tamny, eds., *Certain Philosophical Questions: Newton's Trinity Notebook* (Cambridge: Cambridge University Press, 1983), 253–54, 484–88; Newton, *Opticks*, 124–25, 141–42, 345–48, 353; David Hartley, *Observations on Man* [1749], 2 vols., unchanged reprint ed. (London: J. Johnson, 1791); *1*:192–97; Euler to Bonnet, 18 July 1761, in V.I. Smirnov, et al., eds., *Leonhard Euler: Prisma k uchenym* (Moscow: Akad. nauk USSR, 1963), 29–33.

Scherffer accepted Newton's theory that white light contains an infinitude of unique, homogeneous rays, the Jesuit's theory required the retina to contain infinitely many elements if all the spectral colors were to be seen, a problem Scherffer did not discuss. A decade later, an obscure British dye chemist, George Palmer, suggested (as would Wünsch and Brewster) that white light contains only three homogeneous rays (the traditional painterly primaries of red, yellow, and blue), and that the retina consists of only three types of "particles," "fibers," or "membranes," each of which is capable of being stimulated by only one of the light rays. Equal motion of all three membranes yields the sensation of white; quiescence or minimal motion of all three produces darkness or black; differential motion of the membranes produces sensations of colors. Concerned primarily with explaining color blindness as a defect in one or more of the membranes, Palmer eschewed any interest in exact mechanical details and did not discuss how the frequency or amplitude of relative membrane motions might give rise to the multitudinous phenomena of color vision.[39]

By comparison, Thomas Young postulated a more complex retinal analyzer by dropping the assumption of specific excitability. As David Hargreave has argued, Young, ignorant of Bonnet, Scherffer, and Palmer, intended his speculation solely as a prop for the undulatory theory of light and never sought a comprehensive physiological theory for the variegated phenomena of color vision. As is well known, Young presented his "three-receptor hypothesis" in two versions. In its first appearance, encompassing less than a paragraph of his 1801 Bakerian Lecture, Young described light as a continuum of frequencies and retinal particles as able to vibrate only at single characteristic frequencies. Since every sensitive point on the retina cannot contain an infinitude of vibratory particles, Young proposed that only three types of particles exist, corresponding to the painterly primaries. In his most original move, he further hypothesized that the amplitude of a retinal particle's oscillation is proportional to how closely the frequency of an incoming ray matches the characteristic frequency of the particle. In his 1807

---

39. Charles Bonnet, *Essai analytique sur les facultés de l'âme* [1760], fac. reprint ed. (Geneva: Slatkine Reprints, 1970), 50–2; Scherffer, *Abhandlung*, 36–7, 50–1; Gordon L. Walls, "The G. Palmer Story," *JHMAS* 11 (1956):66–96; and Hargreave, "Young's Theory," 137–45. Palmer's theory, usually attributed to Giros von Gentilly, the pseudonym under which the French version appeared, was not unknown into the nineteenth century. Cf. [Johann Heinrich] V[oigt], "Des Herrn Giros von Gentilly Muthmassungen über die Gesichtsfehler bey Untersuchung der Farben," *Magazin für das Neuste aus der Physik und Naturgeschichte 1* (1781) 2:57–61; Johann Gottfried Voigt, "Beobachtungen und Versuche über farbigtes Licht, Farben und ihre Mischung," *Neues Journal der Physik 3* (1796):235–98, at 245; Muncke, "Gesicht," 1424; and Elie Wartmann, "Mémoire sur le Daltonisme," *Bibliothèque universelle*, n.s. *57* (1845):322–41, *58* (1845):106–32, at 340.

Lectures, Young offered a second account, replacing vibrating particles with three "simple sensations." Incoming homogeneous light stimulates either one simple sensation or two sensations in differing intensities which yield sensations of hues intermediate to the three primaries. Simultaneous stimulation of all three sensations (requiring at least two homogeneous rays) produces white. Young explicitly excluded the simultaneous stimulation of all three sensations by a single homogenous ray, and at one point implied that the eye sees only four uniform colors in the solar spectrum. Young himself discussed no other phenomena of color vision except for noting cryptically in a bibliographical reference to John Dalton's 1798 article on color blindness that the "absence or paralysis of those fibers of the retina, which are calculated to receive red" might account for Dalton's visual defect.[40]

Helmholtz later popularized the view that Young's theory remained "unnoticed until I and Maxwell again directed attention to it."[41] Yet prior to 1852, when Helmholtz first mentioned Young, the three-receptor hypothesis was occasionally discussed as an explanation of red color blindness.[42] Likewise from at least the 1840s on, references to three primary sensations appeared in discussions of subjective colors. And without referring to Young, leading color theorists of the 1840s such as Macedonio Melloni or August Seebeck tried to work out the mathematical details for resonance theories of retinal action.[43] Yet none of these mechanical theories appeared very fruitful or stimulated further experimental explorations of color vision. When Helmholtz turned to color mixing in 1852, he encountered not a field in crisis, such as Brewster's triple spectrum had created for Newton's theory of colors, but a field in disarray. Various methods were employed to "mix

---

40. See Young, "On the Theory of Light and Colours," *Philosophical Transactions* (1802):12–48, at 20–1; idem, "An Account of some Cases of the Production of Colours, not hitherto Described," ibid., 387–97, at 394–95; and idem, *Lectures* 1:437–40, 2:315–16. For a thorough treatment of Young and his British context, see Hargreave, "Young's Theory," 91–185.

41. *Handbuch*, 307.

42. See Muncke, "Gesicht," 1427; William Henry, *Memoirs of the Life and Scientific Researches of John Dalton* (London: Cavendish Society, 1854), 24–7; Victor Szokalski, *Essai sur les sensations des couleurs dans l'état physiologique et pathologique* (Paris: Cousin, 1841), 104; Wartmann, "Mémoire," 128; idem, "Deuxième mémoire sur le Daltonisme ou la dyschromatopsie," *Mémoires de la Société de Physique et d'Histoire Naturelle de Genève* 12 (1849):183–232, at 225–26; Brewster, "Observations on Colour Blindness," *PM*, 3rd ser. 25 (1844):134–41, at 135; and C.G. Theodor Ruete, *Lehrbuch der Ophthalmologie*, 4 parts cont. pag. (Braunschweig: Vieweg, 1845–46), 87.

43. Melloni, "Beobachtungen über die Färbung der Netzhaut und der Krystall-Linse," *AP* 56 (1842):574–87; Seebeck, "Ueber Schwingungen unter der Einwirkung veränderlicher Kräfte," *AP* 62 (1844):289–306; and idem, "Bemerkungen über Resonanz und über Helligkeit der Farben im Spectrum," ibid., 571–76.

colors," not all of which yielded identical results. And color mixers were even unsure whether to expect identical results from different methods.

5. Addition and Subtraction: Helmholtz Reconceptualizes Color Mixing

Helmholtz later recalled that a failed color-mixing experiment had first attracted him to color research. In Königsberg since at least 1850, Helmholtz taught a course every winter semester on "physiology of the senses and generation." He accompanied these lectures with classroom demonstrations, and later remembered his "astonishment" at not observing identical results upon mixing identically colored pigments and lights (yellow + blue).[44] If Helmholtz's memory was correct, this surprise could only have arisen for a young professor lecturing on a topic whose literature he had not mastered. His knowledge of color mixing was only partially increased by the source to which he turned for help, an article on the classification of colors published in 1849 by the Edinburgh professor of physics and former protégé of Brewster, James D. Forbes. Forbes reviewed efforts since Leonardo da Vinci to schematize pigment mixing, but limited his history to the painterly tradition. What caught Helmholtz's attention, however, were two scattered digressions by Forbes. First, the latter repeated Newton's explanation of why a mixture of colored powders does not produce "pure" white as does a mixture of prismatic lights. Second, Forbes argued that, in positing other primaries than the standard red-yellow-blue, some authors confused the question of primaries with that of the objective composition of the solar spectrum. He noted that Mayer considered sunlight to contain only three objectively colored rays. Wollaston in 1802 thought four colored rays comprised the solar spectrum. Thomas Young, continued Forbes (citing Young's *Lectures*), had named the primaries red, green, and violet, a "singular opinion" based on no persuasive evidence, and had supposed that "perfect sensations" of, say, yellow, are produced by mixtures of red and green. And Brewster, whose absorption experiments Forbes loyally praised despite the raging controversies in 1849, considered various colors of the spectrum to be composed of "two, if not three colours each." Forbes thereby

---

44. Geheimes Staatsarchiv Preußischer Kulturbesitz, Merseburg, Kultusministerium, Rep. 76Va Sek. 11 Tit. 7 Nr. 1 Bd. 9, passim; and "Antwortrede [gehalten beim Empfang der Graefe-Medaille zu Heidelberg]," *Bericht über die Achzehnte Versammlung der Ophthalmologischen Gesellschaft* (1886) *1*:43–52, in *VR*[4] 2:311–20, at 317.

offered an incorrect description of Brewster's triple spectrum and an ambiguous statement of Young's hypothesis, not clarifying whether Young's was a theory of color vision or light.[45]

Following Forbes' comments back to the original papers of Wollaston, Young, and Brewster, Helmholtz in his second 1852 paper on color (his *Habilitation* essay) raised a new question about color mixing and presented what he incorrectly thought were the first experiments since Newton of mixing spectral lights.[46] He organized the essay around the question of the meaning of "primary colors," a probe by which he could compare the various mixing traditions. At times repeating nearly verbatim Forbes' history of color mixing, Helmholtz suggested that the notion of "primary colors" (*Grundfarben*) had acquired three different meanings: colors from which all other colors can be compounded; colors corresponding to "objective types of light"; and colors conforming to "primary modes of sensation of the visual nerve fibers," as per Young's theory. Helmholtz explicitly framed his investigation as a test of these agnostic (painterly), physical (Brewster), and physiological (Young) views of the primaries.

Helmholtz described Young's hypothesis in its initial, more mechanical version, offering its most extensive restatement since 1802. Young, he wrote:

> assumed that the particles on the surface of the retina would be capable of vibrations peculiar to themselves, and that at every place particles of three different frequencies would be present beside each other, corresponding to the oscillatory speeds of the three basic colors, violet, green, and red, which stand in relation as 7:6:5. Were a light ray to have a frequency of 5, it would affect only the red-sensitive nerve end; if it were 5½, it would simultaneously excite the red- and green-sensitive, and produce a mixed sensation of yellow.[47]

Helmholtz did not comment on the mechanical or anatomical bases of Young's hypothesis. Nor did he consider whether the hypothesis, which he accurately quoted, could save even the most basic phenomena of color mixing (see p. 226). Instead, he thought his experiments refuted Young's hypothesis.

---

45. J.D. Forbes, "Hints towards a Classification of Colours," *PM*, 3rd ser. *34* (1849):161–78; cf. Hargreave, "Young's Theory," 279–83; and Sherman, *Colour Vision*, 73–9.
46. "Ueber die Theorie der zusammengesetzten Farben," *AP 87* (1852):45–66, in *WA* 2:3–23, at 7, referring to Forbes, "Hints," 164, who wrote that Newton's failed attempts to form white from pairs of colored lights "merit well a careful repetition, which I am not indeed aware that they have ever received."
47. "Zusammengesetzten Farben," 6–7.

Exhibiting the same aptitude for physical optics that he had demonstrated in repeating Brewster's work, Helmholtz arranged for sunlight to pass through a V-shaped slit before falling on a flint prism, which he again borrowed from his Königsberg colleague Neumann. This peculiarly shaped aperture produced two partially overlapping spectra in which each spectral color of one spectrum overlapped every color of the other spectrum (see Figure 5.4). Viewing the V-spectrum directly with a telescope located a suitable distance from the prism, Helmholtz examined separately each pair of superimposed colors. The observations required not color comparisons but absolute color memory. Sensitive to this problem, Helmholtz reported that other witnesses, also "practiced in the judgment of colors," invariably confirmed his observations. He would not be deceived by the "subjective errors of my eyes." Placing thin pieces of paper over one leg of the slit, he varied the intensity but not the hue of each spectral component in the mixture. Such a simple means for prismatic mixing had never before been deployed.

Helmholtz used this apparatus to observe binary mixtures of Newton's seven named colors in all possible combinations and at varied intensities. His "most striking fact, most deviant from earlier views," was that only a single pair (yellow + indigo) gave a pure white. Newton, as noted above, had been unable to mix pure white from pairs of lights, even though the barycentric diagram for color mixing presented in Newton's *Opticks* implied that such complementary binary pairs should exist. And pigment mixers, enthused Helmholtz, had "for a thousand years" produced not white but *green* from yellow and blue. By adding a third slit to his aperture (making a rotated "Z" from the "V"), Helmholtz overlapped three spectra and found many tertiary mixtures which yielded pure white.

From these observations Helmholtz drew three conclusions, two of which he soon abandoned as incorrect. In the one claim that he would not disavow, he distinguished "additive" and "subtractive" mixing to explain why combining lights and pigments are not equivalent. Additive mixing, he argued, occurs when two or more light rays of different wavelengths illuminate the same spot on the retina, either simultaneously or in rapid succession. Subtractive mixing results from the physical combination of different pigments. The former was thus a physiological (or psychological) process, which Helmholtz in 1852 did not explain; the latter was a physical process, which Helmholtz, in the most novel contribution of his 1852 essay, sought to explain. Rather than the Newtonian account, which Forbes and many others had repeated, that pigments decrease the *intensity* of reflected light and thus appear darker or more gray in mixtures, Helmholtz made *hue* the

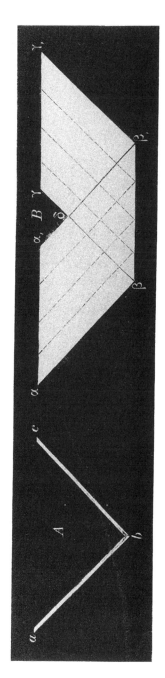

Figure 5.4. Helmholtz's "V-slit" mixed spectra (1852). Reprinted from Handbuch, 393.

key. Drawing on his experience with the ophthalmoscope wherein clear glass plates both transmit and reflect light and on Brücke's studies of "turbid media" whose particles reflect and refract light,[48] Helmholtz described pigment as a series of semi-transparent layers of particles acting as filters to light reflected from the underlying surface. Each particle partly transmits incoming rays to deeper layers, partly reflects light of given wavelengths, and absorbs rays of all other wavelengths. Mixed pigments send back only light not absorbed (subtracted) by either color. Hence, yellow pigment returns red, yellow, and green light; and blue pigment returns green, blue, and violet light. A mixture of yellow and blue pigments absorbs all light except for green. To demonstrate his claim, Helmholtz mixed yellow and blue pigment at the center of a color wheel and painted the circumference in separate sectors of yellow and blue. Upon spinning the wheel, the center looked green but the circumference appeared gray (not white because, as Helmholtz, echoing Newton, noted, intensity is also decreased when light reflects from colored pigments). Helmholtz thereby dramatically supported his new explanation with a simple demonstration on a device that color mixers had used for over a century. With the concepts of additive and subtractive mixing, he provided critical taxonomic principles for ordering and distinguishing the traditions of color mixing.

Less convincingly, as Helmholtz himself would soon admit, he argued that his experiments indicated that a minimum of five "simple colors" were required to produce any spectral hue by paired mixtures. If the primaries were limited to three, these must be red, green, and violet (Young's set, noted Helmholtz) since green could not be made by mixing any other pairs. But mixtures of red + green and green + violet lights did not produce colors as saturated as spectral yellow or blue. Hence the need for five simple colors—red, yellow, green, blue, and violet. Such an empirical conclusion to increase the number of "primaries" violated the assumptions of nearly all the traditions of color mixing discussed above.

Finally, Helmholtz returned to his lead question concerning the meaning of primary colors. He concluded that the physical view—primaries are the only rays composing white light—must be rejected, simply because his other 1852 paper had refuted Brewster's physical theory of the triple spectra. He also considered Young's physiological interpretation untenable. If the sensation of yellow arises only if sensations of red and green are simultaneously excited, then yellow rays

---

48. Ernst Brücke, "Ueber die Farben, welche trübe Medien im auffallenden und durchfallenden Lichte zeigen," *SBW 9* (1852):530–49.

should evoke a sensation identical to the simultaneous effect of red and green rays (assuming flat, partially overlapping response curves for Young's three receptors). Yet Helmholtz's prismatic mixing experiments had demonstrated that this was not the case, for the mixed yellow appeared "duller" than the spectral yellow. By rejecting Brewster's physical and Young's physiological interpretation of primaries, Helmholtz was left with only the agnostic view that "primaries are those from which all other colors are compounded, or at least allow themselves to be compounded."[49] The 1852 paper offered a new concept of additive and subtractive color mixing, but left fully open the question of how the visual system achieves additive mixing.

Unlike his refutation of Brewster's triple spectrum, which met with universal and immediate acceptance, Helmholtz's 1852 paper on color mixing provoked a storm of controversy, beginning with issues of priority. In 1849 the French physicist, Léon Foucault, had combined selected spectral colors with a large achromatic lens; he now challenged Helmholtz's claim to have been the first since Newton to mix colored lights. The Belgian physicist, Plateau, indignantly complained that already in 1829 he had demonstrated that mixing pigments physically and with a spinning disk do not yield identical results.[50] More serious, however, were the theoretical criticisms of Hermann Günther Grassmann and Joseph Grailich, a little-known mathematics teacher in Stettin and a young Viennese university student, respectively. In a brilliantly original essay of 1853, Grassmann sought to apply the vector analysis of his *Ausdehnungslehre* (1844) to the barycentric treatment of color mixing in Newton's *Opticks*. From four "laws" derived from experience and logic, Grassmann proved with what looked like mathematical rigor that every spectral color must have a spectral complementary, as Newton had asserted must be the case but Helmholtz had failed to find. Grassmann politely suggested that altered intensities in Helmholtz's observations would have yielded more complementary pairs, implying that the Königsberg physiologist had not adequately controlled his experimental conditions. Even more fundamentally, Grailich challenged the subjectivity of Helmholtz's procedures. "The eye," he asserted, "is not the highest authority for [determining] the saturation and purity of color tones." Preferring a strictly mechanical

---

49. "Zusammengesetzten Farben," 7.
50. Foucault, "Ueber die Wiedervereinigung der Strahlen des Spectrums zu gleichförmigen Farben," *AP 88* (1853):385–87; and Joseph Plateau, "Reclamation wegen einer Stelle im Aufsatz des Herrn Helmholtz über die Theorie zusammengesetzter Farben," ibid., 172–75.

analysis, Grailich proposed that retinal elements vibrate sympathetically under the influence of incoming undulations in the light aether and combine rays by interference (Newton's theory of retinal action). Since no two sine waves of unequal frequencies (monochromatic rays) can combine to make a third sine wave, Grailich argued that it would be impossible to see a mixture of spectral colors as identical to the sensation of a single monochromatic ray. Such a conclusion clashed starkly with some of Helmholtz's experimental results.[51]

Helmholtz spent the summer of 1853 responding to these challenges, and in January 1855 he published a second essay on color mixing.[52] Admitting that the V-slit produced some overlapped color areas too small for the eye to detect and that the eye could not easily focus on lights of differing refrangibilities, Helmholtz employed another method for mixing colored lights, "similar," he allowed, to Foucault's. Using a large achromatic lens and an adjustable double slit, Helmholtz mixed given pairs of spectral lights at selected intensities on a white screen (see Figure 5.5). He now found seven pairs of complementaries (including all the principal colors of the spectrum except green), the wavelengths of which he measured and plotted by pair. He also compared the relative intensities of the lights of each complementary pair by adjusting the respective slit widths and using his eye as a null instrument to determine when shadows from the two lights appeared equally dark. Such data probably represent the first quantitative measurements of the relative intensities of complementary lights.

Rather than presenting his results as simple empirical facts, as he had done in 1852, Helmholtz in 1855 criticized Grassmann's defense of Newton and proposed his second conceptual innovation for color mixing. Helmholtz had to explain why he had found no spectral complementary to green, thereby violating one of Grassmann's laws. Grassmann's mistake, he argued, was in assuming that purple is a spectral color intermediate between red and violet. No observer since Jean Hassenfratz in 1799 claimed to have seen purple beyond violet in the

---

51. Grassmann, "Zur Theorie der Farbenmischung," *AP 89* (1853):69–84; Grailich, "Beitrag zur Theorie der gemischten Farben," *SBW 12* (1854):783–847, *13* (1854):201–84, at 283; Sherman, *Colour Vision*, 93–111; and Michael J. Crowe, *A History of Vector Analysis* (Notre Dame, Ind.: Notre Dame University Press, 1967), 54–96.

52. "On the Mixture of Homogeneous Colours," *Report of the Twenty-Third Meeting of the British Association for the Advancement of Science; Held at Hull in September 1853* (London: John Murray, 1854) 2:5; "Ueber die Zusammensetzung von Spectralfarben," *AP 94* (1855):1–28, in *WA* 2:45–70. See Koenigsberger *1*:191–92; and Richard L. Kremer, ed., *Letters of Hermann von Helmholtz to His Wife, 1847–1859* (Stuttgart: Steiner, 1990), 116, 134.

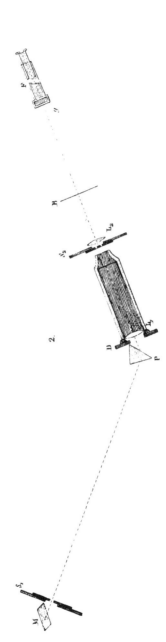

Figure 5.5. Helmholtz's revised mixing experiments (1855). The double slits at S_2 are adjustable both in lateral position and in width, so that pairs of given segments from the solar spectrum can be combined by the achromatic lens L_2. Reprinted from "Ueber die Zusammensetzung von Spectralfarben," AP 94 (1855):1–28, Tafel I, Figure 2, in WA 2:45–70.

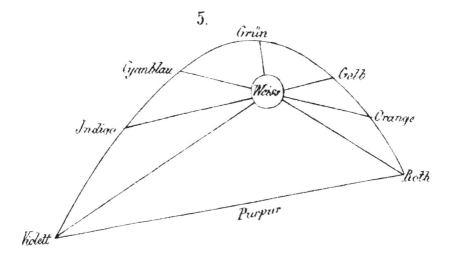

Figure 5.6. Helmholtz's barycentric curve for color mixing (1855). Reprinted from "Zusammensetzung von Spectralfarben," Tafel I, Figure 5.

solar spectrum. Helmholtz himself used two prisms to explore both ends of the spectrum and found no purple. Hence, the curve for a barycentric representation of spectral mixing is not a circle such as Newton had suggested in the *Opticks*, but a triangular curve with the complement of green (purple) "intercalated."[53] The shape of this curve, furthermore, could be determined from the relative intensities of the complementary pairs. Since, for example, the intensity of greenish-blue must be twice that of red to make white, red must be twice as far away from white on the barycentric curve as greenish-blue. By similar reasoning, Helmholtz was able to sketch a curve for how the eye mixes spectral lights to produce white (see Figure 5.6). By criticizing Grassmann, Helmholtz had placed Newton's barycentric construction for spectral mixing on an empirical foundation.[54] Rather than simply classifying colors, as in the painterly tradition (subtractive mixing), Helmholtz provided a graphic and empirical representation of how the human eye mixes colors additively.

53. "On the Mixture," 5.
54. Newton and subsequent users of the barycentric circle assumed that the relative *angular* sectors allotted to each color should correspond to the relative *lengths* of the colors in the prismatic spectrum, an arbitrary if reasonable assumption. Helmholtz replaced this assumption with measurements, although he admitted that "most relations of this drawing are chosen at pleasure, and can make no pretense of exactitude" ("Spectralfarben," 70).

The 1855 paper deserves two final comments. First, by alluding to judgments, Helmholtz revealed the first hint of what by 1860 would become a major rift in his theoretical treatment of color vision. Confirming Dove's observations that two pigment colors which appear equally bright under a given intensity of illumination do not appear equally bright when the illuminating intensity is halved or doubled (the Purkyně shift),[55] Helmholtz noted further that although the relative intensities of two complementary lights appear to vary, the mixed pair still appears white. For example, to produce white light from a mixture of spectral indigo and yellow, the yellow must be four times as intense as the indigo for strong lights but only three times as intense for weak lights. Yet for fixed-slit widths (fixed objective relative intensities), the mixture remained white to the eye when the overall intensity of light entering the apparatus was varied. Helmholtz argued that this resulted not merely from the fact that the eye's ability to differentiate colors decreases with decreasing intensity, but also because "we consider the color of sunlight as normal white under all degrees of brightness."[56] The eye, therefore, judges the color of the "normal white" as unchanged even though the apparent brightness and hue of the components are changing. In the *Handbuch* of 1860, Helmholtz would discuss these effects in terms of "unconscious judgments" (see p. 252).

Second, in the 1855 essay Helmholtz mentioned neither Young nor the problem of the meaning of primary colors which had motivated his 1852 paper on color mixing. Although Helmholtz labeled his work the "physiology of color sensations," he did not consider how the human visual system might produce the phenomena summarized on the barycentric curve. Neither did he report any mixtures of colors other than complementaries. Even though he discussed how Grassmann's four laws could yield a form of Newton's barycentric construction in which the spectral hues are ordered by the placement of three (and no more) arbitrarily selected colors at the vertices of a triangle, Helmholtz in 1855 did not repudiate his 1852 assertion that five primary colors are required to produce all the spectral hues.[57] It appears,

---

55. Dove, "Ueber den Einfluss der Helligkeit einer weissen Beleuchtung auf die relative Intensität verschiedener Farben," *AP 85* (1852):397–408.

56. "Spectralfarben," 63. For another early example in which Helmholtz postulated unconscious judgments, cf. "Ueber die Erklärung des stereoskopischen Erscheinung des Glanzes," *Sitzungsbericht des naturhistorischen Vereins der preussischen Rheinlande und Westphalens 13* (1856):xxxviii–xl, in *WA 3*:4–5.

57. "Spectralfarben," 57, 67–9. See Richard L. Kremer, "From Newton's Color Wheel to the CIE Color Space: Diagrams, Ambiguity, and the Formulation of Modern Theories of Color Vision," forthcoming.

therefore, as if Helmholtz in late 1854 still viewed Young's three-receptor hypothesis with skepticism.

6. Origins of the 'Young-Helmholtz' Theory of Color Vision

Between 1855 and 1860 Helmholtz reversed his position on Young's hypothesis. Rejected in 1855 on empirical grounds, the hypothesis in revised form assumed a central role in Helmholtz's analysis of color vision in the second Part (1860) of the *Handbuch*, which treated "visual sensations." Indeed, the most intriguing and significant conceptual innovation of that Part was Helmholtz's elaboration of Young's hypothesis into a powerful theory of color vision with which he unified a host of previously unrelated phenomena. Yet five years before Helmholtz presented what would soon be called the "Young-Helmholtz" theory, Maxwell had published an identically modified version of Young's hypothesis. The circumstances surrounding Helmholtz's conversion to Young are complex. Helmholtz probably arrived at his version independently of Maxwell; nonetheless, the creation of the Young-Helmholtz theory of color vision again reveals Helmholtz as a master in combining and synthesizing others' ideas, in this case those of Fechner and Müller.

Young's hypothesis appeared midway through the second Part of the *Handbuch*. After reviewing recent work, including his own, on color mixing and reiterating his critique of Brewster's triple spectrum, Helmholtz continued:

> ... it makes no sense whatsoever to speak in objective terms of three fundamental colors.... A reduction of colors to three fundamental colors can only have subjective meaning, can only concern a derivation of color sensations from three fundamental sensations. In this sense Thomas Young correctly comprehended the problem, and in fact his assumption provides an extraordinarily simple and clear overview and explanation of all phenomena of physiological color theory.[58]

Helmholtz outlined Young's "assumption" as follows. First, three types of "nerve fibers" exist in the eye, which produce sensations of red, green, and violet, respectively. Second, the intensity of sensation produced by each fiber depends on the wavelength of the incoming monochromatic light. Third, "it is not thereby excluded, and indeed to explain a host of phenomena it must be assumed, that every spectral

---

58. *Handbuch*, 291.

color excites all types of fibers, but one weakly, the others strongly." Fourth, a simultaneous, "roughly equal" excitation of all three fibers produces the sensation of white. Finally, Helmholtz offered a schematic, hypothetical diagram of response curves for the three fibers as a function of frequency, a powerful, visually persuasive summary of Young's hypothesis. Brewster's objective coterminal triple spectra were thereby internalized as sensations within the visual system, even if Helmholtz did slightly reshape Brewster's equally hypothetical curves (cf. Figures 5.2 and 5.7).[59]

To this presentation of Young's hypothesis Helmholtz added two comments. The colors of the three fundamental sensations, he argued, are "somewhat arbitrary." Any three colors could be chosen, provided they form white when mixed. The only means of determining the actual color sensations would be to investigate color-blind subjects. Furthermore, Helmholtz defended Young's tripling of the number of fibers in the eye, even though, he noted, one might equally well assume that each retinal fiber "could enter three different, independent activities." Since, Helmholtz reasoned, microscopic anatomy has determined neither the number of fibers in the eye nor the function of retinal elements such as rods, cones or ganglia,[60] Young's assumption of independent fibers could not be empirically refuted. Positing such fibers, Helmholtz continued, provides a "more definite perception" of the physiological process and preserves an identity of function for sensory and motor nerves. And with individual fibers, "Young's hypothesis is only a special case of the law of specific sense energies," that is, Müller's assertion that each sensory nerve responds to various types of stimuli (wavelengths of light here) with a single energy or quality of sensation (color here).[61]

Helmholtz's presentation of Young's hypothesis is problematic, as is the question of why he reversed his earlier opposition to Young. Young, as noted above, had explicitly indicated that at most two receptors can be stimulated by an incoming monochromatic ray. By making the response curves coterminal so that any incoming ray stimulates all three receptors, Helmholtz significantly modified Young's hypothesis so that it could account for the decreased saturation of a mixed sensation (Helmholtz's 1852 observation) as well as the fact

---

59. Ibid., 291–92.
60. Helmholtz's reticence here is surprising; the *Handbuch* (19–23) presented up-to-date information on retinal anatomy from Albert Kölliker, *Mikroskopische Anatomie*, 2 vols. (Leipzig: Engelmann, 1850–54), 2/2:648–703, and accepted Kölliker's and Heinrich Müller's conclusion that rods and cones comprise the light-sensitive layer.
61. *Handbuch*, 292.

that the eye can see a full range of hues in the spectrum. Yet the same scientist who in his first public speech had boasted about correcting the "great genius" Newton and refuting Brewster, one of "the most outstanding optical researchers,"[62] now said nothing about what appears to be his most creative innovation to date in color theory. Why did Helmholtz in 1860 refuse to take credit for shaping Young's hypothesis into a powerful explanatory tool?

Hargreave, Sherman, and R. Steven Turner have suggested that Helmholtz could not claim credit for the insight because he borrowed it from James Clerk Maxwell.[63] Since 1849 Maxwell had been experimenting with colors, spurred on by his Edinburgh teacher, James Forbes, whose 1849 paper had also caught Helmholtz's attention. Between 1855 and 1860 Maxwell published two major and two minor essays on color vision, exhibiting a brilliant experimental and mathematical control of the subject. He invented two methods (color top and color box) to mix colored lights so that the eye compared only intensities of grays rather than colors, thereby reducing subjective effects and greatly increasing the quantitative precision of such measurements. In 1855, both Maxwell and Helmholtz independently measured the triangular shape of the barycentric mixing curve for the human eye. Accepting from the beginning Young's hypothesis of "three distinct modes of sensation" (the second, less mechanical version), Maxwell by 1860 *measured* the shapes of the three response curves, shapes which Helmholtz had only guessed. Exhibiting a firm grasp of Grassmann's vector approach, Maxwell proved that only three independent variables are required to characterize any color, and that to specify colors by three modes of sensation is equivalent to specifying them by the three variables of hue, saturation, and intensity.[64] And in 1855, five years before Helmholtz published the Part of the *Handbuch*

---

62. "Ueber die Natur des menschlichen Sinnesempfindungen," *Königsberger naturwissenschaftliche Unterhaltungen 2* (1852) *3*:1–20, in *WA* 2:591–609, at 592, 598.

63. Hargreave, "Young's Theory," 191–92, 299–300; Sherman, *Colour Vision*, 209; and R. Steven Turner, "Fechner, Helmholtz, and Hering on the Interpretation of Simultaneous Contrast," in *G.T. Fechner and Psychology*, eds. Josef Brožek and Horst Gundlach (Passau: Passavia Universitätsverlag, 1988), 137–50, at 141. For a more detailed and even more speculative reconstruction of Helmholtz's conversion to Young's theory, see Kremer, "Newton's Color Wheel."

64. Maxwell, "Experiments on Colour, as perceived by the Eye, with Remarks on Colour-Blindness" [read 19 March 1855], *Transactions of the Royal Society of Edinburgh 21* (1857):275–98, reprinted in W.D. Niven, ed., *Maxwell's Scientific Papers*, 2 vols. (Cambridge: Cambridge University Press, 1890), *1*:126–54, at 131–36. See Sherman, *Colour Vision*, 165–72, for a concise exposition of this argument. Maxwell also showed that Newton's barycentric method of representing color mixtures was equivalent to Young's hypothesis and named Newton as the originator of the three-receptor theory.

dealing with color, Maxwell postulated that Young's response curves must be coterminal.

Like Helmholtz, Maxwell nowhere implied that Young himself had not crafted the hypothesis in this form. Unlike Helmholtz, however, Maxwell considered it necessary to defend the idea of coterminal responses. First, Maxwell argued that the notion is implicit in the barycentric mixing triangle. If every color sensation can be represented by a point inside this triangle, which in turn can be considered as the sum of three "weights" at the vertices, then every color sensation can be viewed as the result of simultaneous excitation of three "elementary colour sensations." Second, Maxwell proposed an analogy between sensation and colored filters. Just as a red glass transmits not only red rays but also rays of other wavelengths at reduced intensities, so too "nerves corresponding to the red sensation are affected chiefly by the red rays, but in some degree also by those of every other part of the spectrum." Müller's doctrine of specific sense energies was not mentioned.[65] A third argument derived from the phenomena of the Purkyně shift. If increasing intensity would saturate the receptors at different rates, one could easily envision three-dimensional response surfaces for the receptors (response as a function of wavelength and intensity) in which at low intensities the violet rays are the last to "go black" and at high intensities the reds are the last to "go white." If a given incoming ray excited only one receptor, however, then no change in hue should be observed upon altering the intensity of that ray. Finally, Maxwell cited Helmholtz's 1852 experiments as proof that any ray simultaneously excites all three sensations. Only if the saturation of sensations produced by monochromatic rays of red, green, and violet were identical to the maximal saturations of the respective receptors would mixtures and monochromatic rays appear equally saturated. Since Helmholtz had shown this was not the case, the sensation of, say, pure spectral red cannot be the "pure sensation" of the red receptor but must be a mixture of sensations from the other receptors as well. Maxwell thus showed how the observations which Helmholtz in 1852 had used to refute Young's hypothesis could undergird the same, if modified to posit coterminal responses.[66]

---

65. Not until 1877 did Maxwell mention Müller; he praised Helmholtz's *Handbuch* and subsequent *Die Lehre von den Tonempfindungen als physiologische Grundlage für die Theorie der Musik* (1863) as "splendid examples" of Müller's law. See Maxwell, "Hermann Ludwig Ferdinand Helmholtz," *Nature* 15 (1876–77):389–91, reprinted in Niven, ed., *Scientific Papers* 2:592–98, at 596.

66. Maxwell, "On the Theory of Colours in Relation to Colour Blindness" [letter dated 4 January 1855], *Transactions of the Royal Scottish Society of Arts 4* (1856):394–400, reprinted in Niven, ed., *Scientific Papers 1*:119–25, at 121–22; and idem, "Experiments," 136, 149–52.

Maxwell's essays on color were regularly abstracted in the *Fortschritte der Physik*, the review journal for which Helmholtz himself wrote.[67] In the *Handbuch* Helmholtz cited the 1855 book on color blindness which contained Maxwell's first paper on color and Young's modified hypothesis. Clearly, Maxwell published a workable version of Young's theory before Helmholtz, and the latter, as he was preparing the color chapters of his *Handbuch*, knew of the Scottish physicist's research. Helmholtz, however, referred only once to Maxwell in the *Handbuch*, noting that Young's theory remained "unnoticed until I and Maxwell again directed attention to it," a statement that is literally correct, but hardly recognizes Maxwell's clear priority in revising and accepting Young's hypothesis.[68] Yet the fact that Maxwell first published the revised Young hypothesis does not prove that Helmholtz simply borrowed Maxwell's innovation without attribution. Indeed, a close reading of several of Helmholtz's minor papers published between 1855 and 1860 suggests, as Turner has recently argued, that Helmholtz might have been prompted to modify and accept Young not only by Maxwell but also by Fechner's work on subjective colors.[69]

In the summer and fall of 1858 Helmholtz delivered two speeches on afterimages, the first fruits of his analysis of the extensive literature on this topic that he would summarize in the second Part of the *Handbuch*.[70] In the first speech Helmholtz indicated that he now conceived of Young's hypothesis in the revised form (coterminal response curves) and considered one type of colored afterimage as evidence supporting the hypothesis. If one views a small red field, for example, on a white ground and suddenly removes the field, an afterimage in the complementary green appears on the white ground. If, however, this afterimage is viewed against a fully saturated green ground, the afterimage appears more saturated than the ground. The ground, equivalent in saturation to prismatic green, is thus not the purest or most saturated color the eye can see. According to the stenographic report of Helmholtz's speech:

> The speaker already had concluded in his earlier works on color mixing that if Thomas Young's theory were correct, i.e., that there are three

---

67. For example, *Fortschritte der Physik 11* (1858):281–84, summarized Maxwell, "Experiments," 1855, and concluded: "The author considers his investigation as a confirmation of Young's theory of color vision."
68. *Handbuch*, 307; Helmholtz left this remark unchanged in *Handbuch²*, 383–84.
69. Turner, "Fechner, Helmholtz, and Hering," 140–41.
70. "Ueber die subjectiven Nachbilder im Auge," *Sitzungsbericht des naturhistorischen Vereins der preussischen Rheinlande und Westphalens 15* (1858):xcviii-c, in *WA* 3:13–5; and "Ueber Nachbilder," *Amtlicher Bericht über die Versammlung deutscher Naturforscher und Aerzte* (1858):225–26.

types of optic nerve fibers, red-sensing, green-sensing, violet-sensing, then the spectral colors would not be the most saturated colors which appear to the sensation of the eye; the investigations described herein were planned to test exactly this point.[71]

In his previous two essays on color mixing, however, Helmholtz had concluded no such thing. As noted above, the 1855 paper had not mentioned Young; the 1852 paper described Young's theory as predicting identical sensations for red and green rays mixed in the eye and for monochromatic yellow rays which excite equally the red and yellow receptors. Although Helmholtz did not describe how Young's postulated receptors would yield such sensations, he must have assumed flat-topped, partially overlapping response curves for the receptors, so that stimulus of a single receptor is the most saturated sensation possible. Hence, in 1858 Helmholtz appears to have reinterpreted his 1852 paper, now claiming that he always had understood Young's theory to require coterminal responses.

Helmholtz framed his 1858 speeches, however, not as tests of Young's hypothesis but as defenses of Fechner's theory of afterimages against Plateau's.[72] These explanations of subjective colors, formulated in the 1830s and still reigning in the 1850s, represented the culmination of two long-standing traditions for dealing with the ever-expanding panoply of subjective color phenomena.[73] One approach postulated an active retina, producing "on its own" (i.e., corresponding to no direct, external stimuli) antagonistic sensations analogous to everyday experiences of contrast, such as the feeling of pleasure which follows the cessation of pain or the sensation of darkness which results upon leaving a brightly illuminated for a darker room. As the Newtonian James Jurin speculated in 1738: "We are therefore not to wonder, that either the whole retina, or that any part of it, after having been intensely affected with one sensation for some space of time, should upon the cessation of the cause that so affected it, immediately be affected with

---

71. "Ueber die subjectiven Nachbilder," 15.
72. Joseph Plateau, "Essai d'une théorie générale comprenant l'ensemble des apparences visuelles qui succèdent à la contemplation des objets colorés, et de celles qui accompagnent cette contemplation," *Nouvelle mémoires de l'Académie Royale des Sciences et des Belles-Lettres de Bruxelles 8* (1834), separately pag.; Gustav Theodor Fechner, "Ueber die subjectiven Complementarfarben," *AP 44* (1838):221–45, 513–35; and idem, "Ueber die subjectiven Nachbilder und Nebenbilder," *AP 50* (1840):193–221, 427–70.
73. All colors, of course, are subjective. By "subjective," "accidental," or "apparent" colors, writers from the mid-eighteenth century on meant those sensations not directly produced by the "usual" action of incoming rays of light on the eye, i.e., sensations dependent solely on refrangibility of the incoming rays. For a concise summary of such phenomena and a bibliography of publications through 1854 on subjective colors, see Hargreave, "Young's Theory," 359–68, 500–6.

a contrary sensation."[74] The second approach, originated several decades later by Buffon and Scherffer, made retinal fatigue the essential mechanism for subjective effects. Rather than acting in the absence of light, the retina when exhausted fails to act in the presence of light. By positing for each color sensation specific retinal elements which, like muscles, could fatigue independently, these theorists could explain a rich variety of subjective visual phenomena which change over time.

Plateau had proposed an active retina oscillating between antagonistic sensations of complementary colors over time and over the space of the visual field, a theory Helmholtz summarily dismissed in 1858, probably because it contradicted Müller's doctrine. Instead, Helmholtz thought his own exploration of afterimages "fully confirmed" Fechner's theory, which in his 1858 speeches Helmholtz reduced to three points. An illuminated retinal field retains a "residual stimulus" for some time after the objective stimulus ceases, thereby generating the afterimage. Once illuminated, retinal elements fatigue and become less sensitive to subsequent external rays until rest (no stimulation) allows recovery. And sensation of the "inner light" of the eye, that small amount of white light always present in the eye, also is affected by the retinal fatigue. Significantly, Helmholtz in the 1858 speeches did not describe how Fechner's mechanisms of fatigue might apply to colorific effects; nor did he fully present Fechner's theory, a speculative attempt to unify the explanation of afterimages (effects over time) and contrast (effects over space on the visual field). Fechner's complete theory of 1840, combining effects of fatigue, inner light, specific irritability of individual retinal fibers (only one wavelength of light stimulates a given retinal element), and a Plateau-like "Gegenwirkung des Organismus" need not be rehearsed here. But one important feature of the theory, not mentioned by Helmholtz, was the claim that a "secondary" complementary sensation accompanies every "primary" color sensation evoked by objective light entering the eye. The fibers excited by the primary rays become immediately fatigued, Fechner speculated, so that they respond less energetically to continuing action of rays of the same wavelength from either outside or inside the eye. Neighboring retinal fibers, specific for other wavelengths, are not fatigued. The sensation from any retinal field containing fibers for all visible wavelengths thus contains the "primary" color sensation mixed with a "secondary" complementary sensation, coming mostly from inner light. Hence, every primary sensation is diluted and less saturated than that which

---

74. James Jurin, "An Essay upon Distinct and Indistinct Vision," in Robert Smith, *A Compleat System of Opticks in Four Books*, 2 vols. (Cambridge: By the author, 1738), 2:115–71, at 170.

would be produced by a hypothetical "pure" sensation. This is exactly the claim made by the modified Young hypothesis, and it has been suggested that Fechner's theory might have led Helmholtz, "in part," to modify Young's hypothesis by 1858.[75]

However, as noted above, Fechner assumed a specific irritability for retinal elements, as had Scherffer seventy-five years earlier; Young assumed a specific energy for the fibers. Both approaches concluded that all color sensations are diluted; but they offered different mechanisms to explain the dilution. Given Helmholtz's incomplete summary of Fechner's theory in his 1858 speeches, it seems more plausible that not Fechner but rather Johannes Müller provided the decisive impulse for Helmholtz to reinterpret Young. Helmholtz later referred to Young's theory as nothing other than a "further specialization" of Müller's doctrine, and recalled that Müller's law (along with the failed classroom demonstrations) had first prompted him to study color.[76] If true, then Helmholtz's "conversion" would have occurred when he realized Young's theory was indeed nothing other than Müller's law, something that he had not understood in 1852.

This realization may have come with Helmholtz's "Kant speech," delivered in February of 1855, before Maxwell's essays were published. Already in his 1852 *Habilitationsrede*, Helmholtz summarized Müller's doctrine: "the peculiarity of light sensation derives not from special characteristics of light but from the special activity of the optic nerve which produces only sensations of the quality of light sensations, regardless of how it is excited."[77] But what about the peculiarity of color sensations? Is color a "quality," like light, which requires its own specific energy (or energies), or is color merely a modification of the light quality? Noting that the same color sensation can be produced by a homogeneous ray or combinations of such rays (one of the results of his 1852 experiments), Helmholtz observed that "which color combinations appear identical" depends not on their "objective relations" but "only on the physiological law of their interaction."[78] As to the nature of this law, Helmholtz in the 1852 speech remained silent. In his 1855 speech, however, he addressed this issue. He now distin-

---

75. Fechner, "Ueber die subjectiven Nachbilder," 428–31; and Turner, "Fechner, Helmholtz, and Hering," 140–41.
76. "Die neueren Fortschritte in der Theorie des Sehens," *Preußische Jahrbücher 21* (1868):149–70, 261–89, 403–34, in *VR⁴ 1*:265–365, at 313; and "Antwortrede Graefe-Medaille," 317.
77. "Sinnesempfindungen," 605.
78. Ibid., 606.

guished, as had his friend Brücke,[79] "qualities" of sensations (light, heat, and sound) from "modifications" thereof (color, temperature, and pitch), and repeated the Newtonian hypothesis that different frequencies of aether oscillations excite different sensations of colors (again without speculating on the physiological processes by which this occurs). But then he added an extraordinary claim: "If light of different colors is mixed, it excites an impression of a new color, a mixed color, which is *always whiter and less saturated* than the simple colors of which it is compounded."[80] How could Helmholtz have arrived at this conclusion? His 1852 experiments showed that mixed rays produced sensations in some cases equal in saturation to the homogeneous prismatic ray of the same color and in some cases less saturated than that ray. His 1855 paper reported only mixtures of complementary rays, which are irrelevant to the above claim. Hence, the short comment in the 1855 speech might have derived from some "physiological law" about how the retina mixes colors, a law which Helmholtz now accepted but did not elaborate.

Might the comment have derived from Fechner's 1840 theory of retinal action? As noted above, Fechner had hypothesized the existence of a secondary, complementary sensation arising from fatigue and the inner light of the eye. Yet two monochromatic rays, fatiguing two elements in the illuminated portion of the retina, would not increase the intensity of the inner light, i.e., reduce the saturation of the mixed sensation over that of the unmixed sensations. The only means of reducing the saturation in a mixture is for more white to be produced than is present when a single monochromatic light illuminates the retina. Precisely such an effect is predicted by Young's hypothesis, if modified so that any monochromatic ray stimulates all three receptors.

Might therefore Helmholtz already by early 1855 have modified Young's hypothesis? And might the decisive spur have come from Müller's doctrine of specific sense energies which Helmholtz had reviewed in his 1852 speech but considered more thoroughly in the Kant speech? If so many different types of stimuli (pressure, electricity, disease, stimulating drink, and even complete darkness) all provoke the eye to produce sensations of light and color, is it not also plausible

---

79. Cf. Ernst Brücke, "Visus" [1846], in *Encyclopädisches Wörterbuch der medicinischen Wissenschaften*, 37 vols. (Berlin: Veit & Comp., 1828–49), *35*:406–53, at 407, 410–11.

80. *Ueber das Sehen des Menschen* (Leipzig: Voss, 1855), in *VR⁴ 1*:87–117, at 98–9 (emphasis added). This claim might derive from a strict reading of Newton's barycentric diagram for color mixing; yet neither Newton nor Helmholtz ever interpreted this diagram so that binary mixtures, even of neighboring hues, would be slightly desaturated. (See Newton, *Opticks*, 132–34; Shapiro, ed., *Optical Papers*, 507–9.)

that any light ray, regardless of frequency, can excite any retinal element? Such reasoning, in conjunction with his 1852 observations of the decreased saturation of mixed lights, might have led Helmholtz to modify Young's hypothesis by postulating coterminal responses. If this speculative reconstruction of Helmholtz's "conversion" to Young's hypothesis is correct, then his puzzling 1858 comment on "earlier works" would refer to the 1855 Kant speech. Further, Helmholtz's persistent refusal to grant Maxwell priority for the modified Young theory would appear less pejorative; and his later description of Young's theory as "nothing more" than Müller's doctrine would describe the path by which Helmholtz came to adopt Young's view.

Young's hypothesis also removes an ambiguity that had always plagued Müller's doctrine concerning the status of color as a "specific energy" of the optic nerve. In the initial formulation of his doctrine (1826), Müller named light, darkness, and color as energies of the "*Sehorgan*," three energies that by Goethe's theory of color, which Müller in 1826 accepted, might be considered continuous. But Müller also allowed that the energy of the eye, for sensations of color, depends on "certain conditions" of the stimuli, a step back toward notions of specific irritability which the doctrine of specific sense energies was designed to replace.[81] Critics such as Alfred W. Volkmann detected this ambiguity in Müller's doctrine, and argued that since individual color sensations are "qualitatively different," Müller's approach would require an infinity of specific energies and hence an equal number of optic nerves.[82] Müller himself may have realized this problem, for in 1838 he suggested a "physiological explanation" of complementary colored afterimages that would reduce the number of "color sensations" to three: "Seeing one of the three primary colors is only one of the three conditions to which the retina, in a condition of excitation, tends; if this condition is artificially excited, then the retina tends in maximum to the complementary color, which thus appears in the afterimage."[83] This extraordinary suggestion, for which Müller cited no source, postulates retinal action identical to the 1807 version of Young's hypothesis! Müller made no attempt to link this suggestion

---

81. Johannes Müller, *Zur vergleichenden Physiologie des Gesichtssinnes des Menschen und der Thiere* (Leipzig: Cnobloch, 1826), 44–5, 404–5.

82. Volkmann, "Nervenphysiologie" [1844], in Rudolph Wagner, ed., *Handwörterbuch der Physiologie mit Rücksicht auf physiologische Pathologie*, 4 vols. (Braunschweig: Vieweg, 1842–53), 2:476–627, at 521–26. Cf. William Woodward, "Hermann Lotze's Critique of Johannes Müller's Doctrine of Specific Sense Energies," *Medical History 19* (1975):147–57.

83. Müller, *Handbuch* 2:368–69.

to his doctrine of specific nerve energies; but by restricting the "condition" of the retina to only three states he could have eliminated the ambiguity concerning the number of "energies" in the visual organ. Volkmann in 1846 and Carl Ludwig in 1852 also considered, but rejected, explanations for afterimages in terms of three independent "specific types of fibers" or three "types of changes" in the retina.[84] Thus even before 1855, Müller's doctrine and Young's hypothesis occasionally intersected in the literature of physiological optics.

It remained for Helmholtz to meld these two concepts into a viable hypothesis for color vision, first somewhat obliquely in his 1855 Kant speech and then explicitly and more completely in the 1860 *Handbuch*. In 1855 Maxwell published a complete description of how Young's modified hypothesis might account for the basic phenomena of color vision. If the above reconstruction of Helmholtz's research trajectory is correct, then his "conversion" to Young's hypothesis in its Müllerian form would have been independent of Maxwell.

7. Extending Young's Hypothesis in the *Handbuch*

Where Helmholtz did go beyond Maxwell was in extending Young's modified hypothesis to account for more phenomena of color vision. The Heidelberg physiologist—Helmholtz left Königsberg in 1855 for Bonn, and then in 1858 left Bonn for Heidelberg—possessed a richer knowledge of the problems and literature of physiological optics than did the Scottish physicist who was interested primarily in the mathematical features of Young's hypothesis. The second Part of Helmholtz's *Handbuch* is not unlike his 1847 essay on the conservation of force: in each, a hypothesis is elevated to a general principle and applied to unify a host of formerly disparate phenomena.[85] Yet as the following examples of color blindness, the Purknyĕ shift, color harmony, and subjective colors show, Helmholtz not only extended Young's hypothesis to ever more features of color vision, but also added auxiliary features to the basic mechanism. For subjective colors, moreover, he was forced to turn from the language of physiology to that of psychology, thereby disrupting the theoretical unity of the second Part of the *Handbuch*. Nonetheless, by bringing Young's hypothesis into so many areas of physiological optics, Helmholtz (and not

---

84. Volkmann, "Sehen" [1846], in Wagner, ed., *Handwörterbuch*, 3/1:265–351, at 314; and Ludwig, *Lehrbuch* 1:234.

85. Cf. Bevilacqua, "Helmholtz's *Ueber die Erhaltung der Kraft*."

Maxwell) earned the eponymous prize from his contemporaries as creator of the "Young-Helmholtz" theory of color vision.

As had Maxwell, Wartmann, Herschel, and Young, Helmholtz noted how easily color blindness might be explained if one assumes that one of the modes of sensation can become dysfunctional. Abandoning the catholicity of most sections of the *Handbuch*, Helmholtz did not rehearse other nineteenth-century explanations for color blindness.[86] Reporting his own cursory investigations of a color-blind student in Heidelberg by means of Maxwell's color top, Helmholtz showed by Grassmann's laws and a barycentric construction that a missing red sensation would produce the vision described by his subject. Citing Seebeck's 1837 division of color blindness into two classes, Helmholtz suggested a loss of red would account for Seebeck's second class, and a loss of green for the first.[87] Revealingly, Helmholtz assumed without comment that all types of color blindness must result from a physical disruption of one or more of Young's three receptors. He assumed, that is, that color blindness is not a selective *inhibitory* process, a type of mechanism which Ewald Hering would later make central to his theory of vision. From the beginning Helmholtz could envision a visual system which acts only by Müller's specific sense energies; in no way could "proto-sensations" interact before becoming sensations in the optic nerve or perceptions in the conscious mind.[88]

In defending Newton against Brewster, Helmholtz had referred to what later would be called the Purkyně shift, the fact that sensations of color depend not only on wavelength but also on the intensity of incoming light. Although this phenomenon had occasionally been discussed since Purkyně first described it in 1825, Helmholtz's former Berlin teacher, Heinrich Dove, was the first to fashion a physiological explanation for it.[89] In 1852 Dove proposed a mechanical law for all

---

86. See "Ueber Farbenblindheit," *Verhandlungen des naturhistorisch-medizinischen Vereins zu Heidelberg 2* (1859–61):1–3, in *WA* 2:346–49; and *Handbuch*, 294–300. For comprehensive reviews, see Wartmann, "Deuxième mémoire"; and Hargreave, "Young's Theory," 302–45.

87. By 1867, after some of his Heidelberg students found ever more complex cases of color blindness, Helmholtz suggested another explanation: perhaps a color-blind eye has differently shaped intensity curves for the three sensations, which would produce a "much greater variability in the appearance of objective colors to the eye" than the simple loss of a receptor (*Handbuch*, 848).

88. Elsewhere in the *Handbuch* (326–27), Helmholtz explicitly rejected Plateau's explanation of irradiation, which had hypothesized interactive lateral effects between a stimulated retinal point and its neighbors.

89. Jan Ev. Purkyně, *Neue Beiträge zur Kenntniss des Sehens in subjectiver Hinsicht* (Berlin: Reimer, 1825), reprinted in idem, *Opera omnia*, 13 vols. (Prague: Spolek Ceskych Lekaru, 1918–85), *1*:57–162, at 118; and Dove, "Einfluss der Helligkeit," 397–99.

sensations, whereby vibratory stimuli on sensory organs produce sensations only if weak movements are repeated rapidly enough. The higher the frequency of a stimulus, the less amplitude is required for sensation to result, and vice versa. Hence, as the intensity of light decreases, the lower frequency colors (red) disappear first. As noted above (see p. 240), Maxwell in 1855 had briefly referred to one half of the Purkyně shift (at high intensities all color sensations become white) to support his modification of Young's hypothesis. By considering the effects of intensity on Young's receptor curves, Helmholtz offered a comprehensive alternative to Dove's mechanical explanation.

Always deferential to his teachers, Helmholtz did not mention Dove's theory in the *Handbuch* and thus avoided having to refute it. To confirm quantitatively Dove's descriptions of the appearance of colors in art galleries at dusk, Helmholtz reported some of his own previous unpublished observations of mixed complementary pairs of spectral colors made with his variable-width, double-slit apparatus of 1855. Just as sunlight appears white at all intensities, so too did Helmholtz's complementary mixtures remain white as he altered the overall intensity of light entering the device. In mixing violet and yellowish green at a low overall intensity, the sensation of white was produced when violet light appeared only one-fifth as intense as the green light. At a higher overall intensity with the same slit widths, the sensation of white remained even as the violet now appeared only one-tenth as intense as the green.[90] Helmholtz briefly suggested how Young's hypothesis might account for such intensity-related effects by postulating "that in the three types of nerves the intensity of sensation is a different function of the [objective] light intensity, so that [for homogeneous violet light] the intensity of sensation in the violet-sensing nerves increases faster and then slower than for the green-sensing nerves, as do those nerves in comparison with the red-sensing nerves."[91] Helmholtz described different functional relations between strengths of sensation and intensity of stimulus for Young's other two receptors. As intensity of the stimulus increases so that each of Young's three receptors is maximally excited, the sensation of white occurs for any wavelength. In his graph of the hypothetical three response curves (see Figure 5.7),

---

90. Helmholtz did not here turn to "unconscious judgments" to explain why the eye keeps seeing white even as the relative apparent brightness of the colors changes. Instead, he vaguely speculated that "sunlight, which we consider by day as normal white, must in different intensities vary its colors similarly as do other white or whitish color mixtures with which we compare it" (*Handbuch*, 319).

91. Ibid., 320.

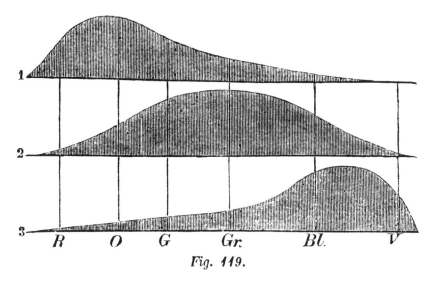

Figure 5.7. Helmholtz's hypothetical response curves for Young's receptors (1860). Reprinted from Handbuch, *291.*

Helmholtz had assumed a constant, middle-range intensity of incoming light. By using Young's theory to account for the Purkyně shift, Helmholtz indicated how a third dimension might be added to those curves, making intensity of sensation a function of both wavelength and intensity of the objective light.[92]

Provoked by Goethe's *Farbenlehre* and more importantly by the work of Michel Chevreul in the 1830s, nineteenth-century artists and scientists had long been seeking empirical "laws of color harmony," often based on analogies with musical tones.[93] In 1852 and 1855, Helmholtz had rejected attempts to explain the "pleasing" effects of colors seen together by analogy to harmonious musical intervals, such as major or minor chords. The numerical ratios of wavelengths for musical relations considered "pleasing" simply did not correspond to

---

92. Helmholtz made no attempt to graph these hypothesized three-dimensional surfaces, although he did sketch hypothetical response curves for two colors, selected because they were complementary, not because they matched Young's receptors (*Handbuch*, 318). Unlike Maxwell, who measured the shapes of Young's response curves, Helmholtz appeared uninterested in exploiting the Purkyně shift to determine empirically the quantitative relations among intensity, wavelength, and color sensation for the human eye.

93. See Chevreul, *De la loi du contraste simultané des couleurs* (Paris: Pitois-Levrault, 1839); and Fr[iedrich] W. Unger, "Ueber die Theorie der Farbenlehre," *AP 87* (1852):121–28.

wavelength ratios for colors considered "meaningful." In the *Handbuch*, however, he proposed another explanation for color harmony. The three colors so admired by Italian painters—red, green, and violet—do not in their relative wavelengths mirror any pleasing musical intervals, but they do exactly match Young's three receptor sensations. Refusing to elaborate, Helmholtz wondered whether in that correspondence might "lie the basis of their aesthetic effect."[94] Unlike his *Die Lehre von den Tonempfindungen als physiologische Grundlage für die Theorie der Musik* (1863), which devoted considerable attention to the aesthetics of music, Helmholtz's *Handbuch* nearly ignored the aesthetic aspects of color.[95]

For color blindness, the Purkyně shift, and color harmony Young's hypothesis served Helmholtz well. Only when considering subjective colors did Helmholtz encounter limits for Young's hypothesis. To explain such colors Helmholtz was forced to supplement Young's mechanisms with hypotheses about mental processes of "judgments" (perhaps borrowed from Fechner or Brücke) and Fechner's idea of retinal fatigue. By attributing some subjective phenomena to strictly physiological mechanisms and others to combined physiological and psychological processes, however, Helmholtz clouded the methodological clarity and simplicity that he had achieved with the modified Young hypothesis in other areas of color vision.

Unlike Fechner, who in 1840 sought a single explanatory program for both afterimages and contrast, Helmholtz separated these two subjective effects. For the former, he combined Young's and Fechner's mechanisms. Simple afterimages, Helmholtz postulated, result from the "persistent stimulus of the retina" after cessation of a primary stimulus, combined with effects of fatigue, the degree of which is proportional to the strength of the primary stimulus. Fechner had hypothesized a separate fatigable receptor for each color; Helmholtz simply transferred this mechanism to Young's three receptors. For the

---

94. "Sinnesempfindungen," 585, 600, 602; "Zusatz" to "Ueber die Messung der Wellenlänge des ultravioletten Lichtes, von E. Esselbach," *Bericht über die zur Bekanntmachung geeigneten Verhandlungen der kgl. Akademie der Wissenschaften zu Berlin* (1855):760–61, in *WA* 2:81–2; and *Handbuch*, 270.

95. In a later series of popular lectures, "Optisches über Malerei" [1871–73], in *VR⁴* 2:93–135, at 117–33, Helmholtz discussed how artists use pure color, intensity, and contrast effects in painting, referring again to Young's color theory (129) to explain why the eye objects to the intermingling of other colors when viewing one of the three "fundamental colors." Yet he again concluded that "[s]olid rules for color harmony, of similar precision and certainty to those for consonance of [musical] tones, have not yet been formulated" (131). Cf. Vogel, "Sensation of Tone," and Gary Hatfield, "Helmholtz and Classicism: The Science of Aesthetics and the Aesthetics of Science," in this volume.

more complex succession of differently colored afterimages which appear after staring at very intense white light or the sun (*Blendungsbilder*), Helmholtz borrowed another explanatory mechanism from Fechner, the "inner light" of the eye. The first three colors in the series of afterimages (bluish-green, blue, and reddish-violet) could be produced by differential rates of fatigue for the persisting primary sensations of Young's receptors. Helmholtz's graph of the persisting intensity over time for Young's three sensations closely follows Fechner's similar diagram for three arbitrary sensations.[96] To explain the final two colors of the sequence (orange and yellowish-green), Helmholtz turned to the action of "inner light" on the fatiguing retina. If retinal elements lose sensitivity for "inner light" as well as for external rays, then the complementary sensations for the former would fade over time just as do the persisting primary sensations. In *Blendungsbilder*, the green afterimage persists the shortest amount of time, which means Young's green receptor would be the first to regain sensitivity for green rays of the "inner light." This green image would mix with the persisting reddish-violet image to create an orange sensation and finally a green sensation as the persisting reddish-violet image disappears. At no point did Helmholtz mention that the various mechanisms he was combining to explain afterimages had all been postulated by Fechner in 1840.

In explaining contrast, however, Helmholtz refused to accept another of Fechner's 1840 mechanisms—antagonistic interactions among lateral retinal elements. Instead, Helmholtz abruptly introduced a new discourse into the *Handbuch*:

> Such phenomena appear to me to be of a completely different type from those previously considered. In general, they can be characterized as cases in which an exact judgment of the reacting color [the subjective sensation] is not possible from a comparison with other inducing colors [the objective sensation]. In such cases we are accustomed to consider such differences, which in conception [*Anschauung*] are perceived as clearer and more certain, to be greater than those which either enter the conception uncertainly or must be judged with help of memory. This is probably a general law of all our perceptions. A person of medium height beside someone very tall appears short, because we see instantaneously that taller persons exist but not that shorter ones also exist. The same person of medium height, beside someone short, would appear tall.[97]

---

96. Cf. *Handbuch*, 372; and Fechner, "Ueber die subjectiven Nachbilder," Figure 2.
97. *Handbuch*, 392–93.

Thus, sensations from different stimuli falling simultaneously on different parts of the retina interact only through psychological processes deriving from judgment, memory, or habit. As in judging human heights, the visual system in comparing color hues (color contrast) or intensities (brightness contrast) exaggerates differences, in inverse proportion to the spatial separation on the retina of the two fields being compared.

Again, this explanation does not lack precedents. Helmholtz made no attempt to explain, either teleologically or with some postulated psychological mechanism, why such judgments of contrast are made. His "rule," merely describing the phenomena, is not unlike Chevreul's 1828 "general law of simultaneous contrast" which stated that "when two colors are viewed together [i.e., side by side], they appear as dissimilar as possible."[98] In 1840 Fechner had allowed that "judgments" affect intensity contrasts and our determination of the sensation of white. When one fixes on a black field against a white ground, the field appears to lighten over time. Accepting the contemporary opinion that black is the absence of sensation, Fechner could not attribute the lightening of black to fatigue. Perhaps, however, the white sensation fatigues and the eye mistakenly judges the change in intensity to have occurred on the black rather than on the white field. Without external comparisons, concluded Fechner, the eye has difficulty in reaching a "decisive judgment" of such relative changes.[99] In 1846 Brücke likewise argued that our "judgments" of white and lightly colored objects can err if no comparison colors are in the visual field. Five years later Brücke elaborated:

> Our entire judgment of color must depend essentially on the perception which we have of neutral gray ... or of pure white. If our sensory memory were absolute and our perception of white were unchangeable, then we would always judge colors correctly. ... But this is not so; because of the imperfections of our sensory memory, we name things white at different times which prove to be colored when seen beside each other.[100]

Attributing subjective colors to judgments was thus well established by 1860.

---

98. Cf. ibid., 393; and Michel Chevreul, "Mémoire sur l'influence que deux couleurs peuvent avoir l'une sur l'autre quand on les voit simultanément," *Mémoires de l'Académie des Sciences 11* (1828):447–520, at 453.

99. Fechner, "Ueber die subjectiven Nachbilder," 203–4; and Turner, "Fechner, Helmholtz and Hering," 141–42.

100. Brücke, "Visus," *35*:415; and idem, "Untersuchungen," 430.

Helmholtz explained additional phenomena of color contrast partly in the language of fatigue but more frequently in the language of unconscious judgments. For example, he repeated Fechner's 1838 observation of colored shadows, in which one color viewed through a blackened tube did not alter in hue as the neighboring color was changed, unlike the case when both colors were in the visual field. No other observation, Helmholtz wrote, shows "more decisively and clearly the influence of judgment on our color determinations."[101] From a series of observations of small white fields viewed against colored grounds, Helmholtz derived another general rule: after a color has filled the entire visual field, a whiter gradation of that color appears white, and an actual white field appears in the complementary color. In such cases, Helmholtz suggested, our very "concept of white" has changed. When we cannot compare a sensation of white simultaneously with another one we "know" to be white, we must "perceive" whether the intensity relations of Young's three sensations have changed. Such a comparison over time is "highly uncertain and imprecise." Hence, we resort to "determinations" which may judge considerably different sets of sensations to be "white." Brücke, as noted above, had made the same point in 1851, only without Young's mechanism for producing the sensation of white. Finally, Helmholtz noted that effects of contrast disappear when the boundary between a white field and a colored ground is marked by black lines. Here, the explanation was teleological. "Because judgment of spatial position, of the bodily independence of an object, is decisive in the determination of its color, it follows that the contrast color arises not through an act of sensation, but through an act of judgment." Helmholtz referred readers to the forthcoming third Part of the *Handbuch* for treatment of "perception of objects."[102]

In the historical overview appended to the *Handbuch*'s chapter on contrast, Helmholtz described himself as the first to separate systematically afterimages and contrast and to discover that "the appearance of contrast colors depends on circumstances determined only by the psychic activities by which visual perceptions are formed."[103] Again, the synthesizer had overestimated his own contribution. Most observers since Chevreul had carefully distinguished these two types of subjective effects,[104] and Fechner in 1840 had considered it meritorious

---

101. *Handbuch*, 393–95, 416; and cf. Turner, "Fechner, Helmholtz, and Hering," 139.
102. *Handbuch*, 396–97, 404, 406.
103. Ibid., 414–15.
104. For example, Müller, *Handbuch* 2:365–75, distinguished three classes of afterimages and two classes of contrast effects.

to find one theoretical discourse to explain both. As Turner has noted, in 1860 Fechner immediately mounted a cogent attack on Helmholtz's bifurcated treatment of contrast and afterimages,[105] and it does seem probable that with only a little ingenuity Helmholtz could have exploited Fechner's flexible mechanisms and unmeasured fatigue rates to explain contrast. Why, then, did Helmholtz in the second Part of the *Handbuch* so boldly and perhaps so unnecessarily disrupt the unity of the theory of color vision he had assembled from the ideas of Young and Fechner by turning from physiological to psychological language?

Helmholtz himself offered few hints for answering this question. The third Part of the *Handbuch* (1867) barely mentioned color contrast even as it fully elaborated Helmholtz's empiricist program of unconscious judgments and spatial perception. The one-paragraph addendum of 1867 updating the 1860 chapter on contrast cited Fechner's critique without comment and did not mention the psychological explanation of contrast. Helmholtz would not again elaborate or defend his theory of contrast against its detractors.[106]

At least several reasons for Helmholtz's bifurcated approach to color sensations seem apparent. First, as noted above, already by the mid-1850s he had begun to formulate his empiricist epistemology, in which judgments and experience, rather than simple mechanical associations, shape *perceptions*. It is thus not surprising that by the late 1850s Helmholtz was willing to employ empiricist language for phenomena at the "lower level" of *sensations* when these appeared inexplicable by Young-Fechner physiological mechanisms. Perhaps more important was Helmholtz's unwavering commitment to a passive "front end" of the sensory neurophysiological system. Helmholtz might accept Fechner's hypothesis of a "persistent stimulus" in afterimages. But he could not accept Plateau's antagonistic retina or Fechner's interactive retinal elements; nor would Helmholtz later tolerate Hering's idea of opponent processes. Although by 1860 histological observations seemed to indicate that networks of neural fibers link the rods and cones in the retina, Helmholtz always considered the "front end" nerves of the visual system to be nothing more than analyzers in Young's and Müller's sense. In an 1868 speech, Helmholtz described each rod and cone as connected to its own nerve fiber. Concerning the fovea, the spot on the human retina of greatest visual discrimination where only cones

---

105. Turner, "Fechner, Helmholtz, and Hering," 145–47.
106. *Handbuch*, 852–53. In the second edition of the *Handbuch*, Helmholtz made only minor changes in the chapter on contrast, but did drop two paragraphs which in the first edition had emphasized the role of judgments (cf. *Handbuch*, 394–95, 417, and *Handbuch²*, 552–53, 566).

are present, Helmholtz noted: "We may assume that from each of these cones an isolated nerve fiber goes through the optic nerve stem to the brain, so that the excitation of each individual cone can also come to sensation isolated from the other [excitations]." Clearly no lateral, front-end interactions would be possible in such a one-fiber, one-sensation view. In the same speech, Helmholtz developed an extended analogy of nerves as telegraph wires, "conducting sensory impressions from the external organs to the brain."[107]

Such a view mirrors the neurophysiological assumption, widely held in the first half of the nineteenth century, that the nervous system consists of two types of elements, those which generate "nervous energy" and those which conduct the same throughout the body.[108] And it conforms to Müller's doctrine of specific nerve energies. Regardless of the nature of its stimulus, a sensory nerve can transmit to the brain a single, peculiar type of sensation, only the intensity of which can vary over time. Fechner's mechanism of retinal fatigue, by which the intensity of sensations vary with time, nicely conforms to Müller's doctrine. But in Helmholtz's view lateral interactions, such as might be required if contrast effects were to be explained physiologically, apparently did not accord with Müller's law. Müller himself had no difficulty accepting lateral interactions across the retina, "by which the qualitative condition of one [retinal particle] influences the condition of another."[109] For Helmholtz, however, such apparent interactions were better explained in psychological terms, as judgments at the level of perception.

8. Helmholtz and Color after 1860

This essay has reviewed Helmholtz's development as a color theorist through 1860. It has argued that his chief contributions were theoretical rather than observational or experimental, and that as a theorist Helmholtz was primarily a synthesizer, creatively combining the hypotheses, concepts, and theories of previous students of color. And it has claimed that despite the bifurcation of color theory in the *Handbuch* into the language and mechanisms of physiology and psychology, Helmholtz nonetheless firmly positioned color as a subjective phenomenon. After

---

107. "Die neueren Fortschritte," VR^4 1:277–78. Cf. $Handbuch^2$, 350.
108. Edwin Clarke and L.S. Jacyna, *Nineteenth-Century Origins of Neuroscientific Concepts* (Berkeley: University of California Press, 1987), 75–7; and Gary Hatfield, *The Natural and the Normative: Theories of Spatial Perception from Kant to Helmholtz* (Cambridge, Mass.: MIT Press, 1990), 172.
109. Müller, *Handbuch* 2:369.

the *Handbuch*, all researchers agreed that theories of color must be theories of color vision.

Yet after the *Handbuch*, Helmholtz himself essentially abandoned active color research. In the 1860s, several of his Heidelberg students published studies relating color blindness to the "Young-Helmholtz" theory of color vision. Some of his students and assistants in the 1870s and 1880s conducted research on color vision in Helmholtz's Berlin physics institute, and went on to become leading sensory physiologists and elaborators of Helmholtz's earlier views. But after 1860 Helmholtz himself published very little on color. During the late 1880s he expanded the coverage of color in the revised edition of his *Handbuch*, incorporating the researches of his students and briefly attacking Hering's alternative "opponent theory" of color vision.[110] Only in 1891 did Helmholtz return to the technical discussion, publishing several relatively unimportant papers linking Fechner's psychophysical law to color vision. Helmholtz thus took little part in the proliferation of research on color vision prompted by his own *Handbuch*.[111]

This essay cannot examine that proliferation, in which the phenomena of color and the postulated mechanisms and theories of the visual system became ever more complex, disordered, and unmanageable. For a variety of philosophical, institutional, and personal reasons, color researchers between 1860 and 1920 simply could not agree on which color experiences are quintessential or on what criteria are appropriate to evaluate hypothetical mechanisms for a psychoneurophysiological system of sensation. The methods by which Helmholtz created order and coherence in the second Part of the *Handbuch* increasingly produced only contradictions and incoherence. It is, therefore, perhaps not surprising that Helmholtz after 1860 drew back from color research. His interests shifted first to spatial perception, the "nativist-empiricist" controversy (in which color played no role), and acoustics, and then with the move to Berlin in 1871 to topics in physics. As the essays in this volume show, throughout his career Helmholtz always sought what he described in 1862 as an "intellectual mastery

---

110. *Handbuch*[2], 376–82.
111. According to Turner, "Paradigms and Productivity," 44, between 1870 and 1885 more works in physiological optics were devoted to color vision than to any other topic.

of nature," a control obtainable through simplicity, classification, and mathematical description.[112] Once after 1871, when asked why he had moved to physics despite his many achievements in physiology, Helmholtz replied that the questions in physiology had become too difficult.[113] Perhaps, too, by 1860 the synthesizing color theorist had reached the limits of his craft.

---

112. "Ueber das Verhältniss der Naturwissenschaften zur Gesammtheit der Wissenschaften," *Heidelberger Universitätsprogramm* (Heidelberg: Mohr, 1862), 3–30, in *VR⁴* 1:157–85, at 183.

113. Quoted in K[arl] Hürthle, "Zum Gedächtniss an Rudolf Heidenhain," *Jahres-Bericht der Schlesischen Gesellschaft für vaterländische Cultur* 76 (1898):15–33, at 22.

6

Sensation of Tone, Perception of Sound, and Empiricism

Helmholtz's Physiological Acoustics

Stephan Vogel

1. Introduction

Helmholtz's studies in physiological acoustics exemplify a primary characteristic of his scientific research as a whole: they simultaneously united several different subjects and employed several different methods of analysis. In his physiological acoustics Helmholtz conducted mathematical analyses of aerial vibrations in tubes; studied the physiological processes in the ear; and discussed problems in musical theory. In so doing, he designed new instruments; performed highly precise experiments; and formulated theories. His research in physiological acoustics, like his research in other areas of science, was of a broad, synthetic nature.

This essay focuses on Helmholtz's work in physiological acoustics. Section 2 briefly indicates the nature and state of pre-Helmholtzian acoustics, in particular paying special attention to the Ohm-Seebeck dispute over the definition of tone. At the heart of this dispute lay the issue of the relationship between sound waves and the characteristics of the perceived sound. As the essay seeks to show, it was precisely around this issue that Helmholtz based most of his acoustical research. In a remarkably short time, as Section 3 indicates, Helmholtz acquired

Acknowledgments: I should like to thank Timothy Lenoir, Simon Schaffer, R. Steven Turner, and William Woodward for stimulating discussions and helpful comments, and, above all, David Cahan for his careful and critical reading of my paper, for many useful suggestions, and for improving my English.

all the essential ideas for his new brand of physiological acoustics; to be sure, it took him several years to elaborate and publish all of his ideas and results, which he saw as constituting a reformation of the field of physiological acoustics.

Section 4, the heart of the essay, argues that Helmholtz formulated four theories in physiological acoustics: a non-linear theory of combination tones; a beat-theory of consonance and dissonance; a theory of tone quality and, as a special application of it, of vowels; and finally, a resonance theory of hearing. In all these areas of physiological acoustics his research method was theory oriented. Moreover, as Section 4 also seeks to show, he treated problems in physiological acoustics in a mathematical manner; this predilection for mathematical analysis of physiological problems was part of his broader program to apply the methods of the physical sciences to physiology. At the same time, Helmholtz also displayed an ability to design new instruments, for example, tuning-fork apparatus and spherical resonators. Appreciating the employment of those instruments, as Section 4 emphasizes, is central to understanding Helmholtz's physiological acoustics, for they embodied exactly his theoretical commitments. In sum, the essential feature of Helmholtz's approach to the study of sensory processes was, therefore, his ability to integrate mathematical, theoretical, and instrumentational elements into a complex yet unified structure. Each element in the structure strengthened the others. As a consequence, even an observation or a hypothesis which, taken alone, remained weak, gained plausibility when connected to the other elements of the larger network.

Helmholtz's research in physiological acoustics was part of his still broader research in sensory physiology and his formulation of a theory of perception. Section 5, this essay's final section, returns to the early 1850s to discuss how Helmholtz developed a new epistemology in conjunction with his ongoing studies in sensory physiology. Numerous scholars of Helmholtz's scientific work have given primacy of place to his philosophical convictions, seeing the development of his scientific theories as an application of his philosophical views. By contrast, Section 5 argues that Helmholtz's empiricist epistemology was developed in tandem with his experimenting on and theorizing about the sensory processes; indeed, his empiricist epistemology eventually developed into an organizing principle for his research in physiological acoustics. The central feature of that epistemology was a division of the sensory process into three parts: the physical, the physiological, and the psychological. Although initially formulated as part of his work in physiological optics, this empiricist epistemology became the basis of Helmholtz's researches in physiological acoustics as well. In fact, the differentiation between the physiological and the psychological part of the sensory process, or what Helmholtz called the differentiation between sensations and perception, became the key to resolving the Ohm-Seebeck controversy.

2. Pre-Helmholtzian Acoustics

Eighteenth-century acoustics was a virtually moribund research field. There were, for example, only two lecture demonstrations on sound: the bell in vacuo to demonstrate that sound depends on a medium for transmission, and the use of small clips on a vibrating string to demonstrate nodal points. Nonetheless, vibrating bodies did play an important role in the development of the calculus: the string, the loaded string, and the fixed and the free rod served as physical models for problems in the theory of differential equations.[1] Towards the end of the eighteenth century, however, acoustical phenomena did become objects of experimental investigation. In particular, in 1787 Ernst Florens Friedrich Chladni gave great momentum to the field of acoustics by performing (at home) an extensive series of experiments on plate vibrations by means of his well-known technique of dust-figures.[2] In 1802, moreover, he published his *Akustik*, the first monograph devoted to experimental acoustics. There he described his own experiments and surveyed the observations and theories of others, thereby producing a systematic account of all known acoustical phenomena. Chladni stressed observation and experiment; he formulated only simple mathematical laws to fit his measurements for vibrating strings and rods; and he compared his empirical laws to those deduced from the theories of leading mathematicians. He could, however, neither report on nor formulate an adequate theory of vibrating plates. Furthermore, along with his highly praised research on vibrating plates, Chladni discovered longitudinal vibrations, and, as an application, he measured the velocity of sound in different materials, including gases. Finally, Chladni exploited every opportunity to argue for acoustics as an independent science. He urged, for example, that it not be subsumed

---

1. Clifford Ambrose Truesdell, *The Rational Mechanics of Flexible or Elastic Bodies, 1638–1788*, Editor's Introduction to *Leonhardi Euler. Opera Omnia*, ser. 2, vols. 10 and 11, in *Leonhardi Euler. Opera Omnia*, ser. 2, vol. 11, part 2 (Zurich: Orell Füssli, 1960).
2. Ernst F.F. Chladni, *Entdeckungen über die Theorie des Klanges* (Leipzig: Weidmann 1787). On Chladni, see Sigalia C. Dostrovsky, "Chladni, Ernst Florenz Friedrich," *DSB*, 3:258–59; and Dieter Ullmann, *Ernst Florens Friedrich Chladni* (Leipzig: B.G. Teubner, 1983), which contains references to the older literature.

under pneumatics, as was then the case, but rather that it be treated as a branch of mechanics.³

Despite Chladni's impressive results, acoustics in fact remained a very minor branch of the emerging discipline of physics. As a count of all papers published in the *Annalen der Physik und Chemie* during the first quarter of the nineteenth century shows, only about two percent concerned acoustics. Most of these dealt with sound propagation, in particular with the dependency of loudness on distance, and most were either by scientists working principally in other fields or by amateurs. Before 1825 or so, not a single German university professor represented acoustics as his major field of research.

Thereafter, however, the field began to emerge as an independent subdiscipline. The leaders of the post-1825 generation of acousticians in Germany included the brothers Ernst Heinrich and Wilhelm Weber, both of whom were greatly influenced by Chladni. In their masterpiece *Wellenlehre, auf Experimente gegründet* (1825), which they dedicated to Chladni, the Weber brothers not only described experiments on the motion of fluids; they also introduced their German readers to the latest mathematical theories of the leading French physicists, especially Siméon Denis Poisson, Pierre-Simon de Laplace, and Augustin-Louis Cauchy.⁴ In France, it was Félix Savart who led the way in acoustical research and who, in contrast to the young Wilhelm Weber, criticized Chladni's results. Apart from the Webers, many of the Germans who advanced acoustics were schoolteachers: for example, Friedrich Strehlke, who worked on vibrating plates; August Roeber, who reviewed publications on acoustics for the *Fortschritte der Physik*; August Seebeck, who wrote articles on acoustics for Heinrich Dove's *Repertorium der Physik* and for the *Annalen* and who was the senior acoustician after Chladni's death in 1827; and, finally, Georg Simon Ohm, who formulated a definition of tone which became central to the entire field and, most pertinent for this essay, to Helmholtz's own acoustical research.

Acoustical research received great impetus in the 1830s and 1840s when an important new acoustical instrument, the siren, became available for research on sound. Invented in different forms by Charles

---

3. See, for instance, Ernst F.F. Chladni, "Beitrag zur Beförderung eines besseren Vortrags der Klanglehre," *Neue Schriften der Gesellschaft der naturforschenden Freunde Berlin 1* (1795):102–24; and idem, *Akustik* (Leipzig: Breitkopf und Härtel, 1802), preface.

4. Ernst Heinrich Weber and Wilhelm Weber, *Wellenlehre, auf Experimente gegründet oder über die Wellen tropfbarer Flüssigkeiten mit Anwendung auf die Schall- und Lichtwellen* (Leipzig: Gerhard Fleischer, 1825)

Caignard de la Tour (1827), Savart (1830), and Friedrich Opelt (1834), and later improved by Dove (1851) and by Helmholtz (1856), the siren was second only to the tuning fork in its importance as a sound generator.[5] All forms of the siren were based on one underlying, common principle: a quick succession of single sounds, it was found, produced the sensation of a continuous sound, the pitch being determined by the frequency of the pulses. Previously, a tone had been regarded as being due to the vibrations of solid bodies (for example, strings, rods, or plates) or air columns (for example, those in organ pipes or wind instruments). As a consequence, the sound wave was considered to be sinusoidal in form. The siren now inspired a new definition of tone; in fact, it gave physicists the occasion to rethink the meaning of a tone. Siren experiments by several scientists now showed that the true causes of the sensation of tone were periodic pulses transmitted to the auditory nerve.[6] The essential feature of this new definition was the reduction of tone to mere periodicity and the elimination of the former assumption about the form of the vibration.

Seebeck performed the most systematic and thorough siren experiments. In 1841 he used the polyphonic siren, which allowed easy control of the phase relation between different tones, in a series of experiments on the interference of two or three tones. He explained his results by introducing a new model of sound: he considered sound to be nothing more than a series of single pulses. Moreover, Seebeck investigated how the ear perceives non-isochronic pulses. For pulses following one another in time intervals $t, t', t, t', \ldots$, where t and t' are two slightly different times, Seebeck could hear two tones, the first corresponding to $t+t'$, the second to its octave $\frac{t+t'}{2}$. He found that the relative intensity depended on the actual difference between t and t'. Seebeck concluded that exact isochronism is not a necessary condition for tone production. His experiments thus led him to claim that a tone was a nearly isochronic series of single pulses of arbitrary form.[7]

---

5. On the history of the siren, see Ernst Robel, *Die Sirenen: Ein Beitrag zur Entwicklungsgeschichte der Akustik*. Teil 2: *Die Arbeiten deutscher Physiker über die Sirene im Zeitraume von 1830–1856*, Programm des Luisenstädtischen Gymnasiums, Beilage (Berlin: Gaertner 1894).
6. See, for instance, Heinrich Ernst Bindseil, *Akustik, mit sorgfältiger Berücksichtigung der neueren Forschung* (Potsdam: Horvarth, 1839), 543; August Röber, "Akustik," in Heinrich Wilhelm Dove and Ludwig Moser, eds., 8 vols. in 4, *Repertorium der Physik 3* (1839):30–1; and Johannes Müller, *Handbuch der Physiologie des Menschen*, 2 vols. (Coblenz: J. Hölscher, 1838), 2:393.
7. August Seebeck, "Beobachtungen über einige Bedingungen der Entstehung von Tönen," *AP 53* (1841):417–36.

Two years later, in 1843, Ohm sought to reinterpret Seebeck's siren experiments; he wanted to rehabilitate the old definition of tone as the basis of acoustics. For Ohm, this definition amounted to three statements: i) to generate a tone of frequency **m**, the pulses in the intervals of length 1/**m** must either be of the form $A\sin 2\pi(mt+p)$, where **A** is the amplitude and **p** the phase, or have this form as a real constituent of the pulses; ii) the phase **p** must be the same in successive intervals; and iii) the amplitude **A** must be of the same sign in successive intervals. To decide whether a sound wave contains a vibration of the form $A\sin 2\pi mt$, Ohm used Fourier analysis, whereby a complex sound vibration is broken up into a harmonic series of sinusoidal vibrations. In order to perform his calculations, Ohm made some rather specific assumptions about the form of the siren pulses. In particular, he took them to be of sinusoidal form, yet extending over only part of the interval between two pulses. From his long and circumstantial calculations he derived a number of results which he used to explain Seebeck's experiments within the framework of his own definition, and he formulated several conclusions about the intensity of the higher harmonics. On the basis of an unnoticed mathematical error, he concluded that, for the case where pulses extend over the entire interval between two pulses, the fundamental frequency must have an infinite amplitude. Although his result was physically impossible, Ohm thought it convincing, for he viewed it as a limiting condition which explained why only the fundamental tone is normally heard whereas the higher harmonics usually remain unnoticed. He did not discuss the dependence of his results on his assumptions about the form of the pulses.[8]

Seebeck rejoined that the problem at issue was not merely one of mathematical method; it was, rather, one about the fundamental principles of acoustics. Seebeck saw acoustics not merely as the theory of sound waves and vibrating bodies, as Chladni had seen it, but also as a theory of sound perception. Accepting Fourier analysis as a valid method for the analysis of sound, Seebeck repeated Ohm's calculations but without making Ohm's specific assumptions about the sound wave's form. For many (but not all) cases he found good agreement between the calculated amplitudes and his own observations. He thus argued that the definition of tone should not be restricted to the

---

8. Georg Simon Ohm, "Ueber die Definition des Tones, nebst daran geknüpfter Theorie der Sirenen und ähnlicher tonbildender Vorrichtungen," *AP 59* (1843):513–65, esp. 518.

sinusoidal form. To further illustrate his point, he considered the case of a harmonic series in which all lower harmonics are absent and only harmonics from the nth onwards are present:

$$f(t) = a_n \cos\frac{2\pi nt}{l} + a_{n+1}\cos\frac{2\pi(n+1)t}{l} + a_{n+2}\cos\frac{2\pi(n+2)t}{l} + \ldots,$$

where l denotes the fundamental period. Seebeck conjectured that f(t) would still produce a tone of the frequency 1/l, but one with a difference in tone quality. He thought that this tone, which he viewed as a combination tone of the higher harmonics, may even be audible when all the single harmonics are too weak to be heard individually. On Ohm's definition of tone, by contrast, no tone would be perceptible in such a case, and certainly none with the frequency 1/l. Although Seebeck did not test his idea experimentally, he found it convincing enough to draw an important conclusion from it: the higher harmonics reinforce the fundamental. Under this assumption Seebeck could give a satisfactory explanation for all siren experiments. He concluded that observation favored his view and refuted Ohm's.[9]

Seebeck's paper disturbed Ohm less because it rejected his definition of tone than because it revealed his calculating errors. While he accepted Seebeck's idea of the reinforcement of the fundamental tone by the higher harmonics, he asked whether or not the 4th, 6th, 8th, ..., harmonics will reinforce the 2nd, and, equally, the 6th, 9th, ..., the 3rd, and so on. All terms in the harmonic series would thereby be reinforced, he argued, hence leading to new difficulties. Ohm thus boldly introduced his own hypothesis: "All contradictions, Seebeck sees, depend on an acoustical illusion. . . . I assume, namely, that our ear involuntarily regards the fundamental tone as stronger than it really is, and the partials fainter than they really are. . . ." Ohm pointed to contrast effects in the case of color perception to illustrate his point. He insisted that his definition of tone included everything needed to explain all sound phenomena.[10]

With that, the Ohm-Seebeck dispute ended; yet it was not resolved. Seebeck "won" insofar as he raised several problems about Ohm's

---

9. August Seebeck "Über die Sirene," *AP 60* (1843):449–81, esp. 474.
10. Georg Simon Ohm, "Noch ein paar Worte über die Definition des Tones," *AP 62* (1844):1–18, on 15.

definition which the latter could not answer.[11] In particular, Seebeck rejected Ohm's definition of tone, arguing that the loudness of the higher harmonics is often not in accord with the results of Fourier analysis. Seebeck also mentioned (and discussed in varying detail) three other problems: variable intensity, combination tones, and tone quality. His objections concerning tone quality and combination tones warrant discussion, for they are highly pertinent to the subsequent development of acoustics, in particular for Helmholtz's acoustical research.

First, Seebeck argued that if a sinusoidal vibration is taken as the only possible form of a simple tone, then only pitch and loudness, not tone quality, can be explained. For Seebeck, who formulated no specific assumptions about the wave form for simple tones, tone quality caused no problems at all: it depended on the form of the vibration. However, having accepted the Fourier series as a valid mathematical representation of that form, Seebeck reformulated his results in the language of harmonic series: the higher harmonics have a considerable influence on tone quality.[12]

The other major problem with Ohm's definition concerned combination tones. According to contemporary theory, combination tones were nothing other than beats of such a frequency that they blend into a continuous tone. On that account, however, they could not be of the sinusoidal form required by Ohm's assumption and, therefore, should be completely inaudible. Ohm's first paper on acoustics in 1839 was actually a short note on combination tones. He reported results about the frequencies of the combination tones without giving any clue as to how he arrived at those results.[13] In 1843, he indicated that he intended to publish a fuller account of those investigations, which he had also based on the assumption that a complex sound is resolved into a series of sinusoidal vibrations. Combination tones remained a problem for Ohm's definition of tone; his inability to solve that problem probably explains why he gave up further research in acoustics.

---

11. Both Ernst Robel, *Die Sirenen: Ein Beitrag zur Entwicklungsgeschichte der Akustik*, Teil 3: *Der Streit über die Definition des Tones*, Programm des Luisenstädtischen Gymnasiums, Beilage (Berlin: Gaertner, 1895), 21, and R. Steven Turner, "The Ohm-Seebeck Dispute, Hermann von Helmholtz, and the Origin of Physiological Acoustics," *BJHS 10* (1977):1–24, on 11, consider Seebeck the winner of the dispute.

12. August Seebeck, "Über die Definition des Tones," *AP 63* (1844):353–68, esp. 364–65; and idem, "Akustik," in Dove and Moser, eds., *Repertorium der Physik 8* (1849) 1–16, on 14–5.

13. Georg Simon Ohm, "Bemerkungen über Combinationstöne und Stösse," *AP 47* (1839):463–66.

3. Helmholtz as Self-conscious Reformer of Physiological Acoustics

Helmholtz began his acoustical research in October 1855, shortly after he had left the University of Königsberg for Bonn. He moved fast: his first order of business was to ask his good friend Emil du Bois-Reymond to have a Berlin instrument maker build him a polyphonic siren.[14] Six months later he wrote du Bois-Reymond that he was preparing a paper on combination tones, one which would also have consequences for the theory of consonance and dissonance in music.[15] Three weeks later he sent the paper to the Berlin Akademie der Wissenschaften, where it was read on 22 May 1856, and a few days thereafter he presented a short paper on this very topic to a meeting of the Niederrheinische Gesellschaft. Later that year he consolidated these preliminary papers into a long article for the *Annalen*.[16] In addition, he made a detailed anatomical study of the ear (in particular of the drum and small bones in the middle ear), research that was closely related to that on combination tones.[17]

From the start Helmholtz planned his studies in physiological acoustics on a grand scale and sought to produce something similar to his work in physiological optics. He wrote du Bois-Reymond in May 1857: "I have gradually collected a considerable amount of material for reforming physiological acoustics, and I am waiting for the instruments to complete it step by step."[18] He envisaged his reformist research program even though he had published only on the theory of combination tones and had yet to advance his other acoustical research plans. Already in January 1857, barely a year after he had begun research in physiological acoustics, he wrote to William Thomson:

> I shall next publish a second part of my acoustical experiments.... The theory of harmony and disharmony and of chords can be derived completely from the investigations on beats of upper partials and of combination tones. I have prepared an apparatus to study tone quality. With

---

14. Helmholtz to du Bois-Reymond, 14 October 1855, in Kirsten, 157.
15. Helmholtz to du Bois-Reymond, 3 May 1856, ibid., 159–61, on 160.
16. "Ueber Combinationstöne," *MB* (22 May 1856):279–85, in *WA* 1:256–62; "Ueber die Combinationstöne oder Tartinischen Töne," *Niederrheinische Sitzungsberichte 13* (1856):75–7, in *WA* 3:7–9; and "Ueber Combinationstöne," *AP 99* (1856):497–540, in *WA* 1:263–302. All references to Helmholtz's research on combination tones are to the third of these three papers.
17. See Helmholtz to Wilhelm Heinrich von Wittich, 21 May 1856, quoted in Koenigsberger *1*:267–68.
18. Helmholtz to du Bois-Reymond, 18 May 1857, in Kirsten, 167–69, on 169.

the help of the hypothesis that each tone of specific pitch is sensed by an individual nerve fiber which is connected at its end to a vibrating pendulum of corresponding frequency—a hypothesis supported by recent anatomical findings—it appears that physiological acoustics will soon receive just as rigorous a mathematical garment as optics.[19]

In this short but historically important statement Helmholtz revealed the essence of his entire research program in physiological acoustics: consonance and dissonance will be explained on the basis of beats. Moreover, the reference to the apparatus for studying tone quality shows that he had already conceived the idea that tone quality depends on the series of higher harmonics, while the reference to the resonance hypothesis, which formed the core of his theory of hearing, shows that he had already conceived this idea, too. Finally, the letter's last sentence shows (again) that Helmholtz saw his studies not merely as a contribution to physiological acoustics but rather as a fundamental change of the field itself.

Although Helmholtz needed only some fourteen months (October 1855–January 1857) to conceive all his fundamental acoustical ideas, he required nearly eight years (1855–63) to develop them in detail and to publish his results. Several reasons account for this slow development. First, Helmholtz was still working on his time-consuming *Handbuch der physiologischen Optik*.[20] Second, he did not confine his scientific interests to one narrow field: in addition to his work on physiological acoustics and optics, he also published papers on muscle physiology and hydrodynamics during this time. Third, academic duties and his move to Heidelberg in 1858, where he first set up a provisional physiological laboratory and then later moved into a new building, interrupted his research. Fourth, he had to wait for the construction of the instruments he needed to perform his planned experiments. The budget for the purchase of scientific instruments was rather low: in Bonn and (at first) in Heidelberg, Helmholtz had only 200 Gulden to spend on instrumentation. Even when money was available, the instrument makers frequently required a notoriously long time to build instruments of new design.[21] Helmholtz's situation improved

---

19. Helmholtz to William Thomson, 16 January 1857, in Kelvin Papers, Glasgow University Library, H 15.
20. *Handbuch*. The first Part appeared in 1856, just after Helmholtz had begun his research in physiological acoustics. The second Part appeared in 1860, while the third Part was delayed until 1867. For a discussion of the *Handbuch*, see the essays by Richard L. Kremer, "Innovation through Synthesis: Helmholtz and Color Research," and R. Steven Turner, "Consensus and Controversy: Helmholtz on the Visual Perception of Space," both in this volume.
21. Helmholtz to du Bois-Reymond, 14 October 1849, in Kirsten, 86–9, on 87–8.

considerably when the King of Bavaria donated 400 Gulden for his acoustical research. He used this money to order the well-known tuning-fork apparatus, which allowed him to synthesize complex sounds from a series of sinusoidal harmonics.[22]

After his initial papers of 1856, Helmholtz promulgated his views through a number of scientific articles, popular public lectures, and a semi-popular book that summarized and extended his and others' findings and that set the stage for much future research in physiological acoustics. In a public lecture in 1857, and again at the Naturforscherversammlung in Karlsruhe in 1858, Helmholtz presented his view that the dissonance between two tones is due to the beats of their higher harmonics. He also used those lectures to argue for his hypothesis that different resonators in the ear decompose complex sound waves in a Fourier-like manner.[23] Along with his papers on tone quality and musical harmony, Helmholtz published two long studies on the theory of pipes. In the first, he dealt with flute-like pipes; in the second, with reed instruments.[24] These papers show a typical characteristic of his research procedure: he frequently took up problems that he encountered in his physiological investigations or in the construction and operation of a scientific instrument as starting points for a mathematical analysis of the given problem. In the case at hand, it was the resonance properties of the tuning-fork apparatus which required clarification. Helmholtz tackled it as a general mathematical problem, which also found its applications in musical instruments.

Helmholtz's major publication in physiological acoustics was, of course, *Die Lehre von den Tonempfindungen als physiologische Grundlage für die Theorie der Musik*.[25] In 1861, as he had already done in 1854, the publisher Friedrich Vieweg asked Helmholtz to write a textbook on medical physics. Helmholtz, who had still not finished his

---

22. Helmholtz to du Bois-Reymond, 15 April 1858, ibid., 182–84, on 184.
23. "Ueber die physikalische Ursache der Harmonie und Disharmonie," *Amtlicher Bericht über die 34. Versammlung deutscher Naturforscher und Aerzte zu Carlsruhe im September 1858*, 157–58. Helmholtz probably first mentioned the resonance hypothesis in a public lecture during the previous year: "Ueber die physiologischen Ursachen der musikalischen Harmonie," in *VR⁴* 1:119–55. However, he substantially revised the lecture for publication, so it is uncertain as to exactly what he said about the resonance theory in 1857.
24. "Theorie der Luftschwingungen in Röhren mit offenen Enden," *JfruaM 57* (1859):1–72, in *WA 1*:303–82; and "Zur Theorie der Zungenpfeifen," *Verhandlungen des naturhistorisch-medicinischen Vereins zu Heidelberg 2* (1861):159–64, in *AP 114* (1861):321–27, and in *WA 1*:388–94.
25. Braunschweig: Friedrich Vieweg, 1863. The present essay uses the English translation of the fourth German edition: *On the Sensations of Tone as a Physiological Basis for the Theory of Music*, trans. and annotated by Alexander J. Ellis (London: Longmans, 1885; reprint New York: Dover Publications, 1954).

Handbuch der physiologischen Optik, was forced to decline. However, in his reply he also shrewdly let Vieweg know that he was working on a manuscript presenting the results of his studies in physiological acoustics. His aim, he said, was to write a popular book for musicians and lovers of music. He added that another publisher, Voss of Leipzig, had expressed interest in publishing the book but that no definite arrangements had yet been made.[26] Vieweg recognized Helmholtz's implicit offer, and immediately sought to get him under contract. In the course of negotiating his royalties, Helmholtz confidently insisted on his demands and stressed his book's originality. He wrote:

> The book's content will be the presentation of the results of a series of acoustical experimental investigations that I have conducted during the last six years. Its aims are: First, to analyze exactly, for the benefit of the physiologist, the ear's different capabilities and to reduce them to a common principle. This I consider to be a substantial advance in physiological acoustics, a science hitherto poorly developed. Second, on that basis it now becomes possible to uncover the actual physiological and physical causes of the differences between harmony and disharmony, thereby grounding the entire theory of musical harmony on natural scientific principles. This part of my work presents completely new material, for everything that has been taught so far about the scientific foundation of harmony has been empty talk.[27]

Here was the voice of the self-avowed reformer of physiological acoustics. His reformation brought him 600 Thaler for the first edition of the *Tonempfindungen*, and 400 Thaler for each subsequent edition.

4. Physiological Acoustics in a Mathematical Garment

When Helmholtz began his research in physiological acoustics, Gustav Hällström's theory of combination tones, including his explanation of higher-order combinations or tones as beat-tones of lower-order combination tones, held sway.[28] However, several acousticians conjectured that those higher-order combination tones might in fact be

---

26. Helmholtz to Friedrich Vieweg, 15 July 1861, Verlagsarchiv Vieweg, Wiesbaden, quoted by permission.
27. Helmholtz to Friedrich Vieweg, 21 November 1861, ibid.
28. Gustav G. Hällström, "Von den Combinationstönen," *AP 24* (1832):438–39. On the early history of combination tones see Arthur T. Jones, "The Discovery of Difference Tones," *American Physics Teacher 3:2* (1935):49–51; and R. Plomp, "Detectability Threshold for Combination Tones," *Journal of the Acoustical Society of America 37:6* (1965):1110–23.

first-order combination tones of the higher harmonics. To study this alternative, experiments with pure tones (that is, tones without higher harmonics) were needed. Helmholtz managed to generate them by coupling a vibrator with weak radiation to a resonator whose upper partials did not coincide with the vibrator's upper partials.[29] In this way only the fundamental tone was reinforced while the higher harmonics were rendered inaudible. Helmholtz experimented with tuning forks as vibrators and with resonance boxes or weighted strings as radiators. His experiments led him to agree with Hällström on the frequencies of the combination tones, but to differ with him on the number of audible combination tones. In all cases where higher harmonics were excluded Helmholtz could hear only the first-order combination tone f_1-f_2. However, the existence of higher-order combination tones could be proved indirectly through the use of a probe-tone which would beat with the expected combination tone. Although Helmholtz confirmed the existence of higher-order combination tones, the important result for him was that they were much fainter than the first-order combination tones. He stressed this fact, since it fit well with the mathematical theory that he developed for combination tones.

Helmholtz was also able to hear a combination tone which no one had previously noticed, a tone corresponding to the frequency $f_1 + f_2$. He called this tone the "summation tone," while for the combination tone f_1-f_2 he introduced the correlative term "difference tone." It is noteworthy that he heard the summation tone only *after* his new theory had predicted its existence. Indeed, he mentioned this fact as an argument in favor of his new theory: not only did it explain the known phenomena as well as the rival beat-theory did; it also predicted a new phenomenon which, in turn, could be observed. Theory ruled observation here.

The summation tones served as only one of three arguments against the old theory of combination tones. Helmholtz also argued that under certain conditions combination tones exist objectively, independently of the ear which might have gathered the beats into a new tone.[30] Moreover, he argued that the old theory conflicted with the law that the ear perceives only tones corresponding to sinusoidal vibrations, a law which Helmholtz considered as confirmed by all other experiments.[31] Helmholtz here turned around one of Seebeck's arguments

---

29. The following presentation of Helmholtz's research on combination tones is based on his "Ueber Combinationstöne."
30. A number of other scientists disputed Helmholtz's claim; see *Sensations*, 527–38, where Helmholtz's translator Ellis summarizes the dispute in an appendix.
31. Ibid., 156.

against Ohm's definition of tone: he argued that it was not the definition that was deficient, but rather the received explanation of combination tones. The latter was based on the assumption that the superposition of sound waves is undisturbed, that is, that the amplitudes of the sound waves add up linearly. However, he noted that this assumption is really an idealization and, like all such idealizations, it served the purpose of making a complicated phenomenon accessible to experimental and theoretical investigation. With his exceptional knowledge of the principles of mechanics, Helmholtz realized that the linearity assumption for physical systems was only a first-order approximation and that it might be worth investigating the non-linear effects. He therefore introduced a quadratic term into the force law:

$$\mathbf{k} = \mathbf{ax} + \mathbf{bx}^2,$$

where $\mathbf{k}$ is the restoring force pulling the vibrating mass $\mathbf{m}$ back to equilibrium, $\mathbf{x}$ is the distance of the mass from the point of equilibrium, and $\mathbf{a}$ and $\mathbf{b}$ are constants. The equation of motion for a point mass $\mathbf{m}$ moving under the influence of two sound waves $A_1 \sin(2\pi f_1 t)$ and $A_2 \sin(2\pi f_2 t + p)$, where A_1 and A_2 denote the amplitudes, f_1 and f_2 the frequencies, and p the phase, is, then:

$$-\mathbf{m}\frac{d^2x}{dt^2} = \mathbf{ax} + \mathbf{bx}^2 + A_1\sin(2\pi f_1 t) + A_2\sin(2\pi f_2 t + p).$$

Helmholtz solved this equation by the standard technique of expanding x into a power series, thus yielding an infinite series of ordinary differential equations. He not only found, as expected, the two primary frequencies $\mathbf{f_1}$ and $\mathbf{f_2}$; he also found, as unexpected, the higher harmonics $\mathbf{nf_1}$ and $\mathbf{mf_2}$, where n and m are natural numbers; the summation tones $\mathbf{nf_1 + mf_2}$; and the difference tones $\mathbf{nf_1 - mf_2}$ and $\mathbf{mf_2 - nf_1}$, $n > m$. The amplitude of the first-order difference tone, he found, is inversely proportional to $(f_1-f_2)^2$, whereas the amplitude of the first-order summation tone is inversely proportional to $(f_1+f_2)^2$. Therefore, the summation tone is much fainter than the difference tone. The amplitudes of the higher-order combination tones are inversely proportional to $(f_1+f_2)^l$ and $(f_1-f_2)^l$, where l denotes the order of the combination tone. This explains why Helmholtz, when reporting his experiments, emphasized the faintness of higher-order combination tones: it was a requirement of his theory. He was apparently untroubled by his assuming a rather special force law, for a different law would have yielded considerably different results.[32]

---

32. Ludimar Hermann, "Zur Theorie der Combinationstöne," *Archiv für die Gesamte Physiologie des Menschen und der Tiere* 49 (1891):499–518, for example, criticized

Helmholtz took as the physical cause of combination tones not so much the non-linearity of the force law as its asymmetry. Of all the ear's parts, the drumskin seemed to him the best candidate for an asymmetrically vibrating body because it is bent inward by the small bones in the middle ear. He therefore conjectured "that this peculiar form of the tympanic membrane conditions the generation of combinational tones."[33] With his asymmetry theory of combination tones Helmholtz had solved one of the problems Seebeck had formulated for Ohm's definition of tone.

Helmholtz next sought to explain tone quality. Seebeck had maintained that the *form* of a sound vibration governs the quality of that sound. Helmholtz reformulated Seebeck's position into Ohm's language. He wrote: "The question then arises: can, and if so, to what extent, the differences of musical quality be reduced to the combination of different partial tones with different intensities in different musical tones?"[34]

To make this problem manageable, Helmholtz limited the domain of phenomena that the theory had to explain. He considered only the steady-state of the tones. Moreover, he abstracted from his phenomena by neglecting accompanying noises like the hissing of air in wind instruments or scratches of the bow in string instruments. For his goal was to give a mathematical explanation of a constructed, idealized, standard case—not a phenomenological description of all the peculiarities, the small case-by-case differences, or the anomalies under special circumstances. In this way, and, parallel to the design of his experimental setup (see below, pp. 274–76), Helmholtz broke with the traditional way of describing and classifying tone qualities and made the problem one of mathematized physiology. His exposition of the theory of the tone quality of musical instruments was essentially grounded in mathematics, a point that is not immediately apparent when looking at the *Tonempfindungen* because Helmholtz relegated all mathematics to the book's appendices or referred his readers to his journal articles for the explanation of details.[35]

Helmholtz's theory because it was based on those specific assumptions. For Rudolph Koenig's and Erich Waetzmann's, as well as Helmholtz's, research on combination tones see also Dieter Ullmann, "Helmholtz-Koenig-Waetzmann und die Natur der Kombinationstöne," *Centaurus* 29 (1986):40–52.

33. *Sensations*, 413.
34. Ibid., 65.
35. See Appendices "V. On the Vibrational Forms of Pianoforte Strings"; "VI. Analysis of the Motion of Violin Strings"; and "VII. On the Theory of Pipes," in ibid., 380–97. Helmholtz had already developed the mathematics of pipes in "Theorie der Luftschwingungen," and "Theorie der Zungenpfeifen."

Before Helmholtz treated the general problem of tone quality, he attempted (1857–60), as a special case, to account for the tonal characteristics of the vowels. In 1857 he stated his basic assumption that vowels differ from one another by their higher harmonics.[36] To make these higher harmonics manifest, he described an experiment in which the loud singing of a vowel into an undamped piano caused the strings to vibrate in correspondence to the vowel's harmonics. In his first attempt to ascertain the loudness of those higher harmonics, he generated the following list of vowels, wherein **1** represents the fundamental harmonic, and **2, 3, 4, 5, 6,** and **7** are the successive harmonics:

A: In addition to **1**, the harmonics **3** and **5** are distinct, while **2, 4,** and **7** are weaker;

O: **3** is somewhat weaker than in A, while **2** and **5** are very weak;

U: Mainly the fundamental, with a weak **3**;

E: Very strong **2**, with the higher harmonics hardly audible;

I: **2** and **3** in relation to the weak fundamental seem characteristic for this vowel, while **5** is also weakly present.[37]

To substantiate his idea by further experiments, Helmholtz constructed a new apparatus that embodied precisely his theoretical assumption. If a sound wave is nothing other than a harmonic series of sinusoidal vibrations, as Helmholtz supposed, then the sound generator should reflects this. The tuning-fork apparatus, consisting of a set of tuning forks each of which is supplied with a corresponding resonator and tuned to produce a harmonic series of pure tones, was Ohm's definition of tone cast, as it were, into wood and brass. This apparatus allowed Helmholtz to synthesize complex sounds at will, and gave him full control over the individual components. The electromagnetically driven tuning forks produced the idealized sounds that Helmholtz had put at center stage: steady state, both in frequency and in loudness; of any desirable duration; and noise-free. Helmholtz used this apparatus to reconstruct and to imitate the vowels. That the sounds he produced resembled not so much spoken but rather sung vowels is scarcely surprising, since this was a necessary consequence of the built-in steadiness of the sounds. In 1859, he lectured on his experimental results

---

36. "Ueber die Vokale," *Archiv für die holländischen Beiträge zur Natur- und Heilkunde I* (1857):354–55, in *WA I*:395–96.

37. Ibid., 355.

before the Bayrische Akademie der Wissenschaften. Although his results were essentially the same as those formulated two years earlier, he now included the minor modification that the characteristic (that is, the loudest) harmonics tended to be higher up the scale by one or two positions.[38]

Helmholtz also addressed the matter of how the situation changed when the fundamental was set to a different frequency. In his apparatus, the frequency was fixed by the tuning forks and could only be shifted one octave up; when shifted, it lost the last of the higher harmonics. Helmholtz cautiously conjectured that "in this case, too, the relation between the harmonics and the fundamental seems to be decisive for the vowel character."[39] In other words, he favored a vowel theory which considered the relative position of the dominant harmonics as the crucial factor.

Along with his synthesizer, Helmholtz also constructed an analyzer consisting of a tuned set of spherical resonators. These resonators were constructed with two openings, one that fit into the ear and one that allowed the sound wave to propagate. The closest of the wave's partials to the resonance frequency was amplified to make it audible. Just as th synthesizer composed sounds on the basis of a set of sinusoidal vibrations, so the analyzer decomposed compound sounds into their sinusoidal components. When analyzing sung vowels with the help of this instrument, Helmholtz detected some deviations from his earlier results with artificially produced vowels. He found that for each vowel harmonics falling into one region of the musical scale were louder than those in other regions.[40]

Helmholtz also used the tuning-fork apparatus to study the role played in tone quality by the phase relations of the higher harmonics to the fundamental. The form of a sound vibration, he knew, is dependent on the phase relation of the single harmonics; conversely, compound vibrations of very different shapes can be constructed from the same set of harmonics. To investigate this matter, Helmholtz made use of a physical property of resonating systems: the phase shift between the oscillation of the driving force and that of the resonator depends on the frequency difference between the two oscillations. By partly covering the opening of the tone synthesizer's resonance boxes, Helmholtz purposely mistuned the resonators and thus introduced a calculable phase shift. He compensated for the unavoidable change in

38. "Ueber die Klangfarbe der Vocale," *AP 108* (1859):280–90, in *WA 1*:397–407.
39. Ibid., 287.
40. Ibid., 288.

loudness, which is also dependent on the frequency difference, by varying the distance between the tuning fork and the resonator. By this arrangement, he kept the intensity of the higher partials constant while changing their phase relation to the fundamental. As a result of his experiments, he concluded that tone quality is not phase dependent.[41]

Helmholtz continued his researches on the nature of vowels, and did so on the basis that higher harmonics falling within a certain region are particularly strong. In 1860, he summarized his results into three points: First, the vowels can be arranged into three series: U-O-A, I-E-A, and Ü-Ö-A, which last, by its characteristics, lies between the first two series. Second, he found that restricting and narrowing the mouth cavity makes the higher harmonics fainter; in general, he found that the higher harmonics are fainter than the lower ones. Third, and as an exception to the second point, those harmonics which fall into the frequency regions corresponding to the resonance frequencies of the mouth's cavity are stronger than expected according to point two. For the first series of vowels these regions are U:f, O:h^1, A:h^2. For the second series, into which Helmholtz included the German vowel Ä, there are three regions: I:f,c^4, E:f^1,g^3, Ä:c^2,e^3. And for the third series two regions can be found, a lower one corresponding to the reinforcement regions of the first series, and an upper one corresponding to the higher regions of vowels in the second series: Ü:f,g^3 (= the lower region of U and I, and the upper of E), and Ö:f^1,e^3 (= the lower region of E and the upper region of Ä). Finally, Helmholtz found that the regions of reinforcement are the same for male and female voices. However, in female voices some of the lower regions fall below the voice range.[42]

The important element in Helmholtz's 1860 exposition of the theory of vowels was the shift in emphasis from the higher harmonics to the reinforcement regions. In contrast to his previous view that the *relative* position of the dominant harmonics determines the characteristics of the different vowels, Helmholtz now considered the *absolute* position of the reinforcement regions to be decisive. The idea that some of the higher harmonics are reinforced by the mouth cavity was already hinted at in his paper of 1859. As his research continued, he came to realize the central role resonance phenomena play. As we shall see presently (pp. 279–81, below), resonance phenomena came to constitute the inner thread connecting most of his acoustical activities from 1858 to 1860.

---

41. Other scientists, including especially Rudolph Koenig, "Ueber den Ursprung der Stösse und Stosstöne bei harmonischen Intervallen," *AP 12* N.F. (1881):335-49, observed a change of tone quality when phases were changed.
42. "Ueber die Klangfarbe."

In the *Tonempfindungen* Helmholtz developed the theory of vowels solely on the basis of the resonance characteristics of the mouth's cavity.[43] Here, too, there were some slight changes in the exact position of the reinforcing regions. As before, the tendency was to raise the upper regions to an even higher position. This reflected the extension of Helmholtz's experimental apparatus. He had more (i.e., higher) tuning forks to synthesize and more resonators tuned to higher frequencies to analyze compound sounds. Moreover, he now for the first time treated the tone quality of musical instruments. The central point was, of course, to show that tone quality in general and the differences of tone quality in individual musical instruments can be explained on the basis of the harmonic series. Helmholtz did precisely this for string instruments—differentiating according to the mode of agitation between plucked (e.g., harp or guitar), hammered (e.g., the piano), and bowed (e.g., the violin) instruments—and for wind instruments—here differentiating between the flutes, brass, and the reed instruments. The main difference between the characteristics of vowels and the tone quality of musical instruments was, in Helmholtz's view, that in the former case it is the absolute position of the higher harmonics which is decisive for the vowel characteristics whereas in the latter case it is the relative position of the strong higher harmonics which governs the tone quality.[44]

The problem of musical harmony was important to Helmholtz from the start of his acoustical research. Already in 1856 he had formulated the idea that the ear perceives continuous movement as consonant and discontinuous movement as dissonant.[45] He soon gave this vague idea substance when, in his public lecture of 1857 and again at the Naturforscherversammlung in Karlsruhe in 1858, he lectured on the physical and physiological causes of musical harmony and explained his beat-theory of consonance and dissonance. On the basis of Ohm's definition of tone, Helmholtz argued that interval tones have to be considered as two series of harmonics. Only very few of those partials will be of exactly the same frequency; the remaining ones, he concluded, will produce more or less distinct beats. If the beating frequency is so high that the single beats cannot be distinguished, then they will blend into a rattling or jarring noise. This unpleasant intermittence, he argued, is the cause of dissonance.[46]

43. *Sensations*, 103–18.
44. Ibid., 69–102.
45. See Helmholtz to du Bois-Reymond, 3 May 1856, in Kirsten, 159–61, on 160; and to Wittich, 21 May 1856, quoted in Koenigsberger *1*:267.
46. "Ueber die physiologischen Ursachen," (1857); and "Ueber die physikalischen Ursache," (1858).

In the *Tonempfindungen*, Helmholtz extensively treated the beat-theory of consonance and dissonance. Two new features merit particular notice. First, in addition to the beats of the higher harmonics, Helmholtz also noted the beats of combination tones as to their influence on the harmonic character of an interval or an accord.[47] On the basis of beating combination tones, he found that intervals of pure tones become dissonant and that concords in the minor mode sound less consonant than concords in the major mode. Second, Helmholtz discussed the influence of tone quality on consonance. Taking the violin as his musical source, he calculated the degree of dissonance on the basis of a simple mathematical relation between the number of beats and the resulting roughness, and constructed a curve showing the degree of roughness for different frequency relations. The minimum of roughness occurred, he found, at precisely those intervals which were traditionally seen as the most consonant; he saw this result as strong evidence in favor of his theory.[48]

Of all his achievements in the reformation of physiological acoustics, Helmholtz seems to have been especially proud of his beat-theory of consonance and dissonance. Even Leonhard Euler, Helmholtz pointed out, had contented himself with metaphysical speculations about the pleasantness of simple relations to the human mind. Where the great mathematician Euler had failed, Helmholtz succeeded—as he did not fail to note: "the results of my investigation," he boasted, "may be said, in one sense, to fill up the gap which Euler's left."[49]

Although in the late 1850s Helmholtz had mentioned the resonance hypothesis and its potential use as a basis for treating physiological acoustics in a mathematical manner similar to that in physiological optics, he only developed this hypothesis into a full-fledged theory in his book of 1863. He introduced the theory in Chapter 6, on the perception of tone qualities.[50] By then he had established the phenomena of higher harmonics and had explained the facts about resonance.

To illustrate the properties of the inner ear as a resonating system, Helmholtz took the piano as a mechanical analogy.[51] In a piano with

47. *Sensations*, 197–211, and see also Appendix XV, 415–18.
48. Although Helmholtz was dealing with psychological concepts—roughness as a characteristic of sensations—he transformed it into a mathematical function. And although he realized that this mathematical relation was chosen rather arbitrarily, he presented his calculations graphically, a point which he considered to be more illuminating than long and circumstantial explanations. See ibid., 192. For a detailed discussion and critique of this calculation see Patrice Bailhache, "Valeur actuelle de l'acoustique musicale de Helmholtz," *Revue d'histoire des sciences* 39:4 (1986):301–24.
49. *Sensations*, 231.
50. Ibid., 119–51.
51. Ibid., 129. Already in the eighteenth century several anatomists had formulated a resonance theory of hearing and had used it or a similar analogy. See Alois Kreidl,

the damper lifted, a complex sound will set those strings into vibration with a frequency corresponding to one of its harmonics. Helmholtz argued that since the inner ear has a structure similar to the series of strings in the piano, a similar function should be expected. After presenting a detailed description of the anatomy of the inner ear, including special reference to the findings of Alfonso Corti,[52] Helmholtz concluded that "the terminations of the auditory nerves [are] everywhere connected with a peculiar auxiliary apparatus, partly elastic, partly firm, which may be put in sympathetic vibration under the influence of external vibration, and will then probably agitate and excite the mass of nerves."[53] He argued for his hypothesis not only by treating the ear as a system of resonators and by discussing their physical properties; but also by demonstrating how the resonance hypothesis can be used to explain pitch perception, pitch discrimination, and tone quality. These points demand individual discussion.

Helmholtz knew, of course, that a resonating system continues to vibrate for some time after the driving force has ceased to act upon it. He applied this principle to the (hypothetical) resonating system of the inner ear and argued that tones could be separated only so long as they did not follow each other too quickly. He therefore sought to determine the limiting speed up to which clear trills can be performed, and found through observation that the limit is not the same throughout the musical scale. He considered his results as further evidence "that there must be different parts of the ear which are set in vibration by tones of different pitch and which receive the sensation of these tones."[54] If, on the other hand, the entire structure of the inner ear were set into vibration by an incoming sound wave, then it would vibrate at its own resonance frequency as soon as the driving sound wave had died off and a tone of that frequency would be heard.

Helmholtz argued that just as a dampened resonator is set into vibration not only by its resonance frequency but also, to a lesser degree, by all neighboring frequencies, so, too, are the resonating fibers in the inner ear. This property of the ear becomes singularly important for understanding consonance and dissonance. A simple tone sets into vibration not only the fiber with the matching resonance frequency but also the neighboring fibers as well. If, now, two tones of nearly the

---

"Zur Geschichte der Hörtheorien," *Archives néerlandaises des sciences exactes et naturelles*, sér. 3c, 7 (1922):502–9.

52. Alfonso Corti, "Recherches sur l'organe de l'ouïe des mammifères," *Zeitschrift für wissenschaftliche Zoologie 3* (1851):109–69.
53. *Sensations*, 142.
54. Ibid., 143–44.

same frequencies are superimposed in the ear, their respective resonance regions will overlap and the fibers in the middle will be set into vibration with varying strength. This, Helmholtz argued, is the physiological cause of beats.

Helmholtz next inquired as to whether enough resonators exist in the ear to accommodate the musical scale. Experiments on differential sensitivity with respect to pitch perception had shown that it is possible to detect a frequency change of 1,000:1,001. This corresponds to about 1/64th of a semitone. Helmholtz knew that anatomists had counted the number of Corti's fibers to be around 3,000. Allowing 200 fibers for the perception of very high and very low tones, where pitch discrimination is quite poor, there remain 2,800 fibers for the 7 octaves used in music, that is, 400 fibers per octave or 33 per semitone. This is only half as many as required by the experiments on pitch discrimination. Helmholtz found a simple explanation for this result: due to the dampening of the resonators, a tone with frequency between the characteristic frequency of two fibers will excite both of them. Pitch discrimination, he concluded, is thus a process which compares the amplitude of two resonating fibers.[55] This also explained why the sensation of tone is continuous rather than discontinuous, as the proposed perception mechanism would suggest. As so often in Helmholtz's theorizing, he simplified the physiological mechanism by means of an implicitly or explicitly postulated psychological process.[56]

Finally, the resonance hypothesis gained additional strength from its ability to explain the perception of different tone qualities. Helmholtz wrote:

> The sensation of different pitch would consequently be a sensation in different nerve fibers. The sensation of a tone quality would depend upon the power of a given compound tone to set in vibration not only those of Corti's arches which correspond to its prime tone, but also a series of other arches, and hence to excite sensation in several different groups of nerve fibers.[57]

With this Helmholtz had achieved what he saw as the essential goal of perception theory: explaining qualitative phenomena by reducing them to quantitative measures.

---

55. Ibid., 147.
56. Cf. Kremer, "Innovation through Synthesis"; Timothy Lenoir, "The Eye as Mathematician: Clinical Practice, Instrumentation, and Helmholtz's Construction of an Empiricist Theory of Vision," this volume; and Turner, "Consensus and Controversy."
57. *Sensations*, 148.

To achieve his reformation of physiological acoustics, Helmholtz had set up a research program which, step by step, incorporated the entire range of issues relating to the production and perception of sound. The result was an organic development of his research program that led to a set of strongly intertwined theories, forming a closely knit structure in which every element was strengthened by its relationship to every other element. Moreover, Helmholtz's fundamental theoretical ideas were paralleled in the principles of the different instruments which he had constructed.

The basis of Helmholtz's research in physiological acoustics had been Ohm's acoustical law. In Ohm's formulation, this law had a twofold purpose: it was a substantive statement about the physical characteristics of sound, and it was also an implicit methodological recommendation for the appropriate mathematical technique to be used in acoustics. Helmholtz used Ohm's law in two further ways: he transformed it into a tool of experimental analysis and he paralleled the mathematics of Fourier analysis with a physiological model of sound perception.

From 1856 on, resonance became the fundamental concept in Helmholtz's research program. The generation of pure tones for research on combination tones involved the reinforcement of sound phenomena by resonance. For the study of vowels, Helmholtz constructed the spherical resonators and the tuning-fork apparatus which contained a set of resonance boxes. To use the sound synthesizer in experiments on tone quality required a mathematical theory of the resonator box. This theory, which found its way into the *Tonempfindungen*'s appendices, was also the starting point for Helmholtz's two papers on aerial vibrations in tubes and in reed pipes. The more Helmholtz became concerned with resonance, both in the design of acoustical apparatus and in theoretical matters, the more the concept came to dominate his theory of vowels as well. Finally, the concept of resonance and the decomposition of a complex sound into a series of harmonics merged in the resonance theory of hearing. Helmholtz wrote of this theory: "it gathers all the various acoustical phenomena with which we are concerned into one sheaf, and gives a clear, intelligible, and evident explanation of all phenomena and their connection."[58]

---

58. Ibid., 227.

5. Physiological Acoustics: A Study in Epistemology

In his lecture "Ueber die Natur der menschlichen Sinnesempfindungen" (1852), Helmholtz inquired into the relationship between our sensations and the objects we perceive. He sought to advance his inquiry on the basis of the law of specific sense energies, which his teacher, Johannes Müller, had previously formulated. Helmholtz characterized the relationship between perception and the perceived object in two statements: First, not all that we perceive as light is actually light, for mechanical and electrical impact can also serve as stimuli for sensations in the eye. Second, since the human eye is sensitive to only a part of the solar spectrum, there is light that we cannot perceive. In particular, Helmholtz pointed to the demonstrated existence (to use modern terminology) of infrared and ultraviolet light. He concluded that, insofar as color is generated by the eye, it is not an objective quality of the objects perceived but rather a subjective quality.[59] He pointedly characterized the relationship between the object and the sensations: "The sensations of light and color are only signs for relations in reality."[60] To clarify this point he used an analogy. Just as there is no similarity between, say, the name of an individual and the actual individual, so there is no similarity between the sensations of an object and the object itself. Yet just as names help us to identify individuals in the world, so sensations inform us, by their identity or their differences, about whether we are dealing with identical or different objects. There is, however, an important limitation to the theory of signs. Unlike linguistic symbolism, the symbolism of the sensory nerves is not arbitrary. Instead, it is based on the structure of the human body and, therefore, is the same for all human beings.

Helmholtz's model of perception as formulated in 1852 was rather simple: the external objects interact with the sense organs through physical forces. The sense organs may function as filters, allowing only forces within a given range to pass and stimulate the sensory nerve. The stimuli then give rise to sensations, the quality of which, according to Müller's law, depends on the sense affected. Helmholtz's model is thus essentially a dualistic one, involving stimulus and sensation. He imported this dualism from Müller's physiology, which differentiated between the living organism and the inorganic material surrounding and interacting with that organism.

---

59. See the analysis in Kremer, "Innovation through Synthesis."
60. "Ueber die Natur der menschlichen Sinnesempfindungen," *Königsberger naturwissenschaftliche Unterhaltungen* 3(1852):1-20, in *WA* 2:591-609, on 608.

By contrast, Helmholtz's position in his 1855 Kant lecture, "Ueber das Sehen des Menschen," was quite different.[61] In the first part of his lecture he compared the eye to a camera obscura by describing the eye's purely optical properties. He called this the physical part of vision. He then discussed the fact that, as a result of physiological processes in the retina, light excites the sensation of light. Here, too, he emphasized Müller's law of specific sense energies, whereby the qualities of the sensations do not depend on the external stimulus but rather on the particular sensory nerve excited. He concluded that human perception depends as much on the characteristics of the senses as on the properties of external objects.[62]

Nonetheless, Helmholtz recognized that to experience a sensation of light is not the same as what we call vision. To perceive, he argued, means to have a conception—Helmholtz used the word "*Vorstellung*"—of an external object. In the transition from sensations to perceptions, sensory illusions may come into play. Helmholtz gave a detailed discussion of different types of sensory illusion and he drew several conclusions, one of which was that perception is subject to learning and practice.[63] Moreover, the fact that we do not normally notice the blind spot in the eye shows, he argued, "that in the unprejudiced use of our senses we take into account only the sensations that give us information about the external world."[64] In short, he saw the sensory processes as highly utilitarian.

The transition from a set of sensations to a perception of an external object forms the third part of the sensory process, one which is neither physical nor physiological but rather, Helmholtz maintained, psychological. Helmholtz gave an initial characterization of these psychological processes by emphasizing the importance of learning in perception. His epistemology was an empiricist epistemology. At the same time, however, he left unanswered the question as to the extent to which these processes were explicable by reference to learning alone and the extent to which "innate ideas" depending on the structural properties of the human organism had to be invoked.[65] Finally, he argued that under normal conditions only the results of the entire process of perception, and not any of the intermediate stages, come into consciousness.[66]

61. *Ueber das Sehen des Menschen* (Leipzig: L. Voss, 1855), and in VR^5 *1*:85–117.
62. *Ueber das Sehen*, 379.
63. Ibid., 386.
64. Ibid., 388.
65. Ibid., 395. On the nativist-empiricist debate see Turner, "Consensus and Controversy."
66. *Ueber das Sehen*, 394.

From this brief characterization of Helmholtz's epistemology it is evident that it had evolved quite substantially in the years between 1852 and 1855. The model of the sensory process postulated in 1852 was a dualistic one; the fundamental feature of the 1855 formulation, by contrast, was the tripartite division of the sensory process into physical, physiological, and psychological parts. This evolution was further reflected in Helmholtz's terminology: in 1852 he spoke only of sensations; in 1855, by contrast, he differentiated between sensations, which he associated with physiological processes, and perception, which was the result of psychological processes.

What tempted Helmholtz to develop this epistemology? In 1853 he had agreed to contribute to Gustav Karsten's gargantuan project to publish an *Allgemeine Encyclopädie der Physik* by writing a handbook on physiological optics. For that handbook Helmholtz had to develop a framework in which he could organize an immense amount of largely unrelated optical phenomena. Such a framework was, in effect, a response to the problem of determining the fundamental characteristics in the process of vision. Helmholtz formulated an epistemology because he required a methodological framework by which he could organize his planned research on physiological optics; and, indeed, the three parts of his *Handbuch der physiologischen Optik*, published in 1856, 1860, and 1867, corresponded exactly to the physical, physiological, and psychological parts of his analysis of vision. That the 1852 model of the sensory process was insufficient became obvious as soon as Helmholtz realized the importance of sensory illusions, for instance in his refutation of Brewster's theory of light. Although in 1852 he still lacked a clear concept of the psychological processes involved, he did use contrast phenomena as a strong argument against Brewster's theory. This was the seed of his elaborate discussion of optical illusions in his 1855 paper.[67]

Yet his work in physiological optics was so time-consuming and his progress so slow, that his new epistemology could not immediately yield a payoff for physiological optics. In 1854, when he started to work on the *Handbuch*, he had hoped to finish it within a year or two;[68] in the event, he required another thirteen years to do so. By then he had long completed his proposed reformation of physiological acoustics by publishing his research results in the *Tonempfindungen*. While Helmholtz clearly formulated his empiricist epistemology in connection with his work in physiological optics, it is equally clear

---

67. For an analysis of this point see Kremer, "Innovation through Synthesis."
68. Helmholtz to Adolph Fick, 4 September 1854, cited in Koenigsberger *1*:266.

that his evolving epistemological stance was also tied to his ongoing research in physiological acoustics.

For example, in the introduction to the *Tonempfindungen* Helmholtz presented his epistemological position in its mature form. (Indeed, this is probably the clearest, most systematic exposition of Helmholtz's theory of perception.) First, he distinguished between physiological and physical acoustics. The latter, he argued, is nothing but a part of the theory of the motion of elastic bodies, just as Chladni had envisaged. As for physiological acoustics, which is the study of the processes that take place within the ear, it can be divided into three parts: first, a physical part, "which treats of the conduction of the motions to which sound is due, from the entrance of the external ear to the expansion of the nerves in the labyrinth of the inner ear"; second, a physiological part, which deals "with the various modes in which the nerves themselves are excited, giving rise to their various sensations"; and third, a psychological part, which seeks "the laws according to which these sensations result in mental images of determinate external objects, that is, in perceptions."[69] Helmholtz explicitly presented this tripartite division of the sensory process to the readers of his *Tonempfindungen* as the epistemological basis of his theories in physiological acoustics.

The relationship between epistemology and acoustical research became evident not only in the *Tonempfindungen*. For all of Helmholtz's physiological acoustics was based on the tripartite division of the sensory process.[70] When he wrote his extended paper of 1856 on combination tones it was not only to report new observations, refined experimental techniques, and a new theory; it was also to demonstrate how his empiricist epistemology served as the methodological framework for studying the problem of sound perception. This new epistemology helped him to surmount the impasse created by the Ohm-Seebeck dispute. He wrote of Seebeck's arguments against Ohm's definition of tone: "I believe ... that the difficulties he found will disappear as soon as one distinguishes here exactly between the auditory nerve's sensory reception and the psychological processes."[71] To help fix this division Helmholtz distinguished the (German) term "*Ton*" (tone), whereby he meant a simple tone corresponding to a sinusoidal

---

69. *Sensations*, 6.
70. On the importance of the tripartite division for Helmholtz's research in physiological acoustics, see also Gary Hatfield, "Helmholtz and the Physiological Foundations of Music Perception," 16th International Congress of the History of Science (Bucharest: Publishing House of the Socialist Republic of Romania, 1981), vol. 4, supplement, pp. 299–307.
71. "Ueber Combinationstöne," 525.

vibration, from the term "*Klang*" (sound), whereby he meant the compound of a fundamental tone with higher harmonics corresponding to a complex vibration which is the sum of several sinusoidal vibrations. Helmholtz's definition of tone was thus the same as Ohm's. By now equating Seebeck's definition with his own definition of sound, he resolved the dispute. What formerly appeared as conflicting positions became, in Helmholtz's framework, both possible and, in fact, necessary stages in the process of tone perception.

In his long paper of 1856 on combination tones, to cite a final example, Helmholtz devoted nearly one-third of that paper to presenting the elements of his empiricist epistemology and to discussing the consequences they had for the formulation of physiological acoustics. Thus, his paper was, in effect, a first test of the potential of this epistemology as a methodological framework for conducting research on perception theory. To deal with the psychological processes operating in sound perception, Helmholtz formulated a set of four statements: First, an apparently simple perception is in fact composed of a multitude of sensations which, however, can be isolated and recovered by using appropriate means. For instance, it is possible to hear the partials of a complex sound by guiding the listener's attention to them or by using resonators. The fact that we hear the partials proves that compounding them into a seemingly simple sensation is "not accomplished by the activity of the nerve but rather only by psychological processes."[72] Second, a set of rules can be given by which the multitude of sensations are compounded into a single perception; Helmholtz called these rules "unconscious inferences." For each of the three characteristics of a tone (pitch, loudness, and tone quality), he formulated a corresponding rule. The pitch of the complex sound is equal to the pitch of the fundamental.[73] The overall loudness is equal to the sum of the loudness of the partials.[74] As for tone quality, he wrote: "According to Ohm's theory we must suppose that differences in timbre of those tones ... originate in the different loudness of their partials."[75] Helmholtz here adumbrated the general strategy he was to employ in subsequent years in order to solve the problem of tone quality. Third, learning is an important element of perception. For instance, the process of recognizing the compound sounds of musical instruments and of the human voice is a learned process.[76] Fourth and

72. Ibid., 527.
73. Ibid., 525.
74. Ibid., 526.
75. Ibid., 525.
76. Ibid., 526. Helmholtz expanded on the role of learning in tone perception in *Sensations*, 101–12.

finally, from the multitude of incoming sensations, an extract or compound is formed which suffices to recognize the external object. The transition from sensations to perceptions normally results in a loss of information on the conscious level. However, this matters little because the lost information gives no advantage towards recognizing the external object. As Helmholtz wrote: "In the normal use of our senses we consider the reception of sensations only insofar as they help us to recognize the objects and events of the external world, and we disregard everything which is unnecessary for doing so. . . ."[77]

In his 1855 Kant speech, Helmholtz had discussed the importance of learning for the correct interpretation of our sensations and the utilitarian character of unconscious inferences. The application of his empiricist epistemology to the case of sound perception resulted in a further development and refinement. In his paper on combination tones of 1856 he introduced still more details of those psychological processes by formulating explicit rules for the transformation of the multitude of simple sensations into a complex perception. Indeed, it seems as if Helmholtz found it easier to apply his ideas on perception to physiological acoustics than to physiological optics. After the first Part of the *Handbuch der physiologischen Optik* was completed in 1856, Helmholtz concentrated his attention on physiological acoustics and he successfully applied his empiricist epistemology to the problem of sound perception. The culminating result, his *Die Lehre von den Tonempfindungen als physiologische Grundlage für die Theorie der Musik*, not only complemented the *Handbuch*; like the latter, the *Tonempfindungen* embodied an intellectually complex and stunning array of philosophical views, methodological approaches, theoretical claims, and empirical results.

---

77. "Ueber Combinationstöne," 526.

Part Two

Physicist

7

Helmholtz's *Ueber die Erhaltung der Kraft*

The Emergence of a Theoretical Physicist

Fabio Bevilacqua

1. Introduction

During the 1840s and early 1850s numerous formulations of the perennial attempt to show the "conversion" among natural phenomena and the "conservation" of something underlying them appeared. Despite a certain persistence of the term "force" (*"Kraft"*) among a few German physicists, by the 1860s the term "energy" was generally adopted, although it did not assume an unequivocal meaning. Before the century closed, several histories of energy conservation were written which raised controversies about priority and about the scientific and philosophical meaning of various formulations of the principle of conservation of energy.[1]

Acknowledgments: Financial support for this research has been provided by the Consiglio Nazionale delle Ricerche. Earlier versions of this paper were presented in Urbino (1989) and Munich (1990), and I thank Enrico Giannetto, Ivor Grattan-Guinness, Rod Home, my wife Leitha Martin, and Stefan Wolff for their comments on the earlier versions. I should also like to thank the editor, David Cahan, for his continuous encouragement and helpful suggestions.

1. See esp. James Clerk Maxwell, *The Theory of Heat* (London: Longmans, Green, 1871); Ernst Mach, *Die Geschichte und die Würzel des Satzes von der Erhaltung der Arbeit* (Prague: Calve, 1872); Balfour Stewart, *The Conservation of Energy*, 2nd ed. (London: H.S. King, 1874); James Clerk Maxwell, *Matter and Motion* (London: Society for Promoting Christian Knowledge, 1876); Moritz Rühlmann, *Vorträge über Geschichte der technischen Mechanik und der theoretischen Maschinenlehre*, 2 vols. (Leipzig: Baumgärtner, 1881–85); J.B. Stallo, *The Concepts and Theories of Modern Physics* (London: Kegan, Paul, Trench, 1882); Ernst Mach, *Die Mechanik in ihrer Entwickelung historisch-kritisch dargestellt* (Leipzig: F.A. Brockhaus, 1883); Max Planck, *Das Princip der Erhaltung der Energie* (Leipzig: B.G. Teubner, 1887); Georg Helm, *Die Lehre von der*

Among modern scholarship the most influential interpretation by far of the appearance of energy conservation between 1830 and 1850 is Thomas S. Kuhn's classic essay, "Energy Conservation as an Example of Simultaneous Discovery," which in turn has stimulated a number of thoughtful alternative approaches to the problem.[2] Kuhn's principal interest was not, however, to write another history of the emergence of the principle of conservation of energy; instead, he sought to identify "the sources of the phenomenon called simultaneous discovery." He argued that between 1832 and 1854 twelve scientists—above all, Julius Robert Mayer, James Prescott Joule, Ludwig Colding, and Hermann von Helmholtz—"grasped for themselves" the essential "elements" of the concept of energy and its conservation, and he asked why these "elements" became accessible at that time, thereby seeking to identify not the innumerable "prerequisites" of the principle of energy conservation but rather only what he called the "trigger factors." In particular, he identified three such factors: the "concern with engines," the "availability of conversion processes," and the "philosophy of nature."[3] For most modern scholars, not the least result of Kuhn's

---

Energie, historisch-kritisch entwickelt (Leipzig: A. Felix, 1887); Ernst Mach, *Die Principien der Wärmelehre, historisch-kritisch entwickelt* (Leipzig: J.A. Barth, 1896); Georg Helm, *Die Energetik nach ihrer geschichtlichen Entwickelung* (Leipzig: Veit, 1898); Henri Poincaré, *La science et l'hypothèse* (Paris: E. Flammarion, 1902); Arthur Erich Haas, *Die Entwicklungsgeschichte des Satzes von der Erhaltung der Kraft* (Vienna: Alfred Hölder, 1909); Wilhelm Ostwald, *Die Energie* (Leipzig: J.A. Barth, 1908); Emile Meyerson, *Identité et réalité* (Paris: F. Alcan, 1908); and Ernst Cassirer, *Substanzbegriff und Funktionsbegriff* (Berlin: Bruno Cassirer, 1910).

2. Thomas Kuhn, "Energy Conservation as an Example of Simultaneous Discovery," in Marshall Clagett, ed., *Critical Problems in the History of Science* (Madison, Wis.: University of Wisconsin Press, 1959), 321–56, reprinted in Kuhn's *The Essential Tension: Selected Studies in Scientific Tradition and Change* (Chicago and London: University of Chicago Press, 1977), 66–104. Other studies on Helmholtz and the conservation of energy include: Yehuda Elkana, "The Conservation of Energy: A Case of Simultaneous Discovery?," *Archives internationales d'histoire des sciences 23* (1970):31–60; idem, "Helmholtz' 'Kraft': An Illustration of Concepts in Flux," *HSPS 2* (1970):263–99; idem, *The Discovery of the Conservation of Energy* (Cambridge, Mass.: Harvard University Press, 1974); Peter Heimann, "Conversion of Forces and the Conservation of Energy," *Centaurus 18* (1974):147–61; P.M. Heimann, "Helmholtz and Kant: The Metaphysical Foundations of *Über die Erhaltung der Kraft*," *SHPS 5* (1974):205–38; Peter Clark, "Elkana on Helmholtz and the Conservation of Energy," *The British Journal for the Philosophy of Science 27* (1976):165–76; Geoffrey Cantor, "William Robert Grove, the Correlation of Forces and the Conservation of Energy," *Centaurus 19* (1976):273–90; P.M. Harman, "Helmholtz: The Principle of the Conservation of Energy," in his *Metaphysics and Natural Philosophy: The Problem of Substance in Classical Physics* (Brighton: Harvester Press, 1982), 105–26; Fabio Bevilacqua, *The Principle of Conservation of Energy and the History of Classical Electromagnetic Theory* (Pavia: La Goliardica Pavese, 1983); and Stephen M. Winters, "Hermann von Helmholtz's Discovery of Force Conservation," Ph.D. dissertation, The Johns Hopkins University, 1985.

3. Kuhn, "Energy Conservation," 66–70, 73.

essay has been to see Helmholtz's essay of 1847, *Ueber die Erhaltung der Kraft. Eine physikalische Abhandlung*, within Kuhn's analytical framework and according his terms.⁴

Kuhn's general analysis of the "trigger factors" contains a number of problems, however. His first factor, the "concern with engines," led him to focus on the concept of work as it related to several traditions, above all that of an older engineering practice.⁵ Yet his assertion about the ineffectiveness of the well-known theoretical engineering tradition dating from Lazare Carnot is surprising and, as we shall see, unjustified.⁶ Moreover, Kuhn neglected the tradition of potential theory, which identified the concept of work with that of potential, and so opened the way for a mathematical expression of "energy" conservation.⁷ Thus, if the concept of work is taken as loosely as Kuhn did, it can be derived from figures ranging from Hero of Alexandria to Leibniz, rather than from the engineering tradition of the eighteenth century, which would then make it more a "prerequisite" than a "trigger factor."⁸ Instead, a more pertinent influence was that derived from

---

4. *Ueber die Erhaltung der Kraft. Eine physikalische Abhandlung* (Berlin: G. Reimer, 1847), in *WA 1*:12–75 (including an appendix [68–75] from 1881). For a facsimile edition see *Über die Erhaltung der Kraft*, transcribed by Christa Kirsten, 2 vols. (Berlin: Akademie Verlag, 1982).

5. Kuhn, "Energy Conservation," 83–4, 86–8, 90.

6. The French theoretical engineering tradition, which Kuhn mentioned but dismissed, has long (and rightly) received special attention from a number of scientists, historians, and philosophers: Rühlmann, *Maschinenlehre*; Helm, *Die Energetik*, 14–5; Haas, *Entwicklungsgeschichte*; Edmund Hoppe, *Histoire de la physique*, trans. Henri Besson (Paris: Payot, 1928), 96, and cf. Kuhn, "Energy Conservation," 86, n.44; Felix Auerbach, "Feld, Potential, Arbeit, Energie und Entropie," in "Grundbegriffe," in *Handbuch der Physik*, ed. Adolph A. Winkelmann, 2nd ed., 6 vols. in 7 (Leipzig: Barth, 1908) *1*:68–91; Ernst Cassirer, *The Problem of Knowledge*, trans. William Woglom and Charles Hendel (New Haven, Conn.: Yale University Press, 1950), 49–50, 72; and John Theodore Merz, *A History of European Thought*, 4 vols. (reprint New York: Dover, 1965) *2*:101. See also Ivor Grattan-Guinness, "Work for the Workers: Advances in Engineering Mechanics and Instruction in France, 1800–1830," *Annals of Science 41* (1984):1–33.

7. M. Norton Wise, "William Thomson's Mathematical Route to Energy Conservation: A Case Study of the Role of Mathematics in Concept Formation," *HSPS 10* (1979):49–83, on 59, notes the lack of mathematical factors in Kuhn's analysis; M. Norton Wise and Crosbie Smith, "Measurement, Work, and Industry in Lord Kelvin's Britain," *HSPS 17* (1986):147–73, on 154, expresses a different view.

8. Erwin Hiebert, "Commentary on the Papers of Thomas S. Kuhn and I. Bernard Cohen," in Marshall Clagett, ed., *Critical Problems in the History of Science* (Madison, Wis.: University of Wisconsin Press, 1959), 391–400, on 394, noted that the concept of work was initially meant to explain the five simple machines in Hero of Alexandria's *Mechanica*. According to Cassirer, it was Leibniz who first connected the concept of work with "energy" conservation; see his *Leibniz' System in seinen wissenschaftlichen Grundlagen* (Marburg: Elwert, 1902); *Substanzbegriff*, 226–48; and *Das Erkenntnisproblem in der Philosophie und Wissenschaft der neueren Zeit*, 4 vols. (vols. 1–3: Berlin:

the technical concept of work linked closely to the emergence of potential theory and *vis viva* conservation.

Kuhn's second factor, the "availability of conversion processes," led Peter Heimann to note that the emphasis on the interconversion of the forces of nature was not specific to the 1830s and 1840s, and thus to argue that it cannot be considered a "trigger factor."[9] Much the same point is essentially true for Kuhn's third "trigger," that of "the philosophy of nature," in particular *Naturphilosophie*. From the Greek atomists onward, numerous metaphysical thinkers posited the unity, uniformity, and homogeneity of natural phenomena, and so contributed to the rise of energy conservation.[10] *Naturphilosophie* has no privileged role in this hoary tradition. To be sure, Kuhn sought to show the relevance of *Naturphilosophie* for Helmholtz's work by referring to a controversial remark of 1882, wherein Helmholtz acknowledged the Kantian influences in his 1847 essay.[11] Yet to note the Kantian roots of the *Naturphilosophen* hardly means that Kant was a *Naturphilosoph*.[12] The methodological role of Kantian, as well as Leibnizian, elements in Helmholtz's 1847 essay should not be confused with ontological commitments typical of *Naturphilosophie*. In sum, Kuhn's three factors and his dismissal of various prerequisites—above all, the dynamical theory of heat and the impossibility of perpetual motion[13]—do not account for the appearance of Helmholtz's *Erhaltung* or the other versions of "energy" conservation.

Moreover, with respect to Helmholtz himself, Kuhn makes a number of historical claims that, as this essay argues, are doubtful: that "Helmholtz fails to notice that body heat may be expended in mechanical work"; that the concept of work from the old mechanical engineering tradition was "all that which is required" and "the most decisive contribution to energy conservation made by the nineteenth-century concern with engines"; that "Helmholtz was not, however,

---

Bruno Cassirer, 1906–20; vol. 4 [in English translation]: 1950, [in German]: Stuttgart: Kohlhammer, 1957; reprinted Darmstadt: Wissenschaftliche Buchgesellschaft, 1973–74), *2*.

9. Heimann, "Conversion of Forces," 147, 159.
10. Haas, *Entwicklungsgeschichte*, esp. 31–45.
11. "Ueber die Erhaltung der Kraft," *WA* 1:68.
12. L. Pearce Williams, "Kant, *Naturphilosophie*, and Scientific Method," in *Foundations of Scientific Method: The Nineteenth Century*, eds. Ronald Giere and Richard Westfall (Bloomington, Ind., and London: Indiana University Press, 1973), 3–22.
13. Kuhn, "Energy Conservation," 101–3. On the role of the dynamical theory in the work of four British scientists see Crosbie Smith, "A New Chart for British Natural Philosophy: The Development of Energy Physics in the Nineteenth Century," *History of Science 16* (1978):231–79; and Donald Moyer, "Energy, Dynamics, Hidden Machinery: Rankine, Thomson and Tait, Maxwell," *SHPS 8* (1977):251–68.

aware of the French theoretical engineering tradition"; that Helmholtz "fails completely to identify $\int \mathbf{p}d\mathbf{p}$ as work or *Arbeitskraft* and instead calls it the 'sum of the tensions' "; that "the dominance of contact theory in Germany" might "account for the rather surprising way in which both Mayer and Helmholtz neglect the battery in their accounts of energy transformations"; and that "Helmholtz was able by 1881 to recognize important Kantian residues" in the *Erhaltung* "that had escaped his earlier censorship" and that this is evidence of the influence of *Naturphilosophie*.[14]

The concerns raised about Kuhn's study suggest that a reassessment of Helmholtz's contributions to energy conservation is in order, and this essay seeks to do so by providing an *explication de text* of Helmholtz's *Ueber die Erhaltung der Kraft*. In so doing, it argues that Helmholtz's work is best seen as the first and primary event in his emergence as a theoretical physicist. The *Erhaltung*, as this essay shows, presented no experimental or mathematical achievements by Helmholtz. Rather, Helmholtz used it to outline an explicit and sophisticated theory and methodology. His theoretical effort resulted in a formulation of the principle of force conservation based on the impossibility of perpetual motion and on the Newtonian model of forces depending only on positions. This allowed him to define two sharply distinguished and main forms of "energy": potential and kinetic. Moreover, his methodology distinguished clearly between theoretical and experimental physics and did so by presenting four hierarchical, interacting levels that articulated and demonstrated force conservation: the physical hypotheses of the impossibility of perpetual motion and of central Newtonian forces (level one); the principle of conservation of force (level two); and various specific empirical laws (level three) and natural phenomena (level four) pertinent to force conservation. With reference to these four levels, Helmholtz drew an explicit distinction between theoretical physics (dealing with deductions from level two to three, that is, with the applications of the principle to empirical laws) and experimental physics (dealing with the inductions from level four to three, that is, from natural phenomena to empirical laws). It is important to note, furthermore, that Helmholtz did not at all identify the use of mathematics as the dividing line between theoretical and experimental physics. Section I of the *Erhaltung* sought to show the equivalence of the two main hypotheses (level one), while Section II sought to deduce from them Helmholtz's version of the principle of

14. Kuhn, "Energy Conservation," 95 (n.68), 84, 90, 88, 73, 100–1, resp.

conservation (level two). The remaining four sections (III-VI) dealt with the interactions between levels two and three, that is, with the application of the principle to the existing empirical laws and with an attempt to derive new empirical laws on theoretical grounds. The *Erhaltung* offered no new experimental data (level four), and it criticized the few available data regarding the mechanical equivalent of heat, not least, as this essay seeks to show, on the basis of a technical mistake in converting units of measurement.

Moreover, the essay argues, *pace* Kuhn, for a series of specific points that have been previously misunderstood or neglected: that prior to 1847 Helmholtz was indeed aware of the interconvertibility of heat and work, although he did lack a numerical equivalent; that the concept of work only became an important contribution to force conservation after it acquired the characteristics of a total differential; that Helmholtz was well aware of the French engineering tradition and that he utilized (without rederiving) the new expression for *vis viva*; that he consciously dropped the new interpretation of "*Arbeit*" in favor of "*Spannkraft*"; that he showed that contact theory was not opposed to force conservation and that, indeed, he dedicated the longest section of the *Erhaltung* to an analysis of batteries; and, finally, that he did not censor but rather reinstated the *Erhaltung*'s philosophical introduction before publication and that Kantian transcendentalism, not *Naturphilosophie*, played the main philosophical role. This essay pays particular attention to the *Erhaltung*'s four-level methodological structure; to the demarcation between theoretical and experimental physics; to the overcoming of both the engineering and the mathematical approaches to the concept of work; to the lack of an experimental determination of a work equivalent of heat and the mistranslation of Joule's values; to the difficult (and at points simply wrong) theory-experiment interplay in applying the principle; and, finally, to the formulation of a lasting methodology and of a subsequently discarded conceptual model of energy.

Much of the immediate background to Helmholtz's essay of 1847 lay in his own experimental work in heat physiology between 1843 and 1846 and in his critical knowledge of the work of his predecessors in this area. Section 2 of this essay briefly treats this topic.[15] Section 3 presents the *Erhaltung*'s methodological structure and its conceptual foundations. Sections 4 and 5 describe and analyze the *Erhaltung*'s concrete results, with Section 4 elaborating the conceptual foundations

---

15. For a detailed analysis of this topic see Kathryn M. Olesko and Frederic L. Holmes's essay, "Experiment, Quantification, and Discovery: Helmholtz's Early Physiological Researches, 1843–50," in this volume.

into the principle of the conservation of force and Section 5 applying the principle to a wide variety of physical topics.

2. On the Borderline between Physiology and Physics: The "Bericht" of 1846

Helmholtz had studied physiology with Johannes Müller in Berlin, where he also formed close and long-lasting friendships with some of Müller's best pupils, above all Emil du Bois-Reymond, Ernst Brücke, and Carl Ludwig. In 1845, these last three, along with several of Gustav Magnus's students (namely, Gustav Karsten, Wilhelm Beetz, Wilhelm Heintz, and Hermann Knoblauch), founded the Berlin Physikalische Gesellschaft. Two years later, the Gesellschaft issued its first installment of the *Fortschritte der Physik*, which included the first of Helmholtz's many "Berichte," or reports reviewing recent research. Physiology was the research context in which Helmholtz published his two papers on force conservation in 1847.[16] So, too, was it the context in which he published a series of five other papers between 1843 and 1848.[17] Although these first works were written in a physiological context,[18] Helmholtz declared in 1882 and again in 1891 that his interest in force conservation did *not* arise from empirical problems in physiology but rather from his teenage inclination to favor the principle of the impossibility of perpetual motion.[19] Moreover, it is well to recall

---

16. "Bericht über 'die Theorie der physiologischen Wärmeerscheinungen' betreffende Arbeiten aus dem Jahre 1845," *Fortschritte der Physik im Jahre 1845 1* (1847):346–55, in *WA 1*:3–11; and "Ueber die Erhaltung der Kraft." See also "Erwiderung auf die Bemerkungen von Hrn. Clausius," *AP 91* (1854):241–60, in *WA 1*:76–93. Unless otherwise noted, all references to the "Bericht" are to the article above.

17. "Ueber das Wesen der Fäulniss und Gährung," *MA* (1843):453–62, in *WA 2*:726–34; "Ueber den Stoffverbrauch bei der Muskelaction," *MA* (1845):72–83, in *WA 2*:735–44; "Wärme, physiologisch," in *Encyklopädisches Handwörterbuch der medicinischen Wissenschaften*, ed. Professoren der medicinischen Facultät zu Berlin, vol.35 (Berlin: Veit, 1846), 523–67, in *WA 2*:680–725; "Ueber die Wärmeentwicklung bei der Muskelaction," *MA* (1848):147–64, in *WA 2*:745–63; and "Bericht über die Theorie der physiologischen Wärmeerscheinungen betreffende Arbeiten aus dem Jahre 1846," *Fortschritte der Physik im Jahre 1846 2* (1848):259–60.

18. See Olesko and Holmes, "Experiment, Quantification, and Discovery." Whether physiological research, apart from constituting the context, should also be seen as the root or one of the roots of Helmholtz's formulation of force conservation, remains uncertain. Richard L. Kremer, "The Thermodynamics of Life and Experimental Physiology, 1770–1880" (Ph.D. diss., Harvard University, 1984), 190–93, perceptively contrasts the "standard" view that research in energy conservation was motivated by physiological problems.

19. "Ueber die Erhaltung der Kraft," *WA 1*:74, and "Erinnerungen," in *VR⁴ 1*:1–21, or 7–11. Leo Koenigsberger accepted this approach (Koenigsberger *1*:1–21, or 12,

that his professional context was that of an army surgeon in Potsdam.[20]

Physiology offered the battleground for the fight over explaining animal heat in terms of the principle of the impossibility of perpetual motion and the consequent refusal to admit vital forces. The acceptance or rejection of vital forces in explaining the origins of animal heat constituted one of the fundamental problems of mid-century German physiology.[21] One of the principals in this fight was Justus von Liebig, who in 1841 had asserted a principle of correlation of forces, that is, of the conversion of forces with constant coefficients—"No force can be generated from nothing...," he averred[22]—and who in 1842 had rejected the idea that vital forces could generate animal heat.[23] Although Helmholtz eventually rejected Liebig's approach to the problem, he was initially much influenced by it.[24] In his first two publications he explicitly followed Liebig by assuming a common origin for mechanical forces and the heat produced by an organism.[25] He asked whether this origin might not be entirely attributed to metabolism, thereby obviating vital forces. More to the point, his discussion in 1846 of the origins of animal heat reveals two elements of his methodological strategy in the *Erhaltung*: First, while he accepted Liebig's

---

50–2); by contrast, Kremer ("Thermodynamics of Life," 237–38) denies the relevance of vitalism in physiological debates and dismisses Helmholtz's autobiographical claims on the role of the principle of the impossibility of perpetual motion.

20. See Arleen Tuchman's essay, "Helmholtz and the German Medical Community," in this volume.

21. Timothy Lenoir, *The Strategy of Life: Teleology and Mechanics in Nineteenth-Century German Biology* (Dordrecht and Boston: D. Reidel, 1982), 195–96, 215–17, 230. For a different interpretation see Kremer, "Thermodynamics of Life," 237–38; and Olesko and Holmes "Experiment, Quantification, and Discovery."

22. Justus Liebig, *Chemische Briefe*, 3rd ed. (Heidelberg: C.F. Winter, 1851), twelfth letter, pp. 116–18, quoted in Helm, *Die Energetik*, 10; Haas, *Entwicklungsgeschichte*, 57; and Kuhn, "Energy Conservation," 95.

23. Kremer, "Thermodynamics of Life," 204–9.

24. Ibid., 238. Liebig has often been credited with being a pioneer in the history of energy conservation; see Planck, *Das Princip der Erhaltung der Energie*, 33; Helm, *Die Energetik*, 10; Haas, *Entwicklungsgeschichte*, 57; Kuhn, "Energy Conservation," 68; Lenoir, *Strategy of Life*, 196; and Kremer "Thermodynamics of Life," 198–215.

25. Koenigsberger and Kremer argue that Helmholtz's first paper, that on fermentation and putrefaction, aimed to support Liebig's antivitalist position but raised confusing conclusions for him (Koenigsberger *1*:53; Kremer, "Thermodynamics of Life," 239–40); by contrast, Lenoir and Yamaguchi argue that Liebig's position stimulated Helmholtz to further research (Lenoir, *Strategy of Life*, 197, and Chûhei Yamaguchi, "On the Formation of Helmholtz' View of Life Processes in His Studies of Fermentation and Muscle Action—in Relation to His Discovery of the Law of Conservation of Energy," *Historia Scientiarum 25* [1983]:29–37). In his second paper, that on metabolism during muscular activity, Helmholtz had inconclusive results about metabolism because he lacked an exact relationship between the muscular action and the heat developed. (Koenigsberger *1*:60; Kremer, "Thermodynamics of Life," 243; Lenoir, *Strategy of Life*, 202; and Olesko and Holmes, "Experiment, Quantification, and Discovery.")

principle of force correlation and his theory of the chemical origin of heat, he also declared that the appropriate conceptual model of heat and the definition of the heat equivalents utilized in the correlation required clarification before a satisfactory application of the principle could be achieved. Second, he saw that Liebig's theoretical determinations did not in fact agree with experimental results.[26]

Helmholtz argued that, assuming a conceptual model of heat as a substance (caloric), the conservation of matter assured that the amount of (latent) heat ingested equalled that emitted by living bodies, and that Hess's law, which claimed that in chemical transformations the order of the intermediate steps was irrelevant for the final emission of heat, gave further support. Although Helmholtz thought that the caloric model of heat was the more useful for refuting the idea of vital forces, he nonetheless used the idea of heat as motion since it was widely spreading throughout the sciences: the recent identification of thermal radiation with light and the generation of heat not ascribable to the liberation of latent heat (for instance, in electrical processes) forced him to accept the model of heat as motion. Yet this model had disadvantages for eliminating vital forces, for the total amount of heat was no longer, as in the caloric model, considered constant, and the production of heat by the action of forces (including, at least in principle, vitalistic ones) now became a possibility.[27] To deny any role to vital forces and to reassert the principle of the impossibility of perpetual motion demanded a solution to the second problem mentioned above: the redefinition of heat equivalents.

Liebig had predicted an amount of animal heat smaller than that measured by Pierre-Louis Dulong and César-Mansuète Despretz. The difference between prediction and measurement might have been explained as due to vital forces. Helmholtz sought to eliminate the discrepancy between theoretical prediction and experimental result so as to eliminate any potential role for vital forces. He attempted to do so by reformulating the terms on both sides of the equation relating the heat ingested and emitted by living bodies: on the one side, the heat of the ingested matter results not from the oxidation of the elements of the food but rather from that of the compounds themselves; on the other, the heat generated in animals occurs not only in the respiratory organs but also in the blood and tissues. Helmholtz's reformulation of this equation gave him a theory-based prediction that satisfied experimental results on the basis of heat as

---

26. "Wärme, physiologisch," *WA* 2:695–700.
27. Ibid., *WA* 2:700.

motion while excluding vital forces. Still, the experimental confirmations remained highly problematic.[28]

Moreover, Helmholtz discussed the mechanical equivalent of heat, even though a determination of that equivalent, and hence a component—the work done by animals—in the theoretical energy balance was lacking.[29] Already in his paper of 1845 on "Muskelaction" he had asserted that the problem was "whether or not the mechanical force and the heat produced in an organism could result entirely from its own metabolism,"[30] and in his paper of 1846 on "Wärme" he noted that one of the differences between the kinetic and the caloric theory of heat was "the determination of the equivalent of heat that can be produced through a given quantity of mechanical or electric force."[31] To be sure, he did not provide a mechanical equivalent of heat in these papers; yet that was due to his lack of reliable experimental data, not adequate conceptualization. Nor did he do so in the *Erhaltung*. There he explained that, given that the amount of work produced by the animals is small in comparison to the heat generated, work can be neglected and the problem of conservation of force in physiology reduced to the problem of whether the combustion and transport of food can produce the same amount of heat as produced by the animals. He added that the results of his own work in the "Wärme" and the "Bericht" of 1846, compared with the Dulong and Despretz's measurements, allowed a positive, if approximate answer.[32]

Helmholtz's first "Bericht," written in October 1846 and published in the *Fortschritte* in 1847, was a key step in his elaboration of a methodological strategy. The paper's outstanding feature was a methodological one: the extension of the correlation principle from physiology to various branches of physics and chemistry. Helmholtz explicitly asserted that heat cannot originate from nothing and he used Liebig's own words to state a principle of correlation of forces.[33] Yet he also criticized that principle's application to animal heat, since Liebig's solution, as already noted, did not correspond to Dulong and Despretz's calorimetric results: the direct measurement of the heat of

---

28. Kremer, "Thermodynamics of Life," 251.
29. "Wärme," *WA* 2:699–700. Kremer argues ("Thermodynamics of Life," 238) that in his physiological research Helmholtz never succeeded in bringing heat and work together operationally.
30. "Muskelaction," *WA* 2:735.
31. "Wärme," *WA* 2:699–700.
32. *Ueber die Erhaltung der Kraft*, 70, in *WA* 1:66. See also Olesko and Holmes, "Experiment, Quantification, and Discovery."
33. "Bericht über 'die Theorie der physiologischen Wärmeerscheinungen,'" in *WA* 1:4, 6.

combustion of the hydrogen and carbon of the food ingested ranged between seventy and ninety percent of the heat generated by the animals.

The use of the correlation principle was not limited to the problem of animal heat. The principle was based on the impossibility of perpetual motion, which, as Helmholtz said, was "logically... completely justified," and had already been used in the "mathematical theories" of Sadi Carnot and Emile Clapeyron (using caloric) and of Franz Neumann (using electrodynamic potential). At the same time, Helmholtz also noted that the principle had yet to receive either full expression or full experimental confirmation. He saw correlation as a more sophisticated expression of the principle of the impossibility of perpetual motion and he immediately used it to provide a series of energy balances based on the model of heat as motion. Using that model and the principle of the "constancy of force-equivalents," he argued that "mechanical, chemical, and electrical forces can always generate a determined equivalent of heat, however complicated the transition from one force to the other." He conceded that empirical evidence for this claim was greatly lacking, yet he thought it nonetheless useful to offer specific theoretical applications of the principle to the heat produced by mechanical, chemical, electrolytic, and electrostatic forces.[34] The case of animal heat now became only the last of five applications of a principle that was becoming increasingly general.

Yet a new difficulty soon appeared: the lack of a mechanical equivalent of heat meant that the most important balance, that between heat and work, could not be written; in fact, the values offered by the theories of Carnot and Clapeyron and of Karl Holtzmann were unacceptable since they were based on the caloric model and since they referred only to the propagation, not the production, of heat. Helmholtz continued to lack experimental data. In October 1846, he still did not know of either Mayer's or Joule's work: "there exist no experiments which can be taken into account for the mechanical forces," he reported.[35] Hence, all the other balances were written as equivalences based on heat units rather than on work units. In the "Bericht," heat, not work, was the unit of measurement common to all the natural phenomena considered. This is a non-trivial difference with the subsequent *Erhaltung*.

In analyzing chemical transformations, Helmholtz employed Hess's law of the constancy of heat production. For electrolytic currents, the

---

34. Ibid., in *WA* 1:6–7.
35. Ibid., in *WA* 1:7.

heat developed in the circuit was seen as equivalent to the electrochemical transformations in the galvanic chain (battery), independently of their order. The circuit's heat θ could be calculated using Ohm's and Lenz's laws (Joule was not mentioned): $θ = J^2Wt$, where J is the current intensity, W the total resistance, and t the time. On the other hand, Helmholtz knew from Faraday's electrolytic law that $θ = AC$, where A is the electrical "difference" of the metals involved and C the quantity of atoms "consumed," that is, that underwent a process of oxidation and reduction. According to the principle of equivalence, the heat produced in the circuit must be equivalent to that which could be produced through the electrochemical transformations in the cells. For static electricity, Helmholtz easily showed that the production of heat by electric discharge followed from Riess's principles; he thus established a balance between the resultant heat on the one side and the product of the quantity of electricity and electrical density (an extant Voltian term for what was later to be known as the tension or potential difference), on the other. Finally, in discussing animal heat Helmholtz identified the latent heat of chemical reactions with the thermal equivalent that could be produced in further reactions. The energy balance had to hold between the latent heat of the ingesta, on the one side, and the heat "provided by the animals" plus the latent heat of the egesta, on the other. He saw that the equivalent on the left side of the balance was no longer the "heat of combustion of carbon and hydrogen but instead that of the food."[36] He reformulated and modified respiratory theory, and so claimed to have eliminated vital forces while satisfying Dulong's and Desprez's experimental results. Liebig's difficulties, he believed, were overcome.

With the "Bericht" of 1846 Helmholtz had acquired a new methodology and was aware of its great generality. The "Bericht" stands on the borderline between physiology and force conservation. In certain fundamental respects Helmholtz's methodology here adumbrated that of the *Erhaltung*: he enunciated a principle, provided a conceptual model of the quantities involved, expressed an equation between the energy terms, and, finally, compared the equation with empirical laws. There is, however, one central difference: despite the application of the equivalence principle to an analysis of several physico-chemical laws, the "Bericht" is still largely dedicated to physiology. In the much longer *Erhaltung*, by contrast, physiology is confined to a few lines of the final section. In the "Bericht," moreover, the equivalence principle

36. Ibid., in *WA I*:8; and Planck, *Princip*, 34.

based on the impossibility of perpetual motion (and on the impossibility of destroying motion) is presented along with a model for many equivalents (the terms of the energy balance); yet the equivalence principle of the "Bericht" is quite different from the mechanical principle of the conservation of force expressed in the *Erhaltung*. Indeed, although Helmholtz did not know of Mayer's and Joule's work, his equivalence principle of 1846 is much closer to their ideas. For it only asserts the numerical equivalence of the effects involved and does not employ the assumption of central Newtonian forces or imply that every effect must have a mechanical interpretation in terms of potential and kinetic energy.[37] Finally, despite his acceptance of the mechanical theory, in the "Bericht" Helmholtz did not discuss the specific determinations of the mechanical equivalent of heat.

In 1847, while still writing the *Erhaltung*, Helmholtz wrote another paper ("Muskelaction") dedicated to physiological problems.[38] There he tried to link the problem of animal heat to that of the mechanical force produced by muscle action. Seeking to demonstrate that heat is produced in the muscle itself, he devised a very sensitive thermocouple which, when linked to an astatic galvanometer and a magnifying coil, could detect differences of temperature in the range of 1000th of a degree centigrade. His thorough experiments on frogs' legs showed that heat is generated directly in the muscle tissue, that its origins are due to chemical processes, and that heat production in the nerves is negligible. He had disposed of vital force on empirical grounds. The role of this experimental research on the sources of animal heat, done simultaneously and immediately after composing of the *Erhaltung*, was particularly germane for Helmholtz's understanding of force conservation: it was, in fact, the only experimental research in this field that he conducted. His understanding and evaluation of the mechanical equivalent of heat is almost certainly connected to this very research.[39] Indeed, as the next section argues, the entire *Erhaltung* only offered a theoretical reinterpretation of known results, but no new experiments. Physiology could not and did not provide a key to Helmholtz's conservation of force: it offered neither theoretical arguments nor experimental evidence for the establishment of the principle of

---

37. Hence one must disagree with Lenoir's claim (*Strategy of Life*, 211) that "the physiology of muscle action laid before Helmholtz all the elements of conservation of energy."
38. "Ueber die Wärmeentwicklung bei der Muskelaction." For a detailed analysis of this paper see Olesko and Holmes, "Experiment, Quantification, and Discovery."
39. Lenoir, *Strategy of Life*, 211; Kremer, "Thermodynamics of Life," 244; and Olesko and Holmes, "Experiment, Quantification, and Discovery."

force correlation. Instead, the rejection of vital force was based on the previously accepted notion of the impossibility of perpetual motion.

3. The *Erhaltung*: Methodological Structure and Conceptual Foundations

While experimenting on the heat produced during muscular action, Helmholtz also worked (October 1846–July 1847) on the *Erhaltung*. He presented the *Erhaltung*'s results to the Physikalische Gesellschaft (apparently with great success) on 23 July 1847. However, Magnus's and, above all, Johann Christian Poggendorff's judgments were less than warm, with the latter refusing to publish it in his *Annalen der Physik* because of its non-experimental character.[40] Helmholtz was forced to turn to a private publisher, Georg Reimer. The final product consisted of an Introduction ("Einleitung"), which is largely methodological and philosophical in character, and six individual sections, the first two of which—"Das Princip von der Erhaltung der lebendigen Kraft" ("The Principle of the Conservation of Living Force") and "Das Princip von der Erhaltung der Kraft" ("The Principle of the Conservation of Force")—are dedicated to formulating the principle, and the following four of which—mechanics, force equivalent of heat, force equivalent of electrostatics and galvanism, and force equivalent of magnetism and electromagnetism—are dedicated to the applications of the principle to their respective fields.

By February 1847, at least, Helmholtz had written a sketch of the *Erhaltung*'s Introduction, which he sent to du Bois-Reymond.[41] It caused him problems: before he presented the *Erhaltung* to the Physikalische Gesellschaft and before he sent it to Magnus, whom he hoped would help him see to its publication in Poggendorff's *Annalen*, he decided to drop the Introduction. Then, following Poggendorff's refusal to publish his essay, Helmholtz, at du Bois-Reymond's request, restored the Introduction, though he altered it in "certain parts" before sending it to Reimer for publication.[42] The alterations are probably what is now the Introduction's opening paragraph.

The Introduction succinctly summarized the *Erhaltung*'s plan. It reveals that the *Erhaltung*'s structure is based on four methodological

---

40. Koenigsberger *1*:68–72. Koenigsberger's claim (ibid., *1*:68) that in the first quarter of 1847 Helmholtz conducted extensive experimentation strictly related to the preparation of the *Erhaltung* is difficult to accept.
41. Ibid., *1*:68.
42. Ibid., *1*:72.

assumptions or considerations: the positing of two "physical assumptions" (the impossibility of perpetual motion and central Newtonian forces) and their equivalence (Section I of the essay); the derivation from these assumptions of a theoretical law, viz., the principle of the conservation of force (Section II); the comparison of this general principle with various empirical laws, and the connection of the principle and those laws to natural phenomena in various fields of physics (Sections III–VI).[43] Unlike most other researchers involved with conservation problems, Helmholtz not only proposed to offer a specific functional formulation of the quantities conserved and their interrelations; he also proposed to derive this "principle" from more general physical assumptions.

Helmholtz's outstanding methodological innovation in the *Erhaltung* was to compare not only empirical laws with natural phenomena but also with a general principle. It is not difficult to understand Magnus's and Poggendorff's concerns about Helmholtz's essay: without presenting any new experimental results, the young physiologist sought to combine two major physical assumptions to a series of empirical laws and phenomena stretching across the entire spectrum of physics. In so doing, he provided one of the first conscious criteria for demarcating theoretical and empirical physics: while the experimental physicist searches for empirical generalizations that might fit experimental data (for example, the laws of light refraction and reflection), the theoretical physicist searches for agreement between a principle and extant empirical laws (the principle's justificatory role) and for the discovery of new theoretical laws (heuristic role).[44] Helmholtz here explicitly posited a long-term task for theoretical physics: empirical laws must agree with principles as well as with experimental data.

The Introduction clearly and consciously shows the methodological control that Helmholtz had achieved over his own research. His methodology had a four-level structure. Two basic physical hypotheses (the impossibility of perpetual motion and central Newtonian forces) constitute the first level; the principle of the conservation of force the second; empirical laws the third; and natural phenomena the fourth. That Helmholtz felt the need to justify his own version of the conservation principle on higher grounds (on which, see below) clearly indicates his awareness of the possibility of alternative formulations of the principle itself. He wanted not only to express a principle, but also to establish a framework and a set of rules by which the principle

---

43. *Ueber die Erhaltung der Kraft*, 1, in *WA 1*:12.
44. *Ueber die Erhaltung der Kraft*, 2, in *WA 1*:13.

could be formulated and used. This distinguished his approach as a major step not only in his own emergence as a theoretical physicist but also in the emergence of theoretical physics as a subdiscipline; at the same time it showed that his version of the principle was not only the application of a (meta)physical assumption but also the application of a sophisticated methodology. The two physical hypotheses (first level) brought together, though not unproblematically, two different but well-known traditions in physics (the Newtonian and analytic mechanics),[45] and thus offered secure grounding for the whole enterprise. Moreover, Helmholtz sought to justify the first level on more abstract grounds: he connected the principle of the impossibility of perpetual motion with the principle of sufficient reason, a regulative condition for the intelligibility of nature, and gave a conceptual explanation of the model of central forces in the Kantian style. Finally, he hinted at a cause-effect relationship embedded in the actual formulation of the conservation principle (second level). This principle in turn required comparison with existing empirical laws (third level), sought to predict new ones and (eventually) to make natural phenomena intelligible (fourth level).

Helmholtz explicitly distinguished between theoretical physics, which dealt with the applications of the principle to empirical laws, that is, with deductions from level two to level three, and experimental physics, which dealt with the inductions from natural phenomena to empirical laws, that is, from level four to level three. The dividing line between theoretical and experimental physics was not the use of mathematics; theoretical physics was no more to be identified with mathematical physics than with experimental physics. Helmholtz's chief and most successful novelty was his stress on the interplay of the second and third levels: after 1847, physical laws (level three) had increasingly to satisfy not only experiments and natural phenomena (level four) but also theoretical principles (level two) as well. Helmholtz saw these principles, whose basic characteristic was to unify the different branches of physical knowledge, as heuristic tools for discovering empirical laws. He thereby helped the new subdiscipline of theoretical physics to emerge.

As noted above, Helmholtz also sought to justify his innovative methodological approach philosophically. The Introduction aimed to show the meaning of the two initial assumptions for the "final" and "true" goal of the physical sciences.[46] In particular, he believed that

45. Elkana, *Discovery of Conservation of Energy*, 49–51.
46. *Ueber die Erhaltung der Kraft*, 1–2, in *WA 1*:12.

physical science should proceed by searching for the "unknown causes" of phenomena and by seeking to understand these phenomena in terms of the law of causality.[47] It is here not certain what he meant by the law of causality ("Gesetz der Causalität"). It seems likely, however, that he was referring to a Kantian concept of cause, a transcendental condition for the interpretation of natural phenomena.[48] On the other hand, Helmholtz also introduced a different meaning of causality, a regulative one, where causality is a precondition for the possibility of scientific knowledge. In this interpretation, the scientist must assume that nature is intelligible, that "every transformation in nature must have a sufficient cause," as Helmholtz wrote.[49] A natural process is intelligible, he held, if it refers to ultimate causes, which act according to a constant law, and thus, if the external conditions are the same (*ceteris paribus*), produce the same effect. To be sure, Helmholtz also asserted, probably with the debate over vital forces in mind, that perhaps some natural processes are not actually intelligible.[50] Some phenomena may belong to a realm of spontaneity and freedom, though this cannot be decided conclusively. Be that as it may, the scientist must assume nature's intelligibility as the departure point for his investigations. Here his second basic physical assumption—the impossibility of perpetual motion—came into play: the impossibility of perpetually providing work without a corresponding compensation limits nature's spontaneity and freedom and offers a physical version of the principle of sufficient cause. The widespread use of the regulative "empirical" causality to unify natural phenomena will lead to the causal link that Helmholtz will establish in Section II between living and tension forces.

As many scholars have noted, the Introduction has a Kantian character.[51] However, the specifics are important, for different parts of Kant's work played different roles here. Both the regulative principle

---

47. Ibid., 2, in *WA 1*:13.
48. For an analysis of the conceptual explanations of "regulative-empirical" and "transcendental" causality in Kant see Gerd Buchdahl, *Kant and the Dynamics of Reason* (Oxford: Blackwell, 1992), passim.
49. *Ueber die Erhaltung der Kraft*, 2, in *WA 1*:13.
50. *Ueber die Erhaltung der Kraft*, 2, in *WA 1*:13.
51. In addition to Helmholtz himself in "Ueber die Erhaltung der Kraft," 68, see, for example, Elkana, "Helmholtz' 'Kraft' "; Heimann, "Helmholtz and Kant"; M. Norton Wise, "German Concepts of Force, Energy, and the Electromagnetic Ether: 1845–1880," in *Conceptions of Ether: Studies in the History of Ether Theories 1740–1900*, eds. G.N. Cantor and M.J.S. Hodge (Cambridge, London, New York: Cambridge University Press, 1981), 269–307; and S.P. Fullinwinder, "Hermann von Helmholtz: The Problem of Kantian Influence," *SHPS 21:1* (1990):41–55.

of empirical causality and the transcendental principle of causality that allow the possibility of scientific knowledge and the lawlikeness of nature have already been noted. In addition, Helmholtz was also preoccupied in the Introduction with the conceptual explanation of a specific physical model, one which tended to show the *possibility* of Newtonian forces and not, at this stage, their inductive validity.[52] The model in question is that of Newtonian central forces dependent on distance alone. Helmholtz presented a detailed conceptual explanation based on the mechanical categories of matter and force in order to show that Newtonian forces can be considered as the ultimate causes of natural phenomena. In so doing, he followed Kant's method as given in the *Metaphysische Anfangsgründe der Naturwissenschaft*. Helmholtz's well-known enunciation of the mechanical worldview is based on the assumption that both matter and force are abstractions and that the first cannot be considered more "real" than the second. He asserted that the problem of finding unchanging, fundamental causes can be interpreted as the problem of finding constant forces; causes and forces were identified with one another. One characteristic of the definition of force is that it is constant over time; bodies with constant forces acting on one another allow only spatial movements and, if the forces of extended bodies are decomposed into forces acting between material points, then the intensity of the forces depends only on the distances. Helmholtz saw this as a direct consequence of the principle of sufficient reason.[53] Thus, if a general application of the principle of "force" conservation allowed the reduction of all natural phenomena to the effects of attractive and repulsive forces whose intensities depended only on distance, then "empirical" causality would match "transcendental" causality and the goal of physical science would be reached: an "intelligible" nature would be "understood."[54] The utopian nature of this conceptual model of forces notwithstanding, it was by no means universally accepted: for example, Wilhelm Weber's electrodynamic law of 1846, based on the alternative assumption of forces dependent on distance, velocity, and acceleration was then gaining widespread recognition. Helmholtz recognized that his (that is, the Newtonian) model was not fully accepted. Moreover, later in his career he began to doubt

52. Heimann, "Helmholtz and Kant," 229.
53. *Ueber die Erhaltung der Kraft*, 5, in *WA* 1:15.
54. Heimann, contra Elkana, shows that the meaning of "Kraft" is always unequivocal in a given context; see Heimann, "Helmholtz and Kant," 207 (n. 10), 209 (n. 14). For a detailed analysis of the philosophical points raised here see Michael Heidelberger's essay, "Force, Law, and Experiment: The Evolution of Helmholtz's Philosophy of Science," in this volume.

or at least redefine certain Kantian categories, including those of causality, that he had announced in the Introduction.[55] By the early 1880s, he abandoned the interpretation of Newtonian forces as final causes, though he still adhered to the regulative and transcendental use of causality.

4. The Two Conceptual Foundations of the *vis viva* Principle and Their Supposed Equivalence

In Section I of the *Erhaltung* Helmholtz sought to demonstrate the equivalence of his two basic assumptions—the impossibility of perpetual motion and central Newtonian forces—by analyzing the *vis viva* principle. He began with a statement of the principle of the impossibility of perpetual motion, namely: "that it is impossible, through any combination of natural bodies, to continually produce a motive force from nothing."[56] He stated that Carnot and Clapeyron had deduced a number of laws from this principle and that his own aim was to introduce the principle into every branch of physics "in the same way" so as to show both its applicability to all instances where laws based on phenomena have already been established (its justificatory role) and its guiding (heuristic) role for future experimental work. The assertion "in the same way" referred to methodology; it did not indicate that the same expression of the principle (as in Carnot and Claypeyron, with their caloric model) was to be applied. Instead, Helmholtz reformulated the principle by utilizing the term "work" ("*Arbeit*") along with the mechanical terms of "force" and "velocity":

> the quantity of work obtained when a system of bodies moves from one position to another under the action of specific forces must be the same as that needed to return the system to the original position, independent of the way, the trajectory, or the velocity of the change.[57]

Hence the term "*Arbeit*" became a function of the system's state (position); it is a total differential: in a closed path work cannot be created or destroyed.

---

55. "Ueber die Erhaltung der Kraft," in *WA* 1:68–70.
56. *Ueber die Erhaltung der Kraft*, 7, in *WA* 1:17.
57. *Ueber die Erhaltung der Kraft*, 8, in *WA* 1:18.

310   Physicist

Helmholtz then equated this innovative concept of work as a function of position to another function of position: the *vis viva*. From Galileo's relation $\mathbf{v} = \sqrt{2\mathbf{gh}}$, where **v** is the final velocity acquired by a body of mass **m** during a fall from height **h** under the acceleration **g**, it follows that the work **mgh** equals the expression $\frac{1}{2}\mathbf{mv}^2$, which also is a function of position. Here Helmholtz again used the term "*Arbeit*" for work and, explicitly following the French engineering definition of "*travail*," in the work-*vis viva* equation gave priority to the concept of work. Indeed, he defined $\frac{1}{2}\mathbf{mv}^2$, not $\mathbf{mv}^2$, as the measure of *vis viva*. In this way, he wrote, it "becomes identical with the quantity of work."[58] He did this, he added, in order "to establish a better agreement with today's customary way of measuring the intensity of forces."[59]

Having equated work and *vis viva*, Helmholtz obtained the "mathematical expression" of the principle of the impossibility of perpetual motion, that is, the law of the conservation of *vis viva*:

> When any arbitrary number of movable point masses moves solely under the influence of forces which they exercise on one another or which are directed vis-à-vis fixed centers, then the sum of the living forces of all the point masses together is the same at every instant of time at which all the points are in the same relative positions with respect to one another and towards the fixed centers at hand, whatever their trajectories and their velocities during the time interval may have been.[60]

The specific meaning of "conservation" here is that the quantity conserved (*vis viva*) occurs at specific positions and not during the process, a definition that echoes Huygens's results for the compound pendulum and Lagrange's definition of *vis viva* conservation.

Helmholtz further sought to show that the principle held only if the forces can be decomposed into central forces of mass points. From

$$d(\mathbf{q}^2) = \frac{d(\mathbf{q}^2)}{d\mathbf{x}}d\mathbf{x} + \frac{d(\mathbf{q}^2)}{d\mathbf{y}}d\mathbf{y} + \frac{d(\mathbf{q}^2)}{d\mathbf{z}}d\mathbf{z},$$

---

58. *Ueber die Erhaltung der Kraft*, 9, in *WA 1*:18. For one indication of Helmholtz's theoretical as opposed to mathematical approach to the concept of work, compare his awareness of the role of that concept to the problems posed in Section V in adopting a definition of "potential in itself" as the equivalent of work.

59. *Ueber die Erhaltung der Kraft*, 9, in *WA 1*:18. Kuhn's suggestion notwithstanding ("Energy Conservation," 88), Helmholtz did not himself rederive the definition of *vis viva*.

60. *Ueber die Erhaltung der Kraft*, 9, in *WA 1*:19.

where **q** is the velocity of a mass point **m** moving under the forces exerted by a fixed system **A**, and where **x**, **y**, and **z** are the Cartesian coordinates, and from

$$d(\mathbf{q}^2) = \frac{2\mathbf{X}}{\mathbf{m}}d\mathbf{x} + \frac{2\mathbf{Y}}{\mathbf{m}}d\mathbf{y} + \frac{2\mathbf{Z}}{\mathbf{m}}d\mathbf{z},$$

where **X**, **Y**, and **Z** are the components of the acting forces and $d\mathbf{q} = \frac{\mathbf{X}}{\mathbf{m}}dt$, Helmholtz, by equating the corresponding components of the second members, incorrectly derived

$$\frac{d(\mathbf{q}^2)}{d\mathbf{x}} = \frac{2\mathbf{X}}{\mathbf{m}}, \frac{d(\mathbf{q}^2)}{d\mathbf{y}} = \frac{2\mathbf{Y}}{\mathbf{m}}, \text{ and } \frac{d(\mathbf{q}^2)}{d\mathbf{z}} = \frac{2\mathbf{Z}}{\mathbf{m}},$$

and from this that the force's magnitude and direction must be a function of the position of **m** and thus of its distance from an attracting point **a**.[61]

Three remarks concerning Helmholtz's synthesis in this section of many elements deriving from different traditions and his introduction of several novelties (sometimes only implicitly and sometimes unsuccessfully) merit attention. First, concerning the formulation of the impossibility of perpetual motion, Helmholtz introduced specific mechanical concepts (work, velocity, and force) that did not belong to the Carnot-Clapeyron expression. He attempted to frame the impossibility of perpetual motion in a mechanical worldview:[62] this was an (implicit) step in his methodological strategy that tended to show that his two initial assumptions belonged to the same conceptual scheme.

Second, Helmholtz reinterpreted the term work (*"Arbeit"*) as a total differential in the (new) expression for the impossibility of perpetual motion. He here united two different traditions in mechanics (analytical mechanics and mechanical engineering) along with the old philosophical principle of "nothing comes out of nothing and nothing is destroyed," which he had already partially used in the "Bericht" of 1846.[63] In the French tradition of mechanical engineering, which as we know was well known to Helmholtz, the term *"travail"* was of cardinal importance: the principle of conservation of *vis viva* became

---

61. *Ueber die Erhaltung der Kraft*, 11–2, in *WA* 1:20–1.
62. On this point, see Helm's sharp criticism (*Die Energetik*, 41).
63. "Bericht über 'die Theorie der physiologischen Wärmeerscheinungen'," in *WA* 1:6.

the principle of transmission of work.[64] Yet while French engineers accepted the impossibility of creating work, they did not exclude the possibility that it could be lost.[65] They were mostly concerned with impact, and so for them work was not a total differential and the concept of potential, which was then only being developed in analytical mechanics, was not generally admitted.[66] On the other hand, in the analytical tradition the quantity that eventually came to be called potential was not meant to be work stored in the system at a certain position. Notwithstanding its formal equivalence, "potential" was understood only as a mathematical function of the positions from which the forces could be derived. In this tradition force by displacement in the direction of force was a total differential but it did not receive a physical interpretation. Helmholtz subtly and skillfully united the two approaches, the concept of work with the function of positions (though not without problems in defining potential).[67] Work could thus no longer be seen as something created or destroyed; instead, it was a state function (of the positions).

Third and finally, Helmholtz's illicit "demonstration" of the equivalence of the two initial assumptions played a vital role in his research program. The generalization of the principle of conservation of *vis viva* into his principle of the conservation of "force," that is, into a principle where the kinetic and positional terms are sharply split, was only possible if the forces depended on distances alone. While this holds for Newton's, Coulomb's, and Ampère's forces, it did not hold for Weber's electrodynamic forces. This was a non-trivial problem for Helmholtz. In 1847, it was impossible to oppose Weber's law on empirical grounds; hence, it was important for Helmholtz to demonstrate that empirical laws admitting forces other than central ones violated the conservation of *vis viva* and the impossibility of perpetual motion. Yet he could not do so, for his demonstration was based on a false assumption: even if the components of the *vis viva* depend on the positions alone, the same does not necessarily follow for the force components. In fact, it was possible to show that forces (like Weber's) depending on velocities and accelerations do not violate the conservation of *vis viva* or the impossibility of perpetual motion. In 1848, Weber showed that his own force admitted a potential, even if a kinetic one. Nonetheless,

---

64. Rühlmann, *Vorträge über Maschinenlehre*; and Haas, *Entwicklungsgeschichte*, 73–83.
65. Haas, *Entwicklungsgeschichte*, 81.
66. Grattan-Guinness, "Work for the Workers," 32.
67. *Ueber die Erhaltung der Kraft*, 38–44, in *WA* 1:41–6.

during the next two decades Helmholtz's approach was astonishingly successful. In the British literature, the point that Weber's force law denied conservation of *vis viva* and energy conservation was maintained by James Clerk Maxwell in 1865, Peter Guthrie Tait and William Thomson in 1867, and Tait in 1868, until, following Helmholtz's own retreat in 1870, it was finally refuted by Maxwell in 1873. By 1882, Helmholtz himself acknowledged that Rudolf Lipschitz had found a flaw in the 1847 demonstration (he still did not mention Rudolf Clausius's criticisms), and he agreed that he had been unable to demonstrate that central Newtonian forces had a privileged status.[68] Based as it was on insecure, not to say faulty, deductive grounds, the validity of Helmholtz's approach had to rely on the inductive side, or its "empirical" success, that is, on its ability to reassess extant results (justificatory power) and disclose new phenomena (heuristic power).

In Section II of the *Erhaltung*, Helmholtz presented his grand generalization: the principle of conservation of *vis viva* became the principle of the conservation of "force." For a point mass **m**, moving with velocity **q** along the path **r** under the action of a central force ϕ, Helmholtz wrote the principle of the conservation of *vis viva* as:[69]

$$\frac{1}{2}md(\mathbf{q}^2) = -\phi d\mathbf{r},$$

or, for **Q** and **q** as the velocities at distances **R** and **r**:

$$\frac{1}{2}m\mathbf{Q}^2 - \frac{1}{2}m\mathbf{q}^2 = -\int_\mathbf{r}^\mathbf{R} \phi d\mathbf{r}. \tag{1}$$

This expression is formally identical to the well-known theorem of *vis viva*-work. The left-hand side of equation (1) represents the variation of the *vis viva* while the right-hand side has the dimension of work

---

68. "Ueber die Erhaltung der Kraft," *WA 1*:70. James Clerk Maxwell, "A Dynamical Theory of the Electromagnetic Field," reprinted in *The Scientific Papers of James Clerk Maxwell*, ed. W.D. Niven, 2 vols. (Cambridge: Cambridge University Press, 1890), *1*:526-37, on 526-27; W. Thomson and P.G. Tait, *Treatise on Natural Philosophy* (Oxford: Clarendon, 1867), 311-12; and P.G. Tait, *Sketch of Thermodynamics* (Edinburgh: Edmonston and Douglas, 1868), 76. See also Carl Neumann's defense of Weber in his *Die Gesetze von Ampère und Weber* (Leipzig: Teubner, 1877), 322-24; and, for a discussion of this, Bevilacqua, *The Principle of Conservation of Energy*, 122-23, 133-36. For Clausius's critique see his "Ueber das mechanische Aequivalent einer elektrischen Entladung und die dabei stattfindende Erwärmung des Leitungsdrahtes," *AP 86* (1852):337-75; idem, "Ueber einige Stellen der Schrift von Helmholtz 'über die Erhaltung der Kraft'," *AP 89* (1853):568-79; and idem, "Ueber einige Stellen der Schrift von Helmholtz 'über die Erhaltung der Kraft', zweite Notiz," *AP 91* (1854):601-4.

69. *Ueber die Erhaltung der Kraft*, 13, in *WA 1*:21-2.

(force by elementary displacement in the direction of the force integrated along a line). In Section I Helmholtz had used the word "*Arbeit*" several times, and so it might be expected that he would mention it again here. Yet he did not. Instead, he boldly reinterpreted equation (1) by redefining the right-hand side not as "*Arbeit*" but rather as "the sum of the tension forces [*Spannkräfte*] between the distances **R** and **r**." The tension force, he explained, was meant explicitly as the conceptual counterpart to the living force ("in contrast to that which mechanics calls living force"), a "force" that attempts to move the point **m** until motion actually occurs. He interpreted the concept geometrically as "the set of all the force intensities acting in the distances between **R** and **r**." If the intensities of ϕ correspond to the ordinates perpendicular to the line of the abscissae connecting the point **m** and the center of force **a**, then the integral represents an area given by the "sum of the infinite abscissae [read: ordinates] lying on it."[70]

His partly unsuccessful effort—the integral is not the sum of the abscissae but rather of the infinitesimal surfaces—to provide a geometrical interpretation of the *Spannkräfte* stressed that the tension forces $\phi d\mathbf{r}$ are also quite different from the Newtonian forces ϕ: they are represented dimensionally by the product of a force by a displacement; exist only when the material point is not in motion; attempt to put it in motion; are "consumed" by the acquired motion (here they should be compared to the constant relation force-matter described in the Introduction); and, finally, while they are a function of distance (two positions), acquire a proper meaning only when summed over a definite interval.

Helmholtz had introduced new theoretical concepts into an old equation. Both the left- and right-hand sides of equation (1) now had a physical theoretical meaning and were connected by an equality holding during a process: a variation of one side equalled a variation in the other. He interpreted the sum of the two sides physically as the conservation of force:

> In all cases in which free material points move under the influence of attractive and repulsive forces whose intensity depends only on the distance, the loss in the amount of the tension force is always equal to the gain in the living force, and the gain in the first is always equal to the loss in the second. *Therefore the sum of the extant tension and living forces is always constant.* We can define our law in this most general form as *the principle of the conservation of force.*[71]

70. *Ueber die Erhaltung der Kraft*, 14, in *WA* 1:22.
71. *Ueber die Erhaltung der Kraft*, 17, in *WA* 1:25.

Helmholtz's meaning of conservation is much different from that of Huygens's. For Huygens, conservation of *vis viva* meant that a system's *vis viva* reacquired the same value when the system reacquired the same positions independently of the trajectories that it followed back to those positions (of course, velocity, and thus *vis viva*, changes during motion). Conservation for Helmholtz meant that force (*Kraft*) is conserved during motion and a variation of *vis viva* corresponded to an opposite variation of tension force.

In addition, Helmholtz also deduced the principle of virtual velocities from the conservation of force: an increase of *vis viva* results only from the consumption of a quantity of tension force. Hence, if there is no consumption of tension forces for every possible direction of motion in the first instant, then a system at rest remains at rest.[72]

Helmholtz had thus achieved three results: the principle of conservation of force implies that the maximum quantity of work obtainable from a system is a determined, finite quantity if the acting forces do not depend on time and velocity; if they do so depend, or if the forces act in directions other than that joining the active material points, the "force" can be gained or lost ad infinitum; and under non-central forces, a system of bodies at rest could be set in motion by the effect of its own internal forces.[73] The hypothesis of central forces depending only on distances was thus basic to Helmholtz's view. Although these three results, as already noted, are not without their problems, Helmholtz's own summation of them did not do himself justice. For he had imparted a real shift in meaning to the well-known old equation (1). In the tradition of analytical mechanics, the stress had been on the conservation of *vis viva*; in the tradition of mechanical engineering on the transmission of work. Helmholtz, by contrast, stressed the equivalence of the two. It was the introduction of the term *Spannkraft* which brought the real shift in meaning: with tension forces we are very far from the concept of work and very close to that of potential energy. Work, which now meant not work done but rather the capacity to do work, now acquired the role of a unit of measurement for a new theoretical concept. As Helmholtz's student Max Planck said: "[h]owever insignificant this interpretation might at first glance seem to be, the perspective that it opens on all fields of physics is nevertheless extraordinarily wide because now the generalization to every natural phenomenon is evident."[74] Helmholtz's formulation of the principle of conservation of "force," Planck further explained, became similar

---

72. *Ueber die Erhaltung der Kraft*, 17–8, in *WA 1*:25.
73. *Ueber die Erhaltung der Kraft*, 19–20, in *WA 1*:26–7.
74. Planck, *Das Princip der Erhaltung der Energie*, 37.

to that of conservation of matter: "force" as matter cannot be increased or diminished, it can only manifest itself in different forms. The two basic forms of "force," *vis viva* and tension force, can appear in many ways: for example, *vis viva* as motion, light, or heat; tension force as elevation of a weight, elastic or electric potential, or chemical difference.[75] It is thus most surprising that modern commentators like Kuhn have considered Helmholtz's grand innovation as a failure and have dismissed it.[76]

Helmholtz's approach to conservation issued from a Leibnizian tradition[77] and, as a few highlights may suggest, unmistakably shows the difference between a theoretical and mathematical approach to physics. First, the duality *lebendige Kraft-Spannkraft* strongly resembles the older *vis viva-vis mortua* duality, as the very terms themselves suggest. While the *vis viva* has almost the same meaning, the *Spannkraft* is quite close to its older Leibnizian counterpart. Yet there are two main differences between Leibniz's and Helmholtz's ideas on positional "energy" which explain the latter's success: Helmholtz provided a formal quantitative expression and he included the Newtonian forces. By properly using the Newtonian concept of force and by fully accepting the inheritance of Newtonian mechanics, Helmholtz provided a formal expression for the second term of the duality that was absent in Leibniz.

A second Leibnizian element, moreover, is that the equality of the two sides of equation (1) no longer meant an analytical identity. Being two independent physical concepts, the equality now implied a causal relationship: the variation in one implied the variation in the other. The equality holds at every instant during a process. This is a Leibnizian concept of conservation. Helmholtz now brought "empirical" causality into play here: not causality as a condition of the possibility of natural laws, but rather causality as a principle which establishes a specific link between different realms of phenomena. The "empirical" causality indicates a quantitative equivalence between phenomena that are qualitatively different (static and dynamic). Like Leibniz, what is conserved in Helmholtz during the process is the specific equivalence between qualitatively different phenomena. A static cause can generate, as an effect, the motion of a body. This motion, which in turn becomes a cause, has the power ("motive force") to produce the effect of returning the body to its initial position. The quantitative equivalence of cause and effect is maintained at every instant of the process: the

---

75. Ibid., 37.
76. Kuhn, "Energy Conservation," 88.
77. Planck, *Das Princip der Erhaltung der Energie*, 35; Koenigsberger *1*:89; and Elkana, *Discovery of Conservation of Energy*, 20 (n. 31).

interchangeability of the initial and final stage is only an exemplification of this principle.

Yet how are two qualitatively different phenomena to be measured so as to establish a quantitative equivalence? A common unit of measurement was obviously needed, and Leibniz had suggested work (to use the eventual designation) as the unit of measurement of all natural phenomena.[78] He recognized the impossibility of continually creating and destroying work (that is, "*ex nihilo*" and "*ad nihilum*") without a corresponding compensation as a necessary condition for guaranteeing the invariability of the chosen unit. Similarly for Helmholtz, work became the common unit of quantitative measurement for different phenomena connected by a causal principle but now mechanically interpreted according to Newton's definition of force and laws of motion. The "*ex nihilo*" and "*ad nihilum*" were now both present: the quantitative aspects of one side of the equation must be the same as those of the other. Work can be neither created nor destroyed. Helmholtz argued that all natural phenomena must be measured by a common unit and interpreted by using only two forms of force (*Kraft*). Here was a grand theoretical unification resulting from a shift in meaning.

A still deeper understanding of Helmholtz's result may emerge by briefly comparing it to Clausius's less philosophically but more mathematically inclined approach. Helmholtz had introduced the concept of *Spannkraft* (potential energy) without discussing that of potential. The peculiarity of his approach lay in his jumping theoretically from the *vis viva* theorem to the conservation of force principle without following the now standard formula of $\mathbf{f} \times \mathbf{ds}$ = work, where work is a total differential or difference of potential. In 1852, Clausius, by contrast, established a different relation between *vis viva* and potential. He started with the *vis viva* theorem and equated the increase in *vis viva* to the quantity of mechanical work produced in the system during the same time.[79] He rejected Helmholtz's "sum of tension forces" (potential energy) and the corresponding interpretation of the conservation principle. For Clausius, work was in most cases a total differential, and thus its integral depended only on the initial and final positions

---

78. Leibniz, "Brevis Demonstratio erroris memorabilis Cartesii et aliorum circa legem naturalem...," in *Acta Eruditorum* (1686):161–63 [Leipzig: Christophori Guntheri, 1686], reprinted in *Leibnizens mathematische Schriften*, ed. C.I. Gerhardt, 7 vols. in 4 (vols. 1–2: Berlin: A. Asher; vols. 3–7: Halle, H.W. Schmidt, 1849–63), 2:117–19; and, more generally, see Ernst Cassirer, *Das Erkenntnisproblem*, 2:165; and idem, *Substanzbegriff*, 226–48.

79. Rudolf Clausius, "Ueber das mechanische Aequivalent einer elektrischen Entladung und die dabei stattfindende Erwärmung des Leitungsdrahtes," *AP* 86 (1852):337–75.

and so was identical with a difference of potential. He explicitly asserted that the potential is work stored in the system. For Clausius, work as total differential and difference of potential were identical concepts. This important statement was a rather different one from Helmholtz's. In the Gauss-Clausius tradition, energy would never acquire the same importance as in Helmholtz's works. The principle of conservation was often to be called, as in the old tradition, the *vis viva* conservation and the only really important requirement was that work be a total differential. This interpretation left open the possibility that forces other than central Newtonian ones could satisfy the conservation principle, if the work done by these forces was, mathematically speaking, a total differential. But in this case the energy terms could not be easily divided into kinetic and positional parts. For central Newtonian forces, as introduced by Helmholtz for example, work was indeed a total differential, and thus his and Clausius's approaches intersected considerably. In Section II of the *Erhaltung*, Helmholtz introduced in a straightforward way the sum of tension forces (potential energy) and the energy (*Kraft*) concept—variation of *vis viva* equals variation of the sum of tension forces, the sum of *vis viva* and tension forces is a constant—as conceptual and physical entities. However, he introduced the physical concept of potential only in Section V, and he did so only with a shaky grasp of that concept.[80]

Helmholtz's result may be further appreciated by briefly comparing it to Mayer's. Despite his acquaintance with Liebig's papers, Helmholtz was, as already noted, surprisingly unaware of Mayer's contribution of 1842 to the *Annalen der Chemie und Pharmacie*. He did not quote Mayer in the *Erhaltung*, and, indeed, he later claimed that in 1847 he had no knowledge of Mayer's work. Although both men used a Leibnizian principle of causality, Helmholtz alone adopted the mechanical conception of nature, the mechanical theory of heat, the central-force hypothesis, and the reduction of all qualitatively different forms of "force" to two basic ones. Mayer, by contrast, rejected all these elements and his expression of conservation of energy was closer to a principle of equivalence, that is, to a correlation principle. On the other hand, Mayer worked out a mechanical equivalent of heat, although, to be sure, he did so through original thinking rather than through original experimentation; Helmholtz, by contrast, as Section 5 shows, did not work out such an equivalent.[81]

---

80. *Ueber die Erhaltung der Kraft*, 39, in *WA 1*:42.
81. "Ueber die Erhaltung der Kraft," *WA 1*:71–3; "Anhang zu dem Vortrag 'Ueber die Wechselwirkung der Naturkräfte und die darauf bezüglichen neuesten Ermittelungen der Physik': Robert Mayer's Priorität," in *VR^5 1*:401–14; "Anhang zu dem Vortrag 'Das

By restricting itself to Newtonian forces, Section II of Helmholtz's *Ueber die Erhaltung der Kraft* sharply distinguished between kinetic and positional terms and fulfilled the far-reaching plan announced in the first lines of the Introduction: the conservation principle just formulated is the "consequence" (level two) of the two basic physical assumptions (level one: impossibility of perpetual motion and central forces). By the end of Section II, Helmholtz had already made impressive achievements: within a few pages he had completely reappraised older traditions in physics by providing an original synthesis of different and previously competing approaches (Newtonian and analytical mechanics, Leibnizian philosophy, mechanical engineering). Nonetheless, he had still only provided a general theoretical framework, one that remained to be filled with specific expressions for the *vis viva* and tension forces that resulted from the interaction of the principle (level two) with the experimental laws (level three) in the various realms of natural phenomena (level four). He demonstrated this interaction between levels two, three, and four in the *Erhaltung*'s remaining four sections through application to an extremely wide variety of empirical laws and natural phenomena. In so doing, the "empty" framework shows the strengths and weaknesses of its justificatory and heuristic power.

5. Applications: Mechanics and the Force-Equivalents of Heat, Electrical Processes, Magnetism, and Electromagnetism

Helmholtz first applied his principle to mechanical theorems (Section III), mostly using known applications of *vis viva* conservation, which is to say that in this short, non-mathematical section he did not deal with specific applications of the concept of tension forces. Instead, he briefly considered the motions (of both celestial and terrestrial bodies) caused by gravitation; the transmission of motion through incompressible solids and fluids when *vis viva* is not lost through friction or inelastic collisions; and the motions of perfectly elastic solid and fluid bodies without internal friction.

Denken in der Medicin,' " VR^5 2:384–86; Planck, *Das Princip der Erhaltung der Energie*, 21–8; Helm, *Die Energetik*, 16–28; Haas, *Entwicklungsgeschichte*, 61–2; Robert Bruce Lindsay, *Julius Robert Mayer. Prophet of Energy* (Oxford and New York: Pergamon, 1973); and Peter Heimann, "Mayer's Concept of 'Force': The 'Axis' of a New Science of Physics," *HSPS* 7 (1976):227–96.

He explicitly noted Fresnel's use of the principle of *vis viva* conservation to derive the laws of light reflection, refraction, and polarization, as well as the application of the principle to interference, thus displaying a broad and deep knowledge of physical problems. His application of the principle of "*Kraft*" suggested that if there is a loss of *vis viva* due to the absorption of elastic, acoustic, or heat waves, then a different kind of quantitatively equivalent "force" must appear.[82] He maintained that heat must be produced by the absorption of heat rays, but asserted that it had not yet been proven experimentally that the amount of heat which disappears from the radiating body reappears in the irradiated one. (Here was a first instance of his predilection in the final four sections of the *Erhaltung* to suggest and outline applications of the principle independently of any experimental confirmations.) While asserting that light absorption can produce heat, light (phosphorescence), and chemical effects, Helmholtz identified light with radiations producing thermal and chemical effects. He remarked, too, on the (small) effects of light and chemical rays on the eye, an indication perhaps of a small value for their heat equivalent.[83] The quantitative relations of the chemical effects produced by light were not well known, and Helmholtz believed that relevant magnitudes were only involved in the case of light absorbed by the green parts of plants.[84]

Even in this short section, one which mostly recalled known results, Helmholtz's method started to reveal its fertility by organizing a very large amount of physical knowledge. Yet its limits also became evident: for example, the difficulty of identifying the "tension forces" here reduced the conservation of "force" to a correlation principle. Moreover, it is evident that the principle can only be heuristically fruitful when its predictions are supported by extant empirical laws in the pertinent fields. To confirm the principle required knowledge of specific coefficients of equivalence, knowledge which Helmholtz lacked. He thus had to confine his efforts to broad theoretical applications based on his surprisingly (for a medical doctor) deep knowledge of the physical literature.

In Section IV Helmholtz turned to the problem of the force-equivalent of heat. Although it might easily be supposed that this section is the *Erhaltung*'s centerpiece, it is not; rather, it is here that Helmholtz's different approach from that of Mayer (unknown to Helmholtz) and Joule become most evident. In contrast to the German physician and the English brewer, Helmholtz did *not* accurately establish the mechanical equivalent; indeed, he does not even seem to have been

---

82. *Ueber die Erhaltung der Kraft*, 23, in *WA* 1:30–1.
83. *Ueber die Erhaltung der Kraft*, 24, in *WA* 1:31.
84. *Ueber die Erhaltung der Kraft*, 25, in *WA* 1:31.

concerned to do so, and the lack of such a determination supports his subsequent claim that the *Erhaltung* sought more to review and synthesize contemporary physical knowledge than to produce original experimental results.[85] Instead, Helmholtz's principal interest in this section lay in a theoretical interpretation of thermal phenomena through his own framework. His approach was rather qualitative, using mathematical formalism only to discuss Clapeyron's and Holtzmann's laws.

He began by looking for actual compensations (equivalents) to an apparent loss of force. He used his principle of the conservation of force to identify the compensation for the loss of living force in inelastic collision and in friction with a supposed increase of tension forces (namely, internal elastic forces) due to the variation of "the molecular constitution of the bodies" and with acoustical, thermal, and electrical effects.[86] He first treated the case of the collision of inelastic bodies, where the loss in *vis viva* reappeared as an increase in the tension forces, as heat and sound. He then treated friction, where there is an increase of the tension forces, heat and electricity. Neglecting molecular effects and electricity, he posed two important questions: does a loss of *vis viva* correspond to an equivalent amount of heat? and how can heat be given a mechanical interpretation?[87] The first question was connected to a "correlation" approach; the second was specific to Helmholtz's program.

Helmholtz quickly disposed of the first question, though without obtaining any definite results. Unaware of Mayer's and Colding's research, he briskly asserted that "perhaps" too few efforts had been dedicated to this issue. He cited only a paper by Joule, recalling his attempts at establishing a mechanical equivalent through the heat produced by the friction of water in narrow tubes and in vessels (the famous paddle-wheel experiment).[88] He reported that Joule's result in the case of narrow tubes was that the heat needed to raise 1 kg of water by 1 °C raises 452 kg to 1 m, and, in the case of vessels, 521 kg. His judgment on Joule's work, the only original experimental determination of the mechanical equivalent of heat cited in the *Erhaltung*, was severe. He thought Joule's measurements inadequate to the "difficulties of the research" and thus false: "probably the figures are too high," he wrote. His criticism of the only empirical evidence corroborating his own theoretical approach is surprising. As a good experimentalist and as one heavily involved in experimental physiological

---

85. "Ueber die Erhaltung der Kraft," in *WA 1*:74.
86. *Ueber die Erhaltung der Kraft*, 26, in *WA 1*:32.
87. *Ueber die Erhaltung der Kraft*, 27, in *WA 1*:32-3.
88. James P. Joule, "On the Existence of an Equivalent Relation between Heat and the Ordinary Forms of Mechanical Power," *PM* ser. 3, *27* (1845):205-7.

research on an intimately related subject, his judgment of inaccuracy about Joule's highly accurate results demands discussion.

In his paper of 1845, Joule had given his results from the paddle-wheel experiments (890 ft-lbs) along with those from previous work: 823 ft-lbs in 1843 from magnetoelectrical experiments, 795 ft-lbs in 1845 from the rarefaction of the air, and 774 ft-lbs from unpublished experiments on the friction of water moving in narrow tubes.[89] Joule averaged the two experiments resulting from the friction of water (890 and 774) to 832 as well the results of all three distinct types of experiments (823, 795, and 832) to 817. Given that the final accepted equivalent value was 778, Joule's averaged results make Helmholtz's criticisms seem excessive, not least since in 1847 Joule's results alone favored Helmholtz's approach.

The explanation for Helmholtz's harsh judgment is threefold. First, the *Erhaltung* was intended as a work in *theoretical* physics. Its origin was largely independent of experimental results. Helmholtz intended it to reinterpret extant knowledge, and so its value was not meant to rest on any doubtful experimental results. As an amateur scientist in 1847, Joule had yet to gain full scientific recognition, and Helmholtz may not have wanted to rely on such a weak ally. Second, although Helmholtz quoted Joule four times in different passages of Section IV, he may have only become aware of Joule's papers during the *Erhaltung*'s final preparation.[90] (Helmholtz did not, for example, quote Joule in the "Bericht," written in October 1846.) Thus, he may not have mastered Joule's results. Third, and perhaps most important, Helmholtz systematically erred in converting British and continental units of measurement (from degrees Fahrenheit, feet, and pounds to degrees Centigrade, kilograms, and meters). In his search for a mechanical equivalent of heat, Joule equated the quantity of heat needed to increase 1 lb of water by 1 °F (latter to be called BTU) with the corresponding work expressed in feet by force-pounds. From his experiments with the friction of water he obtained a mechanical equivalent of 890 ft-lb/BTU for vessels and 774 ft-lb/BTU for narrow tubes.

---

89. Ibid.; and J.P. Joule, "On the Changes of Temperature Produced by the Rarefaction and Condensation of Air," *PM*, ser. 3, *26* (1845):369–83. In 1853 and in 1863, John Tyndall attempted to explain Helmholtz's evaluation of Joule's work by claiming that Helmholtz used Joule's paper of 1843, where the force equivalent varied between 1,040 and 547 ft-lbs. (See J.P. Joule, "On the Caloric Effects of Magneto-Electricity, and on the Mechanical Value of Heat," *PM* ser. 3, *23* [1843]:263–76, 347–55, 435–43, on 438–42; and John Tyndall, "Remarks on the Dynamical Theory of Heat," *PM* ser. 4, *25 [1863]:368–87, on 375–76.) In point of fact, in the passage of the Erhaltung at issue,* Helmholtz did not refer to Joule's 1843 paper but rather to that of 1845.

90. "On the Interaction of Natural Forces and Recent Physical Discoveries Bearing on the Same," *PM*, ser. 4, *11* (1856):489–518, on 499; see also Helmholtz, "Erinnerungen," in *VR⁴ 1*:11; and Koenigsberger *1*:82.

Considering the conversion factors between British and continental units (1 BTU = 0.252 kcal; 1 ft-lb = 0.1382 kgm) to express Joule's results in kgm/kcal, we must multiply them by a factor of 0.5484 (= 0.1382/0.252). Joule's values of 890 and 774 thus correspond to 488 and 424, respectively. The second figure is very close to 778, the definitely accepted value of the mechanical equivalent of heat (in continental units: 778 x 0.5484 = 427 kgm/kcal).

As noted already, Helmholtz's conversion results are different: instead of 488 and 424, he found 521 and 452 for vessels and narrow tubes, respectively. He maintained that the (converted) values of Joule's experiments were too high. Yet the fault lay not in the accuracy of Joule's experiments but rather in Helmholtz's own faulty conversion. The source of his systematic error seems clear enough: the two sets of results above give 521/488.3 = 1.067; and 452/424.6 = 1.065. Now, Helmholtz must have used the *French* foot, a well-known unit of measurement which equals 12.8 inches or 0.3251 meters.[91] And in fact, 12.8/12 = 1.067. Helmholtz's mistake became relevant both for his general evaluation of Joule and for the specific comparison of Joule's results with those of Holtzmann (given in metric units).[92] (See below, p. 326.)

Having misunderstood Joule's experimental results and quickly dismissing the entire unresolved problem of the mechanical equivalent of heat, Helmholtz turned to the second question, a theoretical one which he viewed as far more important: the extent to which heat corresponds to a force equivalent.[93] (Here force equivalents, which are theoretically identifiable energy terms, should not be confused with mechanical equivalents, which are numerical conversion factors.) Helmholtz discussed the caloric theory with explicit reference to the interpretation of Carnot and Clapeyron, for whom the force equivalent was the work produced in the passage of a certain amount of caloric from a higher to a lower temperature. Moreover, he criticized William Henry's and Claude Louis Berthollet's interpretation of the heat produced by friction as a displacement of caloric, and asserted that results from the field of electricity showed that the total amount of a body's heat can actually be increased.[94] He cited experimental evidence, based entirely on electrical research, against the caloric theory and in favor

---

91. On the French foot see H.G. Jerrard and D.B. McNeill, *Dictionary of Scientific Units: Including Dimensionless Numbers and Scale*, 5th ed. (London: Chapman and Hall, 1986), 45.
92. *Ueber die Erhaltung der Kraft*, 27, in *WA 1*:33. Thirty-four years later, in a footnote of 1881 (*WA 1*:33), Helmholtz modified his comments and praised Joule's work, yet he did not acknowledge (or see?) his own conversion mistake of 1847.
93. *Ueber die Erhaltung der Kraft*, 27, in *WA 1*:33.
94. *Ueber die Erhaltung der Kraft*, 28–9, in *WA 1*:34.

of the mechanical. While frictional and voltaic electricity did not give indisputable evidence—since the heat produced could be interpreted as caloric displaced—he nonetheless argued that "we still must explain in a purely mechanical way the production of electrical tensions in two processes [electrical induction and movements of magnets] in which any quantity of heat that can be assumed to be transferred never appears."[95] For electrical induction, he cited the example of an electrophorus used to charge a Leyden jar; for the movements of magnets, that of electromagnetic machines where "heat can be developed ad infinitum." It was only here that Helmholtz recalled Joule's experiments of 1843 and asserted that Joule "endeavored to show directly" that the electromagnetic current produced heat and not cold even in that part which is under the actual action of the magnet (no displacement of caloric is thus conceivable in the electrical circuit).[96] Again, Joule's results played a minor role in the exposition of Helmholtz's ideas.

For Helmholtz, the caloric theory had to be rejected and replaced by the mechanical theory, which allowed heat to be produced indefinitely by mechanical and electrical forces. As noted in Section 2, Helmholtz had already reached this conclusion in the "Bericht." What was new and specific in the *Erhaltung* was the application of the theoretical framework of tension and living forces to the mechanical theory of heat. Again, this was done in a purely conceptual and qualitative way, without mathematical formulation: free heat was now interpreted as the quantity of living force of thermal movement and latent heat as the quantity of tension forces (namely, the elastic forces of atoms). Yet the whole subject remained highly speculative, and Helmholtz was satisfied with "the possibility that thermal phenomena be conceived as motions."[97] Lack of empirical confirmation, not lack of conceptual clarity, as Georg Helm later supposed, fully justified Helmholtz's cautiousness.[98]

Although qualitative, Helmholtz's conceptual scheme was wide-ranging. To conceive of atoms as possessing not only living but also tension forces was a bold step, and his analogy with free and latent heat seems apt, for it allowed an easy reinterpretation of older ideas. The reinterpretation of the heat produced in chemical processes followed: Hess's law, "also partially verified by experience," had been

---

95. *Ueber die Erhaltung der Kraft*, 29, in *WA 1*:34.
96. *Ueber die Erhaltung der Kraft*, 30, in *WA 1*:35, where Joule's results are misdated as 1844.
97. *Ueber die Erhaltung der Kraft*, 31, in *WA 1*:36.
98. Helm, *Die Energetik*, 44.

deduced from the caloric theory. It asserted that "the heat developed in the production of a chemical compound is independent of the order and the intermediate steps of the process."[99] As Helmholtz had shown in the "Bericht," Hess's law agreed with the force-equivalent hypothesis (correlation principle); in the *Erhaltung*, he showed that it can be interpreted in terms of the new concepts of living and tension forces: the heat produced was now considered a living force, generated by the chemical forces of attraction that played the role of tension forces. Helmholtz here implicitly applied the mechanical concept of conservation developed in Section I: the *vis viva* developed between two definite configurations of the system was independent of the trajectory.

The final problem of this section concerned the disappearance of heat; as Helmholtz noted, it had yet to receive much attention. In discussing it, he again displayed his inclination towards a theoretical approach: both the transformations of work into heat and of heat into work were assumed, but as necessary consequences of the principle of the conservation of force and not on the basis of experimental results. Again quoting Joule, Helmholtz asserted that Joule's results on this topic were the only ones available and that they seem "sufficiently reliable." He was referring to Joule's experiment in which compressed air expanding against air pressure cooled itself; however, this did not occur when the air expanded in a vacuum. In the former instance, the compressed air has to exert a mechanical force to overcome the air pressure's resistance; in the latter, it does not. Hence, in the former instance the heat which has disappeared can be equated to the work done and thus a mechanical equivalent can be found (although Helmholtz did not mention any).[100]

Finally, and most puzzlingly, Helmholtz discussed the research of Clapeyron and Holtzmann. He knew that both men had conducted their work on the basis of the caloric theory; in fact, in the "Bericht" he asserted that they dealt more with the propagation than the production of heat. But in the *Erhaltung* he spoke of their research as "tending to find out the force equivalent of heat" and compared their work with his own.[101] He discussed and criticized Clapeyron's approach at length, noting that Clapeyron's law, which assumed the caloric theory, had received empirical support for gases alone, and that for that case it was equivalent to Holtzmann's. The latter, for his part, had assumed that if a certain quantity of heat "enters" into a gas, then it

99. *Ueber die Erhaltung der Kraft*, 32, in *WA* 1:37.
100. *Ueber die Erhaltung der Kraft*, 33, in *WA* 1:37; and Joule, "On the Changes of Temperature."
101. *Ueber die Erhaltung der Kraft*, 33, in *WA* 1:37-8.

produces either an increase in temperature or an expansion. In expansion, the work done allows, according to Helmholtz's account of Holtzmann's work, the possibility of calculating the mechanical equivalent of heat. Using Dulong's values for the specific heats of gases, Holtzmann's equivalent was 374 kgm.[102] Helmholtz warned that this could only be accepted within the framework of the conservation of force if all of the transmitted heat's living force was actually given as work, that is, if the sum of the living and tension forces (or, in the old terminology, the quantity of free and latent heat) of the expanded gas was the same as that of the denser gas at the same temperature. This approach agreed with Joule's as given above, and Helmholtz compared Holtzmann's equivalent of 374 kgm with a series of results by Joule, whom he credited with having actually performed the experiments and not merely having reinterpreted older data. He cited five values from Joule: the two already noted (452 and 521, which, as argued above, should be 424 and 488) derived from the friction of water, and three others (481, 464, and 479, which should be 451, 435, 449). These last three, although cited without a specific reference, were almost certainly taken from Joule's 1845 paper. The first (481/451) referred to the 1843 experiments with an electromagnetic engine; the second (464/435) to the 1845 experiments on air referred noted above; and the third (479/449) to the average mentioned in Joule's 1845 paper. Helmholtz's comparison of Holtzmann's and Joule's results was seriously distorted by the systematic error in his conversion of Joule's units of measurement. Helmholtz concluded with a detailed comparison of the laws of Clapeyron and Holtzmann.[103] Seven years later, in 1854, Clausius, using a mechanical equivalent of 421 kgm, raised a serious objection, asserting that Helmholtz had misunderstood Holtzmann's law wherein the concept of caloric played a role which cannot be eliminated.[104] This criticism was one of the very few that Helmholtz accepted during his long controversy with Clausius.[105]

Helmholtz dedicated the fifth (and longest) section of the *Erhaltung* to applying the law of the conservation of force to static electricity,

---

102. *Ueber die Erhaltung der Kraft*, 35, in *WA* 1:39.
103. *Ueber die Erhaltung der Kraft*, 36–7, in *WA* 1:40–1.
104. Rudolf Clausius, "Ueber die bewegende Kraft der Wärme und die Gesetze, welche sich daraus für die Wärmelehre selbst ableiten lassen," *AP 79* (1850):368–97, 500–24.
105. "Erwiderung auf die Bemerkungen von Hrn. Clausius," *AP 91* (1854):241–60, in *WA* 1:76–93, on 90; and see also Clifford Truesdell, *The Tragicomical History of Thermodynamics. 1822–1854* (New York: Springer, 1980), 162. In an appendix to this section in 1882, Helmholtz discussed the priority problem: "Ueber die Erhaltung der Kraft," *WA* 1:71–4.

galvanism, and thermo-electric currents. Here, too, he displayed an extraordinarily detailed knowledge of the empirical laws of physics. His first application was to Coulomb's law, which, being a strictly central force law involving attractive and repulsive forces, offered the best possible example of how to formulate a sum of tension forces and to equate them with an increase in *vis viva*. Yet difficulties soon ensued: unaware of Green's results, Helmholtz introduced the concept of electric potential. He defined the quantity $-\frac{e_i e_{ii}}{r}$, corresponding to the sum of the tension forces consumed and the living forces acquired in the motion of the two charges e_i and e_{ii} from an infinite distance to the distance r, as the potential of the two electrical charges over the distance r.[106] He then expressed the principle of conservation of force in a new manner, as "the increase of *vis viva* in whichever movement must be considered equal to the difference of the potential at the end of the trajectory with respect to the potential at the beginning."[107] A potential so defined is equivalent (but for the sign) to the modern definition of potential energy. In relating potential and work, for instance in the case of the potential of one body with respect to another, Helmholtz showed a good grasp of their equivalence.[108] Yet his definition of the potential of a body on itself (the sum of the potentials of an electric element of a body with respect to all other such elements of the same body) was problematical: it did not correspond to the work done (the potential was supposed to be twice the work done).[109] Hence, in Helmholtz's approach the two concepts were "independent." To be sure, in the original 1847 edition of the *Erhaltung* there was a final correction (the only one) referring to exactly these problems, which suggests Helmholtz's uncertainties and difficulties with concepts that were then by no means common.[110] He later maintained that his 1847 approach of relating potential and work was basically correct.[111] In any case, he was among the first to interpret and use correctly the "new" mathematical tool of potential.

Although Helmholtz thus explicitly unified the tradition of analytical mechanics (the potential function of Gauss, Hamilton, and Jacobi) and of mechanical engineering (the concept of work), he arrived at the concept of potential not (as did Clausius in his more mathematical

---

106. *Ueber die Erhaltung der Kraft*, 38–9, in *WA* 1:42.
107. *Ueber die Erhaltung der Kraft*, 39, in *WA* 1:42.
108. *Ueber die Erhaltung der Kraft*, 39–40, in *WA* 1:42–3.
109. *Ueber die Erhaltung der Kraft*, 42–3, in *WA* 1:44–5.
110. In the 1882 reprint of "Ueber die Erhaltung der Kraft" in *WA 1* the correction is incorporated in the text (ibid., 45); and cf. ibid., 75.
111. Ibid., *WA* 1:75.

approach of 1852)[112] through the concept of work as a total differential, but rather directly from the concept of the sum of the tension forces. Theoretical rather than mathematical physics lay at the heart of Helmholtz's approach, as is evident from his attempt to clarify the "mathematical" potential through the introduction of physically sound concepts, and not vice versa: he first introduced the "equilibrium surfaces,"[113] later identified with equipotential surfaces, and then the "free tension," later identified (by Helmholtz himself) with the mathematical potential function.[114] As late as 1847, his idea of electric tension was reminiscent of Volta's influential "density of electricity."

Helmholtz sought to apply his *conceptual* framework of living and tension forces to every realm of nature. That his approach differed markedly from that of mathematical physicists, such as Clausius and Bernhard Riemann, as well as from experimentalists, such as Joule, can again be seen in his discussion of galvanism. For Helmholtz, Volta's contact law did not disagree, as Kuhn has claimed,[115] with the impossibility of perpetual motion. Volta's contact tensions were not equivalent to a definite quantity of "force": they did not produce an electrical imbalance but rather originated from such an imbalance. Helmholtz restricted his use of the contact law to first-class conductors (metals) and recognized that second-class conductors conduct only electrolytically. Hence, he interpreted contact force in terms of the attractive and repulsive forces of two metals which remove electrical charges in the contact area from one metal to the other. Equilibrium was reached when an electrical particle, in passing from one metal to another, neither acquired nor lost living force, that is, when the variation of living force from one metal to the other was compensated by an identical variation of tension forces independently of the shape and dimension of the contact surfaces and in agreement with the galvanic series of tensions.[116]

If conservation of force based on central forces once again provided a conceptual explanatory framework for the contact law, it failed to do so for Helmholtz's long analysis of galvanic currents. Here conservation of force was applied to batteries not producing polarization; those producing polarization but not chemical decomposition; and those producing both. It was applied, however, in the non-mechanical

---

112. Clausius, "Ueber das mechanische Aequivalent."
113. *Ueber die Erhaltung der Kraft*, 40–2, in *WA* 1:43–5.
114. "Ueber einige Gesetze der Vertheilung elektrischer Ströme in körperlichen Leitern mit Anwendung auf die thierisch-elektrischen Versuche," *AP 89* (1853):211–33, 352–77, on 224, in *WA* 1:475–519.
115. Kuhn, "Energy Conservation," 73.
116. *Ueber die Erhaltung der Kraft*, 45–7, in *WA* 1:47–9.

sense as equivalence of numerical effects without a reinterpretation in terms of living and tension forces.[117] Helmholtz knew that precise, experimentally confirmed laws existed only for batteries not producing polarization. By using Ohm's, Lenz's, and Joule's laws, Helmholtz gave the amount of heat that must be generated in the circuit to achieve conservation of force. This heat had to be equivalent to the chemical heat developed without electrical effects; the result was that the electromotive forces of the two metals were proportional to the difference of the heat developed by oxidation and by combination with acids.[118] By contrast, Helmholtz discussed batteries producing polarization as well as polarization with chemical decomposition in detail but without applying his conceptual scheme and, because he lacked reliable empirical data, non-quantitatively.[119] Here, too, Helmholtz again cited Joule—this time for his experiments showing the equivalence of chemical and electrical heat[120]—and here, too, he again criticized and judged Joule's results and methods as unreliable, despite their providing evidence for a part of Helmholtz's innovative program.

Once more Helmholtz returned to his main line of thought: a conceptual explanation of electrical movements between metals and fluids through attractive and repulsive forces, in analogy with what he had just done for contact forces. For polarization currents, the two metals attracted (until saturated) positive or negative electrical charges, respectively. For chemical decomposition, there was not a stable equilibrium but instead a continuous process, one whose velocity did not continually increase for the loss of *vis viva* by the heat developed. Helmholtz derived an equivalence between the heat produced (living force) and the consumption of chemical elastic force (tension force). His conservation of force thus helped clarify yet another difficult topic. Finally, Helmholtz discussed thermoelectric currents and the Peltier effect. Without applying the concepts of tension and living forces, he utilized the principle of conservation to derive two consequences (on the heat produced and absorbed at equal [constant] temperatures and on equal currents), yet again complaining that he knew of no experimental measurements.[121]

117. Helm, *Die Energetik*, 44.
118. Given that most of this section (*Ueber die Erhaltung der Kraft*, 48–58 in *WA* I:49–57) deals with a careful analysis of batteries, it is difficult to understand Kuhn's claim ("Energy Conservation," 73) that Helmholtz did not discuss batteries in the *Erhaltung*.
119. *Ueber die Erhaltung der Kraft*, 51–6, in *WA* 1:51–5.
120. J.P. Joule, "On the Heat evolved by Metallic Conductors of Electricity, and in the Cells of a Battery during Electrolysis," *PM* ser. 3, *19* (1841):260–77, on 275; and idem, "On the Electrical Origin of Chemical Heat," ibid., *22* (1843):204–8.
121. *Ueber die Erhaltung der Kraft*, 60, in *WA* 1:58.

In Section VI, the final section, Helmholtz's approach revealed all its strengths and limitations. In treating the force-equivalents for magnetism and electromagnetism, the intrinsic difficulties connected with the formulation and application of the principle of conservation of force are clear enough.[122] In treating magnetism, Helmholtz followed the pattern that he had used for electrostatics: the inverse square law provided an expression for the tension forces. He defined living forces and potentials, both for two bodies and for a body on itself. An interesting application was that of a non-magnetized steel bar brought close to a magnet, then magnetized, and then separated. Here there occurred an expenditure in mechanical work of $-\frac{1}{2}\mathbf{W}$ (again, the potential on itself $\mathbf{W}$ is twice the work) acquired by the magnetized bar. In treating electromagnetism, Helmholtz for the first time showed his mastery of the subject and outlined a research program that he would pursue intermittently during the next forty years. He used not only the well-known laws of Ampère but also the more recent and less-well-known laws of Weber, Lenz, Neumann, and Grassmann. He characterized precisely the approach of Weber's law (which at variance with Ampère's explained electromagnetic induction), and noted how it stood within a conceptual framework that was at odds with his own, for Weber assumed forces depending on velocities and accelerations. Helmholtz pointed out that "until now" it had not been possible to refer Weber's law to central forces.[123]

Both Neumann's and Grassmann's laws, he noted, agreed with Weber's for closed currents, the only laws for which experiments were available. He thus restricted his application of the principle of the conservation of force to closed currents and showed that the "same laws" could be deduced by using the principle.[124] His strategy was clear: lacking a central force law for electromagnetism, he hoped to use the principle to deduce "empirical" laws already deduced on the basis of non-Newtonian hypothetical forces, thereby gaining evidence for the principle's justificatory power and, if new consequences could be successfully predicted, for its heuristic power as well. Yet difficulties soon emerged, as can be seen in the following two cases.

In discussing a system consisting of a magnet moving under the effect of a current, Helmholtz identified the tension forces with those consumed in the current, $\mathbf{aAJdt}$, where $\mathbf{a}$ is the mechanical equivalent

---

122. *Ueber die Erhaltung der Kraft*, 60–9, in *WA* 1:58–65.
123. *Ueber die Erhaltung der Kraft*, 62–3 (quote on 63), in *WA* 1:60–1 (quote on 61).
124. *Ueber die Erhaltung der Kraft*, 64, in *WA* 1:61.

of heat, **A** the electromotive force of a single cell, **J** the current, and dt an infinitely small amount of time; that is, as with his results on galvanism, he identified the tension forces with the heat generated chemically inside a battery. As for the living force, it consisted of two parts: the heat generated in the circuit by the current, aJ^2Wdt, where **W** is the resistance of the circuit; and the living force acquired by the magnet under the effect of the current, $J\dfrac{dV}{dt}dt$, where **V** is the potential of the magnet towards the conductor carrying a unit current. Hence:

$$J = \frac{A - \dfrac{1}{a}\dfrac{dV}{dt}}{W}.$$

Helmholtz interpreted the term $\dfrac{1}{a}\dfrac{dv}{dt}$ as a new electromotive force, namely, that of the induced current. His force was similar to but more precise than Neumann's in that he gave the value 1/**a** for what in Neumann was an undetermined constant.[125]

If this "demonstration" became famous as an instance of the principle's heuristic power,[126] a second case discussed by Helmholtz, that concerning the interactions between two currents, became famous as an instance of a false deduction from the principle. Here Helmholtz simply extended the previous formulation for the tension forces provided by the batteries of the two circuits to A_1J_1 and A_2J_2 and identified the living forces with the heat produced by the current in the two circuits with

$$J_1^2W_1 + J_2^2W_2 + \frac{1}{a}J_1J_2\frac{dV}{dt},$$

where the third term was interpreted as the living force of one circuit under the effect of the current circulating in the other circuit. Although he claimed that his results agreed with Weber's, in point of fact he

---

125. *Ueber die Erhaltung der Kraft*, 64–6, in *WA* 1:62–3.
126. In his *Treatise on Electricity and Magnetism*, Maxwell was still quoting Helmholtz's deduction of the law of induction with praise in 1873, though by the third edition (1891), Maxwell's editor, J.J. Thomson, noted that the law of induction cannot be deduced through the principle of conservation of energy alone. See Maxwell's *A Treatise on Electricity and Magnetism* 2 vols. (Oxford: Clarendon Press, 1873; reprint: New York: Dover, 1954), 2:190–93, on 192.

dismissed two kinds of potentials that indeed exist: the mutual potential of the two currents (electrokinetic energy) and the potential of a current on itself (self-induction).[127]

These two cases, whose conflicting results both derived from Helmholtz's principle, raise a question about its heuristic utility. In the first instance, Helmholtz's deduction was really a reinterpretation of extant knowledge. In the second, he provided neither new predictions nor a rationale for the specific application of the concepts of tension and living forces. His efforts thus show that, lacking experimental data, he could not give an a priori, precise, theoretical deduction of the energy equations.

6. Conclusion

Helmholtz devoted the *Erhaltung*'s short conclusion to physiological problems, his research domain proper. Here, too, his principal problem was that of formulating the force equivalents for the energy balance. For the plant world, he declared that, due to insufficient data, precise application of the principle was impossible; all that could be said was that the stored tension forces were chemical in origin and that the only absorbed living forces were the "chemical rays of sunlight."[128] For the animal world, by contrast, he declared that relatively precise applications of his principle were possible. Summarizing his previous research for the benefit of physicists, he introduced for the first time the concepts of tension and living forces in physiology: animals utilize a certain quantity of chemical tension forces and generate heat and mechanical forces. Yet he thought that the mechanical work done by animals was only a small quantity compared to the heat they produced, and thus that it could be omitted in the equation for force equivalents.[129] On the basis of Dulong and Despretz's experimental work, he believed that the combustion and conversion of nutritive substances generated a quantity of heat equivalent to that produced by animals.[130] Helmholtz's use of the principle in this physiological context shows yet again that his principal aim was to outline, independently of the actual experimental determination of the mechanical equivalent of heat and thus independently of precise experimental corroboration, a general framework in which the principle of the conservation of force would be applicable to the largest possible class of

---

127. Planck, *Das Princip der Erhaltung der Energie*, 47.
128. *Ueber die Erhaltung der Kraft*, 69, in *WA* 1:66.
129. Cf. "Wärme," in *WA* 2:699–700.
130. *Ueber die Erhaltung der Kraft*, 70, in *WA* 1:66.

phenomena. If the exact value of the mechanical equivalent was unknown, that scarcely mattered in this context: in asserting that the work done by animals was a small percentage of the heat produced, Helmholtz showed that he could be satisfied with a gross figure.[131] Indeed, he concluded by making only modest claims for the principle itself, declaring *not* that he had demonstrated the principle but only that it was "not in contradiction to any known fact in natural science but rather that it is confirmed in a remarkable way by a great number of such facts."[132]

In accord with the sophisticated plan that he had outlined at the start of the *Erhaltung*, Helmholtz had completely united the principle of the conservation of force with the pertinent known laws of natural phenomena. At the same time, he was keenly aware that his own extraordinary theoretical efforts lacked experimental confirmation. His goal had been, he said, "to show physicists (with the greatest possible completeness) the theoretical, practical, and heuristic importance of this law [of the conservation of force], whose complete confirmation must indeed be considered as one of physics's main tasks in the near future."[133] Few pieces of scientific literature have ever expressed their goals and results so clearly as did Helmholtz's *Ueber die Erhaltung der Kraft*. It is the first full expression of Helmholtz's ideas on theoretical physics as well as on force. Both from the physical and the methodological point of view, the *Erhaltung* was a masterpiece. Yet it neither enjoyed an immediate success nor did it go uncriticized. Before Helmholtz's career and the nineteenth century ended, Clausius's criticism would lead Helmholtz to abandon his formulation of the principle of the conservation of force and its associated unifying program for physics in favor of a new regulative principle, that of least action.[134]

---

131. *Ueber die Erhaltung der Kraft*, 70, in *WA 1*:66; and see Kremer, "Thermodynamics of Life," 248 (n. 148).
132. *Ueber die Erhaltung der Kraft*, 72, in *WA 1*:67.
133. *Ueber die Erhaltung der Kraft*, 72, in *WA 1*:68.
134. See Günter Bierhalter's essay, "Helmholtz's Mechanical Foundation of Thermodynamics," in this volume.

8

Electrodynamics in Context

Object States, Laboratory Practice, and Anti-Romanticism

Jed Z. Buchwald

1. Introduction

Historians have long known that late nineteenth-century electrodynamics embraced more than the opposing poles of the field and the electric particle. In particular, during the 1870s Hermann von Helmholtz created and developed an alternative electrodynamics, one that historians have increasingly recognized to have provided the framework within which German physicists adapted field theory during the 1890s.[1] From this point of view Helmholtz's theory has appeared rather like a halfway-point between the field, on the one hand, and electric particles governed by distance forces, on the other. It shared with field theory the idea of an ether, and with particle theories the idea that actions are transmitted instantaneously across space.

This essay seeks to go beyond the traditional understanding of Helmholtz's work by uncovering a much broader and more fundamental set of concepts and practices that permeated most of German physics by the mid-1880s. These *Helmholtzian* structures were more general than the particulars of the electrodynamic ether, for they amounted to a new way of doing physics altogether, making contact with the electromagnetic field and even (if necessary) with electric

---

Acknowledgment: I thank the editor, David Cahan, for his many useful comments.

1. Gisela Buchheim, "Hermann von Helmholtz und die klassische Elektrodynamik," *NTM 8* (1971):23–36; and Jed Z. Buchwald, *From Maxwell to Microphysics: Aspects of Electromagnetic Theory in the Last Quarter of the Nineteenth Century* (Chicago and London: The University of Chicago Press, 1985).

particles, but only by transforming them into something quite different from what they had traditionally been to physicists who used them in daily practice. For Helmholtz conceived and developed the notion that physics is the science of objects that interact through system energies that are determined by the physical states of the objects at any given moment. This essay shows that a galaxy of implications for theory and for experiment emerged from Helmholtz's conception of electrodynamics, but that its radically novel structure usually remained below the surface of explicit discourse. It appeared overtly in theoretical discussions only at junctures where the usual terminology failed entirely (such as when Helmholtz's student, Heinrich Hertz, tried to use it to understand the difference between field and particle theories).[2] But in the laboratory, in the *practice* of experimental physics, Helmholtzianism had immediate and fundamental effects.

A number of students of late nineteenth-century philosophy of science have noted that many German-speaking physicists argued vehemently over the issues raised by materialism and atomism, for example, that Ernst Mach's anti-atomistic, sensationalist philosophy and Wilhelm Ostwald's anti-atomistic energetics opposed Ludwig Boltzmann's atomistics.[3] Recognition of the existence of Helmholtzianism should alter these analyses. It provided a *practical* scheme for both theoretical and experimental physics that at its core embodied a way of thinking about physical interactions that could be extended even to the micro-level without simultaneously adopting a fully reductionist atomistics. For, to use ideas that will be discussed below, molecular or even atomic entities could be embraced in the general scheme by treating them as *objects* that can have states and interaction energies. According to this way of thinking neither atoms nor molecules need

---

2. Jed Z. Buchwald, "The Background to Heinrich Hertz's Experiments in Electrodynamics," in Trevor H. Levere and William R. Shea, eds., *Nature, Experiment, and the Sciences: Essays on Galileo and the History of Science in Honour of Stillman Drake* (Dordrecht, Boston, London: Kluwer Academic, 1990), 275–306.

3. Erwin Hiebert, "The Genesis of Mach's Early Views on Atomism," *Boston Studies in the Philosophy of Science*, vol. 6 (Dordrecht: D. Reidel, 1970), 79–106; Martin Klein, *Paul Ehrenfest*. vol. 1: *The Making of a Theoretical Physicist* (Amsterdam and London: North-Holland, 1970); and Andrew D. Wilson, "Hertz, Boltzmann, and Wittgenstein Reconsidered," *SHPS* 20 (1989):245–63, provide illuminating (but considerably different) discussions of Boltzmann's views. See also John T. Blackmore, *Ernst Mach: His Work, Life, and Influence* (Berkeley, Los Angeles, and London: University of California Press, 1972); and Herbert Hörz, "Helmholtz und Boltzmann," in Roman Sexl and John Blackmore, eds., *Ludwig Boltzmann: Internationale Tagung anlässlich des 75. Jahrestages seines Todes 5.-8. September 1981. Ausgewählte Abhandlungen* (= *Ludwig Boltzmann Gesamtausgabe*, Band 8) (Graz: Akademische Druck- und Verlagsanstalt; Braunschweig and Wiesbaden: Friedrich Vieweg & Sohn, 1982), 191–205. On Mach see Hiebert, "Genesis of Mach's Views."

be considered to possess eternally fixed properties; instead, they exist in a state or states that may (or may not) prove to be quantitatively invariant.[4] The long and confusing history and significance of the Mach-Boltzmann-Ostwald triangle may benefit from a study of the actual behavior of practicing physicists during the 1890s and the early 1900s, as they reacted to Helmholtz's immense influence on both the content and the professional structure of the discipline.

This essay concentrates on *Helmholtzianism* proper in order to explain what it was, how it influenced theoretical and laboratory work during the 1870s and the early 1880s, where it originated, and why it so strongly gripped Helmholtz as well as many of his students and assistants during these years. As a consequence, the customary historical categories for discussing Helmholtz's electrodynamics require considerable reinterpretation. Previous discussions of Helmholtz's polarizable ether are usually accurate, as far as they go; but they miss altogether or do not fully perceive the foundation of Helmholtz's electrodynamics in the far more general *Helmholtzian* scheme.[5] Indeed, this essay briefly argues that Helmholtz's full embrace of the polar ether in the late 1870s was more an attempt to rescue Helmholtzianism from laboratory failure than an embrace of a proto-field physics. Though the point goes beyond the scope of this essay, it is also worth noting that the uniquely German understanding of the connection between field physics and microphysical structures that developed during the 1890s owed a great deal to the continuing influence of Helmholtzianism.[6]

Although Helmholtzianism differed fundamentally from its two competitors, the differences could be quite subtle. Accordingly, Section 2 of this essay explains the principal tenets of these three forms of electrodynamics, concentrating on their differences. Section 3 relates the difficulty that many of Helmholtz's contemporaries had in grasping his novel design through the example of an experimental failure that one of them incorrectly claimed to have detected. Section 4 then goes on to provide an example of a quintessential *Helmholtzian* at work in the laboratory, namely, Heinrich Hertz in 1880, long before he became famous for fabricating electric waves. Section 5 considers whether such

---

4. For an instructive example of how this worked see "On the Modern Development of Faraday's Conception of Electricity. The Faraday Lecture," *Journal of the Chemical Society 39* (1881):277–304, in *WA 3*:52–87.

5. A.E. Woodruff, "The Contributions of Hermann von Helmholtz to Electrodynamics," *Isis 59* (1968):300–11, on 308–10 comes closest to detecting the connection in that he recognizes the inextricable link between Helmholtz's potential and his polarizable ether.

6. Buchwald, *From Maxwell to Microphysics*, 133–201.

a thing as a self-identified *Helmholtzian* group, analogous, for example, to the British *Maxwellians*, existed, while Section 6 explores some possible social and cultural roots for Helmholtz's new form of physics.

2. Electrodynamic Interaction circa 1870: Weberean, Maxwellian, and Helmholtzian Approaches

When Helmholtz first published his own account of electrodynamics in 1870 there already existed two major approaches to the subject. The older, and certainly better developed, approach considered electric charge to consist of two kinds of particles, and electric current to consist of the equal and opposite flow of these two kinds of particles *in* a conductor. The German physicist Wilhelm Weber had created this approach in the 1840s on the basis of a hypothesis about the nature of currents first promulgated by his Leipzig colleague Gustav Fechner. By contrast, the newer, and less developed, approach argued that electromagnetic processes occurred primarily in the space *surrounding* electrified bodies. This second system derived from the discoveries and concepts of the Englishman Michael Faraday and, by 1870, had been developed by the Scotsman James Clerk Maxwell. Helmholtz's electrodynamics contrasted markedly with both of these approaches in rather complex ways that highlight its basic precepts. These differences hinged directly upon Helmholtz's image of laboratory processes.

An experimenting physicist seeking to measure charge or current requires two devices: one that generates a sufficiently large effect and one that detects it. Most physicists circa 1870 produced charge by means of an electrostatic machine, a device that operated by friction, and current by means of a battery or an electromagnetic generator. To detect charge they usually employed an electrometer, a device that measured the deflection produced in one charged object by another object against the action of some force, for example, gravity or spring tension. To detect current they usually employed a galvanometer, a device that measured the deflection of a magnet or of another current-bearing body by a given one. In both cases measurement ultimately yielded a force that acted upon the material object in the requisite state of "chargedness" or "current-bearingness." The implications of the implied relationship here between theory and measurement, which at first glance appears to be nothing more than a naive positivist one, require thorough assimilation in order to grasp the sense of the three competing approaches for electrodynamics in their historical context. To do so consider how the three measured forces.

The Fechner-Weber electrodynamic theory postulated a direct transference between the values obtained by laboratory measurements and the physical interactions that produced them. The charged or current-bearing object in effect disappeared and was replaced by a region in which electric particles moved about within a matrix of material particles. The particles exerted specific electrodynamic forces on one another, and each particle responded to a given force with a fixed acceleration in that force's direction. Through some unspecified mechanism the forces between these particles were transferred directly to the particles that formed the bodies in which they occurred, so that the laboratory detector actually measured the net force between the electric particles themselves, which acted directly as sources for the actions whose overall effects were detected. The interactions consequently occurred not between the laboratory objects but rather between entities that subsisted in these objects and that acted upon their constituents. The laboratory detectors were therefore considered to be measuring the end products of the forces between the electric particles.[7]

By contrast, the Faraday-Maxwell electrodynamic theory (or, for short, field theory) had a quite different view of laboratory measurements in that, unlike the Fechner-Weber theory, it did not postulate the direct transference between laboratory measures and physical interactions.[8] Instead of envisioning entities that were different from, yet subsisted in and acted on material objects, field theory dispensed altogether with the objects as electrodynamic entities in their own right and instead introduced something quite different from them. This new entity, the field, subsisted in the space occupied by the objects as well as in that between them. For field theorists, to say that an object was in a "charged" or "current-bearing" condition was actually a shorthand way of referring to the local state of the field, a state that depended upon the presence of matter in the immediate neighborhood. This state eventually affected other states throughout the field, including those at places where other objects might be present. If these bodies also affected the state of the field, then, depending upon whether or not this alters the energy gradient at the body's locus, they might experience a net "force." This net "force" was either one that tended to move them or one that tended to change their state. Consequently, objects with electrodynamic states might rather loosely be said to "interact"

---

7. Wilhelm Weber, "Elektrodynamische Maassbestimmungen. Ueber ein allgemeines Grundgesetz der elektrischen Wirkung," [1846] in *Wilhelm Weber's Werke*, ed. Königliche Gesellschaft der Wissenschaften zu Göttingen, 6 vols. (Berlin: Springer Verlag, 1892–94), vol. 3: *Galvanismus und Elektrodynamik, erster Theil*, ed. Heinrich Weber (1893):25–214.

8. James Clerk Maxwell, *A Treatise on Electricity and Magnetism*, 2 vols. (Oxford: Clarendon Press, 1873).

with one another since each ultimately affected the field at the other's position. Hence, according to field theory the laboratory measure did not emerge from a force-like interaction between entities that comprised the objects, but rather from a connection of an utterly different character between the objects and the state of an entity that comprised nothing but itself: the ether.[9]

Despite the profound difference between the Fechner-Weber and field theories, neither theory conceived laboratory *objects* to interact directly with one another. According to both approaches instruments detected the final results of interactions between things *other than* the laboratory objects themselves. In Fechner-Weber they interacted with electric particles, which in turn interacted with one another. In field theory they were linked only to the local state of the field, and the existence of other laboratory objects at the moment of interaction was considered irrelevant to the process: only the state of the field at the object's locus was thought to activate the detectors. In contrast to both these approaches, Helmholtz in 1870 presented one that differed radically from them in that it refused either to abstract from the laboratory objects, as in Fechner-Weber, or to introduce something entirely different in nature from them, as in field theory. In place of the unitary forces of Fechner-Weber or field theory's duality between the field and object, Helmholtz substituted what may best be called a *taxonomy of interactions*. During the next three decades his approach penetrated far into German experimental practice and theory, in particular profoundly influencing his most brilliant student, Heinrich Hertz.

Although the technical details and implications of Helmholtzian electrodynamics were quite complex, its conceptual foundations can nonetheless be conveyed relatively easily. According to Helmholtz, objects in the laboratory remained entities in their own right. A "charged" object differed from an "uncharged" one by acquiring a condition or state that it did not previously have in relation to other objects that were also charged. Similarly, current-bearing objects had no mutual interactions before they acquired their current state; afterwards, they did. Helmholtz viewed electromagnetic interactions—indeed, all interactions—as instantaneous and bipartite, and he held the nature of the interaction to depend upon the simultaneous *states* of the interacting objects, which themselves were given directly in the laboratory.[10]

---

9. Buchwald, *From Maxwell to Microphysics*, passim.
10. Helmholtz nowhere presented the full shape of his new electrodynamics in a unified way. It can, however, be reconstructed from articles that he wrote between 1870 and 1876: "Ueber die Gesetze der inconstanten elektrischen Ströme in körperlich ausgedehnten Leitern," *Verhandlungen des naturhistorisch-medizinischen Vereins zu Heidelberg* 5 (1870):84–9, in *WA* 1:537–44; "Ueber die Theorie der Elektrodynamik. Erste

The profound implications of Helmholtz's novel view are nicely illustrated by contrasting the three ways in which, *circa* 1870, electrostatic effects could be viewed. According to Fechner-Weber, a charged body was in itself no different from an uncharged one; it merely had more of one of the two kinds of electric particles than the other kind did. Consequently, electrostatic interactions were in fact always present between any two conductors, yet a net force did not necessarily result; moreover, the interactions were always between electric particles. According to field theory, a body remained essentially unchanged by its charged state. However, the field at or within its surface had its state changed as a result of the presence of a second charged body, with an accompanying alteration in the field energy gradient at the first body's locus. This translated into a force upon the object due to the unknown connections that linked the body's state to that of the field. For if an object alters the field at its position, then the field's changed state will propagate from point to point throughout it. This will eventually reach other objects and affect the field's state at their positions, thereby producing forces upon them. Consequently the interaction between field and object at a given moment depends upon other object-field interactions at previous times.[11]

In Helmholtzian electrodynamics, by contrast, there were no electric atoms to transfer force to the laboratory objects and no local field for the bodies to interact with. Rather, the *states of charge* of the bodies at any given instant directly determine their mutual interaction. A charged body, Helmholtz proposed, acts instantly and directly upon another charged body, and vice versa; the charge of the one body does not act upon the charge of the other: the bodies act, and charge is just a shorthand way of specifying the nature of the interaction. To determine what will happen one must begin with an *interaction energy* that is determined by the charged states of the objects at a given moment

---

Abhandlung. Ueber die Bewegungsgleichungen der Elektricität für ruhende leitende Körper," *JfruaM 72* (1870):57–129, in *WA 1*:545–628; "Ueber die Fortpflanzungsgeschwindigkeit der elektrodynamischen Wirkungen," *MB* (1871):292–98, in *WA 1*:629–35; "Ueber die Theorie der Elektrodynamik," *MB* (1872):247–56, in *WA 1*:636–46; "Ueber die Theorie der Elektrodynamik. Zweite Abhandlung. Kritisches," *JfruaM 75* (1873):35–66, in *WA 1*:647–87; "Vergleich des Ampère'schen und Neumann'schen Gesetzes für die elektrodynamischen Kräfte," *MB* (1873):91–104, in *WA 1*:688–701; "Ueber die Theorie der Elektrodynamik. Dritte Abhandlung. Die elektrodynamischen Kräfte in bewegten Leitern," *JfraM 78* (1874):273–324, in *WA 1*:702–62; "Kritisches zur Elektrodynamik," *AP 153* (1874):545–56, in *WA 1*:763–73; "Versuche über die im ungeschlossenen Kreise durch Bewegung inducirten elektromotorischen Kräfte," *AP 158* (1875):87–105 and *MB* (1875):400–15, in *WA 1*:774–90; and "Berichte betreffend Versuche über die elektromagnetische Wirkung elektrischer Convection, ausgeführt von Hrn. Henry A. Rowland," *AP 158* (1876):487–93 and *MB* (1876):211–16, in *WA 1*:791–97.

11. Buchwald, *From Maxwell to Microphysics*, passim.

and by their mutual distance at that moment. If a slight perturbation (or variation) of the objects' states or mutual distance alters this energy, then a corresponding force must come about in order to conserve energy. The force, in other words, is something to be deduced from the system's energy (or, in Helmholtzian terminology, from its potential function). Helmholtz's primary entities are, accordingly, objects, states, and interaction energy; everything else is secondary. Thus for Helmholtz the goal of physics (and not just electrodynamics) was to discover what states objects could have and what the possible forms of interaction energies could be—to establish, that is, a taxonomy of interactions.

Helmholtz's view had radical implications because it meant, for example, that charged objects and current-carrying objects had no necessary interactions with one another, since "chargedness" and "current-bearingness" are different states. If they did interact then a new entry had to be made in the taxonomy, and a new form of interaction energy invented. The instrumental simplicity of Helmholtz's exclusive concentration on the states of laboratory objects consequently exacted a heavy economic toll. This complex point merits further discussion, by example, for it had a major influence on *fin-de-siècle* German physics.

The Helmholtzian scheme, which places strictures on the relationships between objects and which suggests how to probe those relationships, is extraordinarily weak if it is considered to be a *theory* in the usual sense. For unlike many theories, it could not suggest—indeed, it particularly abjured—models. In practice this meant that many questions which Webereans or Maxwellians could answer at once (such as whether a charged object exerted a current-generating action on a conductor) could not be answered a priori by Helmholtzians or at least not without a great deal of argument.[12] On the other hand, Helmholtzians benefited enormously from the comparative freedom from constraint that this very abjuration of models gave them, for they were at liberty to conceive new states and new interaction energies when dealing with novel, unfamiliar laboratory situations. Webereans could not do this at all, since they were firmly restricted by their model; Maxwellians, for their part, could do something similar but more limited than Helmholtzians, since Maxwellians were free to introduce new fields (and field energies) but only under fairly rigorous constraints—for example, that a field which could move a charged object could also generate a current in a conductor. These constraints emerged directly

12. Buchwald, "The Background to Hertz's Experiments."

out of the fundamental Maxwellian requirement that only fields effect changes and that fields are states of something *other than* the interacting objects themselves. Maxwellians did not have a widely accepted model for the ether in the same sense that Webereans had a model for the current, but their scheme specified in rather elaborate detail how to construct fields.[13]

Many physicists trained in Helmholtz's Berlin laboratory used Helmholtzianism to develop a specific way of doing experiments, of approaching both familiar and novel situations. For them new object states were always possible, and if they existed there also had to be corresponding interaction energies. According to *Helmholtzian* principles, such states and energies could only be brought into the open by perturbing an existing configuration, by transforming a static situation into a dynamic one. Consequently Helmholtzian experimenters—for example, Eugen Goldstein and Hertz—were open to novel effects (and knew what to do to manifest them) in ways that Webereans and even Maxwellians were not. However, precisely because Helmholtzians were so thoroughly free they tended not to build connections between effects in the way that Maxwellians did.

The Hall effect provides a good example of the contrast in the three forms of electrodynamics. Maxwellians and Webereans had specific views on the electric current that, for example, led the former to assume that magnetic action will not affect a distribution of current, whereas, according to the latter, it must. When the American physicist Edwin Hall pursued this point shortly after beginning to read Maxwell's *Treatise on Electricity and Magnetism*, he was torn between what he now knew to be Maxwell's position and what he at first thought to be the more intuitive Weberean claim. His initial laboratory practice betrayed this conflict when he cast about for different ways to manipulate a current distribution as he strove to reconcile Maxwellian with Weberean understanding. Suppose that, as Webereans claimed, some action does occur, and suppose further that the general Maxwellian image of a current must nevertheless be maintained. The latter, as Hall rather loosely understood it, required the current in the plate to be treated as though it were incompressible—in which case it cannot be actually deflected, but only pressured. As a result he sought for and found an effect in a particular location, rather than somewhere else; whereas in other locations, which (as Hall's supervisor, Henry Rowland, evidently pointed out) are more naturally suggested by Weberean conceptions, the effect would have escaped his early experiments.[14] Precisely because

---

13. Buchwald, *From Maxwell to Microphysics*, 3–129.
14. Ibid., 78–95.

Hall freely combined concepts that *cannot* be joined consistently he was able first to conceive of the experiment and then to execute it. Theory guided him in designing his experiment, though in a most odd sense, since the theory he used was riddled with internal contradiction.

After he found (to his considerable surprise) that he could detect something—after he discovered the eponymously named Hall effect—the conflict disappeared. It was replaced by a sophisticated, highly constraining Maxwellianism based on the construction of new fields for new effects, a way of thinking which powerfully directed his subsequent laboratory work. The experiment that Hall designed to search for the effect may be said to have embedded aspects of both Maxwellian and Weberean theory, for there were direct isomorphisms between relations that subsisted within each of these two theories and the design of Hall's apparatus. This is hardly surprising since the apparatus was created precisely to look for something implied by one of the two theories (the Weberean) under constraints imposed by the other (the Maxwellian).

Helmholtzians, for their part, did no experiment *like* those of Hall. Indeed, Hall's discovery apparently excited little interest among them, probably because they viewed it as an alteration in electric conductivity rather than as a new interaction between currents. This was Boltzmann's opinion, though he admitted that it was difficult to maintain in the face of other experiments Hall had performed.[15] Boltzmann was not a Helmholtzian, but his view had been common even among some Maxwellians for a time. It would also have been congenial to Helmholtzians for a similar reason—namely, that nothing fundamentally new had to be postulated. Boltzmann's student, Albert von Ettingshausen, did perform early experiments at his instigation (and later ones in conjunction with von Ettingshausen's own student, Walther Nernst).[16] There was only one other, and more radical, way that Helmholtzians could have construed Hall's discovery: they might have interpreted it to be a new interaction between current-bearing objects. Such an interpretation would have shared many features with the Maxwellian one espoused by Hall, Rowland, and many others, but with a signal difference: it would not have signified a new *state* (field) of a mediating object (ether). However, had they taken such an otherwise unwarranted step this would still not have

---

15. Ibid., 99–100.
16. Albert von Ettingshausen, "Bestimmung der absoluten Geschwindigkeit fliessender Elektricität aus dem Hall'schen Phänomen," *SBW 81* (1880):441–52; and Albert von Ettingshausen and Walther Nernst, "Über das Hall'sche Phänomen," ibid., *94* (1886):560–610.

greatly excited them, for the interaction occurred between current-bearing conductors, and there were, in direct consequence, no further results to be obtained from it (for example, links between bodies carrying polarization currents). Only Hendrik Antoon Lorentz outside of the Maxwellian community adopted something like this interpretation, but he apparently placed it within the context of Helmholtz's polar ether.[17] Among many Maxwellians, by contrast, the Hall effect offered the chance to deploy the machinery of field energetics in a striking way, with implications that far exceeded the original discovery. For an effect to capture the attention of Helmholtzians, it had either to stress the experimental boundary for interactions between known states (for example, by reaching to high-frequency currents) or to elicit new, hitherto unknown states (as it did for Goldstein in evacuated tubes [see below, pp. 359–60]) or reveal unexplored interactions between known, but different, states (for example, an interaction between polarization- and conduction-current bearing objects). The Hall effect failed on all three grounds.

In general, Helmholtzianism had a far more limited predictive range than field theory or the Fechner-Weber approach. Yet it was a powerful tool for building an engine for discovery, because it permitted the construction of new states and interactions and because it suggested how to probe for such things. In this sense laboratory practice directly reflected Helmholtzianism. Helmholtzianism entailed unexpected forms of interaction between objects whose state of charge is changing and bodies carrying steady (or varying) currents. The practical effect of these implications was to focus the Helmholtzian experimenter's eye on dynamic, changing situations (and so, for example, to make an experiment like Hall's, which involved unchanging situations, boring). With his potential function, which determines the energy held by a pair of current-carrying objects, Helmholtz had created something that was, on the one hand, non-theoretical in the sense that the field or Fechner-Weber involved theory, yet that was, on the other, capable of yielding novel results. Furthermore, Helmholtz was entirely convinced that his electrodynamic potential embraced every permissible form of electrodynamic interaction. Anything that fit the potential was a possible theory; anything that failed to fit it could not be allowed. He saw the potential as something that stood above all theory, as something that was substantially (though not absolutely) beyond doubt. At the time (1870) he was primarily concerned to establish sets of differential

17. Buchwald, *From Maxwell to Microphysics*, 98–9, 205–9.

equations that follow from the potential function upon which his account was based and to investigate the effects of an arbitrary constant **k** that appears in it. For, Helmholtz believed, both Fechner-Weber and field theory could be embraced by his potential by assigning special values to this disposable constant.[18]

Some physicists soon questioned Helmholtz's potential because it had to emerge, if at all, as an artifact of more fundamental processes required by their often highly detailed *theories*. These critiques divided Helmholtz's attention, which had lately been concentrated on the instabilities that, he was certain, plagued Fechner-Weber. At the same time, he sought to flesh out how the potential could account for processes that seemed difficult for it but easy for Fechner-Weber or even for a primitive field theory. Moreover, he sought to discuss the seemingly unique, new implications of his potential. The experiments Helmholtz designed and stimulated others to perform so as to decide between the implications of the old Ampère law and his new potential law were directly informed by Helmholtzianism. First, they introduced a novel object of investigation (the open-ended circuit) into laboratory work. Second, they stimulated the use of uncommon tools of investigation in electrodynamics, including oscillating currents and electrostatic generators. Helmholtz intended these early experiments primarily to silence questions that arose from the difficult conceptual novelties that he had introduced. Thus, he did not consider such experiments to be connected with *theory*, because they had nothing to do with the disposable constant **k**, where he thought *theory* was localized. Yet the experiments failed to detect the novel implications of the potential, and one of them failed in a particularly disturbing way. Helmholtz now for the first time injected theory into his published discourse. Faced with the possible failure, and hence abandonment, of Helmholtzianism, Helmholtz introduced theory in the form of a polarizable ether into Helmholtzianism. Helmholtzian electrodynamics now acquired a structure similar to that of Fechner-Weber, which required particles, or to that of Maxwellianism, which required the field. What had previously been a question of eliciting the unique features of the potential now became something utterly different. Little wonder that Helmholtz later pressed Hertz to seek effects that depended directly on polarization: for without them not only his polarizable ether, but also something much more fundamental was in doubt: Helmholtzianism itself.[19]

---

18. See, e.g., "Ueber die Bewegungsgleichungen der Elektricität."
19. Jed Z. Buchwald, *The Creation of Scientific Effects. Heinrich Hertz and Electric Waves* (forthcoming), discusses these events in detail.

3. Herwig's Experiment: A Weberean Attempt to Refute Helmholtzianism

The Helmholtzian approach to electrodynamics had direct implications for laboratory practice, for, as already noted, one could only discover the states and interactions that objects possessed by manipulating them. There were no a priori grounds for knowing what would happen: there were no particles and no transmitting ether. Instead, Helmholtzian physics was founded on the assumption that every interaction involved a specific type of energy and that changes in the system's energy translated into various activities. Yet how could the working experimental physicist find these activities or connect the system energy to its actions? The answer, analytically, was through variational principles, which often translated physically into special kinds of experimental procedures: in the Helmholtzian laboratory one took a system and perturbed it so as to discover or highlight the actions that corresponded to its energy. This kind of practice was quite dissimilar to the Weberean physicist's usual procedure. The Weberean Hermann Herwig's attempt in 1874 to demonstrate empirically that Helmholtz's potential must be rejected provides a particularly apposite example of the basic difference.

Like other Webereans, Herwig mentally tended to break devices into separate pieces and to examine what occurs in each piece. The several pieces were not thought to interact, as objects, with one another. Rather, each piece defined a region in which the motions of electric particles occurred: the only proper interactions were thought to be among the particles, not between the regions to which they were confined. Hence objects in the Weberean laboratory were not understood in the same manner as those in the Helmholtzian laboratory, and so the kinds of manipulations and interpretations that suggested themselves tended to be rather different.

To understand what Herwig sought to do first requires understanding an important, indeed central, feature of Helmholtz's electrodynamics. Because it deduced "forces" exclusively from variations in the general potential function, many situations that were quite simple to analyze in the Fechner-Weber approach to electrodynamics or even in field theory became complicated for Helmholtz's electrodynamics. Consider, for example, what occurred when a metal arm was rotated about an axis through one of its ends in a magnetic field perpendicular to its plane of rotation. Both Fechner-Weber and field theory easily implied that the ends of the arm would be charged as a result of the electromagnetic induction produced by its motion. Helmholtz's potential, by contrast, implied that the arm would *not* be charged, since the symmetry of the arm's relationship to the field remained unchanged by its motion, so that no variation in the potential could occur.

Suppose, however, that the arm swept over another, stationary conductor. Then a galvanometer connected between the pinned end of the arm and the stationary conductor would, as was well known, indicate the presence of a current. Fechner-Weber and field theory attributed this current to the very same electromotive force that, they implied, charged the isolated arm. Helmholtzianism, by contrast, argued that the current resulted from the fact that, as the arm swept round, it continuously *altered* the circuit of which it formed a part, and this alteration translated into a variation in the potential. Accordingly, in Helmholtzian electrodynamics effects due to motion arose only at loci where the experimental object was undergoing an actual change in its configuration. Helmholtz called these positions "*Gleitstelle*," or slip-joints.

Herwig built a device in which a massive wire hanger with its ends fixed in position replaced the rotating wire of previous experiments (see Figure 8.1). In Herwig's rather misleading figure the magnet is actually standing vertically. The brass hanger **abc** is pivoted at **a** at the magnet's top end, with its side **bc** also standing in the vertical. In familiar lecture demonstration experiments—though not in this one—the hanger terminated in a vat of mercury, and a circuit was completed through the mercury to the hanger's upper end. It follows from Ampère's law that a hanger free to move will rotate about the magnet's axis as a result of forces exerted primarily on its vertical portion **bc**.[20]

Helmholtz's potential law, it might be thought, could not even accommodate such well-known behavior as this. The current-bearing hanger apparently remains symmetric with respect to the magnet during its rotation, so that the interaction potential cannot change, because it depends only upon the states of the interacting objects and their mutual distances, neither of which apparently changed. As a result, forces could not occur. However, as Helmholtz eventually pointed out in some detail, as the hanger swings round the circuital path formed through the mercury it is constantly changing.[21] This does indeed entail alteration in the potential because the circuit is actually changing during the motion, bringing hitherto unaffected portions under the potential's sway: in effect, the objects themselves change during the motion even though the symmetry of the arrangement remains unaffected

---

20. Hermann Herwig, "Ueber eine Modification des elektromagnetischen Drehversuches," *AP 153* (1874):262–67.
21. "Kritisches zur Elektrodynamik."

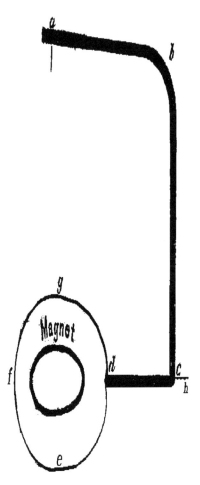

Figure 8.1. Herwig's apparatus for use in refuting Helmholtz's potential.
Source: Hermann Herwig, *"Ueber eine Modification des elektromagnetischen Drehversuches,"* AP 153 (1874):262–67, on 265.

by it. Since this occurred within the mercury (at the slip-joint where the hanger made sliding contact with the liquid), it followed that, according to Helmholtzian electrodynamics, the locus of the action that moved the hanger must lie there as well—and this formed a direct contrast with the requirements of Ampère's law. The difference between the two analyses clearly could not be tested with this sort of apparatus, and indeed Helmholtz argued that it could never be examined as long as all of the current-carrying circuits were closed.

Herwig, who thoroughly understood Weber's electrodynamics, did not so fully and clearly grasp Helmholtz's. He reasoned that the traditional apparatus could probe the difference between Ampère's and Helmholtz's implications for the locus of the action by removing the mercury and, instead, linking the lower end of the hanger **c** rigidly to the end **d** of a fixed wire **cd** located on one side of the vertical magnet. Then (to replace Helmholtz's slip-joints) Herwig formed a thin metal wire **def** which he folded into a semi-circle whose center coincided with the magnet's axis, and which therefore lay in the horizontal plane (that is, in Herwig's diagram **def** and **dgf** are actually perpendicular to the vertical portion **bc** of the wire).

With this configuration the hanger cannot rotate freely; but it can be torqued, and Herwig set out to examine the effect. He at once found that the torque existed, even though Helmholtz's fluid "*Gleitstelle*" had been removed. Herwig was, however, well aware that the fine wire **def** *would* necessarily break its symmetrical position with respect to the magnet's axis since one end of it was tied to the end of the hanger, and the hanger had to move slightly in response to the torque. This naturally entailed a change in the circuit's configuration, which was just the kind of alteration that Helmholtz's account required. Herwig accordingly tried to find some way of demonstrating that the torque on the hanger could not possibly derive from an effect localized in the connecting wire **def**.

To do so he attached another wire, **ch**, to the lower, horizontal part **cd** of the hanger. In a first experiment Herwig insulated **h**, so that the entire hanger bore the current. Here the torque occurred in full measure. He then insulated point **a** of the hanger, and passed the current through **h**, so that only the part **cd** of the hanger carried a current. Here he found that the torque dropped to a tiny fraction of its previous value, and he concluded that the rotational action must occur between the magnet and the vertical part **bc** of the hanger (as Ampère's law, but not Helmholtz's potential, required) since it occurs only when **bc** carried a current.[22]

Helmholtz replied almost immediately.[23] Herwig had wrongly assumed, Helmholtz remarked, that the added wire **ch** is electrodynamically ineffective. Certainly the potential does require that the hanger's vertical part cannot be the locus of any action that occurs, and yet the action vanishes when only **chd** carries a current. But,

---

22. Herwig, "Ueber eine Modification."
23. "Kritisches zur Elektrodynamik," 767–71.

Helmholtz argued, the action does not vanish because the current in **bc** has been removed, as the Ampère law would have it. The situation, he maintained, is rather more complicated than Herwig had assumed, for the second experiment is no more equivalent to the first one from the standpoint of the potential law than it is from that of Ampère's law. With a current flowing from **h** through **c** and on through **def**, Helmholtz noted, a slight displacement of the lower end of the hanger will change the potential of *both* **def** and **ch** with respect to the magnet. Furthermore, the change for the one is opposite to the change for the other, with the result that they are twisted in opposite directions, producing a vanishing net result. Far from demonstrating that **def** cannot be the locus of the twist, Helmholtz concluded, Herwig's experiment is fully consistent with that claim.[24]

Herwig had missed this point because, as a Weberean, he tended to look at his device from within the viewpoint of a highly articulate theory, and so envisioned pieces of wire as objects that bridle the interacting electric particles. The order and placement of the device's parts consequently acted as essential and unalterable constraints on his Weberean analysis: to understand what happened, the regions to which the particles were confined had to be the givens of the problem. Accordingly, Herwig neglected the small deformation that the wire **ch** necessarily experiences when it is torqued, especially since, undeformed, the wire has no potential at all with respect to the magnet. A deformation would simply make the problem much too complicated to solve, for if the particles' loci could not be precisely specified, then their actions could not be estimated.

Helmholtz, by contrast, held a radically different viewpoint. He looked at the device with a decidedly untheoretical eye and saw malleable objects that determined the device's system energies as a function of the distances between their parts. Like the Webereans, he too broke the device up into pieces—albeit into volume elements—but he did not go beyond the pieces. For him the order and placement of the parts were not fixed constraints, since he had actually to deform them in order to calculate the changes in the energies that translate into bodily forces. The Helmholtzian, therefore, tended naturally to think about deforming or mutating a system, whereas the Weberean rarely, if ever, did. This was how the primitive differences between the schemes began to come to life in the laboratory.

By the early 1870s Helmholtz himself certainly found it difficult to think in any other way than his own about electrodynamic interactions.

24. Ibid., 763–73.

In concluding his analysis of Herwig's mistakes (as well as similar ones by Friedrich Zöllner, about whom more presently), Helmholtz remarked:

> The potential law requires one and the same simple, proportionate mathematical expression to encompass the entire, experimentally known realm of electrodynamics, and ponderomotive and electromotive effects; and to the area of the ponderomotive effects it brings the same great simplification and lucidity that the introduction of the idea of the potential brought to the study of electrostatics and magnetism. I myself can bear witness to this, since for thirty years I have applied no other fundamental principle but the potential law and have needed nothing else in order to find my way through the rather labyrinthine problems of electrodynamics.[25]

The practical result of Helmholtzianism's foundation in the potential was to make its electrodynamic laboratory a much more flexible, manipulative place than a Weberean laboratory could ever be. In the latter, one concentrated largely on measuring constants using devices with given, unaltered structures, such as ballistic or standard galvanometers. These pieces of measuring equipment would rarely if ever have been used in conjunction with other devices whose behavior could not be calculated from theory beforehand. Consequently, in the Weberean laboratory apparatus rarely changed in a fundamental way (since that would have destabilized any measurement) and interesting new effects rarely emerged (except as an occasional result of extremely long, elaborate sequences of measurements).[26] In the Helmholtzian laboratory, by contrast, many electrodynamic devices could not be treated as fixed things, because to yield potential effects the Helmholtzian experimenter had to bring about changes in them. Helmholtzian laboratories were places for seeking out unknown phenomena; Weberean laboratories were places for measuring unknown constants.

25. Ibid., 772.
26. David Cahan, "Kohlrausch and Electrolytic Conductivity: Instruments, Institutes, and Scientific Innovation," *Osiris*, Kathryn M. Olesko, ed., 2nd ser. 5 (1989):167–85, discusses Friedrich Kohlrausch's experiments on electrolysis, which provide an excellent example of the extreme concentration on measurement that characterized Weberean laboratory practice. In the late 1870s Kohlrausch correlated immense amounts of elaborately measured data on electrolytic conductivity that eventually enabled him to generate an empirical law for calculating the conductivity of a solution from those of its solutes.

4. A True Helmholtzian: Heinrich Hertz

No more apposite historical example of Helmholtzianism's effect on laboratory practice exists than the young Heinrich Hertz's first experiment in Berlin. Hertz first matriculated at the University of Berlin in October, 1878, the fall after Helmholtz moved from his old institute into his new one, whose quarters included large lecture halls, laboratories designed for introductory instruction and advanced research work, and a library.[27] Under construction since 1873, the new physics institute was and remained the costliest in Germany until after the turn of the century,[28] and its facilities were very well adapted for specialized work. Hertz did not initially intend to sign up for the laboratory; but then he discovered that "one of this year's prize problems more or less falls into my field," to wit electrodynamic measurement. From the very beginning, then, electromagnetism captured Hertz, and almost immediately Helmholtz took an interest in him. "I have already talked about it [a prize problem in electrodynamics put by the Prussian Akademie der Wissenschaften] with Prof. Helmholtz," Hertz wrote his parents, and "he has kindly given me some information about the literature." No doubt Helmholtz included his own articles among "the literature," and Hertz began by "spending most of [his] time in the reading rooms."[29] Hertz quickly recognized the highly competitive environment that he had joined, and he was at first reluctant to commit himself fully to the project "since I may fail."[30] Helmholtz encouraged him: "I reported to Prof. Helmholtz yesterday [November 5] that I had thought over the matter up to a point and would like to begin. He went with me to see the demonstrator and was kind enough to spend another twenty minutes in discussion as to how best to begin and what instruments I should need."[31] On that day Helmholtz began molding Hertz into a Helmholtzian, a process that developed extremely rapidly (as Hertz tackled the prize problem), and was well

---

27. Christa Jungnickel and Russell McCormmach, *Intellectual Mastery of Nature: Theoretical Physics from Ohm to Einstein*, 2 vols. (Chicago and London: The University of Chicago Press, 1986), vol. 2: *The Now Mighty Theoretical Physics 1870–1925*, 28–9.
28. David Cahan, *An Institute for an Empire. The Physikalisch-Technische Reichsanstalt 1871–1918* (Cambridge, New York, New Rochelle: Cambridge University Press, 1989), 21–2.
29. Hertz to his parents, 31 October 1878, in Johanna Hertz, ed., *Heinrich Hertz, Memoirs. Letters. Diaries*, 2nd enl. ed., eds. Mathilde Hertz and Charles Susskind, trans. Lisa Brinner, Mathilde Hertz, and Charles Susskind (San Francisco: San Francisco Press, 1977), 93–4, on 93.
30. Hertz to his parents, 6 November 1878, in ibid., 95–6, on 95.
31. Ibid.

underway if not complete by early February 1879, when Hertz decided that he had finished.

Hertz's first experiment was designed to bring out the inductive relations between objects in current-bearing states. As his device resisted producing sufficiently significant results, Hertz transformed it by stripping away everything that made the inductive links impure (see below). His procedure reflected the experimental craft he had learned in Helmholtz's laboratory. Figure 8.2a is Hertz's own, rather compressed, and difficult drawing of his initial experimental setup.[32] His complicated figure reflects the contemporary state of circuit science: telephony scarcely existed, and electrical engineers had little understanding of the functions of devices like inductors, with the result that circuit science, and attendant standards for quick and unambiguous communication, remained undeveloped. Figure 8.2b, which employs modern conventions, is a redrawing of Figure 8.2a, and is designed to highlight several pertinent features. The device in question is a Wheatstone bridge with double-wound, spiral inductors placed in diagonally opposite branches. ("Double-wound" means that each inductor contained two independent strands of wire, each of which was wrapped around the cylinder in parallel paths. Hertz's experiments concerned the coupling between the two strands in a single spiral.) In his experiment, Hertz used a single commutator—though there appear to be two in Figure 8.2b—to collapse the current in the bridge approximately ten times per second; this produced the inductive actions. The ballistic galvanometer (G') hooked across the bridge measured these actions, and from them Hertz deduced inductive couplings. Hertz sought to determine whether these couplings conformed to a certain expression, about which he was uncertain. He did so by following the Helmholtzian approach, namely, to concentrate on the problematic elements of the device—on those parts of the device that did the interacting—and then to change them.

Figure 8.3 is a rendering of Hertz's altered device (which he never drew). Here the spiral inductors have been removed; there are two circuits instead of one, and the inductive coupling occurred between the circuits as wholes. (In the previous configuration, by contrast, the predominant coupling occurred between the two distinct windings within the far-right spiral.) The Wheatstone bridge was placed on a table; on the floor directly beneath it was a rectangular circuit, which replaced one of the inductors. This transformed experimental setup

---

32. Heinrich Hertz, "Experiments to Determine an Upper Limit to the Kinetic Energy of an Electric Current," in his *Miscellaneous Papers*, trans. D.E. Jones and G.A. Schott (London: Macmillan, 1896), 1–34.

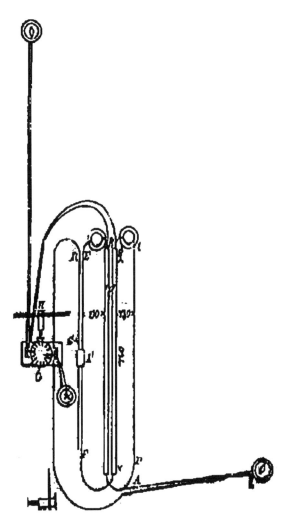

Figure 8.2a. Hertz's original diagram of a Wheatstone bridge with double-wound spiral inductors. Source: *Heinrich Hertz, "Experiments to Determine an Upper Limit to the Kinetic Energy of an Electric Current," in his* Miscellaneous Papers, *trans. D.E. Jones and G.A. Schott (London: Macmillan, 1896), on 22.*

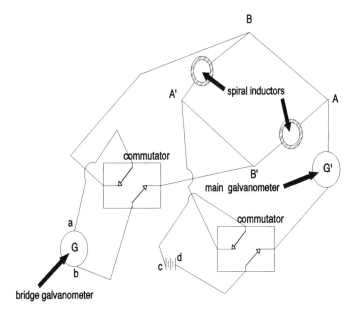

Figure 8.2b. A modern rendering of Hertz's original diagram (see text for discussion).

improved the experiment's accuracy without at all changing the actual measuring devices (the galvanometers). Paradoxically, it did so by reducing the magnitude of the effect since the coupling between the floor and table circuits is vastly smaller than the coupling between the independently wound strands in the spiral inductor.

Hertz had thus transformed his initial device in a most particular way, stripping it of anything that obscured the inductive links between the objects that comprised it. He envisioned the device, in its two major forms, as a set of inductive couplings. In the device's initial configuration the couplings occurred between the separately wound strands in each of the two spirals. In one experiment that probed interactions with small inductances the strands coupled to one another in the same way in both spirals; in another experiment that probed very large inductances, the coupling remained the same in one spiral but was multiplied over 200 times in the other spiral. In Helmholtzian terms, the strands in each spiral formed a system with a characteristic interaction energy. When the weakly coupled system yielded results that were vitiated by the difficulty of calculating the smaller, and measuring the larger, inductance, Hertz created new systems using a pair of rectilinear circuits whose coupling could always be very accurately computed, whether large (for currents in opposite directions in the

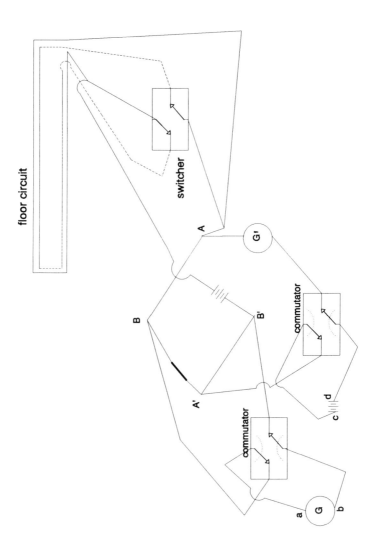

Figure 8.3. A modern rendering of Hertz's apparatus with a floor circuit in place of a spiral inductor (see text for discussion).

circuits) or small (for currents in the same directions in them). Yet the rectilinear arrangement actually reduced the larger inductance and left the smaller one essentially unchanged in magnitude. Its advantage was not to magnify grossly a marginal effect in order to bring it within the ambit of the detectors, but rather to *purify* the inductive couplings in a way that made them accurately computable. By concentrating on the objects that together formed the interacting system Hertz was rather simply and directly able to extract the inductive links that bound its components together.

This archetypically Helmholtzian approach of manipulating coupled objects was rather unnatural, though not inconceivable, to a Weberean. A Weberean performing Hertz's experiment would almost certainly have sought to improve the measuring device's accuracy or the current strength before altering the initial form of the bridge in so radical a manner as Hertz did, because, for him, the wires were the experiment's fixed scaffolding. They contained and directed, but they did not act; there would have been no reason for him to consider the wires directly, much less to alter them. If the Weberean had been interested in calculating what the galvanometers measured, he would have supposed the existence of currents with arbitrary magnitudes and rates of change, and he would then sum up the various forces that move the electric particles through the galvanometer. For Hertz, by contrast, the wires were not scaffolds; they neither directed nor contained. Instead, they were the active entities, and he immediately set about understanding their mutual relations. Precisely because the wires were, for him, electrodynamically alive, a true Helmholtzian like Hertz inevitably altered a structure in order to probe or enhance the systemic effects. In short, in the Weberean laboratory, one built, measured, and then analyzed; in the Helmholtzian, one preferred to build, test, and alter the setup until the device behaved as a system in a satisfactory manner—that is, until it ceased to resist the experimenter's attempts to elicit appropriate effects. This was one aspect of the special craft, the unique experimental culture, that Hertz imbibed in Helmholtz's laboratory, a craft that in the mid-1880s carried him into the heartland of unexplored electrodynamic territory.

5. From Mechanical to Relational Physics

Despite the power and systematic approach of Helmholtzian physics, no self-identified community of Helmholtzian physicists emerged in Germany or at least nothing comparable to that community of Maxwellian physicists which emerged in Britain after 1870. In Britain, students who aimed in the early 1880s to become professional physicists studied from three canonical texts that together presented a set of philosophical viewpoints, physical principles, and experimental techniques, and that together forged for them a unified understanding of field physics. William Thomson's and Peter Guthrie Tait's *Treatise on Natural Philosophy* showed them how to deploy variational principles based on energy densities for material continua. From the *Treatise* British students learned how to derive forces and the resulting motions for continua by placing energy expressions either in Hamilton's principle or in Lagrange's equations; and they learned that the most advanced and interesting subjects in modern dynamics concerned the continuum rather than the interactions of discrete particles. Lord Rayleigh's *Theory of Sound* also explored and applied variational principles, and it treated only continua. Finally, James Clerk Maxwell's own *Treatise on Electricity and Magnetism* showed how to build that subject on a similar basis. By the early 1880s Maxwellians knew these texts thoroughly; they thus had a sense of pursuing physics in a common fashion. Moreover, Maxwellianism yielded new results in its theoretical structure (for example, Poynting's theorem), in laboratory discoveries (for example, the Hall effect), and in unifying disparate phenomena (for example, the Hall, Faraday, and Kerr effects).[33] In short, by the 1880s the Maxwellians evinced a strong sense of intellectual community identity.

Helmholtzians, by contrast, had at best a weak (or at least a comparatively weak) sense of community identity. Several factors account for this. First, Helmholtz produced no textbook. Students could learn his style of physics only by studying his journal articles, attending his lectures in Berlin, or working in his laboratory there. Second, Helmholtz's electrodynamics suffered from a sense of anti-climax: it seemed, at best, an addendum to or a reinterpretation of previously known results. Coupled to the difficult and highly problematic concept of a polarizable ether, Helmholtz's approach to electrodynamics could indeed yield electromagnetic waves, but then Maxwell had already (albeit confusingly) obtained them. Hence many German (as well as French and Dutch) physicists believed that Helmholtz had, in effect, only glossed Maxwell's difficult theory. Nothing novel seemed to follow from Helmholtz's structure: during the mid-1870s Helmholtz himself had worked hard to elicit effects that follow from his approach; yet he and his laboratory failed to do so. On the contrary, he and his collab-

33. Buchwald, *From Maxwell to Microphysics.*

orators' experimental results apparently pointed to the complete equivalence (as they understood it) of Helmholtz's and Maxwell's electrodynamics. In sum, during the 1880s and the first-half of the 1890s there was only a quiet, largely covert Helmholtzianism.

This quiet Helmholtzianism manifested itself in two senses. First, much of the work undertaken in Helmholtz's laboratory during the 1870s and early 1880s was strongly influenced by the general approach and the specific details of his relational electrodynamics. ("Relational" because it was founded on a presumed relation between a pair of objects in given states and at a given distance from one another.) For example, Eugen Goldstein's investigation of electric discharge in partly evacuated glass tubes and Hertz's early experimental work on inductive couplings were both relational in their understanding of electrodynamic phenomena in that they presumed a relation between a pair of objects in given states. In Hertz's case, as already discussed, the relationship was quite overt since it obtained between electric circuits.[34] In Goldstein's case the relationship was more subtle but equally present.

About 1874 Goldstein, following Wilhelm Hittorf and Julius Plücker, undertook a series of experiments with partly evacuated tubes subject to high electric potentials. When the voltage is turned on, a glow is given off by the remaining gas and by the walls of the tube, a glow that could be moved about by magnetic action. The glow does not extend all the way to the cathode; a dark space whose extent is proportional to the tube's evacuation surrounds it. Details aside, the important point about Goldstein's investigations is that, according to Goldstein, the glow does not *directly* involve either conduction or electrolysis. It does not, that is, represent the effects of the transfer of electric charge from point to point throughout the evacuated tube. Instead, the discharge involves a series of successive, localized "ray" generations: where this occurs a particle becomes the source of new rays, in much the same way that the rays originally emanate from the cathode, with these new rays in turn stimulating other particles of the residual gas to produce rays. In Goldstein's words:

> ... each secondary negative bundle [of light produced by processes that begin at the cathode] represents a *motion which, excited at the point of origin of the bundle, is transferred to the surrounding medium*; hence each particle affected [other remarks here as well as elsewhere make clear

---

34. For an example of the effect of Helmholtzianism on Hertz's understanding of electrodynamic theory see Buchwald, "The Background to Hertz's Experiments."

that the particles Goldstein referred to formed the residual gas in the tube—they were not, that is, components of the polarizable ether, which itself transmitted the rays that affected the particles in question], as far as the excitation is propagated, assumes the characteristic form of motion which is produced at the *point of origin* of the rays; whilst a comparison of the discharge at any point with conduction in metals and electrolysis can afford a guide only for the relationships at the point itself.[35]

Goldstein's remarks mirror Helmholtzian tenets. A "motion" (*state*) characterizes the cathode and elicits an interaction between the cathode and the "surrounding medium."[36] This interaction generates "rays" in the medium. These rays bear the stamp of the interaction, and when they strike a material particle of the residual gas a new interaction between the medium and the particle occurs. This interaction endows the latter with the same *state*—the particle "assumes the characteristic form"—that the cathode itself had possessed in its original ray-generating interaction with the medium. The process then repeats itself, only now the particle replaces the cathode as the ray-generator. The particle then couples to the medium and the medium to the particle throughout the path marked by the secondary light. The nature of the process, Goldstein insisted, must be distinguished from conduction. For here, among the particles in the glowing tube of rarefied gas, the last interaction in the sequence bears the stamp of the very first one. Meaning that it has the same *state* as the first particle because its state results from the original cathode state through a causal sequence. Conduction, on the other hand, is a *state* of an object as a whole, so that it makes no sense to think that one part of a conducting object *causes* the state in another part of the same object. In Helmholtzian physics interactions occur between distinct objects, and they always bear the mark of the object *states*. Goldstein's "ray" accordingly represented whatever *state* of the medium interacted with the corresponding *state* of the cathode or the other objects (particles) that were embedded in the medium.

The second way in which Helmholtzianism manifested itself was more subtle than the first, for it spread into nearly every area of physics, and so the nature of its influence warrants greater discussion. It was pointedly articulated by Ludwig Boltzmann, who had also worked for

---

35. Eugen Goldstein, "On the Electric Discharge in Rarefied Gases. Part I," *PM 10* (1880):173–90, on 185.

36. Goldstein understood this medium to be Helmholtz's polarizable ether. As a result, his views went considerably beyond Helmholtzianism proper to embrace higher-order theory as well. A complete account of them would take us far beyond this essay's present purposes.

a time in Helmholtz's laboratory in the 1870s, in an introductory article to the catalogue of the mathematical exhibition, an exhibition held in September 1893 in Munich by the Association of German Mathematicians. Boltzmann's article, "Über die Methoden der theoretischen Physik," contrasted past and present methods of physics. From the late eighteenth century to *circa* 1860, Boltzmann claimed, the method of physics had involved specifying particles and the forces they exert. "The aggregate of these methods was so successful," he declared, "that to explain natural phenomena was defined as the aim of natural science; and what were formerly called the descriptive natural sciences triumphed, when Darwin's hypothesis made it possible not only to describe the various living forms and phenomena, but also to explain them."[37]

According to Boltzmann, the goal of physics, indeed of much of natural science, through at least the 1860s had been the provision of specific mechanisms—for example, Maxwell's repulsive forces for gas particles or Darwin's natural selection—for obtaining observable phenomena. For physics, this meant that the equations governing phenomena were generally derived from, or at least closely associated with, models of the processes that lay behind the observable world. Yet just as this method had begun to penetrate the other natural sciences, Boltzmann argued, "strangely enough Physics made almost exactly at the same time a turn in the opposite direction." Although the old method of model-making was too strongly rooted in mechanics to disappear at once, in newer areas, he remarked,

> the view gained ground that it could not be the object of theory to penetrate the mechanism of Nature, but that, merely starting from the simplest assumptions (that certain magnitudes are linear or other elementary functions), to establish equations as elementary as possible which enable the natural phenomena to be calculated with the closest approximation; as Hertz characteristically says, only to express by bare equations the phenomena directly observed without the variegated garments with which our imagination clothes them.[38]

Boltzmann argued that this method, which differed from the old one of model-making based on particles and forces, differed also from another new one, based on physical analogy, developed by Thomson

---

37. Reprinted in his *Populäre Schriften* (Leipzig: J.A. Barth, 1905), 1–10; and translated as "On the Methods of Theoretical Physics," *Proceedings of the Physical Society of London* 12 (1893):336–45, on 340.
38. Ibid., 340–41.

and Maxwell. The essence of the latter British method, he said, consisted in grouping disparate areas under a single analytical pattern and then using the best understood of the areas to work out the pattern's consequences. But even in Britain, Boltzmann continued, this third method, originally different even from the new one that refused to penetrate the mechanism of nature, gradually merged with it, so that "in Maxwell's [later work] the formulae more and more detach themselves from the model, which process was completed by [Oliver] Heaviside, [John Henry] Poynting, Rowland, Hertz, and [Emil] Cohn." Analytical structure based on analogy gradually came to replace visual representation. As Boltzmann concluded in a beautifully resonant passage:

> The new ideas, however, gradually found entrance into all regions. . . . It was seen indeed that they corresponded better to the spirit of science than the old hypotheses, and were also more convenient for the investigator himself. For the old hypotheses could only be kept up as long as everything just fitted; but now a few failures of agreement did no harm, for it can be no reproach against a mere analogy if it fits rather loosely in some places. Hence the old theories, such as the elastic solid theory of light, the theory of gases, the schemes of chemists for the benzol rings, were now only regarded as mechanical analogies, and philosophy at last generalized Maxwell's ideas in the doctrine that cognition is on the whole nothing else than the discovery of analogies. With this the older scientific method was defined out of the way, and Science now only spoke in parables.[39]

Boltzmann's florid remarks reflected his conviction that his own work in gas statistics was itself rather out of harmony with the present temper of science, even though he himself was thoroughly addicted to mechanical modeling, having avidly pursued it in electrodynamics and thermodynamics. His bracketing of Maxwellians like Poynting and Heaviside with Hertz (on electrodynamics) and Helmholtz (on mechanical analogies to the second law)[40] revealed his conviction that contemporary physics had adopted completely new methods. It did not use particles or force, and it did not investigate models in detail.

---

39. Ibid., 344–45.
40. Boltzmann himself later pursued this topic; see Martin Klein, "Mechanical Explanation at the End of the Nineteenth Century," *Centaurus* 17 (1972):58–82. For a most informative discussion of Boltzmann's understanding of mechanical analogy see idem, *Ehrenfest*, 53–74. Wilson, "Hertz, Boltzmann, and Wittgenstein," discusses differences between Hertz and Boltzmann that hold as well for (and probably derive from) differences between Helmholtz and Boltzmann. See also Engelbert Broda, *Ludwig Boltzmann: Mensch, Physiker, Philosoph* (Vienna: Franz Deuticke, 1955).

Instead, it produced scientific "parables" whose common analytical structure stood in isolated self-fulfillment. In short, the new method was Helmholtz's relational physics. By the early 1890s many German physicists were headed rapidly in that direction.

6. The Roots of Helmholtzianism: Anti-Romanticism

Assuming Boltzmann's account of the new approach to physics to be correct, two sets of obvious and pertinent questions confront the historian. First, where did Helmholtz's relational physics come from and why did it emerge in the late 1860s and early 1870s? Second, why was it so powerfully propagated by him from his professorial chair in Berlin? What did he and many of those who worked in his laboratory see in it that led them to carry it throughout physics, including, as in Hertz's final work, into the heart of mechanics itself? Although definitive answers to these questions cannot be provided here, a sketch of the elements of such an answer may offer some enlightenment.

Helmholtz completed his first studies on a new electrodynamics in the late 1860s, when he was still a professor of physiology at Heidelberg and where his career had come to full bloom. With the death of his one-time teacher at Berlin, Gustav Magnus, a major professional change occurred in Helmholtz's career: in 1871 he left Heidelberg and physiology to assume Magnus's prestigious chair in experimental physics at Berlin. In his inaugural lecture at Berlin and in other addresses published collectively as *Populäre wissenschaftliche Vorträge*, Helmholtz signaled his growing concern for what he perceived as an unfortunate and reactionary trend in some parts of German science. By the second-half of the 1870s Helmholtz apparently saw Weberean physics, particularly in the extreme form advocated by one of its most zealous supporters, the notorious xenophobic and anti-Semitic astrophysicist Friedrich Zöllner, as the embodiment of a cowardly metaphysics that opposed all that he had long embraced, including his view of the German university as a bastion of freedom and of science freed from the influence of a pernicious philosophy.[41]

When Wilhelm Weber first produced his atomistic electrodynamics at Leipzig in the late 1840s, he made no extravagant claims for it, concentrating instead on demonstrating its power to integrate several known electromagnetic effects into a single type of action and on pursuing exact electric measurement in the Gaussian tradition. However,

---

41. For further discussion on this point see David Cahan's essay, "Helmholtz and the Civilizing Power of Science," in this volume.

the basis of Weber's theory—namely, Fechner's hypothesis concerning the electric current—was itself embedded in a much more far-reaching program, one that sought to transform the analytical structure of atoms and forces into something that resonated with metaphysical meaning and that inevitably had powerful social implications. Several historians have recently argued that Weberean electrodynamics appropriated only the apparatus, not the positivistic attitudes, deployed by early nineteenth-century French physicists of atoms and distance forces.[42] They argue that, far from being positivistic, Weberean physics sought to incorporate the imperatives of certain forms of German Idealism while avoiding the mystifications of *Naturphilosophie*. The primary figure in this movement, they claim, was Fechner, who had abandoned physics for psycho-physics after the (temporary) blindness that he suffered had led him to resign his chair at Leipzig, a chair filled thereafter by Weber. Part of Fechner's goal, in this view, was to dematerialize the physics of particles and forces through atomic points. These were unextended centers which pure, immaterial force linked together in pairs. According to this view, matter (atomic points) is not a passive subject of a different, active principle (force); instead, "force" is the manifestation of a living connection between the intrinsically active points. Far from being inert or passive, matter resonates with activity.

Although the extent of Fechner's influence on Weber is debatable, it is certain that Weber's disciple Zöllner, at least, understood atoms and electrodynamic force in a different way from the French physicists of the early nineteenth century. Physical connections had little if any metaphysical character for the French, who had regarded *Naturphilosophie* as simply ridiculous and for whom the central issue was how to create positive science.[43] Some German electrodynamicists, by contrast, apparently invested atoms and force with meaning that went beyond positive science. They eliminated inert matter from the world and populated it instead with centers of vital interaction. In this world the ultimate physical objects are eternally the same, and each is bound to every other object in a bipartite connection that reflects their living

---

42. See Arthur Molella, "Philosophy and Nineteenth-Century Electrodynamics: The Problem of Atomic Action at a Distance," (Ph.D. dissertation, Cornell University, 1972); and M. Norton Wise, "German Concepts of Force, Energy, and the Electromagnetic Ether: 1845–1880," in G.N. Cantor and M.J.S. Hodge, eds., *Conceptions of Ether: Studies in the History of Ether Theories 1740–1900* (Cambridge, London, New York: Cambridge University Press, 1981), 269–307.

43. See, e.g., Maurice Crosland, *The Society of Arcueil: A View of French Science at the Time of Napoleon I* (Cambridge, Mass.: Harvard University Press, 1967), 89–91, who translates (on 90) the following remark from Laplace's *Système du monde*: "The true object of the physical sciences is not the search for primary causes but the search for laws according to which phenomena are produced...."

interaction. This physics was, therefore, fundamentally a relational one, but its relational character did not extend to the objects with which we have to do in the empirical world. For such things were composed of a myriad of atomic points, and the relation subsisted between the points, not the bodies.

Helmholtz's electrodynamics of the 1870s embraced, indeed was explicitly founded on, the axiom that bodies interact with one another through irreducible bipartite relationships. In the Helmholtzian world, as in the Weberean, things did interact in pairs. But Helmholtz's electrodynamics went much further in this vein than Weber's did, and with radically different implications, for Helmholtz's things were not Weber's things. It is entirely possible to dispense with a relational understanding of force and yet still retain the analytical and physical structure of Weberean theory, including his later universal models for all of physics. Whether point atoms are thought to create and to be guided by independent forces, or whether the forces are symbols of an unbreakable unity between points, has no effect at all on the theory's testable consequences.

Helmholtzian physics was utterly different. It rejected atoms in Weber's sense and made no direct use of forces. Instead of building bodies out of invariant and inaccessible entities (Weber's metaphysically pregnant atoms) it built them out of small pieces that were entirely similar to the bodies proper that it took them to be, that is, essentially as they were in the *laboratory* world. Instead of forces linking atoms (or even energy determined by atom pairs), Helmholtz considered the energy of a *system* consisting of a pair of volume elements in given states and at a given distance from one another. A relational understanding of the actions between objects lay at the very core of this enterprise precisely because the relationships were not determined by the invariant natures of the bodies. It was as if Helmholtz had inserted relational physics into phenomena rather than into noumena, and so thereby succeeded in demystifying it.

During the 1850s and 1860s, first at Königsberg, then at Bonn, and finally at Heidelberg, Helmholtz built an extraordinarily influential career through his mastery of new fields of research, in particular physiological acoustics and optics. Moreover, he occasionally, and with great power, produced work in mathematical physics (for example, his 1857 theory of vortex motion). Yet for the most part he did not pursue energy physics. In the late 1860s he changed callings: he began to concentrate on physics proper and to develop a way to free it from the impulsion towards atomic modeling that was present even in his own original formulation of the conservation of forces, a formulation which had indeed deployed atoms and forces.

Very early in his career Helmholtz had probably held a relational view of force,[44] but it was then embedded in an atomistic world. He wrote in 1847:

> Motion is the alteration of the conditions of space. Motion, as a matter of experience, can only appear as a change in the relative position of at least two material bodies. Force, which originates motion, can only be conceived of as referring to the relation of at least two material bodies towards each other; it is therefore to be defined as the endeavor of two masses to alter their positions. *But the forces which two masses exert upon each other must be resolved into those exerted by all their particles upon each other; hence in mechanics we go back to forces exerted by material points.* The relation of one point to another, as regards space, has reference solely to their distance apart: a moving force, therefore, exerted by each upon the other, can only act so as to cause an alteration of their distance, that is, it must be either attractive or repulsive.[45]

Helmholtz thus began with force as the originator of motion, and insisted on its relational (though not necessarily bipartite) character. He then introduced the vector property of force, and from this he leapt to the world of "material points," which enabled him to insist on force's uniquely spatial character. The "material point" consequently had two functions: first, it represented the final stage of material decomposition, and, second, it made force ineluctably spatial. But the point was not in itself a foundation of Helmholtz's argument since he presupposed both the vector character of force as well as its purely spatial dependence. The "material point" provided a uniquely suitable way to embody both characteristics.

For Helmholtz in 1847 the central aspect of a fundamental interaction was its pure dependence on space; it was not a mystical marriage of two entities that transcended in its wholeness the sum of their individualities. The signal importance of spatial relation derived from its prior association with causal invariability. Helmholtz's insistence on invariability may reflect a conviction that natural actions have no teleological elements whatsoever or at least that the proper object of science is causal necessity, which amounted to an outright rejection of Romantic *Naturphilosophie* as well as a highly skeptical attitude

---

44. Wise, "German Concepts," 296.
45. "On the Conservation of Force: A Physical Memoir," trans. John Tyndall, in *Scientific Memoirs, Natural Philosophy*, eds. J. Tyndall and W. Francis (London: Taylor and Francis, 1853), 114–62, on 117. (Emphasis added.)

towards metaphysics of the Hegelian sort.⁴⁶ "The problem of the sciences," Helmholtz remarked well before introducing "material points," was

> to evolve the unknown causes of the processes from the visible actions which they present; ... to comprehend these processes according to the laws of causality. We are justified, and indeed impelled in this proceeding, by the conviction that every change in nature *must* have a sufficient cause. The proximate causes to which we refer phenomena may, in themselves, be either variable or invariable; in the former case the above conviction impels us to seek for causes to account for the change, and thus we proceed until we at length arrive at final causes which are unchangeable, and which therefore must, in all cases where the exterior conditions are the same, produce the same invariable effects. The final aim of the theoretic natural sciences is therefore to discover the ultimate and unchangeable causes of natural phenomena.⁴⁷

Helmholtz designated cause in general as "force," so that he here distinguished between "forces" that are changeable (and therefore not ultimate), and "forces" that are unchangeable (and so final). He developed this as follows: First, he asserted that "bodies with unchangeable forces have been named in science (chemistry) elements." Then he continued: "Let us suppose the universe decomposed into elements possessing unchangeable qualities; the only alteration possible to such a system is an alteration of position, that is motion: *hence* the forces can be only moving forces dependent in their action upon conditions of space."⁴⁸ Helmholtz's unchangeable forces can themselves by definition depend only upon distance, and so he concluded that the only possible effect of such forces, assuming them to attach only to unchangeable bodies, is to alter the distance (because the qualities of such bodies are themselves fixed). The relegation of the effect of an unchangeable "force," that is, an unchangeable cause, to change in position follows from Helmholtz's secondary assertion that unchangeable forces require unchangeable bodies.

Helmholtz's fundamental requirement that ultimate "forces" must be unchangeable, that they must not change explicitly with time, reflected his belief that whatever the ultimate causes are that act on bodies now, they must always have acted, with the same bodies in the same positions, in just this manner in the past, and they must do so

---

46. Timothy Lenoir, *The Strategy of Life: Teleology and Mechanics in Nineteenth Century German Biology* (Dordrecht: D. Reidel, 1982), 197–215, insists on Helmholtz's early dislike for teleology.
47. "On the Conservation," 115.
48. Ibid., emphasis added.

again in the future. In 1847, he used the "material point" as an embodiment of this prior requirement because the point is completely invariant except as regards location. Since, at that time, Helmholtz thought that an unchangeable force should be linked to a similarly unchangeable quality, the material point was a natural—indeed, the only possible—vehicle for him to employ. Had he broken the connection, had he allowed that a purely spatial "force" could affect something other than location, then the material point would not have been essential. He would then have had to allow that they can produce qualitative change as well as change in position. In 1847, Helmholtz did not even conceive that the connection could be broken; by the end of the 1860s, he thought otherwise. Yet to break the connection he had to alter his understanding of physical cause by separating it from moving force and attaching it instead to the much more powerful notion of "energy."

From Helmholtz's metaphysical remarks it follows that he did not discover through misapplied technique that Weber's force law must be rejected in view of the energy principle (misapplied because, as Helmholtz eventually admitted, a force like Weber's that depends on acceleration as well as speed can fit a conservation principle). He knew it even before he developed the principle in technical form because Weberean "force" necessarily involves time through its dependence on speed and acceleration. It is not, in other words, that time-dependent forces necessarily violate the conservation of *vis viva* (though at the time Helmholtz certainly thought they did). Rather, they are a priori unacceptable because they are, by their very nature, variable and so cannot serve as ultimate causes. Helmholtz's rejection of time-dependent "forces" and his formulation of energy conservation both bore the stamp of his deeply felt conviction that nature could only be comprehended through invariable causes. In 1847, he developed a way to remove even implicit variability from "force," and this began to effect a very deep change indeed, one that led in the late 1860s to his new program for physics.

He invented what soon evolved (though not initially in his hands) into the concept of energy, or rather what he at first thought of as the quantitative aspect of "force." Ultimate "force" in Helmholtz's sense had to depend upon space, not time, and the prime example of such a thing is gravitation. Yet even a force like gravity changes over time (albeit only implicitly) as the distances change. Such a "force" is invariable as a cause; yet in another, less important, way it is not. And here Helmholtz apparently detected a puzzle, because he strongly believed (as he put it in 1847) that "it is impossible by any combination whatever of natural bodies, to produce force continually from nothing."[49] The puzzle—that even ultimate "force" does vary, yet that it cannot be produced from nothing—manifests the potency of the concept and the essential ambiguity that accompanies its power. Because, of course, there can be no puzzle here *if* we limit "force," as the analytically minded French *physiciens* did, to signifying whatever equals the product of mass by acceleration. Such a thing is not in any sense conserved. Helmholtz's problem emerged from his insistence on conserving something about force, not just on conserving.

Helmholtz's analytical resolution of the enigma was simple in appearance because he founded it on ultimate (that is, purely spatial) actions between "material points." He identified the integral of such a force over distance with a force quantity, or "tension" as he called it, and at once arrived at the well-known fact that such a thing equals the corresponding change in the sum of half the product of mass by the velocity squared.[50] This tightly bound the notion of conservation to points, which meant that it appeared to contain an internal push towards models, though neither then nor later did Helmholtz himself pursue atomic model-building.

During the next two decades (the 1850s and 1860s) an independent concept, that of energy, was in any case developed, initially by Thomson in Scotland and Rudolf Clausius in Germany, and as it developed energy became separated not only from "force" but from models. It gradually became a thing that in certain respects had its own identity and that could always be traced in fixed amounts through the operations of nature. Then, two decades later (in the 1870s), just when Helmholtz officially changed his calling from physiology to physics by going to Berlin, he grappled anew with the difficult, ambiguous conceptions that had gripped him in his youth, and in the process he developed a new, energy-based physics. The proximate source of that physics—the source that made its formulation possible—was almost certainly Thomson and Tait's *Treatise*, which he and Emil Wertheim translated into German; the first volume was published in 1871. The *Treatise*, as already noted, did not employ atoms and forces. Instead, its structure was molded at the deepest level around the concepts of continuity, energy, and variational principles for mechanics. Their elementary object was not the material point, but the volume element, and their physics concentrated on the energy *densities* that they supposed to be located in these elements. Helmholtz eventually adapted

49. Ibid., 118.
50. Ibid.

these ideas to his own needs, transforming them in fundamental ways, though he never explicitly stated that he was doing so. He used the *Treatise*'s foundation in the differential volume element as the appropriate unit of analysis in his electrodynamics, but he did not adopt its image of an energy spread throughout the volumes.[51] For such an image would not have suited the basis of his thought: his insistence that an interaction must always be understood as a bipartite, spatial process. Helmholtz accordingly invented the concept of an irreducible system energy that is determined by the simultaneous *states* of, and the distance between, a pair of volume elements. In this new way of doing physics there was no vestige of an a priori hypothesis, no hint of anything that might, at the hands of someone like Zöllner, lead to Romantic Idealism.[52]

Helmholtz announced his new method in 1871 in a lecture to the Prussian Akademie der Wissenschaften honoring the memory of his predecessor Magnus. Here he sought to rescue mathematical physics from an Idealist embrace by enunciating a structure for it that avoided the potent imagery of atoms and forces. "To flee into an ideal world," Helmholtz claimed in his Magnus lecture, "is a false resource of transient success; it only facilitates the play of the adversary; and when knowledge only reflects itself, it becomes unsubstantial and empty, or resolves itself into illusions and phrases."[53] That year Helmholtz had found himself confronted with a metaphysics that was undeniably connected to Weber's atoms and forces, and whose principal exponent, Zöllner, had mounted a direct attack on the very source of Helmholtz's new physics, Thomson and Tait's *Treatise*. Ironically, Zöllner had been educated at Berlin under Magnus and Heinrich Dove. He became extraordinary professor of physics at Leipzig in 1866 and, following his invention and deployment of the astrophotometer, in 1872 became ordinary professor of astrophysics there. In 1872 he published a book

---

51. This interpretation of Helmholtz's electrodynamics, and indeed of all his physics after *circa* 1870, is based on a pattern common to all of his work during this period. For full details for his electrodynamics, see Buchwald, *The Creation of Scientific Effects*.

52. The kind of Romantic Idealism represented by, e.g., Arthur Schopenhauer, which Helmholtz particularly detested and which his arch-foe Zöllner ardently embraced, should be distinguished from Idealism considered as the antithesis to Materialism. Helmholtz never embraced Materialism, and he may have been strongly influenced by the non-Romantic Idealism developed by Fichte, on which see the essay by Michael Heidelberger, "Force, Law, and Experiment: The Evolution of Helmholtz's Philosophy of Science," in this volume. Schopenhauer's philosophy powerfully antagonized many physicists associated with Helmholtz. Years later, in 1906, Boltzmann, to cite one example, nearly entitled a talk "Demonstration, that Schopenhauer was a Mindless, Ignorant, Spreader of Nonsense"; see Hörz, "Helmholtz und Boltzmann," 200.

53. "Gustav Magnus. In Memoriam," *Popular Lectures on Scientific Subjects*, 2 vols., trans. E. Atkinson (New York: D. Appleton and Co, 1873 and 1881), 2:1–25, on 14.

entitled *Über die Natur der Cometen. Beiträge zur Geschichte und Theorie der Erkenntnis*. It quickly mired his career in controversy, not least because in its preface Zöllner attacked Thomson and Tait's remark, as given in Helmholtz's translation of their *Treatise*, that Weber's atoms and forces were not only useless but downright pernicious. Zöllner was enraptured by the notion that the mind, in harmony with nature, could by intuition discover its inner workings, and so he insisted that the laboratory could not reveal anything fundamental.[54] This last point was a direct attack on Helmholtz, whose new physics was nothing if not tied to the laboratory. By the end of the decade Zöllner's opposition had evolved into a deep-seated hatred of Helmholtz and his fellow conspirator, Emil du Bois-Reymond. His diatribes against them had vicious overtones of xenophobia, pan-Germanism, and explicit anti-Semitism, culminating in a posthumous pamphlet— he died in 1882—that treated Helmholtz as a dupe of Jewish scientists.[55]

Alongside the *völkisch* cast of Zöllner's ideology was his passionate concern to develop an entire a priori physics based on Weber's electrodynamics. In 1874, through an experiment similar to (but much cruder than) the one that Herwig performed, he attacked what he perceptively recognized as a thoroughly alien approach to Weber's physics, namely, that which Helmholtz was developing on the basis of system energies.[56] In 1874, and again in 1877, Helmholtz felt compelled to strike back. He replied to Zöllner in the preface to the second part of his translation of the *Treatise*, quickly getting to the heart of the matter. He wrote in a tone of unmistakable scorn and anger:

> Judging from what [Zöllner] aims at as his ultimate object, it comes to the same thing as [Arthur] Schopenhauer's Metaphysics. The stars are to "love and hate one another, feel pleasure and displeasure, and to try to move in a way corresponding to these feelings." Indeed, in blurred imitation of the principle of Least Action, Schopenhauer's Pessimism, which declares this world to be indeed the best of possible worlds, but worse than none at all, is formulated as an ostensibly generally applicable principle of the smallest amount of discomfort, and this is proclaimed as the highest law of the world, living as well as lifeless.

---

54. Molella, "Philosophy and Nineteenth-Century Electrodynamics," 199. See Friedrich Zöllner, *Über die Natur der Cometen. Beiträge zur Geschichte und Theorie der Erkenntnis*, 2nd ed. (Leipzig: Wilhelm Engelmann, 1872), v–lxxii.
55. Molella, "Philosophy and Nineteenth-Century Electrodynamics," 212–13; and Friedrich Zöllner, *Beiträge zur deutschen Judenfragen mit akademischen Arabesken als Unterlagen zu einer Reform der deutschen Universitäten*, ed. Moritz Wirth (Leipzig: Oswald Mutze, 1894).
56. Friedrich Zöllner, "Ueber einen elektrodynamischen Versuch," *AP 153* (1874):138–43.

Now, that a man who mentally treads such paths should recognize in the method of Thomson and Tait's book the exact opposite of the right way, or of that which he himself considers such, is natural; that he should seek the ground of the contradiction, not where it is really to be found, but in all conceivable personal weaknesses of his opponents, is quite in keeping with the intolerant manner in which the adherents of metaphysical articles of faith are wont to treat their opponents, in order to conceal from themselves and from the world the weakness of their own position.[57]

In the summer and fall of 1877 Helmholtz twice again replied to Zöllner's vituperative remarks in the *Cometen*. In a popular address entitled "Das Denken in der Medicin," delivered in August to his alma mater, the Friedrich-Wilhelms-Institut in Berlin, he again pointed to the lack of politesse among "metaphysicians," noting that as far as he was concerned "a metaphysical conclusion is either a false conclusion or a concealed experimental conclusion." Two months later he delivered his inaugural address as Rector at Berlin on the topic of academic freedom. He waxed enthusiastic over the great freedom of the German university and, in particular, of the new German Empire where "the most extreme consequences of materialistic metaphysics, the boldest speculations upon the basis of Darwin's theory of evolution, may be taught with as little restraint as the most extreme deification of Papal Infallibility," and where, though "it is forbidden to suspect motives or indulge in abuse of the personal qualities of our opponents, nevertheless there is no obstacle to the discussion of a scientific question in a scientific spirit."[58] The unhindered teaching of materialistic metaphysics was precisely what Zöllner and those of like opinion strongly objected to, and he was, of course, precisely the one who questioned motivations.

Both of these addresses echoed the pointed remarks Helmholtz had made in 1874. Zöllner's insulting comments in the *Cometen* had deeply angered Helmholtz. His xenophobia and his national character assassination combined with his objectionable embrace of Schopenhauer to impel Helmholtz to respond. There were at least two ways to counter Zöllner's influence. On the one hand, his beliefs could be attacked directly. This Helmholtz did on the three occasions noted above. On the other, and far more powerfully, one could provide an alternative,

---

57. "Helmholtz on the Use and Abuse of the Deductive Method in Physical Science," *Nature 11* (1874):149-51, 211-12, on 150.
58. "On Academic Freedom in German Universities," in *Popular Lectures*, 2:237-64, on 255-56. See also Cahan, "Helmholtz and the Civilizing Power of Science."

empirically grounded physics that was ineluctably tied to the laboratory, one that by its very nature could not be founded on a priori considerations. Helmholtz's electrodynamics, with its basis in object states and energetic relations that could be discovered only in the laboratory, had precisely the right character. Accordingly, it seems highly likely (though by no means indisputable) that the extraordinary intensity with which Helmholtz pursued the subject in the mid-1870s at least in part reflected his desire to eliminate the plausibility of Zöllner's Idealistic deductive scheme of Weberean physics by creating situations in the laboratory that Helmholtz, but not Weber, could explain.[59] The constellation of problems that emerged from this heady brew of ideology and technique actually revolved about a single, highly technical point which Helmholtz addressed in nearly all of his papers on electrodynamics during the 1870s: what forces do the ends of open circuits exert? In the event, the experiments to probe the issue had results that required Helmholtz to press his own scheme more deeply, and this opened up a set of problems that had been implicit as early as 1870 and that concerned the connection between Helmholtz's electrodynamics and field theory, problems with which Heinrich Hertz was soon to grapple.

59. In later years, particularly following Hertz's discovery of electric waves, Helmholtz's views changed markedly when he turned to the principle of least action as a foundation for physics. However, even then his understanding of force as a secondary, emergent effect persisted, as is apparent from his comments in *Vorlesungen über Theoretische Physik*, eds. Arthur König, et al., 6 vols. (Leipzig: J. A. Barth, 1897–1903), vol. 1.1: *Einleitung zu den Vorlesungen über Theoretische Physik*, eds. Arthur König and Carl Runge (1903), 15. For a brief discussion of this topic see Buchwald, "The Background to Hertz's Experiments."

9

Helmholtz's Instrumental Role in the Formation of Classical Electrodynamics

Walter Kaiser

1. Introduction

During the last third of the nineteenth century Hermann von Helmholtz led an entire generation of German-speaking physicists to recognize the fecundity and the challenge of James Clerk Maxwell's new electrodynamics. His midwifery in bringing Maxwell's ideas into the German-speaking world and in nurturing the research abilities of young German-speaking physicists, above all his star student Heinrich Hertz as well as Ludwig Boltzmann, noticeably outstripped his own substantive intellectual contributions to electrodynamics. There is much irony in the fact that Hertz's principal scientific achievement—the production and detection of transverse electromagnetic waves—not only provided, as Oliver Heaviside once wrote to Hertz, the "death blow" to the action-at-a-distance physics practiced on the Continent but that it also questioned the predictive capacity of Helmholtz's ideas about elementary electromagnetic interaction and about Helmholtz's own polarization theory.[1] By 1890, as this essay argues, Helmholtz's electrodynamical ideas were in eclipse if not obsolete. The electromagnetic parts of his *Vorlesungen über Theoretische Physik* became essentially outdated even before they were published: both volume

---

Acknowledgments: For stimulating discussions and various types of help I should like to thank Jed Z. Buchwald, David Cahan, and Jörg Meya.
1. Oliver Heaviside to Heinrich Hertz, 13 July 1889, in James G. O'Hara and Willibald Pricha, *Hertz and the Maxwellians* (London: Petrus Peregrinus, 1987), 66–8, on 67.

five, *Vorlesungen über die elektromagnetische Theorie des Lichts* (1897), and volume four, *Vorlesungen über Elektrodynamik und Theorie des Magnetismus* (1907), served far more to honor the respected "*Meister*" than to stimulate further development of his own brand of theoretical physics.[2] His electrodynamics became a kind of interesting historical footnote. By 1890 the new axiomatic interpretation of Maxwell's theory, promulgated by, inter alia, Heaviside, Hertz, and Emil Cohn, represented the forefront of research in electrodynamics.[3] Moreover, it was August Föppl's textbook on Maxwell's electrodynamics, *Einführung in die Maxwell'sche Theorie der Elektricität* (1894), not Helmholtz's own lectures, which became the chief German-language pedagogical source for learning Maxwellian physics—not least because Föppl employed the mathematically appropriate tool of vector analysis.[4] In short, as this essay seeks to show, Helmholtz neither established a fundamental or comprehensive theory of electrodynamics nor did he contribute outstanding experimental results. Instead, he played an instrumental role in the formation of classical electrodynamics: through his critical analysis of the work of others, through creative variations on existing theories, and through his inspiring guidance of his students, he became a major figure in shaping classical, Maxwellian electrodynamics as it emerged during the last third of the nineteenth century.

The essay begins (Section 2) with a brief recapitulation of the background in electrophysiology, thermodynamics, and hydrodynamics that served as Helmholtz's entry vehicles into his analysis of contemporary work in electrodynamics. Section 3 shows that during the 1870s Helmholtz pointed to the shortcomings of the electrodynamic theory promoted by his fellow German Wilhelm Weber and called attention to the potential strengths of the theory established by his British colleague Maxwell. Helmholtz's own theory of the electromagnetic field and its distinctiveness from Maxwell's is the subject of Section 4, while Section 5 treats the experimental testing of the various electrodynamic

---

2. *Vorlesungen über die elektromagnetische Theorie des Lichts*, eds. Arthur König and Carl Runge (Leipzig: Johann Ambrosius Barth, 1897), 27–130; *Vorlesungen über Elektrodynamik und Theorie des Magnetismus*, eds. Otto Krigar-Menzel and Max Laue (Leipzig: Johann Ambrosius Barth, 1907), v, 371–86; and *Dynamik continuirlich verbreiteter Massen*, ed. Otto Krigar-Menzel (Leipzig: Johann Ambrosius Barth, 1902), v–vi, quote on v. These are, respectively, Helmholtz's *Vorlesungen über Theoretische Physik*, vols. 5, 4, and 2.

3. See, e.g., Emil Wiechert, "Grundlagen der Elektrodynamik," in *Festschrift zur Feier der Enthüllung des Gauss-Weber-Denkmals in Göttingen* (Leipzig: B.G. Teubner, 1899), 2–112, esp. 64–78.

4. August Föppl, *Einführung in die Maxwell'sche Theorie der Elektricität* (Leipzig: B.G. Teubner, 1894).

theories by Boltzmann and Hertz. The essay concludes (Section 6) with a brief discussion of the historical meaning of Helmholtz's contributions to electrodynamics for both "classical" and "modern" physics, arguing that as a researcher, teacher, and science organizer Helmholtz assembled a variety of ideas and attracted several individuals who helped clarify and unify "classical" physics.

2. Helmholtz's Route to Electrodynamic Studies

Helmholtz took a circuitous route towards confronting the demanding theoretical problems of electricity. Before taking up systematic electrodynamic studies in the later 1860s, he devoted much of his time to work in electrophysiology, thermodynamics (conservation of energy), and hydrodynamics, fields which helped introduce him to some of the problems and resources that he would later confront and use in his electrodynamic studies.

From the mid-1840s to the mid-1860s Helmholtz worked principally in the new field of electrophysiology. Here he first became acquainted with and copiously used a series of fundamental (and recent) electrodynamic phenomena and electrodynamic laws—above all, Georg Ohm's law of electrical resistance, Michael Faraday's experiments on electromagnetic induction, Gustav Robert Kirchhoff's theory of the distribution of currents in systems of conductors, and several measuring devices (especially Charles Wheatstone's bridge and various galvanometers). These laws and instruments became essential tools in his electrophysiological studies. His masterful experimental discovery and analysis of the propagation of nerve impulses, for example, used electricity as a means to excite nerve action and muscle contraction in a frog's leg. Moreover, and most impressively, he managed to employ a ballistic galvanometer—that is, a multiplicator similar to the apparatus used by Carl Friedrich Gauss and Weber in their early telegraphic work—to measure the extremely brief time taken by a current in the so-called secondary circuit. The current in this secondary circuit switched on simultaneously with the electric setting of a nerve impulse and switched off with the muscle's contraction, thereby indicating that the nerve impulse had reached the muscle. Hence the maximum of deflection of the ballistic galvanometer measuring the current in the secondary circuit was in effect a summation of the charge flow over time. Knowing the nerve length and propagation time allowed Helmholtz to calculate the velocity of a nerve impulse, which turned out to have the unexpectedly small value of approximately 30 meters per second.[5]

Helmholtz's experimental research in the propagation of the nerve impulse probably led him to a general awareness of propagation in time, in the sense that perturbations in physical systems (for example, telegraph wires) could be described physically as propagations in time.[6] Furthermore, his elaborate methods for measuring nerve-impulse propagation, along with the work of his lifelong friend Emil du Bois-Reymond on determining muscle and nerve currents, exemplified the power of a new physiology based solely on physics and chemistry, and thereby challenged the vitalistic approach of physiology during the Romantic period.[7]

Furthermore, Helmholtz's 1847 essay *Ueber die Erhaltung der Kraft* also owed much to his previous and concurrent work in physiology. As is well known, his idea of a general law of conservation of "force"—he did not then use the term "energy"—was stimulated by theoretical and experimental research on the development of heat in muscles. In contrast to his medical colleague Julius Robert Mayer, who never completely overcame his poor training in physics, Helmholtz based his general conservation principle on the law of conservation of mechanical energy. In his essay Helmholtz argued that all physical phenomena can be explained by central ("conservative") forces acting between point masses that depended solely on the distance between them. The treatise, which Helmholtz directed above

---

5. "Messungen über den zeitlichen Verlauf der Zuckung animalischer Muskeln und die Fortpflanzungsgeschwindigkeit der Reizung in den Nerven," *MA* (1850):242–301, in *WA* 2:764–843, esp. 772–79, 821–31; "Messungen über Fortpflanzungsgeschwindigkeit der Reizung in den Nerven. Zweite Reihe," *MA* (1852):199–216, in *WA* 2:844–61; and "Ueber die Methoden, kleinste Zeittheile zu messen, und ihre Anwendung für physiologische Zwecke," *Königsberger naturwissenschaftliche Unterhaltungen 2:2* (Königsberg: Bortäger, 1851), 169–89, in *WA* 2:862–80. For a full analysis of these matters see the essay by Kathryn M. Olesko and Frederic L. Holmes, "Experiment, Quantification, and Discovery: Helmholtz's Early Physiological Researches," in this volume.
6. Koenigsberger *I*:130–31; and Helmholtz to Emil du Bois-Reymond, 11 April 1851, in Kirsten, 111.
7. "Ueber das Wesen der Fäulniss und Gärung," *MA* (1843):453–62, in *WA* 2:726–34; "Ueber den Stoffverbrauch bei der Muskelaction," *MA* (1845):72–83, in *WA* 2:735–44; "Wärme, physiologisch," *Encyklopädisches Handwörterbuch der medicinischen Wissenschaften* (Berlin: Veit, 1846), in *WA* 2:680–725; "Bericht über die Theorie der physiologischen Wärmeerscheinungen für 1845," *Fortschritte der Physik im Jahre 1845 1* (1847):346–55, in *WA* 1:3–11; "Ueber die Wärmeentwicklung bei der Muskelaction," *MA* (1848):147–64, in *WA* 2:745–63; Karl Eduard Rothschuh, *Geschichte der Physiologie* (Berlin, Göttingen, Heidelberg: Springer-Verlag, 1953), 112–33; Gunter Mann, ed., *Naturwissen und Erkenntnis im 19. Jahrhundert: Emil du Bois-Reymond* (Hildesheim: Gerstenberg, 1981); Du Bois-Reymond to Helmholtz, 25 March 1862, in Kirsten, 202; and Timothy Lenoir, "Models and Instruments in the Development of Electrophysiology, 1845–1912," *HSPS 17:1* (1986):1–54, esp. 4–21.

all to physicists, largely contained purely physical, as opposed to physiological, ideas and analyses; moreover, its physical analysis owed much not only to rational mechanics but also to the developing theory of electromagnetic induction.[8]

Where Faraday had given a descriptive explanation of the occurrence of electromagnetic induction phenomena in a wire in terms of "cutting" lines of magnetic force, Continental physicists tackled the problem mathematically, within a framework of action-at-a-distance theories of electrodynamics. In particular, Franz Ernst Neumann, the cofounder of the Königsberg school of theoretical physics, derived a potential for the interaction of linear currents which was based on the action at a distance of infinitely small current elements and which was later interpreted as the magnetic energy issuing from a system of such interacting currents. Carefully restricting his theory to low frequencies, Neumann used his potential to derive a law for the induction of an electromotive force due to a system of two time-variable currents.[9] Taking the time derivative of the potential, with the current in one circuit set equal to unity, Neumann obtained the electromotive force induced by one circuit in the other circuit.

When Helmholtz applied the theory of conservation of "force" to electrical processes, he sought to include all mutual transformations of thermal, electrical, and magnetic energy in a physical system containing a battery, permanent magnets, and electric currents. However, in his derivation of an induction law he concentrated only on systems of linear currents and permanent magnets. He was then unaware of the energy stored in the magnetic field of a single electric current (as opposed to that part of the magnetic energy issuing from the interaction of two linear currents), and so his derivation of an electromagnetic induction law proved valid only for the mutual action of a linear current and a permanent magnet. For this special case he derived the time integral of an additional electromotive force in a circuit from the

---

8. "Wärme, physiologisch," esp. 699–700; "Bericht über die Theorie der physiologischen Wärmeerscheinungen," 6; *Ueber die Erhaltung der Kraft, eine physikalische Abhandlung* (Berlin: G. Reimer, 1847), in *WA* 1:12–75, esp. 12, 61–5. For further analysis of this point see Fabio Bevilacqua's essay, "Helmholtz's *Ueber die Erhaltung der Kraft*: The Emergence of a Theoretical Physicist," in this volume.

9. Franz E. Neumann, *Die mathematischen Gesetze der inducirten elektrischen Ströme*, 1845, reprinted as vol. 10 of Ostwald's Klassiker der exakten Wissenschaften (Leipzig: Wilhelm Engelmann, 1889), 1, 22, 62; idem, *Über ein allgemeines Princip der mathematischen Theorie inducirter elektrischer Ströme*, 1847, ibid., vol. 36 (Leipzig: Wilhelm Engelmann, 1892), 72–5; and Arnold Sommerfeld, *Vorlesungen über Theoretische Physik*, 5th ed., 6 vols. (Leipzig: Akademische Verlagsgesellschaft, 1967), vol. 3: *Elektrodynamik*, 97.

alteration of the potential of a permanent magnet towards an existing current in the circuit.[10] Shortly thereafter, in the early 1850s, William Thomson successfully applied the principle of conservation of energy to a slowly varying system of currents, magnets, and sources.[11]

Helmholtz's theory of hydrodynamics, along with his work on electrophysiology and on conservation of energy, constituted the third of the three major background areas for his future work in electrodynamics. During much of the 1850s and 1860s his abiding interest in electrodynamics remained latent as he devoted himself to work in physiological optics and acoustics, including analysis of the physiological foundations of music. One indication of his continuing interest in electrodynamics in the 1850s is his theory of hydrodynamics, in particular his highly original conservation law for fluid vortices. With his differential equation for the velocity field of a moving fluid,

$$\vec{v} = \nabla P - \text{curl } \vec{A}$$

or

$$\text{curl } \vec{v} = 2\vec{\omega} - \nabla \vec{A}$$

where $\vec{v}$ is the velocity of the fluid, P an arbitrary scalar potential, $\vec{A}$ a kind of vector potential for "hypothetical" magnetic "masses," and $\vec{\omega}$ the angular velocity of a given element. Helmholtz defined the components of the vector $\vec{A}$ analogously to the kind of volume integrals representing the magnetic potential at an external point $\vec{r}_2$ due to magnetic "masses" distributed with density $-\omega(\vec{r}_1)/2\pi$:

$$A_x(\vec{r}_2) = \frac{-1}{2\pi} \int_V \frac{\omega_x(\vec{r}_1)}{|\vec{r}_2 - \vec{r}_1|} d^3\vec{r}_1; \quad A_y(\vec{r}_2) = \ldots$$

This integral over all values of $\omega_x(\vec{r}_1)$ in space describes how the angular velocities at each point—or, in the electrodynamic analogy, the mag-

---

10. "Ueber die Erhaltung der Kraft," 61–5, esp. 62; and Edmund Whittaker, *A History of the Theories of Aether and Electricity*, 2 vols. (reprinted New York: Humanities Press, 1973), *1*:218.

11. William Thomson, "On the Theory of Electromagnetic Induction," [1848], reprinted in his *Mathematical and Physical Papers*, 6 vols. (Cambridge: Cambridge University Press, 1882–1911), *1*:91–2.

netic "mass" densities at each point—formally contribute to the (unobservable) vector potential $\vec{A}$ and eventually yield the (observable) velocity field $\vec{v}$ or the (equally observable) field of magnetic force. For, applying the differential operation "curl" on the vector potential $\vec{A}$ Helmholtz was able to calculate the velocity $\vec{v}_2$ induced in a fluid particle by another rotating particle with angular velocity $\vec{\omega}_1$ at a distance $(\vec{r}_2 - \vec{r}_1)$:

$$\vec{v}_2 = \frac{1}{2\pi} \frac{\vec{\omega}_2 \times (\vec{r}_2 - \vec{r}_1)}{|\vec{r}_2 - \vec{r}_1|^3}.$$

The result was a formula analogous to the expression for the magnetic effect of current densities, which in turn was a part of the electrodynamic force law of Jean-Baptiste Biot and Félix Savart.[12] Although Helmholtz was very cautious with this heuristic transfer of mathematical structures from hydrodynamics to electrodynamics,[13] it is scarcely surprising that Maxwell read with delight Helmholtz's paper on fluid motion in Crelle's *Journal für die reine und angewandte Mathematik* for 1858. Maxwell was particularly impressed to find that Helmholtz had "pointed out that the lines of fluid motion are arranged according to the same laws as the lines of magnetic force, the path of those particles of the fluid which are in a state of rotation." He noted that this was an "additional instance of a physical analogy, the investigation of which may illustrate both electro-magnetism and hydrodynamics."[14] Maxwell's interpretation of Helmholtz's work on fluid vortices involved the use of physically interpreted analogue models, whereas Helmholtz confined himself to a formal understanding of similar mathematical structures in different branches of physics.

3. Maxwell's Electrodynamics and the Weber-Helmholtz Dispute

The advent of Maxwell's electrodynamic field theory (in Britain) also largely coincided with the end of Helmholtz's work as a physiologist. During the 1860s Helmholtz's own electrical research became transformed from a mere auxiliary scientific technique in physiology

---

12. "Ueber Integrale der hydrodynamischen Gleichungen, welche den Wirbelbewegungen entsprechen," *JfruaM* 55 (1858):25–55, in *WA* 1:101–34, on 110–19.
13. Ibid., 103–4.
14. James Clerk Maxwell, "On Physical Lines of Force," in *The Scientific Papers of James Clerk Maxwell*, ed. W.D. Niven, 2 vols. (Cambridge: Cambridge University Press, 1890), 1:451–88, quote on 488.

to the very core of his theoretical thinking in physics. At the time, as Boltzmann later said, only two Continental physicists immediately recognized the importance of Maxwell's theory, namely, Helmholtz, and Josef Stefan, Boltzmann's own teacher.[15] Yet Helmholtz's approach to Maxwell's electrodynamics was by no means simply an outright acceptance of the new theory with its mathematically elegant but physically confusing differential equations representing the contiguous action of electromagnetic forces that resulted in an electromagnetic theory of light. As noted already, Maxwell's physical interpretation of mechanical analogies was not Helmholtz's style of theoretical physics. On the other hand, Helmholtz viewed Maxwell's theory as a striking argument against Weber's proposed law of fundamental electrodynamic interaction. For Maxwell's electrodynamics of a mechanically conceivable ether seemed to provide a means of circumventing the problems of the alleged violation of energy conservation by Weber's electrodynamic law of interacting moving charges, which depended on velocities and accelerations of charge.[16]

Weber's theory constituted the culminating point of a long series of fundamental laws in electrodynamics, one which began with André-Marie Ampère's and Hermann Grassmann's laws for the interaction of infinitely small current elements and went on to include Biot and Savart's formula for the interaction of current elements and particles of magnetic fluids.[17] In their theories of elementary electrodynamic interaction, Ampère and his colleagues began with theoretical models of the microphysical level, which, after integration of the basic equations for complete circuits, furnished the forces acting between electric currents or between electric currents and magnets on the level of experimentally testable setups. Of further fundamental importance were Neumann's interaction potential of linear currents as well as Gauss's attempts to derive the electromagnetic current interaction from the forces of interaction between moving charges.[18] All of these theories

---

15. Ludwig Boltzmann, "Josef Stefan," (1895), in Engelbert Broda, ed., *Ludwig Boltzmann, Populäre Schriften* (reprinted Braunschweig: Vieweg, 1979), 59–66, on 62.

16. "Ueber die Theorie der Elektrodynamik," *MB* (1872):247–56, in *WA* 1:636–46, on 639; and "Ueber die Theorie der Elektrodynamik. Zweite Abhandlung. Kritisches," *JfruaM* 75 (1873):35–66, in *WA* 1:647–87, on 674.

17. André-Marie Ampère, *Théorie mathématique des phénomènes électro-dynamiques uniquement déduite de l'expériment* [1826] (reprinted Paris: Albert Blanchard, 1958); R.A.R. Tricker, *Early Electrodynamics* (Oxford: Pergamon Press, 1965); and, e.g., Walter Kaiser, *Theorien der Elektrodynamik im 19. Jahrhundert* (Hildesheim: Gerstenberg, 1981), and Christine Blondel, *A.-M. Ampère et la création de l'électrodynamique 1820–1827* (Paris: Bibliothèque Nationale, 1982).

18. Carl Friedrich Gauss, "Grundgesetz für alle Wechselwirkungen galvanischer Ströme," in *Carl Friedrich Gauss:Werke*, ed. Königliche Gesellschaft der Wissenschaften

were formally constructed within the context of action-at-a-distance physics; that is, in their formulae for the elementary electrodynamic interaction they neglected the influence of a medium between current elements or moving charges and they neglected the role of time in the propagation of electrodynamic action. So, too, did Weber's elaborate and ambitious theory, which was designed to embrace Coulomb's law for electrostatic interaction between electric charges, Ampère's electrodynamic law for magnetic forces acting between current elements, and Faraday's electromagnetic induction.[19]

However, like Ampère and Gauss, Weber too struggled with the problem of how to conceive the propagation of electrodynamic action.[20] Already in the 1820s, Ampère had pictured the propagation of electrodynamic action as a mechanical stimulation of consecutive layers of the electric fluid, notwithstanding the formal character of his elementary law, which he designed according to the theory of action at a distance. Gauss, for his part, proposed in a letter of 1845 to Weber the idea of a propagation in time of electrodynamic action. In response, Weber agreed that a propagation in time might be the best solution.[21] A year later, Weber conceived of a medium between moving charges that might prove essential for the theory of electrodynamic interaction.[22] Moreover, an additional problem for Weber was the complicated conduction mechanism underlying his theory. Like Gustav Fechner, Weber pictured an electric current as consisting of one (partial) current of positive electricity moving in one direction and another (partial) current of negative electricity moving in the opposite direction. Rudolf Clausius, among others, criticized this dual conduction mechanism in

---

zu Göttingen, 12 vols. (Göttingen: Königliche Gesellschaft der Wissenschaften; Leipzig: Teubner in Komm.; Berlin: Springer, 1863-1929), 5:616-20, esp. 616-17.

19. Wilhelm Weber, "Elektrodynamische Maassbestimmungen. Ueber ein allgemeines Grundgesetz der elektrischen Wirkung," [1846], in *Wilhelm Weber's Werke*, ed. Königliche Gesellschaft der Wissenschaften zu Göttingen, 6 vols. (Berlin: J. Springer, 1892-94), vol. 3, *Galvanismus und Elektrodynamik, erster Theil*, ed. Heinrich Weber (Berlin: J. Springer, 1893), 25-214, esp. 132. See also Karl Heinrich Wiederkehr, "Wilhelm Webers Stellung in der Entwicklung der Elektrizitätslehre" (Ph.D. diss., University of Hamburg, 1960); and M. Norton Wise, "German Concepts of Force, Energy and the Electromagnetic Ether: 1845-1880," in G.N. Cantor and M.J.S. Hodge, eds., *Conceptions of Ether. Studies in the History of Ether Theories 1740-1900* (Cambridge, London, New York: Cambridge University Press, 1981), 269-307, esp. 276-87.

20. Kaiser, *Theorien der Elektrodynamik*, 41-5, 108-12; and Blondel, *Ampère*, 161-64.

21. Carl Friedrich Gauss to Wilhelm Weber, 19 March 1845, in Gauss, *Werke* 5:627-29, esp. 629; Weber to Gauss, 31 March 1845, Niedersächsische Staats- und Universitätsbibliothek Göttingen, Gauss Nachlaß, Nr. 31, 1845, and published in Wiederkehr, "Wilhelm Webers Stellung," 68; see also Kaiser, *Theorien der Elektrodynamik*, 109.

22. Weber, "Elektrodynamische Maassbestimmungen," *Werke*, 3:213-14; see also Kaiser, *Theorien der Elektrodynamik*, 109-12.

Weber's law. As befit the widespread positivistic ideal of creating an "economically" structured physical theory, Clausius suggested starting from a single electrical "fluidum" as a physical analogue model for the electrical current.[23]

Helmholtz, for his part, was deeply concerned about Weber's attempt to explain the fundamental electrodynamic interaction by using moving electricity, for Weber's law contained both the velocities and the accelerations of electrical charge. As early as 1857 Helmholtz challenged Weber's law as the supposed last word on the topic of elementary electrodynamic interaction.[24] He thought Weber's assumption clearly contradicted the principle of conservation of "force" that he had earlier connected with the action of central forces which depend only on the distance of point masses and not on their velocities.[25] His arguments in this electrodynamic context also reveal the distinct change in the status of the principle of conservation of energy: what for a long time had been empirically questionable, or even subject to highly speculative ideas about the unity of forces, had become by the late 1860s a decisive argument for the consistency of a physical theory.[26]

Helmholtz's objections to Weber's electrodynamic law were complex. In evaluating it, he distinguished, on the one hand, between the law's formal structure and how that structure could be embedded in the accepted conservation principles of rational mechanics, and, on the other, the physical contents or experimentally testable range of the law's validity. Although he did not completely renounce his reservations about a force law which depended not only on distances but also on charge velocities and accelerations, by 1870 he did concede that in a formal sense Weber's law was compatible with the energy principle.[27] This formal compatibility meant that Helmholtz could derive Weber's

---

23. Rudolf Clausius, "Ueber die Ableitung eines neuen elektrodynamischen Grundgesetzes," *JfruaM 82* (1877):85–130, esp. 85–6.
24. Helmholtz to Emil du Bois-Reymond, 26 May 1857, in Kirsten, 171–73, on 173. See also "Ueber die Theorie der Elektrodynamik"; and "Ueber die Theorie der Elektrodynamik. Kritisches."
25. "Ueber die Erhaltung der Kraft," 15–7, 19–21.
26. See, e.g., Georg Helm, *Die Energetik nach ihrer geschichtlichen Entwickelung* (Leipzig: Veit, 1898), 48; and Max Planck, *Das Prinzip der Erhaltung der Energie,* 2nd ed. (Leipzig and Berlin: Teubner, 1908), 1–2.
27. "Ueber die Bewegungsgleichungen der Elektricität für ruhende leitende Körper," *JfruaM 72* (1870):57–129, in *WA 1*:545–628, on 553; and "Ueber die Theorie der Elektrodynamik, on 647–49.

force law with the help of a Lagrangian equation from a potential analogous to Weber's potential.[28]

Moreover, Helmholtz demonstrated the use of his law of conservation of energy for fitting Weber's law—as a special case of his own law—into a mathematical description of the conversion of electrostatic and electrodynamic energy into thermal energy.[29] At first glance, Helmholtz argued, Weber's law did not allow for the creation of work from nothing. However, upon closer examination of the time integral in Weber's force law, Helmholtz found that some nonequilibrium states among moving charges in extended conductors do have a smaller energy than those in a state of rest. This led to instabilities in the flow of electricity.[30] As an experimental test case Helmholtz considered the movements of a charged body close to a homogeneously conducting sphere, which, according to Weber's law, would cause unstable movements of electricity on the sphere. Furthermore, Helmholtz constructed a unique physical situation which indicated a violation of the principle of conservation of energy even in the case of two interacting charges Q_1 and Q_2.[31] Neglecting gravitational forces, Helmholtz found that the time integral of Weber's force law yielded, in Helmholtz's formulation,

$$\frac{1}{2c^2}(\frac{\partial r_{12}}{\partial t})^2 = \frac{C' - \dfrac{Q_1Q_2}{r_{12}}}{m_1c^2 - \dfrac{Q_1Q_2}{r_{12}}},$$

where C' is a constant, c a constant of the order of magnitude of the velocity of light, m_1 the mass attached to the moving charge Q_1, r_{12} the distance between the two charges Q_1 and Q_2, and t the time.

In his version of the time integral of Weber's force law, Helmholtz considered all quantities, except c, as completely variable. He thus argued (where the notations, as above, differ slightly from Helmholtz's): "If $Q_1Q_2/r_{12} > m_1c^2 > C'$, then $(\partial r_{12}/\partial t)^2$ is positive and greater than $2c^2$, hence $\partial r_{12}/\partial t$ is a real number. If this term $\partial r_{12}/\partial t$ is positive, r_{12} will increase until $Q_1Q_2/r_{12} = m_1c^2$, hence $\partial r_{12}/\partial t$ increases to in-

---

28. "Ueber die Theorie der Elektrodynamik," 640–43; "Ueber die Theorie der Elektrodynamik. Kritisches," 659–61; Whittaker, *History 1*:203; and Kaiser, *Theorien der Elektrodynamik*, 101–4.
29. "Ueber die Bewegungsgleichungen der Elektricität," 578–81.
30. Ibid., 550–51, 583; and A.E. Woodruff, "The Contributions of Hermann von Helmholtz to Electrodynamics," *Isis 59* (1968):300–11, on 304–5.
31. "Ueber die Bewegungsgleichungen der Elektricität," 553–54.

finity. The same will occur when, in the beginning, $C' > m_1c^2 > Q_1Q_2/r_{12}$ and $\partial r_{12}/\partial t$ is negative."[32] For finite values of m_1c^2 and Q_1Q_2/r_{12} the square of the relative velocities in Weber's potential can obviously become infinite.

In arguing against Weber, Helmholtz introduced the simplest case of two "isolated," moving electrical particles interacting electrodynamically according to Weber's law. Although these moving particles do start with finite kinetic energy, when approaching one another or moving apart they could gain infinite kinetic energy in a certain small but finite distance. Thus a system of interacting moving charges like that described by Weber's law seemed to become unstable, for the time integral, or the equation for the conservation of energy, loses all meaning. Although in the course of his debates with Weber and with Carl Neumann, a mathematical physicist at the University of Leipzig and the son of Franz Neumann, Helmholtz later varied his argument slightly, he stuck to this salient point throughout.[33]

Circa 1870 neither Weber nor Carl Neumann nor anyone else could present experimentally confirmed knowledge of the velocity of light as an upper limit or could give numerical values for molecular forces. The values for molecular radii, calculated approximately by Kirchhoff in 1866, were highly loaded with uncertain, basic assumptions about the kinetic theory of gases (for example, the concept of a mean free path) and thus open to criticism. Accordingly, Weber thought that applying his law to freely moving charges—that is, to elementary particles, not to constituent parts of an electric current—might only provide a hint of the problems of particles interacting within extremely small microphysical distances. Although Weber clearly thought in terms of a molecular structure for matter, he realized that the molecular forces which came into play within those microphysical dimensions necessarily limited any macroscopic electrodynamic force law to a sort of preliminary survey of the field of microphysics. He thus argued in 1871 that Helmholtz's objections, which in the first instance applied to extremely small molecular distances and so were still not within the reach of experimental physics, might not be generally valid.[34] However, in 1873 Helmholtz countered by showing that this operational argument might be incorrect or could at least be avoided by introducing

---

32. Ibid., 553.
33. "Ueber die Theorie der Elektrodynamik," 639–45; "Ueber die Theorie der Elektrodynamik. Kritisches"; "Kritisches zur Elektrodynamik," *AP 153* (1874):545–56, in *WA 1*:763–73.
34. Wilhelm Weber, "Elektrodynamische Maassbestimmungen[,] insbesondere über das Princip der Erhaltung der Energie," [1871], in *Werke* (Berlin: J. Springer, 1894), 4:247–99, esp. 247–49, 268–69, 298.

additional, nonelectric forces. He argued that if, for example, a charged particle moves (even) in a frictional medium, then it might accelerate continuously under the influence of a charged sphere, which could initially be a great distance away.[35] In turn, Carl Neumann, like Weber before him, rejected this argument, questioning its operational meaning. Neumann explicitly denied that Helmholtz had shown the operational meaning for the special microphysical case, wherein Weber's law led to a violation of the principle of conservation of energy. Until Helmholtz could demonstrate how this special case could in fact occur, and until a physical measurement could be obtained, Neumann chose to reject Helmholtz's argument as a test case of Weber's law.[36]

Neither the philosophical nor the physical arguments of the Weber-Helmholtz debate led to a decisive result—perhaps because Weber, Neumann, and Helmholtz were unable to disentangle their dispute from the confusion about the appropriate physical levels (microphysical and macrophysical) to be considered. They could not agree on the limitations of a macroscopic law with regard to the problems arising from microphysical distances. Equally important, Helmholtz's test cases, which actually included the interplay of open currents in general, had not yet been subjected to experimental verification.[37] Moreover, the personal relations between Weber and Helmholtz became troubled in the early 1870s after Helmholtz became the object of a vitriolic polemic instigated by the astrophysicist Friedrich Zöllner, one of Neumann's colleagues at Leipzig.[38] Zöllner, who became a close friend of Weber's during the 1870s, wanted to compromise Helmholtz in his relations with Weber's British critics, namely, Peter Guthrie Tait and William Thomson.

Helmholtz's preference for Maxwell's theory and his dislike of Weber's was continuously nourished by his unceasing doubts about the wisdom of an electrodynamic force law which, unlike a mechanical force law, depended not only on distances but also on velocities and

---

35. "Ueber die Theorie der Elektrodynamik," *WA* 1:647–87, esp. 663–68; and Woodruff, "Contributions," 305.
36. Carl Neumann, "Ueber die gegen das Weber'sche Gesetz erhobenen Einwände," *AP* 155 (1875):211–30, on 226.
37. "Versuche über die im ungeschlossenen Kreise durch Bewegung inducirten elektromotorischen Kräfte," *AP* 158 (1875):87–105, in *WA* 1:774–90.
38. Koenigsberger *1*:148; Woodruff, "Contributions," 305; Friedrich Zöllner, *Ueber die Natur der Cometen. Beiträge zur Geschichte und Theorie der Erkenntnis*s (Leipzig: Engelmann, 1872), v–lxxii, xcvii-c; Helmholtz's preface to William Thomson and Peter Guthrie Tait, *Handbuch der theoretischen Physik*, trans. H. Helmholtz and G. Wertheim, 1 vol. in 2 parts. (Braunschweig: Vieweg, 1871–74), 2:v–xiv; and J. Hamel, "Karl Friedrich Zöllners Tätigkeit als Hochschullehrer an der Universität Leipzig," *NTM* 20 (1983):35–43.

on accelerations as variables.³⁹ Moreover, Helmholtz was also probably reluctant to accept Weber's complicated dual-conduction mechanism, which was an essential part of Weber's derivation of the law of electromagnetic induction and which, in effect, reconciled his electrodynamic law with the Galilean invariance as given in the equations of classical mechanics.⁴⁰

Nonetheless, Helmholtz did more than simply criticize Weber's electrodynamic law and praise Maxwell's theory. For he sought to orientate himself and others in the "pathless wilderness" of competing theories in electrodynamics around 1870;⁴¹ it was in this historical context that he promulgated his own contribution to the ongoing discussion about a fundamental potential for current elements. As already noted, those current potentials were mathematical tools used to derive further equations. Thus, the negative gradient of the potentials (the variation with respect to changing position) furnished laws of *pondero*motive forces, that is, laws of mechanical forces between distant linear currents. The time derivative of the potentials furnished the *electro*motive force induced in systems of time-variant currents. With regard to a fully developed field theory of electrodynamics, those (vector) potentials for currents and (scalar) electrostatic potentials could provide a means to uncouple the field equations in order to derive, for example, separate wave equations for the propagation of electric and magnetic fields. Helmholtz thus established the following equation for a general vector potential U_H:

$$U_H = -\frac{1}{2c^2}\frac{i_1 i_2}{r_{12}}[(1+k)d\vec{s}_1 d\vec{s}_2 + (1-k)\frac{(\vec{r}_{12}\cdot d\vec{s}_1)(\vec{r}_{12}\cdot d\vec{s}_2)}{r_{12}^2}],$$

where $i_1 d\vec{s}_1$ and $i_2 d\vec{s}_2$ are the respective current elements, c a constant of the order of magnitude of the velocity of light, $\vec{r}_{12}$ the distance vector of the current elements, and k a variable factor which could assume the values -1, 0, and $+1$.⁴² Helmholtz tried to adjust his own generalized potential to Neumann's potential (where $k = 1$) and, notwithstanding his own objections, to Weber's potential (where $k = -1$). For $k = 0$, Helmholtz's potential also supposedly embraced a formula

---

39. "Kritisches zur Elektrodynamik," 772; and "Ueber die Theorie der Elektrodynamik. Zweite Abhandlungen. Kritisches," 684–87.
40. Hans-Jürgen Treder, "Helmholtzsche und relativistische Elektrodynamik," *Beiträge zur Geophysik* 79 (1970):401–20, on 414–16; and idem, "Helmholtz' Elektrodynamik und die Beziehungen von Kinematik, Dynamik und Feldtheorie," *WZHUB* 22:3 (1973):327–30, on 328.
41. R. Steven Turner, "Hermann von Helmholtz," *DSB* 6:241–53, on 250.
42. "Ueber die Bewegungsgleichungen," 567.

Maxwell implicitly ("*in verdeckter Form*") used to derive his law for electromagnetic induction.[43] However, what Helmholtz claimed and what often is cited without proof, is far from being obvious in Maxwell's own writings.[44] The only expression comparable to Helmholtz's general electrodynamic potential was that which appeared in Maxwell's derivation of the coefficients of mutual electrodynamic induction. These induction coefficients were a formal tool for introducing the precise geometrical location of interacting currents and for eventually calculating the energy of a system of interacting currents. But Maxwell did not use any explicit formula which can be considered a special form of Helmholtz's expression for a general potential. Maxwell's potential is only comparable to Helmholtz's potential when it is integrated for closed circuits; in this case both expressions are identical. Helmholtz's expression for a generalized potential also agreed with the formal structure of Maxwell's theory insofar as Helmholtz assumed that

$$\text{div} \vec{U}_H = -k\frac{\partial \phi}{\partial t},$$

where ϕ is the electrostatic "potential function of the free electricity."[45] When $k = 0$ this equation is equivalent to a formal condition which (to use modern terminology) is known as the Coulomb gauge, a condition which in fact Maxwell assumed in his presentation of the general equations of the electromagnetic field.[46]

To obtain an idea of the status of Helmholtz's general potential within a more complete theory of electrodynamics, consideration is needed of how his potential functioned in his system of differential equations for the "movement of electricity," equations which did not yet include the influence of dielectric and magnetic properties of matter. When discrete current elements were replaced by continuously distributed current densities, then Helmholtz's potential became primarily a solution to his main differential equation,

$$\Delta \vec{U}_H - (1-k)\nabla \frac{\partial \phi}{\partial t} = -4\pi \vec{j},$$

---

43. Ibid., 548–49.
44. See, e.g., James Clerk Maxwell, "A Dynamical Theory of the Electromagnetic Field," [1864/1865], reprinted in Niven, ed., *Scientific Papers* 1:526–97, on 589–90.
45. "Ueber die Bewegungsgleichungen," 572.
46. Ibid.; Woodruff, "Contributions," 303–4; and Treder, "Helmholtzsche Elektrodynamik," 419.

which summarized the electrodynamic effects of a given current density $\mathbf{j}$ (or one originating from electromagnetic induction) and the electrostatic effects of free electric charges. Yet only by assuming $\mathbf{k} = 0$ or div $\vec{\mathbf{U}}_H = 0$, and only by neglecting $\dfrac{\partial^2 \vec{\mathbf{U}}}{\partial t^2}$, is Helmholtz's equation compatible with modern theory.[47] In fact, Helmholtz's general potential, as Jed Z. Buchwald has argued, had a far more limited predictive capability than Maxwell's field theory and Weber's electrodynamics of moving charged particles. Helmholtz's own understanding of his potential as embracing all possible electrodynamic interactions did, however, open a way for new experiments striving for unknown interaction states of objects carrying time-variant charges and currents.[48]

4. Helmholtz's Theory of the Electromagnetic Field

Helmholtz's criticism of Weber's theory as well as his own proposal of a fundamental electrodynamic law were motivated and nurtured by his attempt to reconcile electrodynamics with the principle of energy conservation. At the same time his efforts were also inseparably linked with his struggle to understand Maxwell's new theory of an electromagnetic field. As early as 1870, while speaking of the power of Maxwell's theory, Helmholtz presented his own contribution to electromagnetic field theory, one which competed with Maxwell's. At the time, the competition was not one of professionals of equal status. Maxwell's standing within the scientific community before 1871, when he became the new Cavendish professor of natural philosophy at Cambridge, was distinctly less than that of Helmholtz. Indeed, in a biographical portrait of Helmholtz in 1876 Maxwell himself referred to the overwhelming, universal scientific work of Helmholtz and called him an "intellectual giant."[49]

Nonetheless, as was the case with Helmholtz's fundamental law for the interaction of current elements within the context of general electrodynamics,[50] his field theory soon fell behind Maxwell's theory, even

---

47. "Ueber die Bewegungsgleichungen," 568–77; John David Jackson, *Classical Electrodynamics* (New York, London, Sidney: John Wiley, 1962), 182; and Harald Stumpf and Wolfgang Schuler, *Elektrodynamik* (Braunschweig: Friedrich Vieweg und Sohn, 1973), 54.
48. See Jed Z. Buchwald's essay, "Electrodynamics in Context: Object States, Laboratory Practice, and Anti-Romanticism," in this volume.
49. James Clerk Maxwell, "Hermann Helmholtz," [1877], in Niven, ed., *Scientific Papers*, 2:592–98, on 598.
50. See, e.g., his *Vorlesungen über Theoretische Physik*, vol. 4: *Vorlesungen über Elektrodynamik und Theorie des Magnetismus*, 341.

as his approach had an enormous impact on the process of reception of Maxwell's theory by Continental physicists. As with the impact of Helmholtz's fundamental law, it was his approach to electrodynamic field equations—that is, its formal generality—that attracted the attention of Continental physicists.[51] Indeed, Buchwald has even gone so far as to argue that, except for Föppl's textbook of 1894, all Continental attempts to understand Maxwell's theory did so only "as a limiting case of Helmholtz's" theory.[52]

Generality came into Helmholtz's theory, in the first place, through an adaptive factor $\vec{k}$ in his generalized vector potential $\vec{U}_H$. This generalized potential $\vec{U}_H$ contributed not only to the various discussions of the fundamental laws for the elementary electrodynamic interaction but also to the inner foundation of Helmholtz's own differential equations for "moving electricity" and eventually to his development of a complete alternative to Maxwell's field theory, including the influence of matter on electrodynamic action.[53] Thus this generalized potential $\vec{U}_H$ appeared again in a number of places in Helmholtz's elaborate electrodynamics, for example, in his formula linking a magnetic potential within a diamagnetic medium and the magnetic effects of current systems with the magnetization $\vec{M}$:

$$\vec{M} = \phi(-\nabla \vec{U}_M + \frac{1}{c}\nabla \times \vec{U}_H),$$

where ϕ is the magnetic susceptibility and $\vec{U}_M$ the magnetic potential.[54]

Helmholtz used this equation to unite all the possible sources of the electric field.[55] Consequently, he assumed that induced electromotive forces may be produced by time changes of either the vector potential $\vec{U}$ due to time-variant current systems or the magnetic potential due to the changing magnetization of matter. He termed the important source for an "electric momentum" within a dielectric medium the "dielectric polarization." Like Faraday's notion of "induction," in Helmholtz's theory "polarization" described the displacement of charges within a dielectric medium. The electric momentum

---

51. See e.g., Ludwig Boltzmann, *Vorlesungen über Maxwells Theorie der Elektricität und des Lichtes*, 2 vols. (Leipzig: Johann Ambrosius Barth, 1891–93; reprinted Braunschweig, Wiesbaden: Friedrich Vieweg, 1982), 2:133–40.
52. Jed Z. Buchwald, *From Maxwell to Microphysics. Aspects of Electromagnetic Theory in the Last Quarter of the Nineteenth Century* (Chicago and London: The University of Chicago Press, 1985), 177–86, esp. 177.
53. "Ueber die Bewegungsgleichungen," 556–57, 572–75, 611–29; Woodruff, "Contributions," 303–4; and Buchwald, *From Maxwell to Microphysics*, 178–79.
54. "Ueber die Bewegungsggleichungen," 617–19.
55. Ibid., 614–20.

density due to polarization was proportional to the electrostatic field within the medium:

$$\vec{P} = -\chi_e \nabla \phi_T,$$

where χ_e is the dielectric susceptibility of the medium and ϕ_T the scalar potential related to the total electromotive force in the electrically polarizable medium: $\vec{E}_T = -\nabla \phi_T$. The total electromotive force $\vec{E}_T$, on the other hand, originates from a superposition of a force $\vec{E}_E$ which creates the polarization, and an oppositely directed force $\vec{E}_P$, which results from the polarization charges produced by the action of $\vec{E}_E$.[56]

Although Helmholtz's and Maxwell's systems of field equations contained suggestive formal relationships and common, empirically testable consequences, there were still marked differences in their underlying physical concepts. According to Buchwald, Maxwell's basic physical concept was the displacement (wherever it originates) in an elastic ether filling all space.[57] This displacement breaks down continuously in a conductor, which causes an ordinary conduction current, and, in turn, produces heat. In vacua and in dielectric media the elastic displacement may remain more stable. Moreover, due to different decay rates of the state of displacement in different media, Maxwell's displacement takes on different values in different media, and, consequently, the divergence of the displacement field may not vanish at the boundary of different media. Maxwell interpreted precisely this nonvanishing divergence of the displacement field as the electric charge. Hence, in Maxwell's electrodynamics (unlike in Continental theories) charge was not an entity in its own physical right but rather a derivative quantity.[58] Thus in Maxwell's theory a current did not arise primarily from moving charges or from a flow of charged particles but rather from a time change of displacement. Together with the other differential equations of Maxwell's theory, especially the induction law, Maxwell's extension of Ampère's law by a displacement current (along with the requirements $\nabla \vec{B} = 0$, $\vec{j} = \rho = 0$ in unbounded, nonconducting media) gave rise to wave equations, the solutions to which led to the prediction of transverse electromagnetic waves propagating in the ether.

56. Ibid., 612; and Buchwald, *From Maxwell to Microphysics*, 179.
57. Buchwald, *From Maxwell to Microphysics*, 20–40.
58. Ibid., 28; Maxwell, "On Physical Lines of Force," in Niven, ed., *Scientific Papers 1*:451–513, on 491; and idem, "On a Method of Making a Direct Comparison of Electrostatic with Electromagnetic Force," [1868], in Niven, ed., *Scientific Papers 2*:125–43, on 139.

Helmholtz, by contrast, did not think primarily let alone exclusively in terms of fields. Instead, like a good Continental theorist he began with moving charges as an ordinary current in a conductor; these current elements act upon one another instantaneously at a distance. However, in responding to Maxwell's theory, Helmholtz concentrated in his theory of polarization on the local reaction of dielectric matter (and, as a tentative extrapolation, of the ether) upon electrical forces, in the sense that in the particles of dielectric matter and in the particles of a dielectric ether (in an otherwise empty space) opposite charges are separated. He called this separation of charges in the particles of a dielectric the "separation of the electricities" ("*Scheidung der Elektricitäten*") or "dielectric polarization" or, in abbreviated form, the "distribution."[59] Hence the original electromotive forces, which act at a distance, are altered by a contiguously acting state of polarization within a dielectric.

In Helmholtz's theory this polarization gave rise to polarization charges which, together with ordinary conduction charges, represented Helmholtz's so-called free charges. First, this density of free charges was subject to the formulation of a continuity equation for charge and current that linked the time change of charges and the (spatial) divergence of a current.[60] The more important consequence of the dominant role of polarization in Helmholtz's theory was that, unlike in Maxwell's theory of a general time-variant displacement, the current density in Helmholtz's theory was a sum of an ordinary conduction current and a nonconducting term originating from the time change of polarization $\frac{\partial \vec{P}}{\partial t}$.[61]

Finally, central to understanding the different physical concepts used by Maxwell and Helmholtz as the basis for their electrodynamic theories are the different meanings attached to "polarization" and "total electrical field" within a dielectric medium. In Helmholtz's theory of polarization the field of electromotive forces within the dielectric contained not only electrostatic fields, originating from a superposition of an original field and the field of the polarization charges, but also electric fields due to electromagnetic induction.[62] In Maxwell's theory, by contrast, the field of electromotive forces ("electromotive intensi-

---

59. "Ueber die Bewegungsgleichungen," 612.
60. Ibid., 611–16.
61. Ibid., 616.
62. Ibid., 572–73, 612; and Buchwald, *From Maxwell to Microphysics*, 179–81.

ties")[63] may create a field of "electric displacement." The latter is subject to variation within dielectric matter due to the varying influence of such matter on the mechanically conceived state of displacement. Thus, the nonconducting part of the current in Maxwell's theory is solely the displacement current $\frac{1}{c}\frac{\partial}{\partial t}\vec{D}$, whereas in Helmholtz's theory it is $\frac{1}{c}\frac{\partial}{\partial t}(\vec{D}-\vec{E}_{ind})$.[64]

Still, an important result of Helmholtz's field theory, which, like the basic interaction potential of currents, was designed to create a greater variety of experimentally testable predictions, was the derivation of the wave equations. Here Helmholtz still concentrated on the standard quantities of a dielectric and diamagnetic medium, namely, polarization and magnetization. He started with a differential equation for the polarization which summed up all the electrodynamic and electrostatic forces polarizing a medium:

$$\frac{\vec{P}}{\chi_e} = -\nabla\phi - \frac{1}{c^2}\frac{\partial \vec{U}_H}{\partial t} + \frac{1}{c}\frac{\partial}{\partial t}\nabla \times \int \frac{\vec{M} d^3\vec{r}'}{|\vec{r}-\vec{r}'|}.$$

Combining his new expression for polarizable media with that for magnetizable media (see p. 390), and thus uncoupling his field equations, Helmholtz arrived at two separate wave equations for $\vec{P}$ and $\vec{M}$ in a homogenous dielectric and diamagnetic medium:[65]

$$\Delta\vec{M} = 4\pi\chi_e(1+4\pi\phi)\frac{1}{c^2}\frac{\partial^2\vec{M}}{\partial t^2},$$

and

$$\Delta\vec{P} = 4\pi\chi_e(1+4\pi\phi)\frac{1}{c^2}\frac{\partial^2\vec{P}}{\partial t^2} + [1 - \frac{(1+4\pi\phi)(1+4\pi\chi_e)}{k}]\nabla\nabla\vec{P}.$$

Thus Helmholtz's electromagnetic field theory predicted the occurrence of transverse waves of polarization and magnetization propagating in polarizable and magnetizable media.

Unlike Maxwell's differential equations with their single solution for transverse waves, Helmholtz's theory also indicated the existence

---

63. James Clerk Maxwell, *A Treatise on Electricity and Magnetism*, 3rd ed., 2 vols. (Oxford: Clarendon Press, 1891; reprinted New York: Dover, 1954), 2:252-53 (§ 608-11).
64. Buchwald, *From Maxwell to Microphysics*, 181.
65. "Ueber die Bewegungsgleichungen," 556-58, 614-15, 619, 625-26.

of longitudinal waves of electric polarization. Referring to the known solutions of the analogue wave equations for the propagation of mechanical perturbations in an elastic solid, Helmholtz concluded that the perturbations of polarization in a dielectric matter travel with different velocities for transversal (v_t) and longitudinal (v_l) waves:

$$v_t = \frac{c}{\sqrt{4\pi\chi_e(1+4\pi\phi)}}; \quad v_l = c\sqrt{\frac{1+4\pi\chi_e}{4\pi\chi_e k}}.$$

However, in order to present Maxwell's theory as a limiting case of his own theory, Helmholtz, by simply equating the notorious factor **k** to 0, allowed the longitudinal waves to disappear. For the case of **k** = 0 the velocity of the longitudinal waves becomes infinite. In order to formally reconcile the wave equations of Maxwell's theory of light with his own wave equations, Helmholtz further assumed a highly polarizable matter in air and a highly polarizable ether ("*Lichtäther*") in empty space ("*Weltraum*") (χ_e and ϕ equal to infinity in the Maxwellian limit). However, this polarization in turn inevitably tended to compensate the electromotive force engendering the polarization charges. Thus, in effect, Helmholtz's polarization model tended to lose its physical consistency. In 1892, Hertz complained how in Helmholtz's theory for the Maxwellian limit the electrostatic forces acting at a distance almost disappeared completely. Yet Hertz offered no other consolation for a Helmholtzian theorist than to face the inconsistency of, on the one hand, forces acting at a distance and engendering polarization, and, on the other, polarization compensating these forces.[66]

As already suggested, Helmholtz's measurements of the propagation of nerve impulses almost certainly made him keenly aware of the necessity of general concepts of propagation in time. More specifically, he sought to clarify the validity of the different approaches in electrodynamics, namely, Weber's instantaneous action at a distance of moving charges, Carl Neumann's potentials propagating in time, and Maxwell's theory of contiguous action with its consequence of electromagnetic waves travelling in the electromagnetic ether. Both Neumann's potential theory and Maxwell's field theory assumed that the velocity of propagation was of the order of magnitude of the

---

66. Heinrich Hertz, *Gesammelte Werke*, ed. Philipp Lenard, 3 vols. (Leipzig: Johann Ambrosius Barth, 1894–95), vol. 2: *Untersuchungen über die Ausbreitung der elektrischen Kraft*, 2nd ed. (1894), 25–31, esp. 27; and Buchwald, *From Maxwell to Microphysics*, 190–93.

velocity of light. Hence, immediately after the publication of his theory of wave propagation in dielectric and diamagnetic media Helmholtz began conducting experimental work on the propagation of perturbations in oscillating electrodynamical systems. Measuring only the time lag of electromagnetic induction effects in induction coils at some distance apart, he obtained the rather disappointing result that the velocity of propagation must be greater than 314,400 meters per second.[67] In a letter to his parents in 1887 Hertz noted this meager result of a lower limit of the propagation velocity: Helmholtz's measurements were especially dissatisfying because Hertz's own experimental results suggested a velocity exceeding even that of light.[68]

5. Experimental Tests of Maxwell's Theory: Boltzmann, Hertz, and Helmholtz

Helmholtz contributed to the reception of Maxwell's theory not only by noting its intellectual strength, praising its elegant equations, and suggesting experiments in its favor. With his intimate if not unmatched knowledge of both Continental action-at-a-distance theories and Maxwell's contiguous-action theory, Helmholtz was also highly sensitive to the possibilities of either uniting these two theories or reformulating them into a more general theory or demonstrating, through a decisive experiment, the superiority of one theory over the other. One opportunity for so doing emerged in 1871–72 with the appearance of Boltzmann in Berlin, who came from Vienna to work under Helmholtz in his Berlin physical institute in order to undertake measurements of the specific inductive capacity of insulating substances and so to test the experimental foundations of Maxwell's electromagnetic theory of light. He did so, too, in order partially to balance his education as a purely theoretical physicist by doing some experimental work, while also wanting to come into personal contact with Helmholtz, the prominent holder of the most important chair of physics at a German university.[69] At first Boltzmann and, ironically, Helmholtz misinterpreted

---

67. "Ueber die Fortpflanzungsgeschwindigkeit der elektrodynamischen Wirkungen," *MB* (1871):292–98, in *WA 1*:629–35, on 635.
68. Hertz to his parents, 23 December 1887 and 1 January 1888, in *Erinnerungen, Briefe, Tagebücher*, ed. Johanna Hertz (Leipzig: Akademische Verlagsgesellschaft, 1927), 183–86.
69. Boltzmann to Josef Stefan (?), 2 February 1872, Staatsbibliothek Preussischer Kulturbesitz, Berlin, Handschriftenabteilung, Slg. Darmstaedter F 1 e 1871 (3) L. Boltzmann.

Boltzmann's measurements of the specific inductive capacity as compared with the indices of the refraction of light. They were uncertain about the well-known Maxwellian relation $\mathbf{n} = \sqrt{\varepsilon}$, where $\mathbf{n}$ is the index of refraction and ε is the specific inductive capacity, which is related to the dielectric susceptibility according to $\vec{E} = \vec{D} - 4\pi\vec{P} = \varepsilon\vec{E} - 4\pi\chi_e\vec{E}$, $\varepsilon = 1 + 4\pi\chi_e$. Because they assumed $\mathbf{n} = \varepsilon$, Boltzmann at first thought that he had completely disproved Maxwell's electromagnetic theory of light. In point of fact, Boltzmann's data fit quite nicely into the relation $\mathbf{n} = \sqrt{\varepsilon}$.[70]

Experimental proofs sometimes depend on lucky circumstances; that was certainly the case with Boltzmann's measurements of specific inductive capacity. For Boltzmann mostly measured the specific inductive capacity of nonpolar (for example, organic) substances, where Maxwell's relation indeed holds. Had he measured polar substances (for example, water molecules with their high dipole moment which yield a high inductive capacity), he would have been confronted by numerous deviations from Maxwell's law, deviations that show values of the specific inductive capacity far from $\mathbf{n}^2$, which is of the order of magnitude of 1. Although Boltzmann then found impressive experimental evidence for Maxwell's electromagnetic theory of light, the latent experimental problems of polar substances, of the dispersion of light in general, and of metal optics showed clear indications of the shortcomings of Maxwell's electrodynamics, which assumed both the ether and matter to be purely continuous and which treated both the specific inductive capacity and the specific resistance as independent of the frequency of the electromagnetic oscillations of light.

With the completion of his experimental work, Boltzmann left Berlin and returned to Vienna in 1872. His place in Helmholtz's institute was (in effect) taken up in the late 1870s and early 1880s by Hertz. In a prize question for the Prussian Akademie der Wissenschaften for the year 1879, Helmholtz called for experimental proof of the electrodynamic effect of increasing or diminishing states of polarization in dielectric media.[71] The question reflected the transformation Maxwell's theory of a general displacement current had undergone in Helmholtz's mind. More to the point, Helmholtz's own objective in setting this

---

70. Koenigsberger 2:201; and Walter Kaiser, Introduction, in Boltzmann, *Vorlesungen* (reprint), on 12*-13*.
71. Adolf Harnack, *Geschichte der Königlichen Preussischen Akademie der Wissenschaften zu Berlin*, 3 vols. (Berlin: Reichsdruckerei, 1900), vol. 2: *Urkunden und Actenstücke zur Geschichte der Königlich Preussischen Akademie der Wissenschaften*, 617, No. 229.

question was to stimulate his student Hertz to work on experimentally testing the basic hypotheses of Maxwell's electrodynamics. As Helmholtz solemnly put it in the preface to Hertz's posthumously published *Die Prinzipien der Mechanik*, he felt from the very beginning of their relationship that in Hertz he had his most promising pupil and one who might promulgate his own scientific ideas.[72] Nevertheless, for a number of years Hertz sensed a marked reserve in Helmholtz's behavior. Helmholtz did not interfere in the development of Hertz's early scientific career, for example, in Hertz's plan to habilitate at the University of Kiel; indeed, he did all he could to help Hertz. Yet only when Helmholtz received Hertz's first promising paper on the propagation of electrodynamic action under the influence of isolating substances did Helmholtz's tone of voice become truly friendly.[73]

The objective of a direct experimental test of Maxwell's basic assumption of a displacement current had already been discussed by Maxwell himself.[74] Experiments had been undertaken by Nikolai Schiller in 1875 and, especially, by Henry A. Rowland, who worked upon his own proposal in Helmholtz's institute in 1876. Thanks to Helmholtz, Rowland's measurements of (small) magnetic effects of polarization charges moving at high speeds with rotating disks were published in 1876. Contrary to Helmholtz's expectations, however, the results were not decisive. Helmholtz conceded that the effect could either be interpreted within the framework of Weber's theory (as a magnetic effect of moving charges attached to ponderable masses) or in the sense of Maxwell's theory of displacement current or in the sense of his own theory of polarization currents. In the last case, Helmholtz considered the magnetic effect of time-varying polarization as due to air vortices engendered by the rotating disks.[75]

Hertz's work shows a clear indication of the lasting influence of Helmholtz's prize question of 1879: for example, as already noted, Hertz himself had tried to measure the influence of a massive dielectric attached to the emitting dipole on the propagation of electromagnetic

---

72. Preface by Helmholtz to Heinrich Hertz, *Werke*, vol. 3: *Die Prinzipien der Mechanik*, 2nd ed. (Leipzig: Johann Ambrosius Barth, 1910), xiii–xxviii, on xiv–xvi; and vol. 2: Hertz, *Untersuchungen über die Ausbreitung der elektrischen Kraft* (Leipzig: Johann Ambrosius Barth, 1894), 1.

73. Hertz to his father, 1 March 1883, in J. Hertz, ed., *Erinnerungen*, 135–37; Helmholtz to Hertz (postcard), 7 November 1887, in ibid., 180.

74. Salvo d'Agostino, "Hertz's Researches on Electromagnetic Waves," *HSPS 6* (1975):261–323, on 271.

75. "Bericht betreffend Versuche über die elektromagnetische Wirkung elektrischer Convection, ausgeführt von Hrn. Henry A.Rowland," *AP 158* (1876):487–93, in *WA 1*:791–97, on 796–97.

action. Moreover, referring to Wilhelm Conrad Röntgen's experiments on the magnetic effect of time-varying states of polarization in a dielectric, Hertz himself was very clear about the relationship of the prize question, of the Rowland and Röntgen experiments, and of his own intermediate experiments in 1887. Here Hertz concentrated on the alleged effect of polarizable dielectric masses on the propagation of electromagnetic action of high frequency. Although this propagation was still not identified as wave propagation, it is hardly surprising that Helmholtz acknowledged the receipt of Hertz's manuscript, which he helped prepare for publication in the *Sitzungsberichte* of the Prussian Akademie, with an enthusiastic "Bravo!!"[76]

Notwithstanding the remaining problem of a direct experimental confirmation, Hertz and Helmholtz believed that Hertz's electromagnetic waves meant the scientific breakthrough of Maxwell's theory and its consequent transverse waves. In the introduction to his *Untersuchungen über die Ausbreitung der elektrischen Kraft*, Hertz claimed that the "aim of his experiments was the test of the fundamental hypotheses of the Faraday-Maxwell theory and that the result of the experiments was the confirmation of the fundamental hypotheses of this theory." Historical evidence for this statement can be found as early as 1884 in his theoretical studies on the competing theories, where Hertz concentrated on Maxwell's electrodynamics. Although Hertz avoided the field approach in 1884, instead using formally retarded potentials to derive Maxwell's equation, he in fact already favored Maxwell's theory. And although his derivation was hardly a logically consistent one, for the first time Hertz's paper of 1884 showed the highly symmetrical form of Maxwell's equation in vacuum.[77] Like Maxwell's idea of a primarily existing state of displacement in the ether, Hertz stated that the field of electromotive force "is an entity which does exist autonomously in space and which does exist independently from the means of its creation."[78]

Hence Hertz's later remarks, where he politely praised the influence of his teacher Helmholtz, are best understood as a general indication

---

76. H. Hertz, "Ueber Inductionserscheinungen, hervorgerufen durch die elektrischen Vorgänge in Isolatoren," [1887], in Hertz, *Untersuchungen*, 102–14, esp. 108–13; Hertz to Helmholtz, 5 November 1887, in J. Hertz, ed., *Erinnerungen*, 179–80; and Helmholtz to Hertz, 7 November 1887.

77. D'Agostino, "Hertz's Researches," 284–96; and Kaiser, *Theorien der Elektrodynamik*, 164–75.

78. H. Hertz, "Über die Beziehungen zwischen den Maxwell'schen elektrodynamischen Grundgleichungen und den Grundgleichungen der gegnerischen Elektrodynamik," [1884], in *Werke*, vol. 1: *Schriften vermischten Inhalts* (Leipzig: Johann Ambrosius Barth, 1895), 295–314, esp. 296.

of the stimulating scientific atmosphere in Helmholtz's Berlin institute and of Hertz's general gratitude towards Helmholtz. The closest contact between Helmholtz's theory of dielectric polarization and Hertz's own thinking can be seen in Hertz's previously mentioned intermediate experiments of 1887 aimed at detecting an influence of polarizable matter on the propagation of high-frequency electrodynamic action. Hertz even apologized for not having mentioned Helmholtz in his papers on electromagnetic waves.[79]

The years from 1886 to 1888 witnessed the triumph, at least in Germany, of Maxwell's electrodynamics: Hertz discharged electricity over a spark gap and thereby found a method for producing and detecting free electromagnetic waves. As discussed above, one typical Hertzian move was to measure the influence of a dielectric in the vicinity of the radiating dipole on the propagation of electromagnetic action.[80] Most physicists eventually interpreted Hertz's experiments on electromagnetic waves as the definitive confirmation of Maxwell's prediction of wave solutions for his system of field equations. In particular, Hertz's results on quasi-optical experiments, namely, the reflection, polarization, and interference of electromagnetic waves, were so striking that Hertz, in a lecture before the sixty-second Versammlung deutscher Naturforscher und Ärzte in 1889 at Heidelberg, could speak of having almost united the fields of electricity and of light.[81] As he later stated: "The aim of these experiments was to confirm the basic hypotheses of Maxwell's theory, and the outcome of these experiments was a confirmation of the basic hypotheses of Maxwell's theory."[82] Oliver Heaviside wrote that the discovery of electromagnetic waves was the "death blow" to the old theories of action at a distance.[83] Nonetheless, despite his final experimental successes, Hertz's research on electromagnetic waves did not confront the basic assumptions of Maxwell's (or Helmholtz's) theory, namely, to show the displacement (or polarization) current directly by experiment. Hertz found an experimental device to test the wave solutions, which Maxwell and Helmholtz could derive from their fundamental field equations.

In December 1888 Helmholtz wrote Hertz how "happy" he was about his erstwhile student's latest achievements ("*Thaten*")—he was

---

79. Hertz to Helmholtz, 24 February 1892, in J. Hertz, ed., *Erinnerungen*, 240–41.
80. D'Agostino, "Hertz's Researches," 301.
81. Heinrich Hertz, "Ueber die Beziehung zwischen Licht und Elektricität" (1889), in his *Schriften*, 339–54, on 352–53.
82. Hertz, *Untersuchungen*, 21.
83. Heaviside to Hertz, 13 July 1889, reprinted in O'Hara and Pricha, *Hertz and the Maxwellians*, 66–8, on 67.

apparently referring to Hertz's quasi-optical experiments. Yet Helmholtz's letter also betrayed an undertone of self-disappointment: "These are matters that I gnawed around at for years, looking for an opening whereby one could get at them. . . ."[84] There is no little irony in the fact that Hertz's principal experimental achievement—the production and detection of polarizable, transverse electromagnetic waves—favored the predictions of Maxwell's theory and questioned the predictive capacity of Helmholtz's theory of dielectric polarization. And it was Hertz, too, who renounced Helmholtz's idea of a modified theory of action at a distance. Accordingly, Helmholtz, in his evaluation of Hertz's experimental work, referred solely to Faraday and Maxwell as the builders of the theoretical basis of Hertz's experiments on electromagnetic waves.[85]

In his last years Helmholtz again tackled a basic problem in physics which, like the invariance principles, belongs to the metatheoretical level. He ventured to apply the principle of least action to the equations of electrodynamics. This last piece of work in electrodynamics was an attempt to transfer the structural principles of rational mechanics to Maxwell's dynamics of the electromagnetic field. In Bernhard Riemann's and others' theories of action at a distance the derivation of force laws from Lagrangians (representing all the energies of systems of currents or moving charges) was already a well-known theoretical method.[86] Maxwell, by contrast, had treated current systems with the help of the principle of least action, using formally electric currents as generalized coordinates.[87] Helmholtz, however, wanted to find a Lagrangian from which one could derive Maxwell's differential equations of the electromagnetic field. Put the other way around, he wanted to unite the entire set of Maxwell's fundamental differential equations into one single equation governed by the principle of least action. Helmholtz's expression for a Lagrangian of the electromagnetic field included electrostatic energy. But more significant was his attempt to propose a so-called kinetic potential, one which still contained polarization and magnetization along with their time variations.[88] In the hands of Boltzmann, Hendrik Antoon Lorentz, Karl Schwarzschild,

---

84. Helmholtz to Hertz, 15 December 1888, copy in Deutsches Museum, Munich, Handschriftenabteilung, No. 3110.
85. Helmholtz's preface to Hertz, *Die Prinzipien der Mechanik*, xx–xxiv.
86. Bernhard Riemann, *Schwere, Elektricität und Magnetismus, Nach den Vorlesungen von Bernhard Riemann vom Sommersemester 1861*, ed. Karl Hattendorff (Hannover: Carl Rümpler, 1876), 316–18, 323–27.
87. Maxwell, *Treatise* 2:223–28 (§ 578–84).
88. "Das Princip der kleinsten Wirkung in der Elektrodynamik," *AP 47* (1892):1–26, in *WA* 3:476–504, on 485–90.

and Gustav Mie, this approach became a powerful tool for an elegant representation of field equations in electrodynamics and in the theory of relativity.[89]

6. Conclusion

Helmholtz's substantive results in electrodynamics were far less successful than either Maxwell's theory of the electromagnetic field or Hertz's experimental confirmation and theoretical clarification of Maxwell's theory or Lorentz's synthesis of particles and fields in his electron theory.[90] Helmholtz's importance in shaping classical electrodynamics lay more in stimulating others than in creating his own theory; more in attracting them to his institute and providing them with experimental means than in conducting experiments himself; more in mediating and criticizing than in determining. In short, his role in the formation of classical electrodynamics was an instrumental one. As such, he helped synthesize "classical" physics.

Yet at the same time his electrodynamics, even as it became increasingly obsolete after 1890, contained elements that others would pursue to help shape "modern" physics. For one, his polarization theory, with its consequence of longitudinal waves, was used as a theoretical basis in experimental areas beyond dielectric media. In particular, cathode rays were interpreted as longitudinal waves. Helmholtz himself was inclined to think of these rays as longitudinal waves in the ether.[91] And again in 1895, barely a year after Helmholtz's death, when Röntgen found a new type of radiation, the X-rays, the longitudinal ether waves appeared: in his preliminary announcement to the Physikalisch-Medicinische Gesellschaft in Würzburg Röntgen suggested interpreting his discovery as experimental proof of the predicted longitudinal undulations in the ether.[92] For another, Helmholtz's work in electrodynamics contained vague ideas foreshadowing relativity: for

---

89. See, e.g., Wolfgang Pauli, "Raelitivitätstheorie," in *Encyklopädie der mathematischen Wissenschaften. Mit Einschluss ihrer Anwendungen*, eds., H. Burkhardt, et al., 6 vols. in several parts (Leipzig: B.G. Teubner, 1896–1935), *5:2: Physik*, ed. Arnold Sommerfeld (1920), 539–775, on 642–43, 755; and Günter Bierhalter's essay, "Helmholtz's Mechanical Foundation of Thermodynamics," in this volume.

90. Hendrik Antoon Lorentz, "La théorie électromagnétique de Maxwell et son application aux corps mouvants," *Archives néerlandaises des sciences exactes et naturelles* 25, sér. 1 (1892):363–552, on 475–97. For Helmholtz's influence on Lorentz see Ole Knudsen, "Electric Displacement and the Development of Optics after Maxwell," *Centaurus* 22 (1978):53–60.

91. Koenigsberger 2:305.

92. Wilhelm Conrad Röntgen, "Ueber eine neue Art von Strahlen," *Sitzungsberichte der physikalisch-medicinischen Gesellschaft in Würzburg 137* (1895):132–41, on 140–41.

example, in his tentative introduction of retardation effects in his basic law for electrodynamic interaction. Along with the contemporary work of Riemann, Enrico Betti, Ludvig Valentin Lorenz, and Tullio Levi-Civita on retarded potentials, which took into account the finite velocity of propagation of electromagnetic action, the overall thrust of Helmholtz's electrodynamics clearly constituted within the electrodynamic arena the continuing erosion of the Newtonian physical worldview.[93] Whether or not Helmholtz's generalized vector potential for the interaction of distant current elements was a conscious abandonment of the structures of classical mechanics, it nonetheless shattered the predominant position of electrodynamic theories like Weber's. Ideas foreshadowing relativity can also be seen in Helmholtz's attempt to find an appropriate Lagrangian for the electromagnetic field, and, though this did not appear directly in his electrodynamics, in his stress on the empirical character of the axioms of geometry.[94]

Yet even as he helped point the way to the future, Helmholtz remained firmly rooted in nineteenth-century traditions. With his criticism of theories of action at a distance, with his formulation of a generalized potential for the fundamental electrodynamic interaction, and with his own generalized field equations he stimulated the reception and eventual establishment of Maxwellian electrodynamics. Together with his work on hydrodynamics and on thermodynamics, his electrodynamics greatly influenced the formation of classical theoretical physics as a whole. Yet unlike his British colleagues and unlike Boltzmann, Helmholtz always remained very cautious with hypotheses concerning the microphysical level. In that respect, at least, his work accurately reflected the epistemological atmosphere in physics in the German-speaking countries during the second half of the nineteenth century.

93. Treder, "Helmholtzsche Elektrodynamik"; idem, "Helmholtzsche Elektrodynamik und die Beziehungen von Kinematik"; W. Kaiser, "Die zeitliche Ausbreitung von Potentialen in der Elektrodynamik," *Gesnerus 35* (1978):279–317; Wiechert, "Grundlagen," 78–9; and Tullio Levi-Civita, "Sulla reductibilità delle equazioni elettrodinamiche di Helmholtz alla forma Hertziana," *Il Nuovo Cimento*, 6th ser. (1897):93–108.

94. On this last point see Robert DiSalle's essay, "Helmholtz's Empiricist Philosophy of Mathematics: Between Laws of Perception and Laws of Nature," in this volume.

10

Between Physics and Chemistry

Helmholtz's Route to a Theory of Chemical Thermodynamics

Helge Kragh

1. Introduction

Mid-nineteenth-century chemistry was almost a purely experimental and classificatory science. In contrast to physics, it largely lacked theoretical foundations and showed little progress in supplying such foundations. Nearly all chemists merely collected data and analyzed specific compounds—an empiricist trend reinforced by the emergence in the 1830s of the powerful new subdiscipline of organic chemistry. In 1873, the distinguished chemist and historian of chemistry, Hermann Kopp, confessed that in "chemistry, no theory has yet been developed which attempts to derive the results of experience as necessary consequences of a definite principle. The theoretical principles of chemistry are still only applicable to more or less special cases in practical chemistry. As yet we have no comprehensive picture of the connections between these special cases or of the image which we construct from them, however we imagine these connections."[1] Most chemists agreed.

Nonetheless, some progress had been made, particularly in the fields of chemical equilibrium and rates of reaction, electrochemistry, and thermochemistry.[2] The theoretical foundation for much of this work

---

Acknowledgments: I would like to thank Pat Munday for comments on this paper, and David Cahan for extensive criticism and many helpful suggestions.
 1. Hermann Kopp, *Die Entwicklung der Chemie in der neueren Zeit* (Munich: R. Oldenburg, 1873), 844.
 2. For brief descriptions of these developments, see, e.g., Aaron J. Ihde, *The Development of Modern Chemistry* (New York: Dover, 1984), 391–417; and A.J.B. Rob-

was the concept of "chemical force," or "affinity," the nature of which was, however, obscure. It was at first widely assumed that the affinity of chemical substances would find its explanation in terms of gravitational or electrical forces, or perhaps by means of mechanical models of molecules; yet nothing came of these speculations. Then with the establishment of the principle of energy conservation around 1850 the heat evolved in chemical processes seemed to offer a quantitative measure of affinity and so to provide a theoretical basis for chemical change. Chemists applied the thermal theory of affinity to chemical reactions and used it to study chemical equilibria and electrochemical processes. For these processes it was generally assumed that the electrical heat produced by the electromotive force of the cell equalled the chemical heat, in which case electrochemistry could presumably be understood in terms of thermochemistry.

According to classical thermochemistry, as founded by Julius Thomsen and Marcellin Berthelot, the heat evolved in a chemical reaction was the true measure of its affinity.[3] The theory rested on the Thomsen-Berthelot principle that all chemical changes were accompanied by heat production and that the actual process which occurred was the one in which the most heat was produced. This principle, formulated in slightly different versions by Thomsen in 1854 and by Berthelot in 1864, became the controversial foundation of a research program that lasted for two decades. It was criticized from the start on theoretical and empirical grounds; by 1880, it was widely recognized that the thermal theory of affinity needed replacement. Its successor, as physicists more than chemists recognized, would have to rely on the second law of thermodynamics. Although this law, along with the associated concept of entropy, had been introduced in the 1850s (by Rudolf Clausius, William Thomson [Lord Kelvin], William Rankine, and others), it had diffused only slowly to chemistry. The new theory, as it emerged during the 1880s, transformed the methodology of theoretical chemistry without constituting a revolutionary break with its past.

The traditional conceptual objects of chemistry were atoms and molecules; yet many chemists argued that such objects served only a heuristic purpose and had no real existence. Positivistically oriented

---

ertson, "Physical Chemistry," in Colin A. Russell, ed., *Recent Developments in the History of Chemistry* (London: The Royal Society of Chemistry, 1985), 153–78.

3. For this theory and references to the literature, see Helge Kragh, "Julius Thomsen and Classical Thermochemistry," *BJHS* 17 (1984):255–72; and R.G.A. Dolby, "Thermochemistry versus Thermodynamics: The Nineteenth-century Controversy," *History of Science* 22 (1984):375–400.

chemists maintained that a realistic interpretation of atoms and molecules was "metaphysical" and that chemistry should build solely on measurable quantities, such as equivalent weights.[4] Such philosophical niceties notwithstanding, in practice much of chemistry was firmly based on the assumption of atoms and molecules. Yet during the latter part of the century many chemists and physicists thought that this assumption blocked the path to a proper theoretical understanding of chemical phenomena; Berthelot, Josiah Willard Gibbs, Pierre Duhem, Wilhelm Ostwald, and Georg Helm, among others, assumed a general theory would be neutral with respect to hypotheses about the constitution of matter, and they regarded hydro- and thermodynamics as methodological models for future chemical theory.

The transformation of chemistry from a largely static and experimental into a dynamic and theoretical science was produced not by run-of-the-mill chemists but by physicists or chemists with a strong background in physics. Hermann von Helmholtz was one such important figure in that transformation; others were Gibbs, Ostwald, August Horstmann, Svante Arrhenius, and Jacobus Henricus van't Hoff. Although Helmholtz was not a chemist, his contributions to the fields of electrochemistry and chemical thermodynamics secured him a distinguished place in the history of chemistry.[5] That reputation rested, above all, on his thermodynamic explanation (in 1882) of chemical changes, an explanation that provided chemistry with a solid theoretical foundation. Building on earlier insights gained by himself and other scientists, Helmholtz proved that affinity was not given by the heat evolved in a chemical reaction but rather by the maximum work produced when the reaction was carried out reversibly. Characteristically, Helmholtz's demonstration was the result of abstract physical reasoning and not of chemical experimentation. His work of 1882–83 was convincing evidence that advanced physical theory had a fruitful role to play even in the traditionally empirical science of chemistry. Although organic chemists continued to distrust the increasing mathematization of chemistry, by 1890 theoretical chemistry had undeniably evolved into an important and useful subdiscipline. Helmholtz

---

4. See David M. Knight, *Atoms and Elements* (London: Hutchison, 1967); and Helge Kragh, "Julius Thomsen and 19th-Century Speculations on the Complexity of Atoms," *Annals of Science* 39 (1982):37–60.

5. For appreciations of Helmholtz's work in chemistry, see Wilhelm Ostwald, *Große Männer* (Leipzig: Akademische Verlagsgesellschaft, 1909), 265–311; Walther Nernst, "Die elektrochemischen Arbeiten von Helmholtz," *Die Naturwissenschaften* 9 (1921):699–702; and Karl Heinig, "Einige Bemerkungen zu Helmholtz's wissenschaftlichem Wirken und der Chemie des 19. Jahrhunderts," *WZHUB* 22 (1973):357–61. See also the references to Section 4 of the present essay, especially notes 68–81.

was neither the sole nor the most important contributor to this development; but his thermodynamic theory of 1882–83 was the pioneering work on which much of the new theoretical chemistry rested.

Helmholtz's route to chemical thermodynamics was long and tortuous; it was only after many years' occupation with energetics[6] and electrochemistry—disrupted by a long period of work in physiology—that he realized how to incorporate fully the laws of thermodynamics into chemistry. For Helmholtz, as for nineteenth-century science in general, energetics, thermodynamics, and electrochemistry were inseparably bound. In order to understand his chemical thermodynamics, and thereby to appreciate better the methodological transformation of chemistry in the late nineteenth century, this essay first examines his earlier work in electrochemistry and related areas. This is the subject of Section 2, which surveys Helmholtz's early work in physiological chemistry (1843–45), the chemical parts of his 1847 treatise on energy conservation, his work on galvanic polarization and concentration cells (1872–77), and, finally, his analysis of the atomic nature of electricity (1880–81). Apart from their importance as prologue to Helmholtz's theory of 1882, these contributions also had considerable importance in their own right and were widely appreciated by contemporary chemists. In his theory of chemical thermodynamics, the subject of Section 3 below, Helmholtz introduced the fundamental concept of free energy; he thereby eliminated the old thermal notion of affinity. His work of 1882–83 proved important for the new field of physical chemistry which emerged a few years later. Yet Helmholtz's support of the physical chemistry of Ostwald, van't Hoff, and Arrhenius was not wholehearted; indeed, he was skeptical of the ionic theory of dissociation, a cornerstone of the new theory. Section 4 examines the relation of Helmholtz's work in chemical thermodynamics to the new physical chemistry and evaluates the influence of Helmholtz's work on chemistry as a whole.

2. Energetics and Electrochemistry

Helmholtz's initiation into the sciences of his day included a solid education in chemistry. As a medical student at Berlin's Königlich medizinisch-chirurgisches Friedrich-Wilhelms-Institut, he attended Eilhard Mitscherlich's lectures on chemistry; yet he apparently found

---

6. Here and elsewhere in this essay, "energetics" refers to the science of energy and its transformations, and not to the viewpoint developed in the 1890s by the energeticist school of Ostwald, Helm, and others.

the lectures boring and the laboratory exercises tedious.[7] He passed his preliminary examinations in December 1839 and received his highest mark (*"vorzüglich gut"*) in chemistry.[8] He was apparently not interested in chemistry for its own sake, yet he realized that it would be useful to him for his work in physics and physiology. Decades later, in the course of giving advice to his son Robert, Helmholtz revealed his own attitude towards chemistry: "Chemistry can really interest only those who, in order to gain firsthand experience of the circumstances, have themselves made experiments in it .... After all, a certain amount of chemical knowledge is essential for progress in the (other) sciences."[9]

It was the physiological aspects of chemistry that interested the young Helmholtz. His first publications dealt with physiological chemistry and animal heat, a fact that reflected, on the one hand, the influence of Justus von Liebig—Helmholtz rather sardonically (and privately) anointed him "the king of chemists"[10]—and, on the other, the tension between physiology and chemistry in explaining organic processes. In his very first publication (1843), Helmholtz confirmed the observation of Theodor Schwann and others of the vegetable nature of yeast. Most chemists, according to Helmholtz, considered this controversial conclusion to be based on "physiological phantasies."[11] The burning issue of the day in chemistry and physiology was that of "vitalism" versus "mechanism." The vitalists believed that processes in organic nature were governed not only by the general laws of physics

---

7. David Cahan, ed., *Letters of Hermann von Helmholtz to His Parents: The Medical Education of a German Scientist, 1837–1846* (Stuttgart: Franz Steiner Verlag, 1993), Letter 6 (5 November 1838), Letter 9 (5 May 1839), and Letter 12 (11 July 1839), 48–52, 56–60, 65–6, resp.
8. Ibid., Letter 16 (11 December 1839), 74–5. In physics, psychology, zoology, and botany, Helmholtz received a *"sehr gut,"* and in mineralogy a *"ziemlich gut."* (Ibid.)
9. Hermann von Helmholtz to Robert von Helmholtz, 4 July 1880, in Ellen von Siemens-Helmholtz, ed., *Anna von Helmholtz. Ein Lebensbild in Briefen*, 2 vols. (Berlin: Verlag für Kulturpolitik, 1929), *1*:248–50, quote on 249. Helmholtz advised Robert to attend a course in inorganic chemical analysis in Robert Bunsen's laboratory. (Ibid.)
10. Helmholtz to Olga von Helmholtz, 10 August 1851, in Richard L. Kremer, ed., *Letters of Hermann von Helmholtz to His Wife, 1847–1859* (Stuttgart: Franz Steiner, 1990), 53–60, quote on 56.
11. "Ueber das Wesen der Fäulniss und Gährung," *MA* (1843):453–62, in *WA* 2:726–34, quote on 727. For the vexing question of "vitalism" in the works of Liebig and Helmholtz, see Yehuda Elkana, *The Discovery of the Conservation of Energy* (London: Hutchison, 1974), 103–11; Timothy Lenoir, *The Strategy of Life. Teleology and Mechanics in Nineteenth Century German Biology* (Dordrecht: Reidel, 1982), 197–215; and Richard Lynn Kremer, "The Thermodynamics of Life and Experimental Physiology, 1770–1880," (Ph.D. diss., Harvard University, 1984), chap. 6: "Animal Heat and Energy Conservation, 1837–1847." For Helmholtz's early work in physiology see the essay by Kathryn M. Olesko and Frederic L. Holmes, "Experiment, Quantification, and Discovery: Helmholtz's Early Physiological Researches, 1843–50," in this volume.

and chemistry, but also by the purposeful action of a special "vital force." Extreme mechanists or reductionists, by contrast, argued that biological phenomena could be fully reduced to physical-chemical processes. Although Helmholtz sympathized with Liebig's version of antivitalism, his conclusions gave only ambiguous support to "the greatest of living chemists,"[12] as he called Liebig. Helmholtz showed experimentally that putrefaction and fermentation could not be due to either very hot air or electrolytic oxygen, which led him to conclude, with Liebig, that putrefaction could be maintained independently of vital processes. But he also argued, against Liebig, that fermentation was essentially a different type of organic process, namely "a form of putrefaction bound to and modified by the presence of an organism."[13]

Whereas Helmholtz's first paper was not concerned with the problem of the origin and transformation of organic ("animal") heat, his second was.[14] In critically reviewing the problem of animal heat as given in Liebig's "Über die thierische Wärme," he demonstrated a sure knowledge of the still immature science of thermochemistry and discussed various experimental determinations of the heat of combustion of carbon, a topic of evident importance to the problem of animal heat and energy conservation in the organic world. He criticized results—of César Despretz, Pierre Dulong, Pierre Favre, and Johann Silbermann—that were not based on the mechanical theory of heat. That criticism indicated Helmholtz's growing conviction of the unity of natural forces and pointed toward his idea of the conservation of force, published two years later (1847). "In considering heat as motion," he wrote in 1845, "we must first suppose that mechanical, electrical, and chemical forces can produce only a definite equivalent of work, however complicated the manner of transformation of one force into another may be."[15]

The chemical content of Helmholtz's epochmaking study of 1847— *Ueber die Erhaltung der Kraft*—included two areas to which Helmholtz would later greatly contribute: thermochemistry and electrochemistry. In Section IV of his study, he dealt with the heat formed by chemical

---

12. Helmholtz to Olga von Helmholtz, 10 August 1851, on 56.
13. "Fäulniss und Gährung," 437.
14. "Bericht über die Theorie der physiologischen Wärmeerscheinungen für 1845," in *Fortschritte der Physik im Jahre 1845* (Berlin, 1847), 346–55, in *WA 1*:3–11.
15. Ibid., *1*:7. Although Helmholtz here expressed the idea of the unity of forces, he did not articulate it into the principle of energy conservation. (See Thomas S. Kuhn, "Energy Conservation as an Example of Simultaneous Discovery," in Kuhn's *The Essential Tension: Selected Studies in Scientific Tradition and Change* [Chicago and London: The University of Chicago Press, 1977], 66–104, esp. 95.)

processes. Referring to Germain Henri Hess's recently stated law that the amount of heat evolved in a chemical process was independent of the steps through which the reaction occurred, Helmholtz argued that this result simply reflected the principle of force conservation. Hess had originally derived his law on the assumption that heat was a substance; Helmholtz showed that this assumption was unnecessary. "According to our way of viewing the subject," he wrote, "the quantity of heat developed by chemical processes would be the quantity of *vis viva* produced by the chemical attractions, and in this case the above law [Hess's] would be the expression for the principle of the conservation of force."[16] Helmholtz then went on to address the earlier works of Benoit-Pierre-Emile Clapeyron and Karl Holtzmann (1834 and 1845, respectively) which were based on Sadi Carnot's theory of thermal work and the associated caloric notion of heat. Holtzmann and Clapeyron had derived different equations for the conversion coefficient between heat and mechanical force; Helmholtz showed that, when seen from the point of view of conservation of force, these equations were identical.

In Section V of *Ueber die Erhaltung der Kraft* Helmholtz discussed polarization, contact electricity, and various kinds of galvanic cells from the point of view of conservation of force. Using the simple case of a Daniell cell or a battery consisting of such cells, he argued that the net (electrical) heat produced arose from the difference in the chemical heats evolved at the two electrodes. Denoting the heat produced by one equivalent of the electropositive metal by its oxidation and dissolution a_z and the corresponding heat absorbed at the electronegative metal a_c, he found the electromotive force $\mathscr{E}$ of the cell to be given by $\mathscr{E} = a_z - a_c$, which implied that the electromotive force was fully reducible to thermochemical quantities. James Prescott Joule had reached a similar conclusion in 1841 using proportionality instead of equality; and William Thomson in 1851 confirmed the equality of chemical and electric heats.[17] This general idea that in a galvanic cell chemical energy was completely transformed into electric energy was subsequently often referred to as the "Thomson-Helmholtz rule."

In his 1847 masterpiece, Helmholtz assumed that each kind of matter had a specific affinity for electricity, positive or negative, and that

---

16. "Ueber die Erhaltung der Kraft. Eine physikalische Abhandlung," *WA* *1*:12–75, first published in book form (Berlin: G. Reimer, 1847). For an analysis of this work see the essay by Fabio Bevilacqua, "Helmholtz's *Ueber die Erhaltung der Kraft*: The Emergence of a Theoretical Physicist," in this volume.

17. See Wilhelm Ostwald, *Electrochemistry. History and Theory*, 2 vols. (Leipzig: Veit & Co., 1896; reprinted New Delhi: American Publishing Co., 1980), 2:743–66.

during electrolysis work was done in separating atoms from their electrical charges rather than in dissociating the atoms. He referred repeatedly to "electrical particles" and composite atoms with internal degrees of freedom and imagined that "portions of the composite atom of a liquid are endowed with different powers of attraction for electricities." According to Helmholtz, during electrolysis these portions of atoms were separated on the electrodes to which they deposited their electricity. "We can therefore consider," he wrote, "that in chemical compounds the atoms are also associated with equivalent quantities of electricity $\pm E$. These quantities of electricity are just as equal for all atoms as the stoichiometric equivalents of ponderable substances in different compounds."[18] Thus, as early as 1847 Helmholtz considered electricity to be corpuscular in nature, a view which was then far from common. In sum, Helmholtz's early works show that already by 1847 he had a solid understanding of thermochemical and electrochemical matters. He used that understanding to advance his larger scientific program.

Despite his early, mid-century achievements in electrochemistry, Helmholtz refrained from further electrochemical studies for a full quarter of a century and instead devoted himself principally to physiological investigations. With his appointment in 1871 as the new professor of experimental physics at the University of Berlin, he devoted all his efforts to physical enquiries. In 1872–73, he resumed his electrochemical studies with an investigation of galvanic polarization in a Daniell element connected to an electrolytic cell with platinum electrodes.[19] It was well known that in such a system the "polarizing current" from the Daniell element declined rapidly but without ever entirely vanishing; for small currents no decomposition of water was observed. When the electrolytic system was disconnected, the platinum cell would decompose water and produce a "depolarizing" current that soon became imperceptible. (Helmholtz introduced the terms *polarizing* and *depolarizing currents*.) To explain this phenomenon, Helmholtz characteristically focused on its apparent contradiction with the principle of energy conservation (as the conservation of "force" had by then become known). He asked: "On what does the apparently unlimited duration of the polarizing current depend? In a series specified as above (that is, a Daniell element connected to a platinum cell), if no other changes occur in it, then the electrolytic conduction in the

---

18. "Ueber die Erhaltung der Kraft," *WA 1*:54.
19. "Ueber die galvanische Polarisation des Platin," *Halle, Zeitschrift für die gesammten Naturwissenschaften* 6 (1872):186–88; and "Ueber galvanische Polarisation in gasfreien Flüssigkeiten," *AP 150* (1873):483–95, in *WA 1*:823–34.

liquids, as predicted by Faraday's law, cannot come about without violating the law of the conservation of energy."[20] Helmholtz saw that any explanation had to agree with Faraday's law as well as with the law of energy conservation. He suggested that the platinum cell acted like an imperfect capacitor, an analogy that James Clerk Maxwell had also suggested about this time.[21] Helmholtz argued that the mechanism occurring in the platinum cell capacitor was what he called an electrolytic convection: in galvanic processes the gases (hydrogen and oxygen) would not necessarily evolve as bubbles, but might instead move within the liquid or penetrate the surface layer of the platinum plates. In this way he explained the puzzling fact that a very small electromotive force could cause a current without water being decomposed.

Helmholtz got the idea that hydrogen—and, to a much lesser extent, oxygen—can diffuse into certain metals, a process known as occlusion, from Thomas Graham, who in 1858 had demonstrated the effect for palladium.[22] In 1872-73, Helmholtz provided partial experimental support for his suggestion of electrolytic convection currents and the role of occlusion effects in platinum. In 1876, the American physicist Elihu Root, who worked in Helmholtz's laboratory in Berlin, supplied definitive experimental proof of the correctness of Helmholtz's theory.[23]

By 1873, then, Helmholtz had broadened the spectrum of processes by which electrochemical action could occur. Nonetheless, he still adhered to the thermal notion of affinity, according to which the heat of combustion of hydrogen provided a theoretical limit below which the decomposition of water could not occur. He stated this view explicitly, and it is only from this viewpoint that the case of a Daniell cell connected to a platinum electrolytic cell constituted a puzzle in need of explanation: from the viewpoint of the second law of thermodynamics—still not incorporated into chemistry—electrochemical processes that consumed more heat than supplied electrically would not necessarily conflict with the law of conservation of energy.[24] Although Helmholtz thus based his theory on false premises—the thermal theory

---

20. "Ueber gasfreien Flüssigkeiten," *WA 1*:824.
21. James Clerk Maxwell, *A Treatise on Electricity and Magnetism*, 2 vols. (Oxford: Clarendon Press, 1873), *1*:322.
22. Thomas Graham, "On the Occlusion of Hydrogen Gas by Metals," *Proceedings of the Royal Society 16* (1868):422-27.
23. Elihu Root, "Versuche über die Durchdringung des Platina mit elektrolytischen Gasen," *MB* (1876):217-20. See also Helmholtz's "Bericht über Versuche des Hrn. Dr. E. Root aus Boston, die Durchdringung des Platins mit elektrolytischen Gasen betreffend," *AP 159* (1876):416-20, in *WA 1*:835-40.
24. See Ostwald, *Electrochemistry 2*:975-77.

of affinity—his conclusion concerning convection currents remained valid even after 1882 when the modern understanding of the relationship between heat and affinity became established.

The single most important background element for Helmholtz's thermodynamic theory of 1882–83 issued from his work of 1877 on concentration cells, that is, two identical metals immersed in two different concentrations of the same salt (for example, one copper plate in a more concentrated solution of copper sulphate, another copper plate in a less concentrated solution of the same salt). Concentration cells had been studied by Antoine-César Becquerel around 1830, but until Helmholtz investigated such cells, they had played only an insignificant role in the development of electrochemistry.[25]

Helmholtz argued that a concentration cell could be understood as consisting of two processes: First, as an electric process in which ions were transferred from one part of the electrolyte to the other, corresponding to the transport of water through a semipermeable membrane; this process would give rise to an electric current and tend to dilute the more concentrated solution. Second, as an evaporation process in which the difference in concentration would be equalized by evaporation of water from the diluted solution. Helmholtz showed that if certain secondary processes were neglected, both primary processes were reversible. It then followed from the second law of thermodynamics, which Helmholtz here applied to chemical reactions for the first time, that the energies involved would be equal. Hence there existed a relationship that depended only on the terminal states of the system, which related the vapor pressure and the electromotive force. "We can ... apply Carnot and Clausius's law to the reversible processes," Helmholtz wrote. "Since the temperature of all bodies taking part in the process should be equal and constant, no heat can be converted into work, and, through the reversible processes, no work can be converted into heat. The sum of the work gained and lost must, therefore, taken by itself, be equal to nil, as must also the sum of the heat withdrawn and supplied." Converting this idea into formulae, Helmholtz calculated the electric energy and the mechanical work gained by evaporation. Equating them, he found the general relationship

$$\mathscr{E} = \mathbf{P}_k - \mathbf{P}_a = \int_k^a q(1 - \frac{1}{n}) \frac{dW}{dp} dp.$$

---

25. Antoine-César Becquerel, "De l'électricité dégagée dans les actions chimiques," *Annales de Chimie* 35 (1827):113–40.

Here **P** is the potential (**a** and **k** refer to the anode and cathode, respectively), **n** the transport number of the cation (defined as the fraction of the electrolytic current carried by cations), **q** the mass of water that combines with one electrolytic equivalent of the salt, **W** the work of evaporation, and **p** the vapor pressure.[26]

As Helmholtz remarked, "this equation shows the existence of an electromotive force, the magnitude of which depends only on the concentration at both electrodes, and not on the distribution of concentrated and diluted layers within the liquid."[27] Aiming to calculate the dependency, Helmholtz noted that the water pressure would be subject to the Boyle-Mariotte law as well as to Adolph Wüllner's finding of 1858 that the change in vapor pressure was directly proportional to the salt concentration.[28] In Helmholtz's formulation, Wüllner's result was that

$$P = b\left(\frac{1}{q_0} - \frac{1}{q}\right),$$

where **q** is the "dilution" of the electrolyte (the inverse of the concentration), q_0 the dilution of pure water, and **b** a constant that depends only on the kind of salt dissolved. Introducing these results in the above equation, Helmholtz obtained the final result, namely,

$$\mathscr{E} = P_k - P_a = bV\left(1 - \frac{1}{n}\right) \log\left(\frac{q_a}{q_k}\right),$$

where **V** denotes the volume of one gram of vapor in equilibrium with the water. In terms of concentrations **c**, the result was

$$\mathscr{E} = bV\left(1 - \frac{1}{n}\right) \log\left(\frac{c_k}{c_a}\right).$$

In 1878 Helmholtz had his student James Moser experimentally test his equation; Moser found it to be approximately valid.[29] A decade

---

26. "Ueber galvanische Ströme, verursacht durch Concentrations-Unterschiede; Folgerungen aus der mechanischen Wärmetheorie," *MB* (1877):713–26, quote on 717, in *WA 1*:840–54. See also Ollin J. Drennan, "Electrolytic Solution Theory. Foundations of Modern Thermodynamical Considerations" (Ph.D. diss., University of Wisconsin-Madison, 1961), 92–101.
27. "Concentrations-Unterschiede," 719.
28. Adolph Wüllner, "Ueber die Spannkraft des Dampfes von wässerigen Salzlösungen," *AP 103* (1858):529–62.
29. James Moser, "Galvanische Ströme zwischen verschieden concentrirten Lösungen desselben Körpers und deren Spannungsreihen," *AP 3* (1878):216–18.

later, in 1888, Walther Nernst perfected Helmholtz's equation into his celebrated Nernst equation in physical chemistry.[30]

Helmholtz's equation contained no term corresponding to the evolution of heat occurring during the dilution of the electrolyte. Hence it contradicted the thermal theory of affinity and the thermoelectric orthodoxy to which Helmholtz had subscribed since 1847. "Such contradictions could not remain concealed from someone like Helmholtz," Ostwald wrote in 1898. They forced Helmholtz, Ostwald believed, to "reexamine his premises."[31] Before considering this reexamination, which led to his theory of chemical thermodynamics (Section 3), still another of his influential contributions to theoretical chemistry must be considered: the atomic nature of electricity and Faraday's laws of electrolysis.

In 1880 and 1881, Helmholtz reanalyzed the issue of the atomic nature of electricity and its relevance to Faraday's laws of electrolysis. First, in the course of a study of polarizing currents, he largely reiterated his atomistic view of 1847, according to which chemical electricity was of a corpuscular nature and attached to the valencies of the atoms.[32] Then in early April of 1881, as part of his Faraday Lecture before London's Chemical Society, he greatly elaborated this view.

The president of the society was Helmholtz's close friend Henry Roscoe, with whom Helmholtz had spent several days in Manchester before going to London.[33] Roscoe introduced the Faraday lecturer as being "eminent as an anatomist, as a physiologist, as a physicist, as a mathematician, and as a philosopher, (and) we chemists are now about to claim him as our own."[34] Yet Helmholtz, lecturing before some 1,100 listeners, including "notabilities of all kinds, and also many women,"[35] was acutely aware of being a physicist among chemists. "I am not sufficiently acquainted with chemistry," he confessed, "to be confident that I have given the right interpretation." He diplomatically declared that he wanted to imitate the great Faraday, thus making it unnecessary for him "to speculate about the real nature of that which we call a quantity of positive or negative electricity." Yet in fact he used most of his lecture to argue confidently for the atomic hypothesis

---

30. Walther Nernst, "Zur Kinetik der in Lösung befindlichen Körper," *ZPC 2* (1888):613–37.
31. Ostwald, *Electrochemistry* 2:983.
32. "Ueber Bewegungsströme am polarisirten Platina," *AP 11* (1880):737–59, in *WA 1*:899–921.
33. Henry Roscoe, *The Life and Experiences of Sir Henry Enfield Roscoe* (London: Macmillan, 1906), 89–95.
34. Ibid., 89.
35. Siemens-Helmholtz, ed., *Anna von Helmholtz 1*:257. See also Koenigsberger 2:278, 284; and Roscoe, *Life and Experiences*, 90–1, which includes three letters from Helmholtz to Roscoe concerning the Faraday Lecture.

of electricity. The atoms of modern chemistry may be hypothetical, he said, but they are also indispensable, and they must form the basis of any theory of electrolysis. His "startling result," as he called the atomic interpretation of Faraday's laws, was that electricity was not continuous, but divided into definite elementary portions which behave like "atoms of electricity." He argued that ions were associated with one or more of these quanta which were given off only at the electrodes, transforming the ions into neutral atoms. He stressed that his "atoms of electricity" were independent of the substance with which they combined and that they acted as if they were real particles associated with the chemical atoms.[36] The Faraday lecture proved a great success. Afterwards, Helmholtz went to Cambridge where he met with Arthur Schuster, who had worked in Helmholtz's Berlin laboratory in 1874,[37] and Lord Rayleigh (John William Strutt), and where he received an honorary Doctor of Laws degree. From Cambridge he went to Dublin and from there to Edinburgh and Glasgow to visit his good friend William Thomson.[38]

In his original formulation of the electrolytic laws, Faraday had not conceived of electrolysis as a transfer of electrical particles. The particles of an electrolyte, according to Faraday, were subject to a rapidly changing electric strain that produced a transfer of energy (power), not matter, between the electrodes.[39] Later developments, by Alexander Williamson (1852), Clausius (1857), and others, led to an incomplete form of dissociation hypothesis.[40] According to the Clausius-Williamson hypothesis, dissociation products appeared during electrolysis only at the poles and only a tiny amount of the electrolyte was dissociated. Although Helmholtz did not clearly state his view on dissociation, he apparently accepted the Clausius-Williamson hypothesis and associated ions with electrolysis, not with a dissociation process caused by the interaction between electrolyte and solvent.

---

36. "On the Modern Development of Faraday's Conception of Electricity," *Journal of the Chemical Society (London) 39* (1881):277–304, quote on 303, in *WA 3*:52–87. See also Stanley M. Guralnick, "The Contexts of Faraday's Electrochemical Laws," *Isis 70* (1979):59–75.
37. Arthur Schuster, *Biographical Fragments* (London: Macmillan, 1932), 64–5.
38. Koenigsberger 2:284–85; and Siemens-Helmholtz, ed., *Anna von Helmholtz* 1:254–57.
39. L. Pearce Williams, *Michael Faraday* (New York: Basic Books, 1965), 240–57; and Trevor H. Levere, "Faraday, Electrochemistry, and Natural Philosophy," in George Dubpernell and J.H. Westbrook, eds., *Selected Topics in the History of Electrochemistry* (Princeton: The Electrochemical Society, 1978), 150–61.
40. Alexander Williamson, "On the Constitution of Salts," *Journal of the Chemical Society (London) 4* (1852):350–55; and Rudolf Clausius, "Ueber die Electricitätsleitung in Elektrolyten," *AP 101* (1857):338–60. See also Ostwald, *Electrochemistry 2*:864–80; and James R. Partington, *A History of Chemistry*, 4 vols. (London: MacMillan, 1964), 4:672.

Helmholtz's atom of electricity was not, as is sometimes argued, the electron, an elementary particle yet to be discovered.[41] Quanta of electricity figured in the electrical theories of Ottaviano Mossotti, Wilhelm Weber, Clausius, Bernhard Riemann, and others, and were popular among speculative German scientists. Maxwell, in 1873, had also entertained the idea of "molecules of electricity" in electrolysis. But Maxwell spoke only of molecular charges in a general, ambiguous way, not of an elementary charge independent of the chemical substance.[42] More to the point, in 1874 George Johnstone Stoney had introduced the "electrine"—he later rechristened it the "electron"—as a unit quantity of electricity and made many of the same points that Helmholtz would make seven years later. Helmholtz did not refer to Stoney's work, which was first published in 1881, and Stoney later claimed priority for the idea of atomic electricity.[43]

The significance of Helmholtz's suggestion thus lay neither in its anticipation of the electron nor in the notion of electrical quanta. Rather, it lay in its impact on the physical and, in particular, chemical ideas of electricity. The Faraday lecture was instrumental in reviving the hypothesis of atomic electricity at a time when the notion was threatened by the continuity conception of Maxwellian theory.[44] In language familiar to chemists, Helmholtz indicated the chemical significance of the idea of atomistic electricity. Coming at the right time and place, his address to the Chemical Society supplied powerful rhetoric for the idea and suggested its usefulness in electrolysis and related areas. As van't Hoff, one of the founders of the new physical chemistry, put it in 1891, Helmholtz's Faraday lecture opened the eyes of chemists, stimulating them to take up a more serious study of electricity.[45]

---

41. L. Marton and C. Marton, "Evolution of the Concept of the Elementary Charge," *Advances in Electronics and Electron Physics 50* (1980):449-72.

42. Maxwell, *Treatise 1*:381.

43. George Johnstone Stoney, "On the Physical Units of Nature," *PM 11* (1881):381-90, on 384, paper read to the British Association for the Advancement of Science meeting in Belfast in 1874. For the priority claim, see Stoney, "Of the 'Electron', or Atom of Electricity," *PM 38* (1894):418-20. Stoney's priority claim was triggered by a paper by the German physicist Hermann Ebert, who hailed Helmholtz's lecture as a great advance in the electrochemical theory of affinity. (Hermann Ebert, "Heat of Dissociation according to the Electrochemical Theory," *PM 38* [1894]:332-36.) See also Franz Richarz, "On Helmholtz's Electrochemical Theory," *PM 39* (1895):529-31.

44. Edmund Whittaker, *A History of the Theories of Aether and Electricity*, 2 vols. (London: Thomas Nelson, 1958), *1*:354; and Jed Z. Buchwald, *From Maxwell to Microphysics: Aspects of Electromagnetic Theory in the Last Quarter of the Nineteenth Century* (Chicago and London: The University of Chicago Press, 1985), 23-9, 177-86.

45. See Ernst Cohen, *Jacobus Henricus van't Hoff. Sein Leben und Wirken* (Leipzig: Akademische Verlagsgesellschaft, 1912), 299. Similar appraisals appeared in Matthew M. Pattison Muir, *A History of Chemical Theories and Laws* (New York: Wiley, 1907),

Indeed, Helmholtz himself pointed to the problem of chemical combination as a prime area for applying his ideas. Identifying units of electricity with units of chemical affinity implied, he said, the idea of "compounds in which every unit of affinity of every atom is connected with one and only one other unit of another atom." This, he pointed out, "is the modern chemical theory of quantivalence, comprising all the saturated compounds."[46] Whereas Stoney had only associated the valencies of electrolytes with electrical quanta, Helmholtz argued that the valencies of non-electrolytes were also charged with the same atoms of electricity. Although Helmholtz did not develop the chemical consequences of this view, he did apply it to other fields; for example, in one of his last research papers, he based his dispersion theory of light on what he called "the electrochemical theory."[47]

3. Chemical Thermodynamics

Although by the late 1870s many physicists had rejected the thermochemical theory of affinity, it still enjoyed support from the majority of chemists. Physicists and physically inclined chemists, including Gibbs, Rayleigh, Carl Neumann, and Henri Etienne Sainte-Claire Deville, judged it to be theoretically primitive and argued that a new, proper chemical thermodynamics was needed. They pointed out, for example, that the Thomsen-Berthelot theory rested only on the first law of thermodynamics, while it ignored the chemical implications of the second law.

Chemical thermodynamics, as a general and consistent physical theory of the energies involved in chemical change, was largely founded by Helmholtz in an influential memoir of 1882. It was only with this paper, and especially with its subsequent elaboration by van't Hoff in 1884, that chemists discovered thermodynamics.[48] Yet Helmholtz was-

---

333–37; and in Emil Fischer, "Zum Gedächtnis von Hermann von Helmholtz," memorial lecture at the Deutsche Chemische Gesellschaft on 15 October 1894, printed in Emil Fischer, *Gesammelte Werke*, ed. M. Bergmann, 8 vols. (Berlin: Springer, 1906–24), 8:848–55.

46. "Faraday's Conception of Electricity," 303.

47. "Elektromagnetische Theorie der Farbenzerstreuung," *AP 48* (1893):389–405. This study is analyzed in Buchwald, *From Maxwell to Microphysics*, 237–39. In the early 1890s, Helmholtz analyzed (in unpublished notes) the chemical consequences of replacing his dualistic theory of electricity, which assumed the existence of both positively and negatively charged electrical atoms, with a unitary theory of the type suggested by Franz Richarz. (See Koenigsberger 2:279–84; and Franz Richarz, "Ueber die elektrischen Kräfte der Atome," *Sitzungsberichte der Niederrheinischen Gesellschaft für Natur- und Heilkunde zu Bonn* [1891]:18–32.)

48. Jacobus H. van't Hoff, *Etudes de Dynamique Chimique* (Amsterdam: F. Muller, 1884), 186–209.

not the first to apply the second law of thermodynamics to chemistry. The first important step was taken by August Horstmann, who attended some of Helmholtz's lectures in Heidelberg and who incorporated Clausius's thermodynamic theory into chemistry. In important studies of 1869 and 1873, Horstmann applied the concept of entropy to equilibrium processes and derived various thermodynamic relations for chemical processes.[49] Slightly later, Gibbs independently began research in chemical thermodynamics; his publications made even less of an impact than did Horstmann's.[50] When, in 1875, Rayleigh criticized the chemists for not recognizing the importance of the second law and its incompatibility with the thermal theory of affinity, he was unaware of the work of Horstmann.[51] Similarly, even though Gibbs had sent a reprint of his paper to Helmholtz, the latter took no notice of it.[52] (Curiously, Gibbs, Horstmann, and Helmholtz were all in Heidelberg in 1868, yet were not in contact with one another.)

Helmholtz's inspiration to develop a chemical thermodynamics thus did not issue from the works of his predecessors in the field, but rather from his own work in electrochemistry. In 1882 he stated that his theory of chemical energy was inspired by his earlier work on the connection between chemical changes and electromotive forces.[53] He did not mention Thomsen or Berthelot explicitly, but he did refer to the thermal theory of affinity as "the older view, which I myself defended in my earlier works." This view, he now argued, had only limited validity. Instead of focusing on heat alone one should distinguish between a part of the energy that appeared only as heat and a part that could be freely converted into other kinds of work. This latter part Helmholtz called "free energy."

Helmholtz's concept of free energy started with Clausius's definition of entropy S and its relationship to heat Q, absolute temperature T, and internal energy U, namely

---

49. August F. Horstmann, "Dampfspannung und Verdampfungswärme des Salmiaks," *Berichte der deutschen chemischen Gesellschaft 2* (1869):137–40; and idem, "Zur Theorie der Dissociation," *Annalen der Chemie und Pharmacie 170* (1873):192–210. See H.A.M. Snelders, "Dissociation, Darwinism, and Entropy," *Janus 64* (1977):51–75.
50. Josiah W. Gibbs, "On the Equilibrium of Heterogeneous Substances," *Transactions of the Connecticut Academy 3* (1875–78):108–248, 343–524.
51. Lord Rayleigh, "On the Dissipation of Energy," *Nature 11* (1875):454–55.
52. For Gibbs's mailing list of reprints, see Lynde Phelps Wheeler, *Josiah Willard Gibbs: The History of a Great Mind* (New Haven, Conn.: Yale University Press, 1951), 238–39.
53. "Die Thermodynamik chemischer Vorgänge," *SB* (1882):22–39, in *WA* 2:958–78.

$$dS = \frac{dQ}{T}, \qquad (1)$$

and

$$\frac{\partial S}{\partial T} = \frac{1}{T}\frac{\partial U}{\partial T}, \qquad (2)$$

which are valid for reversible processes. Equation (1) states that a very small change in the heat of a system at constant temperature determines the change in entropy; equation (2) is derived by combining this definition of entropy with the law of conservation of energy. Helmholtz relied on Clausius's mechanical theory of heat, a theory which probably also inspired him to separate the internal energy into a free and a bound part. (Clausius had earlier suggested a separation between kinetic and potential energies.)[54] Extending Clausius's treatment, Helmholtz considered generalized mechanical systems described by the temperature and an arbitrary, finite number of other parameters p_α, such as volume and electrical charge. The first law could then be stated in the general form

$$JdQ = \frac{\partial U}{\partial T}dT + \sum_\alpha (\frac{\partial U}{\partial p_\alpha} + P_\alpha)\, dp_\alpha, \qquad (3)$$

where J is the mechanical equivalent of heat and the product $P_\alpha dp_\alpha$ the external work produced by the change in p_α. (P_α may be a pressure or an electric potential difference.) The first term represents the heat energy, which is decomposed into an internal energy and the work done by the system. Rewriting equation (1) as

$$dS = \frac{\partial S}{\partial T}dT + \sum_\alpha \frac{\partial S}{\partial p_\alpha}dp_\alpha, \qquad (4)$$

equations (3) and (4) could be combined so that, after some manipulation, Helmholtz obtained the result

$$P_\alpha = -\frac{\partial}{\partial p_\alpha}(U - JTS), \qquad (5)$$

---

54. Rudolf Clausius, "Über die Anwendung des Satzes von der Äquivalenz der Verwandlungen auf die innere Arbeit," *AP 116* (1862):73–112.

where the quantity in the square brackets represented the free energy,

$$F = U - JTS. \tag{6}$$

Helmholtz introduced the concept of free energy as analogous to the potential energy in mechanics, an analogy which is justified by the equation

$$\mathbf{P}_\alpha = -\frac{\partial F}{\partial \mathbf{p}_\alpha}. \tag{7}$$

Since U is the total energy, $JTS = U-F$ signified what Helmholtz called the "bound energy." It now became easy to express the energy and entropy in terms of the differential quotient of the free energy. Equation (6) yielded

$$\frac{\partial F}{\partial T} = \frac{\partial U}{\partial T} - JS - JT\frac{\partial S}{\partial T},$$

which, in combination with equation (2), led to the results

$$\frac{\partial F}{\partial T} = -JS \tag{8}$$

and

$$T\frac{\partial F}{\partial T} = F - U, \tag{9}$$

where all parameters other than T are kept constant. Equation (9) gives the amount of energy that can be "freed" in a reversible and isothermal process and converted to mechanical work.[55]

In order to relate thermodynamic results to chemical processes, Helmholtz argued that isothermal changes can only proceed spontaneously in the direction corresponding to a decrease in free energy. The condition for chemical stability could then be stated as

$$\delta F \geq \emptyset,$$

and was not determined by the heat production. It followed immediately from equation (9) that heat-absorbing (endothermic) reactions

---

55. For a clear exposition of Helmholtz's theory, see Georg Helm, *Die Energetik nach ihrer geschichtlichen Entwickelung* (Leipzig: Veit & Co., 1898), 175-89. See also Koenigsberger 2:361-79; and Drennan, "Electrolytic Solution Theory," 131-52.

might spontaneously occur. The existence of such processes was incompatible with the thermal theory of affinity, but according to Helmholtz they would take place only if $T(\partial F/\partial T)$ was larger than F; in that case, the internal energy would be negative, corresponding to an absorption of heat.

In general, Helmholtz pointed out, any external work would take place at the expense of free energy while heat production would result in a loss of bound energy only—a result which is not valid for adiabatic processes. If the temperature of a chemical system increased, free energy would be converted into bound energy. This again led Helmholtz to conclude that the work performed by the free energies of a system determined the direction of chemical change. The work would be simply the difference in free energy between the initial and final states,

$$W = F_i - F_f.$$

The heat produced, on the other hand, would be

$$JQ = U_i - U_f,$$

which was the quantity erroneously measured as affinity in classical thermochemistry. "Given the unlimited validity of Clausius's law," Helmholtz wrote, "it would then be the value of the free energy, not that of the total energy resulting from heat production, which determines in which sense the chemical affinity can be active."[56] Furthermore, Helmholtz distinguished between what he called "ordered" and "disordered" motions, the latter being processes in which "the motion of each individual particle does not necessarily have any similarity at all to the motions of its neighbors." In this vague sense, he sought to "describe the amount of entropy as the measure of disorder."[57] Although this sounds like a definition of Ludwig Boltzmann's probabilistic theory of entropy, it was not. Helmholtz's thermodynamics remained solidly based on Clausius's mechanical theory, even though Boltzmann in 1884 cited "famous scientists, such as Helmholtz" in support of his theory.[58] The importance of Helmholtz's 1882 memoir derived from the concept of free energy and its use by means of equation (9). This equation, known today as the "Gibbs-Helmholtz equation," became the cornerstone of chemical thermodynamics. Although

---

56. "Die Thermodynamik," 23.
57. Ibid., 33.
58. Quoted in Stephen Brush, *The Kind of Motion We Call Heat*, 2 vols. (Amsterdam: North Holland, 1976), 2:633.

corresponding results had previously been found or anticipated by other scientists, Helmholtz obtained his results independently. Thermodynamic functions similar to free energy had been obtained by the French physicist François Massieu in 1869, by Gibbs in 1876, and by Kelvin in 1879; yet, unlike Helmholtz, none of these physicists explored the chemical significance of their functions.[59] Moreover, the Gibbs-Helmholtz equation, as is well known, was not established by Gibbs, who only discussed what he called "Helmholtz's equation" in 1887.[60] It, or an equation similar to it, was first derived in 1855 by Kelvin, who failed to relate it to chemical changes; in 1872 Horstmann obtained it independently and applied it to dissociation processes.[61] Helmholtz was unaware of Kelvin's work, a surprising fact given that Kelvin and Helmholtz were close friends and regular correspondents. (They had first met in 1855, the year Kelvin completed his paper.) It was only with Helmholtz that the equation got a prominent position in physical chemistry, and it was Helmholtz's version of it that van't Hoff and Nernst further explored. Moreover, it was only with Helmholtz's work that the equation received experimental support, first provided in 1884 by Siegfried Czapski who worked under Helmholtz's guidance.[62] Finally, later in 1882 and then again in 1883 Helmholtz extended his approach to chemical thermodynamics. He applied his theory to a number of chemical problems, including electrochemistry and the calculation of heats of dilution, and found the theory's predictions to agree well with measurements. By now aware of the works of his predecessors, he gave full credit to the earlier findings of Rayleigh, Massieu, and Gibbs. For some unknown reason he neglected to

---

59. For the entangled history of the thermodynamic potentials, see James R. Partington, *An Advanced Treatise on Physical Chemistry*, 6 vols. (London: Longmans, Green & Co., 1949), *1*:182–87; and W. Lash Miller, "The Method of Willard Gibbs in Chemical Thermodynamics," *Chemical Reviews 1* (1925):293–344, on 314–16.

60. Josiah W. Gibbs, "Electrochemical Thermodynamics (letter to Oliver Lodge of 21 November 1887)," *Report of the British Association for the Advancement of Science* (1888):343–46.

61. William Thomson, "On the Thermo-elastic and Thermo-magnetic Properties of Matter," *Quarterly Journal of Mathematics 1* (1857):57–77; and August Horstmann, "Ueber den zweiten Hauptsatz der mechanischen Wärmetheorie und dessen Anwendung auf einige Zersetzungserscheinungen," *Annalen der Chemie und Pharmacie Supplementband 8* (1872):112–33.

62. Siegfried Czapski, "Ueber die thermische Veränderlichkeit der elektromotorischen Kraft galvanischer Elemente und ihre Beziehung zur freien Energie derselben," *AP 21* (1884):209–43. The early experiments of Czapski and others showed only qualitative agreement with Helmholtz's theory. In 1888, Nernst explained the reasons for the deviations between theory and measurements; by 1889, Helmholtz's equation was fully verified. (Nernst, "Zur Kinetik"; and "Die elektromotorische Wirksamkeit der Ionen," *ZPC 4* [1889]:129–81.)

acknowledge his compatriot Horstmann, who more than anyone else had pioneered chemical thermodynamics.[63]

Helmholtz's chemical thermodynamics and the related work of Gibbs, Duhem, van't Hoff, and others signified the death of classical thermochemistry. Although the limited validity of the Thomsen-Berthelot principle had now been clearly demonstrated, classical thermochemists, such as Berthelot, continued to maintain the principle for several years and hesitated to give up the thermal theory of affinity. As for Helmholtz, he did not further pursue his thermodynamic researches of 1882-83. Instead, he used this research as an important starting point for his work on monocyclic systems and the principle of least action.[64] In 1887 he briefly returned to chemistry: in an experimental examination of the electrolysis of water he confirmed the predictions of his thermodynamic theory of 1882.[65] Around the same time Helmholtz contemplated bringing his writings on chemical thermodynamics into a more definite and broader form: he began writing a manuscript entitled "Thermodynamische Betrachtungen über chemische Vorgänge"; it remained unfinished.[66] Although Helmholtz did no further research in chemical thermodynamics, he did disseminate the theory in a coherent and instructive manner through his regular lectures at the University of Berlin. There he presented a full account of the concept of free energy and its use in chemical and galvanic processes.[67]

63. "Zur Thermodynamik chemischer Vorgänge (zweiter Beitrag)," *SB* (1882):825-36, in *WA* 2:979-92; and "Zur Thermodynamik chemischer Vorgänge (dritter Beitrag)," *SB* (1883):647-65, in *WA* 3:92-114. These two works, together with Helmholtz's "Die Thermodynamik," and "Concentrations-Unterschiede," were reprinted in Max Planck, ed., *Abhandlungen zur Thermodynamik von H. Helmholtz* (Leipzig: Engelmann, 1902) in Ostwald's series *Klassiker der Exakten Wissenschaften*. Planck corrected a number of printing and calculational errors which appeared both in Helmholtz's original papers and in *WA* 2 and *WA* 3.

64. "Studien zur Statik monocyklischer Systeme," *SB*, 1884, 159-77, 311-18, 755-59, in *WA* 3:119-42, 163-78. See Martin J. Klein, "Mechanical Explanation at the End of the Nineteenth Century," *Centaurus 17* (1972):58-82; and Günter Bierhalter's essay, "Helmholtz's Mechanical Foundation of Thermodynamics," in this volume.

65. "Weitere Untersuchungen, die Elektrolyse des Wassers betreffend," *SB* (1887):749-57, in *WA* 2:267-81.

66. Koenigsberger 2:361-79. In 1893, another theoretical physicist, Max Planck, completed a project similar to that begun by Helmholtz: see Max Planck, *Grundriß der Thermochemie* (Breslau: E. Trewendt, 1893).

67. *Vorlesungen über Theoretische Physik*, eds., Arthur König, et al., 6 vols. (Leipzig: J.A. Barth, 1897-1907), vol. 6: *Vorlesungen über Theorie der Wärme*, ed., Franz Richarz (1903), 267-337. Although first published in 1903, volume 6 originated from a combination of Helmholtz's lecture notebook of 1890, a stenographic lecture notebook of 1893, and lecture notes from the early 1880s taken by his student (and subsequent editor of volume 6), Franz Richarz. (Ibid., v-vi.)

4. Helmholtz and the "New Era in Chemistry"

Helmholtz's contributions to various branches of chemistry—physiological chemistry, electrochemistry, and chemical thermodynamics—were fundamental and highly regarded by chemists. In 1891, on the occasion of his seventieth birthday, August Wilhelm von Hofmann, then the *doyen* of German chemists, praised Helmholtz as having done pathbreaking work in chemistry. Hofmann mentioned Helmholtz's early works on physiological chemistry and energy conservation, and, in particular, his contributions to chemical thermodynamics. These, Hofmann said, "must be characterized as the beginning of a new era in chemical research ... [and to] have thrown new light on entire areas [of chemistry]." Helmholtz, with false modesty, replied that his contributions to chemistry had been merely "amateur efforts."[68] A year later the Deutsche Chemische Gesellschaft in Berlin elected him an honorary member, thereby signifying the chemical community's appreciation of the work of the great chemical amateur. Similar appreciations were contained in memorial lectures and reminiscences by Nernst, Ostwald, Roscoe, Emil Fischer, Henry Armstrong, A.G. Vernon Harcourt, and Edward Frankland.[69] His friend Roscoe wrote: "Helmholtz was certainly, taking him altogether, the most wonderful man I ever knew, and his character was as charming and simple and his heart as kind as his intellect was great."[70] Yet such homage tells us nothing about the actual reception of Helmholtz's works among chemists. What role did his works in electrochemistry and thermodynamics play in the "new era in chemistry" and, in particular, in the development of the new physical chemistry?[71]

Most major treatises on theoretical chemistry in the period between 1880 and 1910 included references to Helmholtz's contributions to galvanic polarization, concentration cells, and thermodynamics.[72]

---

68. *Ansprachen und Reden gehalten bei der am 2. November 1891 zu Ehren von Hermann von Helmholtz veranstalteten Feier* (Berlin: Hirschwald, 1892), 40–1, quotation on 41. See also Koenigsberger 2:279.
69. Nernst, "Elektrochemischen Arbeiten von Helmholtz"; Ostwald, *Große Männer*, 265–311; Roscoe, *Life and Experiences*, 89–95; and Fischer, "Gedächtnis." For the British chemists, see *Journal of the Chemical Society (Proceedings)* 12 (1896):25–9.
70. Roscoe, *Life and Experiences*, 91.
71. Harry C. Jones, *A New Era in Chemistry* (New York: Van Nostrand, 1913). The "new era" refers to the emergence of theoretical and atomic chemistry, a broader concept than physical chemistry proper.
72. E.g., J.J. Thomson, *Applications of Dynamics to Physics and Chemistry* (London: Macmillan, 1888), 95–9; Matthew M. Pattison Muir, *A Treatise on the Principles of*

However, in most cases Helmholtz's works were mentioned only in conjunction with Gibbs's and often with more space being devoted to the latter's contributions. By the late 1880s, many chemists had come to recognize the value of Gibbs's theory; they preferred his more general thermodynamics to Helmholtz's. Van't Hoff's pioneering *Etudes* of 1884 made use of Helmholtz's theory; but later works by van't Hoff, as well as those by his countrymen Johannes van der Waals and Johannes van Laar, built on Gibbs's thermodynamics.[73]

Helmholtz's chemical works exerted a direct and important influence on French theoretical chemistry in general and on Duhem in particular. Under the guidance of his teacher Jules Moutier, Duhem studied Helmholtz's theory of chemical thermodynamics shortly after its publication in 1882–83. Duhem was fascinated by the notion of free energy, which he developed into a more rigorous theory of chemical potentials in his doctoral dissertation.[74] He later cited Helmholtz's thermodynamics as the pioneering work of the new chemistry and wrote that it "vividly had attracted attention to the role which thermodynamics should play in chemical mechanics."[75] Part of the reason for Duhem's enthusiastic response to Helmholtz's work was that Duhem recognized that it undermined Berthelot's thermal theory of affinity which still held great authority in France. Other French chemists made use of Helmholtz's concept of free energy and the Gibbs-Helmholtz equation in their efforts to establish an alternative to Berthelot's orthodoxy. Both Paul Sabatier and Henri Le Chatelier, for example, were directly influenced by Helmholtz's work.[76]

Chemistry, 2nd ed. (Cambridge: Cambridge University Press, 1889), 382–83; Albert Ladenburg, *Vorträge über die Entwicklungsgeschichte der Chemie*, 3rd ed. (Braunschweig: Vieweg, 1902), 322–24; Harry C. Jones, *The Elements of Physical Chemistry*, 3rd ed. (New York: MacMillan, 1907), 344–49; van't Hoff, *Etudes*, 203–7; Max Planck, *Thermochemie*, 175; and the works by Nernst (notes 78–9), Duhem (notes 74–5), Planck (note 91), and Le Chatelier (note 76) mentioned below.

73. Jacobus H. van't Hoff, *Lectures on Theoretical and Physical Chemistry*, 3 vols. (London: Arnold, 1898-1900); idem, *Physical Chemistry in the Service of the Sciences* (Chicago: University of Chicago Press, 1903); Johannes van der Waals, *Lehrbuch der Thermodynamik* (Amsterdam: Mass & Van Suchten, 1908); and Johannes J. van Laar, *Die Thermodynamik in der Chemie* (Leipzig: Engelmann, 1893).

74. Pierre Duhem, *Le Potentiel Thermodynamique* (Paris: Hermann, 1886); and Jules Moutier, "Recherches de M. Helmholtz sur l'origine de la chaleur voltaique," *La lumière électrique 13* (1884):281–86, 331–34. Helmholtz's "Die Thermodynamik" appeared in French translation in *Journal de Physique 3* (1884):396–414.

75. Pierre Duhem, *Introduction à la Mécanique Chimique* (Lille: Gand, 1893), 113. See also Duhem, "Thermochimie," *Revue des Questions Scientifiques 12* (1897):361–92, on 368; and Stanley L. Jaki, *Uneasy Genius: The Life and Work of Pierre Duhem* (The Hague: Martinus Nijhoff, 1984), 333, where Helmholtz is referred to as Duhem's "hero."

76. Paul Sabatier, "Essai critique de la thermochimie," *Mémoires de l'Académie des Sciences Toulouse*, sér. 8, *10* (1888):289–300; and Henri Le Chatelier, *Recherches Expérimentales et Théoretiques sur les Equilibres Chimiques* (Paris: Dunod, 1888).

Helmholtz's impact on German chemistry owed much to Ostwald and Nernst, two of the leaders of the new physical chemistry. Ostwald developed aspects of Helmholtz's electrochemistry and Nernst took up both his electrochemical and thermodynamical works.[77] Fascinated by the Gibbs-Helmholtz equation, which he always used in the form given by Helmholtz, Nernst thought that it contained "in a general manner all that the laws of thermodynamics can teach concerning chemical processes."[78] In Helmholtz's theory the free energy was only determined within an additive constant, and attempts by Le Chatelier and Theodore W. Richards to establish an absolute value for the free energy failed. This problem—the integration of the Gibbs-Helmholtz equation so as to provide the thermodynamical quantities with absolute values—furnished the immediate background for Nernst's famous third law of thermodynamics in 1906.[79]

While Helmholtz's work thus exerted considerable influence on French and German chemists, it does not appear to have had a similar impact on British and American chemistry. The lack of impact may simply reflect the fact that British chemists were either hostile to or uninterested in the new trends in theoretical chemistry. In his address to the British Association in 1884, Rayleigh praised Helmholtz's "series of masterly papers" on chemical thermodynamics and pointed out that the full application of thermodynamics to chemical processes was the result of the work of physicists such as Helmholtz and Gibbs.[80] But Rayleigh was himself a physicist, and his views were not representative of the attitude of British chemists. Armstrong and other British chemists disapproved of the increasing mathematization of chemistry and the abstract thermodynamic reasoning which characterized, for example, Helmholtz's works. The Helmholtz memorial meeting of the Chemical Society illustrates the resistance of British chemistry to the new trends in theory: while Helmholtz's chemical thermodynamics was barely mentioned, the vortex atomic theory—which had its roots

---

77. Wilhelm Ostwald, "Studien über Kontaktelektrizität," *ZPC 1* (1887):583–610; Nernst, "Zur Kinetik"; and idem, "Elektromotorische Wirksamkeit."

78. Walther Nernst, *Experimental and Theoretical Applications of Thermodynamics to Chemistry* (New Haven, Conn.: Yale University Press, 1903), 3. See also Nernst, *Theoretische Chemie* (Stuttgart: F. Encke, 1893), 17, 537, and 556–68.

79. Walther Nernst, "Über die Berechnung chemischer Gleichgewichte aus thermischen Messungen," *Nachrichten von der Gesellschaft der Wissenschaften zu Göttingen* (1), (1906), 1–40. See also Nernst, *The New Heat Theorem*, 2nd ed. (New York: E. P. Dutton, 1926), 2–10; and Erwin Hiebert, "Nernst, Hermann Walther," in *DSB*, Supplement I, 432–53.

80. Lord Rayleigh, "Presidential Address," *Report of the British Association for the Advancement of Science, Montreal Meeting, 1884* (London: John Murray, 1885), 1–23, on 11.

in Helmholtz's hydrodynamical theory of 1858 but to which Helmholtz himself never contributed—was discussed at length. George Francis FitzGerald, giving the Helmholtz memorial lecture, used the occasion to criticize physical chemistry and to argue that thermodynamics had only limited chemical applicability.[81]

The new physical chemistry was established as a powerful research program in the 1880s, primarily through the efforts of Arrhenius, van't Hoff, Ostwald, and Nernst.[82] It rested on two foundations, chemical thermodynamics and the theory of solutions, of which the latter included the theory of ionic dissociation and the extension of Avogadro's law to solutions. More than thermodynamics, the theory of solutions was the distinguishing feature of physical chemistry. Helmholtz never contributed to this area and he failed to incorporate current knowledge of chemical equilibria and reaction kinetics into his works. As Nernst pointed out, this neglect made important parts of Helmholtz's contributions incomplete and less attractive to chemists like van't Hoff and Arrhenius.[83] In his regular lecture course on theoretical physics, Helmholtz barely referred to the theory of solutions; he devoted only half an hour to the subject, and that at the very end of the semester. He knew that solution theory "really formed the foundation of the more recent development of theoretical thermochemistry," but he claimed that he lacked the time to present it.[84] Yet more than lack of time was at play here. As Helmholtz made clear in his lectures during the summer semester of 1893, he had little confidence in the chemists' conception of osmotic pressure. He argued that van't Hoff's theory of osmosis was based on a rough analogy (with gases) which restricted its validity to highly diluted solutions. Rather than follow the chemists in giving high priority to the theory of osmotic pressure, he wanted

---

81. *Journal of the Chemical Society (Proceedings)* 12 (1896):25–9; and George Francis FitzGerald, "Helmholtz Memorial Lecture," *Journal of the Chemical Society (Transactions)* 69 (1896):885–912. See also Arthur W. Rücker, "Hermann von Helmholtz," *Annual Report of the Smithsonian Institution . . . 1894* (Washington, D.C.: Government Printing Office, 1896), 709–18. For the vortex atomic theory, see Robert H. Silliman, "William Thomson: Smoke Rings and Nineteenth-Century Atomism," *Isis* 54 (1963):461–74.

82. See Robert S. Root-Bernstein, "The Ionists: Founding Physical Chemistry, 1872–1890," (Ph.D. diss., Princeton University, 1980); John Servos, *Physical Chemistry from Ostwald to Pauling: The Making of a Science in America* (Princeton, N.J.: Princeton University Press, 1990), 3–45; and Keith J. Laidler, "Chemical Kinetics and the Origins of Physical Chemistry," *Archive for History of Exact Sciences* 32 (1985):43–75. For Helmholtz's contributions, see L.S. Polak and I.I. Solov'ev, "On the History of Physical Chemistry. The Studies of Helmholtz in the Field of Physical Chemistry," *Voprosy Istorii Estestvoznaniia i Tekhniki* 8 (1959):48–56 (in Russian).

83. Nernst, "Elektrochemischen Arbeiten von Helmholtz," 702.

84. *Vorlesungen* 6:309.

to deduce the properties of solutions from the concept of free energy, which "has a much more definite and general meaning than the osmotic pressure, such as it is defined by the chemists."[85] Helmholtz's lack of interest in solution theory, and the fact that to some extent his thermodynamics was overshadowed by Gibbs's system, may explain his relatively modest influence on physical chemistry. Although Helmholtz was frequently cited in the first volumes of the *Zeitschrift für physikalische Chemie*, he was apparently not a very important figure in the new physical chemistry.[86]

Nonetheless, Helmholtz was basically sympathetic and keenly interested in the new field of physical chemistry. Arrhenius wrote Ostwald in 1888 of his first meeting with Helmholtz, whose status in German science was then even greater than that of Liebig's forty years earlier. "I found H. at home, and after I had waited for about five minutes, I stood before the great man. He received me most kindly, which is said to be an extraordinary distinction. He was very much interested in the rational development of chemistry. Then we talked about the historical development of the theory of affinity and about the simple application of the laws of entropy and gases to solutions."[87] Three years later, Helmholtz supported the recommendation that physical chemistry be included in the curriculum at the University of Berlin. Yet he had reservations about some of the new notions on which physical chemistry built. "Thermochemistry," he said, "seems in fact determined to be the mechanics of chemical forces, since we have learned to distinguish between that part of heat evolution belonging to the heat capacity, and that part belonging to the production of heat by chemical forces. However, the importance and meaning of the newest works of this school, disguised in the hypothesis of osmotic pressure, are not easily recognized."[88]

Physical chemistry built on Arrhenius's hypothesis of *electrolytic* dissociation, which the Swedish chemist had first proposed in his dissertation of 1884 and which he later, in 1887, developed into a more

---

85. Ibid., 325–26. According to the editor, Franz Richarz, the quoted remarks were taken "almost literally" from the stenographic report of Helmholtz's lecture.

86. The first five volumes of the *Zeitschrift für physikalische Chemie* (1887–90) contained thirty references to Helmholtz, who was cited by Ostwald, Gibbs, Nernst, Richarz, and others. In Root-Bernstein's comprehensive "The Ionists," in Servos's *Physical Chemistry*, and in Jones's *A New Era*, Helmholtz appears only as a secondary figure. Jones, a former student of van't Hoff's and Ostwald's, indeed praised Helmholtz as "one of the broadest and most profound men of science who has ever lived" (85), but he referred only once to his works.

87. Arrhenius to Ostwald, 4 August 1888, in Hans-Georg Körber, ed., *Aus dem wissenschaftlichen Briefwechsel Wilhelm Ostwalds*, 2 vols. (Berlin: Akademie-Verlag, 1969), 2:48.

88. Report to the Physikalisch-mathematische Klasse of the Prussian Akademie der Wissenschaften, discussed at its meeting on 12 November 1891. Quoted in ibid., 2:xviii.

general theory of *ionic* dissociation.[89] Arrhenius's theory rested on the notion of atomistic electricity, as argued by Helmholtz in his Faraday lecture. Nonetheless, Arrhenius based his theory on the idea of ionic dissociation in aqueous solutions not subject to electrolysis; it thus went much further than Helmholtz's idea. Arrhenius owed little, if anything, to Helmholtz, who, for his part, did not wholeheartedly accept the theory of ionic dissociation. In a letter of 1891, Helmholtz expressed reservations about Arrhenius's theory, which, he wrote, implied "some arbitrary assumptions which do not seem to me to be proven." And he showed his low opinion of traditional chemistry:

> The chemists ... make use of this hypothesis [of ionic dissociation] in order to form a clear conception of the processes, and they must be allowed to do this after their fashion, since the whole extraordinarily comprehensive system of organic chemistry has developed in the most irrational manner, always linked with sensory images that could not possibly be legitimate in the form in which they are represented. There is a sound core in this whole movement, the application of thermodynamics to chemistry, which is much purer in [Max] Planck's work. But thermodynamic laws in their abstract form can only be grasped by rigorously schooled mathematicians, and are accordingly scarcely accessible to the people who want to do experiments on solutions and their vapor tensions, freezing points, heats of solution, &c.[90]

This was the voice of a theoretical physicist, not a chemist.

As the above remarks indicate, Helmholtz thought highly of Planck's contributions to thermodynamics. In several studies during the 1880s, Planck, a former student of Helmholtz's, treated chemistry from the standpoint of entropy considerations.[91] Helmholtz recognized the similarity between Planck's approach and his own; indeed, he saw Planck's chemical thermodynamics as the completion of his own theory of 1882. In his nomination of Planck for an ordinary membership

---

89. Svante Arrhenius, "Recherches sur la conductibilité galvanique des electrolytes," *Bihang Kungliga Svenska Vetenskaps-Akademiens Handlingar 8*, no. 13 (1884):1–63, and no. 14 (1884):1–89; and idem, "Über die Dissociation der im Wasser gelösten Stoffe," *ZPC 1* (1887):631–48. For the development of Arrhenius's concept of dissociation, see Root-Bernstein, "The Ionists," and Robert S. Bernstein, "Svante Arrhenius and Ionic Dissociation: A Reevaluation," in Dubpernell and Westbrook, eds., *Electrochemistry*, 201–12.

90. Quoted in Koenigsberger 2:298. See also Cohen, *Van't Hoff*, 314; and, for Armstrong's interpretation of the letter, R.G.A. Dolby, "Debates over the Theory of Solution: A Study of Dissent in Physical Chemistry in the English-speaking World in the Late Nineteenth and Early Twentieth Centuries," *HSPS 7* (1978):297–404, on 389.

91. Max Planck, "Über das Prinzip der Vermehrung der Entropie," *AP 30* (1887): 562–82; *31* (1887):189–203; and *32* (1887):462–503. For Planck as a physical chemist, see Root-Bernstein, "The Ionists," 417–73.

in the Berlin Akademie der Wissenschaften in 1894, he called particular attention to his younger colleague's work in physical chemistry. He praised Planck's chemical thermodynamics for not resting on special assumptions about atoms and ions and for being more complete and exact than previous theories.[92] For his part, Planck appreciated Helmholtz's "pioneering start of the development of pure thermodynamics, that is, the development of those theories of heat which disregard special kinetic hypotheses and confine themselves to the application of both of its two main laws."[93] Planck and other German scientists saw Helmholtz's thermodynamics as a step toward the realization of the energeticist program of basing physical science on purely macroscopic quantities such as energy.[94]

5. Conclusion

Throughout his scientific life Helmholtz remained first and foremost a physicist (and physiologist) rather than a chemist. He was little interested in traditional chemistry, which, because of its lack of theoretical foundations, did not appeal to him. Yet he had considerable knowledge of the field and contributed to areas bordering between physics and chemistry. His chemical studies certainly occupied a less prominent role in his career than did his works in physics and physiology, yet they were of prime importance for the development of chemistry. His thermodynamical theory of chemical change provided a general basis for those relationships—electrical and energetical— which were independent of chemical qualities; it proved the fertility of thermodynamics in chemistry by establishing free energy as the true measure of chemical affinity. Although Helmholtz's works formed an important background for the transformation of chemistry that took place during the last decades of the nineteenth century, Helmholtz never joined forces with the new physical chemistry. The last scientist to make significant contributions to such diverse fields as chemistry, physics, and physiology, Helmholtz had ceased doing chemical research by 1884 and he only followed the new developments from a

---

92. See Helmholtz's address to the Berlin Akademie, reprinted in Christa Kirsten and Hans-Günther Körber, eds., *Physiker über Physiker: Wahlvorschläge zur Aufnahme von Physikern in die Berliner Akademie 1870 bis 1929 von Hermann v. Helmholtz bis Erwin Schrödinger* (Berlin: Akademie-Verlag, 1975), 125–26.
93. Planck ed., *Abhandlungen zur Thermodynamik*, 73. See also Planck, *Thermochemie*, 175.
94. Max Planck, *Treatise on Thermodynamics* (Leipzig: Engelmann, 1898; and London: Longmans, Green & Co., 1903), ix–x; and Helm, *Energetik*, 175.

distance. Increasing specialization within subdisciplines, such as physical chemistry, made it difficult for him to remain abreast of such fields during his later years. Moreover, he regarded central parts of physical chemistry—the ionic theory of solutions and the theory of osmotic pressure—with skepticism. To his mind, the new physical chemistry was too much chemistry and too little physics.

11

Helmholtz's Mechanical Foundation of Thermodynamics*

Günter Bierhalter

1. Introduction

From the seventeenth to the turn of the twentieth century, mechanics held center stage in the physical sciences. It did so not only by constituting the paradigm for physical thought in general but also by in good measure forming the basis of other physical disciplines. This point is especially true for post-1850 thermodynamics. As physicists created and absorbed the first two laws of thermodynamics, they came to understand the first law as the extension of the law of energy conservation to mechanics while remaining skeptical about the status and meaning of the second law, that concerning entropy, because its mechanical foundations were uncertain. For instance, in a lecture before the Hungarian Academy of Sciences in December 1871, the physicist Coloman Szily maintained that only physical theories based on mechanical principles could be counted as successful. Since the second law then lacked such a mechanical foundation, Szily asserted that physicists distrusted it; indeed, some even denied it altogether.[1] On the other hand, this very lack of such foundations strongly motivated a number of physicists to provide a mechanical deduction for the entropy law. Among these highly motivated physicists was Hermann von

---

I should like to thank David Cahan for his comments and criticisms.
*Translated by David Cahan.
1. Coloman Szily, "Das Hamilton'sche Princip und der zweite Hauptsatz der mechanischen Wärmetheorie," *AP* 145 (1872):295–302, esp. 295–96.

Helmholtz, who devoted much of his intellectual energies during the final decade of his life (1884–94) to the fundamental questions of mechanics, and, in so doing, played a particularly important (if not always decisive) role in physicists' attempts to provide a mechanical foundation to the second law. Accordingly, analysis of his various investigations in mechanics, in particular his studies on the properties of monocyclic systems and their relation to the theory of heat, forms the principal object of this essay.

This essay argues that Helmholtz's views on the mechanical foundations of thermodynamics are best understood through the presentation and analysis of three fundamental theses:

i. All processes in nature satisfy the law of conservation of energy.
ii. All processes in nature satisfy a set of differential equations derived from Hamilton's principle of least action.
iii. The motion of heat consists in the stationary, hidden motion of molecules.

As the essay argues, these theses lay at the heart of Helmholtz's mechanical interpretation of thermodynamics and, ultimately, of his Helmholtz's *mechanical view of nature*. Section 2 of this essay investigates how Helmholtz used these theses in 1884 to advance an interpretation of the thermodynamics of reversible processes. Section 3 then turns to one of the roots of Helmholtz's ideas in the work of Ludwig Boltzmann, and shows how the latter refined and criticized Helmholtz's analysis of monocycles. Section 4 again uses these theses, only now to investigate Helmholtz's contributions to the topic of irreversible processes. Finally, Section 5, the epilogue, briefly touches on some aspects of Helmholtz's physical legacy.

2. Helmholtz's Mechanical Interpretation of the Thermodynamics of Reversible Processes

Helmholtz's work on the mechanical interpretation of thermodynamics appeared chiefly in two essays, both published in 1884.[2] He

---

2. "Principien der Statik monocyklischer Systeme," *JfruaM 97* (1884):111–40, 317–36, in *WA 3*:142–62, 179–202. To a far lesser extent, pertinent work on monocyclic systems also appeared in volume 6 of Helmholtz's *Vorlesungen über Theoretische Physik*, eds. Arthur König, et al., 6 vols. (Leipzig: J.A. Barth, 1897–1907), vol. 6: *Vorlesungen über Theorie der Wärme*, ed. Franz Richarz (1903), 338–71. Although first published in 1903, volume 6 originated from a combination of Helmholtz's lecture notebook of 1890,

considered a special type of stationary motion occurring in monocyclic systems. In such systems one or more cyclical stationary motions occurs. He called the first type a simple monocyclic motion; the second a compound monocyclic motion. For the latter, he assumed that the different motions depend on only *one* parameter. As examples of simple monocyclic systems he cited a rotating top and a liquid enclosed within a tube such that the liquid rotated back into its original position. He considered the motion of both the top and the liquid as stationary, for each time that a particle left its position, another particle assumed that position with the same state of motion.[3]

Helmholtz's principal interest in monocyclic systems lay in the fact that they display the essential properties of thermal motion. He assumed that heat consisted in the motion of molecules, whereby each particle changes its type of motion, that is, that over time it assumes different positions and velocities. Because of the tremendously large number of particles involved, all possible states of motion are represented at any point in time; thus, a stationary motion necessarily occurs. Moreover, since each particle in its thermal motion displays (in successive time intervals) distinct velocities, thermal theory required calculation by average values. In monocyclic systems, by contrast, Helmholtz showed that such temporal variations do not occur. Thus, an arbitrarily chosen mass particle of a frictionless top rotating on its axis continuously maintains a constant velocity. The same point holds for a liquid particle circulating in a ringlike tube with the same cross-section throughout. Helmholtz found that in monocycles he could calculate with individual values. Although he acknowledged that the molecular thermal motion is not strongly monocyclic, he did argue that it approximates quite closely that of monocyclic systems. He thus viewed his studies on monocyclic systems as *analogous* to thermodynamic systems.[4]

Helmholtz's ideas had their roots in the kinetic theory of gases, not merely as it developed in the mid-nineteenth century but also as it had originally taken shape in the eighteenth. Various theorists—from Jakob Hermann, Leonhard Euler, and Daniel Bernoulli in the eighteenth century to John Herapath, August Krönig, and Rudolf Clausius in the nineteenth—all proceeded from the assumption that heat consists in the continuous motion of a very large number of the smallest components of matter. In particular, they presumed, largely without further

---

a stenographic lecture notebook of 1893, and lecture notes from the early 1880s taken by his student (and subsequent editor of volume 6), Franz Richarz. (Ibid., v–vi.)
   3. "Principien," 111.
   4. Ibid., 111–12.

justification, that all particle velocities were constant and equal. Helmholtz used this approach insofar as he, too, connected heat with the motion of particles. Moreover, like the older model of the kinetic theory, his monocyclic systems also had no temporal change in velocity.[5]

In 1859, however, James Clerk Maxwell recognized that the assumption of uniform particle velocities was false. The collisions between the molecules, he argued, produced different particle velocities. He found through probability considerations that the velocity distribution is represented by a function so constituted that the particle velocities coalesce around a mean value, with very large and very small velocities appearing only quite infrequently. This so-called Maxwell velocity distribution law provided a statistical structure for the kinetic theory of gases, and so Maxwell could calculate average values of functions of particle velocities.[6] Although Helmholtz knew about these average values, he did not include them in his model. Thus, the monocycle analogies for heat motion were more closely related to ideas which had been developed in the structure of the kinetic theory *before* 1859. As Section 4 below shows, Helmholtz's initial decision not to take a probabilistic approach to monocyclic systems brought him a number of problems.

Notwithstanding its usefulness for Helmholtz, the kinetic theory of gases sought answers to rather detailed questions, for example, to that concerning the velocity of gas molecules and the ideal gas law. By contrast, in his 1884 study on monocyclic systems Helmholtz addressed a much larger and more fundamental problem: he wanted to proffer a mechanical understanding of the two laws of thermodynamics. With this purpose in mind, he began with a recapitulation of these laws as well as a general characterization of monocyclic systems. He pointed out that if one adds to a system the heat dQ and the external work $\sum A da$, then the system's energy E is raised by an amount

$$dE = dQ + \sum A da, \qquad (1)$$

where A are the forces by which the system reacts to a change of coordinates of work or parameter a. One such work coordinate is the volume; the corresponding force is the negative pressure. According

---

5. Clifford Truesdell, "Early Kinetic Theory of Gases," *Archive for History of Exact Sciences* 15 (1975):1–66.
6. James Clerk Maxwell, "Illustrations of the Dynamical Theory of Gases. Part I. On the Motions and Collisions of Perfectly Elastic Spheres," in *The Scientific Papers of James Clerk Maxwell*, ed. W.D. Niven, 2 vols. (Cambridge: Cambridge University Press, 1890), *1*:377–409, on 380–90.

to the second law, as first formulated by Rudolf Clausius, the heat added is related to the change of a state function S:

$$dQ = \theta dS, \qquad (2)$$

where θ designates the absolute temperature at which the heat added occurs and S the entropy. Although the first law always holds for all processes, equation (2) is conditioned by the fact that the heat added to the system must occur in such a way that the temperature θ and the parameters **a** change only very slowly so that the system passes through a succession of equilibrium states. Among the pertinent states here are those in which the system, in the absence of any external influence, continually remains unchanged. The ability to pass through all changes of state in the opposite direction constitutes so-called reversible processes, since they can be reversed at any point in time. Finally, as for the energy E and the entropy S, they are state functions: they depend only on the temperature θ and the coordinates of work **a**. By contrast, dQ and $\Sigma A d\mathbf{a}$ depend on the type and way heat and work are added.[7]

Relation (2) drew Helmholtz's special attention, for it displays a peculiarity. The heat equilibrium of two bodies is indicated by the equality of their temperatures. Every function of temperature also exhibits this property. The different thermometric scales—mercury, alcohol or gas—provide examples of such different functions of θ. On the other hand, Helmholtz also realized that equation (2) could be rewritten in the form

$$dQ = \delta d\sigma, \qquad (3)$$

where σ is an arbitrary function of the entropy S, and δ a function of θ *and* the working coordinates **a**. Thus δ could not characterize the heat equilibrium. Yet since θ is an *arbitrary* function of the entropy, Helmholtz knew that there exist *arbitrarily many relations* having the structure of equation (3). Hence, in his attempt to develop a mechanical interpretation of the second law he knew that he could not proceed immediately to a mechanical analogue of (2); instead, he hoped to do so through an equation of the form (3). He thus required another condition, one which included the state of the heat equilibrium. He began with the well-known fact that a system showing the same temperature throughout may be constructed out of two partial systems, **1** and **2**. Thus, for the heat added

$$dQ = dQ_1 + dQ_2 = \theta dS_1 + \theta dS_2 = \theta d(S_1+S_2) = \theta dS. \qquad (4)$$

---

7. "Principien," 112–17.

This meant that, assuming temperature equality, the same equation (2) is valid for the entire system, just as it is valid for the two partial systems taken individually.[8] Helmholtz thus faced the problem of finding a mechanical counterpart for equation (4). He had come a bit closer to his goal of understanding *all* natural processes according to mechanical categories, and so the entire system of phenomenological thermodynamics might thus be built upon the foundation of the two laws.

Along with the centrality of the two laws, Helmholtz's basic ideas on monocycles depended on Lagrange's equations, which he in turn derived from Hamilton's principle of least action. In particular, he began with Lagrange's equation in the form:[9]

$$\mathbf{P} = -\frac{\partial L}{\partial \mathbf{q}} + \frac{d}{dt}\frac{\partial L}{\partial \dot{\mathbf{q}}}, \quad (5)$$

where **P** is the force acting on the generalized coordinate **q**, $\dot{\mathbf{q}} = \frac{d\mathbf{q}}{dt}$ the generalized velocity, L the Lagrangian function (that is, the difference between the kinetic energy T and the potential energy U):

$$L = T - U. \quad (6)$$

The potential energy U is a function of the coordinates, and T has the structure:

$$T = \sum \frac{1}{2}\mathbf{p}\dot{\mathbf{q}}, \quad (7)$$

where **p** is a generalized momentum:

$$\mathbf{p} = \frac{\partial L}{\partial \dot{\mathbf{q}}}. \quad (8)$$

Now there exist as many equations (5) as there are coordinates available.[10]

---

8. Ibid., 113–15.
9. Ibid., 118. In 1884 Helmholtz presented the Lagrangian equations in complete form, that is, he did not at all seek to derive them from Hamilton's principle. In the 1903 presentation, however, the Lagrangian equations are fully explicated: their derivation and analysis take up over one-fourth of the pages devoted to explaining the monocycle analogies.
10. Ibid. Helmholtz actually designated these symbols differently. Here **q**, $\dot{\mathbf{q}}$, and **p** are the coordinates, velocities, and momenta, respectively; Helmholtz used the symbols **p**, **q**, and **s** for these respective quantities. Thus, here $\frac{d\mathbf{p}}{dt}$ designates a force, while for Helmholtz it designated a velocity. Moreover, Helmholtz did not calculate using the Lagrangian function L, but rather with its negative value, that is, with the difference

Helmholtz adapted equation (5) to the relations in a monocylic system by introducing *one* rapidly changing coordinate q_b. This coordinate represents stationary motion which, in the thermodynamic case, consists of the heat motion of molecules. Since immediately after a particle leaves its position a similarly constituted particle with the same state of motion takes its place, it followed that, notwithstanding the perpetual motion, neither the potential nor the kinetic energy depended on q_b. Because these two quantities of energy also determine the Lagrangian, L is also independent of the cyclic coordinate $\frac{\partial L}{\partial q_b} = 0$. Thus, Helmholtz simplified (5) to:

$$\mathbf{P}_b = \frac{d}{dt}\frac{\partial L}{\partial \dot{q}_b} = \dot{\mathbf{p}}_b, \tag{9}$$

while simultaneously taking (8) into account.[11]

Although q_b did not occur in L, it nonetheless followed according to (9) that the appropriate cyclical velocity $\dot{q}_b$ did enter into L. This seemingly contradictory state of affairs becomes understandable when one considers a sphere rotating on its axis in the Earth's gravitational field as a special type of monocycle. Here q_b equals the angle of rotation, $\dot{q}_b$ the angular velocity. For the sphere's kinetic and potential energy, and thus for its Lagrangian, the angle of rotation through which the sphere may have already passed is completely irrelevant. Correspondingly, the angle of rotation q_b also does not enter into L. The kinetic energy (and thus L) depends only on the angular velocity $\dot{q}_b$. Again correspondingly, the same holds for the molecular motion of gas particles: the molecular positions corresponding to q_b have no influence on the thermodynamic properties of the gas. By contrast, however, the molecular velocities corresponding to $\dot{q}_b$ do indeed determine its temperature.

Helmholtz also characterized a monocycle by another group of coordinates q_a, which, in comparison to q_b, vary slowly. In particular, he

---

between the potential and kinetic energy, to which he applied the symbol H. For the kinetic energy he used the symbol L, doubtless in association with the concept of "living force" ("*lebendige Kraft*"). His rather arbitrary symbolism led to misunderstandings between mathematicians and physicists; see Felix Klein, *Vorlesungen über die Entwicklung der Mathematik im 19. Jahrhundert*, 2 vols., Teil 1, eds. R. Courant and O. Neugebauer (Berlin: J. Springer, 1926), *1*:205. This essay avoids his (confusing) notation.

11. "Principien," 119.

assumed that the $\dot{q}_a$ are vanishingly small and thus do not contribute to the kinetic energy. Hence the $\dot{q}_a$ do not enter into L, and so the terms with $\dfrac{\partial L}{\partial \dot{q}_a}$ can be neglected in equation (5). Under this assumption Helmholtz obtained from equation (5) the special relation:

$$\mathbf{P}_a = -\frac{\partial L}{\partial q_a}. \tag{10}$$

As already noted, one such slowly changing coordinate in thermodynamics is the volume.[12]

Under the condition that the $\dot{q}_a$ do not contribute to the kinetic energy, T thus depends solely on the cyclic velocity $\dot{q}_b$, and so equation (7) simplifies to:

$$T = \frac{1}{2} p_b \dot{q}_b. \tag{11}$$

Helmholtz here considered the following thermodynamic state of affairs: if one imagines a gas enclosed in a container with a moveable piston, then one can neglect the piston's kinetic energy which moves the slowly expanding gas forward.[13] The kinetic energy is accordingly composed of only that part which depends on the molecular thermal motion and is included through the cyclic velocity $\dot{q}_b$.

Having laid the groundwork for his analysis of monocyclic systems, Helmholtz then turned to the case where q_b and the q_a are varied. He stipulated that, *analogous* to thermodynamically reversible processes, these variations occur only very slowly so that the monocycle will pass through a series of equilibrium states.[14] He thus sought to develop a *statics* of monocyclic systems, and in this sense he saw the laws of reversible processes as laws of statics. He found the cyclic work dQ_b, which is added to a monocycle by varying the cyclic coordinate under the influence of the force $\mathbf{P}_b$, by multiplying equation (9) by dq_b; a bit of manipulation gave him:[15]

$$dQ_b = \dot{q}_b dp_b. \tag{12}$$

The work dQ causes an increase in cyclic velocity, just as in thermodynamic processes the heat added leads to increased molecular

12. Ibid., 120.
13. *Vorlesungen* 6:349.
14. "Principien," 115.
15. Ibid., 119.

velocities. Helmholtz thus built the work conditioned by the forces $\mathbf{P}_a$ from the sum of the individual $\mathbf{P}_a d\mathbf{q}$. This work constituted the analogue to the external thermodynamic work. Since the total energy dE added to the monocycle is composed of these two different works, Helmholtz finally got:[16]

$$dE = dQ_b + \sum \mathbf{P}_a d\mathbf{q}_a. \qquad (13)$$

The mathematical structure of equations (13) and (12) recalls the thermodynamic laws (2.1) and (2.2), respectively, where $\dot{q}_b$ and p_b replace the temperature and the entropy.[17] As we shall see, Helmholtz gradually endowed the mathematical equations (12) and (13) with physical content.[18]

Having recapitulated Helmholtz's route to establishing the fundamental equations (12) and (13), analysis of the meaning of theses i–iii is now possible. In accord with thesis (ii), Helmholtz began with the differential equations (5) of mechanics. He refined these to equation (9) by introducing the coordinate q_b whereby he took into account the stationary movement as expressed in thesis (iii). By introducing the coordinates q_a he got from equation (5) the forces expressed in equation (10), forces which correspond to external thermodynamic forces. In this way he obtained the energy balance of equation (13) in the sense of thesis (i), which represents the analogy to the thermodynamic statement that an increase of energy in the general case can occur in two ways, namely, in the form of heat and work. With equation (12) Helmholtz thus found a representation for the cyclic work corresponding to the second law.

However, in seeking to obtain mechanical analogies for thermodynamics, Helmholtz could not permit all kinds of energy exchange to occur in monocyclic systems. In 1903, Franz Richarz gave a nice example that illustrated the problem that confronted Helmholtz.[19] Richarz imagined a special monocycle, that of a rotating top on whose axis of rotation one winds up (from some definite moment in time) a

---

16. Ibid., 122.
17. The negative Lagrangian corresponds to the free energy of thermodynamics. See ibid., 116, 122.
18. For a more formally mathematical treatment of Helmholtz's work here see Günter Bierhalter, "Zu Hermann v. Helmholtzens mechanischer Grundlegung der Wärmelehre aus dem Jahre 1884," *Archive for History of Exact Sciences* 25 (1981):71–84; and idem, "Wie erfolgreich waren die im 19. Jahrhundert betriebenen Versuche einer mechanischen Grundlegung des zweiten Hauptsatzes der Thermodynamik?," ibid., 37 (1987):77–99.
19. *Vorlesungen* 6:370 (n1).

thread attached to an elastic spiral spring. In so doing, the top's total rotational kinetic energy is transformed into potential energy, since the thread pulls on the spring and increases its length. Now, there exists no analogous thermodynamic process: there is no process which could cause only the total kinetic energy of heat motion to be withdrawn from a body and be transformed by, for example, the raising of a weight into potential energy, thus causing the body to cool to absolute zero. There is no way of intervening directly in the thermal motion as can occur in the rotational. To assure that his monocycle analogies really described the essential nature of heat, Helmholtz thus had to exclude such energy transmissions as those which occur by means of a winding thread. In particular, he required that whenever cyclic work dQ_b is added to a monocycle this must already have been stored as cyclic energy in another monocycle.[20] Only by this limitation could he use monocyclic systems to give a suitable picture of heat transfer between bodies.

Helmholtz used the example of a frictionless top to illustrate the concept of hidden motion as expressed in thesis (iii).[21] If such a top rotates with its tip touching a smooth surface, then no force increases or decreases its rotational velocity. However, the top does react to forces which try to displace its rotational axis with motions which it could not achieve without rotating. Therefore, Helmholtz imagined the top as being enclosed within a shell so that it could not be seen, and he constructed a system with a hidden motion that could not be directly affected. A system based on hidden motion, he argued, behaves completely differently than one without such motion. In his view, therefore, the behavior of warm bodies only becomes understandable on the assumption that within their interior there occurs a hidden motion inaccessible to our immediate perception.

By asserting such limiting conditions as those just described, Helmholtz found that monocyclic motion cannot be directly influenced. He thus created a mechanical *analogy* to thermal motion: for the latter, too, no direct influence is possible since there is no way to influence every individual particle directly. All manipulations necessarily influence all molecules in a given spatial area. This human limitation provided Helmholtz with the reason why we cannot directly influence thermal motion as the laws of thermodynamics would otherwise allow.[22]

---

20. Ibid., 350.
21. Ibid., 354.
22. "Principien," 136. See also *Vorlesungen* 6:370.

This situation can be formulated in another way: very many molecules are always necessarily involved in all manipulations because the quantities of any substance on which thermodynamic investigations are made always contain a tremendously large number of particles. This state of affairs led Maxwell to attack the problem of thermal motion by means of probabilistic methods and to determine the average values of the functions of particle velocities describing a system's thermal behavior. He could thus show, for example, that the temperature is proportional to the system's mean kinetic energy.[23] By contrast, Helmholtz treated the case of a large number of particles slightly differently: he explained the special properties of warm bodies with his model of a hidden, not directly influenced, and stationary motion. Where Maxwell in effect departed from classical mechanics by enlisting new probabilistic considerations into his analysis, Helmholtz, by contrast, remained rooted in the tradition of classical mechanics; he employed no probabilistic calculations in his model's mathematical formulation.

The concept of hidden motion also conditioned another aspect of Helmholtz's analysis: in the mathematical formulation of equations (9) – (13) he never employed particle-based atomistic concepts such as particle position, particle velocity, and numbers of particles, since such quantities are, according to his theory, indeed hidden from human perception and so impossible to include within a mathematical formulation. In Helmholtz's theory such quantities are, instead, hidden parameters. On this rather technical point he stood closer to energeticists like Wilhelm Ostwald and Georg Helm, not to mention the positivism of Ernst Mach, all of whom held atoms and molecules, which we cannot experience directly, for a pure fiction. Nonetheless, despite the seeming similarity of Helmholtz's concept of hidden motion to energetic and positivistic concepts, Helmholtz cannot be counted amongst the energeticists or positivists. Indeed, he belongs most decisively to the atomists. For example, he interwove molecular theories of heat into his lectures on thermodynamics,[24] and he based his mechanical analogies for thermodynamics on atomistic pictures.

One final issue concerning Helmholtz's mechanical interpretation of the thermodynamics of reversible processes requires treatment: Is the cyclic velocity $\dot{q}_b$ in equation (12) identical with the temperature θ, and thus is the cyclic momentum P_b identical with the entropy?[25] This question touches on the problem which Helmholtz engaged mathematically in equation (3): since in his theory there occurred only *one*

23. Maxwell, "Illustrations of the Dynamical Theory of Gases."
24. *Vorlesungen* 6:v–vi.
25. Ibid., 352–53.

cyclic velocity, he imagined that there was also only one type of thermal motion. For a body which is in thermal equilibrium throughout its parts, there is only *one* type of temperature. Yet Helmholtz argued that the cyclic velocity does not have the same physical meaning for all monocylic systems.[26] In the case of a rotating sphere, $\dot{q}_b$ equals the angular velocity; in the case of a liquid circulating inside a closed tube, by contrast, $\dot{q}_b$ is the quantity of liquid flowing through the tube's cross-section in unit time. Helmholtz had therefore to replace q_b in equation (12) with a quantity which was the same for all monocyclic systems and which possessed a definite physical meaning, namely, kinetic energy.[27] He thus rewrote equation (12) with the value of the kinetic energy from (11):

$$dQ_b = TdS_b, \qquad (14)$$

where the entropy S_b of the monocycle is given by

$$S_b = 2\log p_b. \qquad (15)$$

Here he was struck by the fact that the kinetic energy in equation (14) assumed the same position for the cyclic work as the temperature in the thermodynamic case in equation (2). As we shall see, the step to (14) became important for Helmholtz for yet other reasons.

At the same time, equation (15) displays no noticeable similarity with the entropy function of an ideal gas, for example. That function is dependent on the temperature and the volume **V** with the constants C_v and **R** given by:

$$S = C_v \log \theta + R \log V. \qquad (16)$$

To compare equations (15) and (16), Helmholtz employed a simple mathematical reformulation.[28] He used equation (11) to rewrite relation (15) as

$$S_b = \log T + \log \frac{p_b}{\dot{q}_b}. \qquad (17)$$

Here the kinetic energy corresponds to the temperature in (16) and $\frac{p_b}{\dot{q}_b}$ represents the moment of inertia in a rotational motion. Like the

---

26. Ibid., 363; and "Principien," 119–20.
27. "Principien," 123.
28. Ibid.

gas volume, this too depends on the spatial dimensions of the system in question.

3. Boltzmann, Helmholtz, and Monocycles

Helmholtz's idea of using the kinetic energy as the counterpart to the temperature came from work by Boltzmann in 1866 and Clausius in 1871, a point that he did not hesitate to acknowledge.[29] Of the two papers, Boltzmann's alone merits discussion, for it is noticeably clearer than Clausius's and its critique led Helmholtz to refine his own ideas.

In his paper of 1866, Boltzmann assumed that particles undergo a periodic thermal motion and that, after a time $t = t_0 + i$, all particles with the same velocities return to the same positions as at time $t = t_0$, where i is the period of thermal motion. Furthermore, he equated the temperature with the mean kinetic energy $<T_0>$ of a particle over the period i, and conceived the heat added in such a way that a quantum of energy ε is added to each atom yielding an average increase of kinetic energy and an average work performed:

$$\varepsilon = <\delta T_0> - <X\delta x + Y\delta y + Z\delta z>, \qquad (18)$$

where X, Y, and Z designate the components of force in the direction of the coordinate axes x, y, and z. Boltzmann found the heat added to the total (particle) system as equivalent to the sum of the εs formed over all atoms. Under these assumptions he calculated the relation for the entropy as:

$$S = \sum 2\log(i<T_0>), \qquad (19)$$

where again the sum is taken over all atoms.[30]

---

29. Ibid., 129. Ludwig Boltzmann, "Über die mechanische Bedeutung des zweiten Hauptsatzes der Wärmetheorie," *SBW 53* (1866):195-220, reprinted in *Wissenschaftliche Abhandlungen von Ludwig Boltzmann*, ed. Fritz Hasenöhrl, 3 vols. (Leipzig: J.A. Barth, 1909), *1*:9-33; and Rudolf Clausius, "Ueber die Zurückführung des zweiten Hauptsatzes der mechanischen Wärmetheorie auf allgemeine mechanische Principien," *AP 142* (1871):433-61. For an analysis of Boltzmann on monocycles see the excellent article by Martin J. Klein, "Boltzmann, Monocycles, and Mechanical Explanation," in *Philosophical Foundations of Science* (= Boston Studies in the Philosophy of Science, vol. 11), eds. Raymond J. Seeger and Robert S. Cohen (Dordrecht and Boston: D. Reidel, 1974), 155-75, which is in good part identical to idem, "Mechanical Explanation at the End of the Nineteenth Century," *Centaurus 17* (1972):58-82, and which have served as models for the present essay.

30. Boltzmann, "Über die mechanische Bedeutung," 23-8.

Boltzmann's approach gave Helmholtz the idea of viewing the kinetic energy as the measure of temperature. To be sure, there was an important difference between their approaches: whereas Boltzmann used the mean kinetic energy of a *single* particle as the measure, Helmholtz used the *total kinetic energy of a monocycle*. Indeed, Boltzmann's approach had several shortcomings, and Helmholtz's made several improvements on it. First, Boltzmann used only particle-based quantities in equation (18); in particular, he nowhere stated exactly how the quantity of energy ε, and hence the heat, was to be added, although he did vaguely mention that the quantity of energy ε should be added to each particle.[31] However, that implied the inadmissible assumption that, like macroscopic bodies, the particles can be seen directly. In contrast to Helmholtz, Boltzmann had no concept of hidden motion prohibiting direct influence. Accordingly, he recognized no distinction of coordinates in the sense of Helmholtz's q_b and q_a. Further, it followed that Boltzmann could not build the energy added from the different energy forms of heat and external work, for he did not include the concepts of total energy E and external work. Nor did he explain why he attributed a periodic thermal motion to the particles. Indeed, he introduced this ad hoc hypothesis solely to maintain the simple relation (19), and in 1872, for example, he dropped the hypothesis altogether.[32] By contrast, Helmholtz could justify physically his hypothesis of stationary thermal motion. Finally, although Helmholtz was doubtless familiar with Boltzmann's statistical approach, in 1884 he only cited Boltzmann's first, and still quite deficient, treatise on heat theory. The reason for this is clear: during the last decade of his life Helmholtz sought to secure a *mechanical* understanding of physics, and Boltzmann's 1866 treatise sought to provide a *purely mechanical* justification of the second law.[33] Boltzmann's approach was further congenial to Helmholtz in that entropy for Boltzmann is determined mechanically by the (Leibnizian) quantity of action $i<T_0>$; and, as

---

31. Cf. Martin J. Klein, "Maxwell, His Demon, and the Second Law of Thermodynamics," *American Scientist* 58 (1970):84–97, on 90.

32. Ludwig Boltzmann, "Weitere Studien über das Wärmegleichgewicht unter Gasmolekülen," *SBW* 66 (1872):275–370, reprinted in idem, *Wissenschaftliche Abhandlungen 1*:316–402.

33. For an extensive treatment of the mechanical derivations of the second law see George H. Bryan, "Researches Relating to the Connection of the Second Law with Dynamical Principles," *Report of the Sixty-First Meeting of the British Association for the Advancement of Science held at Cardiff in August 1891* (London: John Murray, 1891):85–122. Bryan also wrote "Allgemeine Grundlegung der Thermodynamik," *Encyklopädie der mathematischen Wissenschaften mit Einschluss ihrer Anwendungen*, eds. Akademie der Wissenschaften zu Göttingen, et al., 6 vols. (Leipzig: B.G. Teubner, 1898/1904–1935), vol. 5: *Physik*, ed. A. Sommerfeld, Teil 1 (Leipzig: B.G. Teubner, 1903–21), where on pp. 146–60 he further analyzed the ideas of Boltzmann, Clausius, and Helmholtz.

we shall see, in 1887 Helmholtz investigated the physical meaning of the principle of action.

Helmholtz knew that if the properties of monocyclic systems were to properly represent the properties of thermal bodies, then monocycles must also be able to couple, just as thermal bodies of equal temperature may combine. Monocyclic systems should thus also show a state analogous to heat equilibrium. To demonstrate such coupling, Helmholtz imagined two identical rotating bodies coupled at their axes at a point in time when both exhibit the same angular velocity and kinetic energy. The two bodies thus rotate with the same angular velocity. If any force changes the angular velocity of one body, then the angular velocity of the other must also change in exactly the same way. In particular, Helmholtz found for the added cyclic work a relation analogous to equation (4), wherein the kinetic energy replaced the temperature. He called such a coupling an isomoric coupling.[34]

According to Helmholtz, kinetic energy in monocycles plays the same role as temperature in thermal bodies. Yet, as Richarz later pointed out, there was one important difference. While two bodies of different mass can exhibit the same temperature throughout, the kinetic energies of their thermal motions, in contrast to the temperature, have different values. Now Helmholtz knew that the kinetic energies are proportional to the masses of the respective systems. In order to address this difficulty in monocyclic systems, he referred all equations containing the kinetic energy to a mass unit. Hence, he did not, as he believed, assume precisely Boltzmann's concept of temperature; rather, he chose as the counterpart to temperature the total kinetic energy of a monocycle. Although Helmholtz knew and recognized Boltzmann's critique of this shortcoming of his presentation, for some unknown reason he did not correct it.[35]

Helmholtz dedicated most of his second essay of 1884 to extending his theorems for simple monocyclic systems to complex systems where *several* $\dot{q}_b$ and p_b appear yet depend on a common parameter. He assumed that complex monocyclic motions could be achieved by means of actual mechanical apparatus, composed, for example, of several rotating bodies. Gear wheels, cords, and so on could be used to achieve connections between these bodies in order to produce manifold and varying relations between the $\dot{q}_b$. He imagined that all processes occurring in a complex monocycle do so at various velocities: if a move-

---

34. *Vorlesungen* 6:367–69; and "Principien," 134–38. For a critical treatment of isomoric coupling see Bierhalter, "Zu Helmholtzens mechanischer Grundlegung," 80–1; and idem, "Wie erfolgreich waren," 85–8.

35. *Vorlesungen* 6:366 (n1).

ment originally takes a time t_0, then one could, by varying the forces, have the movement occur in time $t = nt_0$; the period t is thus, according to the choice of **n**, smaller or larger than t_0. Under this assumption Helmholtz found that the $\dot{q}_b$ and the p_b always remained proportional to the common parameter **n**. For such a complex monocycle the cyclic work and the kinetic energy is the sum of the expressions (12) and (11). Helmholtz argued that for this case the cyclic work dQ_b can also be given in the form of equation (14); the entropy of the complex monocycle is determined by the parameter **n**:

$$S_b = 2\log n. \qquad (20)$$

With this expression, he believed that he had shown that theorem (14) is valid for *all* monocyclic systems, simple and complex.[36]

However, from the physical point of view Helmholtz's analysis contained a serious shortcoming: what is the exact meaning of **n**? Helmholtz could not answer this question. He merely noted that Boltzmann and Clausius, in their works of 1866 and 1871, had perhaps determined **n**, which there was connected with the quantities of action $i < T_0 >$.[37] In Helmholtz's own work, however, **n** had no physical meaning: it is a pure number which simply takes into account how fast or slow a motion occurs.

Helmholtz's work immediately aroused Boltzmann's interest. Before the year had ended, Boltzmann published a study, one of the most important on the subject, of the physics of monocyclic systems. He not only extended Helmholtz's point of view; he went well beyond it by explaining more exactly why monocyclic systems show some properties of warm bodies.[38]

In the introduction to his paper Boltzmann noted that Helmholtz had only formulated *analogies* to the second law of thermodynamics.[39] No *mechanical proof* of the second law was possible, Boltzmann said, since one has neither exact knowledge of the nature of atoms nor the

---

36. "Principien," 123, 317–19, 322–23.
37. Ibid., 325.
38. Ludwig Boltzmann, "Über die Eigenschaften monozyklischer und anderer damit verwandter Systeme," *JfruaM 98* (1884–85):68–94, reprinted in idem, *Wissenschaftliche Abhandlungen 3*:122–52.
39. Clausius had missed this very point; see Klein, "Mechanical Explanation at the End of the Nineteenth Century," esp. 66–7. Klein's essay goes beyond the scope of the present essay in that it shows how monocyclic systems *generally* were called upon in the late nineteenth century to explain natural laws.

exact conditions under which thermal motion occurs.[40] Boltzmann's declaration essentially marked a change in his own point of view. Previously, for example in his work of 1866, he had advertised his ideas as proofs of the second law.[41] By 1884, however, he had become more cautious. He now departed from the older, naive idea of the kinetic theory of gases, according to which atoms are *structureless* mass points. More to the point, Boltzmann eliminated a painfully noticeable defect in Helmholtz's work. Helmholtz had cited only rotating bodies and liquids circulating within closed tubes as examples of monocyclic systems. Boltzmann now expanded the number of such concrete cases and developed them mathematically. He showed, for example, that Saturn's rings represent a monocyclic system just as do particle currents which are reflected elastically at a definite angle from a container's walls. Moreover, he treated such systems mathematically, and so elucidated the physical meaning to be ascribed to Helmholtz's $\dot{q}_b$ and p_b for these special cases.[42] In short, he showed physical realities where Helmholtz could only show mathematical formalism.

Moreover, Boltzmann went on to show how his theorems, which he developed on the basis of his new statistical mechanics, fit into the scheme of the monocycle analogies.[43] Here he referred principally to his work of 1871, wherein he had sought to prove the second law by using an expanded form of Maxwell's law of velocity distribution.[44] At the same time, he also discussed the relation between his earlier views of 1866 and Helmholtz's theory.[45]

One of Boltzmann's central points in 1884 was to consider ergodic systems, by which he meant systems composed of elements whose coordinates and velocities assume (over time) all possible values consistent with the fixed total energy of each system. He now expanded his 1871 presentation of ergodic systems by giving them for generalized coordinates. He determined the number of individual elements whose coordinates were included between the limits q_1 and $q_1 + dq_1, \ldots, q_g$ and $q_g + dq_g$, where g indicates the number of coordinates distin-

---

40. Boltzmann, "Über die Eigenschaften monozyklischer Systeme," 122.
41. Boltzmann, "Über die mechanische Bedeutung," 23 (esp. the title to Section IV: "Beweis des zweiten Hauptsatzes der mechanischen Wärmetheorie").
42. Boltzmann, "Über die Eigenschaften monozyklischer Systeme," 123–30.
43. Ibid., 131–37.
44. Ludwig Boltzmann, "Analytischer Beweis des zweiten Hauptsatzes der mechanischen Wärmetheorie aus den Sätzen über das Gleichgewicht der lebendigen Kraft," *SBW 63* (1871):712–32, reprinted in idem, *Wissenschaftliche Abhandlungen 1*:288–308.
45. Boltzmann, "Über die Eigenschaften monozyklischer Systeme," 139.

guishing the ergodic system. From this Boltzmann developed a new general expression for the added heat δQ to an ergodic system:

$$\delta Q = \frac{2T}{g}\delta\log\int\ldots\int \Delta^{-\frac{1}{2}}\psi^{\frac{g}{2}}dq_1\ldots dq_g, \qquad (21)$$

where the q_g—in the special case of a monatomic gas—are rectangular coordinates of the particles, ψ the total kinetic energy, and Δ is dependent on the mass of the atoms. He proved that from equation (21) the entropy function (16) of a monatomic ideal gas followed. An important property of equation (21) is that its validity does *not* depend on the number of coordinates distinguishing an ergodic system. It is thus even valid if only *one* coordinate and *one* related velocity occur. In particular, this holds for the case of a monocyclic system formed by a rotating solid body. Here Boltzmann equated the single q to the angle of rotation, where ψ is the rotational kinetic energy, and Δ the reciprocal moment of inertia. Under these assumptions he reduced equation (21) to the simpler relation (15) for the entropy S_b of a monocycle. Moreover, he also verified this relation by means of equation (21) for Helmholtz's example of a liquid flowing inside a closed tube.[46]

Boltzmann's paper of 1884 constituted still another advance over that of Helmholtz's: where Helmholtz had only proved that monocyclic systems show analogies to warm bodies, Boltzmann explained why these analogies existed. The reason lay precisely in the general validity of equation (21) for ergodic systems.[47] Boltzmann showed that the formula (21) was thus also valid for complex monocyclic systems displaying an ergodic character.[48] His analysis allowed retention of Helmholtz's theorem (14) that for complex monocyclic systems, as well as simple systems, the kinetic energy can replace the temperature. Nonetheless, Boltzmann also showed that equation (14) is, despite Helmholtz's claim, not possible for monocycles exhibiting non-ergodic characteristics.[49] In this case, relation (21) no longer exists, and so the kinetic energy no longer corresponds to the temperature, a point that he further justified by means of a special mechanical model that demonstrated his typical ability to think in terms of concrete pictures.[50]

---

46. Ibid., 134–38.
47. On the meaning of ergodic systems in Boltzmann's 1884 essay see Stephen G. Brush, "Foundations of Statistical Mechanics 1845-1915," *Archive for History of Exact Sciences 4* (1967):145–83, esp. 169.
48. Boltzmann, "Über die Eigenschaften monozyklischer Systeme," 140–41.
49. See Klein, "Mechanical Explanation," 71.
50. Boltzmann, "Über die Eigenschaften monozyklischer Systeme," 141–52.

Later, in 1885 and 1886, Boltzmann published two additional studies on monocyclic systems and deepened his analysis still further.[51]

In sum, Helmholtz's attempt to prove that *all* monocyclic systems show analogies to warm bodies ran aground against Boltzmann's critique. Only *simple* monocyclic systems always possess this property. His mechanical analogies did not show, as he had hoped, far-reaching agreement with thermodynamics. Boltzmann's critique perhaps explains why there was no mention of complex monocyclic systems in Helmholtz's posthumous *Vorlesungen über Theorie der Wärme* (1903). Be that as it may, Boltzmann's critique certainly brought the best out of Helmholtz's ideas on monocycles, combining his own ideas with those of Helmholtz's so as to deepen understanding of the monocycle analogies in a way that proved impossible for Helmholtz to do alone.

4. Helmholtz's Contributions to Understanding Irreversible Processes

In 1884, Helmholtz did not examine the thermodynamics of irreversible processes. He did, however, assume that his readers were already familiar with this topic, and he returned to it in 1886 as well as in his lectures and research in the early 1890s.

Helmholtz knew that, according to the second law, only those processes can occur in (adiabatically) isolated systems which increase the total entropy. Such a system does not exchange any heat energy dQ with its environment. For example: let a colder body with temperature θ_1 and a warmer body with temperature θ_2, where $\theta_2 > \theta_1$, be in thermal contact. If a shell impermeable to heat surrounds both bodies, then they can exchange heat with one another but not with their environment. The system is thus adiabatically isolated from the outside. The heat flows from the warmer to the colder body, and indeed such that in unit time the warmer body delivers the heat $-\dot{Q}$ at temperature θ_2, while the colder body, on the other hand, receives the heat $+\dot{Q}$ at the temperature θ_1 per unit time. The change in total entropy per unit time is:

$$\Delta \dot{S} = \frac{-\dot{Q}}{\theta_2} + \frac{+\dot{Q}}{\theta_1} = +\dot{Q}\frac{\theta_2 - \theta_1}{\theta_1 \theta_2} > \emptyset. \qquad (22)$$

---

51. Ludwig Boltzmann, "Über einige Fälle, wo die lebendige Kraft nicht integrirender Nenner des Differentials der zugeführten Energie ist," *SBW* 92 (1885):853–75, reprinted in idem, *Wissenschaftliche Abhandlungen 3*:153–75; and idem, "Neuer Beweis

As required by the second law, the entropy increases. The process described by equation (22) continues until both bodies have reached the same temperature: $\theta_1 = \theta_2$. No further entropy increase occurs; the entropy has reached its maximum value. Conversely, the second law prohibits the reverse process; in complete agreement with experience, heat does not flow from a colder to a warmer body. Equation (22) thus describes a natural process, one which is irreversible.

In 1876, Josef Loschmidt noted that there exists a strange discord between the fundamental equations of mechanics and the unidirectional behavior of many physical processes. Loschmidt pointed out that, in principle, these fundamental equations permitted the *reverse* of *every* process to occur.[52] Graphically speaking, this meant that if at a definite moment in time the directions of all particle velocities are reversed, then the system reverts through all its previous states, including those with lower entropy. Hence, mechanics seemed to allow processes to occur that the second law prohibited. Helmholtz faced a dilemma: he wanted to interpret all natural events—including irreversible processes—mechanically (or by mechanical pictures). Yet how could he make such an interpretation with a theory which recognized only reversible processes?

In his investigation of 1886 on Hamilton's principle, Helmholtz analyzed the physical meaning of the principle of least action by employing a particularly useful form of the principle first given in 1835 by William Rowan Hamilton.[53] From this principle Hamilton had derived both the Lagrangian equations of motion and his eponymously named canonical equations of dynamics. These equations formed an extremely versatile tool for treating the most diverse mechanical problems. In particular, for thermodynamics Hamilton's equations proved essential for developing the kinetic theory of gases into the new field of statistical mechanics. In 1886, Helmholtz sought to develop an expanded version of Hamilton's principle. His goal was to generalize the

---

eines von Helmholtz aufgestellten Theorems betreffend die Eigenschaften monozyklischer Systeme," *Göttinger Nachrichten* (1886):209, reprinted in idem, *Wissenschaftliche Abhandlungen* 3:176–81.

52. Josef Loschmidt, "Über den Zustand des Wärmegleichgewichtes eines Systems von Körpern mit Rücksicht auf die Schwerkraft," *SBW 73* (1876):128–42, on 139. For a survey of Loschmidt's considerations on the fundamental reversibility of mechanical processes see Martin J. Klein, "The Development of Boltzmann's Statistical Ideas," *Acta Physica Austriaca Supplement 10* (1973):53–106, esp. 71–2.

53. "Über die physikalische Bedeutung des Princips der kleinsten Wirkung," *JfruaM 100* (1886):137–66, 213–22, and in *WA* 3:202–48; and Leo Koenigsberger, *Helmholtz's Untersuchungen über die Grundlagen der Mathematik und Mechanik* (Leipzig: Teubner, 1896), 44. For clarity's sake, this essay refers to Helmholtz's analysis of this problem in his *Vorlesungen*.

principle so that it might serve as the foundation for all fields of physics.

In the past, Hamilton's principle had been demonstrated by means of Newton's laws of motion and, with respect to conservative forces, the presumed law of energy conservation. Helmholtz now took exactly the opposite tack: he used the principle itself as his starting point and placed it at the top of his entire deductive structure. Moreover, he took into account his own *general* law of energy conservation. In other words, in addition to kinetic and potential energy he now also included non-mechanical forms of energy, for example, that characteristic for thermal or electrical phenomena. In this way he extended the principle of least action so as to derive the equations of motion. On the basis of this generalized principle of least action he not only rederived the Newtonian, Lagrangian, and Hamiltonian equations of motion but also the differential equations representing thermal and electrodynamic forces.[54] In short, he expanded Hamilton's principle by means of the law of conservation of energy such that it was no longer restricted to its original range of validity, namely, the assumed conservative forces. And he included in his new, expanded mechanical concept such phenomena as heat and electricity, phenomena which until now had usually been considered as beyond the pale of mechanics.

Helmholtz first considered the kinetic energy. He found that if all coordinates, all motions, and the changes of all variable states are completely known, then the kinetic energy depended quadratically on the velocities $\dot{q}$. This property of kinetic energy, according to Helmholtz, is nonetheless lost if one does not or need not know some of these quantities since these do not occur in the expression for kinetic energy. This happens when variables are eliminated, for example, when rigid connections exist between point systems.[55] Helmholtz thus treated the problem as if certain distances between points were invariant or as if no forces worked towards their change. This allowed him to set up conditional equations between the coordinates which served to eliminate some of them.

A still far more important case was the following. Helmholtz found that for cyclical systems the kinetic energy does not remain a quadratic function of $\dot{q}$ if there does not exist forces which act on the cyclic coordinates. Thus, according to equation (9),

$$\mathbf{P_b} = \dot{\mathbf{p}}_b = 0. \tag{23}$$

---

54. Koenigsberger, *Helmholtz's Untersuchungen*, 44–6.
55. *Vorlesungen* 6:353–54.

This equation need not hold for *all* cyclical coordinates, only for *one or more* of them. For the special case of a monocyclic system, there exists only one relation of the type (23). From (23) Helmholtz concluded that the cyclical momentum is a time-independent constant c. He then considered a system described by a finite number of coordinates q and velocities q̇ which are not subject to any conditions. The velocities q̇ thus need not be vanishingly small, as Helmholtz had required for the q̇$_a$; the q̇ thus contribute to the kinetic energy. The general Lagrangian equations (5) are then valid. Moreover, Helmholtz assumed that *one* cyclical coordinate was available to satisfy condition (23). He showed that if the cyclic momentum is a constant, the cyclic velocity can be expressed by the remaining coordinates q and velocities q̇. In this way he could eliminate the cyclical velocity q̇$_b$ in the expression for the kinetic energy T. In addition to the quadratic velocity terms q̇^{2}, there then also appears in T the linear terms q̇. Helmholtz explained that in this case the Lagrangian L also contains linear velocity terms, since the kinetic energy enters into L. To indicate the Lagrangian altered by the elimination, Helmholtz introduced a special symbol, here $\mathscr{L}$. He emphasized that $\mathscr{L}$ has the same value as L; only its mathematical form is different. He then used the Lagrangian equations to prove that

$$\mathbf{P} = -\frac{\partial}{\partial \mathbf{q}}(\mathscr{L} - c\dot{q}_b) + \frac{d}{dt}\frac{\partial}{\partial \dot{\mathbf{q}}}(\mathscr{L} - c\dot{q}_b), \qquad (24)$$

where there exist as many equations of motion for the system as coordinates q are available.[56]

Helmholtz explained the meaning of (24) as follows: In place of L in the Lagrangian equations (5), the function $\mathscr{L} - c\dot{q}_b$ appears in (24), which, concerning $\mathscr{L}$, contains linear q̇ as well as quadratic terms. Moreover, he pointed out that the q̇$_b$ also depended linearly on the remaining velocities q̇. To make a series of changes of state run backwards, the directions of the velocities q̇ are reversed—mathematically stated, their signs are reversed. In this case the quadratic terms do not change $(+\dot{q})^2 = (-\dot{q})^2$, although the linear q̇ terms do. Hence the reverse motion gives another function $\mathscr{L} - c\dot{q}_b$, and with it also other differential equations (24). Helmholtz thus found that the process cannot run in the same way backwards as it does forwards; the process

---

56. Ibid., 355–56. Equation (24) follows from Helmholtz's calculations on p. 356, if one substitutes L for H and, further, instead of relation 17 used there the Lagrangian equations (5), expressed in standard formulation, are used. On Helmholtz's use of mathematical symbols see also the explanation in note 10.

is irreversible. According to Helmholtz, the movement could only be reversed, if, in addition to the sign changes of $\dot{q}$, the signs of the eliminated $\dot{q}_b$ are also changed.[57] This key point shows that Helmholtz in no way doubted the fundamental reversibility of mechanical processes as formulated by Loschmidt. Irreversibility was rather based on the impossibility of our directly influencing the individual particles. The cyclic velocity $\dot{q}_b$ encompasses the molecular motion which we cannot directly influence; hence we cannot reverse its direction. Thus we again meet thesis (iii) on hidden motion, which in Helmholtz's interpretation of irreversibility takes on a special meaning.

The significance of the idea of a hidden motion not directly subject to influence becomes even more clear in light of the so-called Maxwell demon. This creature, conceived by Maxwell in 1867, possesses special powers.[58] Maxwell imagined a gas-filled container divided by a partition wall into two chambers, **A** and **B**. The gas in **A** is at a higher temperature than that in **B**; from the molecular point of view this means that the gas molecules in **A** possess a greater mean kinetic energy than those in **B**. Maxwell employed his velocity distribution function, according to which all possible particle velocities occur, and imagined a creature ("a finite being") that could recognize all the molecular paths and velocities but that could do nothing other than open and close a small hole in the partition wall. Now the demon is instructed to open the hole to allow a molecule to exit from **A** and to pass into **B** only if its velocity corresponds to a kinetic energy smaller than the mean kinetic energy of the molecules in **B**. Conversely, the demon can allow a molecule from **B** to enter **A** only if its kinetic energy is greater than the mean kinetic energy of the molecules in **A**. Finally, the demon is supposed to alternately execute these two processes so that the number of particles in each chamber remains unchanged. Maxwell realized that the result of this process was to increase the mean kinetic energy, and thus the temperature, of the molecules in **A** and to decrease that in **B**. Hence, the warmer system **A** becomes warmer and the cooler system **B** cooler. The entire process reverses that described by (22); in contradiction to the second law, this process leads to a decrease in entropy.

From the very beginning Helmholtz had conceived his interpretation of thermodynamics such that there was no place for a creature like Maxwell's demon. According to thesis (iii), where the thermal motion represents a *hidden* motion, no demon capable of recognizing the molecular paths and velocities is imaginable. Helmholtz's theory

57. Ibid., 357–58.
58. The presentation given here follows Klein, "Maxwell, His Demon," 85–6. For the original see Cargill Gilston Knott, *Life and Scientific Work of Peter Guthrie Tait* ... (Cambridge: Cambridge University Press, 1911), 213–14.

excluded the very possibility of a demon ordering molecular events so as to obviate the second law.[59]

Boltzmann was not Helmholtz's only critic. In 1889, Henri Poincaré responded to Helmholtz's presentation of 1886 on the meaning of the linear $\dot{q}$ terms with a four-page yet penetrating paper analyzing the issue of whether the second law could be deduced on a purely mechanical basis. He argued that, on the basis of mechanics alone, no function S having the unidirectional properties of entropy increase expressed, for example, by equation (22) could be established. He thus held Helmholtz's analysis as unfit for understanding irreversible processes.[60] Yet Helmholtz never claimed to have found a *proof* of the theorem of entropy increase; indeed, on the basis of his model, such a proof was impossible. The theorem of entropy increase holds for an (adiabatically) isolated system, and he formed the mechanical analogy to such a system through the condition (23): the absence of a force P_b implied that the system can neither take up nor give off cyclical work dQ_b. However, in contrast to the thermodynamic case, the value of the monocyclic entropy S_b shows no increase. For S_b is given by the logarithm of the cyclic momentum, and according to equation (23) this equals a time-independent constant c:

$$S_b = 2\log c = \text{constant}. \qquad (25)$$

That is, the monocyclic entropy maintains a constant value. Thus, while Poincaré's statement that the theorem of entropy increase cannot be proved mechanically was quite correct, it could not be applied against Helmholtz's argument since Helmholtz never claimed to have proven such a theorem.

Nearly two decades later, in 1906, Poincaré's paper occasioned Paul Ehrenfest to reanalyze the problem of irreversibility. Ehrenfest claimed that Helmholtz sought neither to give a mechanical analogy nor even to explain irreversibility, a point which, he said, Poincaré had misunderstood. Ehrenfest stressed that, according to Helmholtz, a hidden motion can be reversed, if the sign of $\dot{q}_b$ is also reversed, as discussed above.[61] Yet Ehrenfest's view was only partially correct: it was right insofar as Helmholtz did not try to prove the theorem of entropy

---

59. Klein, "Mechanical Explanation," 81 (n40).
60. Henri Poincaré, "Sur les tentatives d'explication mécanique des principes de la thermodynamique," *Comptes rendus de l'Académie des sciences* 108 (1889):550–53. Poincaré referred to Helmholtz's essay "Über die physikalische Bedeutung des Princips der kleinsten Wirkung," *JfruaM 100*:146–47 (esp.), in *WA* 3:214–15.
61. Paul Ehrenfest, "Bemerkungen zur Abhandlung des Hrn. Reisner: 'Anwendung der Statik und Dynamik monozyklischer Systeme auf die Elastizitätstheorie,' " *AP 19* (1906):210–14, esp. 214.

increase but wrong insofar as Helmholtz did—of course not in its entire generality—treat the problem of irreversibility, a problem that Helmholtz sought to illuminate solely through analysis of monocyclic systems.

If Helmholtz gave no proof of the theorem of entropy increase, then which aspects of irreversibility could he clarify? His thesis (iii) concerning the hidden, stationary motion, not directly subject to influence, shows that the entropy increase in an isolated system constitutes a natural law which cannot be affected by human means, even though, according to the laws of mechanics, every process is reversible. This insight was the kernel of Helmholtz's contribution to understanding the problem of irreversibility.[62] Still, the monocycle analogies could only give an *incomplete* image of the laws of thermodynamics. Moreover, the Helmholtzian model provided no structured arrangement of matter; it did not, for example, state that molecules are built from atoms. Hence, Helmholtz could not, for example, give values of the specific heats of matter. Such values could only be given by the kinetic theory of gases or, subsequently, statistical mechanics. Notwithstanding his influence on Boltzmann, who developed statistical mechanics to prove the theorem of entropy increase,[63] Helmholtz's influence on the development of statistical mechanics was rather marginal.

5. Epilogue: Helmholtz's Mechanical Heritage

If, to take the long historical view, Helmholtz had inherited his general research program of a mechanical interpretation of nature from Newton and the other mechanical philosophers of the Scientific Revolution along with many general (mechanical) laws from Jean d'Alembert, Joseph Louis Lagrange, Pierre Louis Maupertuis, and the other leading rational mechanists of the eighteenth century, then he also passed on that heritage to the generation of students who studied physics during the last third of the nineteenth century, not least to his own students Heinrich Hertz and Max Planck. Helmholtz had confronted the issue of how mechanics itself might be most simply constructed so that every possible movement could be understood analytically as due to a system of laws that had no recourse to experience yet which disallowed any movement excluded from the viewpoint of human experience.[64] It was an issue which both Hertz and Planck took with the utmost seriousness.

62. *Vorlesungen* 6:358 (n1).
63. Boltzmann, "Weitere Studien über das Wärmegleichgewicht."
64. Koenigsberger, *Helmholtz's Untersuchungen*, 42.

At the center of Helmholtz's mechanical investigations had stood the concept of action and, with it, the principle of least action. In 1893, Helmholtz spoke to the mathematician Felix Klein about the origin of his mechanical view of physics. The two men spent a good deal of time together during the course of a voyage across the Atlantic in order to participate in the World Exhibition at Chicago. Helmholtz told Klein that the transference of his mechanical ideas and viewpoint had begun with his work of 1847 on energy conservation and that it had been completely natural for him to extend his work to other physical processes, for example, the generalizations of mechanics, including Hamilton's principle of least action.[65] As Helmholtz's point suggests, mechanical principles constituted overarching principles of physical research for him. Nonetheless, it was only in the final decade of his life that he pursued the complete, systematic working out of his general mechanical point of view in physics.

Both Hertz and Planck welcomed Helmholtz's mechanical ideas and sought to continue the traditional program of a mechanical view of nature. Hertz, for one, considered it an advantage that the Newtonian concept of force had now lost its dominating role in physics and that, in its place, the central position of the concept of energy was now stressed. Yet Hertz also recognized some problems confronting this new viewpoint. In particular, he found that Hamilton's principle expressed an extremely complicated reality, one conceivable only mathematically. For Hertz, the complication lay in the fact that, according to Hamilton's principle, both a system's past (initial position) and its future (final position) determine its motion. The principle seemed to express a peculiar teleological tendency for natural processes. Hamilton's principle certainly did not correspond to Hertz's ideal of the simplest law possible, by means of which mechanics should be developed. Hertz thus constructed his own analytic mechanics on the basis of the following law. One can imagine that at a certain point in time, all connections between individual particles are dissolved; it followed that they will disperse themselves over space rectilinearly and with constant velocities. Since, however, this cannot really happen, the mass points remain as close as possible to that idealized motion (Hertz's principle). Helmholtz, for his part, conceded a great heuristic value to the picture outlined by Hertz: he thought that it might be useful for discovering the general character of natural forces.[66] As for

65. Klein, *Vorlesungen*, 226.
66. Koenigsberger, *Helmholtz's Untersuchungen*, 50. See also Hertz's *Gesammelte Werke*, ed. Philipp Lenard, 3 vols. (Leipzig: Johann Ambrosius Barth, 1894–95), vol. 3: *Die Prinzipien der Mechanik in neuem Zusammenhange dargestellt* (1894), 162–70. For the reasons why Hertz favored his "Grundgesetz" over Hamilton's principle, see ibid., 27–8. In his Foreword to Hertz's *Mechanik*, Helmholtz wrote in 1894, during his

Planck, he too was inspired by Helmholtz's insistence on the centrality of the principle of least action for reconstructing the mechanical foundation of physics. He used the principle as one of the fundamental pillars in his various attempts to establish a world picture.[67]

What Planck and others so valued in Helmholtz's physics was his goal of achieving a *unified view of all natural forces* along with the means, such as the principle of least action, of reaching that goal. In an appreciation of Helmholtz's accomplishments in physics, Planck found it difficult simply to list all of the fields in which Helmholtz had made contributions—Planck thought that crystal physics alone was the only branch of physics in which Helmholtz had published no study.[68] As Planck knew, Helmholtz's ability to work in virtually all branches of physics lay in his use of general principles aimed at illuminating those specific branches.

Felix Klein pointed out that as a scientist Helmholtz operated neither like Charles Darwin, who sought to bring living nature under order, nor like Michael Faraday, who discovered new physical phenomena, nor was he a mathematician who pursued mathematics for its own sake. Klein believed that all approaches and results interested Helmholtz only insofar as they illuminated the larger context of nature. For Klein, Helmholtz's strength as a physicist lay less in creative phantasy than in his ability to think conceptually.[69] This characteristic of Helmholtz appeared in especially marked measure during the last decade of his life as he sought to use the *generalized* theorems of energy conservation and the principle of least action to achieve a total physical world picture. Even as he failed to achieve that goal, he inspired others to make it their own.

---

final days, that, in attempts to give mechanical explanations of natural events, he himself always felt most secure with the differential equations derived from Hamilton's principle (thesis ii). At the same time, he raised no principled objections that other approaches could also lead to a mechanical understanding of nature. (See Helmholtz's "Vorwort," ibid., xiii–xxviii, esp. xxvii–xxviii.) Although Koenigsberger did not especially emphasize the point, this "Vorwort" served him as an important source in his analysis, as his formulations, which are in part taken literally from Helmholtz, prove.

67. Max Planck, "Das Prinzip der kleinsten Wirkung," in his *Physikalische Abhandlungen und Vorträge*, 3 vols. (Braunschweig: Vieweg, 1958), *3*:91–101.

68. Max Planck, "Helmholtz's Leistungen auf dem Gebiete der theoretischen Physik," *Allgemeine Deutsche Biographie 51* (1906):470–72, esp. 470.

69. Klein, *Vorlesungen*, 224.

Part Three

Philosopher

12

Force, Law, and Experiment

The Evolution of Helmholtz's
Philosophy of Science

Michael Heidelberger

1. Introduction

In Germany and beyond, Hermann von Helmholtz became a leading figure in shaping philosophy of science during the second half of the nineteenth century. In the mid-1840s, he helped establish the program of physiological reductionism. In the late 1860s, as his views further evolved, he came to share much common ground with contemporary antimetaphysical currents.[1] Scholars have long recognized

---

Acknowledgment: I am grateful to David Cahan for helping me translate this article into English.

1. Paul F. Cranefield, "The Organic Physics of 1847 and the Biophysics of Today," *JHMAS* 12 (1957):407–23; Everett Mendelsohn, "The Biological Sciences in the Nineteenth Century: Some Problems and Sources," *History of Science* 3 (1964):39–59, esp. 44–8; Paul F. Cranefield, "The Philosophical and Cultural Interests of the Biophysics Movement of 1847," *JHMAS* 21 (1966):1–7; Timothy Lenoir, *The Strategy of Life: Teleology and Mechanics in Nineteenth Century German Biology* (Dordrecht: Reidel, 1982), chap. 5: "Worlds in Collision"; idem, "Social Interests and the Organic Physics of 1847," in *Science in Reflection*, ed. Edna Ullmann-Margalit (Dordrecht: Kluwer, 1988), 169–91, esp. 171; William Coleman, *Biology in the Nineteenth Century: Problems of Form, Function, and Transformation* (New York: Wiley, 1971), 150–54; Frederick Gregory, *Scientific Materialism in Nineteenth Century Germany* (Dordrecht: Reidel, 1977), 145–88; Yehuda Elkana, *The Discovery of the Conservation of Energy* (Cambridge, Mass.: Harvard University Press, 1974), chaps. 4 and 7, and p. 129; Herbert Hörz and Siegfried Wollgast, "Einleitung," to *Hermann von Helmholtz. Philosophische Vorträge und Aufsätze* (Berlin: Akademie-Verlag, 1971), v–lxxix; Hörz and Wollgast, "Hermann von Helmholtz und Emil du Bois-Reymond. Wissenschaftsgeschichtliche Einordnung in die naturwissenschaftlichen und philosophischen Bewegungen ihrer Zeit," in Kirsten, 11–64. See also David H. Galaty, "The Philosophical Basis of Mid-Nineteenth Century

that Helmholtz's philosophical views had strong connections to Immanuel Kant's epistemology and that his views contributed much to the rise of neo-Kantianism. For most scholars, the significance of Helmholtz's own contributions to epistemology and philosophy of science lay precisely in his empirical criticism of Kant's a priori presuppositions.[2]

Yet as this essay seeks to show, contrary to the accepted scholarly view it was neither Kant alone nor his epistemology that decisively influenced Helmholtz's philosophy of science. Instead, Helmholtz's philosophy and methodology of science were far more influenced by Kant's metaphysics of nature, by the idealism of Johann Gottlieb Fichte, and by Michael Faraday's views on the nature of force and matter. To appreciate fully Helmholtz's philosophical views, above all his well-known empiricist outlook, also requires an understanding of the influence of Fichte's philosophy on Helmholtz,[3] as well as that of Faraday and other British physicists.

Section 2 of this essay investigates the precise influence of Kant's metaphysics of nature on Helmholtz. Through Kant's influence, Helmholtz came to advocate a metaphysical realism with respect to the ontology of scientific theories. According to this view, science deals ultimately with a reality that is inaccessible to our senses. At the same time, Helmholtz complemented his realism with a metaphysical epistemology. He believed that the fundamental scientific concepts of force and matter can also be justified metaphysically. As Section 3

German Reductionism," *JHMAS* 29 (1974):295–316; and Owsei Temkin, "Materialism in French and German Physiology in the early Nineteenth Century," *Bulletin of the History of Medicine* 20 (1946):322–27.

2. Hörz and Wollgast, "Einleitung," xxvii–xxix, xxxiv–xxxv.

3. The influence of Kant's *Metaphysische Anfangsgründe der Naturwissenschaft* on Helmholtz is stressed by Peter M. Heimann, "Helmholtz and Kant: The Metaphysical Foundations of 'Über die Erhaltung der Kraft'," *SHPS* 5 (1974):205–38; Peter M. Harman, *Metaphysics and Natural Philosophy: The Problem of Substance in Elastical Physics* (Brighton: Harvester, 1982), chap. 6; and idem, "The Foundations of Mechanical Physics in the Nineteenth Century: Helmholtz and Maxwell," in *Zum Wandel des Naturverständnisses*, eds. Clemens Burrichter, et al. (Paderborn: Schöningh, 1987), 35–46. R. Steven Turner is apparently the only scholar to have discussed Helmholtz's intellectual debt to Fichte; see his "Hermann von Helmholtz and the Empiricist Vision," *JHBS 13* (1977):48–58, on 56–7. Charles A. Culotta sought to identify a romantic inclination even in German biophysics; his argument, however, seems too general and unconvincing. See his "German Biophysics, Objective Knowledge, and Romanticism," *HSPS 4* (1974):3–38. Helmholtz's place in contemporary philosophy is discussed by Klaus Christian Köhnke, *Entstehung und Aufstieg des Neukantianismus* (Frankfurt: Suhrkamp, 1986), 151–57; and by Herbert Schnädelbach, *Philosophie in Deutschland 1831-1933* (Frankfurt: Suhrkamp, 1983), 108–14, 132–33. For the general reception of Fichte in nineteenth-century philosophy see Köhnke, *Entstehung und Aufstieg*, 179–211.

shows, under the influence of Michael Faraday and other British physicists, during the late 1860s Helmholtz developed an empiricist epistemology which intermingled metaphysical realism with positivist elements without, however, fundamentally disturbing his realist position. As a result, the concept of force for Helmholtz was no longer a metaphysically real abstraction but rather the manifestation of stable, lawful relations between phenomena.

Notwithstanding the influence of Kant and Faraday, Section 4 of this essay seeks to show that the inner core of Helmholtz's philosophy of science had its roots in Fichte's philosophy. In particular, Fichte's philosophy of action provided Helmholtz with the appropriate means to develop his scientific methodology, a methodology perhaps best labelled as "experimental interactionism." From Fichte Helmholtz appropriated the view that our consciousness comes to shape its conception of the outer world through the limitations we experience in our practical actions. Only by actively interfering with the world of external objects can we interpret our sensations as due to external causes and thereby distinguish them from the free acts of thinking inside our consciousness. The originality of Helmholtz's epistemology thus lay in the role he assigned to experiment in the process of constructing reality.

As a consequence, Helmholtz, as Section 5 of this essay argues, was far more indebted to German idealism than is commonly supposed. To be sure, his philosophical views were the antithesis of those of philosophers like Schelling and Hegel. Nonetheless, faithfully following his Fichtean heritage Helmholtz sought to distinctly separate mind from matter. Viewed from this angle, Helmholtz's views had little to do with materialism and much with idealism.

2. Metaphysical Realism I: Matter and Force as Necessary Forms of the Concepts of Science

In the Introduction ("Einleitung") to his essay *Ueber die Erhaltung der Kraft* (1847), Helmholtz presented the first public statement of his metaphysical realism.[4] He had originally written a philosophical preface to his essay; however, he decided to rid it of anything "that smelled

---

4. For Helmholtz's derivation of the energy conservation principle here see Fabio Bevilacqua's essay "Helmholtz's *Ueber die Erhaltung der Kraft*: The Emergence of a Theoretical Physicist," in this volume; Max Planck, *Das Prinzip der Erhaltung der Energie*, 3rd ed. (Leipzig: Teubner, 1913), 38–56; Heimann, "Helmholtz and Kant";

of philosophy," as he said. Yet in the end he followed the advice of his good friend Emil du Bois-Reymond and indeed included the philosophical Introduction.[5] It provides much insight into Helmholtz's early philosophical position.

The Introduction's fundamental idea is that natural phenomena are the result of the activity of hidden causes. Helmholtz identified these causes as the attractive and repulsive forces emanating from matter. He sought to prove that these forces are central forces depending only on the distance of the material objects involved, not on their motion. According to this schema, nature is a material system whose changes are due to inner, conservative forces.[6]

The aim of science, therefore, is to grasp the unobservable activity of matter behind the observable phenomena. "The phenomena of nature," Helmholtz declared, "are to be reduced to movements of bits of matter with unalterable moving forces that depend only on their spatial relations."[7] A few years later, he restated this view more precisely: "A complete explanation of a natural phenomenon requires its reduction to the ultimate natural forces that supply its foundation and that are active in it."[8] He believed that neither matter nor (constant) force were mere concepts or fictions "which do not correspond to anything real."[9] On the contrary, forces really exist and stand behind, so to speak, the phenomena. In short, Helmholtz was a *realist* in relation to force and matter.

---

and Harman, "Helmholtz and Maxwell." On other contemporary attempts to derive the energy principle see Peter M. Heimann, "Conversion of Forces and the Conservation of Energy," *Centaurus 18* (1973):147–61; Thomas S. Kuhn, "Energy Conservation as an Example of Simultaneous Discovery," in *Critical Problems in the History of Science*, ed. Marshall Clagett (Madison, Wis.: University of Wisconsin Press, 1959), 321–56, reprinted in Kuhn's *The Essential Tension: Selected Studies in Scientific Tradition and Change* (Chicago and London: The University of Chicago Press, 1977), 66–104; and David Gooding, "Metaphysics versus Measurement: The Conversion and Conservation of Force in Faraday's Physics," *Annals of Science 37* (1980):1–29.

5. Koenigsberger *1*:68, 72; see also Helmholtz to du Bois-Reymond, 12 February, 10 April, and 21 July 1847, and du Bois-Reymond to Helmholtz, 4 and 6 August 1847, in Kirsten, 78–9, 79–81, 81–2, 82–3, 84, resp.

6. On Helmholtz's concept of central forces see M. Norton Wise, "German Concepts of Force, Energy, and the Electromagnetic Ether: 1845–1880," in *Conceptions of Ether: Studies in the History of Ether Theories, 1740–1900*, eds. G.N. Cantor and M.J.S. Hodge (Cambridge: Cambridge University Press, 1981), 269–307, on 295–301. On nineteenth-century theories of matter in general see the still valuable account by Friedrich Albert Lange, *Geschichte des Materialismus und Kritik seiner Bedeutung in der Gegenwart*, 2nd ed., 2 vols. (Iserlohn: Baedeker, 1875; reprinted Frankfurt: Suhrkamp, 1974), 628–65; on Helmholtz's theory of matter see ibid., esp. 655–56, 661–63.

7. *Ueber die Erhaltung der Kraft. Eine physikalische Abhandlung* (Berlin: G. Reimer, 1847), in *WA 1*:12–68, on 15.

8. "Ueber Goethe's naturwissenschaftliche Arbeiten," [1853], VR^5 *1*:25–45, on 40.

9. "Erhaltung," *WA 1*:14.

However, he also ascribed a *metaphysical* status to matter and force: neither matter nor its forces appear, that is, they are not given in experience. He argued that we can only know them indirectly by inferring their nature from their observable effects:

> Since we can never perceive the forces per se but only their effects, we have to leave the realm of the senses in every explanation of natural phenomena and [instead] turn to unobservable objects that are determined only by concepts.[10]

Helmholtz contrasted his position with Goethe's notion of scientific explanation, which he ascribed to the *Naturphilosophie* of Schelling and Hegel.[11] In Goethe's understanding, science dealt exclusively with phenomena and their mutual interrelations. That understanding, Helmholtz explained,

> demands for the investigation of physical affairs an arrangement of the observations in such a way that one fact always explains the other one. This procedure is supposed to give insight into the interrelations without leaving the realm of sense perception. This demand is highly attractive, yet intrinsically and radically false.[12]

He believed that the physicist cannot be content with direct sensory perceptions in the same way that the poet can. He has to go beyond,

> into a world of invisible atoms, [where there are] movements and attractive and repulsive forces that intricately operate on each other in a lawlike yet scarcely comprehensible confusion. To the physicist, the sensory impression is not an irrevocable authority.[13]

Nonetheless, Helmholtz did not undervalue or denigrate efforts to establish lawful connections between observable phenomena. Such efforts, he maintained, can lead to a comprehensive formulation of the observable processes in terms of "general rules" and "general generic concepts" that form the "experimental part" of the sciences.[14] It would,

---

10. "Goethe," *VR^5* 1:40.
11. Ibid., 35. Cf. "Ueber die Natur der menschlichen Sinnesempfindungen," *Königsberger naturwissenschaftliche Unterhaltungen 3* (1852):1–20, in *WA* 2:591–609, on 601; and *Ueber das Sehen des Menschen* (Leipzig: L. Voss, 1855), in *VR^5* 1:85–117, on 99. On the relation of Helmholtz to Goethe see Jeffrey Barnouw, "Goethe and Helmholtz: Science and Sensation," in *Goethe and the Sciences: A Reappraisal*, eds. F. Amrine, F.J. Zucker and H. Wheeler (Dordrecht: Reidel, 1987), 45–82.
12. "Goethe," *VR^5* 1:40.
13. Ibid., 40, 44.
14. "Erhaltung," *WA* 1:13.

however, be fallacious to believe that one had explained the phenomena or recognized the "spiritual content" of nature after the (mere) discovery of those rules and concepts. A true understanding can only be gained in the "theoretical part" of science, he claimed, when the "unknown causes of the processes have been found from their visible effects."[15] The experimental part thus deals with provisional descriptions of circumstances that are to be represented in the theory as the effects of fundamental, yet hidden causes. In subdividing the sciences into experimental and theoretical parts, Helmholtz followed Kant, who distinguished between an "empirical rule" as a subjective connection of perceptions and an objective "law" that comprised the necessity and the universal validity of the connection.[16]

In attempting to justify the central force principle Helmholtz tried to follow Kant's arguments as set out in the *Metaphysische Anfangsgründe der Naturwissenschaft*.[17] In the preface to that work, Kant declared that only such knowledge properly merits the title of a true science when it "treats its object entirely according to a priori principles."[18] Kant argued that science, as a "doctrine of material objects [*Körperlehre*]," has to begin with a "complete analysis of the concept of matter in general" before it can conduct its proper business.[19] Such an analysis cannot refer to any empirical knowledge but has instead to attend to the content of our concept of matter as "separated" from "special experiences" and as related to "the pure intuitions in space and time." Physics in particular could only treat its objects mathematically after it had first obtained the "principles of the *construction* of concepts that belong to the possibility of matter in general."[20] These "principles of the necessity of what belongs to the *existence* of an object" are to be gained in a "metaphysics of nature."[21]

---

15. Ibid.
16. Immanuel Kant, *Prolegomena zu einer künftigen Metaphysik, die als Wissenschaft wird auftreten können* (Riga: Hartknoch, 1783), § 29.
17. See Heimann, "Helmholtz and Kant," esp. 220–32, for a detailed comparison of Kant's and Helmholtz's conceptions; see also Harman, "Helmholtz and Maxwell," 40–1, and Harman, *Metaphysics*, 118–26. On Kant's *Metaphysische Anfangsgründe der Naturwissenschaft* see Peter Plaas, *Kants Theorie der Naturwissenschaft* (Göttingen: Vandenhoeck, 1965); Hansgeorg Hoppe, *Kants Theorie der Physik* (Frankfurt: Klostermann, 1969); Lothar Schäfer, *Kants Metaphysik der Natur* (Berlin: de Gruyter, 1966); and, most recently, Brigitte Falkenburg, *Die Form der Materie. Zur Metaphysik der Natur bei Kant und Hegel* (Frankfurt: Athenäum, 1987), Teil I. The reception of Kant's theory of matter by his contemporaries is treated by Martin Carrier, "Kants Theorie der Materie und ihre Wirkung auf die zeitgenössische Chemie," *Kantstudien 81* (1990):170–210.
18. Immanuel Kant, *Metaphysische Anfangsgründe der Naturwissenschaft* (Riga: Hartknoch, 1786), A v.
19. Ibid., A xii.
20. Ibid.
21. Ibid., A viii.

Kant further argued that we cannot and should not expect of this metaphysics that it will in any way extend our empirical knowledge of corporeal objects as they appear to us. To lay claim to such knowledge would be to invite illusion and fictitious metaphysics. The existence of matter can only be determined empirically and never in an a priori manner. Yet Kant also argued that a metaphysical construction can alone show how matter can be a possible object of experience in reality. The specific attributes of matter thereby brought to light can claim to be universally valid and necessary. Thus, Kant's metaphysics did not pretend to transcend experience. Instead, it sought to elucidate the attributes belonging to matter prior to experience that make it possible for us to experience matter at all. In short, he aimed to clarify our concept of sensible matter.

In following Kant's view, Helmholtz began with the question of how the "step into the conceptual realm" can be realized if we "want to ascend to the causes of natural phenomena" in physics.[22] In the Introduction to *Ueber die Erhaltung der Kraft*, he claimed that in considering "objects of the external world" science has to abstract its concepts from experience in a twofold way. The first abstraction concerned objects "in their mere presence, apart from their effects on other objects or on our sense organs."[23] It leads to the concept of matter. For Kant, this concept signifies "what is movable, in separation from everything else that exists in space besides it."[24] Helmholtz discerned two attributes in this "separated concept" of matter: "its spatial distribution and its quantity (mass) that is posited as eternally unalterable."[25] Matter here is *posited*, not experienced, that is, it is conceptually determined by our understanding prior to any experience, thereby enabling experience. In itself, matter is everywhere similarly constituted. The only change that it can undergo is spatial change. However, Helmholtz claimed that there are major differences in the way matter is endowed with force:

> Matter as such does not exhibit qualitative differences. If we talk of different matters we posit their diversity as resulting from the diversity of their effects, i.e., their forces. The only change that matter as such can therefore undergo is a spatial one, i.e., motion.[26]

Here, too, Helmholtz again made it plain that one has to deal with the metaphysical attributes of corporeal matter in the Kantian sense:

---

22. "Goethe," VR^5 1:40.
23. "Erhaltung," WA 1:14.
24. Kant, *Metaphysische Anfangsgründe*, A 42, Erklärung 5.
25. "Erhaltung," WA 1:14.
26. Ibid.

human understanding posits the concept of matter prior to all experience in such a way that qualitative differences of experienced matter can only be the effect of different forces.

Insofar as Helmholtz followed the transcendental disposition of Kant's view, he succumbed to the temptation to transcend the Kantian limits and to conceive of the "realm of concepts" as a domain of real, yet experientially inaccessible objects. Kant called his procedure metaphysical only because it dealt with the question of under what attributes matter has to be conceived of prior to any experience so as to become an object of experience and, secondly, how these attributes can be justified. For Kant, the metaphysics of nature has to do only with the peculiar manner of this justification; for Helmholtz, it deals with a realm of real objects beyond all empirically accessible phenomena. Helmholtz seems to have confused the metaphysical determination of concepts with the determination of the thing-in-itself.

In his metaphysical preliminaries to *Ueber die Erhaltung der Kraft* Helmholtz concluded that spatial distribution and mass were the fundamental a priori attributes of matter. In order to make it conceivable that matter can actually be experienced—in order, as Helmholtz said, that we can "apply" the concept of matter "in reality"—we have, by a second process of abstraction, again to provide matter with an attribute "from which we wanted to abstract in the first place, namely, its ability to exert influences." Such an ability is what Helmholtz meant by "force." Therefore, for Helmholtz the concept of force is a metaphysical concept that does not stem from experience. On the contrary, it is formed in abstraction from sense experience in order to make experience possible. We generally infer the existence of a force from the effects which matter causes in our sense organs. The world of objects is only accessible to us because of these effects: "The objects of nature are, however, not without effects," he wrote. "We arrive at knowing them only through the influences they exert on our sense organs by inferring an operating agent from the effects."[27] Still, for every concrete, individual case it can only be discovered by empirical means if and how a force is active.

Helmholtz's view of force led to the consequence that all empirical qualities of objects are dispositional in nature: "All qualities of natural objects manifest themselves only when we put them into reciprocal action with other natural objects or with our sense organs."[28] In this, he followed the traditional doctrine of corpuscularism as found, for

27. Ibid.
28. *Handbuch*, 444.

example, in the writings of John Locke: sensible qualities of natural objects are caused in other objects and in our senses by the forces (powers) that an object has due to the special internal constitution of its primary qualities.[29]

Helmholtz's concept of matter dealt with the same attributes of matter which Kant determined as a priori. Yet Helmholtz altered Kant's view of their mutual relations with each other. In the *Metaphysische Anfangsgründe der Naturwissenschaft* Kant treated the theory of matter in four main parts which themselves derive from the four categories of the understanding: that movement is the "fundamental attribute" of matter, that matter fills a part of space, that it exerts a moving force, and that it is an object of our experience. Kant only briefly justified why "movement had to be the fundamental attribute of a something that is supposed to be the object of external senses"; for these senses can only be affected through movement. "The understanding, too, reduces to it all other predicates of matter that belong to its nature."[30] For Kant, the concept of movability contained precisely that combination of the determination of space and time that every object must have in order to become accessible to the external senses.

For Helmholtz, the fundamental attributes of matter at rest are not only (as already noted) its spatial distribution and mass, but also that it is endowed with forces. The latter can be inferred from the experience that there are changes in the world, that is, that matter can move. The fact that matter can influence other matter, that it has a certain force, explains why matter can be an object of experience and can change through motion. Although this sounds very much like Kant's conception of matter, the difference is that it makes the concept of force play the role that movability played for Kant. On the one hand, Helmholtz's conception made it impossible to reduce, as Kant did in his dynamics, the property of filling space to attractive and repulsive forces. Where Kant held matter to fill space by its moving force, Helmholtz argued that it was the other way around: only what fills space can exert force. On the other hand, Helmholtz revalued the concept of force, for force is now the cause and source of all changes, that is, of all movements and of the effects of matter on our senses. Helmholtz could not follow Kant in seeing force as the cause of matter. He did, however, accept that empirical knowledge in physics was only possible through understanding forces. Helmholtz's concept of matter, by contrast, seems

---

29. John Locke, *An Essay Concerning Human Understanding* (London: Basset, 1690), II, viii, 23.
30. Kant, *Metaphysische Anfangsgründe*, A xx.

curiously pale. The only positive definition he gives of matter is that it acts by means of its force.

Helmholtz sought to clarify the nature of force and the way it acts. He posited that any change in nature is caused by either a variable or a constant force. The fundamental forces "behind" the appearances have to be unchangeable over time, that is, they have to produce the same effects under the same circumstances. If they themselves could change, then other, more basic forces would have to be imagined to account for this change. Thus, in the final analysis Helmholtz thought of matter as consisting of parts which differ only in the indestructible and unchangeable forces they possess. These are the chemical elements.

If the fundamental forces are unchangeable, Helmholtz further argued, then any real change in the world is caused by movement. Any other alteration would presuppose a variable force. If pieces of matter possessing the same force have different effects, then a change can only depend on the distance between the agent and the place of its effect:

> If we consider . . . the universe as consisting of elements possessing ineradicable qualities [i.e., forces], the only possible changes in such a system are spatial, i.e., movements. The action of the forces can only be modified by external spatial circumstances. That means that the forces are nothing but moving forces whose action is determined by their spatial relations.[31]

Helmholtz defined motion of a body as a change in relation to another body, not in relation to "homogeneous empty space." Force as the cause of such changes can "only be inferred from the relation which at least two bodies have to one another."[32] Kant likewise held that the motion of a body can only be determined in relation to another body and that an absolute movement is therefore impossible.[33] Strictly speaking, Helmholtz argued that one cannot talk of a force between two masses but only "between all its parts in relation to each other." The real seats of forces are therefore not bodies at all, but rather points. "Mechanics," he wrote, "thus reduces to the forces of material points, i.e., of the points of space that are filled with matter."[34] Forces acting between points can alter only the distance between them and nothing

---

31. "Erhaltung," *WA* 1:15.
32. Ibid.; Wise, "German Concepts," 296, mistranslates "*erschlossen*" as "predicated" instead of "inferred," and therefore draws the erroneous conclusion that to Helmholtz "forces existed as relations."
33. Kant, *Metaphysische Anfangsgründe*, A 154; see also ibid. A 139–40, A 3, A 15. Kant's denial of absolute motion does not imply the denial of pure or absolute space; see, e.g., ibid., A 1; and Falkenburg, *Form der Materie*, 72–5.
34. "Erhaltung," *WA* 1:15.

else; they are attractive or repulsive according to whether two points move towards or recede from each other.

Thus Helmholtz's Introduction concluded that all physical phenomena are the result of motions of material bodies acting on each other through central forces. To understand nature means to "reduce natural phenomena to unalterable forces of attraction and repulsion whose intensity is dependent upon the distance."[35] As a corollary of his deduction of the central force principle, he further concluded that matter and force depend on one another and that it would be logical nonsense to think that they can exist independently of one another: If the existence of matter were separable from the existence of force, we would not be able to know anything of matter since it could not have any effect on other matter or on our senses. Similarly, the existence of force as separate from matter is equally inconceivable. In this case, force would be something "that, at the same time, should exist and yet should not exist since anything that exists we call matter."[36] Hence, all matter possesses force and all force has to have a material seat.

At the end of the Introduction to *Ueber die Erhaltung der Kraft* Helmholtz remarked that the task of theoretical science would be fulfilled

> if, at some time, the reduction of all phenomena to elementary forces were completed, and if, at the same time, this reduction were shown to be the only possible one that the phenomena permit. The necessary conceptual form for understanding nature would then be provided and objective truth would have to be ascribed to it.[37]

Although he did not say how far away the science of his day was from this goal, the spirit of his Introduction makes it highly probable that he considered it to be close at hand. In any case, his concluding words are, in effect, a confession that the necessity and objective truth of the necessary conceptual form have yet to be completely established.

Helmholtz's biographer Leo Koenigsberger has provided some of Helmholtz's draft philosophical notes which he wrote down several years prior to the publication of *Ueber die Erhaltung der Kraft*. Helmholtz ultimately included essential parts of these draft notes in his Introduction. In this text, he contended that science has to provide the concepts "from which the individual, distinct empirical phenomena can be deduced." Otherwise, science could only offer "an ordered

---

35. Ibid., 16.
36. Ibid., 14.
37. Ibid., 17.

survey of everything that is empirical," thereby restricting itself to a mere description of nature and to experimental physics.[38] The "general concepts of nature," he wrote, are "derived from the possibility of any knowledge of nature [*Naturanschauung*]" and are constituted "on the basis" of it. They constitute the "general and necessary form" which makes science conceivable in the first place. "The certainty" of the general or pure sciences thus attained is "absolute." We cannot, however, obtain from them an "empirical fact or law"; "they can only serve as a norm for our explanations."[39] From this point of view, Helmholtz's principle of the conservation of force appears as a metaphysical theorem of pure science in the Kantian sense of serving as a constraint on empirical explanations. This early text shows even more clearly that Helmholtz was led by metaphysical considerations and that Kant's *Metaphysische Anfangsgründe der Naturwissenschaft* was the model for his procedure.

Yet seen in the wider context of the scientific goals of his time, Helmholtz's object was a metaphysical justification of the Laplacian program in physics.[40] Laplace and his school sought to establish "that the phenomena of nature can be reduced in the final analysis to actions *ad distans* from molecule to molecule, and that the considerations of these actions must serve as the basis of the mathematical theory of these phenomena."[41] Helmholtz sought to show that the metaphysical concept of matter already necessarily contains the physics of point-atoms with attractive and repulsive forces depending solely on the distance and that such a physics is enough to explain *all* natural phenomena. It is therefore not surprising that physicists in Helmholtz's circle who were committed to an experimental, Baconian approach to nature as opposed to a theoretical program did not like Helmholtz's idea very much. In particular, Gustav Magnus and Johann Christian Poggendorff regarded *Ueber die Erhaltung der Kraft* as filled with "speculations" which, while stimulating, could lead, as Koenigsberger later

---

38. Koenigsberger 2:126–27.
39. Ibid.
40. Wise, "German Concepts," 296. On the Laplacian program see Robert Fox, "The Rise and Fall of Laplacian Physics," *HSPS* 4 (1974):89–136; Ivor Grattan-Guinness, "From Laplacian Physics to Mathematical Physics, 1805–1827," in Burrichter, et al., eds., *Zum Wandel des Naturverständnisses*, 11–34; and Ivor Grattan-Guinness, *Convolutions in French Mathematics, 1800–1840*, 3 vols. (Basel: Birkhäuser, 1990), vol. 1: *The Settings*, chap. 7.
41. Pierre-Simon Laplace, "Note" to "Sur les mouvements de la lumière dans les milieux diaphanes," *Mémoires de la classe des sciences mathématiques et physiques de l'Institut de France 10* (1809, published 1810):326–42, on 325; and in *Oeuvres complètes de Laplace*, ed, Académie des sciences, 14 vols. in 15 (Paris: Gauthier-Villars, 1878–1912; reprinted Hildesheim: Olms, 1966), *12*:286–98, as cited in Grattan-Guinness, *Convolutions 1*:440.

put it, to the revival of the "phantasies of Hegelian *Naturphilosophie*."[42] Significantly, only the mathematician Carl Gustav Jacobi seems to have grasped the relationship between Helmholtz's views and those of the great French masters of eighteenth-century rational mechanics.[43]

In consciously using Kant's *Metaphysische Anfangsgründe der Naturwissenschaft* as his model and in allotting to the concept of force such a prominent position, Helmholtz also seems to have revived the old, seemingly dead controversy between atomists and dynamicists. Where dynamicists followed Kant in deducing the "specific diversity of matter" from the "moving forces of attraction and repulsion,"[44] atomists looked to the movement of atoms in the void. Most physicists of Helmholtz's day were (by far) atomists. The few dynamicists left concentrated their efforts on an attempt to explain light, matter, electricity, and magnetism through interacting forces without assuming the existence of imponderable fluids, as the atomists did.[45]

Helmholtz, for his part, proved to be an atomist only insofar as he assumed that talk of forces is legitimate only if there are material points that possess forces.[46] He did, however, reveal some inclination towards dynamicism through his denial that matter without force could be experienced and so claim the complete reduction of all phenomena to attractive and repulsive forces. To reduce the various appearances of matter exclusively to forces is alien to classical atomism, which, traditionally, explained the different appearances of matter to different distributions of atoms in a given volume. For Helmholtz, in contrast to classical atomism, matter could not be experienced directly; rather, as noted above, it played the same role as the thing-in-itself in Kant's philosophy. We can know of matter only that it exists, moves, and constitutes the seat of force. In contrast to atomism, matter in Helmholtz's system serves merely as a dummy category, introduced for formal reasons and legitimized only through force.

3. Metaphysical Realism II: Forces as Substrates of the Laws of Nature

Despite a series of alternatives or potential objections to his metaphysical realism of 1847, Helmholtz long held to it. In 1855, William

---

42. Koenigsberger *1*:69–79, 85.
43. Ibid., 80; and "Erinnerungen," [1891], *VR⁵ 1*:1–21, on 11.
44. Kant, *Metaphysische Anfangsgründe*, A 101.
45. Walter Kaiser, "Zur Struktur wissenschaftlicher Kontroversen," (unpublished habilitation thesis, Department of Mathematics, University of Mainz, 1984), chap. 2.
46. "Erhaltung," *WA 1*:24.

Rankine proposed to reconstruct physics as a "science of energetics" that could do without the "hypothetical suppositions of atoms, forces etc. hitherto existing." Helmholtz made it absolutely clear that, as he wrote in a review of Rankine's proposal, he did "not share the basic philosophical views from which Mr. Rankine departs."[47] In 1862, he repeated that a true understanding of the complicated processes of nature requires "abstractions from the sensible appearance."[48] And in 1867, he again wrote that the concepts of matter and force "are only abstract ways of viewing the same natural objects in different respects. It is precisely for this reason, however, that neither matter nor force can be directly perceived. They can only be inferred causes of the facts experienced."[49] As before, he continued to maintain that the attributes of objects as we experience them are seen as effects exerted on us by forces issuing from unknown matter. And finally, in 1869, at the annual meeting of the Gesellschaft Deutscher Naturforscher und Ärzte at Innsbruck, he presented his now time-honored metaphysical realist position: if the only change of elementary substances is change of spatial position, then all changes in the world are caused by motion. "If motion, however, is the basic change underlying all the alterations in the world, then all the elementary forces are moving forces. The final goal of the sciences is thus to find all the movements and driving forces supplying the foundation of all other change. In other words, the final goal of the sciences is to reduce [everything] to mechanics."[50]

However, Helmholtz's Innsbruck speech also revealed the first signs of a significant and deep change in his view of force. Until that speech, force had played the fundamental role in his philosophy of science; now it was replaced by the concept of natural law. Forces, Helmholtz had argued up until now, constituted the true reality behind the appearances, and to grasp the appearances meant to find the unalterable moving forces behind them. According to his revised view, however, forces are nothing other than the epitome of lawful relations among the appearances. Forces, he now said, concerned the classifications of facts, namely, those facts whose regular connection is not a matter of convention. In notes appended to a reprinting in 1881 of his *Ueber die Erhaltung der Kraft* for the first volume of his *Wissenschaftliche*

---

47. "[Review of] W.J.M. Rankine, *Outline of the Science of Energetics*," *Fortschritte der Physik im Jahre 1855 11* (1858):365.
48. "Ueber das Verhältniss der Naturwissenschaften zur Gesammtheit der Wissenschaft," [1862], *VR⁵* 1:157–85, on 165.
49. *Handbuch*, 454.
50. "Ueber das Ziel und die Fortschritte der Naturwissenschaft," *VR⁵* 1:367–98, on 379.

Abhandlungen, he claimed that the causal principle, that is, the assumption of causally effective forces behind the appearances, is "indeed nothing other . . . than the supposition of the lawlikeness [*Gesetzlichkeit*] of all natural phenomena."[51] This reinterpretation denuded Helmholtz's concept of force of its previously metaphysical character. Previously, forces (and matter as their carrier) formed the ontological fundament, inaccessible to the senses. Now they amounted to nothing more than a tautological reshuffle of lawlike relations of directly experienced facts; they were dissolved into relations of perceptions.

Not all classifications of appearances, Helmholtz added, had the right to represent a concept of force. To be sure, in order to facilitate recall of our experiences we often subsume them under general rules which are nothing more than arbitrarily chosen instruments, for example, when we classify plants or animals under generic names relative to specific attributes. By contrast, true laws of nature have the property of presenting themselves to us as an "alien power" not at our disposal. "If we are sure of the conditions under which the law has to operate," he wrote,

> then we must experience the effect with a necessity compelling the objects of the external world as well as our perceptions. This must happen without arbitrariness, without choice, without our interference. Thus, law confronts us as an objective power and it is for this reason that we call it *force*.[52]

He explained that when we claim that a force is present, we mean that a certain lawful connection exists objectively in the world of the objects themselves, not subjectively in our consciousness. To say that a force operates on a body is not to say anything about the fundamental entities acting behind the appearances; rather, it is to say that under certain circumstances there is a certain acceleration of the body. To refer to the force of the refraction of light, for example, is to refer to the objective law of light refraction. Helmholtz called this the "factual sense" of our concept of force: "We can compare this factual sense with the facts and verify it. By inserting the abstract concept of force we only add that this law is not an arbitrary invention but rather a compelling law of the phenomena."[53]

---

51. "Zusätze (1881)," appended to the reprint of "Ueber die Erhaltung der Kraft," *WA* 1:68–75, on 68.
52. "Ziel und Fortschritte," *VR*^{5} *1*:375.
53. Ibid., 376.

After his Innsbruck speech of 1869, Helmholtz steadily intensified his tendency to instrumentally reinterpret part of his earlier realistic ontology. In the programmatic vision of 1847, to complete the sciences meant to reduce all phenomena to irreducible forces. Nearly a half-century later, by 1893, the concept of force had shrunk to mean "in a certain sense an empty abstraction." When we use it in formulating a law we are including, he now said, something "that can easily be regarded as a hypothetical element and that, strictly speaking, is not given by the facts."[54] "As a matter of fact, what we mean by force is nothing other—at least nothing other that still has a factual sense—than that which is contained in a mere description of the phenomena."[55] As far as his interpretation of the concept of force was concerned, Helmholtz now professed himself a positivist. His interpretation scarcely differed from similar proposals by Gustav Theodor Fechner in 1855, by Ernst Mach in 1872, and by Gustav Robert Kirchhoff in 1876.[56]

Helmholtz's revised view of force implied a far-reaching transformation in his view of the logical role played by the fundamental concepts of a scientific theory. In the *Ueber die Erhaltung der Kraft* the true object of the sciences had to be posited as a metaphysical necessity, abstracted from experience, in order that science could be founded on it and sense perception reduced to it. After 1869, by contrast, our knowledge of the true object of science becomes a distant goal that we can approach through investigating the empirical phenomena but that can never be completely grasped. Previously, the concept of force was the fundament on which science was to be erected; now it was a mere hypothesis that could, to be sure, be progressively corroborated but

---

54. *Einleitung zu den Vorlesungen über theoretische Physik*, eds. Arthur König and Carl Runge (Leipzig: Barth, 1903) (= Band I, Abtheilung 1 of the *Vorlesungen über theoretische Physik*, eds. Arthur König et al., 6 vols. [vol. 1-4 and 6: Leipzig: Barth, 1897-1907; and vol. 5: Hamburg and Leipzig: Voss]), 12.
55. Ibid., 11.
56. Gustav Theodor Fechner, *Ueber die physikalische und philosophische Atomenlehre* (Leipzig: Hermann Mendelssohn,1855), 90, 107, 113, 179-80; Ernst Mach, *Die Geschichte und die Wurzel des Satzes von der Erhaltung der Arbeit* (Prague: Calve, 1872; reprint of 2nd ed. [1909] Amsterdam: Bonset, 1969), 37, 45-6, 53-4; and Gustav Robert Kirchhoff, *Vorlesungen über mathematische Physik. Mechanik* (Leipzig: Teubner, 1876), 1. On Mach see Michael Heidelberger, "Fechner und Mach zum Atomismus in der Physik," in *Die geschichtliche Perspektive in den Disziplinen der Wissenschaftsforschung*, eds. Hans Poser and Clemens Burrichter (= *TUB-Dokumentation Kongresse und Tagungen*) (Berlin: Technische Universität, 1988), Heft 39, pp. 75-112. See also Gert König, "Der Wissenschaftsbegriff bei Helmholtz und Mach," *Beiträge zur Entwicklung der Wissenschaftstheorie im 19. Jahrhundert*, ed. Alwin Diemer (Meisenheim: Hain, 1968), 90-114 for instructive comparison of Helmholtz's concept of science with that of Mach and others.

that could never be exhaustively confirmed let alone proven necessary. Where originally the scientific object was a metaphysically necessary presupposition of science, now it became the hypothetical summation of lawful relations, a virtual limit to an inaccessibly distant ideal.

Along with his transformation of the concept of force, Helmholtz also developed his empirical criticism of the basic concepts of geometry. If the concept of force cannot be justified metaphysically, then the alleged metaphysical status of geometrical concepts was also questionable. The geometrical axioms, like the axioms of mechanics, are "indeed assertions ... that can be observationally tested and that could, if incorrect, eventually also be refuted.... Yet thereby all possibility of giving a metaphysical foundation to science is abolished, a metaphysical foundation that Kant deeply believed in."[57]

Although Helmholtz's revamped concept of force imparted a strong positivist flavor to his philosophy of science, he did not renounce his metaphysics of realism or abandon realism altogether. In fact, he retained many metaphysical and realistic elements from his earlier mechanistic view. Unlike Fechner and Mach, he did not identify or reduce the objects of science to the sum of their appearances. On the contrary, he continued to speak of "the hidden and immutable ground of the phenomena" that "lies behind the change of appearances and acts upon us";[58] the only difference was that now this ground is no longer grasped by a priori concepts of force and matter but is rather to be sought in the concept of natural law. He conceived the transitory "as the mode of appearance of the unalterable, the law."[59] In both his earlier and his later views, reality as the ground of our perceptions remained hidden from our senses and could only be guessed at or indirectly inferred. To be sure, Helmholtz did renounce any hope for a metaphysical founding of the basic concepts of science and he did come to favor an empirical procedure. Yet, as before, the ultimate reference of the newly founded concepts remains hidden from our sensuous experience. The laws of nature provide us with only a dim reflection, only a limited, partial aspect of reality. Helmholtz thus insistently maintained a metaphysical ontology of the sciences even as he rejected any metaphysical argument in favor of it.

The more precise, the more general, and the more complete the antecedent condition of a law is formulated, the closer we are to the

---

57. Koenigsberger *1*:141–42. On Helmholtz's empirical critique of geometry see Robert DiSalle's essay, "Helmholtz's Empiricist Philosophy of Mathematics: Between Laws of Perception and Laws of Nature," in this volume.

58. *Vorlesungen 1:1*:16; and "Die Thatsachen in der Wahrnehmung," [1878], *VR*⁵ *2*:213–47, 387–406, on 241.

59. Koenigsberger *2*:345.

realm of the true reality that causes the consequences which the law asserts. "That which behind the changing phenomena remains quantitatively invariable," he declared, we call "the actual" or "substance" or "the cause."[60] We can never perceive this substance per se, only "the lawlike," that is, the stable relations between the appearances. Only through laborious work can a faint presentiment of the real, of the "power confronting us," be developed out of the lawlike; yet it can never be grasped completely. No matter how successful our science may be, we can never be sure that we have reached the final and inalterable basis of the phenomena. "The real, or Kant's 'thing in itself' " cannot be represented in positive terms. We can only approach it gradually through extending our "acquaintance with the lawlike order in the realm of the actual."[61]

Thus, in contrast to positivism, Helmholtz regarded laws as possessing a certain degree of proximity to reality. "The more details are regulated by a law as a [scientific] investigation proceeds," he declared, "the more are the hypothetical elements removed from it."[62] For a positivist, by contrast, all laws are equally hypothetical and provisional. The relations of appearances which they embody do not mirror an ontological order. Helmholtz, however, claimed that the amount of reality a law contains depends on its degree of generality. This contrast with the positivists helps explain the strangely tortuous note he appended to Kirchhoff's famous definition of the task of mechanics. Kirchhoff required from mechanics that it "describe in *the most complete and simplest manner* [possible] the motions that occur in nature."[63] Helmholtz appended to Kirchhoff's statement the comment "that the most complete and simplest description can only be given in such a way that one expresses the laws that form the basis of the phenomena."[64] This means that only such a description can be true (is the most complete and the simplest) if it deals with the causes *behind* the phenomena that do not themselves appear. It was of course precisely Kirchhoff's aim to exclude such a domain from mechanics.

---

60. "Die neuere Entwickelung von Faraday's Ideen über Elektricität," [1881], *VR⁵* 2:249–91, on 262 (authorized English translation in *WA* 3:52–87, on 60, slightly amended); and Helmholtz, "Thatsachen," *VR⁵* 1:240–41.
61. "Thatsachen," *VR⁵* 1:242; cf. also 243.
62. *Vorlesungen 1:1*:19.
63. Kirchhoff, *Mechanik*, 1. See also König, "Wissenschaftsbegriff," 96–7. Robert von Helmholtz, "Gustav Robert Kirchhoff," *Deutsche Rundschau 54* (1888):232–45, on 243–45, gives a brief but perceptive comparison of Kirchhoff's doctrine of the simplest description with that of his father.
64. *Vorlesungen 1:1*:13.

A further difference between Helmholtz and the positivists also manifested itself in his reservation towards the idea of simplifying physics by avoiding abstract concepts such as force. It is true, he conceded, that one can formulate Newton's second law without the concept of force such that "it only deals with observable facts."[65] Yet the use of the concept of force "allows a much briefer linguistic expression ... than the description that is developed in conditional clauses" of the lawlike connections of the phenomena, and so it is a most desirable concept.[66]

In 1871, looking back upon his time at the University of Berlin under Magnus during the mid-1840s, Helmholtz noted that he and some of his companions pursued science in a way that

> did not yet separate experiential facts from what was mere verbal definition and hypothesis. One tried to declare the unclear mixture in these elements as axioms of metaphysical necessity and, moreover, to claim their consequences as a similar kind of necessity.[67]

Here he hardly seemed to mask self-criticism of his philosophical views in the Introduction to *Ueber die Erhaltung der Kraft*.

In the same speech, Helmholtz also mentioned some early nineteenth-century theories of matter—and thus, implicitly, his own theory of 1847—as examples of bungled metaphysics. Atomism, he declared, tried to "deduce the foundations of theoretical physics from purely hypothetical suppositions of the constitution of atoms."[68] If one wants to start from experience instead of from hypotheses, as the sciences now demand, one should begin with the volume elements given in experience and the effective laws that exist between them. These laws, "freed from the fortuitousness of the form of the bodies," can only be found empirically, he explained. In any case, theoretical physics "is subjugated to control by experience in exactly the same way ..., as is the case with so-called experimental physics."[69] One can no longer justify any essential difference between the two fields.

To say that Helmholtz changed his view of the nature of force and matter around 1869 is not to say that he was free from problems of his metaphysical realism before or that he could not have imagined a

---

65. "Goethe's Vorahnungen kommender naturwissenschaftlicher Ideen," [1892], *VR⁵* 2:335–61, on 353.
66. Ibid., 354.
67. "Zum Gedächtniss an Gustav Magnus," [1871], *VR⁵* 2:33–51, on 45.
68. Ibid.
69. Ibid., 46.

concrete alternative. For example, du Bois-Reymond had already shown in the Introduction to his *Untersuchungen über tierische Elektrizität* (1848) that one can treat the question of reducing all phenomena to forces in a much more relaxed and pragmatic way, without being forced to introduce an entire metaphysics. He thought talk of force as the cause of movement was anthropomorphic. Nothing is gained if one imagines a hand, so to speak, "that shoves the inert matter silently before itself" or, worse still, the arms of a polyp "with which the particles of matter clasp each other, mutually seizing hold of one another."[70] Forces can only be regarded admissible as causes if one recognizes that there is "nothing actually at the basis" of force, that force "does not possess any reality." The existence of forces is, he said, a mere "fiction."[71] It was precisely Helmholtz's tendency to confound the metaphysical justification of concepts with the proof of a metaphysical ontology that du Bois-Reymond criticized. As to the question of "what then remains if neither forces nor matter possess any reality," du Bois-Reymond answered that one had to restrict oneself to the appearances: "Instead of spinning round in sterile speculations or cutting the knot with the sword of self-delusion, we prefer, therefore, to keep to the appearance of the objects as they are."[72] It was thus not the task of science to know the causes of the movements but rather their laws. "For us, force is the measure of movement, not its cause."[73] As a consequence, du Bois-Reymond no longer spoke of the reduction of phenomena to their causes but rather of "the causal connection of the natural appearances" which we are supposed to imagine "under the mathematical picture of dependency, of function."[74]

Helmholtz came to change his philosophy of science, as he noted on several occasions, in response to the views of Faraday, William Thomson, James Clark Maxwell, and other British physicists. "It was mainly the influence of Faraday," Helmholtz wrote, "that occasioned mathematical physics to the mentioned progress," that is, away from

---

70. Emil du Bois-Reymond, "Über die Lebenskraft. Aus der Vorrede zu den 'Untersuchungen über tierische Elektrizität' vom März 1848," *Reden von Emil du Bois-Reymond*, ed. Estelle du Bois-Reymond, 4th ed., 2 vols. (Leipzig: Veit, 1912), 1:1–26, on 14.
71. Ibid., 6, 13.
72. Ibid., 14.
73. Ibid., 15.
74. Ibid., 6. On du Bois-Reymond's concepts of matter and force see Res Jost, "Das Wesen von Materie und Kraft. Emil du Bois-Reymonds Weltmodell," *Vierteljahrsschrift der Naturforschenden Gesellschaft in Zürich* 128 (1983):145–65.

hypothetical, metaphysical assumptions and towards direct perceptions.[75] Faraday's

> principal aim was to express in his new conceptions only facts, with the least possible use of hypothetical substances and forces. This was really an advance in general scientific method, destined to purify science from the last remnants of metaphysics.[76]

In particular, Faraday could explain the realm of electromagnetic and electrochemical phenomena by experimentally developed "abstractions" that were completely different from the traditional view that used central forces, material points or imponderable fluids. In comparison to Faraday's new concepts, Helmholtz said, the concepts of force and matter of the earlier physics were "void of intuitive content [*anschauungsleer*]," "transcendental, unimaginable abstractions."[77]

In the light of Faraday's views, Helmholtz also revised his opinion about Goethe. To be sure, he did not change his negative appraisal of Goethe's theory of color. In contrast to his Goethe speech of 1853, however, Helmholtz later came to agree with Goethe that he "did not want to be confused by metaphysical visions" in his investigation of science.[78] In refusing to accept the abstract concepts of force and matter, "physics has nowadays [i.e., in 1892] completely taken the roads upon which Goethe wanted to lead it."[79] Indeed, Helmholtz now even identified Goethe's method of looking for the *Urphänomen* of a phenomenon with Kirchhoff's maxim to describe the phenomena in the simplest way possible.

From Faraday's general antimetaphysical views Helmholtz deduced two particular consequences that led him to change his philosophy of science: First, from a theoretical point of view, the abstractions which Faraday's electromagnetic theory presupposed had exactly the same

---

75. "Goethe's Vorahnungen," *VR^5* 2:352. For Helmholtz's remarks on the influence of Faraday and other British physicists on his changing philosophy of science see "Vorrede zur Übersetzung," in John Tyndall, *Faraday und seine Entdeckungen. Eine Gedenkschrift*, authorized trans., ed. H. Helmholtz (Braunschweig: Vieweg, 1870), v–xi; "Gustav Magnus," *VR^5* 2:47; "Nachschrift (1875)," [to "Ueber Goethe's naturwissenschaftliche Arbeiten"], *VR^5* 1:46–7; "Faraday's Ideen," *VR^5* 2:passim; "Heinrich Hertz. Vorwort zu dessen Principien der Mechanik," [1894], *VR^5* 2:363–78, on 370–72; and Helmholtz, *Vorlesungen* 1:1:12–3. Cf. also David Cahan's essay "Helmholtz and the Civilizing Power of Science," (esp. 588–89), in this volume.
76. "Faraday's Ideen," *VR^5* 2:252 (*WA* 3:53).
77. "Goethe's Vorahnungen," *VR^5* 2:351.
78. "Nachschrift (1875)," *VR^5* 1:46. For Helmholtz's opinions of 1853 on Goethe see above, 465.
79. "Goethe's Vorahnungen," *VR^5* 2:351.

legitimacy with those which Helmholtz had always favored, while from an experiential point of view they were at least equivalent if nonetheless fundamentally different. Second, Faraday's theory worked without using the concepts that played the major supporting metaphysical role in Helmholtz's theory.

From the first point Helmholtz concluded that physics could no longer claim any necessity for its foundational abstractions. If there are equally legitimate, yet radically diverging alternatives for the experientially independent presuppositions of a physical theory, then none of these alternatives can be necessary a priori. Yet if there is neither necessity nor universality then all presuppositions lose their claim to a higher validity or absolute certainty. They all have to be measured against experience. If, for example, there is another, equally legitimate way to do electrodynamics than by assuming the existence of atoms, material points, and conservative forces acting at a distance, then no reduction of the phenomena to basic elements would any longer be absolute and necessary. On the contrary, this shows that the abstractions of physics are themselves also of a hypothetical origin.

From the second point it followed that physics can, in principle, altogether dispense with hypotheses about its fundamental objects. It is true, Helmholtz conceded, that all hypotheses are admissible which agree with experience and that science must discuss them "in order to retain a fully comprehensive view of the possible attempts at explanation."[80] Yet they are superfluous, at least in principle, since one could imagine a physics that operated without any abstract, hypothetical ideas. However, as long as the reality behind the phenomena is not completely understood, hypotheses are very advantageous for experimental physics: "He alone can fruitfully experiment," he declared, "who has a penetrating knowledge of theory and can accordingly pose and investigate the right questions."[81] What Helmholtz previously took as an a priori presupposition for physical knowledge he now characterized as a mere instrument for the psychology of research.

4. Experimental Interactionism and Its Fichtean Roots

Why did Helmholtz try to reconcile Faraday's "principles of scientific method" with the entirely antithetical idea that behind the visible phenomena there is the hidden domain of the real which causes

---

80. "Thatsachen," VR^5 2:239.
81. "Gustav Magnus," 46.

them? The answer to this question is best understood by examining another of his convictions, namely, the eminent role he attributed to experiment in the process of knowledge formation. Indeed, this latter conviction appears to have been an even deeper one than that of metaphysical realism; its roots, as this section argues, lay in Fichte's idealism. The special function that experiments had to fulfill for Helmholtz is especially relevant for the way he constructed his theory of perception and geometry. If Helmholtz had taken the "easy" route and merely embraced positivism wholeheartedly, he would have had to cut his ties to Fichte's philosophy and renounce his doctrine of the priority of experiment. This in turn would have forced him to radically change his theory of perception and his theory of geometry. He would have lost too much.

Helmholtz declared that we learn to read the signs of nature "by comparing them with the effect of our movements and the changes which we produce by these movements in the external world."[82] He thus advocated an *experimental interactionism*. The only way to recognize external causes was by intervening in the world and by willfully changing its circumstances. This view was directly relevant for the kind of experience that makes science progress. For Helmholtz, active experiment is much to be preferred to passive observation:

> Relatively few but well performed experiments are enough to allow me to see the original causal conditions of an event with greater certainty than a millionfold observations by which I could not arbitrarily vary the conditions.... We learn how to make reliable judgments of the causes of our sense perceptions only when we place, through our own will, our sense organs into different perspectives to the objects. Such experimenting happens from early youth onwards.[83]

Thus the first and foremost function of an experiment for Helmholtz is to find causes. The function of testing hypotheses is only secondary. For him experiment is far more an *ars inveniendi* than an *ars demonstrandi*.

Helmholtz's strong preference for intervention as the method of cognition derived from his view of human knowledge in general, not

---

82. "Die neueren Fortschritte in der Theorie des Sehen," [1868], *VR^5* 1:265–365, on 354.
83. *Handbuch*, 451–52. See also "Theorie des Sehens," *VR^5* 1:355; and "Goethe's Vorahnungen," *VR^5* 2:359. For background to this discussion, see the analyses of Helmholtz's theory of perception by Timothy Lenoir, "The Eye as Mathematician: Clinical Practice, Instrumentation, and Helmholtz's Construction of an Empiricist Theory of Vision," and R. Steven Turner, "Consensus and Controversy: Helmholtz on the Visual Perception of Space," both in this volume.

just from considerations limited to scientific methodology. He believed that our experience of the external world was due to the formation of knowledge through the voluntary, intervening activity of man. Experience arises only through our being confronted with the possibilities and limits of our actions. Knowledge only arises through our surveying what is and what is not possible through our technical, practical manipulation.

His thesis had two implications: First, it meant that sensations without active manipulation are not enough for acquiring experience. Only by comparing given sensations with those resulting from our actions can we interpret them and thus truly have experiences. Only through our active interventions do we get, so to speak, acquainted with our sensations. Sensations without active experience do not inform us about whether they have their origin in the inner or the outer world. By themselves they are worthless. Second, his thesis meant that we are directly familiar with *our own* will to change something. As concerns our own will and action, we do not require recourse to a special method so as to know ourselves, as we do with any other content of experience.

Although Helmholtz often stated his view that sensations are interpreted through the consciousness of one's own will and actions, the best version of his view is his address of 1878 celebrating the founding of the University of Berlin: "Die Tatsachen in der Wahrnehmung." There, in his foremost epistemological statement, he treated the problem of how those sensations that represent external reality can be discovered and how they can be distinguished from those that belong "to the mind's own workings."[84]

In his address, Helmholtz noted the great merit in Kant's doctrine that "any content we may represent" must necessarily be absorbed into the forms of intuition and thinking if that content was to become a representation. The roots of this doctrine, he argued, were to be found in Locke's distinction between primary and secondary qualities. "Regarding the qualities of sensation," he said, "Locke had already established a claim for the share which our corporeal and mental make-up has in the manner in which things appear to us."[85] Moreover, he maintained that the investigations of his teacher, the physiologist Johannes Müller, showed that even the qualities of the sensations were a mere form, given in advance, of intuition, a point that was, of course, scarcely reconcilable with Kant's actual view.[86] Helmholtz regarded

---

84. "Thatsachen," *VR^5* 2:218.
85. Ibid., 219.
86. On Helmholtz's misunderstanding of and deviation from Kant's original views see, e.g., Ernst Laas, *Idealismus und Positivismus*, 3 vols. (Berlin: Weidmann, 1884),

Kant's and Müller's views as the culminating points of a tradition for which the sensory perceptions are necessarily subjective. Kant and Müller, each in his own way, made it clear that the course by which natural objects affect our sense organs "must naturally and always depend on the peculiarities of both the affective body and the body being affected."[87]

Now, he continued, Kant extended the realm of the forms of intuition far beyond sensory perception: "He spoke not only of the qualities of sensations as given by the peculiarities of our intuitive faculty but also of space and time."[88] Helmholtz wanted to show that Kant exaggerated here. He believed that the dividing line between the objective and subjective portions of experience had to be drawn differently. It is true that the sensory qualities are subjective, he argued, but space and time are results of thought processes and not forms of intuition. "I believe that the most essential advance in our time is the resolution of the concept of intuition into the elementary thought processes, a resolution not yet to be found in Kant," he declared.[89] This statement indicates, on the one hand, that the influence of Kant's epistemology on Helmholtz was weaker than is often thought. On the other, it clearly suggests that with his methodological doctrine of experimental interactionism he owed much more, as we shall see presently, to the philosophical views of the individual who first set himself the task of resolving the intuition into thought: Johann Gottlieb Fichte. As Helmholtz said in his Kant speech of 1855: Fichte's "presentation of sense perception is in the most exact agreement with the conclusions which the physiology of the senses later drew from the facts of experience."[90]

If even the perception of external objects is affected by the activity of thought, Helmholtz asked, how can we distinguish the sensation of external objects from inner sensations? He answered this question by appealing to the consciousness given by our actions when we move our body. If, for example, I follow the voluntary impulse to walk to and fro, then certain sensations change in a lawlike way. Exactly those sensations change that come from the objects in space. Other contents

---

vol. 3: *Idealistische und positivistische Erkenntnisstheorie*, 572–97; and Alois Riehl, "Hermann von Helmholtz in seinem Verhältnis zu Kant," *Kantstudien 9* (1904):261–85.
  87. *Handbuch*, 444.
  88. "Thatsachen," VR^5 2:223.
  89. Ibid., 244.
  90. "Ueber das Sehen," VR^5 1:89.

of experience—such as recollections, intentions, wishes, and emotions—do not change in the course of this action or at least do not change systematically but only contingently. External objects are therefore such things that change if we move our body in executing certain voluntary impulses. Helmholtz wrote:

> We do not merely have alternating sense impressions which come upon us without our doing anything about it. Rather, during our own continuing activity we observe and so become acquainted with the *enduring existence* of a lawlike relationship between our innervations and the realization of the various impressions from the current range of presentables [i.e., sensations]. Each of our voluntary movements, whereby we modify the manner of appearance of objects, is to be regarded as an experiment through which we test whether we have correctly apprehended the lawlike behavior of the appearance before us, i.e., correctly apprehended the latter's presupposed enduring existence in a specific spatial arrangement.[91]

Some changes of our sensations depend on our will, others on the nature of the sensed objects. Someone who had no previous experiences could change his sensations at will through "motoric impulses." He could find out by testing these voluntary impulses how he could put himself into a former state of sensation by taking the same spatial relation to the objects as before.

Helmholtz further maintained that we can discover those changes of sensations that do not have a psychic connection with a preceding voluntary impulse, that is, that are not directly the content of the act of volition but rather its learned effect. We learn that we can make certain sensations recur by making appropriate movements, for example those needed to gain a certain view of a table. "In this way an experientially grounded opinion arises in us that our movements are the reason for the changing views of the table and that this [table], whether we see it directly or not, can indeed be seen by us as soon as we want to."[92] Other states of ours do not lead to changing views of the table. If I am delighted or if I have a certain wish, the sense perceptions that I have of the table do not change thereby. In time, we can quickly distinguish the two cases. "Those alterations which we can produce and revoke by conscious impulses of the will," Helmholtz concluded, "are distinct from those which are not consequences of such impulses and cannot be eliminated by them."[93]

---

91. "Thatsachen," VR^5 2:237.
92. *Handbuch*, 452.
93. "Thatsachen," VR^5 2:226–27.

All of Helmholtz's considerations presupposed that we possess self-consciousness, that is, a consciousness of an individual's own consciousness which is directly given and known to itself without mediation. Self-consciousness, Helmholtz held, is the seat of the free impulses of our will to act, a will not determined by physical causes of the physical process of action.[94] He further supposed that all additional knowledge that we acquire beyond our self-consciousness, that is, knowledge of the external world, is gained *from judgments about actions of self-consciousness*. Knowledge is thus nothing but the judgment of one's own state of self-consciousness and its changes. He wrote:

> The chief reason ... why the power of any experiment [whereby we modify the manner of appearance of objects] meant to convince us is so much greater than that of observing a process going on without our assistance is that with the experiment the chain of causes runs through our own self-consciousness. We are acquainted with one member of [the chain of] these causes—the impulse of our will—from inner intuition, and know by what motives it came about. From this, as from an initial member known to us and at a point in time known to us, there then begins to act that chain of physical causes which terminates in the outcome of the experiment.[95]

Through the comparison of the impulses of my will with my sensations that arise after the performance of the impulse I become able to discern what I can change at will from what cannot be so changed. "Fichte's appropriate expression for this is that the 'I' is faced with a 'not-I' which demands recognition."[96] For Helmholtz, as for Fichte, this discrimination lies at the bottom of distinguishing the spatially determined external world from the world of inner intuition, of self-consciousness.

To sum up Helmholtz's complex argument, the genesis of knowledge takes the following threefold order: First, we learn to distinguish between the lawlike effects of our actions from those that do not follow any laws. Second, this in turn leads to the distinction between those sensations that self-consciousness can change and those that it alone cannot so change. Third and finally, the difference between the changeable and the unchangeable forces induces self-consciousness to confront its own inner world of consciousness with the outer world of an alien power. As Helmholtz wrote:

94. Ibid., 237.
95. Ibid.
96. Ibid., 227; see also ibid., 238, 219, 241, and *Vorlesungen 1:1*:14.

It is clear ... that a division between what is thought and what is actual does not become possible until we know how to make the division between what the "I" can and cannot alter. This does not become possible, however, until we discern what lawlike consequences the impulses of our will have at the given time. The lawlike is therefore the essential presupposition for the character of the actual.[97]

The essential elements of Helmholtz's experimental interactionism are to be found in Fichte.[98] Fichte developed a theory of self-consciousness, of the "I" or "Ego" as he called it. Self-consciousness, he maintained, comes through a primary act of self-affirmation of the ego:

The I *posits itself*, and it *is* through this mere positing by itself. ... It is simultaneously the acting and the product of action; the efficacious and that which is produced by activity; activity and action are one and the same; and the *I am* is therefore the expression of an active action [*Tathandlung*].[99]

Fichte here claimed that our self-consciousness cannot be brought about from the outside in the way that the consciousness one has of an object is to be regarded as externally caused. It is consciousness itself through which self-consciousness (the ego) is brought about. The activity of self-consciousness is produced through the action of its own production.

Fichte did, however, maintain that we also have conscious appearances in our consciousness that do not appear as consciousness itself; these belong to the "non-ego" or the "not-I," to use his idiosyncratic expression. A non-ego is opposed to consciousness. But its opposition occurs inside consciousness, for otherwise consciousness would not be conscious of something opposed to itself. We posit something in our consciousness that is not self-consciousness. This positing is also a primary activity of the I. Objects are therefore appearances that are

---

97. "Thatsachen," *VR⁵* 2:241–42.
98. For Fichte see Walter Schulz, *Johann Gottlieb Fichte, Vernunft und Freiheit* (Pfullingen: Neske, 1962); Ernst Cassirer, *Das Erkenntnisproblem in der Philosophie und Wissenschaft der neueren Zeit*, 4 vols. (vols. 1–3: Berlin: Bruno Cassirer, 1906–20; vol. 4 [in English translation]: 1950, [in German]: Stuttgart: Kohlhammer, 1957; reprinted Darmstadt: Wissenschaftliche Buchgesellschaft, 1973–74), *3*:126–216; and Heinz Heimsoeth, *Fichte* (Munich: Reinhardt, 1923). For Fichte's influence on Helmholtz see also Turner, "Helmholtz and the Empiricist Vision," 56–7.
99. Johann Gottlieb Fichte, *Grundlage der gesammten Wissenschaftslehre* (Jena and Leipzig: Gabler, 1794), in *Johann Gottlieb Fichte's sämmtliche Werke*, ed. Immanuel Hermann Fichte, 1st partition, 8 vols. (Berlin: Veit, 1845–46; reprinted Berlin: de Gruyter, 1965 and 1973), *1*:83–328, on 96.

regarded from our (empirical) consciousness as something that comes from the outside, although it is produced by consciousness itself.

This view had consequences for the concept of perception. Fichte developed his theory of perception most clearly in his book *Die Bestimmung des Menschen* (1800). He thought that if we establish the presence of an (external) object we do so not through a direct perception of the object but rather through a judgment of our own sensations. He wrote:

> The perception of the objects arises from the perception of my own state and is conditioned by it, not the other way around. I discern objects only by discerning my own states. ... I sense inside myself, not in the object, because I am myself and not the object; therefore I sense only myself and my own state, not the state of the object. If there is a consciousness of the object, at least it cannot be sensation or perception; that much is clear.[100]

Like Fichte, Helmholtz did not see in sensation a direct relation to the object. The only thing, he wrote, that "we can directly perceive [are] the excitation of the nerves, that is, the effects, never the external objects." By observing these effects we obtain "the idea of a cause of this effect" through an inference.[101]

Fichte called all immediate awareness of a sensation "feeling" and distinguished it from "intuition." In intuition one reflects upon one's sensation, one observes oneself, so to speak, in dealing with one's own sensations. The consciousness of an external object results from an intuition of one's own sensation and, at the same time, "through thinking according to the principle of [sufficient] reason," that is, according to the causal law. "You not only feel your own state, you also think it; yet it [your state] does not give you a complete thought; you are compelled to add something to it, a reason outside of you, a foreign force."[102] The existence of an external force is thus inferred from knowledge of one's own inner state. Fichte wrote:

> Any awareness of the object outside me is determined through the clear, precise awareness of my own state. In a conscious act like this there is always an inference from something well-founded inside me to a foundation outside of me.[103]

---

100. Johann Gottlieb Fichte, *Die Bestimmung des Menschen* (Berlin: Voss, 1800), in his *Werke* 2:165–319, on 203–4.
101. *Handbuch*, 430; see also "Theorie des Sehens," VR^5 1:296.
102. Fichte, *Bestimmung*, 238.
103. Ibid., 233; see also ibid., 218.

As empirical creatures we are unaware that we ourselves draw this inference. The awareness of objects is a conscious act "not known as such" of the production of the idea of a force outside of consciousness.

Fichte further held that consciousness does not seize the external force directly but rather only through its manifestations. What exists in the external object independently of consciousness is given neither in the sensations nor in the intuitions, but only in thought. Force that is thought in this way is the carrier of all attributes of objects; it eternally remains the same through all changes of appearance.[104]

Yet how did Fichte think the object as existing outside fit together with the nature of the ego as action? In resolving the external object into mere ideas of the consciousness, he confronted the problem of how the reality of the I itself, as the carrier of ideas, could be understood. The I itself cannot be a mere idea. Just as the external world is not directly given in perception, so the I is not given itself in the ideas it has of itself. Since consciousness is, however, a reality, it must be given in another way than through ideas. Fichte solved this problem by claiming that the certainty of the ego, the way it is present to itself, is affected in its action. The I carries in itself the "drive to an absolute and independent self-activity."[105] My own free action gives me the practical certainty that I am actual myself.

Fichte further held that the external world becomes real for me only through my action. Through positing the non-ego, he claimed, I produce a boundary for myself which makes it possible that I can be conscious of myself as someone who acts; I produce a realm in which I can develop my activity. Without such a boundary which stands in opposition to me I could not become conscious of myself as an acting being. Thus for Fichte the external world is the totality of all experiences that limit my activity and interference and that at the same time makes it possible in the first place. "The Nature in which I have to act is ... formed by my own laws of thought. ... It never expresses anything other than the circumstances and relations of myself to myself."[106] Only through my actions can I establish knowledge of causality, that is, the world of external objects.

For his definition of natural science, Helmholtz adopted Fichte's distinction between the free activity of the subject and the object which is thought as an alien thought. Sometime prior to 1847, Helmholtz wrote:

104. Ibid., 236–39.
105. Ibid., 249.
106. Ibid., 258.

The object of science is that content of our ideas which we do not perceive as something produced by the self-activity of our conceptual power [*Vorstellungsvermögen*]; i.e., that which is perceived as real.[107]

Later, in 1855, he referred to the "self-conscious acts of our will and thinking" as "free," and he contrasted them to the processes in the world of external objects which constitute "necessary effects of sufficient causes."[108] The objective circumstances of causes and effects in the world of objects limits the possibilities of our actions. The task of science, Helmholtz held, is to represent nature as the totality of activities that are possible to us.

Helmholtz's father Ferdinand, who was an enthusiastic admirer of Fichte,[109] recognized clearly that such a program stood in complete agreement with Fichte's philosophy. Commenting on his son's habilitation speech in a letter to him of 1852, he wrote:

It almost seems to me as if with this mathematical and experimenting way of investigation . . . a new and slow, but secure course for philosophy has started which will precisely delimit in this way at least the objective substrate of all knowledge and exhibit it undoubtedly clear in its essence. It will thereby in some future day confirm and make evident Fichte's doctrine of the ego as the only possible way of philosophizing.[110]

According to his biographer Leo Koenigsberger, Helmholtz explicitly confirmed "that he indeed had the intention, as his father realized, to give an empirical representation of Fichte's fundamental view of sense perception."[111] In short, Helmholtz took from Fichte the idea of "the resolution of the concept of intuition into the elementary processes of thought" that was still absent in Kant.[112]

5. Helmholtz's Idealism

Notwithstanding the Fichtean elements within Helmholtz's thought, one might maintain that Helmholtz differed deeply from Fichte in

---

107. As quoted in Koenigsberger *2*:126.
108. "Ueber das Sehen," *VR⁵ 1*:116.
109. For Ferdinand Helmholtz's admiration of Fichte see Helmholtz, "Erinnerungen," *VR⁵ 1*:17, and Koenigsberger *1*:168, 242, 244, 284. For the relation of Ferdinand Helmholtz with Fichte's son, the philosopher Immanuel Hermann Fichte, see ibid., 7, 161, 331–32, and also Köhnke, *Neukantianismus*, 151–57.
110. As quoted in Koenigsberger *1*:168.
111. Ibid., 169.
112. Ibid.

taking a final stand against idealism. For Fichte, the consciousness of an external object is a product of our conceptual power; for Helmholtz, by contrast, the objective existence of external objects is, as he put it, "an excellently serviceable and precise hypothesis."[113] Fichte thus appears as the speculative idealist and Helmholtz (at first glance) as the realist and the sober scientist, averse to all metaphysics.[114] Moreover, Helmholtz tried to apply his eventual criticism of a metaphysical justification of force, matter, space, and time to Fichte's idealism, and to reinterpret it in an empirical and realist manner. As argued above, the existence of a true alternative to his metaphysical realism forced him to conclude that the apparent necessities of thought are only hypotheses that either will or will not be confirmed by experimental science. In "Die Tatsachen in der Wahrnehmung" he similarly argued that Fichte's idealist conception of the genesis of external objects was not a necessary one, for it can be interpreted realistically without changing its inner core. Both the extreme idealism that regards the world as a web of purely psychic interrelations and the realism that acknowledges the material world outside us as objectively existing are only metaphysical hypotheses. In Helmholtz's view, the advantage of realism is that it "trusts the testimony of ordinary self-observation" and is the simplest hypothesis we can form: "it has been tested and confirmed in an extraordinarily wide range of application."[115]

If Helmholtz prudently preferred realism and remained indifferent to the idealistic, metaphysical superstructure of Fichte's resolution of the external object into elementary thought processes, it was, nonetheless, impossible for him to be rid of this idealist attitude merely per fiat. In conceiving of external perception as perception of one's own inner state (of nervous excitations conducted to the brain) Helmholtz, no matter how much he denied it, remained an idealist.[116] For as Kant noted, it is the mark of idealism to believe "that the sole immediate experience is the inner one from which one only *infers* external objects."[117]

It might be objected that it is quite implausible to call Helmholtz an idealist since that philosophical stance agrees neither with his me-

---

113. "Thatsachen," VR^5 2:244.
114. Cf., e.g., Koenigsberger's view: Koenigsberger *1*:169.
115. "Thatsachen," VR^5 2:238–39. See also "Das Denken in der Medicin," [1877], VR^5 2:165–90, on 186.
116. For Helmholtz's idealism see also Köhnke, *Neukantianismus*, 414–16, 151–57.
117. Immanuel Kant, *Kritik der reinen Vernunft*, 2nd ed. (Riga: Hartknoch, 1787), B 276.

chanistic reductionism nor with his antimetaphysical and realist pronouncements. Yet the contradiction is only an apparent one. For Helmholtz's reductionism did not reach beyond physiology. He maintained only that all inorganic and organic phenomena are reducible to attractive and repulsive forces; he did not claim that the mind itself was subject to reductionism. Had he done so, his idealism would indeed have come into conflict with his mechanistic worldview. In point of fact, Helmholtz never advocated such a reductive materialism; he never claimed that mental processes or states are reducible to the mechanics of atoms. Instead, he argued just the other way around: that the experience of freedom of the will that we gain through our actions evidently proves that there are exceptions to the causal law. He did not here mean, however, that the causal law was refuted, for "the conclusions drawn from it do not pertain to real experience but only to the understanding of it."[118] If we want to understand anything in nature, he argued, we have "to start from the supposition that it is intelligible," that is, we have to presuppose the law of causality.[119]

Thus, Helmholtz's physiological reductionism strictly separated mind and its principles from the world of matter. The evil for Helmholtz was the mingling of the science of the necessary processes in nature with insights into the free, spontaneous self-activity of the mind. Between the not-I and the I there was a sharp difference. In this, Helmholtz further followed Fichte, who held that "intelligence [i.e., mind or consciousness] and object are therefore directly opposed to each other: they lie in two different worlds between which there is no bridge."[120]

Hence, Helmholtz's true adversary was less vitalism than the various versions of a "philosophy of identity," a philosophy that proclaimed the "laws of the mind to be nothing but laws of [external] reality," and so believed that it was possible to dispense with external experience in seeking the laws of nature. For Helmholtz to believe that nature could be reconstructed a priori in pure thinking was of course an untenable metaphysics.[121] It is noteworthy that he did not initially

---

118. *Handbuch*, 454. For Helmholtz's doctrine of freedom of the will see Laas, *Erkenntnisstheorie*, 576–77.
119. "Ueber die Erhaltung," *WA* 1:13; and see Heimann, "Helmholtz and Kant," 217.
120. Johann Gottlieb Fichte, *Erste Einleitung in die Wissenschaftslehre* [1797], in his *Werke*, 1st partition, 1:435.
121. "Ueber das Sehen," VR^5 1:99. See also ibid., 89, 162, and "Gustav Magnus," VR^5 2:43.

raise his charge against the epigones of the "romantic school" as a charge of metaphysics; that came only after he had freed himself from his earlier, metaphysical view of mechanics.[122]

Seen from this point of view, Helmholtz's mechanistic biophysics, like that of his fellow "organic physicists," had much less in common with materialism than did the "hypothesis of identity." Although the reductive materialists and the *Naturphilosophen* who followed Schelling differed radically in their explanatory principles, they did agree upon a naturalism for which the laws of nature are sufficient to explain the operations of the mind. For Helmholtz, in contrast to both groups, mind and nature are strictly separated so that no inferences from one to the other are possible. His contemporaries well realized that this stance was an idealist one. It was for this reason that he so bristled when critics, like Ewald Hering, of his theory of sense perception charged him with being a "spiritualist." In irritated response, Helmholtz characterized his two groups of adversaries as belonging to one and the same camp of dogmatic metaphysicians hostile to experience: the nativists, who sought to impose all those things by "true intuition a priori," for which they had no experiential evidence; and the intolerant reductive materialists, who brushed aside unpleasant facts in order to save their system.[123] He saw the epigones of the philosophy of identity and the materialists as together in an unholy alliance.

6. Conclusion

This essay has argued that the essential key to understanding Helmholtz's philosophy of science lies in appreciating the influence of Fichte's idealism on Helmholtz. Recognizing the importance of Fichte's philosophy of action for Helmholtz helps clarify (at least) two major features of Helmholtz's philosophy of science: first, the high value he placed on experiment and his playing it off against mere observation; and second, his need to infer external reality from knowledge of one's own self-consciousness. For Helmholtz, physical thought was indeed to be constructed out of the experiences we make as acting beings. We gain an understanding of external effects only through our

---

122. Helmholtz apparently first reproached metaphysics during the course of his polemical exchange with the astrophysicist Friedrich Zöllner; see "Ueber das Streben nach Popularisirung der Wissenschaft. Vorrede zur Uebersetzung von Tyndall's 'Fragments of Science' 1874," *VR⁵* 2:422–34, on 432.
123. "Denken in der Medicin," *VR⁵* 2:186–87.

actions. Reality for him is thus the sum of all effects produced by an unknown substance which limits our free action.

The resolution of the concept of force under the influence of Faraday and others into observable relations did not essentially change these views. On the contrary: towards the end of his life Helmholtz came to believe that, although it is true that all talk about forces in nature can be reformulated in terms of observational descriptions of lawlike relations, only those descriptions are to be regarded as truly *lawlike* (that is, as describing *the actual)* which can be reformulated as propositions about the operation of a force. A law for Helmholtz has to state the conditions under which certain observable effects arise, independently of our will. Both elements, the antecedent conditions and their effects, are equally important and necessary, as he had already stressed in *Ueber die Erhaltung der Kraft.*

A law formulated without describing the material conditions of its efficacy is no law at all: "The law of an effect presupposes conditions under which it comes to efficacy. A force that is detached from matter would be the objectification of a law which lacks the conditions of its effectiveness."[124] This is also the reason why he rejected the assumption of a vital force independent of objects: "A force without matter does not make any sense; this would correspond to a law that talked of changes where there are no objects that could be changed. Such a law would contradict and repeal itself."[125]

Conversely, Helmholtz believed that it was a similar if opposite mistake to formulate a law by describing only the conditions under which it might hold without, however, referring to any change thereby effected. To talk thus of matter without force, he held, was equally senseless, "for such material objects could not be subjugated to changes, because changes always presuppose the existence of a force."[126] All change in nature assumes a mass which is set into motion by another mass. Notwithstanding his concessions to Kirchhoff's positivism, Helmholtz's concept of law thus remained committed to interactionism and to mechanistic, metaphysical realism.

This may perhaps also explain why Helmholtz could not fully understand Faraday's and Maxwell's idea of the electromagnetic field.[127]

---

124. "Zusätze (1881)," *WA 1*:68.
125. *Vorlesungen 1:1*:15. On the vital force see also ibid., 16.
126. Ibid., 15.
127. See Jed Buchwald's essay, "Electrodynamics in Context: Object States, Laboratory Practice, and Anti-Romanticism," in this volume; and Wise, "German Concepts," 298–301.

In Maxwell's laws, the connection between bodies is no longer formulated as the causal efficacy of a physical force but rather as a mathematical, functional relationship between the values of a function and its differential quotients. The concept of the field made not only the concept of action at a distance obsolete; it also challenged deep-seated ideas of force and causality. For Helmholtz, the field must have appeared as the hypostatization of a force which, like the vital force, was supposed to be effective without a material substance being attached to it. As such, it was antithetical to his entire way of thinking.

Nonetheless, he sought as best he could to adjust his conception of central forces to Maxwell's electromagnetic ideas. In so doing, he assumed that two bodies in a vacuum not only influence one another directly through their central forces but also through the dielectric, the ether. The smallest elements of the space-filling ether are, according to him, polarized by the actions of two bodies and act at a distance. Throughout his scientific life he persisted in taking the acting body as, to use the words of his student Heinrich Hertz, "simultaneously seat and origin of force."[128] The greatest concession that Helmholtz could make to Maxwell's theory was to recognize the influence exerted by the medium's polarized particles.

Late in his career, Helmholtz developed doubts about whether his principle of the conservation of energy really fulfilled the criteria that he had demanded of a natural law. Strictly speaking, the energy conservation principle is best regarded as a constraint on effective laws, not as a law itself. It says nothing about how these constrained processes actually operate in detail. To remedy this defect, Helmholtz tried to find a still more general law into which he could embed the conservation principle. Near the end of his life, he came to believe that the principle of least action provided just such a law. He wrote:

> The law of least action is closely connected with the energy law and completes it. For the last-named law specifies only a single magnitude, the quantum of energy which must be constant in the course of every natural process. It does not specify how, after all, the process will come to pass between its initial and its end states. In order to ascertain this for systems of ponderable bodies one needs recourse to the general equations of motion. Now, it is precisely this that is determined by the law of least action.[129]

---

128. Heinrich Hertz, "Einleitende Übersicht," *Untersuchungen über die Ausbreitung der elektrischen Kraft*, 3rd ed. (Leipzig: Barth, 1914), 1–31, on 24. (= vol. 2 of *Gesammelte Werke*, ed. Philipp Lenard, 1st ed., 4 vols. [Leipzig: Barth, 1894–95]).

129. "Rede über die Entdeckungsgeschichte des Princips der kleinsten Action," [1887], in Adolf Harnack, *Geschichte der Königlich Preussischen Akademie der Wissen-*

The privileged position that Helmholtz granted to the principle of least action was later strongly advocated by another of his students, Max Planck. For Planck, this principle "takes on the highest position among all physical laws" and it comes closest of all to the goal of subsuming "all natural phenomena that have been and will be observed."[130] Planck believed so strongly in the principle that not even Einstein's theory of relativity could shake his belief in it; indeed, the latter strengthened his belief in the importance of the least action principle. Planck thus adopted an essential part of Helmholtz's philosophy of science: the principle of least action, freed of its mechanistic interpretation, served him as an abstract substitute for what Helmholtz had called the real, the substance, the objective behind the appearances. Thus substituted, Planck carried Helmholtz's philosophy of science into the twentieth century.

---

schaften zu Berlin, 3 vols. (Berlin: Reichsdruckerei, 1900), vol. 2: *Urkunden und Aktenstücke*, 282–96, on 284. On Helmholtz's treatment of the principle of least action see Heimann, "Helmholtz and Kant," 237–38; and Günter Bierhalter's essay, "Helmholtz's Mechanical Foundation of Thermodynamics," in this volume.

130. Max Planck, "Das Prinzip der kleinsten Wirkung," [1915], in his *Vorträge und Erinnerungen*, 5th ed. (Stuttgart: Hirzel, 1949; reprinted Darmstadt: Wissenschaftliche Buchgesellschaft 1983), 95–105, on 104, 95. See also idem "Die Stellung der neueren Physik zur mechanischen Naturanschauung," [1910], in ibid., 52–68, on 65–6. On Planck's high regard for the principle of least action see Stanley Goldberg, "Max Planck's Philosophy of Nature and His Elaboration of the Special Theory of Relativity," *HSPS* 7 (1976):125–60, on 148–49.

13

Helmholtz's Empiricist Philosophy of Mathematics

Between Laws of Perception and Laws of Nature

Robert DiSalle

1. Introduction

Helmholtz articulated his empiricist philosophy of science in its clearest form in two related controversies concerning space: the empiricist-nativist debate over the visual perception of spatial relations, and a debate with Kantian philosophers over the epistemological status of non-Euclidean geometry. In the first controversy, Helmholtz defended the empirical theory of space-perception, according to which distances in perceptual space are not given in sensation, but instead are *inferred* from sensation by habits acquired through experience. The theory's most important specific claim was that visual perception of depth occurred not, as Helmholtz's opponents believed, directly, but rather indirectly, through the mediation of inferences from two-dimensional sensations. This claim was the core of Helmholtz's dispute with "nativist" psychologists, above all with Ewald Hering, who sought experimental evidence of immediate "intuitions" of distance.[1] In the second controversy, Helmholtz argued that geometry in general derives from physical measurement rather than from a priori features (the Kantian "form") of our spatial intuition. Thus, Euclidean geometry represented merely one possible outcome of our spatial measurements, and the choice between it and various non-Euclidean geometries was necessarily an empirical choice. Helmholtz supported this position

---

1. For analysis of Helmholtz's dispute with Hering see R. Steven Turner's essay, "Consensus and Controversy: Helmholtz on the Visual Perception of Space," in this volume.

with the historic mathematical result that the general metrical form for three-dimensional spaces of constant curvature can actually be deduced from the chief presupposition of ordinary measurement, viz., the free mobility of rigid bodies. A striking, and characteristic, feature of Helmholtz's part in both controversies was his effort to transform general philosophical positions regarding space into specific experimental and mathematical questions.

For at least two reasons, Helmholtz often spoke of a deep connection between his research in space-perception and his mathematical work. First, his interest in the foundations of geometry, and subsequently of mathematics in general, originally developed out of his investigation of visual space.[2] Second, and more broadly, his stands against innate spatial awareness and a priori geometrical knowledge were founded alike on the same empiricist epistemology. Thus the real subject-matter of his geometrical investigations, in contrast to those of mathematicians who worked on similar problems in a more abstract context— for example, Bernhard Riemann and Sophus Lie—was always the "real" space in which humans learn to make perceptual judgments.[3] This essay examines the character of the connection between these two fields, seeking to illuminate some important general methodological and philosophical aspects of Helmholtz's thought. It first sketches (Section 2) the nativist background against which Helmholtz set his empiricist view and analyzes nativism's neo-Kantian philosophical elements.[4] It then turns (Section 3) to a brief discussion of those aspects of Helmholtz's work on perception that touched most closely on the foundations of geometry. Finally, and most extensively, it considers (Section 4) the geometrical writings themselves, both in relation to perception theory and in relation to Helmholtz's empiricist approach to mathematics as a whole.

Above all, this essay aims to clarify Helmholtz's view of the nature of the connection between mathematics and the world of experience. For while Helmholtz certainly understood geometry to be an empirical

---

2. On Helmholtz's mathematical approach to his spatial theory see Timothy Lenoir's essay, "The Eye as Mathematician: Clinical Practice, Instrumentation, and Helmholtz's Construction of an Empiricist Theory of Vision," in this volume.

3. See, for example, Joan Richards, "The Evolution of Empiricism: Hermann von Helmholtz and the Foundations of Geometry," *The British Journal for the Philosophy of Science 28* (1977):235–53; Howard Stein, "Some Philosophical Prehistory of General Relativity," in John Earman, Clark Glymour, and John Stachel, eds., *Foundations of Space-Time Theories* (Minnesota Studies in the Philosophy of Science, vol. 8, Minneapolis: University of Minnesota Press, 1977), 3–49; and Roberto Torretti, *Philosophy of Geometry from Riemann to Poincaré* (Dordrecht, Boston, Lancaster: Reidel, 1978), esp. chap. 3.

4. For a full discussion of the nativist-empiricist issue see Turner's "Consensus and Controversy."

science in some sense, he also recognized its status as a formal deductive structure that stands independently of its intuitive or sensory content. Moreover, he argued that his empiricist view of geometry could be extended to arithmetic as well; for just as geometry is related to the form of intuition of space, so the arithmetical axioms "are in the corresponding relation to the form of intuition of time."[5] Yet Helmholtz did not portray arithmetic as the empirical science of time in the sense in which geometry could be called the science of space. Thus, this essay argues that to the extent that Helmholtz's empiricism constituted a coherent view of mathematics as a whole, it was more complicated and subtle than the bald claim that mathematical truths are empirical truths. By examining the origins of this view in Helmholtz's work on perception, we shall see that it was, rather, a theory of the connection between mathematical and empirical truths—an account of the interaction between the psychological processes underlying mathematics and the *laws* governing the objective world.

2. Kant and the Intuitionists on Spatial Perception

Helmholtz and his contemporaries described the nativist or "intuitionist" theories of space perception as those that "followed the lead" of Kant,[6] insofar as they invoked an innate "space of intuition" to explain particular sensations of distance. This recalled Kant's claim that the observed order of nature reflects cognitive features of the observer rather than the nature of things in themselves, and in particular that the a priori "form" of intuition determines the character of perceived space and time. For the experimental psychology of sensation and perception, Johannes Müller's law of specific nerve energies suggested an analogous focus on the perceiver instead of the objects perceived: according to Müller's law, the character of sensory information depends not on the nature of the stimulus, but rather on the nature of the nerve stimulated. Regarding spatial perception, the intuitionist view held that the salient feature of the nerves stimulated—those of the skin and the retina—was that they are themselves objects extended in space, and so their impressions retain the quality of spatial extension.[7] Thus, nativism granted a kind of a priori status to the

---

5. "Zählen und Messen, erkenntnisstheoretische betrachtet," in *Philosophische Aufsätze. Eduard Zeller zu seinem fünfzigjährigen Doctorjubiläum gewidmet* (Leipzig: Fues' Verlag, 1887), 17–52, in *WA* 3:356–391, on 357.
6. *Handbuch*[3] 3:33.
7. "Die neueren Fortschritte in der Theorie des Sehens," in *VR*[4] 2:265–365, on 331–32.

spatial arrangement of sensation yet at the same time explicated what Kant called the "form of intuition" through the anatomy of the human sensory apparatus.

But if we ask, what *is* the space of intuition?—or, more precisely, what does it mean to say that spatial relations are "given in intuition"?—we see that the connection between nativism and Kantianism was not completely straightforward. Nativism distinguished itself from Kantianism by its account of the nature and the amount of spatial information supposedly captured just by the physiology of the eyes and the optic nerves. Hering's theory, for example, linked the structure of visual space directly to the structure of the retina by a kind of "retinal geometry." The values of certain "spatial feelings" played the role of coordinates: feelings of height and breadth are measured directly by reference to the lines of latitude and longitude of the retinae, while the feeling for depth arises from these together with binocular convergence. Thus, taking the midpoint between the two eyes as the center, symmetrically opposite parts of the retinae have the same depth value. Hering argued that:

> as in geometry the position of a point is determined by its spatial relation to three coordinates, so likewise the location of a point in a visual image is determined by the three simple spatial feelings that it produces in the sensorium, although these three are united in one mixed feeling.[8]

Yet while Hering clearly provided a physiological interpretation of the claim that spatial relations are "given in intuition," he did not propose a theory of a priori knowledge. Like other nativists, Hering did not generally maintain that the spatial feelings accurately represented objective spatial relations. Rather, he recognized that the perceiver had to *learn* the correspondences between his subjective feelings and objective space. (Indeed, Helmholtz criticized Hering's theory as "superfluous" precisely because the supposed spatial feelings had to be supplanted by ideas derived from experience, so that the former were neither necessary nor useful.)[9] Thus, there is no conception of true distance until we learn to orient ourselves in space, and this is possible only through experience. The nativist theory supposed, then, that spatial learning was the comparison of one three-dimensional space with

---

8. Ewald Hering, *Beiträge zur Physiologie: Zur Lehre vom Ortsinn der Netzhaut* (Leipzig: Engelmann, 1861–64), Part 5, *Vom binoculären Tiefsehen* (1864), 325. By invoking a kind of implicit geometrical calculation, Hering's analysis of depth-perception recalls René Descartes's *Optics*, in *Discourse on Method, Optics, Geometry, and Meteorology*, trans. Paul J. Olscamp (Indianapolis: Bobbs-Merrill, 1965), 63–173, esp. 104–6.

9. *Handbuch*³ 3:17.

another: the space of intuition given by the geometry of "spatial feelings" with the space of real objects to which the perceiver gradually adjusts.

This supposition indicates why the intuitionist theory of perception was not a Kantian theory of the "form of intuition." First, it did not attempt to capture the *formal* aspect of the Kantian form: the latter was not part of the actual content of experience, but rather represented a set of formal or structural (in the geometrical sense) *conditions* "underlying" possible experience. As Kant wrote:

> Space is not an empirical concept derived from outer experiences.... Space is a necessary *a priori* representation, which underlies all outer intuitions. We can never represent to ourselves the absence of space, though we can quite well think of it as empty of objects. It must therefore be regarded as the condition of the possibility of appearances, and not as a determination dependent upon them.[10]

Like nativism, Kant's theory implied that experience is (in some sense) originally spatial, inasmuch as objects necessarily appear to the senses in some spatial arrangement. But it also implied—and this is the second distinction—that space involves a set of laws governing the possible course of experience. Kant understood this aspect of space through the connection between the "form of intuition" and the foundations of geometry: geometrical propositions are based on the construction of geometrical figures, and the "form of intuition" grounds the apparent evidence—the "intuitive" validity—of these constructions. He explained:

> The mathematics of space (geometry) is based upon [the] successive synthesis of the productive imagination in the generation of figures. This is the basis of the axioms which formulate the conditions of sensible *a priori* intuition under which alone the schema of a pure concept of outer intuition can arise—for instance, that between two points only one straight line is possible, or that two straight lines cannot enclose a space, etc.[11]

Kant thus argued that Euclid's axioms are "transcendental necessities" because he thought that the conditions under which the Euclidean constructions are possible are exactly the conditions of the possibility of experience of extended bodies in space; the axioms act as a priori

---

10. Immanuel Kant, *The Critique of Pure Reason* [1787], trans. Norman Kemp Smith (London: MacMillan, 1929), 68.
11. Ibid., 199.

constraints on the relations among bodies that the imagination can represent to itself, both simultaneously and successively.

Since the Kantian spatial form of intuition thus concerned the organization rather than the content of our experience of external things—and, analogously, the temporal form of intuition concerned the sequential ordering of our inner experience—Hering and others were not nativists in quite the same sense as Kant. Kant himself anticipated the distinction as follows: The hypothesis that the concepts of space and time are "connate"

> is not to be rashly admitted, since, in appealing to a first cause, it opens a path for that lazy philosophy which declares further research to be in vain. Both concepts are without doubt acquired, as abstracted, not indeed from the sensing of objects (for sensation gives the matter, not the form, of human apprehension), but from the action of the mind in coordinating its sensa according to unchanging laws. . . . Nothing here is connate save the law of the mind, according to which it combines in a fixed manner the sensa produced in it by the presence of the object.[12]

Kant's critique of nativism, like Helmholtz's, recognized that nativism takes the concept of space as a first cause, incapable of further explanation; it thereby forecloses research into what is most crucial to spatial intuition, namely, "the action of the mind in coordinating its sensa." In sharp contrast to Helmholtz, however, Kant asserted that the "unchanging laws" by which sensa are combined included the postulates of Euclidean geometry for outer experience and the axioms of arithmetic for inner experience, and that both sciences are therefore valid a priori. Accordingly, by the mid-nineteenth century the Kantian view of spatial intuition was best represented not by nativists like Hering, but by thinkers like Rudolf Hermann Lotze. According to Lotze, the content of spatial experience came from "local signs," or the characteristics of sensations that depend on the place in the nervous system (e.g., a point on the retina) at which they occur; but the form that content takes—the spatial arrangement—had to be provided by the action of the mind, which Lotze thought "had neither the capability nor the disposition for spatial intuitions," but was "by its nature driven to the spatial unfolding of its intensive contents."[13] Helmholtz himself later adopted the theory of local signs, but not the notion that they are organized according to a priori principles.

12. Kant, "On the Form and Principles of the Sensible and Intelligible World" (Inaugural Dissertation, 1770), in Lewis White Beck, ed., *Kant's Latin Writings*, trans. John Handyside (New York: Peter Lang, 1986), 145–92, on 171–72.
13. Rudolf Hermann Lotze, *Medicinische Psychologie* (Leipzig: Weidmann, 1852), 335.

3. Helmholtz on the Construction of Visual Space

Helmholtz opposed both the nativistic account of the spatial content of sensation and the Kantian account of the spatial form of sensation, above all the belief that spatial form is constrained by Euclidean geometry. As he repeatedly and consistently stressed, however, his empirical theory borrowed certain Kantian principles that he thought were indispensable for explaining how a conception of space could be acquired. The connection between Kant and Helmholtz has occasioned considerable philosophical debate, most particularly concerning the fact that his theory of spatial learning took for granted something like Kant's a priori category of causality. In his most important works on perception from 1855 onward,[14] Helmholtz asserted that all learning would be impossible without the prior assumption that like effects always followed from like causes; he called this a "purely logical law" which "can never be overthrown by any possible experience."[15] Some have interpreted such Kantian statements as conflicting with Helmholtz's empiricist principles;[16] others, minimizing Helmholtz's Kantianism, suggest that he regarded the law of causality as an inductive inference rather than as something known a priori.[17] Both views have some merit and yet at the same time mislead. To understand why this is so, a reconsideration of Helmholtz's relation to Kant is in order. This reconsideration ought to clarify three central issues: what did Helmholtz mean by a conception of visual space?; how could such a conception have some of the formal characteristics attributed to it by Kant and Lotze, and yet be gradually acquired from experience?; and in what way did this study of visual space suggest to Helmholtz an approach to the philosophy of mathematics?

Helmholtz himself pointedly denied that the law of causality was acquired by induction; like Kant, he thought that inductive reasoning necessarily presupposes that law. Indeed, he even noted that much

---

14. For example, "Über das Sehen des Menschen," (1855) in *VR⁴* *1*:87–117, on 115–16; *Handbuch³* *3*:30–1; and "Die Thatsachen in der Wahrnehmung," (1878), in *VR⁴* *2*:215–47, on 243–44. For an analysis of the formation of Helmholtz's empiricist theory of vision see Lenoir, "The Eye as Mathematician."

15. *Handbuch³* *3*:30.

16. For example, Alois Riehl, "Hermann von Helmholtz in seinem Verhältnis zu Kant," *Kantstudien* *9* (1904):261–85.

17. For example, Benno Erdmann, "Die philosophischen Grundlagen von Helmholtz' Wahrnehmungstheorie," *Abhandlungen der Preussischen Akademie der Wissenschaften, Philosophisch-historische Klasse* (1921):pp. 1–45. S.P. Fullinwider emphasizes Helmholtz's lifelong adherence to Kant on this point, but nonetheless mentions "the *fact* that Helmholtz described causality as an inductive inference." See "Hermann von Helmholtz: The Problem of Kantian Influence," *SHPS 21* (1990):41–55, on 54.

empirical evidence seemed to contradict the law, especially the intuitive psychological evidence that everyone has, from choices made in daily life, that the human will is free from causal necessity.[18] His repeated Kantian-like statements notwithstanding, he appeared to view causality as an empirical law, at least to some readers, because he did not assume that the law of sufficient reason is *valid a priori*. Instead of a principle that necessarily regulates all possible experience, the law, for Helmholtz, was an "adaptation of our laws of thought to the laws of nature" that "does not have to be complete or exact." He wrote:

> Our intellectual capacities are connected to the capacities of a corporeal organ, the brain, just as the ability to see is connected to the eye. Human understanding works wonderfully well in the world and brings it under a strict law of causation. *Whether it necessarily must be able to control whatever exists or can happen in the world—it seems to me that there is no guarantee for that.*[19]

Thus, while Helmholtz thought that causality was assumed prior to all experience, he clearly did not regard it as transcendental precisely in Kant's sense, as a condition of the possibility of experience to which all experience must therefore conform. Instead, his position resembled David Hume's account of causality: causality is based on a kind of trust in the regularity of nature, a trust that is always repaid but could never be justified by reason or empirical evidence—above all a trust whose utility is contingent upon the regular stability of the external world. "Thus the law of sufficient reason is nothing other than the drive of our understanding to bring all of our perceptions under its control, and is not a natural law."[20] Helmholtz's perceptual work convinced him that this "drive" to fit particular sensations under general conceptions did not require the aid of specific physiological mechanisms or innate ideas to explain human perceptual learning. At the same time, this conviction was also partly methodological: he was dismayed by what he called "the frightful hypothesis-industry" concerning nervous mechanisms. "I believe it is important," he wrote in 1868, "to open people's eyes to how many unnecessary hypotheses they are making, and would rather risk exaggerating the opposing view than remain in the present rut."[21]

Thus, Helmholtz reinterpreted the Kantian a priori as something that Kant never intended it to be: a species of psychological adaptation

---

18. *Handbuch³* 3:30.
19. Ibid., 22–3 (emphasis added).
20. Ibid., 31.
21. Helmholtz to Franz Donders, quoted in Koenigsberger 2:87–8.

to regularities in the external world. This reinterpretation in turn enabled him to redefine the process of learning about visual space, too, as a process of learning about external regularities. For Helmholtz, visual perceptions were not merely sensations, but "ideas [*Vorstellungen*] of the existence, the form, and the place of external objects" that were "formed out of" sensations.[22] He therefore thought that the seemingly immediate awareness we have of spatial relations arises from our constantly learning the regularities governing the dispositions of bodies. Those relations cannot be simply given in intuition in the manner suggested by nativist theories, he argued, because they involve not merely sense-impressions, but also patterns of expectations about *series* of impressions. He noted, for example, that knowing the shape of an object involves knowing how it would appear from arbitrarily many different points of view; knowing its distance involves knowing (among other things) how its apparent size would change as one approached it.[23] Thus Helmholtz analyzed the notion of spatial intuition into associations of given, remembered, and anticipated impressions, all guided by the assumption that these impressions will succeed one another according to some empirical laws.

Helmholtz's analysis simultaneously drew upon an important insight of Kant's and yet departed from Kant in an essential way. Kant's insight concerned the link between spatial perception and the mathematics of space: as Kant said, geometry rests on "the successive synthesis of the productive imagination in the generation of figures." In other words, the evidence of the Euclidean postulates depended on the intuitive clarity of the constructions that they propose—to draw a line from any point to any point, to continue a line indefinitely in either direction, and, in particular, given a line and a point not on the line, to draw exactly one line through the point that does not meet the first line (Euclid's postulate of parallel lines). And these constructions arose not from immediate sensation, but from "successive synthesis," from a series of intuitions. So for both Kant and Helmholtz, spatial knowledge concerned operations we can perform and patterns we can expect to observe.

According to Kant, however, these synthetic operations yielded "apodeictically certain" propositions, that is, propositions "bound up with the consciousness of their necessity."[24] Since such propositions could never be so certain or exact if they were contingent upon features of things in themselves, they had to be determined a priori by the

22. *Handbuch*[3] 3:3.
23. See, e.g., ibid., 21, 28.
24. Kant, *Critique*, 70.

form of our intuition. Helmholtz asserted, by contrast, that spatial intuitions are contingent, yet they seem self-evident and exact because they exploit some of the most fundamental regularities governing the physical world, namely, the laws of motion of bodies and light. For the process of learning the shape or distance of a thing involves turning its various sides to one's view, moving toward or away from it, or comparing it to something whose shape or size is known. Moreover, in moving toward an object we are looking at, Helmholtz argued, the viewer follows the straight lines by which light from the object reaches our eyes. When in geometry we imagine a line to be produced in a given direction, we mentally represent the uniform direction of a light ray by analogy to ordinary experience. Geometry therefore corresponds to intuition, and its evidence and precision rest on the evidence and precision of the underlying physical laws.

Thus the key to organizing experience into its spatial "form" lies in acquired associations between visual and motor sensations. Helmholtz did not believe, however, that these associations simply link visual and tactile sensations, in the way that a written letter is linked with the sound it denotes. Instead, they arise from the *exploration* of the visual field through the motions of the body and its parts, and they depend on our finding regularities governing the disposition of the body. Helmholtz found examples of this exploration in the most primitive human experience: infants learn by trial and error to correlate sight and motion by reaching for objects that they see or extending a hand to block light from reaching their eyes. "We can therefore determine directly through such motions," Helmholtz wrote, "the direction in the visual field where the object is located, and we learn directly to connect the particular local signs of the experience with the place in the visual field where the object belongs."[25] Where Lotze had contended that the organization of local signs obeyed a priori constraints, Helmholtz argued that the signs would eventually be tacitly understood by inferences from these primitive experiences. Moreover, in learning to reach for or to hide from view the objects we see, Helmholtz emphasized that we are learning basic patterns in the behavior of light; optical illusion experiments in which the normal behavior of light is tampered with reveal our reliance on these regularities. He defined normal conditions of observation as those not only in which "light rays must pass in straight lines from luminous points to the cornea," but also in which "we use our eyes as they must be used in order to have the clearest and most easily distinguishable images."[26]

25. *Handbuch³* 3:139.
26. "Die neueren Fortschritte," *VR⁴* 2:357.

For example, if the expected paths of light rays reaching the eye are altered by some refracting medium—for example a pair of prismatic spectacles—the field of view will be shifted or distorted accordingly, and illusions will result, for the sense of touch and the sense of bodily motion will be disturbed from their usual correspondence.

According to Helmholtz's theory, however, we could adjust to the anomalous circumstances using the same processes by which we acquired our original knowledge of spatial relations. "The possibility of [spatial] learning depends," he wrote, "on our having the movable parts of our own body in our visual field," and also on our ability to move our eyes and therefore to vary the positions of other objects in the field.[27] These conditions allowed for a primitive sort of geometrical measurement: our bodily parts serve as standards of size which we always assume to be unchanging, and "an unvarying retinal image, which shifts its position on the retina with the turning of the eye, is like a compass which we move about on a drawing in order to judge which distances are equal or unequal."[28] In both cases, the chief requirement for the objects we manipulate—that which qualifies them as means of probing spatial relations—is their assumed capacity to be moved without change of form; thus, any apparent change can be regarded as a change of distance. We easily acquire the conception of such motions by voluntary movements of the eyes and limbs. For Helmholtz, then, the parts of our bodies that we move at will serve as measuring instruments whose behavior reveals regularities in our environment.

4. From Rigid Motions to the Mathematics of Objective Space

We can now see some of the features that distinguished Helmholtz's empiricism from the position of classical empiricists like Bishop George Berkeley. Helmholtz did not reduce the meaning of visual local signs to tactile sensations, and he did not grant epistemological primacy to tactile space. In fact, he argued that neither the sense of sight nor that of touch could give us direct access to objective relations; he claimed instead that "both senses, which work at the same task[,] . . . complete each other in a very fortunate way."[29] He did not mean by this that sight is necessary in order to form an adequate conception

---

27. Ibid., 335.
28. Ibid.
29. Ibid., 329.

of space, since he frequently emphasized that the blind could do so. Rather, he meant to indicate the fundamental nature of spatial relations and of space itself. He explained that we could recognize the movements of hand and eye as purely spatial changes because spatial relations, unlike all other real relations between substances, are "changeable relations between substances that do not depend on their quality or quantity."[30] Thus, their empirical content derived entirely from the range of possible dispositions of rigid and manipulable bodies and images. For Helmholtz, then, space was entirely characterized by the conditions in which such manipulations were possible and by the constraints upon the motions they represented. Tactile space and visual space were both mere clues, whose interpretation had to be learned, to the structure of objective space, which is the space characterized by the laws of bodily motion and light propagation.

Helmholtz began his formal work on geometry in 1866, when he was finishing the third part of his *Handbuch der physiologischen Optik*; not surprisingly, his geometrical writings show an intimate connection with the account of perceptual space outlined above. His papers, "Über die tatsächlichen Grundlagen der Geometrie" (1866) and "Über die Thatsachen, die der Geometrie zum Grunde liegen" (1868),[31] presented a mathematical answer to Kant's question: if the apparent certainty of geometry rests upon the intuitive validity of geometrical constructions, what is the foundation for the soundness of these intuitions? Helmholtz answered that exact geometrical measurement relied ultimately on the concept of congruence, and that the determination of congruence necessarily involved an empirical element imported from outside of geometry, namely, the motion of rigid bodies. For congruent geometrical figures can be moved into coincidence with one another and this motion must be assumed to leave their form and dimensions unchanged. Helmholtz pointed out that

> we measure the distance between points by applying to them the compass, rule, or chain. We measure angles by bringing the divided circle or theodolite to the vertex of the angle.... Thus, all our geometric measurements are based on the assumption that our supposedly rigid measuring instruments really are bodies of unvarying form, or at least that they undergo only the small changes we know of as arising from vari-

---

30. Ibid., 356.
31. "Über die tatsächlichen Grundlagen der Geometrie," *Verhandlungen des naturhistorisch-medicinischen Vereins zu Heidelberg* 4 (1866):197–202, in *WA* 2:610–17; and "Über die Thatsachen, die der Geometrie zum Grunde liegen," *Nachrichten von der königlichen Gesellschaft der Wissenschaften zu Göttingen* 9 (1868):193–221, in *WA* 2:618–39.

ations of temperature or from gravity acting differently at different places.[32]

Helmholtz evidently thought that the step from spatial perception to geometry depended on the fact that both exploited the same objective constraints on the motions of bodies.

Helmholtz presented his basic approach and results in "Über die tatsächlichen Grundlagen der Geometrie" in 1866; while he was still working on their mathematical formulation in 1868, he learned of Bernhard Riemann's habilitation lecture, "Über die Hypothesen, die der Geometrie zu Grunde liegen."[33] In April 1868, he wrote to Ernst Schering in Göttingen to request a copy of Riemann's lecture, and explained:

> I myself have been occupied with the same subject for the last two years in connection with my investigations in physiological optics, but the work is not yet completed and published, because I still hope to be able to generalize a few points. In particular I cannot make everything as general for three dimensions as I can for two. Now I see . . . that Riemann came to exactly the same results as I did.[34]

While Riemann's and Helmholtz's approaches and conclusions actually differed in important ways, they did agree on several fundamental points. First, they both presented a generalized notion of space as a manifold of several dimensions, defined independently of the metrical relations associated with geometrical space; they pointed out that the metrical relations were not defined on such a manifold naturally, but only by way of special assumptions about the comparison of spatial magnitudes. Second, they both asserted that only spaces of uniform curvature allow for the free motion of rigid bodies without change of form or dimension. Third, they found that the measurement of distances in any such space is given by a generalized form of the Pythagorean Theorem (that is, by a quadratic equation in the coordinate differentials of the space). For the most part, then, Helmholtz independently arrived at a version, restricted to three dimensions, of Riemann's general treatment of metric spaces of constant curvature.[35]

---

32. "Über den Ursprung und die Bedeutung der geometrischen Axiome," in *VR*⁴ 2:1–31, on 23.
33. This lecture, delivered in 1854 to the philosophical faculty of Göttingen, was first published in 1867; see *Abhandlungen der Königlichen Gesellschaft der Wissenschaften zu Göttingen 13* (1867):133–52.
34. Helmholtz to Ernst Schering, 21 April 1868, quoted in Koenigsberger 2:138.
35. For further discussion of the relation between Helmholtz and Riemann, see Stein, "Some Philosophical Prehistory," 21–3; Torretti, *Philosophy of Geometry*, chap. 3; and Bertrand Russell, *An Essay on the Foundations of Geometry* (Cambridge: Cambridge

Riemann's more general treatment notwithstanding, Helmholtz decided to complete and publish his own work because it had a strikingly unique starting point. He sought to answer the question:

> What must be the nature of a magnitude of several dimensions if rigid bodies (i.e., bodies of unvarying relative measurements) are to be able to move everywhere in it continuously, monodromously, and as freely as bodies move in actual space?[36]

The answer proposed in his 1866 paper was derived for the three-dimensional case in "Über die Thatsachen, die der Geometrie zum Grunde liegen." He began by proposing new geometrical postulates that contained some explicit mathematical definitions, first, of space as a magnitude of several dimensions, and, second, of the set of physical facts (in idealized form) that permit the geometrical treatment of space. He presented four "Hypotheses which lie at the foundation of the investigation." First, space of **n** dimensions is an **n**-fold extended manifold, that is, its element, the point, is specified by **n** independently, continuously, and differentiably variable quantities. Second, there exist movable but internally rigid point systems (bodies); this assumption, he noted, "is necessary so that the comparison of spatial magnitudes by congruence can be carried out." To avoid presupposing any method of measurement, he stated the assumption analytically: "Between the 2**n** coordinates of any point-pair belonging to a rigid body, there is an equation independent of the body's motion that is the same for all congruent point-pairs."[37] Third, rigid bodies are capable of completely free motion, so that the points of any one can be carried into the place of any other without changing their internal relations. This assumption made congruence independent of place. Helmholtz here remarked that in the visual field, retinal images enjoy a restricted kind of mobility on the retina, thus underscoring the continuity between his mathematical hypotheses and his hypotheses about perceptual learning.[38] Fourth and finally, the relation of congruence is not affected when a body is transposed into itself: that is, a rigid point-system, rotated about any axis, eventually returns exactly to the place from which it started. (This is the "Axiom of Monodromy"; later, Sophus Lie, in his

---

University Press, 1897; reprint, New York: Dover, 1956), chap. 1, esp. sections 16–26.

36. Helmholtz to Schering, 21 April 1868, quoted in Koenigsberger 2:138. "Monodromously" derives from "monodromy," the property that a body's dimensions are unchanged by a rotation about any axis.
37. "Über die Thatsachen, die der Geometrie zum Grunde liegen," 622.
38. Ibid., 624.

work generalizing Helmholtz's results from the group-theoretic standpoint, showed it to be unnecessary in three or more dimensions, as Helmholtz's results could be derived from the first three postulates alone.)[39] Thus, Helmholtz offered only one rather weak assumption about space, specifying only its topological and differentiable structure, and he proceeded to derive all further structure, especially metrical structure, from the three postulates about motion.

Helmholtz first applied these postulates to the case of a vanishingly small body (an infinitesimally separated point-pair) in a three-dimensional space; he gave a proof for what Riemann had taken as a hypothesis—namely, that the distance between the points must be expressed by some particular case of the generalized Pythagorean quadratic form, now known as a Riemannian metric. He also derived Riemann's claim that if we assume the rigid mobility in space of bodies of finite size—to which the ordinary bodies of our experience are approximations—then the space must be one of constant curvature, of which Euclidean space is but a special case. If we add to the postulates given that space is three-dimensional and infinitely extended, then, Helmholtz concluded, we obtain "the sufficient foundations for the development of the theory of space."[40] By this claim he meant that these postulates narrow down the generality of the first four to the case of the space of our experience, which therefore would have to be either Euclidean or of constant negative curvature (since the requirement of infinite extension would exclude spherical spaces). It followed that only empirical measurements could determine which of the two alternatives was correct. All of these results showed, he added, that "the independence of congruence from place, from the orientation of congruent spatial figures, and from the path on which they are transposed into one another, is the fact on which the measurability of space is based."[41]

Perhaps because he was convinced that (approximately or sufficiently) rigid bodies really existed, he did not try to match Riemann's generality; except as a mere mathematical possibility, he did not seriously consider the possibility of spaces where his postulates would not hold.[42] For the physiological roots of Helmholtz's investigations

---

39. Sophus Lie, *Über die Grundlagen der Geometrie* (Darmstadt: Wissenschaftliche Buchgesellschaft, 1967), originally published in *Berichte über die Abhandlungen der Königlichen Sächsischen Gesellschaft der Wissenschaften in Leipzig, mathematisch-physische Klasse* 42 (1890): 284–321, 335–418. See also Torretti, *Philosophy of Geometry*, 176–79; Russell, *Essay*, 46–50; and J.L. Coolidge, *A History of Geometrical Methods* (Oxford: Clarendon Press, 1940), 78–80.
40. "Über die Thatsachen, die der Geometrie zum Grunde liegen," 638–39.
41. Ibid., 639.
42. One of the interesting mathematical possibilities that Helmholtz did mention was the case of a plane in which one coordinate is imaginary, so that the distance (an

led him to concerns that Riemann would not have shared, yet for which Riemann's postulates accomplished two important things. First, they showed that geometry could have a consistent and sufficient foundation in principles borrowed from physical experience and more general than those that would determine it to be Euclidean or non-Euclidean. Second, by giving mathematical reconstructions of the physical conditions that allow for the comparison of distances, they showed from the mathematical side what Helmholtz had already argued from the psychological side: that geometrical constructions have an empirical basis, and that we can therefore understand "how geometry is possible as a science" without invoking the synthetic a priori judgments that had answered that question for Kant.

There were thus two reasons for Helmholtz's claim to have presented "the *facts* which lie at the foundations of geometry," whereas Riemann had more circumspectly written "on the *hypotheses*." First, Helmholtz's acceptance of ordinary congruence as the necessary foundation of geometry, rather than merely as its historical origin and its present basis, admittedly showed a certain narrowness of view and complacency, at least in comparison with Riemann. This acceptance seemed to imply that the Riemannian character of the metric and the condition of constant curvature had been shown to be factual rather than hypothetical, and that other possibilities could be excluded.[43] The second reason was more compelling: since we are in fact able to treat certain bodies as effectively rigid, it is simply a fact, and not an apodeictic certainty, that geometry is possible as a science. Indeed, in the *Handbuch* Helmholtz had remarked that "if there were no rigid bodies, our geometrical abilities would remain undeveloped and unused, just as the material eye would not help us in a world in which no light existed."[44] Some scholars have interpreted this and similar remarks to mean that the rigid body was a "transcendental concept" for Helmholtz in Kant's sense, a concept "constitutive of physical experience."[45] If that were true, Helmholtz would have been wrong to suppose that this concept could serve as a *factual* foundation for geometry.[46] Since Helmholtz was questioning only our "geometrical abilities," however, and

---

"indefinite" metric) is equal to x^2-y^2 and the line connecting all points equidistant from a given point is a hyperbola rather than a circle. ("Über die Thatsachen, die der Geometrie zum Grunde liegen," 638.) This, of course, is the two-dimensional space corresponding to the Minkowski spacetime of special relativity.

43. See, for example, Torretti, *Philosophy of Geometry*, 167–68; and idem, *Relativity and Geometry* (Oxford: Pergamon Press, 1983), 238–40; see also Stein, "Some Philosophical Prehistory," 22–3.
44. *Handbuch*³ 3:22.
45. Torretti, *Philosophy of Geometry*, 168.
46. Hugo Dingler, "H. Helmholtz und die Grundlagen der Geometrie," *Zeitschrift für Physik* 90 (1934):348–54.

not the possibility of experience in general, the Kantian interpretation seems unnecessary. Instead, in conformity with his theory of perception, he restricted the subject matter of geometry to the conditions on the rigid mobility of measuring instruments. He simply took it as a fact about the laws governing the external world—not as a constraint on our cognitive faculties—that ordinary measurement is possible.

In his well-known popular lecture of 1870, "Über den Ursprung und die Bedeutung der geometrischen Axiome," Helmholtz brought together his most important results in both mathematics and sense perception and confronted his major philosophical opponents. Philosophers who upheld the privileged status of Euclidean geometry commonly invoked a special "space of intuition" governed by Euclid's axioms, and claimed that non-Euclidean geometries, even if they were consistent and of mathematical interest, could only be "conceptual abstractions" with no bearing on the space of intuition.[47] As Helmholtz's researches showed, however, geometry never came directly to the senses, since all knowledge of spatial relations depended on the accumulation of inferences based on the lawlike behavior of bodies. It followed that any geometry, including Euclid's, was only a conceptual abstraction until it could be connected with physical principles concerning the motions of bodies and light. Helmholtz concluded from this that the philosopher's demand that we be able to "visualize" non-Euclidean spaces was beside the point. Instead, the real challenge lay in Kant's claim that the "successive synthesis" of intuitions, both in geometrical construction and in ordinary perception, was somehow constrained a priori to follow the Euclidean pattern.

Helmholtz implicitly took up this challenge through his proffered definition of "visualizable": a thing is visualizable if "one could depict the series of sense impressions one would have if such a thing happened in a particular case."[48] If we represent to ourselves "a connected series of facts" indicating a non-Euclidean geometry, then we are simply imagining dispositions of bodies and light rays according to laws other than those we are accustomed to. The possibility of doing so, Helmholtz argued, meant that Kant must have been mistaken to say that Euclidean geometry necessarily constrains the organization of visual appearances. And the possibility of such anomalous dispositions of light and bodies, along with our capacity to discover the new laws of their behavior, was precisely what Helmholtz's examples of distorting

---

47. Torretti, in *Philosophy of Geometry*, 285–87, describes the "uproar of Boeotians," or the reaction against non-Euclidean geometries by philosophers as diverse as Lotze and Wilhelm Wundt.
48. "Über den Ursprung," in *VR⁴* 2:8.

lenses showed. If light travels on the shortest lines of any space, Helmholtz contended, then to imagine experience in a non-Euclidean space is to imagine objects as they would appear to us if the light they reflected reached our eyes by curved paths. For example, we could approximate the experiences of a pseudospherical space by wearing concave lenses, with a correspondingly negative focal length, which would make remote objects appear nearer than they normally appear to be. Standing still we would notice nothing odd; but if we moved about, we would find that objects dilate as we approach them and seem to come nearer than they actually do—that is, the visual impression would lead us to expect to encounter the object sooner than we actually do. Helmholtz continued:

> But after [we] went about a little, the illusion would vanish, and in spite of the false images [we] would then judge distances correctly. We have every reason to suppose that what happens in a few hours to anyone beginning to wear spectacles would soon enough happen in pseudospherical space.[49]

The example was compelling because, again, it illustrated the principle that the geometrical structure of space amounts to no more than laws governing certain kinds of motion and that learning that structure is only learning what patterns of motion to expect.

The Helmholtzian way of understanding the structure of space changed the subject of arguments about the foundations of geometry, since it implied that geometrical axioms "do not speak of spatial relations alone, but also, at the same time, of the mechanical behavior of our most fixed bodies during motion."[50] Failure to recognize the changed nature of the subject became, in fact, an important element in the philosophical criticism of Helmholtz's ideas. For example, J.P.N. Land, in "Kant's Space and Modern Mathematics" (1877), took Kant's position on the status of Euclidean geometry, and he objected to Helmholtz's claim that non-Euclidean spaces are visualizable. He contended that in Helmholtz's examples

> the most characteristic features of the thing we are trying to imagine must be done away with, and all we are able to grasp with our intuition is a translation of that thing into something else.[51]

---

49. Ibid., 27.
50. Ibid., 30.
51. J.P.N. Land, "Kant's Space and Modern Mathematics," *Mind, A Quarterly Review* 2 (1877):38–46, quote on 42.

He argued, in other words, that non-Euclidean geometry was representable only by constructions ("translations," in the linguistic sense) in Euclidean space. He thought that we could imagine only Euclidean space, "which remains the same from whatever point we inspect it," and "not any space in which motion implies flattening or change of form of any kind."[52] Two decades later, Bertrand Russell, in his *Essay on the Foundations of Geometry* (1897), noted Land's obvious mistake: as Helmholtz and Riemann had shown, motion without change of form is possible in any space of uniform curvature. Russell agreed with Land, however, that Helmholtz's prescription for visualization, "though it enables us to *describe* our new space, does not enable us to *imagine* it, in the sense of calling up images of the way things would look in it."[53] He further agreed that space is *essentially* uniform, so that what Helmholtz thought of as the empirical *fact* of free mobility simply followed from the concept of space; space could not be conceived of as presenting resistance to the motion of bodies, on account of the relativity of position.[54] (So Russell was at that time a kind of Kantian apriorist, but, in contrast to Land, he gave a priori status only to the common features of spaces of constant curvature.)

Helmholtz found that criticisms of this sort ignored his arguments about the origins of geometrical knowledge. As he replied to Land, since our intuitions are not of space itself, but rather of the dispositions of bodies and light, the same kinds of learning and practice were necessary to acquire a conception either of Euclidean or of non-Euclidean space.[55] Therefore Euclidean space was easiest to visualize only insofar as space really was approximately Euclidean and our habits were actually developed as adaptations to its laws. Moreover, given the source of our knowledge of the laws of space, nothing about its mere concept could entail free mobility; the uniformity and homogeneity of space, from Helmholtz's point of view, first derive their empirical meaning from the supposition of free mobility. Uniformity and non-uniformity alike would be evidenced only by patterns of the motions of bodies; the relativity of position is not essential to the concept of space, but rather is a fact about actual laws of motion. It did not matter to Helmholtz's understanding of space whether the external world was "really" spatial or whether spatiality was only a feature of our intuition or

---

52. Ibid.
53. Russell, *Essay*, 73.
54. Ibid., 151.
55. "Über den Ursprung und Sinn der geometrischen Sätze; Antwort gegen Herrn Professor Land," in *WA* 2:640–60, on 645; English translation in *Mind 10* (1878):212–24.

sensory apparatus; he demanded only that "there must exist some other relations, or complexes of relations, which determine at what place in space an object appears to us."[56] He called these relations "topogenous factors," and he held that we need to know only that there is a lawlike connection between these and what we perceive as spatial relations—that "the coming about of spatially different perceptions presupposes a difference in the topogenous factors."[57] So the laws which our mind represents as the structure of space are just the natural laws governing the topogenous factors.

The foregoing account of Helmholtz's philosophy of geometry explains his belief that his general approach could apply to mathematics as a whole. He took up this larger issue in 1887, in "Zählen und Messen, erkenntnisstheoretische betrachtet." He explained that his empirical theory,

> if it no longer recognizes the axioms of geometry as unprovable propositions and in no need of proof [because the form of intuition of space has been empirically analyzed], must also justify itself regarding the origins of the axioms of arithmetic, which stand in the corresponding relation to the form of intuition of time.[58]

It might indeed be "going too far," as Hermann Weyl later remarked, "to claim that arithmetic is the science of time in the same sense that geometry is the science of space,"[59] or to claim that the truth of arithmetical propositions could depend upon observations of external conditions in the same way that geometrical propositions do. But Helmholtz did not go so far. In order to show that arithmetic followed the empiricist pattern of geometry, he did not try to show that arithmetical propositions could be tested empirically—it is, after all, a formal, deductive structure that "investigates which different ways of combining [arithmetical] symbols (calculative operations) lead to the same result."[60] Instead, he argued that, like geometry, it arose from the application of some basic, "intuitive" operation to empirical circumstances in order to find, test, and exploit lawlike features of the external world.

In this case the basic operation was not rigid motion in external space, but the internal psychological operation of counting; and in

---

56. Ibid., 657.
57. Ibid.
58. "Zählen und Messen," 357.
59. Hermann Weyl, *Philosophy of Mathematics and Natural Science* (Princeton: Princeton University Press, 1949), 39.
60. "Zählen und Messen," 359.

place of spatial congruence, the objective relation to be explored was physical "likeness" in general. Helmholtz explained his viewpoint thus:

> I consider arithmetic, or the theory of pure numbers, as a method constructed on purely psychological facts, through which the logically consistent application of a symbol system (the numbers) of unlimited extent and an unlimited possibility of refinement is taught.... [B]y means of this symbolic system of the numbers we give descriptions of the relations of real objects, descriptions which, where they are applicable, can reach any required degree of exactitude....[61]

The empirical question corresponding to that of "the meaning of the axioms of geometry" then became: "what is the objective sense of our expressing the relations of real objects as magnitudes by using designated [*benannte*] numbers; and under what conditions can we do this?"[62] Helmholtz explicated these questions with two others: what does it mean to say that two objects in a certain relation are alike? and what sort of physical relation must objects have if we are to be able to treat their properties as comparable magnitudes? In considering these questions he did not reach any striking new results, as he had in his geometrical investigations, but a brief look at his approach will clarify his view of the empirical significance of arithmetic.

Since the psychological facts mentioned concerned the process of counting, Helmholtz predictably based the concept of number on the concept of ordinal number. Independently of the familiar philosophical controversy over whether ordinal or cardinal number is conceptually prior,[63] the choice of the former obviously suited Helmholtz's aim to provide a *genetic* account of arithmetic. "Counting," he explained, "is a method based on the fact that we are able to retain in our memory the sequence in which acts of consciousness succeed one another."[64] This fact enabled our "forefathers" to establish a "lawlike series" of symbols, the natural numbers, as a way of determining the cardinal number of any set of objects. Its foundation lay, he added, in the deeper psychological fact that there is an unambiguous distinction between immediately present acts of consciousness and remembered past ones, "an opposition related to the form of intuition of time." "In this sense ordering in the time sequence is the inevitable form of our inner intuition."[65] Helmholtz did not attempt to relate this "form of intuition"

---

61. Ibid.
62. Ibid., 359–60.
63. See, for example, Weyl, *Philosophy of Mathematics*, 34–5.
64. "Zählen und Messen," 360.
65. Ibid., 361.

explicitly to the principle of mathematical induction, but rather took the principle for granted, as is apparent whenever he extends a relation between a number and its successor to the entire series with the phrase "and so on without limit."[66]

These foundations sufficed for Helmholtz's purpose, which was to prove "the series of axioms necessary for the foundation of arithmetic for the concepts of number and sum from which we started out, concepts taken only from inner intuition."[67] His models for this undertaking were the attempts by Hermann and Robert Grassmann to derive arithmetic from a few axioms of equality.[68] He then turned to the empirical problem, which was how we designate numbers representing physical magnitudes in order to compare those magnitudes. To solve this, he attempted to show that "physical likeness" obeyed the axioms for arithmetical equality already established; for example, physical likeness is transitive, so that like physical magnitudes can be substituted for one another. He emphasized, however, that this conformity to the axioms does not provide an empirical test of the latter. "We should not wonder if the axioms of arithmetic justify themselves in the course of nature," he remarked, since "we recognize as addition only those physical connections that satisfy the axioms of addition."[69] Thus, Helmholtz's empiricism did not seek to reduce mathematical truth to empirical truth. Instead, it described arithmetic as the theory of a process, "based on psychological facts," that identifies and systematizes lawlike relations in nature, "reducing the varied manifold of things and changes before us to quantitative relations."[70] For Helmholtz, the truth of arithmetic remained a matter of internal consistency, but its objective significance lay in the correspondence it provided between a subjective operation and objective states of affairs.

5. Conclusion

In 1878, Helmholtz wrote: "I believe that the most essential advance of recent times is the resolution of the concept of intuition into the elementary processes of thought."[71] Kant, he continued, did

---

66. For example, ibid., 365.
67. Ibid., 373.
68. Ibid., 357. See Hermann Grassmann, *Die Wissenschaft der extensiven Grösse oder die Ausdehnungslehre* (Leipzig: Wigand, 1844), and Robert Grassmann, *Die Formenlehre oder Mathematik* (Stettin: published privately by the author, 1872); and see also Michael Crowe, *A History of Vector Analysis* (Notre Dame: University of Notre Dame Press, 1967; reprint, New York: Dover Publications, 1985), chap. 3.
69. "Zählen und Messen," 384.
70. Ibid., 391.
71. "Die Thatsachen in der Wahrnehmung," VR^4 2:244.

not pursue this analysis sufficiently, since he took geometrical intuition to be not further reducible, and indeed to be a condition of possible experience. In analyzing the course of Helmholtz's further pursuit of this reduction and some of the scientific and philosophical concerns that played a part in it, this essay has sought to illuminate the intimate connection between the sensory and the mathematical approaches to space that lay at the core of Helmholtz's, as at Kant's, efforts. It has also sought to show why, in spite of Helmholtz's psychologistic interpretation of Kant's view of causal necessity, Helmholtz did not fully adopt a Humean position. Hume, Helmholtz wrote, was skeptical enough about the empirical world to end with "denying all possibility of objective knowledge."[72] For Helmholtz, by contrast, the empirical world revealed deep relationships in the world of objective law, and human awareness of spatial movement and temporal succession made it possible to express those laws through mathematical structures.

The connection between perception and mathematics reveals something of Helmholtz's metaphysics, inasmuch as it exhibits his conception that space consists of a certain kind of lawlike ordering of the phenomena of motion. More precisely, perhaps, it reveals why certain traditional metaphysical questions about space—questions concerning what kind of "thing" space is, whether it is substance, accident, or relation, and so on—could be ignored in researches like Helmholtz's. For Helmholtz reduced questions about the nature of space to those which bore on the unfolding of natural processes and which suggested experimental results. This achievement makes it easy to understand why subsequent philosophers in the logical empiricist tradition not only adopted Helmholtz's analysis of the empirical content of geometry, but also looked back on him as a philosophical and methodological forebear.[73] Moreover, it shows that Helmholtz, like Riemann, was an obvious forebear to modern spacetime physics, which continues to use the physical laws governing bodies and light to explore the structure of space and time.[74] In general relativity the

---

72. *Handbuch*³ *3*:32.

73. See, for example, Moritz Schlick, "Helmholtz als Erkenntnistheoretiker," in Emil Warburg, Max Rubner, and Moritz Schlick, *Helmholtz als Physiker, Physiologe, und Philosoph* (Berlin: C.F. Müllersche Hofbuchhandlung, 1922), 29–40.

74. See, for example, Hermann Weyl, *Mathematische Analyse des Raumproblemes* (Berlin: Springer, 1923). More recently, Jürgen Ehlers, Felix Pirani, and Alfred Schild derived the metric of general-relativistic spacetime from assumptions about falling bodies and light rays; they remarked that their method "has some similarity to Helmholtz's derivation of the metrics of spaces of constant curvature." (See their "The Geometry of Free Fall and Light Propagation," in L. O'Raifeartaigh, ed., *General Relativity* [Oxford: Oxford University Press, 1972], 63–84, quote on 65.)

fundamental motion of a body is gravitational free-fall, which together with light propagation reveals the variable curvature of spacetime as determined by the influence of matter and energy. Even if relativity theory has left behind Helmholtz's assumptions about rigid motion and uniform curvature, it has done so by following his program for treating geometry as an empirical science.

14

Helmholtz and Classicism

The Science of Aesthetics and the Aesthetics of Science

Gary Hatfield

1. Introduction

An avid musician and an erstwhile teacher of anatomy to artists, Hermann von Helmholtz made a lasting contribution to music theory through his *Die Lehre von den Tonempfindungen als physiologische Grundlage für die Theorie der Musik* (1863) and he applied his optical expertise to the art of painting in lectures during the early 1870s. Although he largely eschewed the subject of aesthetics, neither developing nor espousing a full-scale aesthetic theory, he nonetheless brought his physiological and psychological researches to bear on selected aesthetic topics. Indeed, he advertised his *Tonempfindungen* as an attempt to connect physical and physiological acoustics with musical science and aesthetics, even though he himself restricted the extent of the connection.[1] More generally, in his lectures on Goethe, in his 1862 address to the assembled faculties of the University of Heidelberg, and in other

---

Acknowledgments: Earlier versions of parts of this paper were presented at a lecture series on science and the humanities celebrating the opening of the John Crerar Science Library of the University of Chicago (May, 1985) and at the meeting of the History of Science Society in Gainesville (October, 1989); I benefited from comments on both occasions. Parts of this essay develop points first raised in Gary Hatfield, *The Natural and the Normative: Theories of Spatial Perception from Kant to Helmholtz* (Cambridge, Mass.: MIT Press, 1990).

1. *Die Lehre von den Tonempfindungen als physiologische Grundlage für die Theorie der Musik* (Braunschweig: Vieweg, 1863), 1–8, 357–59. On the social and cultural context of Helmholtz's remarks about the arts, see David Cahan's essay, "Helmholtz and the Civilizing Power of Science," in this volume.

lectures, he compared the thought processes of artist and scientist. Ultimately, he sought to explain the artist's mental processes by appeal to general psychological laws, the very laws he saw operative in both scientific inference and the unconscious inferences of everyday perception.

The nineteenth century brought a highly accelerated separation of the natural sciences from one another and from other domains of knowledge, and Helmholtz's writings on the arts are thus of interest not least for what they reveal about his conception of the relation between natural science and other intellectual activities. In nineteenth-century aesthetic writings there was disagreement over whether aesthetics could itself be a science (*Wissenschaft*). Immanuel Kant had said it could not; others, including Johann Friedrich Herbart and his follower Robert Zimmermann, said it could.[2] Although Helmholtz probably was not closely acquainted with the literature on this topic, he knew of and discussed the issue.[3] Further, he had practical acquaintance with the issue of whether aesthetic questions could be answered by *natural* science, inasmuch as he attempted to apply the findings of natural science to music and painting. Perhaps as a result of his work on music, he recognized strict boundaries to natural scientific explanations of the arts, boundaries which precluded his advance into serious aesthetics through physiological and psychological research. He codified these boundaries as part of a general distinction between *Naturwissenschaften* and *Geisteswissenschaften*. He stands out as an early non-Hegelian and non-Romantic advocate of the position that aesthetic principles are historically conditioned. Having rejected Hegel's historicist metaphysics, he argued that historical subject-matters must be approached empirically, not metaphysically, and that the appropriate empirical techniques for doing so were those of the *Geisteswissenschaften*, which differ in kind from those of the natural sciences.

Helmholtz's writings on the arts reveal not only his conception of the cognitive status of aesthetic research; they also embody a conception of the cognitive activity of the artist in comparison with that of

---

2. Immanuel Kant, *Critique of Judgment*, trans. Werner S. Pluhar (Indianapolis, IN: Hackett Publishing Company, 1987), 172; Johann Friedrich Herbart, *Lehrbuch zur Einleitung in die Philosophie* (Königsberg: Unzer, 1813), 72–3; and Robert Zimmermann, "Zur Reform der Aesthetik als exakter Wissenschaft," *Zeitschrift für exacte Philosophie* 2 (1862):309–58.
3. *Tonempfindungen*, 357–59; see also "Ueber das Verhältniss der Naturwissenschaften zur Gesammtheit der Wissenschaften. Akademische Festrede," (1862), in *VR*[4] 1:157–85. The only aestheticians mentioned by Helmholtz are Friedrich Theodor Vischer and Eduard Hanslick.

the scientist. In so doing, they again show Helmholtz taking a position on a matter that was the subject of ongoing dispute within aesthetics, and again it may be doubted whether Helmholtz had detailed acquaintance with this dispute. Nonetheless, his comparative account of the thought processes of artist and scientist is of double interest. First, it highlights a fundamental development in his own thought—specifically, in his psychological theory of unconscious inference—and in so doing exemplifies what may be termed his "classicist" aesthetics of scientific explanation. Second, his psychological theory of "artistic intuition" reveals his own naturalistic solution to one of the central aesthetic problems of his century: the relation between the paired contrasts of thought and feeling or of understanding and imagination, contrasts which reflected the division between (neo-)Classicist and Romantic aesthetics. Aestheticians had long disputed whether aesthetic judgments are based primarily in sensation and feeling, or in intellectual apprehension, or in Kant's proposed noncognitive relation between imagination and understanding.[4] Helmholtz's position cut across the dispute by emphasizing the intellectual content of artistic and aesthetic judgment in a manner consistent with a classical aesthetics; at the same time, he effectively reduced intellect to imagination. He thus proposed a means of achieving a goal of many post-Kantian philosophers of art: a reconciliation between imagination and understanding. Earlier Classicists equated aesthetic judgments with the understanding's apprehension of an ideal type in nature (or in an artist's idealized imitation of nature). A prominent Romantic tendency emphasized the expression of feeling (contrasted with understanding) through individual imagination. Hegelian aestheticians saw the artistic imagination as giving concrete realization to the "Idea," an object of Reason. Helmholtz gave imagination the dominant role in artistic and aesthetic cognition, but he viewed it as a filter for the lawlike regularity of an ideal type, the operation of which served to reveal general truths.

This essay first pursues these themes through two case studies: an examination of Helmholtz's application of sensory physiology and psychology to music (Section 2) and to painting (Section 3), respectively. Consideration of these concrete cases leads to an analysis of Helmholtz's account of the methodology of aesthetics, more specifically to

---

4. For a general survey of eighteenth- and nineteenth-century aesthetics, see Monroe Beardsley, *Aesthetics from Classical Greece to the Present* (University, AL: University of Alabama Press, 1975), chaps. 8–10. The terms "classical" and "classicist," like the term "romantic," have a variety of uses with varied connotations; on these terms as used herein, see nn. 79 and 87.

his formulation of the distinction between *Geisteswissenschaften* and *Naturwissenschaften* (Section 4). The essay then turns to examine the development of Helmholtz's comparative account of the thought processes of artist and scientist. This development parallels the development of his psychological theory of judgment into his famous theory of unconscious inference (Section 5); comparison of Helmholtz's two Goethe lectures shows that his changed conception of unconscious inference led him to posit a closer relation between artist and scientist than he had at first (Section 6). The latter development, as this essay argues in Section 7, exemplifies his classical "aesthetics of science."

2. The Physiology and Psychology of Music Perception

Helmholtz ventured to apply natural science to the arts in his monumental *Tonempfindungen* in the hope of illuminating the relation between vibrating musical instruments and their effects upon the listener. Music must of course be experienced to be appreciated, and human auditory experience depends on the way the ears are affected by sound. Helmholtz was, however, circumspect in his claims of the extent to which natural science could explain the effects of music upon the listener. How is one to treat the subtleties of feeling that works of music arouse when agreement is lacking, especially when historical and national differences are taken into account, about which works arouse which feelings? Helmholtz simplified his task by attempting to account solely for the experience of musical tones in their simplest melodic and harmonic relations, thereby abandoning any attempt to account for musical expression or to analyze whole compositions aesthetically. Pitch, tone quality, consonance and dissonance, and the development of musical scales are the chief topics of his work, and he believed that all except the last could be approached within the confines of natural science.[5] Systematic investigation of the limits Helmholtz set on natural-scientific explanation in aesthetics must be saved for Section 4. This section restricts its focus to the physiological and psychological aspects of Helmholtz's work on auditory perception and its application to music.

By the early 1860s, when Helmholtz wrote his *Tonempfindungen*, many basic facts pertaining to the physics of sound were known, including the relation between frequency and pitch and the fact that the

---

5. *Tonempfindungen*, 1–8, 357–59.

sounds produced by musical instruments typically are complex, consisting of a fundamental and overtones (or upper partials).[6] In the early 1840s the physicist Georg Simon Ohm applied Fourier analysis to sound waves, contending that musical sounds are analyzed by the ear into sine wave vibrations, each of which gives rise to an appropriate sensation of tone. Helmholtz, using resonators and other sympathetically vibrating bodies, provided numerous demonstrations that typical musical sounds actually do contain upper partials that a listener can experience under appropriate conditions. This led him to accept Ohm's law, which he stated as follows: "the human ear senses pendular vibrations alone as simple tones, and resolves all other periodic motions of the air into a series of pendular vibrations, sensing the simple tones which correspond with these simple vibrations."[7] In his celebrated resonance theory of hearing, Helmholtz posited that the inner ear contains a large number of resonating receptors—he accepted an estimate of 3,000—each tuned to vibrate maximally at a different frequency across the range of audible sounds.[8]

The resonance theory provided a physical analysis of incoming sound waves, but what of the physiological and psychological upshot? That is, how do these analyzed sound waves affect the nervous system, and how are they related to conscious experience? Helmholtz contended that each cochlear resonator is attached to a single nerve fiber which, when the resonator vibrates, produces the sensation of a single tone of an appropriate pitch. In this manner, he extended the law of specific nerve energies to the sensation of pitch. According to this law, as stated by Helmholtz's revered teacher Johannes Müller, the fact that stimulation of the various sense organs gives rise in each case to experience of a specific phenomenal character—light and color in the case of vision, sound in audition, and so on—is to be explained by the peculiar (and, as yet, unknown) characteristics of the nerves leading

---

6. A.J. Ellis, the English translator of Helmholtz's *Tonempfindungen*, translated the German word *"Oberton"* as "upper partial tone," and contended that *"Oberton"* was short for *"Oberpartialton"* (See *On the Sensations of Tone as a Physiological Basis for the Theory of Music*, 4th ed., trans. A.J. Ellis [London: Longmans Green, 1912; reprint New York: Dover, 1954], 25n). Helmholtz's usage included *"Oberton," "harmonischer Oberton,"* and *"Partialton."*

7. *Tonempfindungen*, 97; the whole of chapter 4 of the *Tonempfindungen* is devoted to Ohm's law and its support. On Helmholtz's work in physical and physiological acoustics, see Stephan Vogel's essay, "Sensation of Tone, Perception of Sound, and Empiricism: Helmholtz's Physiological Acoustics," in this volume; see also R. Steven Turner, "The Ohm-Seebeck Dispute, Hermann von Helmholtz, and the Origins of Physiological Acoustics," *BJHS* 10 (1977):1–24.

8. *Tonempfindungen*, 219. In the third edition (225–31), Helmholtz designated the basilar membrane as the chief resonating organ, thus relegating the organs of Corti to secondary status.

away from each sense organ or of the termination of those nerves in the brain. Earlier, Helmholtz had adapted this principle to color vision by positing three kinds of optic nerve fiber, each with its own specific nerve energy, corresponding to the three primary colors; the colors we experience are combinations of color sensations from these three kinds of nerve.[9] In extending the principle to hearing, he increased the number of specific energies, positing for each cochlear resonator a separate nerve fiber with its own specific nerve energy that, when excited, produces the sensation of a tone of a particular pitch.[10] In this fashion he carried through his Ohmian analysis from the physical, to the physiological, to the psychological levels, positing a discrete, pure sensation for each simple pendular vibration in the physical stimulus, mediated by cochlear resonators attached to discrete nerve fibers.

The postulation of a discrete, pure sensation for each cochlear nerve fiber bridged the gap between the physiological and the mental. However, Helmholtz did not conceive the posited sensations as mere neural activity, but instead as mental events possessing their own phenomenal character. He spoke of them as being recognized, identified, and compared, in accordance with their status as mental events. Yet as individual partial tones they—like Gottfried Wilhelm Leibniz's *petite perceptions*, Ernst Heinrich Weber's "pure sensations," and Hermann Lotze's "unconscious mental states"—typically go unnoticed.[11] We experience a musical note sounded by a single instrument as single, and not as a chord consisting of the fundamental and upper partials. The postulation of single sensations raised two questions: Why should discrete sensations be postulated in the first place? And, given that they have been postulated, how does the unity of musical notes as experienced arise?

Others before Helmholtz had faced these questions. In the eighteenth century it was known that the presence and strength of various

---

9. On Helmholtz's color theory, see Richard L. Kremer's essay, "Innovation through Synthesis: Helmholtz and Color Research," in this volume. For a critical discussion of Helmholtz's assumption that Müller's law of specific nerve energies was Kantian in inspiration see Gary Hatfield, *The Natural and the Normative: Theories of Spatial Perception from Kant to Helmholtz* (Cambridge, Mass.: MIT Press, 1990), 155–56.
10. *Tonempfindungen*, 219–23.
11. Ibid., 552–59; previously (ibid., 101), he distinguished "pure sensation" ("*reine Empfindung*") from "perception" ("*Wahrnehmung*"). In the fourth edition of the *Tonempfindungen* (Braunschweig: Vieweg, 1877), 107, he recast the distinction in Leibnizean terminology, as one between those sensations that are perceived (*percipirt*) and those that are apperceived (*appercipirt*). E.H. Weber, "Der Tastsinn und das Gemeingefühl," in R. Wagner, ed., *Handwörterbuch der Physiologie*, 4 vols. (Braunschweig: Vieweg, 1842-53), 3:481–543; and R.H. Lotze, *Medizinische Psychologie oder Physiologie der Seele* (Leipzig: Weidmann, 1852), 179–80.

upper partials distinguished notes of the same fundamental pitch when played on different instruments, thereby giving each instrument its distinctive tone quality. In the decades prior to Helmholtz's own work, there was disagreement about whether these upper partials produce distinct sensations, as Helmholtz later maintained, or directly produce an experience of the appropriate tone.[12] Given his postulation of distinct sensations for the fundamental and upper partials, Helmholtz needed to explain not only why the discrete sensations fuse into a single perceived note, but also how they do so in such a way that when several instruments are played simultaneously the multitude of pure sensations are fused into discriminably distinct tones corresponding to each instrument. His explanation appealed to the commonsense generalization that the fundamental and partials from single instruments are always found together (perhaps with other sounds peculiar to the instrument, such as the scraping of the bow), whereas they are found in the presence of other instruments only on occasion. He argued that we combine the partials of single instruments as a result of habitually finding them together, and that they somehow then fuse into a phenomenally single note, the quality of which depends upon the strength and number of the partials actually entering into the mix.[13] The actually distinct partial tones are singly unnoticed because they are combined by unnoticed mental operations into a phenomenally simple (though actually complex) tone.

Although Helmholtz thus explained how the partial tones are combined given that they are originally distinct, what he really needed was justification for his initial posit of distinct sensations corresponding to each partial tone. For although the existence of partials in physical wave pulses typically was not in question, the existence of discrete sensations for each partial tone was.

In the *Tonempfindungen*, Helmholtz offered a problematic account of the basis for postulating discrete sensations. On the one hand, he repeatedly averred that his position could be justified independently of physiological hypotheses. Thus, in the Introduction he asserted

> that the following investigation deals only with the analysis of actually existing sensations—that the physical methods of observation are almost

---

12. H.F. Cohen, *Quantifying Music* (Dordrecht and Boston: Reidel, 1984), 102–3, 234–43; Carl Stumpf, *Tonpsychologie*, 2 vols. (Leipzig: Hirzel, 1883–90), 2:17–20; Turner, "Ohm-Seebeck Dispute"; and *Tonempfindungen*, chap. 4.

13. *Tonempfindungen*, 1st–3rd eds., 105; and "Ueber die physiologischen Ursachen der musikalischen Harmonie," (1857), in *VR*[4] *1*:119–55, on 147. Helmholtz dropped this explanation from the fourth edition of the *Tonempfindungen*; see Stumpf, *Tonpsychologie* 2:70–81, for a discussion.

solely meant to facilitate and assure the work of this analysis and check its completeness—and that this analysis of the sensations would suffice to furnish all the results required for musical theory, even independently of my physiological hypothesis concerning the mechanisms of hearing.[14]

He construed his task as one of establishing psychophysical mappings, and claimed that this could be accomplished without appeal to physiology. In the preface to the third edition of the *Tonempfindungen* (1870), he responded to the charge that his approach was "too coarsely mechanical" by again denying any essential role for his physiological account of the origin of auditory sensations.[15] On the other hand, he readily admitted that partials are difficult to experience for untrained observers without the aid of resonators or other special apparatus.[16] It was Helmholtz's general view that we are in the habit of disregarding sensations, which are correlated with the state of the nervous system, in favor of the perceptions formed by combinations of sensations, since these perceptions are correlated with and hence informative about external objects.[17]

Yet Helmholtz did not regard it as impossible to isolate partial tones within conscious perception. The solvent needed to resolve individual partials out of compound perceptions was the proper direction of attention, which could serve to resolve the upper partials out of a compound tone. Helmholtz accounted for the difficulty in observing partials by appealing to the difference between the naive listener, who has not learned to direct his or her attention properly, and the trained listener, who, perhaps with the aid of resonators, has learned to direct his or her attention to partials in such a way that they can be singly discriminated. Accordingly, upper partial tones can be experienced. In Helmholtz's view, perceptions produced by directed attention simply reveal the primitive elements of sensation that underlie our ordinary perceptions, and these latter, though phenomenally simple, are in actual fact complexes of sensations.[18]

It is precisely at this point that Helmholtz's claim to be able to conduct his analysis of "actual sensations" independently of his physiological hypotheses becomes suspect. In the isolation of partials

---

14. *Tonempfindungen*, 9.
15. *Tonempfindungen*, 3rd ed., ix.
16. *Tonempfindungen*, 98–101; and "Musikalische Harmonie," 145–49.
17. *Handbuch*³ 3:7–10.
18. *Tonempfindungen*, 84–9, 100–12. In the fourth edition of the *Tonempfindungen* Helmholtz abbreviated his discussion of attention (4th ed., 106–11); he retained his commitment to unnoticed sensations that can be brought to consciousness through the proper directing of attention (ibid., 111).

through trained introspection, the experience of the tone itself is altered. In one of Helmholtz's demonstrations, a simple tone sounding like the *oo* in *too* is combined with a tone an octave higher to produce, when the two tones fuse, a sound like the *oa* in *toad*.[19] Helmholtz reported that through directing his attention to the upper partial, he could isolate the higher tone, with the consequence that he heard both it and the lower tone, the latter sounding again like the *oo* in *too*. Demonstrations such as this constituted the entire empirical case for the existence of simple sensations corresponding to each partial tone. Yet as later critics observed, the demonstration did not prove the existence of simple sensations, but only showed that in some circumstances a compound wave form produced the experience of two tones, whereas in other cases it produced the experience of one tone.[20]

Helmholtz's confidence that partial tones exist and are uncovered through the analytic solvent of attention rested upon his conception of the relation between neural activity and sensation. Along with a host of other investigators, he accepted the one-fiber/one-sensation doctrine, according to which simple sensations are the inalterable product of the activity of single neural fibers; they are thus the primitive elements of mental life. By contrast, any psychological state that can be altered by attention, such as a compound tone, is a derivative product of the combination of sensations. The resolvability of the *oa* sound (as above) could thus be taken as evidence that it was compound; *ipso facto*, the products of resolution are the elements from which the compound has been composed.[21] This conception of the relation between

---

19. *Tonempfindungen*, 109–10. The English-language version of this example is from *On the Sensation*, trans. Ellis, 60–1.

20. For one instance of criticism, see Wolfgang Köhler, *Gestalt Psychology* (New York: Liveright, 1929), 124. Helmholtz also described a set of demonstrations to show that the sounding of strings and other bodies is properly analyzed physically (and mathematically) into a fundamental and upper partials, and he described demonstrations which show that if the upper partials are absent on one occasion and present on another, then, all else being equal, different sounding tones result (*Tonempfindungen*, 89–96). But this again shows only that different complex wave forms sound different; the fact that the complex wave forms are subject to Fourier analysis was admitted by the opponents of Ohm's law (ibid., 100–1); the matter in question was whether individual sensations are produced only by simple pendular motions or are produced by more complex wave forms.

21. *Tonempfindungen*, 219–20; and "Musikalische Harmonie," VR^4 1:139–40. He did not here state the assumption that inalterability through directed attention is the mark of a truly simple sensation; but he did state such an assumption in *Handbuch*³ 3:12–3. He explained the production of simple sensations by postulating that nerve fibers are insulated, and he compared such fibers to telegraph wires; see "Die neueren Fortschritte in der Theorie des Sehens," in VR^4 1:265–365, on 295–98. On the central role of the one-fiber/one-sensation hypothesis in Helmholtz's sensory physiology and psychology, see Hatfield, *The Natural and the Normative*, 172, 184, 196.

neural activity and sensation was disputed by contemporary sensory physiologists, most notably by Ewald Hering in his theory of binocular vision.[22] Be that as it may, Helmholtz's own arguments for the existence of simple sensations ultimately rested upon the one-fiber/one-sensation doctrine. Here, at least, he gave primacy to physiology in resolving a controversy about the psychology of sensation.[23]

When treating the aesthetic qualities of musical tones, Helmholtz attempted to extend physiological explanation as far as it could take him. The second part of the *Tonempfindungen* examined the basis of harmonic relations and of disturbances to them; Helmholtz's account of consonance and dissonance was the centerpiece. In his explanation of consonance he rejected earlier psychological explanations in favor of a physiological one. By the eighteenth century the Pythagorean discovery that strings whose lengths form simple ratios give consonant sounds when plucked had been formulated in terms of the ratios between pitch numbers, and hence could be applied to all sounding bodies (not just strings).[24] Helmholtz reported that the predominant account of musical consonance in the eighteenth century was that proposed by Leonhard Euler in 1739, according to which the human mind takes a special pleasure in the orderliness of simple ratios of pitch numbers. In Euler's account, the mind actually perceives the pitch numbers of tones from separate instruments and discerns the commensurability or incommensurability of their ratios. A feeling of pleasure then results from those unconscious comparisons in which commensurability is discerned.[25]

Helmholtz rejected Euler's explanation. He did not object to the appeal to an unanalyzed sense of pleasure in simplicity and order, nor to Euler's positing that mental acts of comparison and pleasure-taking were unconscious. Rather, he objected that Euler provided no account of how the mind actually comes to recognize and compare the simple ratios of pitch numbers. He found the work of Euler's contemporaries, Jean-Philippe Rameau and Jean d'Alembert, more to his liking, for

---

22. Ewald Hering, *Beiträge zur Physiologie* (Leipzig: Engelmann, 1861–64), 323, and *Zur Lehre vom Lichtsinne*, 2nd ed. (Vienna: C. Gerolds Sohn, 1878), 2–3.

23. *Tonempfindungen*, 99–100; see also "Ueber das Sehen des Menschen," in *VR⁴* 1:85–117, on 111, and "Ueber das Verhältniss," *VR⁴* 1:184. For a review of the psychological and physiological arguments for positing discrete pure sensations, see Stumpf, *Tonpsychologie* 2:1–17, 70–99.

24. The generalization, which built upon seventeenth-century work, was made explicit in the work of Joseph Sauveur, Leonhard Euler, Jean-Philippe Rameau, and Jean d'Alembert; see Cohen, *Quantifying Music*, 231–38.

25. *Tonempfindungen*, 347–51. Leonhard Euler, *Tentamen novae theoriae musicae* (St. Petersburg: St. Petersburg Academy of Sciences, 1739; reprint New York: Broude, 1968), chap. 2, secs. 5,7, 16–20; chap. 4, secs. 2–3.

they emphasized the physical basis of consonance in various patterns of matching wave peaks.[26] In the 1850s, Friedrich Theodor Vischer, in a work on musical aesthetics cited by Helmholtz, postulated that consonance results from the soul's (unconscious) perception of regularity in the relations among consonant tones, while dissonance results from the perception of irregularity in those relations.[27] Yet none of these writers provided what Helmholtz wanted: an explanation of why these matching patterns should yield consonance. Helmholtz saw himself not so much as rejecting the earlier conception outright, as correcting and completing it by explaining how the relations among consonant tones are perceived.

In Helmholtz's account, consonance results from the absence of beats and dissonance from their presence. In the case of dissonant notes, for which the ratio of frequencies is not a pair of small integers, the fundamentals and harmonics of the two notes combine to produce beats (fluctuations in intensity that are too slow to be heard as a pitch); Helmholtz considered this intermittency to be irritating, and hence to result in dissonance. The waveforms of two such notes may also combine to produce combinational tones, which, if audible, sound like an out-of-tune bass accompaniment. Consonance, by contrast, occurs with tones whose pitch numbers form simple ratios because the minima of the waveforms stand in a simple and uniform relation to one another and hence do not produce varying interference; furthermore, some of the upper partials of consonant tones coincide. Helmholtz argued that the most perfect consonance occurs with the octave, in which the frequencies stand in a ratio of two to one and there are numerous correspondences between the upper partials of the two notes. In a fifth, such as C and G, the frequencies are related in a ratio of two to three, and there are numerous coincident upper partials. According to Helmholtz, consonance of simultaneously sounding notes is not explained psychologically, in terms of the mind's pleasure in simple ratios, but physically and physiologically, in terms of the absence of varying interference. He thus met his own objection to Euler by showing how the mind could perceive not the ratios themselves, but rather the physical effects of those ratios, manifested as continuous and intermittent stimulation of the auditory nerves.[28]

---

26. *Tonempfindungen*, 351–54.
27. Friedrich Theodor Vischer, *Aesthetik oder Wissenschaft des Schönen*, 3 vols. (Leipzig and Stuttgart: Carl Mäcken, 1846–57), 3:847, 855, 857–58, 881.
28. *Tonempfindungen*, 340–46; and "Musikalische Harmonie," VR^4 *1*:149–55. Helmholtz's theory attracted numerous critics; for a contemporary review, see Carl

While Helmholtz's account of consonance relied on the physical interaction of sound waves, his account of the aesthetics of simple melodic relations was thoroughly psychological in character and very much in the spirit of Euler's account of consonance. He adopted the commonplace that the aesthetic effects of a work of art depend upon the orderly relations of the parts to the whole—an orderly relation that is not so simple and obvious as to be readily made the object of conscious reflection, but one which must be intuitively grasped. The problem for the theoretician, as Helmholtz put it, "is to understand how regularity can be apprehended by intuition without being consciously felt to exist." He regarded this task as central to aesthetics, for, as he enjoined, "this unconsciousness of regularity is not a mere accident in the effect of the beautiful on our mind, which may indifferently exist or not; it is, on the contrary, most clearly, prominently, and essentially important."[29] Although Helmholtz forsook full-scale aesthetic analysis in the *Tonempfindungen*, as in all of his writings on the arts, he considered commentary on the general conditions for aesthetic judgment to be within his purview as a sensory physiologist; for he believed that aesthetic intuition depends on the same sort of unconscious mental operations as occur in ordinary sense perception—operations that he had already described in a preliminary way in his lecture on vision of 1855. In fact, he later acknowledged that aestheticians had discussed more fully than others the unconscious judgments that, according to Helmholtz, underlie both aesthetic pleasure and ordinary sensory perception.[30] In the case of aesthetic pleasure, he considered the relevant perceptions to be the unconscious grasping

---

Stumpf, *Konsonanz und Dissonanz* (Beiträge zur Akustik und Musikwissenschaft) (Leipzig: Barth, 1898), 1–19.

29. *Tonempfindungen*, 554; see also "Musikalische Harmonie," *VR⁴* 1:154. Earlier, Kant had observed that "even though the purposiveness in a product of fine art is intentional, it must still not seem intentional. . . . there must be no hint that the rule was hovering before the artist's eyes and putting fetters on his mental powers." (*Critique of Judgment*, sec. 45, p. 174)

30. In 1855, Helmholtz discussed unconscious judgments of vision without using the term "*unbewusster Schluss*" ("Ueber das Sehen," *VR⁴* 1:110, 112); in 1868, he referred to unnamed writers on aesthetics ("Die neuern Fortschritte," *VR⁴* 1:361). Aestheticians regularly posited an unconscious perception of order; besides Vischer (see n. 27, above), Eduard Hanslick, the other aesthetic writer whom Helmholtz mentioned directly (*Tonempfindungen*, 2, 386), discussed the unconscious judgments that underlie aesthetic intuition in his *Vom Musikalisch-Schönen: Ein Beitrag zur Revision der Aesthetik der Tonkunst* (Leipzig: Weigel, 1854): "Vorstellen und Urteilen, letzteres natürlich mit solcher Schnelligkeit, daß die einzelnen Vorgänge uns gar nicht zum Bewusstsein kommen"; Dietmar Strauss, ed., *Eduard Hanslick: Vom Musikalisch-Schönen*, 2 vols. (Mainz and New York: Schott, 1990), *1*:28.

of "regularity, connection, and order," or of "order, connection, and equilibrium."[31] These descriptions contain a tacit aesthetic, one that will receive further analysis below (see Section 7). Here we may consider why Helmholtz believed that his postulation of sensations corresponding to upper partial tones by itself contributed to musical aesthetics.

Helmholtz maintained that the aesthetic effect of simple melodic relations depends upon an unconscious sensing of resemblance, which he compared to that which allows us to detect the resemblance between the faces of members of a family even if we do not consciously attend to specific similarities.[32] In the case of melody, the unconsciously detected resemblances are between the fundamental and upper partials of various notes played in succession. If two notes form a fifth, the second partial of the second note is identical to the third partial of the first note. Similar relations hold for the octave, the fourth, and the other consonant intervals. Helmholtz's contribution here was the explanation of how such similarities could be grasped without being subject to conscious thought. His theoretical posit of discrete sensations for the fundamental and harmonic upper partials bore the explanatory burden. In his analysis, the feeling of melodic relationship is founded on the sensation of identical partial tones in successive compound tones.[33] He thus explained the aesthetics of simple melodic relations in terms of unconscious processes similar to those posited by Euler to explain both harmony and melody, except that in his account the material for unconscious comparison consists of pure sensations rather than the actual periodic motions themselves. In effect, Helmholtz extended Euler's account by postulating a physiological mechanism that yields sensations through which the mind could become aware of and compare the common elements in a succession of musical tones.

In Helmholtz's analysis of the perception of harmonic and melodic relations, the physiology of the sensation of upper partial tones led the way. Properly psychological processes, such as those involved in grasping the regularity in artistic compositions, played by far the smaller role. These facts are not, however, indicative of Helmholtz's general attitude about the relation of physiology to psychology; rather, they

---

31. *Tonempfindungen*, 554. Euler, *Tentamen novae theoriae*, chap. 5, secs. 1–3, also ascribed pleasure in successions to a perception of order, but again, an order among actual pitch numbers.
32. *Tonempfindungen*, 558–59.
33. *Tonempfindungen*, 555–60; 3rd ed., 566–67.

reflect his finding that in aesthetics natural science could deal effectively with the immediate sensational effects of art works, but that it could not derive "laws" for the analysis of the aesthetic effects of whole works of art, effects that are mediated by psychological processes. Which is not to say that Helmholtz denied that there are natural laws in psychology; as Section 4 shows, in his view the problem lay not with the inadequacy of psychology, but with the fact that aesthetic principles are historically conditioned and so require a methodology that does not rest all explanations on universal laws. Although Helmholtz emphasized physiology in his work on music, he did not in general grant physiology predominance over psychology. In fact, he considered physiological investigation to be neither necessary nor sufficient for securing the foundations of psychology. Quite the opposite. He argued that psychology possessed its own laws, independent of physiology, and he impugned the program of physiological reduction for its unfounded claims and its tendency toward materialist metaphysics, a metaphysics he rejected.[34] We shall return to his postulation of autonomous psychological laws in Sections 5 and 7, below.

3. The Relation of Painting to Visual Science

In a series of lectures delivered in Berlin, Düsseldorf, and Cologne between 1871 and 1873, Helmholtz brought his research in physiological optics to bear on painting, the mode of artistic experience that he considered, after music, to be the most closely conditioned by sensory physiology.[35] He discussed the basic elements of painting—form, color, and brightness—less systematically than he had treated musical tones and their relations, as befit the subject matter. He distinguished painting—and other arts, such as sculpture and poetry—from music on the grounds that whereas the latter deals in pure sensations and does not aim to represent nature, painting seeks to portray external objects. He averred that in painting and sculpture, the sensory qualities of the works of art are not their fundamental artistic effect, which consists instead in the images such works produce in the "imagination" of the spectator—with the exception of color in painting, the effects of which he compared to pure sensations of tone.[36] He compared the painter to

---

34. *Handbuch*[3] 3:432–33. For further discussion see Hatfield, *The Natural and the Normative*, 192–93, 232–33, and, more generally, Frederick Gregory, *Scientific Materialism in Nineteenth Century Germany* (Boston: Reidel, 1977).
35. "Optisches über Malerei," *VR*[4] 2:93–135, esp. 96.
36. *Tonempfindungen*, 2–4.

a translator who transforms imperfect nature into a thing of beauty, producing a "refined fidelity to nature." Although this "refinement," as he described it, might consist simply in presenting light and color that do not hurt the eyes (as they would if nature were merely *copied*, including the brilliance of sun at high noon), he also suggested that painting should aim to "join for us in vivid intuition the collected features of an ideal type."[37] This aim touches upon what Helmholtz called "the ultimate aims and purposes of art."[38]

As in his analysis of music, Helmholtz did not claim that his physiological approach would allow him to evaluate whether individual works of art met such aims. Indeed, he explained that he was concerned "not with the ultimate aims and purposes of art, but only with discussing the effects of the elemental media with which the artist works."[39] He thus limited his discussion to explicating the relations between the material techniques used in painting and their perceptual effects, relations which he discussed under four headings: form, variations in brightness, color, and color harmony. Under the heading of form he explained the techniques artists use to convey impressions of depth and distance; under variations in brightness he explained how and why the range of brightness differences used in paintings differ from naturally occurring variations out of doors; and under color and color harmony he discussed color contrast and the use of pure, saturated color and of complementaries. As in his study of music, those parts of his lectures that were not devoted to pure sensory physiology were largely limited to explaining established artistic practices, though despite his own warning he did proffer some "tips" to the working artist.

As suggested above, Helmholtz's remarks on color bore the strongest analogy with his work on the sensations of tone, although, as he observed, the relations among color sensations do not show the high degree of formal structure found among musical tones. His remarks on color took the form of a loosely structured set of observations, rather than, say, a theory of color harmony. He used sensory physiology to explain the experiences associated with painting in the case of successive color contrast (previously discussed by Michel Eugène Chevreul), in which gazing at one color affects the quality of subsequently seen colors. This occurs in the production of afterimages. Helmholtz's explanation drew upon his rehabilitated version of Thomas Young's color theory, according to which there are three types of color sensitive

37. "Malerei," *VR*[4] 2:127, 134, 135.
38. Ibid., *VR*[4] 2:97.
39. Ibid.

elements in the retina. Contrast phenomena and afterimages are explained through the fatiguing of one type of element, causing an area of the visual field to manifest the complementary tint. Thus, if an area of bright color in a painting were fixated for a period of time, the color of subsequently fixated areas would be affected.[40]

Helmholtz observed that under the normal conditions of illumination in which paintings are observed contrast effects for adjacent areas will not arise; hence, these "subjective" effects that arise for viewers in many ordinary circumstances must be introduced "objectively" into the painting. He noted that painters and draftsmen, when depicting a plain, uniformly lighted surface, brighten it where it meets a dark surface and lighten it where it meets a dark one; similarly, they give gray surfaces a yellowish tint where such surfaces meet blue and a reddish tint where such surfaces meet green.[41] Furthermore, by considering changes in sensitivity to brightness and to various colors in the spectrum in daylight and moonlight viewing, Helmholtz could explain the artistic practice of rendering daylight and moonlit scenes. The sensitivity to differences between moderately bright and very bright objects diminishes as the overall illumination is increased toward dazzling brightness. Similarly, in very weak illumination, the eye is less able to discriminate between moderately bright and the darkest objects, since the level of illumination puts all of these objects at the very threshold of sensory discrimination. Moreover, in daylight the relative sensitivity to red and yellow is higher than to blue, whereas the reverse is the case in moonlight. For this reason, Helmholtz observed, artists rendering a daylight scene paint moderately bright objects nearly as bright as the brightest objects, and make yellow tints predominate. Conversely, in rendering a moonlit scene, they paint moderately bright objects almost as dark as the darkest objects, and make bluish tints predominate.[42] Of course, these techniques were not universally employed. Helmholtz realized this, and in general he was skeptical about the possibility of finding general formulae or recipes for art.

Historically, the strongest relationship between optics and painting pertained to the rendering of depth and distance on the two-dimensional picture plane. Since the discovery of the retinal image, students of optics had compared the sensory impression received in the eye

---

40. Ibid., VR^4 2:120–22. M.E. Chevreul, *The Laws of Contrast of Colour*, new ed., trans. John Spanton (London: Routledge, 1868), secs. 8–15. For analysis of Helmholtz's work in color theory see Kremer, "Innovation through Synthesis."
41. "Malerei," VR^4 2:123; and Chevreul, *Laws of Contrast*, secs. 320–23.
42. "Optisches über Malerei," VR^4 2:107–17.

with a picture or drawing. Helmholtz himself, in the portion of his *Handbuch der physiologischen Optik* devoted to the psychology of distance perception, made use of this comparison. Now, in his lecture on painting, he engaged in turnabout, and used his knowledge of visual science to explain various aspects of the pictorial problem of representing depth. For beyond simply reviewing the means for rendering depth and distance that students of optics typically illustrated by referring to painter's techniques—including the use of perspective, of objects of known size to cue distance, of shading and shadows to render depth, and of aerial perspective—he used his scientific knowledge to explain two aspects of painting. First, he elaborated the analogies and disanalogies between paintings and retinal images. Both are perspective images, in which the three-dimensional world is rendered in two dimensions. But there is a crucial difference: we normally view the world with two eyes and with head and body mobile. As Charles Wheatstone had shown with his stereoscope, the slight disparity between the two retinal images is a potent indicator of depth relations. Further, the difference between the successively sampled images of a moving observer constitute a second powerful "cue" to depth relations, known as motion perspective. Neither of these is available in a painting, which presents a single, flat, stationary image. Indeed, the surface and edges of the painting yield binocular disparities indicating a flat object, which is also indicated by the lack of motion perspective. These factors must be overcome if a painting is to yield a vivid impression of depth.

The second aspect of the representation of depth and distance in painting that Helmholtz discussed at length was "aerial perspective"—the dulling of distant colors, the fuzziness of distant contours, and the blueness of background mountains. Artists had long used aerial perspective. Helmholtz examined the atmospheric conditions that lead to subjective effects in actual landscapes similar to those represented by the artist. Having given a detailed account of the atmospheric conditions favorable to the creation of aerial perspective, he gave a "tip" to the artist:

> The high, clear landscapes of mountains, which so often lead mountain climbers to underestimate the distance and size of mountain peaks, are difficult to realize in a painting. Views looking upward to the mountains from valleys, seas, or plains, in which the aerial light is faintly but clearly shown, are far better. Not only do such perspectives allow the various distances and sizes of objects to stand out; they also help to produce an artistic unity of coloration [in a painting].[43]

---

43. Ibid., *VR⁴* 2:106.

Here the visual scientist gives the artist some advice on choice of viewpoint.

Despite his disavowal of treating the "ultimate aims" of art, Helmholtz in fact did have a conception of such aims and this conception conditioned his discussion of the portrayal of depth and distance in painting. At the beginning of his discussion, he argued that even if art does not merely copy nature but rather idealizes its subjects, nonetheless it must command techniques for accurate representation. He asked rhetorically: "Even though a painting may present idealized types, must it not still convey at least the true images of the natural objects it portrays?"[44] This it does, he answered, by presenting accurate perspective constructions and by using the other means for creating depth and distance reviewed above. At one point in his exposition he allowed that "concern for the clarity of what is presented is, of course, seemingly only a subordinate consideration in relation to the ideal goals of art."[45] But in the concluding paragraphs he argued that "sensory intelligibility" should be given greater weight in aesthetics than was usual. He said that this conclusion has "forced itself upon me as I have sought to investigate the physiological aspects of the effects of works of art." But, in fact, he explicated it by appealing directly to the highest goals of art. Explicitly addressing the question of what a work of art, "in the highest sense," should effect, he answered that "it should capture and excite our attention, it should arouse a host of slumbering connections among representations and thereby arouse the feelings connected with them into tireless play and direct them to a common goal; in this manner it will join for us in vivid intuition the collected features of an ideal type, which lie scattered in our memory in detached fragments, overgrown by the wild briars of accident." In order to achieve this effect the work must exhibit the "intelligibility" Helmholtz so touted. And he now declared that his investigations are, after all, closely related to the highest goals of art, speculating that artistic beauty is in the end

> founded upon a sense of the graceful, harmonious, and lively current of our representations which, in spite of rich variation, flows toward a common point and brings to a more complete intuition lawfulness hitherto obscured, allowing us to gaze into the ultimate depths of the sensation of our own soul.[46]

---

44. Ibid., VR^4 2:98.
45. Ibid., VR^4 2:101.
46. Ibid., VR^4 2:134–35.

Disregarding the emotional enthusiasm Helmholtz often exhibited at the end of public lectures, this passage may well hold a key to his most fundamental conception of the purpose of art, a conception that itself depends upon his conception of the working of the artist's and the viewer's minds. The elements of his conception of art are that art should bring us to "vivid intuition" of an ideal type, and that the feeling we get from the beautiful reveals a hitherto concealed lawfulness and reveals something of our own soul. As Section 6 below argues, in fact the lawfulness revealed lies not only in the intuited object, but also in the mind's power to extract the ideal type from experience, and here Helmholtz tacitly invoked a greater unity between art and science than that which he had spoken of to his audiences, a unity between the purpose and modes of cognition of artist and scientist. According to Helmholtz, art and science are two ways of getting at the same thing: universal truth. Helmholtz had not, in fact, always felt this way, and in order to understand the position that he had come to by 1870, we need to examine the development of his attitude toward the cognitive processes of both artist and scientist. It will, however, be useful to approach his attitude toward artistic cognition by first examining his conception of the cognitive status of the investigation of the arts and artistic style themselves.

4. Studying Aesthetics: *Geisteswissenschaften* versus *Naturwissenschaften*

Post-Kantian aestheticians disputed whether aesthetics was in fact a science. Kant had denied that it could be, explaining that there could be only *critique* of the beautiful, but no science of it, just as there was no fine science, but only fine art; as he explained it, art is the product of genius and genius neither produces a scientific description of its product nor is itself subject to scientific analysis. Others disagreed, arguing that the beautiful in art can be the subject of scientific (*wissenschaftlich*) study, even if the creative act itself is not. Thus, the Hegelian Vischer defined aesthetics as "the science [*Wissenschaft*] of the beautiful," and Zimmermann, who later appreciatively cited Helmholtz's *Tonempfindungen*, argued in 1862 that aesthetics should be made into an exact science, explaining that the rest of philosophy had abandoned the error of metaphysical idealism, and aesthetics should follow suit.[47]

---

47. Kant, *Critique of Judgment*, sec. 44, p. 172; Vischer, *Aesthetik 1*:3; and Zimmermann, "Zur Reform," 309–10. Zimmermann mentioned Helmholtz favorably in his *Aesthetik*, 2 vols. (Vienna: Wilhelm Braumüller, 1858–65), 2:ix.

Helmholtz first formulated the problem of the relation between aesthetics and science in terms of the relation between aesthetic studies and natural scientific investigation. When in his writings on art he began to discuss the expressive quality of art and the arousal of feelings in the spectator, he spoke of reaching the limits of natural science. His language in such passages is strikingly similar over a period of fifteen years: he spoke of reaching the "border" of natural science and of being commanded to stop, or of wishing to stay within its comfortable confines, or of crossing the border only with difficulty. A striking example of his being halted at the border occurred near the end of his 1857 lecture on the physiology of harmony. He compared the rhythmic crashing of ocean waves with the flow of a musical composition:

> Whereas in the sea, blind physical forces alone are at work and the final impression on the spectator's mind is nothing but solitude, in a musical work of art the movement follows the outflow of the artist's own emotions. Now gently gliding, now gracefully leaping, now violently stirred, penetrated by or laboriously contending with the natural expression of passion, the stream of sound, in primitive vivacity, bears over into the hearer's soul unimagined moods which the artist has overheard from his own, and finally raises him to that repose of everlasting beauty of which God has allowed but few of his elect favorites to be the heralds.[48]

Although Helmholtz's audience, upon hearing this passage, may have expected that he would continue with an analysis of the aesthetics of musical expression, Helmholtz summarily closed his lecture with the statement: "Here however are the borders of natural science, and I am commanded to stop." He used similar language at the end of the *Tonempfindungen*; having observed that he had not entered far into musical aesthetics proper because he had not treated the theory of rhythm, or forms of composition, or the means of musical expression, he explained that he preferred to stay on the "soil" of natural science where he was "at home."[49] And at the beginning of the collected lectures on painting, he observed that although many in the audience may be "well traveled" in "the beautiful lands of art," he was like a traveler who had made his entrance over "steep and stony bordering mountains," from which, however, he had been able to obtain a good overview.[50]

---

48. "Musikalische Harmonie," *VR⁴* 1:155. Helmholtz here seems to take the usual position that the aesthetic function of music is to express emotion; in the *Tonempfindungen* (2–3), however, he sided with Hanslick against the "false standpoint of exaggerated sentimentality" of much musical criticism and endorsed Hanslick's view that the content of music is motion, not emotion.
49. *Tonempfindungen*, 560.
50. "Malerei," *VR⁴* 2:95.

In fact, however, only to the extent that Helmholtz remained a natural scientist was he precluded, in his view, from entering into properly aesthetic inquiries. For although he did not think that the physiological approach he took in these writings could go beyond explaining the pure sensations produced by artistic materials, he was ready to allow that the study of aesthetic principles could be *wissenschaftlich*. But it would belong to the historical sciences, or, as he also put it, to the *Geisteswissenschaften*. It was only by undertaking historical studies that he could achieve systematic results pertaining to aesthetic principles. Thus, in the *Tonempfindungen*, as he passed from the study of the physiology and psychology of isolated or simultaneous tones to the study of musical scales and tonic relations, he noted that the character of his investigation must change. He wrote:

> Up to this point our investigation has been of a purely natural-scientific type. We have analyzed the sensations of hearing, and we have investigated the physical and physiological bases for the phenomena discovered—partial tones, combinational tones, and beats. In this whole field we have dealt solely with natural phenomena, which must always present themselves in a purely mechanical fashion and without choice to all living beings whose ears are on the same anatomical plan as are ours. In such a field, where mechanical necessity reigns and all free will is excluded, science is rightfully called upon to establish constant laws of phenomena and to demonstrate strictly a strict connection between cause and effect.[51]

He continued, contrasting the previous investigation with the topic at hand:

> Because in this third part of our inquiry we turn primarily to music and wish to furnish a satisfactory foundation for the elementary rules of musical composition, here we tread on new ground, which is no longer purely natural-scientific, even if the knowledge which we have gained of the nature of hearing will still find numerous applications. We pass on to a problem which by its very nature belongs to the domain of aesthetics. When we spoke previously, in the theory of consonance, of the agreeable and the disagreeable, we considered only the immediate impression made on the senses when an isolated combination of sounds strikes the ear, without regard to artistic contrasts and means of expression; we considered only sensuous pleasure, not aesthetic beauty. The two must be kept strictly apart, even if the first is an important means for attaining the second.[52]

---

51. *Tonempfindungen*, 357.
52. Ibid., 358. The contrast between sensory and aesthetic pleasure was commonplace in German aesthetics after Kant.

Sensuous pleasure may be fixed by the laws of nature, by the physiology of sensation, but aesthetic pleasure is not. It depends on taste, which is historically and culturally conditioned. As Helmholtz put it for the aesthetics of music, "the system of scales, modes, and harmonic tissues does not rest upon inalterable natural laws, but is the result of aesthetical principles, which have already undergone change, and will undergo further change, with the progressive development of humanity."[53]

In sum, Helmholtz argued that the aesthetic aspects of art cannot be reached by the methods of the natural sciences; instead, they must be studied through what he termed historical and aesthetical inquiry. That is to say, distinct modes of inquiry are involved in investigating the material elements of the arts as opposed to their aesthetic and stylistic aspects. These distinct modes involve the search for strict causal laws on the one hand and the determination of principles of style through the historical investigation of the various periods of art and national or cultural artistic traditions on the other.

Some of Helmholtz's contemporaries would have drawn different conclusions from the fact that aesthetic principles must be studied as they develop through history. Vischer agreed that aesthetic judgments change over time, but he regarded this change as the unfolding of concrete manifestations of the "Idea" in history. From his Hegelian vantage point, the foundational discipline for studying historical developments of this sort was metaphysics: such study begins with an understanding that the Absolute Idea manifests itself in various ways over the course of history.[54] John Stuart Mill, who agreed with Helmholtz in classifying historical subject-matters among the "moral sciences" (a term translated into German as *Geisteswissenschaften*), would not have concluded that, therefore, the methods of natural science are inapplicable. In fact, Mill argued that the moral sciences should copy the methods of natural science.[55]

Helmholtz disagreed. In a lecture entitled "Ueber das Verhältniss der Naturwissenschaften zur Gesammtheit der Wissenschaften,"

---

53. Ibid.: (In later editions Helmholtz qualified the passage to read "does not rest solely upon ... but is, at least in part," etc. [3rd ed., 370]). Although Helmholtz considered aesthetic principles to be historically variant, he repeated the standard Kantian point about the objectivity of taste, namely: "that we do not understand delight in the beautiful as something individual, but as a lawful accordance with the nature of our mind, appears by our expecting and requiring from every other healthy human intellect the same acknowledgement of the beautiful that we ourselves render to it" (*Tonempfindungen*, 553).
54. Vischer, *Aesthetik 1*:3, 47–53. According to Vischer, the world spirit progresses from religion to art to philosophy.
55. J.S. Mill, *System of Logic* (London: Parker, 1843), Book VI, chapter 1, section 1; chapter, 10, section 7.

which he presented to the assembled faculties of the University of Heidelberg in 1862 as prorector, he distinguished the *Naturwissenschaften* and the *Geisteswissenschaften* in a manner that accords with the two sorts of investigations he described in the *Tonempfindungen*. Although he may have appropriated the terminological distinction between the *Natur-* and *Geisteswissenschaften* from the German version of Mill's *Logic*,[56] he did not adopt Mill's construal of the relation between the two areas of study. And while he credited Hegel with advancing the *Geisteswissenschaften*, he rejected the metaphysics of the Absolute Idea with its allegedly a priori derivation of the course of history; moreover, he blamed Hegel for attempting to extend his (ostensibly) a priori approach from the domain of *Geist* into the natural sciences, with disastrous results that unnecessarily alienated natural scientists from philosophy.[57] In opposition to Mill and Hegel, Helmholtz contrasted the natural and moral sciences in terms of both their subject matter and their characteristic modes of thought. The natural sciences seek universal categories and laws, which may be taxonomic, as in natural history, but which nonetheless tend to be exceptionless; their method is deductive. Physics and mathematics are the paradigms. The moral sciences, by contrast, deal with human institutions and products; their method is inductive, and they cannot achieve the certainty of the natural sciences. Philology and history are paradigms.[58]

Enemy of metaphysics that he was, Helmholtz rejected appeals to a priori principles in both the moral and natural sciences. But even though he urged, against the Hegelians, that the *Geisteswissenschaften* must rest their claims upon experience, he did not, like Mill, contend that they should mimic the methods of natural science. He believed that real methodological differences separate the two areas of investigation:

> As empirical investigation of facts has again come to the fore in the other sciences, the opposition between them and the natural sciences

---

56. The term "*Geisteswissenschaften*" appeared as a translation for "moral sciences" in *Die inductive Logik*, trans. J. Schiel (Braunschweig: Vieweg, 1849), as the title to Book VI, and passim. (Later editions bear the title *System der inductiven und deductiven Logik*.) Although it is commonly stated that Schiel's translation introduced the term into the German language, such is not the case: see Alwin Diemer, "Die Differenzierung der Wissenschaften in die Natur- und die Geisteswissenschaften und die Begründung der Geisteswissenschaften als Wissenschaften," in *Beiträge zur Entwicklung der Wissenschaftstheorie im 19. Jahrhundert*, ed. A. Diemer (Meisenheim am Glan: Anton Hain, 1968), 174–223, esp. 187.

57. "Ueber das Verhältniss," *VR⁴* *1*:163–65.

58. Ibid., *1*:167–78. In his first lecture on Goethe (see Section 6, below), Helmholtz had emphasized the role of deduction in mathematical physics as opposed to natural history.

has become less marked. Nonetheless, even though this opposition was expressed in an excessively stark form through the influence of the previously named philosophy [Hegel's], it is grounded in the nature of things and must be acknowledged. It depends partly upon the kinds of intellectual processes characteristic of the previously named disciplines and partly, as the names *Natur-* and *Geisteswissenschaften* indicate, on the subjects of which they treat.[59]

The contrast he drew between the intellectual activities characteristic of the two approaches is of special interest, for he framed this contrast as one between *logical* and *aesthetic* induction, or between conscious formulation of exceptionless laws, and inferences drawn with a sensitive touch from a mass of information held in memory and not reducible to exceptionless laws. In explicating the modes of thought of the *Natur-* and *Geisteswissenschaften*, Helmholtz gave an extended portrayal of his conception of the mental activity appropriate to artist and natural scientist.

His portrayal in 1862 of the mode of cognition appropriate to the natural sciences emphasized the conscious formation of general categories and laws. As an example of what he termed a "logical induction," Helmholtz described a natural scientist arriving at a generalization from a mass of facts, such as the generalization that all mammals have lungs.[60] Once such an induction has been established and codified it can serve as the basis for a deduction—for example, from the observation that something is a mammal, to the conclusion that it must have lungs. Helmholtz emphasized the exceptionless character of the laws in natural science and the deductive manner in which explanations could proceed.

By contrast, the ability to reach conclusions based on large amounts of data but without the benefit of explicitly formulable rules is characteristic of the intellectual activities found in the moral sciences. The jurist, the philologist, and the historian must bring enormous amounts of information to bear on single cases. They must be able to see what fits and what does not, which generalizations should be applied and which should not. This ability requires a certain "psychological instinct" or "feel"—"*Tactgefühl*" is Helmholtz's word—that results from a process of unconscious reasoning. This feel for what is relevant, this ability to grasp what is essential in a large body of information or in

---

59. Ibid., *1*:165. The present argument thus disagrees with Hans-Georg Gadamer, *Truth and Method*, 2nd ed. (New York: Crossroad, 1991), 3–6, who describes Helmholtz's position in the 1862 lecture as agreeing with Mill's position.

60. "Ueber das Verhältniss," VR^4 *1*:169–71. Helmholtz realized that induction could not be reduced to a strict logical principle.

the flux of everyday experience,[61] is what Helmholtz termed "artistic induction." He justified this ascription as follows:

> In opposition to *logical induction*, which operates only with clearly defined universal propositions, we shall call this kind of reasoning *artistic induction*, because it is present in the highest degree in the most exceptional works of art. It is an essential part of artistic talent to be able to reproduce—by words, forms, colors, or musical tones—the characteristic external indications of a person's character and mood, and to grasp, by a kind of instinctive intuition, how our mental states must unfold, without being led in such cases by any definable rule.[62]

The moral sciences, dealing as they do with human actions and motives, make particular use of this "artistic" induction.

In his essay of 1862, then, Helmholtz assimilated the characteristic thought processes of the moral sciences to those of the arts and set them both apart from the natural sciences. In subsequent years, his conception of the relations among the thought processes of the scientist and the artist changed. Although there is no evidence that his attitude toward the distinction between the natural sciences and the *Geisteswissenschaften* altered, his attitude toward the relation between the psychology of scientific inference and "artistic induction" underwent a complete turnabout—and this in the space of a few years. By the end of the decade he stressed the similarity between the intellectual activity of the natural scientist and that of the artist, for he now emphasized the role of what he had earlier called "artistic induction" in the thought processes of the natural scientist.[63] One factor in this change, especially after about 1870, came in his downplaying the importance of deduction in natural science—except for the presentation of results already obtained—and his corresponding emphasis on the importance of inductive intuitions in the work of active natural scientists; this change perhaps resulted from his own deeper involvement in physical research

---

61. Helmholtz observed that "memory alone is insufficient without the general ability to discover the essential similarities, and without a finely and copiously developed intuition of human motives" (ibid., *1*:172–73); cf. Gadamer, *Truth and Method*, 15–7. While Gadamer may be correct in asserting that Helmholtz underdescribed the type of memory involved, he is incorrect in describing Helmholtz as treating this use of memory as a "mere faculty," for Helmholtz himself (*VR⁴ 1*:168–69, 172–73) made Gadamer's point that "memory must be cultivated" and, contrary to what Gadamer implies, he did not simply "presuppose" the concept of *Bildung* but instead explicitly invoked it (ibid., *1*:166).
62. "Ueber das Verhältniss," *VR⁴ 1*:171.
63. "Die neueren Fortschritte in der Theorie des Sehens" (1868), *VR⁴ 1*:265–365, on 361.

after 1871, when he was called to the chair of experimental physics at Berlin. A second factor, perhaps more fundamental, was the further development of his theory of the psychology of unconscious inference in perception.

5. The Psychology of Unconscious Inference, 1855–66

According to the mature Helmholtz, the intellectual processes involved in both science and art are one with the intellectual processes that underlie perception. Although Helmholtz had, from the mid-1850s, held the doctrine that unconscious judgments underlie perception, and had in 1862 alluded to a connection between artistic induction and unconscious perceptual inferences, a connection he also mentioned in the *Tonempfindungen*, at the time of these earlier works he had not yet developed his account of the psychology of such inferences. His mature theory developed in the period shortly after 1862, as he completed work on the third part of his *Handbuch der physiologischen Optik*.[64] This development in his thought led him to reconceive the relation between the thought processes of artist and scientist. At the same time, it also led him to reconceive the relation between the faculties of understanding and imagination, which he had sharply distinguished in earlier work.[65] On the one hand, he claimed to have shown that perceptual intuition results from thought processes akin to those of conscious thought.[66] On the other, he identified the thought processes of perception with those of both artistic experience and scientific induction, explaining the role of "understanding" in all three by appeal to the faculties of imagination and memory.

Helmholtz's theory of unconscious inference developed in connection with a physiological hypothesis he had held from early on. We have become acquainted with the key physiological assumption in the discussion of music perception. Basically, it is the idea that there is a one-to-one correspondence between individual sensory nerve fibers and unitary sensations that vary only in quality and intensity. The stimulation of 100 nerve fibers produces 100 discrete sensations. It is the task of unconscious inferences to put these sensations together into a representation of an external object.

---

64. For a fuller defense of the following account of the development of Helmholtz's theory see Hatfield, *The Natural and the Normative*, 195–218.
65. "Ueber das Sehen," *VR⁴* 1:115–16.
66. "Die Thatsachen in der Wahrnehmung" (1878), *VR⁴* 2:213–47, on 233.

Helmholtz's mature doctrine of unconscious inference can best be illustrated by considering the case of vision. The one-fiber/one-sensation doctrine as applied to vision implies that each fiber in the optic nerve produces a single sensation that varies only in color and intensity. According to Helmholtz, sensations do not, originally, convey spatial meaning; but since we in fact do make spatial discriminations visually (and since he maintained that we have no source of knowledge of the external world except sensation), sensations must represent spatial location in some manner. To explain the basis of localization, Helmholtz adopted the theory of "local signs," according to which each fiber gives a special quality to the sensation that it produces, which labels the sensation as belonging to that fiber.[67] Since the fibers remain in a fixed relation to one another, it should be possible to assign a location (or a direction in space) to each sensation on the basis of its attendant local sign. But this requires interpreting the local sign. As Helmholtz never tired of repeating, his primary conclusion with regard to sense perception was "that sensory impressions are only signs [*Zeichen*] for the properties of the external world, the meaning of which must be learned through experience."[68]

The interpretation of signs occurs through a process of induction. Over repeated experience, a number of laws or rules for localizing sensations are established—presumably one such rule for each local sign. These rules are acquired by a kind of trial-and-error learning. Take the example of a local sign associated with a fiber on the right side of the eye. This fiber would typically be stimulated by light coming from the left side of the visual field. The location of the light would be determined either through the sense of touch or by means of the muscle movements required to turn the eye so that it fixates the light. After repeated sensations, a rule would be learned assigning a direction

---

67. *Handbuch*[3] 3:130, 154, 192; and "Die neueren Fortschritte," *VR*[4] *1*:332. On the theory of local signs, originally so-named by Lotze, see Theodule Ribot, *German Psychology of Today*, trans. James Mark Baldwin (New York: Charles Scribner's Sons, 1886), chap. 3; and Hatfield, *The Natural and the Normative*, 157–62, 174–77. William R. Woodward, "From Association to Gestalt: The Fate of Hermann Lotze's Theory of Spatial Perception, 1846–1920," *Isis 69* (1978):572–82, traces the reception and ultimate rejection of Lotze's theory.

68. "Erinnerungen" (1891), *VR*[4] *1*:1–21, on 17. In his first statement of the doctrine, Helmholtz spoke of "*Symbole*" rather than "*Zeichen*"; see his "Ueber die Natur der menschlichen Sinnesempfindungen" (1852), *WA* 2:591–609, on 608, and also "Ueber Goethe's naturwissenschaftliche Arbeiten" (1853), *VR*[4] *1*:23–45, on 42. He used "*Zeichen*" in "Musikalische Harmonie" (1858), *VR*[4] *1*:146; *Handbuch*[3] 3:22; "Die neueren Fortschritte," *VR*[4] *1*:319; "Ueber das Ziel und Fortschritte der Naturwissenschaften" (1869), *VR*[4] *1*:367–98, on 393; "Wahrnehmung," *VR*[4] 2:222; and "Goethes Vorahnungen kommender naturwissenschaftlicher Ideen," (1892), *VR*[4] 2:335–61, on 357.

to the left to the visual sensation accompanying the local sign associated with the fiber. Once the rule has been learned it then serves as the major premise in unconscious inferences. Suppose that a blue patch occurs to the left. The following would take place: The fiber on the right would be stimulated, giving rise to a sensation of blue plus an attendant local sign. This would serve as the minor premise, the major premise being the rule localizing the sensation to the left on the basis of the local sign. The "conclusion" of the inference would be the conscious experience of a blue patch on the left side of the visual field.[69] If, now, we think of a myriad of such micro-localizations being combined into a single, wildly disjunctive inference—one that draws upon rules for local signs plus rules for interpreting other cues to distance such as binocular disparity—the result will be our ordinary experience of a world spatially articulated in three dimensions.

According to the doctrine of unconscious inference, then, our ordinary sensory experience depends upon the prior learning of rules for relating our sensations to one another. Helmholtz saw a strong connection between these inductive inferences and those by which the scientist grasps the lawlike in nature. He stated this parallel nowhere more forcefully than in his 1878 address on "Die Thatsachen in der Wahrnehmung," which stands as the major statement of his theory of knowledge. There he spoke of the process of learning the rules that guide our perceptual inferences as a matter of gaining an acquaintance "with the enduring existence of a lawlike relationship" between visual sensations and our voluntary movements. He continued:

> Each of our voluntary movements, whereby we modify the manner of appearance of the objects, is to be regarded as an experiment through which we test whether we have correctly apprehended the lawlike behavior of the appearance before us, i.e., correctly apprehended the latter's presupposed enduring existence in a specific spatial arrangement.[70]

Helmholtz then went on to develop a corresponding account of the scientist as someone who seeks to grasp the lawlike in the phenomena by means of experiment, explicitly linking the scientist's grasp of law with the thought processes underlying perception.

---

69. *Handbuch*³ 3:4–5, 129–40; "Die neueren Fortschritte," *VR*⁴ *1*:329–33, 353–58; and "Ueber das Sehen," *VR*⁴ *1*:100–1.
70. "Wahrnehmung," *VR*⁴ 2:237. Translation of "The Facts in Perception," in *Hermann von Helmholtz. Epistemological Writings*, eds. Robert S. Cohen and Yehuda Elkana, trans. Malcolm F. Lowe (Dordrecht and Boston: D. Reidel, 1977), 115–63, on 136.

In both the 1878 address and the corresponding discussion in the third part of the *Handbuch*, there is an apparent tension between the programmatic description of unconscious inference and the subsequent accounts of actual perceptual processes. In Section 26 of the *Handbuch*, Helmholtz attributed an active role to the perceiver in the formation of judgments regarding the causes of sensations, whereas in the succeeding sections he seems to refer only to passive associational processes. This associational account is behind Helmholtz's explanation of the mapping of local signs into visual directions, according to which the facts of the perception of direction can be explained "by the known capacities of sensory memory."[71] He ascribed the process of matching local sign with visual direction to the merely passive collection of associations in memory, rather than to an active process of judgment; for although movement of the eyes and body might be required in order to produce the paired sensations requisite for establishing a mapping, he assigned the process of learning to sensory memory, pure and simple. This tension is only heightened when we recall Helmholtz's repeated claims that in his account of space perception, he had "resolved the concept of intuition into the elementary processes of thought."[72]

Although there exists an apparent tension between the active inferential account of Section 26 and the subsequent references to bare memory, Helmholtz himself did not believe that the two accounts were in conflict; he considered them identical. He thought that the psychology of unconscious inference could be explicated wholly in terms of the ordinary functioning of memory and association, a view he attributed to Mill.[73] The general rules for visual localization are formed using the same kind of inductive process as underlies the discovery of scientific generalizations. In each case, the general law arises out of particular experiences as a result of a tendency for the irregular and the variable to be "washed out" in the associative process of memory. No active process is required in the subsequent application of an associatively based rule; given a representation of the antecedent, the associated consequent simply comes to life. In Helmholtz's account the active role of the perceiver is restricted to the "testing" of associatively formed rules of localization through motor activity. As he made clear in the concluding Section 33 of the *Handbuch*, he considered the laws of association to be the fundamental laws of psychology,

---

71. *Handbuch*³ 3:135.
72. "Wahrnehmungen," *VR*⁴ 2:244. See also ibid., 391; "Das Denken in der Medizin" (1877), *VR*⁴ 2:165–90, on 186; and "Goethes Vorahnungen," *VR*⁴ 2:338.
73. *Handbuch*³ 3:23–4.

from which psychological explanations, including those of the inferential processes in spatial perception, would proceed.[74]

6. Disparity and Unity between Artist and Scientist: Helmholtz's Two Goethe Lectures

Armed with his account of the psychology of unconscious inference, Helmholtz ventured further into aesthetics than he had heretofore. He did so not as a physiologist, but rather as a psychologist characterizing the thought-processes of the artist. Indeed, we have seen that already in 1863, in the *Tonempfindungen*, he placed certain cognitive conditions on the aesthetically pleasing in music, including unstudied conformity to a rule. In his treatment of music he did not formulate a general statement of any such rule. But we have seen that in the case of painting, a form of art he considered to differ from music by being essentially directed toward the portrayal of external objects, he was willing to state the end toward which art should be directed: it should "join for us in vivid intuition the collected features of an ideal type, which lie scattered in our memory in detached fragments, overgrown by the wild briars of accident."[75] In a subsequent lecture on Goethe as scientist, he elaborated this conception of the artist's aim, arguing that in seeking the ideal type the artist pursues the same goal as the scientist: universal or general truth. In so arguing, he broke with the Kantian and post-Kantian aesthetic tradition, according to which so-called "theoretical truth"—truth about the external world—is irrelevant to the aesthetic value of art.[76] Helmholtz made truth fundamental, at least in painting and other nature-depictive arts. It was his unified conception of their intellectual processes that led him to conceive scientist and artist as companions in the search for truth.

The development of this dimension of Helmholtz's thought may be brought into relief by comparing his two addresses on Goethe, written forty years apart. In the first, delivered in 1853, Helmholtz attempted to explain Goethe's achievements as a natural historian by showing how his abilities as a poet carried over to his grasp of natural phenomena.[77] Yet he also emphasized a dividing line in Goethe's abilities.

---

74. Ibid., *3*:439; see also ibid., *3*:24, and "Die neueren Fortschritte," *VR⁴ 1*:353.
75. "Malerei," *VR⁴ 2*:134.
76. Zimmermann, *Aesthetik 2*:11; Kant, *Critique of Judgment*, sec. 60 (p. 355), made truthful rendering a condition of fine art, but distinguished it from art itself. Hegelian aesthetics, by contrast, conceived art as expressing truth: Vischer, *Aesthetik 1*:88–92, 172–78.
77. "Ueber Goethe's naturwissenschaftliche Arbeiten," *VR⁴ 1*:23–45.

The poet was successful in areas such as plant and animal morphology, in which he could treat nature as a work of art. Enlisting his artistic sensibilities, he was enabled to grasp manifest though subtle kinships of form or shape. By contrast, Goethe failed in his theory of color, Helmholtz argued, because success in that area requires the ability to apply abstract, precise mathematical laws under artificial conditions, a requirement that conflicted with the poet's direct, intuitive approach to natural phenomena.[78] In the terminology of the 1862 essay, the poet succeeded in the first area because of his aptitude for artistic induction; he failed in the second because he was not suited to abstract and precise logical induction. In his description of Goethe's failings, Helmholtz in effect endorsed one conception of the division between science and art current in the late eighteenth and early nineteenth centuries, according to which the artist, guided by sensibility and imagination, forsakes the abstract thought of the scientist, who is guided by the faculties of reason and understanding.[79]

In the second address, delivered in 1892, Helmholtz revised his treatment of Goethe in accordance with the change in his position on the relation between scientific cognition and the associative processes of "artistic induction."[80] He now presented art as a second way, besides science, of grasping the lawful in phenomena. He rejected what he termed the common philosophical conception that artistic intuition stands in opposition to thought proper.[81] On his view, artistic intuition acts like a filter for the invariant in nature: the lawful or the ideal is captured in memory, and the variable falls by the wayside. In its intuitive grasp of lawful regularity, art is like sensory perception and scientific inference. In all three cases the mind retains, out of the flux of experience, only what is regular and invariant.[82] The artist's imagination, on Helmholtz's view, reflects the associative laws of thought; through these it grasps the lawful in nature, and this grasp informs the artist's productions. The audience to the artist's works can then find in it a truth that they can recognize from their own experience, but one which the artist has distilled and presented in a clear and forceful manner. Inasmuch as each seeks to make comprehensible what is lawful in nature, the artist has the same aim as the scientist: "Art as well

---

78. Ibid., *VR⁴* 1:34–41.
79. On the Romantic tendency to favor imagination over understanding, see M.H. Abrams, *Natural Supernaturalism: Tradition and Revolution in Romantic Literature* (New York: Norton, 1971), 344–47, and Beardsley, *Aesthetics*, 247–59.
80. "Goethes Vorahnungen kommender naturwissenschaftlicher Ideen," *VR⁴* 2:335–61.
81. Ibid., *VR⁴* 2:341.
82. Ibid., *VR⁴* 2:339–44.

as science can represent and convey the truth." The truth that the artist grasps and conveys is that of the ideal type; the artist "transforms" individual cases, but does not deviate "from the lawfulness of the type."[83] Artist and scientist are not identical in their activities; as he had in 1862, Helmholtz again emphasized that artistic memory is sensitive to detail. But he now saw a bond between artist and scientist even here, for he observed that in the initial stages of scientific work the scientist follows "intuition" based on an acquaintance with the particulars of the phenomena.[84] In the thought of both artist and scientist, variant details fall away as the regular or the lawful emerges.

Helmholtz suggested in his second Goethe lecture, as he had at the end of his lecture on painting, that in proposing that art represents types he was verging on a theory of the nature of beauty in art. He did not state such a theory; but it is apparent what he had in mind. He was tempted to equate the perception of beauty with the grasping of the lawful, the regular, the ideal, discerned amid the flux and variation of the phenomena. Since he equated the lawful with the actual and the true, he was in effect tempted to equate beauty with truth. Helmholtz was drawn toward the aesthetics of classicism, in which beauty arises in the perception of the ideal type amidst the imperfections of the particular.

7. Helmholtz's Classical Aesthetic of Art and of Science

If we were simply to recall the passage on musical expression at the end of his lecture on musical harmony, and to place it beside another early passage in which he spoke of art as resulting from "excited feeling,"[85] we might conclude that Helmholtz had taken over the Romantic conception of art as the expression of inner feelings—where the more particular these feelings are to the artist, the better. But while Helmholtz did occasionally use the language of the Romantics, his

---

83. Ibid., *VR⁴* 2:344–45.
84. Ibid., *VR⁴* 2:348. Helmholtz made the same point in "Das Denken in der Medizin," *VR⁴* 2:184–85. Furthermore, in his 1869 lecture on progress in natural science, he urged the importance of memory and of sensitivity to minute differences of detail in natural scientific research, and in his foreword to the German translation of John Tyndall's *Fragments of Science* he elaborated Tyndall's distinction between mathematical and experimental science by assimilating the psychological processes of the experimentalist to those of the artist; see "Ueber das Ziel," *VR⁴* 1:371, and "Ueber das Streben nach Popularisirung der Wissenschaft," *VR⁴* 2:422–34, on 430.
85. "Musikalische Harmonie," *VR⁴* 1:155; and "Ueber Goethe's naturwissenschaftliche Arbeiten," *VR⁴* 1:34.

mature conception of art is better termed classical than Romantic. For although he did think of art as expressing the intuitions of the artist, he thought that such intuitions should express not the particular feelings of the artist but rather the ideal type. Helmholtz himself distinguished his account of artistic imagination, couched in terms of "the laws of memory," from the Romantic understanding of artistic creation as "the free play of fancy," an understanding which, he asserted, could not account for the "consistency" (*Folgerichtigkeit*) found in even the greatest works of art.[86]

The ascription of a classical aesthetic to Helmholtz could mean two different things: it could mean that his taste in art ran toward the classical, and it could mean that his theory of aesthetics was in agreement with neo-Classical rather than Romantic theories.[87] In fact, both ascriptions are correct. The evidence available suggests that Helmholtz's taste in music and art ran toward the classical. In letters to his parents from the late 1830s, he expressed his preference for Mozart and Beethoven over more recent composers.[88] And his architectural tastes, as evinced in the style of his home in Charlottenburg and in the construction of the Berlin Physikalisches Institut and the Physikalisch-Technische Reichsanstalt, ran toward the neo-Classical.[89] As for his theory of aesthetics, we have reviewed what little textual evidence there is, and the indication is clear: Helmholtz favored the equation of beauty with the rendering of an ideal type.

This same classical taste is also expressed in Helmholtz's attitude toward scientific theory. Although he rarely voiced judgments that could be described as "aesthetic" in respect to specific theory-choices in his technical work,[90] his taste ran toward comprehensive theories

---

86. "Goethes Vorahnungen," VR^4 2:344.

87. On the various senses of "classicism," see Michael Greenhalgh, *The Classical Tradition in Art* (New York: Harper & Row, 1978), 11–5. On the neo-Classical aesthetic of the ideal type, in both neo-Platonic and Aristotelian (empiricist) versions, see Erwin Panofsky, *Idea: A Concept in Art Theory*, trans. Joseph J.S. Peake (New York: Harper & Row, 1968), chap. 6. "Classical" and "Romantic" were terms of art used in nineteenth-century aesthetic works: see, e.g., Vischer, *Aesthetik 1*:33; and Zimmermann, *Aesthetik 2*:89–101.

88. Koenigsberger *1*:23, 27, 29, 33; and David Cahan, ed., *Letters of Hermann von Helmholtz to His Parents: The Medical Education of a German Scientist, 1837–1846* (Stuttgart: Franz Steiner, 1993), Letters 5, 6, 7, 13, pp. 43–8, 48–52, 53–4, 67–70, resp.

89. David Cahan, *An Institute for an Empire: The Physikalisch-Technische Reichsanstalt 1871–1918* (Cambridge, New York, New Rochelle: Cambridge University Press, 1989), 93, 96–7, 101.

90. In his Faraday lecture Helmholtz mentioned that the hypothesis of electrical dualism (positive and negative polarity) was preferable because it was more "convenient" (*geschickter*) than the unitary theory; see "Die neuere Entwicklung von Faraday's Ideen über Elektricität," VR^4 2:249–91, on 263. More often, he spoke of achieving the

that bring unity to a wide range of phenomena by subsuming under universal laws the actions of simple elements varying in only a few dimensions. This mode of explanation, familiar from classical mechanics, was Helmholtz's expressed explanatory ideal. In his memoir of 1847 he characterized the ideal of physical explanation as follows: "natural phenomena should be traced back to the movements of material objects which possess inalterable motive forces that are dependent only on spatial relations."[91] This ideal also describes his preferred mode of explanation in psychology, as is evident in his appeal to the laws of association in his theory of unconscious inference. Associative laws are laws for relating sensations in accordance with their elementary properties, including their qualitative character and intensity and their temporal relations. Reflection on Helmholtz's explanatory practice in psychology suggests the following ideal of psychological explanation: mental phenomena should be traced back to the association of atomic sensations according to inalterable laws that are dependent only on their quality, intensity, and temporal relations. Helmholtz's "unconscious inferences" occur through the lawful interaction of simple sensations, these interactions being determined by the qualitative characteristics of the sensations as these fall under the laws of association. Helmholtz was, in fact, opposed to reductionism and materialism with respect to mental processes; but he maintained that the same scheme of explanation, in terms of simple elements and laws, has application in both physics and psychology.[92] As he announced at regular intervals throughout his scientific career, an assumption of the "intelligibility" or "comprehensibility" of nature drove his postulation of this explanatory ideal.[93]

---

simplest expression of a law, although this sort of economy did not have to reflect a principle of theory choice, since it could be sought within a single theory; see, for example, *Vorlesungen über theoretische Physik*, eds. Arthur König, et al., 6 vols. (Leipzig: J.A. Barth, 1897–1907), vol. 1.2: *Vorlesungen über die Dynamik discreter Massenpunkten*, ed. Otto Krigar-Menzel (1898), 309.

91. *Ueber die Erhaltung der Kraft. Eine physikalische Abhandlung* (Berlin: G. Reimer, 1847), 4–5, in *WA 1*:12–68, on 15.

92. Although Helmholtz and his companions in the biophysics movement of 1847 were united in their opposition to the postulation of "vital forces" to explain physiological processes in the organism, and although they maintained that only physics and chemistry were needed to account for physiological processes such as digestion and respiration, they did not extend their reductionism to include the processes of thought; see Gregory, *Scientific Materialism*, 148–51, and Hatfield, *The Natural and the Normative*, 181–82, 193–94.

93. "Ueber die Erhaltung der Kraft," *WA 1*:13; *Handbuch³ 3*:31; "Wahrnehmung," *VR⁴ 2*:243; and *Vorlesungen 1.1*:17. The German words Helmholtz used are forms of *Begriff* and *begreifen*. He equated the assumption of comprehensibility with acceptance of the law of cause; his attitude toward the status of these commitments changed over time, on which see Hatfield, *The Natural and Normative*, 211–18.

Helmholtz's *taste* in science and art was "classic" in the usual sense of that word; that is, he had a taste for simplicity, regularity, harmony, and order.[94] His *theory* of taste in art and science, to the extent that he expressed one, also fit the standard definition, inasmuch as he expressed a conception of art according to which art should present an ideal type. This theory is a version of classicism also in its rejection of the view that art ought simply to copy nature—art is to present a "refined" or "ennobled" version of nature—and in its assertion that in so doing, art conveys a deeper truth than ordinary sense perception. His classical taste (at least in science) and his classical theory of taste, as this essay proposes by way of conclusion, were related symbiotically, each strengthening the other.

Consider again Helmholtz's one expression of a conception of artistic beauty, offered near the end of his lecture on painting. There he speculated that artistic beauty is founded on "a sense of the graceful, harmonious, and lively current of our representations," which "brings to a more complete intuition lawfulness hitherto obscured," and in so doing allows us "to gaze into the ultimate depths of the sensation of our own soul."[95] What is the relation between the lawfulness portrayed in the work of art and the attendant feeling from the depths of our own souls? By now we know that the lawfulness is that of the ideal type. But we have also seen that the ideal type is perceived by means of the lawful operation of the associative mechanism of memory, which serves to filter variable reality and to preserve the invariant and lawful. Helmholtz's unstated theory of beauty may depend on an implied connection among the ideal as object of art, the intellectual process of "artistic intuition" that allows the artist to apprehend the ideal, and a corresponding process that guides the viewer's response. That is, perhaps pleasure in the beautiful arises when a work of art occasions, in the unconscious mental processes of the viewer, a harmonious flow of representations. This account would explain why the ideal type is perceived as beautiful: because in creating a work the artist distills the invariant from the flux of representations, viewing the work can bring a correspondent order to representations scattered in the viewer's memory. And it would further explain how viewing the ideal type could stimulate in the viewer a "sensation" of his or her own soul: a work of art can reveal to him or her a lawfulness previously concealed

---

94. *Tonempfindungen*, 554–55; and "Malerei," *VR⁴* 2:135. Helmholtz drew a connection between artistic sensibility and the search for simple laws in science in his "Wahrnehmung," *VR⁴* 2:242, and in "Goethes Vorahnungen," *VR⁴* 2:352. On classicism, see Greenhalgh, *Classical Tradition*, 11–5.

95. "Malerei," *VR⁴* 2:135.

only if it resonates with his or her own patterns of (lawfully governed) associations. The "sensation" in his or her soul is the sensation of the mind's own lawfulness, as exhibited in these lawfully flowing representations.

This conception of the perception of artistic beauty, when applied to Helmholtz's taste in scientific theories, might explain his penchant for seeking simple laws from variant phenomena. His attraction to simple, universal laws led him to posit the laws of association as the laws of mental life. With the laws of association in hand, he constructed his theory of the unconscious inferences underlying sensory intuition. He also applied the theory in explanations of scientific and artistic cognition. He then came to posit the same aesthetic aim for art and for science: to find the lawful, to discover the ideal within the variant.

Helmholtz's brand of classicism has been ably captured by Russell McCormmach in *Night Thoughts of a Classical Physicist*. McCormmach creates a composite classical physicist who, after attending a play, begins to reflect on the relation between literary classics and physics:

> The Classical Physicist knew it might be fanciful, but he saw a bond between physics and the classics. Classics penetrated to the spirit of ancient thought and life, and physics penetrated to the essence of nature. Classics was premised on the similarity of ancient minds and modern minds, and physics was premised on the similarity of mental and natural processes. It was a bit simplistic, he realized.[96]

Helmholtz held mental and natural processes to be similar inasmuch as both follow laws having classical form—universal laws regulating the interaction among simple elements, whether these elements be material particles or pure sensations. But he also posited a deeper relation between the mind's form of conception and the lawfulness of nature itself. He treated the mind as a natural filter for the lawful, as able to grasp or comprehend nature only to the extent that nature is lawful, and indeed as driven to seek such lawfulness. Helmholtz did not here propose a pre-established harmony between mind and world—he rejected such a harmony out of hand.[97] Rather, he asserted that because the mind can only comprehend a nature that is lawful, any

---

96. Russell McCormmach, *Night Thoughts of a Classical Physicist* (Cambridge, Mass.: Harvard University Press, 1982), 133.

97. He considered nativist theories of spatial perception to be committed to such harmony, a position that reasserted but did not explain the facts; see *Handbuch³* 3:17–8; "Die neueren Fortschritte," *VR⁴* 1:332–33; and Hatfield, *The Natural and the Normative*, 190–91.

nature that we do in fact comprehend will, for this reason, be governed by law.[98]

In construing the imagination as a faculty that grasps the lawful in both art and science, Helmholtz advanced another "classicist" cause: his proposal implied a means for bridging the divide between imagination and understanding, and between art and science, that had been upheld and celebrated in some Romantic circles. But this is not to suggest that he acted with this intent in mind, or that his choice of psychological theory was guided by a pre-existent theory of aesthetics. Indeed, it would seem that his theory of psychology led him to his aesthetic theory. More than likely, Helmholtz's standard of comprehensibility in science served as the model for his understanding of the aesthetic of the ideal type, as his second lecture on Goethe strongly suggests. In Helmholtz's eyes, the classical aesthetic was an aesthetic of comprehensibility and universal truth. However, it was only after he came to associate the drive for comprehensibility with art that he described it as an aesthetic at all. Arrested though he had been in his natural-scientific study of artistic aesthetics, he finally ventured an aesthetics for art after having come to conceive the task of art on analogy with that of science. Though he conceded that his scientific treatment of aesthetics was unable to explicate the "highest aims of art," once he had postulated an identity between artistic and scientific cognition, he found the aim of art to be the same as that of science: the search for universal truths. A bit simplistic, we realize.

---

98. *Handbuch*³ 3:31.

15

Helmholtz and the Civilizing Power of Science

David Cahan

1. Introduction

Between 1853 and 1892 Hermann von Helmholtz delivered approximately twenty-five popular addresses on the nature, purposes, and results of science, on the optimum institutional and political conditions for scientific advance, on the relations of science and civilization, and on the work of various scientists. These lectures and speeches, which eventually appeared in a two-volume collection entitled *Vorträge und Reden* and which by 1906 had reached the fifth and final edition, constitute the counterpart and natural complement to his three-volume *Wissenschaftliche Abhandlungen*, upon which they sometimes drew and to which they in part owed their existence.[1] They

---

Acknowledgments: Earlier versions of parts of this essay were presented at the annual meeting of the History of Science Society in Cincinnati, 28 December 1988; to the Boston Colloquium for the Philosophy and History of Science, Boston University, 4 April 1989; and at the Science Museum, London, 22 November 1989. I thank these audiences, as well as Mitchell G. Ash, Richard L. Kremer, and R. Steven Turner for many useful comments and criticisms.

1. Helmholtz's original set of popular lectures and speeches appeared in two small volumes (*Hefte*) as *Populäre wissenschaftliche Vorträge* (Braunschweig: Friedrich Vieweg und Sohn, 1865, 1870). A second edition appeared as three small volumes (*Hefte*) (Braunschweig: Vieweg, 1875). The success of these volumes led Helmholtz and his publisher to bring out a third edition under the title *Vorträge und Reden* (Braunschweig: Vieweg, 1884), which included a number of his additional lectures and speeches between 1876 and 1884 and which appeared in two substantial volumes. Following his death in 1894, Helmholtz's wife Anna and his publisher brought out a fourth edition (Braunschweig: Vieweg, 1896), which included additional lectures and speeches from 1884 to 1894.

are science popularized; their audience was largely the German and European political and social elite, on the one hand, and the German middle classes—the *Besitzbürgertum* and the *Bildungsbürgertum*—on the other. Broadly conceived, the lectures and speeches attempted, at times implicitly and at times explicitly, to enlighten Europe's, and above all Germany's, rulers and leaders about the potential benefits of science and technology for making a modern society and economy and for broadening the educated populace beyond its relatively narrow literary culture. The lectures and speeches sought to make both the elite and the middle classes aware of the intellectual pleasures and, as Helmholtz saw it, the practical (and positive) results of pure science. In so doing, at points they became polemical if not ideological. Furthermore, they sought in effect to educate Helmholtz's scientific colleagues; they implicitly challenged the value of the accelerating specialization of science and the formation of modern scientific disciplines that occurred during Helmholtz's lifetime. In short, the overall purpose of Helmholtz's popular lectures and speeches was less to teach science *per se* than to teach what science was about, to show what Helmholtz viewed as the interconnections of its parts, to relate what science can and cannot do, and to polemicize in its favor. In so doing, Helmholtz became Germany's ambassador of science—foreign as well as domestic—during the second-half of the nineteenth century.

Helmholtz's lectures and writings were occasional pieces: while each had a clear subject and theme, Helmholtz neither designed nor published them with any one unifying, systematic theme in mind.[2] Nonetheless, this essay argues that his diverse addresses contain an unarticulated, yet central and recurring theme. That theme may perhaps be best characterized as the civilizing power of science, by which is meant the intellectual understanding, and the socioeconomic and political power and control that humanity, in Helmholtz's view, derives from the pursuit of science. More particularly, this essay argues that Helmholtz's popular writings contain a complex of four intimately related categories that constitute the civilizing power of science: First, Helmholtz believed that science gives us the capacity to understand the natural world and humanity's place in it (Section 2). Second, he believed that science enables us to command and control the natural world (Section 3). Third, he believed that science provides the foundations for aesthetic life (Section 4). Fourth and finally, he believed

---

Finally, a fifth edition (Braunschweig: Vieweg, 1903) appeared, but it is simply a reprint of the fourth.

2. "Vorrede zum ersten Bande der dritten Auflage 1884," *VR⁵ I*:vii–xiii, on viii.

that the scientific enterprise could help unite individuals into a well-knit social community and help bind them to the larger polity of the nation-state (Section 5). In brief, these categories are about truth, power, and beauty. In the Helmholtzian vision, science enabled humanity to understand its world, to enrich itself materially and aesthetically, and to unite individual human beings for the common social good.

By no means were all of Helmholtz's views original with him. This essay does not claim that Helmholtz was the font of all wisdom on the nature of science and its effects, nor even that all of his views went undisputed. Any number of nineteenth-century German *Kulturphilosophen* and their foreign counterparts held and enunciated some (if not many) of Helmholtz's views; indeed, although this essay can only touch upon it, in large measure his views were essentially those of that amorphous and multifaceted (if not contradictory) nineteenth-century phenomenon known as German liberalism. Nonetheless, during the second-half of the nineteenth century no other German scientist, and few if any scientists elsewhere, equalled Helmholtz's broad range of creative scientific accomplishments in both the physical and the life sciences; as a consequence, few if any scientists or philosophers matched his overall intellectual authority and power. Already by 1870, Prussia's Kultusminister, Heinrich von Mühler, reportedly considered him to be "almost as great a *savant* [*Gelehrter*] as [Alexander von] Humboldt."[3] As this remark suggests, during the second-half of the nineteenth century Helmholtz succeeded Humboldt as Germany's premier spokesman of science, just as he in turn was succeeded by his student Max Planck during the first-half of the twentieth century.[4] Moreover, Helmholtz addressed the issue of science and civilization far more frequently and extensively than nearly all scientists before and after him. Although he was a true German *Kulturträger* (or "culture bearer") and arguably a mandarin member of the German academic elite—and so in part a spokesman for traditional and widespread views among the German cultural elite—he was also one of the leading figures in shaping and modernizing that elite's understanding about the relations of scientific, socioeconomic, and political life. Indeed, as this essay briefly argues in conclusion (Section 6), Helmholtz's views

---

3. Emil du Bois-Reymond to Helmholtz, 14 June 1870, in Kirsten, 240–41, on 241.
4. On Humboldt, see, e.g., Kurt-R. Biermann, "Humboldt, Friedrich Wilhelm Heinrich Alexander von," *DSB* 6 (1972):549–55; and on Planck, see J.L. Heilbron, *The Dilemmas of an Upright Man: Max Planck as Spokesman for German Science* (Berkeley, Los Angeles, London: University of California Press, 1986).

represented an extension into the late nineteenth century of certain Enlightenment values; as such, they also represented an uncritical attitude towards the scientific enterprise.

2. Science as the Capacity to Understand the Natural World and Humanity's Place in It

Helmholtz's belief in science's capacity to understand the natural world and humanity's place in it forms the foundation for the other three categories that constitute the theme of the civilizing power of science. His belief issued in part from the wide scope and great complexity of the German term *Wissenschaft* and its philosophically charged implications. For nineteenth-century German academics, *Wissenschaft* meant, in the first instance, the systematic pursuit of knowledge in all possible fields, including not only the totality of natural and social sciences—as the English-speaking world understood (and understands) the term "science"—but the humanities as well. Second, *Wissenschaft* was, at least in theory, less a set of results than a means to a higher, an ethical realm, that of *Bildung* or the cultivation of the individual personality. The term *Wissenschaft* thus had both a cultural and a cognitive dimension, and the individual who became *gebildet* through *Wissenschaft* was at once a person of systematic knowledge and higher culture. The individual became, in a word, a *Kulturträger*.[5]

For Helmholtz and his fellow *Kulturträger* the *Wissenschaften* split into two (more or less) fundamental branches: the human or moral sciences (*Geisteswissenschaften*) and the natural sciences.[6] The human sciences were those subjects that owed their origin to humanity's moral nature, that were directly concerned with the human mind and institutions. These subjects were based, Helmholtz believed, on the descriptive facts of humanity's past and present.[7] Though not a *Geisteswissenschaftler* himself, Helmholtz deeply and sincerely appreciated the *Geisteswissenschaften*. That appreciation originated in his culturally invigorating Potsdam, where he was born in 1821; in his home,

---

5. For good English-language treatments of these concepts see, e.g., John Theodore Merz, *A History of European Thought in the Nineteenth Century*, 4 vols. ([1904–12], reprinted New York: Dover, 1965), *1*:168–73, 202–6, 223–24; and Fritz K. Ringer, *The Decline of the German Mandarins: The German Academic Community 1890-1933* (Cambridge, Mass.: Harvard University Press, 1969), 21, 83–4, 86–7, 102–4, 110–11.

6. For an analysis of this distinction in the light of Helmholtz's aesthetics see Section 4 of Gary Hatfield's essay, "Helmholtz and Classicism: The Science of Aesthetics and the Aesthetics of Science," in this volume.

7. "Über das Verhältnis der Naturwissenschaften zur Gesamtheit der Wissenschaft," *VR*[5] *1*:157–85, on 166, 171–75.

where his parents stressed the importance of learning in general and of music and art in particular; and in his local gymnasium, where his father was a professor and where young Hermann studied classical, Semitic, and modern languages (Greek, Latin, Hebrew, Arabic, English, French, German and Italian) along with other parts of the humanities, the natural sciences, and mathematics.[8] Encouraged by his home and school, Helmholtz became a lifelong lover of literature ancient and modern. From his teenage years onward, he became an habitual visitor of Europe's theaters, museums and cultural monuments. Notwithstanding his professional preoccupation with the natural sciences, he maintained a strong interest in human culture. He declared:

> It is man in the various aspects of his intellectual activity that captivates us. Every great deed which history tells us about, every powerful passion which art represents, every portrayal of manners, of civic affairs, of the culture of peoples of distant lands or of remote times, seizes and interests us, even if we do not know them in any exact scientific context.[9]

He esteemed the human sciences and numbered several *Geisteswissenschaftler*, including the historians Theodor Mommsen and Heinrich von Treitschke, among his closest friends. He valued not only such fields as classical literature, which embodied "the treasures of intellectual [*geistiger*] culture bequeathed by antiquity," but also even the most mundane scholarly tools and products, including scientific catalogues, dictionaries, indices, natural histories, and so on, works which he characterized as forming the "basic stock of humanity's scientific principal, with whose interest we trade; one could compare them with capital invested in different realms."[10]

Yet that investment could only produce limited profits. For Helmholtz argued that although the human sciences dealt with facts they did not deal with laws, and thus they could not produce universal conclusions. Instead, their conclusions were more like commandments

---

8. "Erinnerungen," *VR*[5] 1:1–21, on 6–8, 16–7; Helmholtz's testimony in *Verhandlungen über Fragen des höheren Unterrichts* (Berlin: Wilhelm Hertz, 1891), 205–6; Ernst Kusch, "C.G.J. Jacobi und Helmholtz auf dem Gymnasium. Beitrag zur Geschichte des Victoria-Gymnasiums zu Potsdam," *Beilage des Schulprogramms des Victoria-Gymnasiums zu Potsdam* (Potsdam: Kramer'sche Buchdruckerei, 1896), 1–43, on 24–30; Koenigsberger *1*:11–21; and Gustav Uhlig, "Das Abiturientenzeugnis von H. Helmholtz," *Das humanistische Gymnasium 23* (1912):40–1.
9. "Ueber die Erhaltung der Kraft," *VR*[5] 1:187–229, on 190.
10. "Ueber die akademische Freiheit der deutschen Universitäten," *VR*[5] 2:191–212, on 195 (quote); "Ueber das Ziel und die Fortschritte der Naturwissenschaft," *VR*[5] *1*:367–98, on 369–70; and "Ueber das Verhältniss der Naturwissenschaften zur Gesammtheit der Wissenschaft," *VR*[5] *1*:168 (quote).

("*Gebote*") based on facts about humanity, and so they required "a delicate and well-formed insight into the human soul." These commandments "are given by an alien authority so that [humanity will] believe and act according to moral and legal relations." For Helmholtz, such commandments had a lower epistemological but a higher moral value than the laws of the natural sciences, and they contributed directly to a person's *Bildung*.[11]

Like the human sciences, Helmholtz claimed that the natural sciences were also based on facts—in this case facts which originated in inert matter, not in human life and history. Although most of his own creative scientific work was often quite theoretical and mathematical in nature, he nonetheless claimed that experiment formed the basis of natural science.[12] He characterized "the natural sciences, in the strict sense," as "the laborious and protracted task of collecting experimental facts."[13] He applauded, for example, Michael Faraday's aim "to express in his theoretical views only observable and observed facts while carefully avoiding mixing in any hypothetical elements." Faraday's procedure was really "an essential advance in the principles of scientific methodology," Helmholtz thought, one aiming "to liberate natural science from the last remnants of metaphysics." He judged Faraday a characteristic man of "high intelligence" precisely because he avoided making theoretical speculations before he had the facts to point the way.[14] By contrast, he thought that the work of his compatriot Julius Robert Mayer on the conservation of force during the early 1840s had failed to gain recognition from the scientific community precisely because it was too speculative and without a sufficient empirical basis— and this in an age reacting strongly against Hegelian metaphysics.[15]

Yet neither in theory nor in practice was Helmholtz a positivist. For Helmholtz, facts and theories existed in a dynamic relationship, mutually feeding into and supporting one another. On the one hand,

---

11. "Ueber das Verhältniss der Naturwissenschaften zur Gesammtheit der Wissenschaft," *VR⁵* 1:172–73.
12. "Das Denken in der Medicin," *VR⁵* 2:165–90, on 180, 183–84.
13. "Ueber den Ursprung und die Bedeutung der geometrischen Axiome," *VR⁵* 2:1–31, on 3.
14. "Die neuere Entwickelung von Faraday's Ideen über Elektricität," *VR⁵* 2:249–91, on 252, 291. On the influence of Faraday on Helmholtz's philosophy of science see Michael Heidelberger's essay, "Force, Law, and Experiment: The Evolution of Helmholtz's Philosophy of Science," in this volume.
15. "Anhang zu dem Vortrag 'Ueber die Wechselwirkung der Naturkräfte'. Robert Mayer's Priorität," *VR⁵*, 1:401–14, on 408–9. Ironically, these were the very grounds on which Johann Christian Poggendorff, the editor of the *Annalen der Physik*, rejected Helmholtz's essay, "Ueber die Erhaltung der Kraft. Eine physikalische Abhandlung," in 1847. (Koenigsberger *1*:70–1.)

the mere recognition of facts was not enough, for true "science only begins when its law and its causes are unveiled."[16] On the other, scientists

> only feel certain about the causes of the phenomena in question when the very same forces can also be experimentally demonstrated in our laboratories. The non-experimental sciences have not acquainted us with even a single new force.[17]

This relationship between facts and theories was hierarchical as well as dynamic. Helmholtz scorned the "false ideal" of "a biased and erroneously defined reverence for deductive methods." Philosophers (above all *Naturphilosophen* in general and followers of Hegel and Arthur Schopenhauer in particular), medical men prior to 1850, and too many scientists in generations past "believed they could deduce before they had established their general theorems by induction."[18] Natural science had become "great" due to the "inductive method," not "the windy speculations of a so-called 'deductive method'." To succeed, experimentalists needed "a penetrating knowledge of theory" just as theorists needed "broad, practical experience in experiment."[19]

For Helmholtz the decisive difference between the human and natural sciences lay not in their similar factual bases but rather in natural science's use of "logical induction" as its method for transforming facts about inert matter into laws of nature. He readily granted that mere facts were not science, arguing that facts "only get theoretical or practical value when they acquaint us with the law of a series of uniformly recurring phenomena."[20] He wrote:

---

16. "Ueber das Verhältniss der Naturwissenschaften zur Gesammtheit der Wissenschaft," VR^5 1:169.
17. "Die neueren Fortschritte in der Theorie des Sehens," VR^5 1:265-365, on 355-56.
18. "Das Denken in der Medicin," VR^5 2:170 (quote), 175 (quote), 180. As one of the leading epistemologists, philosophers of science, and aestheticians, and as a leading figure of the neo-Kantian movement of the second-half of the nineteenth century, Helmholtz was hardly an "enemy" of philosophy. However, he had nothing but scorn for metaphysics in general—including both the materialist and the idealist varieties—and Hegelians and *Naturphilosophen* in particular. (Ibid., VR^5 2:173, 184, 188-90; "Die neuere Entwickelung von Faraday's Ideen," VR^5 2:252; "Ueber das Sehen des Menschen," VR^5 1:85-117, on 88-9, 116-7; "Ueber das Verhältniss der Naturwissenschaften zur Gesammtheit der Wissenschaft," VR^5 1:162-65; and "Anhang zu dem Vortrag 'Ueber die Wechselwirkung der Naturkräfte'," VR^5 1:402, 408-10.)
19. "Ueber die Entstehung des Planetensystems," VR^5 2:53-91, on 56; and "Zum Gedächtniss an Gustav Magnus," VR^5 2:33-51, on 46, for the respective quotes.
20. "Ueber das Verhältniss der Naturwissenschaften zur Gesammtheit der Wissenschaft," VR^5 1:171-72; and "Ueber das Ziel und die Fortschritte der Naturwissenschaft," VR^5 1:367-98, on 374, for the respective quotes.

With respect to nature, there is no doubt that we are dealing with a completely rigorous causal nexus, one which admits no exceptions. We are thus further required to continue working until we have discovered laws containing no exceptions. We may not rest before then; only in this form does our knowledge preserve the victorious power over space, time, and nature's forces.[21]

Nature's character was one of necessity; it was an external order that operated completely independently of a person's perceptions or interventions. Its laws never changed.[22]

Helmholtz repeatedly declared that the natural scientist's task, the aim of all natural science, was to discover, under a given set of conditions, the unvarying laws of nature, "to find the laws of facts." To do so was to achieve understanding of nature or at least of a part of it. For Helmholtz, scientists had an "intellectual interest" in nature precisely because and only when they discovered laws that revealed the connections between multifarious physical phenomena.[23] These connections, these regulators of factual phenomena, in a word, these laws, brought understanding.[24] As he once put the point in a lapidary manner: "physical-mechanical laws are like telescopes of our mind's [*geistigen*] eye; they penetrate into the deepest night of the past and the future."[25]

Helmholtz's understanding of the nature of scientific "law" was not merely metaphorical, however. When he spoke of a "law of nature" he meant, first, a generalized conceptual structure that represented a repeatedly occurring natural process. He believed that laws were "discover[ed] in the facts," and that they required repeated testing through observation and experiment under repeatedly different conditions. Discovering laws was thus not a logical but a factual or a factual-cum-conceptual business.[26] A fully tested, true law admitted no exceptions; in the final analysis, a true law was an "objective power" otherwise known as a "force." He wrote:

> If we can make sure that the conditions obtain under which the law operates, then we must see the result follow without arbitrariness, without choice, without our assistance, with a necessity which compels the

---

21. "Ueber das Verhältniss der Naturwissenschaften zur Gesammtheit der Wissenschaft," *VR⁵* 1:178.
22. "Ueber die Entstehung des Planetensystems," *VR⁵* 2:58.
23. "Zum Gedächtniss an Magnus," *VR⁵* 2:47; and "Ueber die Erhaltung der Kraft," *VR⁵* 1:190, for the respective quotes.
24. "Ueber das Ziel und die Fortschritte der Naturwissenschaft," *VR⁵* 1:375.
25. "Ueber die Wechselwirkung der Naturkräfte und die darauf bezüglichen neuesten Ermittelungen der Physik," *VR⁵* 1:48–83, on 80.
26. "Ueber das Ziel und die Fortschritte der Naturwissenschaft," *VR⁵* 1:375.

things of the external world as well as our perception. The law then presents itself to us as an objective power, and accordingly we call it force.[27]

Thus for Helmholtz understanding natural phenomena ultimately meant determining the laws that governed them which in turn meant "the forces ... which are the causes of the phenomena." True laws were causal laws operating in the natural world "independent of our thought and will."[28] They required constant testing. And although he did not believe that natural scientists could ever discover "unconditional truth," he had little doubt that they could attain laws of "such a high degree of probability that they [the laws] are practically equal to certainty."[29]

Helmholtz considered the discovery of a true law of nature as the fullest measure of scientific progress. As the premier examples of such progress he repeatedly cited the law of conservation of force (or energy) and Darwin's theory of evolution by natural selection.[30] Such laws showed, he said, that science had progressed not merely in understanding this or that part of nature, but in understanding nature's overall structure and causal processes. He presented his audiences with an optimistic view of humanity's ability to comprehend nature.[31]

He believed that the laws of thermodynamics and of organic adaptation allowed humanity to comprehend its place in nature and its history, that they gave license, at least in public talks to lay audiences, to infer broad cosmological and biological conclusions about nature and humanity. The law of conservation of force—generalized and applied to several domains of physics by Helmholtz in 1847 and, thanks largely to the work of William Thomson in the 1860s, latter rechristened and reunderstood as the law of conservation of energy—showed that nature had a fixed storehouse of force, one that in algebraic sum neither increased nor decreased, so that "the quantity of effective force in inorganic nature is just as eternal and unchangeable as the quantity

27. Ibid., *VR*[5] *1*:375–76.
28. Ibid., *VR*[5] *1*:377.
29. "Das Denken in der Medicin," *VR*[5] *2*:184.
30. Helmholtz referred to Newton's "discovery of the law of universal gravitation and its consequences [as] the most impressive performance which the logical power of the human mind has ever performed." ("Ueber das Verhältniss der Naturwissenschaften zur Gesammtheit der Wissenschaft," *VR*[5] *1*:176–77.) Yet he mentioned it only *en passant*, perhaps because he and his audiences were more interested in the newly discovered laws of thermodynamics and organic adaptability. Moreover, he also doubted that the life sciences could achieve the same degree of rigor in their laws as those of the physical sciences. ("Antwortrede, gehalten beim Empfang der Graefe-Medaille," *VR*[5] *2*:311–20, on 318.)
31. "Ueber das Ziel und die Fortschritte der Naturwissenschaft," *VR*[5] *1*:377, 384, 387, 390, 395–96.

of matter."[32] This new law of natural philosophy, Helmholtz rightly maintained, was an "achievement of very general interest." He used the law to claim an interconnection between all natural phenomena. There was, he said,

> a new universal, natural law which rules the action of all natural forces in their mutual relations to one another and that has an equally great meaning for our theoretical views of natural processes just as it is of importance for technical applications.[33]

This was no parochial, merely terrestrial law. Its universality allowed its possessor "to understand something of the economy of the universe with respect to the supply of effective force."[34] The law represented power, and those who understood it were accordingly powerful.

Moreover, the law gave humans a sense of standards and of measures. Helmholtz argued that since change was fundamentally motion, it followed that the basic forces of nature were forces of motion, all of which were "of the same nature." Hence all such forces were measurable "according to the same standard, namely, the standard of the mechanical forces." It was precisely this, he argued, that the law of conservation of forces demonstrated.[35] The law showed "that all nature's forces are measurable according to the same mechanical standard, and that, with respect to the performance of work, all are equivalent to pure forces of motion."[36] A universal law brought universal standards to humanity just as it brought it intellectual and practical power.

Although Helmholtz spoke far more frequently about the first law of thermodynamics—he devoted two entire lectures to it and often mentioned it *en passant* in other talks—than about the second, he nonetheless used the latter, in the form of an eternal increase of entropy, to speculate about the origins and destiny of the world and human life. He employed the new laws of thermodynamics to supplant older religious and anthropological speculations and imaginations, and to go beyond both older biblical calculations of humanity's supposed sudden appearance some 6,000 years ago as well as the newer estimates of contemporary geologists that the Earth's age was so many unknown millions of years, to make the startling and seemingly rigorous calculation that the sun, that "dispenser of all life, the ultimate source

---

32. "Ueber die Wechselwirkung der Naturkräfte und die darauf bezüglichen neuesten Ermittelungen der Physik," VR^5 *1*:48–83, on 65.
33. Ibid., VR^5 *1*:51.
34. Ibid., VR^5 *1*:68.
35. "Ueber das Ziel und die Fortschritte der Naturwissenschaft," VR^5 *1*:379.
36. Ibid., VR^5 *1*:383.

of all that happens on Earth," must be at least twenty-two million years old and that it would expire in some seventeen million years.[37] As for the origins of human life, he speculated, as had others before him, that it may have been due to extraterrestrial phenomena; in 1871, he suggested that some form of life may have originated from germs long ago emitted from comets or meteors and that over the course of time those germs had evolved into human life as we now know it.[38] He considered the thought of a thermodynamic death "an insult" against the universe's creative, evolutionary forces: "what disturbs our moral feeling at the thought of a future, even though very remote, destruction of all living creation on this Earth," he wrote, "is principally the question of whether all life may only be a purposeless sport that will ultimately submit to destruction by brute force."[39] Still, he argued that humanity itself would never experience a thermodynamic death because long before thermodynamic forces could take effect meteorological and geological forces would "bring about the last day of the human race." He concluded: "The history of man to date is thus only but a short ripple in the ocean of time."[40] Natural science showed, he believed, that humanity is

> but a speck of dust on the Earth, which itself is but a speck of dust in the immense cosmos. The duration of our race to date, if we follow it far beyond written history back into the era of the lake dwellings or the mammoths, is but a moment compared to the prehistory of our planet when living creatures dwelled upon it, whose strange and sinister remains gaze at us from their ancient graves. Moreover, the duration of the human race disappears still more in comparison to the enormous periods of time during which worlds were formed, as they will indeed continue to form if our sun is extinguished and our Earth is either grown cold or is united with the glowing central body of our system.[41]

Such speculations, whether based on the scientific laws of modern thermodynamics or not, must have given little solace to his listeners and readers.

---

37. "Ueber die Entstehung des Planetensystems," *VR⁵* 2:79 (quote), 80, 82–3, 88–9.
38. Ibid., *VR⁵* 2:89; and for the general scientific context and Helmholtz's immediate predecessors on this topic see Michael J. Crowe, *The Extraterrestrial Life Debate 1750–1900. The Idea of a Plurality of Worlds from Kant to Lowell* (Cambridge, New York, and New Rochelle: Cambridge University Press, 1988), 359–406, esp. 404–5; and John G. Burke, *Cosmic Debris: Meteorites in History* (Berkeley, Los Angeles, London: University of California Press, 1986), 167–70.
39. "Ueber die Entstehung des Planetensystems," *VR⁵* 2:88, 89–90, for the respective quotes.
40. "Ueber die Wechselwirkung der Naturkräfte und die darauf bezüglichen neuesten Ermittelungen der Physik," *VR⁵* 1:83.
41. "Ueber die Entstehung des Planetensystems," *VR⁵* 2:88–89.

Nineteenth-century science taught humanity not only about its possible origins and fate. After 1859, it also taught, in the form of Darwin's theory of evolution by natural selection, about "the fundamental principles of the theory of life," one that Helmholtz esteemed as "an entirely new interpretation of organic adaptation."[42] He believed that Darwin's theory helped rid science of "the interference of free intelligence in the course of natural processes" and that, along with his own law of conservation of force, by which he had analyzed the human body as a heat-consuming and work-producing machine, it had also rid science of so-called vitalistic forces. One no longer needed, Helmholtz claimed, vitalism or "an act of supernatural intelligence" to explain the origin of species; instead, there were Darwin's "great and bold ideas."[43]

The difference between humans and other animals, Helmholtz believed, was that the former alone could think. The capacity to think was the source of their power in general and of their "superiority" over animals, who supposedly lived by "individual experience" instead of by the transmission of knowledge from one generation to another, which in turn allowed civilization to originate and flourish.[44] Not biblical dicta, but rather reason, above all in its manifestation as rigorous scientific thought, made humanity the Earth's ruler. "Under the light of Darwin's great thought," Helmholtz philosophized,

> we begin to see that not merely pleasure and joy, but also pain, struggle, and death are the powerful means by which nature has developed its more delicate and more complete forms of life. And we men know that in our intelligence, civic order, and morality we are living off the inheritance which our forefathers gained for us through their work, struggle, and the courage to sacrifice, and that what we achieve in the same way will ennoble the life of our posterity. Thus the individual who works, even if in a modest position and in a narrow sphere of activity, for the ideal purposes of humanity can fearlessly bear the thought that the thread of his own consciousness will some day break.[45]

Thus, in the Helmholtzian vision the grand laws of thermodynamics and evolution taught humanity of its and the universe's origin, course, and fate. They gave humanity a distinct place in the evolutionary

---

42. "Ueber das Ziel und die Fortschritte der Naturwissenschaft," VR^5 *1*:387.
43. Ibid., VR^5 *1*:387, and "Die neueren Fortschritte in der Theorie des Sehens," VR^5 *1*:293, for the respective quotes.
44. "Das Denken in der Medicin," VR^5 *2*:171.
45. "Ueber die Entstehung des Planetensystems," VR^5 *2*:90; and cf. "Antwortrede, gehalten beim Empfang der Graefe-Medaille," VR^5 *2*:318.

history of nature; and they gave human beings, or at least they gave Helmholtz, existential solace and the conviction that work alone was their (his) calling. They gave meaning to human history—but a science-based meaning.[46]

Although Helmholtz argued that the ability to produce laws gave the natural sciences a greater epistemological value over the human sciences, he did not think that the former were superior to the latter. Rather, he thought that the two domains complemented one another. He perceived and desired a harmony of interests, if not a strong interaction, between them. He thus opposed contemporary efforts to reorganize the traditional four faculties of the German university—theology, law, and the medical and philosophical faculties—into separate institutions; on both material and ideal grounds, he said, the "connection" of the natural and human sciences was "essential." Both types of sciences sought to civilize the individual: the human sciences contributed "directly" to the individual's cultivation, his *Bildung*, by enriching intellectual life, while the natural sciences did so "indirectly," by freeing humanity from "material restraints." Yet while recognizing that the traditional emphasis in education had been overwhelmingly devoted to the human sciences, he argued that an education dominated by the human sciences was incomplete and that the natural sciences should receive an increased emphasis in order to complement fully the human sciences. "A complete education of individual man," he wrote, "as well as of nations, is no longer possible without a combination of the traditional literary-logical studies and the new natural scientific direction." Together these sciences could "establish the supremacy of intelligence over the world."[47] *Wissenschaft* in the Helmholtzian vision gave humanity intellectual mastery of the world.

3. Science as the Means to Command and Control the Natural World

The second of the four categories that constitute the theme of the civilizing power of science in Helmholtz's addresses—science as command and control of the natural world—emerges from the first. In

---

46. Cf. Helmholtz's remarks in "Ueber die Wechselwirkung der Naturkräfte und die darauf bezüglichen neuesten Ermittelungen der Physik," *VR⁵* 1:83.

47. "Ueber das Verhältniss der Naturwissenschaften zur Gesammtheit der Wissenschaft," *VR⁵* 1:166 (quote), 178–79, 183 (quote); and "Ueber das Streben nach Popularisirung der Wissenschaft. Vorrede zu der Uebersetzung von Tyndall's 'Fragments of Science'," *VR⁵* 2:422–34, on 425 (penultimate quote).

Helmholtz's view, understanding nature's forces was the precondition for their use. He maintained this hierarchical view of the relationship of science and practice not only with respect to, say, physics and engineering. He similarly claimed that "the physician, the statesman, the lawyer, the clergyman, and the teacher have to be able to build upon the knowledge of mental processes if they want to acquire a true scientific basis for their practical work."[48] Indeed, even "the practical ends of the businessman who, in part unintentionally through daily experience and in part intentionally through the study of science, must rely for support on acquired knowledge of natural laws," fell within the compass of his view of the relationship of science and practice.[49]

Helmholtz argued that the scientist's mission was either to extend or to apply knowledge; in either case, the scientist served a "practical" end. He did *not* believe in simply increasing knowledge for its own sake. "Although the sciences arouse and train the subtlest powers of the mind," he wrote, "the individual who studied simply for the sake of knowing will surely find no true fulfillment to his purpose on Earth." That purpose, he maintained, was to act or to work:

> Action alone gives man a worthy existence, and thus his goal must be either the practical application of the known or the increase of science itself. For the latter, too, is an act for the progress of humanity.[50]

Whether the scientist chose to extend or directly apply knowledge, his ultimate aim was "to subjugate nature's reasonless forces for the moral purposes of humanity."[51]

For Helmholtz, intellectual progress in science entailed material and social progress in the world at large. The scientist who achieved understanding of the forces of nature simultaneously achieved mastery over nature. In good Lutheran language worthy of his Brandenburgian roots, he declared that the progress of the physical sciences had "taught us to exploit the forces of nature surrounding us for our own use and to subject them to our will." That explained why the modern nation-state supported science, he said.[52] To cite an example: in the preface to the first edition of *Die Lehre von den Tonempfindungen, als physiologische Grundlage für die Theorie der Musik* (1863), Helmholtz

---

48. "Das Denken in der Medicin," *VR*^{5} 2:189.
49. "Ueber das Ziel und die Fortschritte der Naturwissenschaft," *VR*^{5} 1:395.
50. "Ueber das Verhältniss der Naturwissenschaften zur Gesammtheit der Wissenschaft," *VR*^{5} 1:180.
51. "Ueber das Ziel und die Fortschritte der Naturwissenschaft," *VR*^{5} 1:372.
52. "Ueber das Verhältniss der Naturwissenschaften zur Gesammtheit der Wissenschaft," *VR*^{5} 1:177

expressed publicly his "debt of gratitude" to his immediate benefactors for their support of his research in physiological acoustics:

> The following investigations could not have been accomplished without the construction of new instruments, which did not enter into the inventory of a Physiological Institute, and which far exceeded in cost the usual resources of a German philosopher. The means for obtaining them have come to me from unusual sources. The apparatus for the artificial construction of vowels . . . I owe to the munificence of his Majesty King Maximilian of Bavaria, to whom German science is indebted, on so many of its fields, for everready sympathy and assistance. For the construction of my Harmonium in perfectly natural intonation, . . . , I was able to use the Sommering prize which had been awarded me by the Senckenberg Physical Society (*die Senckenbergische naturforschende Gesellschaft*) at Frankfurt-on-the-Main.[53]

Such support made Helmholtz into a natural ally of the state. The latter, for its part, supported science for its own self-preservation. In 1862, as William I struggled with his recalcitrant parliament to increase the size of the Prussian army, Helmholtz selected an apt metaphor to make his point:

> In fact, the men of science form a type of organized army. They seek the best for the entire nation, and almost always at its order and its cost to increase knowledge, which in turn serves to promote industry, to increase wealth, to enhance the beauty of life, and to improve political organization and the moral development of individuals.[54]

To Helmholtz and many of his fellow academics in post-1848 Germany, the scientist soldiered for his state and society.

Helmholtz's professional accomplishments and the course of Germany's economic development during the nineteenth century gave conviction and authority to his pronouncements on the applications of science. Helmholtz came of professional age in the early 1840s, just as Germany's industrial revolution began. He thus experienced firsthand both pre-industrial and industrial ways of life. Just as he began medical school in 1838 in Berlin, the Berlin-Potsdam railroad was established; he soon exploited the expanding German railway network to travel quickly around Germany and Europe.[55] (Like others of his

---

53. *On the Sensations of Tone as a Physiological Basis for the Theory of Music*, trans. by Alexander J. Ellis from the second edition (reprint New York: Dover, 1954), vi.

54. "Ueber das Verhältniss der Naturwissenschaften zur Gesammtheit der Wissenschaft," *VR^5 1*:181.

55. David Cahan, ed., *Letters of Hermann von Helmholtz to His Parents: The Medical Education of a German Scientist, 1837–1846* (Stuttgart: Franz Steiner Verlag, 1993),

day, he saw that the steam engine had been essential to "the great development of industry which has distinguished our century before all others.")[56] After graduating from medical school in 1842, he spent the next five years gaining firsthand medical experience as an army doctor in Berlin and Potsdam.[57] In late 1850, barely two years after he had begun his academic career at Königsberg, he invented the ophthalmoscope. His invention, he later insisted, was due to a combination of good training in mathematics and physics, on the one hand, and "luck," on the other, since he had always wanted to be a physicist but felt compelled for financial reasons to study medicine. In the event, he combined his knowledge of geometrical optics and the physiology of the eye to illuminate the previously invisible human retina. Indeed, the very circumstances of his invention occurred in a practical context: he had to prepare lectures on the physiology of the eye. "The ophthalmoscope," he later averred, "was more a discovery than an invention."[58] Be that as it may, it revolutionized ophthalmology in that it allowed unprecedented diagnosis—and, in time, treatment—of the living human retina. The young Helmholtz's name—he was scarcely thirty years of age at the time—spread quickly throughout the academic and public worlds. At a stroke he achieved career advancement and worldly recognition, thereby embodying his doctrine of the practical benefits of science. Though he was neither a businessman nor an industrialist, he appreciated the value of property: in December 1850, prior to his publication of the principles of the ophthalmoscope, he carefully staked his claim to the ophthalmoscope as "my property."[59] Indeed, for Helmholtz nature itself had property value: "The owner of a mill," he later declared, "claims the gravity of descending water or the energy [*lebendige Kraft*] of the wind rushing by as his property. These parts of the general supply of energy are what give his property its principal value." He spoke, too, of "the immense riches of ever-changing meteorological, climatic, geological, and organic processes on our earth."[60] Here was the voice of the *Besitzbürger* rather than the *Bildungsbürger*. During the 1870s and 1880s, as he devoted his efforts almost exclu-

---

Letters 5 and 7, pp. 43–8, 53–4, resp., and Richard L. Kremer, ed., *Letters of Hermann von Helmholtz to His Wife 1847–1859* (Stuttgart: Franz Steiner Verlag, 1990), xix–xxi.
  56. "Ueber die Erhaltung der Kraft," VR^5 1:211.
  57. Arleen Tuchman, "Helmholtz and the German Medical Community," this volume.
  58. "Antwortrede Graefe-Medaille," VR^5 2:314–16 (quotes on 314).
  59. Koenigsberger 3:142 ff. and 1:133–35.
  60. "Ueber die Wechselwirkung der Naturkräfte und die darauf bezüglichen neuesten Ermittelungen der Physik," VR^5 1:65, 78, for the respective quotes.

sively to physics (especially to problems in electromagnetism), he became peripherally involved with the booming German electrical industry and, particularly in his capacity as founding president of the new Physikalisch-Technische Reichsanstalt, with the construction and improvement of precision instrumentation and electrical standards, both of which were vital to industrial and academic life.[61] In 1881 he represented Germany at the Paris congress on electrical standards, and shortly thereafter he reported on the results to the Elektrotechnischer Verein in Berlin.[62] At the Columbia World Exhibition in Chicago in 1893, he again headed, now in his capacity as president of the Reichsanstalt, the German delegation to an electrical congress devoted to setting international electrical standards.[63] He fully recognized the importance of such standards to the fast-growing electrical industry:

> The electrical engineering industry has gradually developed to the point that it now consumes a tremendous amount of capital and represents an extraordinarily lively industry. Under these circumstances it cannot fail to happen that controversial issues concerning the industry will come before the courts, and the necessity will be felt to regulate these issues legally, namely, to determine the units of measurement to which one can refer to in such decisions.[64]

He saw that academic science and industry benefited one another, be it through the application of high theory, the creation and use of expensive precision instrumentation, or the establishment of legally binding physical standards.[65] For Helmholtz, talk of applied science was not abstract talk: his daily professional and social life was pervaded by the practical use of knowledge about anatomical, physiological, chemical, mechanical, electromagnetic, optical, and thermal phenomena.

The closely entwined courses of science, technology, and the economy during the nineteenth century demonstrated, Helmholtz believed, the living truth behind Francis Bacon's claim that "[k]nowledge is

---

61. David Cahan, *An Institute for an Empire: The Physikalisch-Technische Reichsanstalt 1871–1918* (Cambridge, New York, New Rochelle: Cambridge University Press, 1989), chap. 3.
62. "Ueber die elektrischen Maasseinheiten nach den Berathungen des elektrischen Congresses, versammelt zu Paris 1881," VR^5 2:293–309.
63. Cahan, *An Institute for an Empire*, 122.
64. "Ueber die elektrischen Maasseinheiten," VR^5 2:295.
65. On physical standards and instrumentation, see ibid., VR^5 2:296; and Cahan, *An Institute for an Empire*, 27, 33–4, 92, 105.

power." In his prorectoral address at Heidelberg in 1862, he further announced that humanity "teaches the natural forces of the inorganic world to serve the needs of human life and the aims of the human spirit [*Geistes*]."[66] Against Goethe and others who repudiated the mechanical interpretation of nature adhered to by most scientists (not least by Helmholtz himself) since the seventeenth century, he argued that in order to govern machinery one had to understand its full, detailed mechanical operations—not to ignore them, as did Goethe.[67] Furthermore, unlike the neo-humanist mandarins who fought a rearguard battle against the forces of industrialization and modernization, he prized machinery and industry. He welcomed the introduction of labor-saving devices such as steam engines and textile machinery; new transportation and telecommunication systems like the railroad, the telegraph, and the telephone (shortly after the latter's invention he published a paper that sought to explain sound transmission in telephones);[68] and the new sources and systems of electrical power and illumination.

Nonetheless, although Helmholtz welcomed industrial products and systems, and indeed polemicized for the benefits of applied science, he was no technocrat. He argued that to translate knowledge into economic and political power meant far more than merely accumulating capital and introducing new technology. It also required individual freedom and discipline as well as social justice and organization. States that sought to modernize had "to remove restrictions on industry and to concede a due voice in their council to the political interests of the industrious middle classes." He argued that, in addition to the necessities of capital and technology, one should recognize that

> the political and legal organization of the state and the moral discipline of the individual are what establish the superiority of the cultured [*gebildeten*] over the uncultured nations; and the latter, insofar as they do not know how to acquire culture, are headed towards unavoidable destruction. Here everything affects everything else. Where there is no certain legal condition and where the interests of the majority of the people cannot make themselves felt in an orderly manner, the development of national wealth and of the power which depends upon it is also impossible. Moreover, only he who has learned under just laws to develop the sense of honor of an independent man will be able to become

---

66. "Ueber das Verhältniss der Naturwissenschaften zur Gesammtheit der Wissenschaft," *VR*[5] *1*:180.
67. "Ueber Goethe's naturwissenschaftliche Arbeiten," *VR*[5] *1*:23–47, on 45.
68. "Telephon und Klangfarbe," *MB* (1878):488–500, reprinted in *AP* 5:448–60, and in *WA* *1*:463–74.

a good soldier—certainly not the slave who has submitted to the caprice of an arbitrary ruler.

To build and maintain the modern nation-state, he argued, required not only "the development of the natural sciences and their technical application" but also the "political, legal, and moral sciences, and of all the associated historical and philological studies." That was why "the cultivated peoples of Europe" were now devoting so much money to education and scientific research.[69]

While he welcomed applied science and argued for a broad understanding of the components necessary to create a modern nation-state, he also warned repeatedly against pursuing science for immediate economic ends. He believed that science's economic benefits would eventually emerge and best contribute to human civilization when scientists were expected only to pursue understanding of nature's forces. Only those ignorant of scientific life, he declared, sought "direct use." As examples, he claimed that Galileo's study of the pendulum resulted in improved time measurement and improved chronometers, and that Luigi Galvani's pure scientific work on animal electricity early in the century later led to practical results, in this case to electrical wires traversing Europe carrying news, as he dramatically put it, "from Madrid to St. Petersburg with the speed of lightning."[70] Here, too, he might well have cited his own motives in inventing the ophthalmoscope. "Whoever in the pursuit of the sciences," he proclaimed, "hunts after immediate practical uses can be rather certain that he will hunt in vain. All that science can strive after is complete knowledge and complete understanding of the rule of natural and intellectual forces."[71]

4. Science as the Foundation of Aesthetic Culture

Science's ability to provide the foundations for aesthetic life is the third of the four categories constituting the theme of the civilizing power of science. Between the mid-1850s and mid-1870s, as this section indicates in the broadest outline only, Helmholtz provided novel

---

69. "Ueber das Verhältniss der Naturwissenschaften zur Gesammtheit der Wissenschaft," VR^5 *1*:180–81. In lectures in 1862 and 1874 he again expressed, if more succinctly, essentially the same thoughts. ("Ueber die Erhaltung der Kraft," VR^5 *1*:189; and "Ueber das Streben nach Popularisirung der Wissenschaft. Vorrede zur Uebersetzung von Tyndall's 'Fragments of Science'," VR 2^5:422–23.)

70. "Ueber das Verhältniss der Naturwissenschaften zur Gesammtheit der Wissenschaft," VR^5 *1*:181–82.

71. Ibid., VR^5 *1*:182.

explanations of the scientific roots of painting and music, thereby showing how natural science formed the basis of aesthetic culture.[72] His views on the relations of science and art, like those on science and society, were broad and complex. While he unquestionably believed that painting and music, along with the other fine arts, rested on scientific foundations, he also maintained that art is never, even in principle, fully susceptible to scientific analysis. Still, he believed that "art, like science, can represent and convey truth," and that art therefore constituted a legitimate object of scientific inquiry.[73]

He thought the scientific analysis of art a supremely worthy goal, not least for the enhanced aesthetic pleasure which it brought. He argued that scientists or artists, like Goethe, who believed that scientific "dissection"—the term is Helmholtz's—or analysis destroyed the work of art were mistaken.[74] They failed to understand the nature of science, which, since it concerns the inner workings of phenomena, requires the scientist to invoke abstractions in order "to discover the levers, cords, and pulleys which work behind the scenes and control them" and to "want to ascend to the causes of natural phenomena," thereby aiding his search for laws.[75] Against Goethe and other artists who denied the possibility or desirability of analyzing artistic phenomena scientifically, he declared that:

> we cannot triumph over the mechanism of matter by denying it, but rather only by subordinating it to the purposes of our moral intelligence [*Geistes*]. Though it may also disturb the poetic contemplation of nature, we must become acquainted with its levers and pulleys in order to govern them according to our own will. Therein lies the great meaning of physical research for the culture of human civilization and therein is its complete justification founded.[76]

Indeed, he went so far as to claim "that the artist's work can only succeed when he has a good knowledge of the lawlike behavior of the phenomena represented and their effect on the listener or viewer."[77]

---

72. For a full discussion of Helmholtz's work in physiological acoustics and aesthetics see the respective essays by Stephan Vogel, "Sensation of Tone, Perception of Sound, and Empiricism: Helmholtz's Physiological Acoustics," and Gary Hatfield, "Helmholtz and Classicism," in this volume.
73. "Goethe's Vorahnungen kommender naturwissenschaftlicher Ideen," VR^5 2:335–61, on 344–45.
74. "Ueber Goethe's naturwissenschaftliche Arbeiten," VR^5 1:38.
75. Ibid., VR^5 1:44, 40, for the respective quotes.
76. Ibid., VR^5 1:45.
77. "Goethe's Vorahnungen kommender naturwissenschaftlicher Ideen," VR^5 2:345.

To be sure, he readily acknowledged that some artistic truths or representations were of an entirely different order from scientific truths or representations. Art represented "another way than that of science to acquire insight into the complicated mechanisms of nature and of the human spirit."[78] And he believed that scientists had little or no understanding of "artistic intuition" or the spiritual sources of art.[79] Yet he also stressed that such intuitions depended on experience, and to that extent they necessarily "fall within the domain of thought" and were subject (in principle) to scientific analysis.[80] "Artistic intuitions," like their counterparts "sensuous intuitions" in the non-artistic world, "are drawn from experience," and thus can be "conceived as lawlike."[81]

Helmholtz's interest in art ranked second only to his interest in natural science. Particularly in the realm of music, his theoretical judgments were based in good measure on his own experience. His aesthetic ideas issued both from his own participation in art and from his natural scientific knowledge. His parents fostered art in their Potsdam home. As a boy he was given piano lessons, and even as late as his first year of medical school in Berlin his parents admonished him not to discontinue playing the piano; in point of fact, he needed no admonishment: he took his piano with him to Berlin. Later in life, he numbered the violinist Joachim Jacobi among his friends, and occasionally visited Richard Wagner. He acted in plays at home with his family and friends, and he read literature aloud with them, too. He regularly attended the theater, symphonies, opera, and art museums. To be sure, he (almost) never himself painted, and he made a point of noting this fact when he lectured on painting, declaring that he lacked all "practical experience in the execution of art" and that his knowledge derived exclusively from his physiological research. Still, he had firsthand contact with painters, starting with his sister and continuing with his eldest daughter. During the 1840s he belonged to the Potsdam Kunstverein and was a good friend of its leading figure, Wilhelm Puhlmann, himself a painter and friend of many painters. Moreover, he spent one year (1847–48) as an anatomy instructor at the Berlin Kunstakademie, and so further experienced firsthand some of the practical problems of painters. He later developed friendships with a number of painters, including Adolph von Menzel, one of Germany's foremost painters during the second-half of the nineteenth century. Altogether, these

78. Ibid., VR^5 2:339.
79. Ibid., VR^5 2:341.
80. Ibid.
81. Ibid., VR^5 2:344.

artistic activities and experiences brought him relaxation, inspiration, and a preoccupation with gaining a natural scientific understanding of painting and music.[82]

During the second-half of the century, Helmholtz became one of the leading students of physical and physiological optics, and the leading proponent of an empiricist theory of human vision. He used his scientific and philosophical understanding of optics and perception—the details of which go beyond this essay[83]—to explain how, under given circumstances and within the inherent limitations of painting (i.e., the two-dimensional canvas, the finite range of the painter's colors, and the anatomical and physiological nature of the human eye) form, shading, color, and harmony of color are best rendered and function, and thus how a painting is perceived or, as he put it, made intelligible.[84]

Helmholtz believed that the best painting "ennobled" nature by highlighting the latter's "essential" features, thus allowing the viewer to contemplate nature "more serenely and continuously" than he or she could do by viewing it directly. He considered painters not transcribers but rather translators or interpreters of nature, for they translated nature into an "altered scale" that allowed the viewer to exclude the inessential as well as that which in nature fatigues the eye (e.g., dazzling sunlight). A well-executed painting, he argued, stimulates in us "a rich host of slumbering conceptions, and thereby awakens related feelings in easy play and directs them towards a common end." It thus allowed the viewer to focus on particular objects as imperfect embodiments

> of an ideal type, whose individual fragments, overgrown by the wild thicket of accident, lie scattered in our memory.... It seems that we can only explain the frequent power of art over reality in the human

---

82. E.g., Koenigsberger *1*:1–21, 44, 107–10, 221; Cahan, ed., *Letters*, Letters 1, 5–10, 13, 20, 31, 33–4, 35–8, 43–8, 48–52, 53–4, 55–6, 56–60, 60–3, 67–70, 82, 95, 96–7, 98–9, resp.; Kremer, ed., *Letters*, xi, xiii, xv, xviii–xix, 4–5, 10–1, 14, 17, 23, 36–8, 53, 97, 99, 102–3, 195, 199; Wundt, *Erlebtes und Erkanntes*, 2nd ed. (Stuttgart: Alfred Kröner, 1921), 158–59; "Optisches über Malerei," VR^5 2:93–135, on 95 (quote); Ellen von Siemens-Helmholtz, ed., *Anna von Helmholtz. Ein Lebensbild in Briefen*, 2 vols. (Berlin: Verlag für Kulturpolitik, 1929), *1*:116, 158, 165, 178, 180, 183–85, 193, 204–5, 225, 231, 250–51, 258–59, 281, 291–92, 309, and 2:36, 38, 44; and Jürgen Kuczynski, "Das Haus Helmholtz," in his *Gelehrtenbiographien* (= *Studien zu einer Geschichte der Gesellschaftswissenschaften*, vol. 6) (Berlin: Akademie-Verlag, 1977), 103–24.

83. See the essays in this volume by Timothy Lenoir, "The Eye as Mathematician: Clinical Practice, Instrumentation, and Helmholtz's Construction of an Empiricist Theory of Vision"; R. Steven Turner, "Consensus and Controversy: Helmholtz on the Visual Perception of Space"; and Richard L. Kremer, "Innovation through Synthesis: Helmholtz and Color Research."

84. "Optisches über Malerei," VR^5 2:97, 110–11, 133. On the limitations see esp. ibid., 96–7, 107–8, 110–15, 117–20, 125–29, 132–33.

soul [*Gemüth*] by the fact that reality always mixes something disturbing, distracting, and offensive into its impressions, while art can let all the elements gather unrestrained for the intended impression. The power of this impression will doubtless be greater the deeper, the finer, and the richer is the natural truth of the sensuous impression which is supposed to call forth the series of images and the affects bound with them.[85]

This is what Helmholtz meant when he spoke of a painting being intelligible. In 1863, he took the point even further, declaring that art teaches us that within the subconscious human mind

> there slumbers a germ of order that is capable of rich intellectual cultivation, and we learn to recognise and admire in the work of art, though draughted in unimportant material, the picture of a similar arrangement of the universe, governed by law and reason in all its parts. The contemplation of a real work of art awakens our confidence in the originally healthy nature of the human mind, when uncribbed, unharassed, unobscured, and unfalsified.[86]

In effect, Helmholtz considered art to be, among other things, a kind of psychoanalysis and an indicator of mental well-being.

Helmholtz's understanding of music rested both on his scientific theories and experiments and on his own piano training and performance. First in his popular lecture of 1857 on the physiological causes of harmony in music and then in his long, dense, and highly scientific and scholarly musicological masterpiece, *Die Lehre von den Tonempfindungen, als physiologische Grundlage für die Theorie der Musik* (1863), he employed both scientific research on physical and physiological acoustics and scholarly research on the history of music to establish the scientific foundations of music.[87] In the *Tonempfindungen* he presented his own and systematized and synthesized the work of others' investigations on the physics of musical waves and instrumentation, the anatomy and physiology of the human voice and ear, and the perception of sound. He described and explained the physical, anatomical, and physiological aspects of the elements

---

85. Ibid., *VR^5* 2:127, 134–35, for the respective quotes.
86. *On the Sensations*, 367. Cf. "Optisches über Malerei," *VR^5* 2:135.
87. Under the imprint of Friedrich Vieweg und Sohn in Braunschweig, the first edition of 1863 was followed by a second (1865), a third revised (1870), a fourth revised (1877), a fifth revised (1896), and a sixth revised (1913) edition. It was perhaps not by chance that Helmholtz gave his lecture on the physiological causes of harmony in music in Bonn, the hometown of Beethoven, whom Helmholtz regarded as "the most powerful among the heroes of musical art." ("Ueber die physiologischen Ursachen der musikalischen Harmonie," *VR^5* 1:119–55, on 121.)

of music, above all upper partial tones and quality of tone, combinational tones and beats, consonance and dissonance, and scales and tonality. Moreover, and most unexpectedly for a work on music, his book contained many drawings and employed the use of acoustical and other scientific instruments, diagrams representing wave phenomena and the anatomy of the ear, tables presenting various observational data on sound and musical phenomena, and, in a long series of appendices, mathematical equations (including the use of Fourier analysis and differential equations) to represent various sound-wave phenomena pertinent to music. In Helmholtz's scientific analysis, the playing of musical instruments and the sounding of human voices, not to mention "the rustling of dresses," required that even the concert hall or ballroom be reconceived as "a variegated multitude of intersecting wave-systems."[88] Perhaps most surprising for a work that presented a scientific (that is, structural or transcendental) analysis of music, Helmholtz also insisted upon the importance of changing musical styles and historical traditions, a point consonant with his overall empirical approach to understanding natural phenomena. Here he put to good use his gymnasium training in ancient and Semitic, as well as modern, languages, a training that helped sensitize him to the musical achievements (and limitations) of non-European peoples and older civilizations. Though he considered modern European music to be by far more developed than that of (and historically dependent on) non-European and pre-1600 European music, the *Tonempfindungen* was anything but an ethnocentric study.[89] Altogether, the scientific and musical worlds had never seen anything like Helmholtz's book. Prior to his lecture of 1857 and the first edition of his book in 1863, "music, more than any other art," he wrote, "has to date evaded scientific treatment." Echoing his recent (1853) critique of Goethe as a scientist, he noted how "it appears, at first sight, as if in music ... those individuals who reject the critical 'dissection of its pleasures' are still in the right."[90] Immediately upon publication the *Tonempfindungen* became the premier musicological study, and Helmholtz became "in the right." His

---

88. Ibid., VR^5 1:135.
89. *On the Sensations*, vii, 1, 5, 14, 190, 196, 228–29, 234–35, 237, 240–43, 249, 251, 255, 257–58, 260–72, 278–86, 288, 364.
90. "Ueber die physiologischen Ursachen der musikalischen Harmonie," VR^5 1:121. Cf. his remark: "And to my metaphysico-esthetical opponents I must reply, that I cannot think I have undervalued the artistic emotions of the human mind in the Theory of Melodic Construction, by endeavouring to establish the physiological facts on which esthetic feeling is based." (*On the Sensations*, vii.) See also ibid., 6.

work revolutionized musicology and remained the standard, comprehensive analysis until the early twentieth century.[91]

Helmholtz's overall musicological aim was to relate "*physical and physiological acoustics*," on the one hand, to "*musical science and aesthetics* on the other."[92] He considered musical pleasure the purest of all artistic pleasures because its sources were tones only, not the images of external objects that poetry and painting produced; music had a closer relationship to "pure sensation" than did poetry or painting.[93] The most elementary musical phenomena were, he stressed,

> natural phenomena, which present themselves mechanically, without any choice, to all living beings whose ears are constructed on the same anatomical plan as our own. In such a field, where necessity is paramount and nothing is arbitrary, science is rightfully called upon to establish constant laws of phenomena, and to demonstrate strictly a strict connection between cause and effect.[94]

At the elementary level, he argued, all "musical consequences" followed logically from and ipso facto simultaneously confirmed the laws and results of physical and physiological acoustics.[95]

Helmholtz had no doubt that there existed some sort of still undiscovered "nature of artistic beauty" and that it was the business of aesthetics to discover that nature "in its unconscious reasonableness."[96] The difficulty was that reason, and its laws, lay in the unconscious part of the mind, both the artist's and the viewer's or listener's. Indeed, he considered it an essential function of all good art that the viewer or listener never consciously recognized the law or "intelligence" behind the work of art in question, even when he or she sought to do so. Instead, "regularity . . . [is] apprehended by intuition without being consciously felt to exist." In a great work of art we merely sense—and it is "essential" that we merely sense—the design without ever

---

91. Sigalia Dostrovsky, James F. Bell, and C. Truesdell, "Physics of Music," in *The New Grove Dictionary of Music and Musicians*, ed. Stanley Sadie, 20 vols. (London, Washington, D.C., and Hong Kong: Macmillan, 1980), *14*:664–77, on 671–74; and Natasha Spender and Rosamund Shuter-Dyson, "Psychology of Music," ibid., *15*:338–427, on 389–90.
92. *On the Sensations*, 1. Helmholtz's lecture "Ueber die physiologischen Ursachen der musikalischen Harmonie," had essentially the same purpose (cf. ibid., VR^5 *1*:122).
93. *On the Sensations*, 2–3 (quote); and "Ueber die physiologischen Ursachen der musikalischen Harmonie," VR^5 *1*:121.
94. *On the Sensations*, 234.
95. Ibid., 5.
96. "Ueber die physiologischen Ursachen der musikalischen Harmonie," VR^5 *1*:154.

becoming fully conscious of the law or regularity behind it; thereby, we are excited aesthetically.[97] For the art excites our emotional, not our cognitive self. Comparing the flow of musical tone to the waves of the sea, Helmholtz declared that

> in the musical work of art the movement follows the currents of the artist's inspired soul. Now softly flowing this way, now charmingly skipping, now powerfully excited, convulsed or worked by the natural sounds of passion, the flow of sounds transmits unimagined moods into the listener's soul in primitive animation which the artist has sensed in his own and by which he finally raises him up to the peace of immortal beauty which the godhead has allowed only a few of his select favorites among men to herald.[98]

The result was twofold: on the one hand, great music uplifts morally and aesthetically; on the other, as he cautioned his listeners and readers, musical aesthetics was only partly susceptible to musicological analysis.

Helmholtz thus claimed only to have set the foundations upon which more complex aspects of musical composition and appreciation must rest; he did not think that he had fully explained, or that anyone could in principle fully explain, scientifically all aspects of musical experience. He emphasized in particular that he had dealt only with the elements of music and had only barely entered into musical aesthetics, and not at all into "the theory of rhythm, forms of composition, and means of musical expression."[99] He wrote (and the emphasis is his):

> Hence it follows,—and the proposition is not even now sufficiently present to the minds of our musical theoreticians and historians—*that the system of Scales, Modes, and Harmonic Tissues does not rest solely upon inalterable natural laws, but is also, at least partly, the result of aesthetical principles, which have already changed, and will still further change, with the progressive development of humanity.*[100]

At the same time, he argued that the elements of music were not arbitrary and that a fully developed musical style exhibited rules and

---

97. *On the Sensations*, 366–67.
98. "Ueber die physiologischen Ursachen der musikalischen Harmonie," *VR⁵ 1*:155.
99. *On the Sensations*, vii, 4, 234–35, 371 (quote); and "Ueber die physiologischen Ursachen der musikalischen Harmonie," *VR⁵ 1*:155.
100. *On the Sensations*, 235.

systematic structures; it was the business of musical science, he said, to unveil and to analyze those rules and structures, "to discover the motors, whether psychological or technical," at work in the "artistic process," with physics analyzing the technical motor and aesthetics the psychological.[101] As for the more advanced topics, above all the "psychical motives" at play in music, he left to others the attempt "to explain the wonders of great works of art, and to learn the utterances and actions of the various affections of the mind," preferring instead to "remain on the safe ground of natural philosophy, in which I am at home."[102]

Helmholtz thus viewed art as an admixture of natural phenomena subject to scientific analysis and laws, on the one hand, and ineffable expressions of human emotions or artistic spirit, on the other. Scientists and artists, he held, shared many similarities; for example, he considered a scientist's discovery of a new law to be "of the same type as the highest performances of artistic insight," and he stressed that "[t]he true artist and the true scientific investigator know that great works are produced only through hard work."[103] For Helmholtz, science and art shared much common ground; their procedures for finding and relating "insight into the intricate works of nature and the human spirit" were complementary rather than antithetical.[104] In both science and art, the first, but

[o]nly the first creative thought, which must proceed verbal expression, is formed and must emerge in both types of activity in exactly the same way. In fact, that can initially always occur only in a way analogous to artistic insight—as a presentiment of new lawlike behavior. Such a creative thought consists in the discovery of a previously unknown similarity in the way in which certain phenomena repeat themselves in a series of typically congruent cases.[105]

The difference between science and art, when he chose to simplify the matter, was that science concerned the describable, art "the indescribable."[106]

---

101. Ibid., 235.
102. Ibid., 371.
103. "Das Denken in der Medicin," VR^5 2:184.
104. "Goethe's Vorahnungen kommender naturwissenschaftlicher Ideen," VR^5 2:339; cf. ibid., 344–45.
105. Ibid., VR^5 2:348.
106. Ibid., VR^5 2:348 and 358 (quote). For an extended treatment of the theme of science and art in Helmholtz's thought see Hatfield, "Helmholtz and Classicism."

In his scientific analyses of the foundations of aesthetic life Helmholtz claimed to have done no more than to have given some of the means towards discovering the laws and principles of aesthetics.[107] As he said of his musicological results, "we are here still treading the lower walks of art, and are not dealing with the expression of deep psychological problems."[108] Musical science provided tools in the quest to grasp the underlying reasonableness, and hence morality, behind the art. Although Helmholtz recognized their limits, he thought that visual and musical science aided us in the moral act of contemplating art. For painting and music, two preeminent manifestations of civilized life, Helmholtz indicated how an understanding of the scientific foundations of art enhanced both the creative powers of the artist and the ability of the viewer or listener to deepen his or her experience and understanding of art.

5. The Social and Political Function of Science: Education and National Unity

Helmholtz's fourth and final unspoken theme constituting the civilizing power of science envisioned the scientific enterprise as strengthening community bonds and helping to bind individuals (or at least the German peoples) to the nation-state. At the centerpiece of this vision of the social and political function of science lay Helmholtz's understanding of the contemporary German university system—both its strengths and its weaknesses.

Helmholtz regarded the German university system as the principal institutional means for promoting science and, ultimately, social peace and prosperity, and he thought it decidedly superior to the French and English systems. Its alleged superiority lay, above all, in its capacity to advance research, which Helmholtz, like virtually all German professors, in turn considered the *sine qua non* of good teaching. "Whoever wants to impart to his listeners the full conviction of the truth of his theories," he asserted, "must above all else know from his own experience how one gets and does not get such conviction." The teacher, like the researcher, needed to "have conquered new domains."[109] The signal failure of the French and English university systems, he thought, was that they did not incorporate research into their

---

107. "Ueber die physiologischen Ursachen der musikalischen Harmonie," *VR⁵* 1:155; "Optisches über Malerei," *VR⁵* 2:97; and *On the Sensations*, 366.
108. *On the Sensations*, 368.
109. "Ueber die akademische Freiheit der deutschen Universitäten," *VR⁵* 2:204.

institutional missions; as a consequence, French and English university scientists lacked state support for research and an atmosphere of intellectual freedom. Like many others, he viewed German universities as a model for the world: "the eyes of the civilized world are upon them," he declared.[110]

Helmholtz, who lacked firsthand knowledge of the French higher educational system and who had only limited personal acquaintanceships with Frenchmen, held French universities to be strongly anti-historical and highly rationalistic in nature; as a consequence, they had become "institutes of pure instruction" with "rigid" programs and courses of instruction, replete with examinations, and "completely separated" from French research institutions like the Collège de France, the Jardin des Plantes, and the Ecole des Etudes Supérieures. Students were required to attend lectures and to pay high fees for doing so; the government used the income from student fees to pay staff salaries yet provided little money to maintain, let alone expand, French universities. The teaching staffs, Helmholtz asserted, simply taught the known—and so "require only good receptive talents"—and were self-complacent. To be sure, he judged the system to function well "at giving students of modest talent sufficient knowledge of routine matters in their profession." Yet neither faculty nor students faced intellectual challenges and doubt; one group, he said, transmitted the known, the other received it—"*in verba magistri.*" While he thought that the talents, energy, and ambitions of individual Frenchmen partially corrected the faults of the French system, he nonetheless maintained that French universities did not contribute to "the progress of science."[111] He judged them to be intellectually stagnant and filled with mediocrity.

The English system, of which Helmholtz had some firsthand knowledge, fared more favorably in his eyes than the French. He excluded the Scottish universities and the recently established colleges and universities in London and in the provinces from his judgments, saying

---

110. Ibid., *VR^5* 2:194 (quote), 211-12. While Helmholtz's account of the strengths of the German university system might easily be charged as self-serving and chauvinistic, it is essentially confirmed by recent work of Charles E. McClelland, one of the closest students of the nature and development of the German university system from the Enlightenment to the early 20th century. See McClelland's " 'To Live for Science': Ideals and Realities at the University of Berlin," 181-97, esp. 188, 190-91, 194, in Thomas Bender, ed., *The University and the City: From Medieval Origins to the Present* (New York and Oxford: Oxford University Press, 1988); and idem, "Republics Within the Empire: The Universities," 169-80 in Jack R. Dukes and Joachim Remak, eds., *Another Germany: A Reconsideration of the Imperial Era* (Boulder and London: Westview Press, 1988).

111. "Ueber die akademische Freiheit der deutschen Universitäten," *VR^5* 2:199-200.

simply that they were more like those of Germany and Holland. The English system thus meant Oxford and Cambridge, which he characterized, at least in their pre-reform phase,[112] as largely clerical training schools with well-endowed colleges and a fellow system, and as institutions that catered to and furthered "the style and customs of England's well-to-do classes." He admired how Oxford and Cambridge cultivated in their students the classical languages and literatures that led to a strong sense of antiquity and, in turn, an enviable command over the modern English language, just as he admired how they cultivated the bodily training and health of their students. Yet he thought the methods and range of instruction at Oxford and Cambridge, with their heavy emphasis on highly specialized examinations, to be comparable to German gymnasia, not universities. And while he prized the excellent qualities of Oxford and Cambridge fellows and acknowledged the potential inherent in their unfettered time to conduct research, he claimed that they offered "so little ... for science" because "during their student years they do not come in sufficient contact with the living spirit of research so as later to continue their work from their own interest and by their own inspiration." He believed that the failure lay in Oxford and Cambridge as institutions, not in the abilities of Englishmen.[113] Time and again he gladly acknowledged how gifted he thought the British were, and how much he owed to individual British scientists such as Michael Faraday, William Thomson, and James Clerk Maxwell. "I owe a great deal to England for my own intellectual education," he wrote in 1867. "Grown up among the traditions of high-flown metaphysics, I have learned to value the reality

---

112. In the 1880 English translation of Helmholtz's lecture, the translator, E. Atkinson, took "exception" to Helmholtz's views on Oxford and Cambridge. He wrote: "These statements were a fair representation of the impression produced on the mind of a foreigner by a state of things which no longer exists in those Universities, at least to the same extent." The post-1854 reforms, he claimed, produced significant changes at Oxford and Cambridge. "Hence, in respect of this article ['On Academic Freedom'], I have availed myself of the liberty granted by Professor Helmholtz, and have altogether omitted some passages, and have slightly modified others, which would convey an erroneous impression of the present state of things. I have also on these points consulted members of the University on whose judgment I think I can rely." (*Popular Lectures on Scientific Subjects*, trans. E. Atkinson, 2nd ser. [London, New York, and Bombay: Longmans, Green, 1903], v–vi.) Yet as the English chemist Henry Roscoe, who knew the German academic scene well, indicated, there were indeed native British scientists who shared Helmholtz's views on the shortcomings of scientific education and research at Oxford and Cambridge. (See Henry E. Roscoe, "Science Education in Germany. I. The German University System," *Nature 1* [1869]:157–59.)

113. "Ueber die akademische Freiheit der deutschen Universitäten," VR^5 2:196–99, quotes on 196 and 198.

of facts in opposition to theoretical probabilities by the great example of English science."¹¹⁴ Yet the intellectual strengths of individual Englishmen could not offset the institutional weaknesses of Oxford and Cambridge.

In stark contrast to his assessment of the intellectual poverty of the French and English university systems, Helmholtz presented the German system as a dynamic set of institutions fostering scientific research and, hence, scientific progress. He believed that German universities owed their strength and success in scientific research to a combination of state financial support and political independence.¹¹⁵ During the eighteenth and early nineteenth centuries the financial poverty and political impotency of German universities had led them, he argued, to accept state financial aid and political control. While he conceded that in the past "ruthless use" of such control had occasionally occurred, he nonetheless maintained that the individual German states (*Länder*) had in modern times by and large not interfered with their universities' decision-making procedures. The post-Napoleonic German universities, he said, had "preserved a far greater nucleus of their internal freedom, and indeed the most valuable part of this freedom, than in scrupulously conservative England and in France with its wild hunt for freedom."¹¹⁶ In 1877, in apparent disregard of Bismarck's *Kulturkampf* against German Catholics and attacks on social democracy, Helmholtz felt warranted to claim that "the advanced political freedom of the new German Empire had brought a cure" against interference by the state, society, and religious authorities in scientific matters. He proudly declared that German universities, in contrast to French and English universities, permitted the teaching of "the most extreme consequences of materialistic metaphysics, the boldest speculations on the basis of Darwin's theory of evolution . . . as well as the most extreme deification of papal infallibility." He claimed that German universities, unlike their French and English counterparts, presented "no obstacle of any sort to discussion in a scientific spirit of any scientific question." Moreover, he praised the scientific freedom

---

114. Helmholtz to du Bois-Reymond, 15 May 1864, in Kirsten, 208–9, on 209; Helmholtz to Alexander Williamson, 1867, reprinted in part in George Haines IV, *German Influence upon English Education and Science, 1800–1866* (New London, Conn.: Connecticut College, 1957), 16–7 (quote), citing William A. Tilden, *Famous Chemists. The Men and Their Work* (London: G. Routledge, 1930), 239. Cf. also Wundt, *Erlebtes und Erkanntes*, 158–59.

115. "Ueber die akademische Freiheit der deutschen Universitäten," VR^5 2:196, 200–1.

116. Ibid., VR^5 2:200–1 (quotes), 205–6, 210.

supposedly accompanying the unique German position of *Privatdozent* or private lecturer; while he acknowledged that the *Privatdozent* lacked control over the resources of a scientific institute and the right to give examinations, he claimed that the *Privatdozent* nonetheless held "exactly the same" "legal rights" as an ordinary professor and was often enough promoted from being a professor's assistant to being his competitor.[117] Few *Privatdozenten* would have given such a positive account of their situation.

The spirit of freedom that Helmholtz detected within the German university had two well-known dimensions: *Lernfreiheit* and *Lehrfreiheit*. *Lernfreiheit*, he explained, permitted German students, again supposedly in contrast to their French and English counterparts, to study whatever, wherever, with whomever, and by whatever means they chose. He claimed that the German student,

> freed from working for other interests, can live exclusively for the task of striving after the best and noblest which the human race to date has been able to attain in knowledge and in point of view, bound closely together in friendly rivalry with a large number of similarly striving fellow students and in daily intellectual intercourse with teachers from whom he learns how independent minds think.[118]

Lehrfreiheit complemented this rosy picture: it permitted German university teachers to pursue their scientific interests as they saw fit. Only in times of "heated party struggles" did governments override faculty recommendations, Helmholtz asserted.[119] Since German university teachers were also researchers, they had every material interest, he explained, to try and attract German students in as large a number as possible to their lecture halls and laboratories.[120] "The entire organization of our universities," he proclaimed, "is thus pervaded by respect for free, independent conviction, which is more strongly imprinted in the Germans than in their Aryan relatives the Romanic and Celtic tribes. The latter place more weight on political and practical motives." In contrast to their German cousins, the Romanic and Celtic professoriates supposedly "withhold research ideas from investigating such

---

117. Ibid., *VR⁵* 2:205–6 (quotes), 207. Cf. McClelland, "Republics Within the Empire," for a similar but modern assessment of the German universities' "genuine academic freedom," their relative political independence yet strong support from the state, and their meritocratically based openness to all who could afford to attend (esp. 170–75, 178, quote on 170).

118. "Ueber die akademische Freiheit der deutschen Universitäten," *VR⁵* 2:201–3 (quote on 202–3).

119. Ibid., *VR⁵* 2:207.

120. Ibid., *VR⁵* 2:207–8.

theories that seem to them undiscussable as concerns the necessary foundation of their political, social, and religious organization. They find it completely justified not to allow their young men to look beyond the boundary which they themselves are unwilling to go beyond."[121] German scientists showed no such timidity, he averred; they were

> ... more fearless than others of the consequences of the entire and full truth. In England and France, too, there are excellent researchers who can work with full energy and proper understanding of scientific methods; yet to date they have almost always bent before social and ecclesiastical prejudices, and if they wanted to express their convictions openly they could do so only at the expense of their social influence and effectiveness.[122]

In Helmholtz's view, therefore, it was thus largely social and political constraints, not intellectual ability, that had (supposedly) held back French and English science. The "work-loving, modest, and morally austere" German people, by contrast, supposedly displayed the necessary "boldness" to "look the truth full in the face."[123] In turn, the freedom-bearing professoriate, in Helmholtz's view, had shown by its scientific accomplishments and work ethic that its members were worthy servants of the German state and society.[124]

Helmholtz believed that German university research flourished not only because of internal intellectual freedom and non-interference on the part of the state, but also because the state educational ministries provided the necessary financial support to conduct research. To discover the fact-based laws that he held as the goal of all science meant that scientists needed to observe and experiment, which meant in effect that, given the socioeconomic and political realities of nineteenth-century Germany, they needed the state as the material provider of laboratories, experimental equipment and instrumentation, staff, and all the other components that constituted a scientific institute.[125] With the state's material help, German scientists could discover the facts upon which laws are based and repeatedly test established laws under new circumstances. The search for laws of nature led to expenses that

---

121. Ibid., *VR⁵* 2:209–10 (quotes). And cf. "Ueber die Entstehung des Planetensystems," *VR⁵* 2:78, where he refers to "our forefathers of the Aryan race in India and Persia."
122. "Ueber das Ziel und die Fortschritte der Naturwissenschaft," *VR⁵* 1:397.
123. Ibid., *VR⁵* 1:397.
124. "Ueber die akademische Freiheit der deutschen Universitäten," *VR⁵* 2:203.
125. Cf. David Cahan, "The Institutional Revolution in German Physics, 1865–1914," *HSPS* 15:2 (1985):1–65.

could be met, Helmholtz believed, only by the state. In his view, university scientists became debtors to the state. As he told the general assembly of the Deutsche Gesellschaft der Naturforscher und Ärzte at Innsbruck in 1869:

> We rejoice to find among us a large number of participants from the nation's cultivated classes. We see influential statesmen among us. They are all interested in our work; they expect further progress in civilization from us, further victories over nature's forces. They are the ones who make the external means for our work available, and therefore they are justified in asking after the results of this work.[126]

Over two decades later, at the public celebration of his seventieth birthday in 1891, he again declared: "The state, which provided my livelihood, scientific equipment, and a good portion of my free time, thus had, in my view, a right to demand that I communicate, in appropriate form, freely and completely to my fellow citizens all that I had discovered with its help."[127] The communication of scientific results was simply a small and partial price that university scientists paid for intellectual freedom and financial support.

The other, and perhaps not so small, part was institutional and social isolation. Protection from external interference and the prosecution of advanced, esoteric knowledge comprehensible to only a handful of fellow specialists meant that German scientists had relatively limited contact with their fellow, non-academic Germans. Insularity from the surrounding German society was the obverse of the German scientist's freedom and capacity to do research within the German university system; as nineteenth-century German science became institutionalized within the universities, it also became insular. Helmholtz thought that the great institutional problem confronting German science was its relation to the non-scientific and non-academic worlds. Here lay the great weakness of the dynamic German university system. "On the Continent [in comparison to England]," he wrote Alexander Williamson in 1867, "the conditions of life for scientific men have been different; the great part of them belonged always to a peculiar class, more isolated from other classes of men, more connected by its interests and occupations."[128] German university science produced a relatively large and excellent cadre of scientists, but one essentially without contact with the educated public and without interest in educating

---

126. "Ueber das Ziel und die Fortschritte der Naturwissenschaft," VR^5 1:373.
127. "Erinnerungen," VR^5 1:18.
128. Helmholtz to Williamson, quoted in Haines, *German Influence*, 16-7.

the public on matters scientific. The freedom that permitted specialized scientific research to flourish simultaneously permitted scientists to neglect contacts with their neo-humanist colleagues within the universities and with the educated world beyond. The point is pertinently illustrated by a request of the undersecretary of state in the Prussian Kultusministerium, which in 1869 was hoping to win Helmholtz for Bonn—"well-nigh *coûte que coûte*," as the secretary told Helmholtz's intimate friend Emil du Bois-Reymond—for a copy of Helmholtz's popular lecture "Ueber die Wechselwirkung der Naturkräfte." Du Bois-Reymond sent him all of Helmholtz's popular lectures.[129] How else could ministry officials possibly understand Helmholtz's work?

To overcome or at least alleviate the weakness of the German university system, Helmholtz sought to promote the popularization of science. He thought popular science could build bridges to the German elite both within and beyond the university, and so help surmount the main institutional problem of German science. His own popular scientific lectures and essays did just that; in time, they became an essential aspect of his own ever-increasing public responsibilities and persona as Germany's first man of science.

There were also other motives behind his popularization of science. For one, it brought income and helped promote his own career. In 1864–65, for example, he lectured on the conservation of force and gave lectures in Karlsruhe and in London largely to earn money—some of which he then used to pay for new apparatus for his Heidelberg physiological institute. Yet money was not the only motivation. He also did so for "some better uses from such lectures," presumably meaning public enlightenment. Finally, his lecturing in London, at least, provided him with "a cheap and advantageous way of coming into close contact with London scientists [*Gelehrten*] and English circumstances; the stay in England in general [is] like a sort of spiritual treatment at a watering-place: it shakes up the mind's activity whenever it [the mind] begins to fall asleep in comfortable southern Germany."[130] As Helmholtz popularized science, he also popularized himself.

To help popularize science in Germany, Helmholtz supervised the translation of a number of John Tyndall's works into German, including *Faraday as a Discoverer* (1870), *Heat as a Mode of Motion* (1871), *Lectures on Sound* (1874), and *Fragments of Science* (1874). Despite feeling "overwhelmed by other official and scientific work,"

---

129. Du Bois-Reymond to Helmholtz, 22 October 1868, in Kirsten, 231.
130. Helmholtz to du Bois-Reymond, 26 February 1864, in ibid., 207–8, on 207.

he found time to help bring Tyndall's books—he was speaking in particular about *Fragments of Science*—before the educated German public, which sought instruction in science, he said, "not merely" for purposes of "entertainment or empty, barren curiosity but rather from a well-justified intellectual need." That need issued partly from a desire to understand the "very considerable influence" that the natural sciences now exerted "on the formation of the social, industrial, and political life of civilized nations," and partly from something "much deeper . . . and further reaching," "namely, its influence on the course of the intellectual progress of humanity."[131]

In his long preface to Tyndall's *Fragments*, Helmholtz argued that the rise of the natural sciences had created "a schism in the intellectual formation of modern humanity" and that the sciences now deserved equal treatment in the curriculum alongside such traditional subjects as ancient languages, philosophy, and history.[132] Traditional education, unlike experimental natural science, had failed to make new discoveries and to grasp the real world through observation and experience.[133] "The natural sciences . . .," he said, "are a new and essential element in human education, an element of lasting meaning for all future development in education." Well-rounded men and nations, he argued, required instruction in both traditional "literary-logical studies and the new natural sciences."[134] The great institutional problem here was less that of educating the young than of reeducating the old, those who, as he put it, currently "direct our state, teach our children, and maintain upright respect for the moral order."[135] These educators "have to be encouraged or pressured through public opinion from those classes among all people, men and women, who are competent to judge."[136] For Helmholtz, the problem became one of overcoming "a type of barrier between men of science and laymen," with the latter needing to learn about science for both intellectual and practical reasons.[137] He thought that English scientists had developed the popularization of science into a fine art, and that Tyndall was its master.[138] He hoped that Tyndall's book would prove an important source for enlightening those Germans—above all, the cultivated (*gebildete*) elite—who had no

---

131. "Ueber das Streben nach Popularisirung der Wissenschaft," VR^5 2:422–23.
132. Ibid., VR^5 2:423–28 (quote on 423).
133. Ibid., VR^5 2:424–25.
134. Ibid., VR^5 2:425.
135. Ibid., VR^5 2:425–26.
136. Ibid., VR^5 2:426.
137. Ibid., VR^5 2:428.
138. Ibid., VR^5 2:428–30.

training in the natural sciences but who wanted to know or should know about the nature, methods, and aims of modern natural science.[139] To popularize science meant to win popular favor for the scientific enterprise.

In the Helmholtzian vision, popularizing science could perform two other closely related functions: it could help rid the German peoples of mystical, obscurantist beliefs and unite them into a more integrated social and political whole. From the very start of his career, and probably already in his youth, he believed that science could effectively oppose the anti-rational forces of mysticism and spiritualism. One of the unstated aims of his 1847 essay on the conservation of force, for example, was to rid science (and in turn the non-scientific world) of the vague, mystical notion of "*Lebenskräfte.*"[140] Knowledge of "precisely defined, exceptionless ruling laws," like that on force conservation, served not only humanity's theoretical understanding of the world and its practical needs; it also "liberated" humans, he said, "from everything dark, everything mystical," for example, from belief in the soul's immortality.[141] It could help dispel them of the illusions brought by metaphysical systems, either those of the "spiritualists," who claim that a human is "a being who towers over the standard of the rest of nature," or that of the "materialists," who claim that a man or woman "through his [or her] thinking is absolute lord of the world." He considered himself and his generation as the standard-bearers against the former, and the coming generation as that against the latter.[142]

On a less philosophical level, the spread of scientific knowledge could help dispel the contemporary widespread obscurantist beliefs in spiritual media, hobgoblins, table-moving, and so on.[143] He believed that a better understanding of the anatomy and physiology of visual

139. Ibid., *VR⁵* 2:425–27.
140. Cf. Helmholtz, "Ueber das Ziel und die Fortschritte der Naturwissenschaft," *VR⁵* 1:385–86, where he states that the vitalistic theory of life "directly contradicts the law of the conservation of force."
141. "Antwortrede, gehalten beim Empfang der Graefe-Medaille," *VR* 2:318 (quotes); and Helmholtz to du Bois-Reymond, 22 May 1853, in Kirsten, 141–42, on 142 (on the specious proof of the soul's immortality).
142. "Das Denken in der Medicin," *VR⁵* 2:183 (quotes) and 186; cf. "Anhang zu dem Vortrag 'Ueber die Wechselwirkung der Naturkräfte'," *VR⁵* 1:402–3 and 408–10, where he noted that his critical evaluation of the work of Mayer, whose overly rationalistic, insufficiently empirical, and speculative nature he deplored, had in part led to his unpopularity with speculative metaphysicians.
143. Helmholtz to du Bois-Reymond, 22 May 1853, 14 July 1857, and 15 May 1864, all in Kirsten, on 142, 173–74, and 209, resp.
144. "Ueber die physiologischen Ursachen der musikalischen Harmonie," *VR⁵* 1:146.

perception, for example, could enlighten individuals suffering from eye inflammation who often had "hypochondriacal ideas."[144] Medical science, he thought, could enlighten humanity about the human body and the appropriate types of treatment just as the laws of planetary motion and Newton's law of gravitation had helped humanity eliminate "astrological superstitions."[145] Medical instrumentation, for example, revealed new and more "parasitical organisms" that replaced "mystical disease-entities"[146]—just as his own hypothetical, extraterrestrial germs were used to provide a new, scientific account for humanity's origins and development.

Beyond these specific social benefits, Helmholtz believed that science, as an increasingly larger and more important part of German culture, could help overcome German-speaking Central Europe's social and political divisions. While the establishment of the Prusso-German Zollverein in the late 1820s and 1830s and the Reich in 1871 had led to increased economic and political unity, there continued to exist social, political, and religious tensions between northern and southern Germans. Moreover, the "takeoff" of German industry in the 1840s brought the emergence of an essentially new class, the urban proletariat, and allowed an older class, the bourgeoisie, to expand noticeably. The result was increased alienation of these social classes from one another and from the nation-state. At the same time, proletariat and bourgeoisie together threatened the monarchical, conservative forces and stood in opposition to the principal economic source of the monarchy's and aristocracy's power: large-scale agriculture. The deep and unprecedented socioeconomic changes that began occurring in *Vormärz* Germany later manifested themselves in a series of political disturbances: the revolutions of 1848, the Constitutional Crisis and wars against Denmark and Austria during the 1860s, the Franco-Prussian War, as well as in the rise of political anti-Semitism, the *Kulturkampf* against Catholics, and the anti-socialist laws of the 1870s and 1880s.

The primary bond uniting the various Germanic peoples of Central Europe before 1914 was thus less that of economics or politics than of culture. From the early nineteenth century on, the educated elite sought to substitute a *Kulturstaat* or *Kulturnation* for the non-existent social and political unity or to supplement what little genuine unity there was; they sought to use *Kultur* to overcome or alleviate antag-

---

145. "Ueber das Verhältniss der Naturwissenschaften zur Gesammtheit der Wissenschaften," *VR⁵ 1*:177.

146. "Das Denken in der Medicin," *VR⁵* 2:182.

onistic social and political forces. The gymnasia and universities greatly fostered German unity and nationalism.[147] Yet by mid-century the cultural values of the rapidly expanding class of industrial capitalists, who prized technological and social change, began to challenge the values of the neo-humanist elite: study of the Greek past and the cultural and ethical growth of the individual (*Bildung*) by means of *Wissenschaft* offered little to most members of the bourgeoisie and not much more to most of the aristocracy. The neo-humanist rulers of the gymnasia and the universities felt increasingly threatened by the values of modern industrial society. *Besitzbürgertum* confronted *Bildungsbürgertum*. Moreover, the neo-humanist pedagogues were challenged from within as well as from without: after 1825 the number of teachers of natural science began to increase, and after 1860 the rise of university scientific laboratories and institutes meant greater institutional power for the natural scientists who constituted a potential threat to the neo-humanist rule of the universities.

New cultural leadership was needed, and the four editions of Helmholtz's *Vorträge und Reden* that appeared during his lifetime were, in effect, meant to show the divided German elite the importance, indeed necessity, of science for helping to overcome the divisions and tensions within Germany.[148] As he said in 1891:

> Science and art are now the only bonds of peace remaining among civilized nations. Their continued cultivation is a common aim of all, aspired to through the common work of all, for the common advantage of all. A great and holy work![149]

To teach Germany's neo-humanist elite, its new business and industrial leaders, and its new middle classes in general about the nature, methods, and goals of science was to strengthen bonds of community between them and other groups. Scientific values and interests repre-

---

147. McClelland, "Republics Within the Empire," 177–79; and idem, "The Wise Man's Burden: The Role of Academicians in Imperial German Culture," 45–69 in *Essays on Culture and Society in Modern Germany*, eds. Gary D. Stark and Karl Bede Lackner (College Station, Texas: Texas A & M University Press, 1982), on 45–6, 50.

148. Nonetheless, Helmholtz supported instruction in ancient languages and literature (above all, Greek) at the gymnasia, along with other dimensions of the neo-humanist curriculum (e.g., the study of German and mathematics), precisely because he believed such instruction to be excellent training for the future study of the natural sciences. The problem with the educated (*gebildete*) elite, he thought, was that they were overly rationalistic in their thought and did not know how to recognize and handle facts sufficiently. (Helmholtz testimony in *Verhandlungen über Fragen des höheren Unterrichts*, 202–5, 208–9.)

149. "Erinnerungen," *VR^5 1*:4.

sented values and interests that all groups could benefit from and unite around. Popular science provided one means to help integrate diverse social orders or classes—not least the insular and isolated German university scientists—into one united society. By teaching established scientific laws and by establishing a universal set of physical units and standards—more generally, by constructing a rational world order—science created a common ground for all Germans. It thereby strengthened community bonds, helped defuse antagonistic class forces, and helped align the socially centrifugal, particularistic ethnic forces within Germany into one national community. Helmholtz's great friend Heinrich von Treitschke, the nationalist historian and editor of the *Preussische Jahrbücher*, recognized the necessity of instructing Germany's leaders in the natural sciences. He wrote to a colleague in 1867:

> We will never become a *revue des deux mondes* if we do not on occasion also have something to say about the natural sciences—and then only from a classical pen. Here, fortunately, I've had luck. Helmholtz, one of the first, if not the very first, authority in physiology, and at the same time a nimble and skilled pen, has promised me 3 long essays ... for the end of Dec., wherein he wants to summarize the results of his investigations on the sense of sight. That would thus be something really important for Feb., March and April—[.][150]

The result was Helmholtz's long three-part essay "Die neueren Fortschritte in der Theorie des Sehens."[151] Germany's statesmen, Helmholtz declared, looked to their scientists for "further progress in civilization, further victories over the forces of nature."[152] It was one of the purposes of science, he proclaimed, "to improve political organization and the moral development of individuals."[153] Science was, moreover, "also a center for strengthening the sense of cohesion with the great fatherland."[154] Helmholtz hoped that science would help construct an alliance between otherwise hostile social and political forces and help modernize and liberalize the essentially illiberal Prusso-German state in which he had grown up and lived. He sought to extend the alliance between German university scientists and the state in a

---

150. Treitschke to Wilhelm Wehrenpfenning, 20 November 1867, in *Heinrich von Treitschkes Briefe*, ed. Max Cornicelius, 2nd ed., 3 vols. (Leipzig: S. Hirzel, 1914–20), 3:192.
151. VR^5 *1*:265–366.
152. "Ueber das Ziel und die Fortschritte der Naturwissenschaft," VR^5 *1*:373.
153. "Ueber das Verhältniss der Naturwissenschaften zur Gesammtheit der Wissenschaft," VR^5 *1*:181.
154. "Ueber das Ziel und die Fortschritte der Naturwissenschaft," VR^5 *1*:398.

way that would allow the state and society to be governed rationally by laws, not arbitrarily by men. The ultimate social and political function of the addresses published as *Vorträge und Reden* was thus to help educate the German elite and so, in effect, transform and integrate it into a new ruling class that would lead Germany to prosperity, social peace, and political unity. With views that at points bordered on the ideological, Helmholtz became a leading German mandarin.

6. Conclusion

Cultural historians believe that the European Enlightenment ended in the 1780s, and historians of modern Germany believe that the German Enlightenment was all too short and limited in its scope and effects, resulting in a society that continued to maintain authoritarian and illiberal values well into the twentieth century.[155] Whether generally true or not, Helmholtz represents a counter-example to these beliefs. His vision of science and of its relations with society embodied Enlightenment and "liberal" values which he bore and effectively proselytized for into the 1890s. Yet his Enlightenment and his "liberalism"—more precisely, his conservative liberalism—were in large measure that of Prussia-Germany with its pietistic notions of schooling and education, its stress on duty, its strong respect for the monarchical state and for political change from above, and its conflicting advocacy of free enterprise and a strong nation-state.[156] Helmholtz carried the Enlightenment views of Kant (not uncritically) and of Alexander von Humboldt into the post-1848 world, just as he sought to help rid that world of Hegelian metaphysical idealism and of Goethian romantic science and to help modernize German science and society. He was one of the last great figures of the *Aufklärung*.

Helmholtz believed, as this essay has sought to show, that the natural world was a rational place, one governed by laws of nature which humanity, through science, was capable of discovering. He believed, too, that for science to advance scientists had to complement their theorizing about nature with a constant, never-ending search for new empirical facts by which they simultaneously discovered and tested their laws. And finally, he believed that science also had utilitarian

---

155. On German academic illiberalism, see Konrad H. Jarausch, *Students, Society, and Politics in Imperial Germany: The Rise of Academic Illiberalism* (Princeton, N.J.: Princeton University Press, 1982).

156. Helmholtz's socioeconomic and political views and attitudes are strikingly consonant with those described by James J. Sheehan in his *German Liberalism in the Nineteenth Century* (Chicago and London: The University of Chicago Press, 1978).

purposes, that it should be used by humanity to benefit its physical and mental health, to develop its educational systems and its economy, to deepen its appreciation of the fine arts, and to help achieve just social and political organizations. Reason, empiricism, and utility were the watchwords of Helmholtz's world view just as they were those of the eighteenth-century *philosophes*. The four categories of analysis that constitute what this essay has called the civilizing power of science represent an Enlightenment understanding of the world. His vision of science and its effects was that of a latter-day *philosophe prussien*: he believed that science civilized humanity and its world.

As such, his vision had its shortcomings. Like the *philosophes* and most of their nineteenth-century successors, Helmholtz virtually never considered, at least in his published and unpublished writings, that there may be no truth about the world for science to discover, just as he, and they, virtually never thought that, should such truth exist, science may not be the route to it. His insistence on the supreme importance of laws as the means of understanding a supposedly unchanging natural order and on humanity's ability to grasp such laws gave human beings a demigod status. The laws of conservation of energy and of Darwinian evolution became the new commandments by which the individual could supposedly understand humanity's origins and destiny. For Helmholtz, as for many others in the nineteenth century, science in effect replaced religion as humanity's guide to understanding the world and its fate. And as a never-ending enterprise, science for Helmholtz continually provided humans with something noble to do in life—to work at discovering the laws of nature—and with solace that their work had meaning. The mature scientific researcher, Helmholtz declared, saw that

> the entire intellectual world of civilized humanity comes before him as a continuous, constantly developing whole, whose duration seems eternal in comparison to the brevity of the individual's life. He sees himself, with his tiny contributions to the development of science, placed in the service of an eternal, holy cause. And so his work itself becomes sacred.[157]

Here, to paraphrase Carl Becker, was the heavenly city of nineteenth-century science.

As a good Enlightenment and "liberal" figure Helmholtz failed to see that science might have harmful as well as ameliorative consequences and he did not sufficiently appreciate that the relations of the

---

157. "Erinnerungen," *VR⁵* 1:19–20.

state and science could be negative as well as positive. While he warmly advocated scientists soldiering for the (German) state, he showed no awareness, in public at least, that such soldiering might affect the scientists' choice of research problems and personnel, and that it might lead them to adopt some of the state's interests and values at the cost of losing their own independent ones. Moreover, he did not see that science, or at least science-based technology, might possibly be a force for increased intolerance, intellectual uncertainty, unwanted cultural and social change, and so on. He did not see that the supposed advances and benefits of science and technology may themselves lead, as Freud later put it, to discontents.[158] Nor did he see that science may not be the sole source of technological invention and economic growth or that technology and values may at times be the driving forces behind scientific and technological change. Moreover, as a great beneficiary of the German university system he naturally played up its strong points. While quick to criticize the French and the English for the shortcomings of their systems, he failed to see the deleterious effects of the hierarchical, authoritarian nature of German universities, just as he failed to recognize their discrimination against Catholics, Jews, social democrats, and women. He stressed the internal freedom that the state allowed the German universities, not the political price paid by German academic scientists and others for that limited freedom. And finally, if his readiness to advocate trust and cooperation between university science and the state was understandable and relatively harmless in the nineteenth century, when the use of science and science-based technology for aggressive military and repressive political purposes was still limited, by the twentieth century such advocacy would at best be naive. In short, he did not see that science might have anti-civilizing as well as civilizing powers.

158. Sigmund Freud, *Civilization and its Discontents* (New York: Norton, 1961), 34–5.

Bibliography

Bibliographical listings of Helmholtz's published works, many of which have been translated into English and other foreign languages, are already available in several versions, and thus another such listing would be redundant. See, especially, "Titelverzeichniss sämmtlicher Veröffentlichungen von Hermann von Helmholtz," in his *Wissenschaftliche Abhandlungen* 3:607–36; Royal Society of London, comp., *Catalogue of Scientific Papers*, 19 vols. (London and Cambridge: C.J. Clay and Sons; George Edward Eyre and William Spottiswoode; and Cambridge University Press, 1867–1925), *3*:270–71; *7*:946–47; *10*:188–89; *12*:324; and *15*:747–48; and Russell Kahl's "Bibliography," in his edition of *Selected Writings of Hermann von Helmholtz*, 430–42. See also the Chronology to the present volume. In addition to the reprinting of (most of) his papers in *WA*, Helmholtz's other published work includes the *Handbuch der physiologischen Optik*; *Die Lehre von den Tonempfindungen als physiologische Grundlage für die Theorie der Musik*; *Vorlesungen über theoretische Physik*; and *Vorträge und Reden*. Except for the *Vorlesungen*, these items appeared in two or more editions and printings.

By contrast, Helmholtz's unpublished writings, above all his correspondence, have yet to be fully located and inventoried. It is worth noting, however, that Helmholtz's (relatively few) extant manuscripts are located in the Archiv of the Akademie der Wissenschaften, Berlin, which also contains some 1,600 letters to Helmholtz, and that some additional manuscripts are also to be found in the Deutsches Museum, Munich. Other substantial collections of Helmholtz's correspondence are located in the Staatsbibliothek Preußischer Kulturbesitz, Handschriftenabteilung, Berlin; the Geheimes Staatsarchiv, Stiftung Preußischer Kulturbesitz, Berlin-Dahlem; the Mathematisches Institut, University of Bonn; Houghton Library, Harvard University, Cambridge, Mass.; Badisches Generallandesarchiv, Karlsruhe; The Royal Institution, London; and the archive of Helmholtz's publisher, Friedrich Vieweg & Sohn, Wiesbaden. Helmholtz's remaining known correspondence is principally located in archives scattered throughout Germany; and, to a lesser extent, in Britain and elsewhere in Europe as well as in the United States. The edited works listed below by David Cahan, Christa Kirsten et al., Richard L. Kremer, Winfried Scharlau, and Ellen von Siemens-Helmholtz as well as Johanna Hertz's collection of Heinrich Hertz materials, Herbert Hörz and Andreas Laass's volume on Ludwig Boltzmann, Leo Koenigsberger's three-volume biography of Helmholtz, and Silvanus Thompson's two-volume biography of William Thomson contain published versions of substantial portions of Helmholtz's correspondence.

The present Bibliography lists all known articles and books devoted solely to Helmholtz (exclusive of anonymous, trivial obituary notices); other secondary sources that discuss Helmholtz's life or work as part of some other context; and, as noted above, published versions of his correspondence.

Ahrens, Wilhelm. "Kleinere Geschichten von Astronomen und Mathematikern, von Meteorologen und Physikern." *Das Weltall* 26:9 (1927):137–40.

Albring, Werner. "Gedanken von Helmholtz über schöpferische Impulse und über das Zusammenwirken verschiedener Wissenschaftszweige." In Scheel, ed., *Gedanken von Helmholtz über schöpferische Impulse* (1972):7–23.

Alexander, Karl-Friedrich. "Zum Begriff der Kraft bei Helmholtz." In Scheel, ed., *Gedanken von Helmholtz über schöpferische Impulse* (1972):65–8.

Aliotta, Antonio. *The Idealistic Reaction against Science.* Trans. Agnes McCaskill. London: Macmillan, 1914.

Anon. "Hermann von Helmholtz. Zu seinem 70. Geburtstage gewidmet von einem seiner Schüler." *Die Gartenlaube* (1891):604–6.

———. "The Late Professor Helmholtz." *The Dublin Journal of Medical Science* 98 (1894):459–61.

———. "Prof. Von Helmholtz." *The Electrical Review* 35 (1894):319–20.

———. "Hermann von Helmholtz." *Nature* 50 (1894):479–80.

———. "Hermann Helmholtz." *Wiener Medizinische Wochenschrift* 44 (1894):1646–47.

———. [Essay review of six books by and about Hermann von Helmholtz]. *The Edinburgh Review* 192 (1900):382–404.

———. "Scientific Literature. Hermann von Helmholtz." *Popular Science Monthly* 63 (1903):186–87.

———. "Intimes aus dem Leben von Hermann von Helmholtz." *Die Gartenlaube* (1903):487-89.
———. "Ein Helmholtz-Archiv." *Gaea. Natur und Leben 41:1* (1905):1-3.
———. "The Progress of Science. Hermann von Helmholtz." *Popular Science Monthly 71* (1907):283-85.
———. "Cotugno, non Helmholtz fu l'ideatore della teoria sull'audizione." *Rivisita di Storia critica della Scienze Mediche e Naturali: Organo ufficiale della Società Italiana di Storia della Scienze Mediche e Naturali 10* (1919):143-45.
Ansprachen und Reden gehalten bei der am 2. November 1891 zu ehren von Hermann von Helmholtz veranstalteten Feier, nebst einem Verzeichnisse der überreichten Diplome und Ernennungen, sowie der Adressen und Glückwunschschreiben. Berlin: Hirschwald, 1892.
Arbeitsgemeinschaft der Vier. "Der Weg zur Farbe: Von Newton bis Ostwald." *Fachblatt der Maler: Zeitschrift zur Förderung der Handwerklichen Wertarbeit in Farbe Form und Raum 3:1* (1927):36-9.
Archibald, Thomas. " 'Eine sinnreiche Hypothese': Aspects of Action-at-a-Distance Electromagnetic Theory, 1820-1880." Ph.D. diss., University of Toronto, 1987.
———. "Energy and the Mathematization of Electrodynamics in Germany, 1845-1875." *Archives internationales d'histoire des sciences 39* (1989):276-308.
Ash, Mitchell Graham. "The Emergence of Gestalt Theory: Experimental Psychology in Germany 1890-1920." Ph.D. diss., Harvard University, 1982.
Auerbach, Felix. "Hermann Helmholtz und die wissenschaftlichen Grundlagen der Musik." *Nord und Süd 19* (1881):217-44.
Bailhache, Patrice. "Valeur actuelle de l'acoustique musicale de Helmholtz." *Revue d'histoire des sciences 39:4* (1986):301-24.
Banasiewicz, Wiesław. "Der materialistische Charakter der erkenntnistheoretischen Auffassungen von Hermann von Helmholtz." *WZHUB 22:3* (1973):303-5.
Bancroft, Wilder D. "Helmholtz and Nernst." *Journal of Physical Chemistry 42* (1938):687-91.
Barnouw, Jeffrey. "Goethe and Helmholtz: Science and Sensation." Frederick Amrine, Francis J. Zucker, and Harvey Wheeler, eds. *Goethe and the Sciences: A Reappraisal.* Dordrecht, Boston, Lancaster: D. Reidel, 1987. Pp. 45-82.
Bebel, Dietrich. "Die physikalischen Arbeiten von Hermann von Helmholtz." *WZHUB 22:3* (1973):319-26.
Bell, James F. "Helmholtz, Hermann (Ludwig Ferdinand) von." Stanley Sadie, ed. *The New Grove Dictionary of Music and Musicians.* 20 vols. London, Washington, D.C., and Hong Kong: Macmillan, 1980. 8:466-67.
Belloni, Luigi. "Hermann von Helmholtz und Franz Boll." *Medizinhistorisches Journal 17:1/2* (1982):129-37.
Berkson, William. *Fields of Force: The Development of a World View from Faraday to Einstein.* New York: John Wiley, 1974.

Bernfeld, Siegfried. "Freud's Earliest Theories and the School of Helmholtz." *The Psychoanalytic Quarterly 13* (1944):341-62.
———. "Freud's Scientific Beginnings." *Imago 6* (1949):163-96.
Bernhardt, Hannelore. "Helmholtz' Stellung zum 2. Hauptsatz der Thermodynamik." *WZHUB 22:3* (1973):341-44.
Bernhardt, Karl-Heinz. "Der Beitrag Hermann von Helmholtz' zur Physik der Atmosphäre." *WZHUB 23:3* (1973):331-40.
Bernstein, Julius. "Hermann von Helmholtz †." *Naturwissenschaftliche Rundschau 10:6* (1895):73-9.
———. "Hermann von Helmholtz." Friedrich von Weech and U. Krieger, eds. *Badische Biographien.* Heidelberg: Carl Winter, 1906. 5:281-294.
Bevilacqua, Fabio. *The Principle of Conservation of Energy and the History of Classical Electromagnetic Theory.* Pavia: LaGoliardica Pavese, 1983.
———. "H. Hertz's Experiments and the Shift towards Contiguous Propagation in the early Nineties." *Rivista di Storia della Scienza 1* (1984):239-56.
Bezold, Wilhelm von. "[Glückwunschadresse an Hrn. H. von Helmholtz]." *Verhandlungen der Physikalischen Gesellschaft zu Berlin 10* (1891):81-2.
———. *Hermann von Helmholtz. Gedächtnisrede gehalten in der Singakademie zu Berlin am 14. Dezember 1894.* Leipzig: J.A. Barth, 1895.
Bierhalter, Günter. "Zu Hermann von Helmholtzens mechanischer Grundlegung der Wärmelehre aus dem Jahre 1884." *Archive for History of Exact Sciences 25:1* (1981):71-84.
———. "Die v. Helmholtzschen Monozykel-Analogien zur Thermodynamik und das Clausiussche Disgregationskonzept." *Archive for History of Exact Sciences 29:1* (1983):95-100.
———. "Wie erfolgreich waren die im 19. Jahrhundert betriebenen Versuche einer mechanischen Grundlegung des zweiten Hauptsatzes der Thermodynamik?" *Archive for History of Exact Sciences 37:1* (1987):77-99.
Blackmore, John T. *Ernst Mach: His Work, Life, and Influence.* Berkeley, Los Angeles, London: University of California Press, 1972.
———. [Essay Review of Hermann von Helmholtz, *Epistemological Writings*, eds. Robert S. Cohen and Yehuda Elkana, trans. Malcolm F. Lowe (Dordrecht and Boston: D. Reidel, 1977)]. *Annals of Science 35* (1978):427-31.
Blasius, W. "Die Bestimmung der Leitungsgeschwindigkeit im Nerven durch Hermann v. Helmholtz am Beginn der naturwissenschaftlichen Ära der Neurophysiologie." Karl Rothschuh, ed. *Von Boerhaave bis Berger: Die Entwicklung der kontinentalen Physiologie im 18. und 19. Jahrhundert mit besonderer Berücksichtigung der Neurophysiologie* [= *Medizin in Geschichte und Kultur*, Band 5]. Stuttgart: Gustav Fischer, 1974. Pp. 71-84.
Böhm, Karl. [Review of H. von Helmholtz, *Einleitung zu den Vorlesungen über theoretische Physik.*] *Physikalische Zeitschrift 5* (1904):140-43.
Boring, Edwin G. *Sensation and Perception in the History of Experimental Psychology.* New York: Appleton-Century-Crofts, 1942.
———. *A History of Experimental Psychology.* 2nd ed. New York: Appleton-Century-Crofts, 1950.

Braun-Artaria, Rosalie. "Anna von Helmholtz. Ein Erinnerungsblatt." *Münchener Allgemeine Zeitung.* Wissenschaftliche Beilage. No. 285. 14 December 1899; reprinted Berlin: n.p., 1899.

———. *Von berühmten Zeitgenossen. Lebenserinnerungen einer Siebzigerin.* Munich: C.H. Beck, 1918.

Brazier, Mary A.B. *A History of Neurophysiology in the 19th Century.* New York: Raven Press, 1988.

Breger, Herbert. *Die Natur als arbeitende Maschine. Zur Entstehung des Energiebegriffs in der Physik 1840–1850.* Frankfurt and New York: Campus Verlag, 1982.

Bringmann, Wolfgang G., Gottfried Bringmann, and David Cottrell. "Helmholtz und Wundt an der Heidelberger Universität 1858–1871." *Heidelberger Jahrbücher 20* (1976):79–88.

Brossmer, Karl. "Hermann von Helmholtz (1821–1894)." *Ruperto-Carola. Mitteilungen der Vereinigung der Freunde der Studentenschaft der Universität Heidelberg e.V.* 24 (1958):147–48.

Brown, Theodore M. "Resource Letter EEC-1 on the Evolution of Energy Concepts from Galileo to Helmholtz." *American Journal of Physics 33* (1965):759–65.

Br[üche, Ernst]. "Helmholtz und Rutherford." *Physikalische Blätter 4* (1948):21–2.

———. "Aus der Vergangenheit der Physikalischen Gesellschaft." *Physikalische Blätter 16* (1960):499–505; 616–21; *17* (1961):27–33; 120–27; 225–32; 400–10.

Brush, Stephen G. *The Kind of Motion We Call Heat: A History of the Kinetic Theory of Gases in the 19th Century.* 2 vols. Vol. 1: *Physics and the Atomists.* vol. 2: *Statistical Physics and Irreversible Processes.* Amsterdam, New York, Oxford: North Holland, 1976.

Buchheim, Gisela. "Zur Geschichte der Bestimmung von Dielektrizitätskonstanten." *NTM 3* (1966):38–45.

———. "Zur Geschichte der Elektrodynamik: Briefe Ludwig Boltzmanns an Hermann von Helmholtz." *NTM 5* (1968):125–31.

———. "Hermann von Helmholtz und die klassische Elektrodynamik." *NTM 8* (1971):23–36.

———. "Die Entwicklung des elektrischen Meßwesens und die Gründung der Physikalisch-Technischen Reichsanstalt." *NTM 14* (1977):16–32.

Buchwald, Jed Z. *From Maxwell to Microphysics: Aspects of Electromagnetic Theory in the Last Quarter of the Nineteenth Century.* Chicago and London: The University of Chicago Press, 1985.

———. "The Background to Heinrich Hertz's Experiments in Electrodynamics." Trevor H. Levere and William R. Shea, eds. *Nature, Experiment, and the Sciences: Essays on Galileo and the History of Science in Honour of Stillman Drake.* (= Boston Studies in the Philosophy of Science, vol. 120). Dordrecht, Boston, London: Kluwer Academic 1990. Pp. 275–306.

Bunsen, Marie von. *Zur Erinnerung an Frau Anna von Helmholtz.* N.p.: n.p., 10 December 1899.

Burchfield, Joe D. *Lord Kelvin and the Age of the Earth.* New York: Science History Publications, 1975.
Cahan, David. "Werner Siemens and the Origin of the Physikalisch-Technische Reichsanstalt, 1872–1887." *HSPS 12:2* (1982):253–83.
———. *An Institute for an Empire: The Physikalisch-Technische Reichsanstalt 1871–1918.* Cambridge, New York, and New Rochelle: Cambridge University Press, 1989.
———., ed. *Letters of Hermann von Helmholtz to His Parents: The Medical Education of a German Scientist, 1837–1846.* Stuttgart: Franz Steiner Verlag 1993.
Čapek, Milič. "Ernst Mach's Biological Theory of Knowledge." *Synthese 18* (1968):171–91.
Cappelletti, Vincenzo. "Helmholtz heute." *Studia Leibniziana 1* (1969):253–62.
———. "Dinamica della Scienza e Immagini del Mondo." *Veltro 30* (1986):441–52.
Carazza, B., and G.P. Guidetti. "Helmholtz, la 'Legge di Ohm', e il Problema dell'armonia." *Giornale di Fisica 30* (1989):207–14.
Carhart, Henry S. "Professor von Helmholtz." *The Electrical World 24* (1894):542–43.
Cassirer, Ernst. *The Problem of Knowledge: Philosophy, Science, and History since Hegel.* Trans. William H. Woglom and Charles W. Hendel. New Haven, Conn.: Yale University Press, 1950.
———. *Determinism and Indeterminism in Modern Physics. Historical and Systematic Studies of the Problem of Causality.* Trans. O. Theodor Benfey. New Haven: Yale University Press, 1956.
Chalmers, A.F. "The Electromagnetic Theory of J.C. Maxwell and Some Aspects of Its Subsequent Development." D. Phil. diss., University of London, 1971.
Chesnokova, S.A. "Herman Helmholtz (On the 150th Anniversary of His Birth)." [In Russian.] *Fiziologicheskii Zhurnal SSSR im. I.M. Sechenova 57* (1971):1871–72.
Cierpka, Rudolf. *Das Gesetz von der Erhaltung der Energie in seiner Auswirkung auf die Physiologie in Deutschland.* (Inaugural diss., University of Berlin, 1941.) Berlin: Franz Linke, 1941.
Clark, Peter. "Elkana on Helmholtz and the Conservation of Energy." *The British Journal for the Philosophy of Science 27* (1976):165–76.
Classen, J. *Vorlesungen über modernen Naturphilosophen (Du Bois-Reymond, F.A. Lange, Haeckel, Ostwald, Mach, Helmholtz, Boltzmann, Poincaré und Kant).* Hamburg: C. Boysen, 1908.
Cogan, Robert. "Reconceiving Theory: The Analysis of Tone Color." *College Music Symposium 15* (1975):52–69.
Cohen, Robert S., and Yehuda Elkana, eds. *Hermann von Helmholtz. Epistemological Writings.* Trans. Malcolm F. Lowe. (= Boston Studies in the Philosophy of Science, vol. 37). Dordrecht and Boston: D. Reidel, 1977.
Conrat, Friedrich. *Helmholtz' Verhältnis zur Psychologie.* (Inaugural diss., Bonn University.) Halle: Ehrhardt Karras, 1903.

———. *Hermann von Helmholtz' psychologische Anschauungen.* (= *Abhandlungen zur Philosophie und ihrer Geschichte*, Heft 18). Halle: Max Niemeyer, 1904.

Cranefield, Paul F. "The Organic Physics of 1847 and the Biophysics of Today." *JHMAS* 12 (1957):407–23.

———. "The Nineteenth-Century Prelude to Modern Biophysics." Henry Quastler and Harold J. Morowitz, eds. *Proceedings of the First National Biophysics Conference. Columbus, Ohio, March 4–6, 1957.* New Haven, Conn.: Yale University Press, 1959. Pp. 19–26.

———. "The Philosophical and Cultural Interests of the Biophysics Movement of 1847." *JHMAS* 21 (1966):1–7.

———. "Freud and the 'School of Helmholtz'." *Gesnerus* 23 (1966):35–9.

Cremer. "Hermann v. Helmholtz †." *Münchener Medicinische Wochenschrift* 41 (1894):912–13.

Crew, Henry. "Helmholtz on the Doctrine of Energy." *Journal of the Optical Society of America* 6 (1922):312–26.

Crombie, A.C. "Helmholtz." *Scientific American* 198:3 (1958):94–102.

Cuddy, Lola L., and M.G. Wiebe. "Editors' Introduction: Music and the Experimental Sciences." *The Humanities Association Review* 30:1/2 (1979):1–10.

Culotta, Charles A. "German Biophysics, Objective Knowledge, and Romanticism." *HSPS* 4 (1974):3–38.

Cyon, E. de. "Les bases naturelles de la géometrie d'Euclide." *Revue philosophique* 52 (1901):1–30.

———. *Das Ohrlabyrinth als Organ der mathematischen Sinne für Raum und Zeit.* Berlin: Julius Springer, 1908.

D'Agostino, Salvo. "Hertz e Helmholtz sulle Onde Elettromagnetiche." *Scientia 106* (1971):623–36, followed by an English translation: "Hertz and Helmholtz on Electromagnetic Waves." Pp. 637–48.

———. "Hertz's Researches on Electromagnetic Waves." *HSPS* 6 (1975):261–323.

———. "Scienza e Cultura nella Germania dell'800: Le Basi Filosofiche della Fisica-matematica nell'Opera di Helmholtz." *Cultura e Scuola* 25:99 (1986):265–81.

Dahlhaus, Carl. "Hermann von Helmholtz und der Wissenschaftscharakter der Musiktheorie." Frieder Zaminer, ed. *Über Musiktheorie. Referate der Arbeitstagung 1970 in Berlin.* Cologne: Arno Volk, 1970. Pp. 49–58.

Dale, Peter Allan. *In Pursuit of a Scientific Culture: Science, Art, and Society in the Victorian Age.* Madison, Wis. and London: The University of Wisconsin Press, 1989.

Dannemann, Friedrich. *Die Naturwissenschaften in Ihrer Entwicklung und in Ihrem Zusammenhange.* 2nd ed. 4 vols. Vol. 4: *Das Emporblühen der modernen Naturwissenschaften seit der Entdeckung des Energieprinzips.* Leipzig: Wilhelm Engelmann, 1920–23.

Deeson, Eric. "Hermann von Helmholtz, 1821–1894." *New Scientist and Science Journal 51* (1971):584–86.

Deltete, Robert John. "The Energetics Controversy in Late Nineteenth-Century Germany: Helm, Ostwald and Their Critics." Ph.D. diss., Yale University, 1983.

Deutsche Physikalische Gesellschaft. "Festsitzung zur Feier des 90 jährigen Bestehens der physikalischen Gesellschaft zu Berlin am 25. January 1935 im großen Hörsaal des Physikalischen Instituts der Universität." *Verhandlungen der Deutschen Physikalischen Gesellschafte 16:1* (1935):1–17.

Diepgen, Paul. *Geschichte der Medizin. Die historische Entwicklung der Heilkunde und des ärztlichen Lebens.* 2 vols. in 3. Berlin: Walter de Gruyter, 1949–55. *2:2*

Dilthey, Wilhelm. "Anna von Helmholtz." *Deutsche Rundschau 102* (1900):226–35.

Dingler, Hugo. "Über den Zirkel in der empirischen Begründung der Geometrie." *Kant-Studien 30* (1925):310–30.

———. "H. Helmholtz und die Grundlagen der Geometrie." *Zeitschrift für Physik 90* (1934):348–54.

———. "Nochmals 'H. Helmholtz und die Grundlagen der Geometrie.' " *Zeitschrift für Physik 94* (1935):674–76.

Dostrovsky, Sigalia, James F. Bell, and C. Truesdell. "Physics of Music." Stanley Sadie, ed. *The New Grove Dictionary of Music and Musicians.* 20 vols. London, Washington, D.C., and Hong Kong: Macmillan, 1980. *14*:644–77.

Drenkhahn, R. "Liebigs und Helmholtzs Aeußerungen über den Wunderglauben." *Münchener Medizinische Wochenschrift 85* (1938):66.

Dreyer, Ernst Adolf. *Friedr. Vieweg & Sohn in 150 Jahren deutscher Geistesgeschichte 1786–1936.* Braunschweig: Friedr. Vieweg & Sohn, [1936].

Du Bois-Reymond, Emil. "Ansprache an Se. Exzellenz Hrn. von Helmholtz zur Feier seines fünfzigjährigen Doktorjubiläums in der Gesamtsitzung der Akademie der Wissenschaften am 3. November 1892." *SBB.* 1892. Pp. 905–9; reprinted in *Reden von Emil Du Bois-Reymond.* Estelle Du Bois-Reymond, ed. 2 vols. Leipzig: Veit, 1912. *2*:643–48.

———. "Gedächtnisrede auf Hermann von Helmholtz." *Abhandlungen der Königlichen Preussischen Akademie der Wissenschaften zu Berlin.* "Gedächtnisreden II." 1896. Pp. 5–50; reprinted in *Reden von Emil Du Bois-Reymond.* Estelle Du Bois-Reymond, ed. 2 vols. Leipzig: Veit, 1912. *2*:516–70; and in Kirsten and Körber, eds., *Physiker über Physiker II,* 68–99.

———. "Hermann von Helmholtz." *Revue scientifique* 4th ser. *8* (1897):321–28 and 360–67. [French translation of previous item.]

Dugas, René. *A History of Mechanics.* Trans. J.R. Maddox. Neuchâtel: Editions du Griffon; New York: Central Book, 1955.

Duhem, Pierre-Marie-Maurice. *The Evolution of Mechanics.* Trans. Michael Cole. Alphen aan den Rijn and Germantown, MD: Sijthoff & Noordhoff, 1980.

Dühring, Eugen. *Kritische Geschichte der allgemeinen Principien der Mechanik.* Berlin: Theobald Grieben, 1873; 2nd rev. and enl. ed. Leipzig: Fues's Verlag, 1877; 3rd rev. and enl. ed. Leipzig: Fues's Verlag, 1887.

———. *Robert Mayer. Der Galilei des neunzehnten Jahrhunderts. Eine Einführung in seine Leistungen und Schicksale.* Chemnitz: Ernst Schmeitzner, 1880.
———. *Sache, Leben und Feinde. Als Hauptwerk und Schlüssel zu seinen sämmtlichen Schriften.* H. Reuther: Karlsruhe und Leipzig, 1882; 2nd enl. ed., Leipzig: C.G. Naumann, 1903.
Ebeling, Werner, and Dieter Hoffmann. "The Berlin School of Thermodynamics Founded by Helmholtz and Clausius." *European Journal of Physics* 12 (1991):1-9.
Ebert, Hermann. *Hermann von Helmholtz.* Stuttgart: Wissenschaftliche Verlagsgesellschaft, 1949.
Ebstein, Erich. "Über die Fortpflanzungsgeschwindigkeit der Nervenreizung." *Janus 11* (1906):322-23.
———, ed. *Ärzte-Briefe aus vier Jahrhunderten.* Berlin: Julius Springer, 1920.
E-e. "Hermann von Helmholtz. Zum 31. August 1891." *Tägliche Rundschau. Unterhaltungs-Beilage. 11:201*, Sonnabend, 29 August 1891. Pp. 801-2; *11:202*, Sonntag, 30 August 1891. Pp. 805-7.
Einstein, Albert. [Review of *Hermann von Helmholtz. Zwei Vorträge über Goethe.* W. König, ed. (Braunschweig: Vieweg, 1917).] *Die Naturwissenschaften 5* (1917):675.
Elbogen, Paul, ed. *Liebster Sohn . . . liebe Eltern. Briefe berühmter Deutscher.* Hamburg: Rowohlt, [1956].
Elkana, Yehuda. "Helmholtz' 'Kraft': An Illustration of Concepts in Flux." *HSPS 2* (1970):263-98.
———. "The Conservation of Energy: A Case of Simultaneous Discovery?" *Archives internationales d'histoire des sciences 23* (1970):31-60.
———. *The Discovery of the Conservation of Energy.* Cambridge, Mass.: Harvard University Press, 1974.
Engelking, E. "Hermann von Helmholtz in seiner Bedeutung für die Augenheilkunde." *Berichte der Deutschen Ophthalmologischen Gesellschaft 56* (1950):12-30.
———, ed. *Dokumente zur Erfindung des Augenspiegels durch Hermann von Helmholtz im Jahre 1850.* Munich: J.F. Bergmann, 1950.
Engelmann, Theodor W. *Gedächtnisrede auf Hermann von Helmholtz. Gehalten am 28. September 1894 in der Aula der Universität Utrecht.* Leipzig: Wilhelm Engelmann, 1894.
Engels, Friedrich. *Anti-Dühring.* (= *Karl Marx-Friedrich Engels Werke*, Band 20.) Berlin: Dietz, 1962.
———. *Dialektik der Natur.* (= *Karl Marx-Friedrich Engels Werke*, Band 20). Berlin: Dietz, 1962.
Epstein, Sigmund Stefan. "Hermann von Helmholtz als Mensch und Gelehrter." *Deutsche Revue 21:2* (1896):31-41, 192-202, 328-39; reprinted Stuttgart, Leipzig, Berlin: Deutscher Verlags-Anstalt, 1896.
Erdmann, Benno. *Die Axiome der Geometrie. Eine philosophische Untersuchung der Riemann-Helmholtz'schen Raumtheorie.* Leipzig: Voss, 1877.
———. "Die philosophischen Grundlagen von Helmholtz' Wahrnehmungstheorie. Kritische erläutert." (= *Abhandlungen der Preussischen Akademie*

der Wissenschaften, Philosophisch-historische Klasse, 1921, Nr. 1). Berlin: Verlag der Akademie der Wissenschaften, 1921.

Erggelet, H. "Hermann v. Helmholtz und die Augenheilkunde." *Die Naturwissenschaften* 9 (1921):967–72.

Ernst, K. "Die historische Entwicklung der Stimulationselektromyographie. Ein Beitrag zum 75. Todestag von Hermann von Helmholtz." *Zeitschrift für ärztliche Fortbildung 63:22* (1969):1200–4.

Esser, Albert. "Zur Geschichte der Erfindung des Augenspiegels." *Klinische Monatsblätter für Augenheilkunde 116* (1950):1–14.

Falkenhagen, Hans. *Die Naturwissenschaft in Lebensbildern großer Forscher.* Zurich: Shirzel, 1949.

Fancher, Raymond. *Pioneers of Psychology.* New York: W.W. Norton, 1979.

Fischer, Emil. "Hermann von Helmholtz." *Berichte der Deutschen Chemischen Gesellschaft 27* (1894):2643–52.

———. *Aus meinem Leben.* Reprint. Berlin, Heidelberg, and New York: Springer-Verlag, 1987.

Fischer, Ernst. "A Historical Note on Helmholtz's Theory of Hearing." *Annals of Medical History* n.s. *8* (1936):357–58.

FitzGerald, George Francis. "Helmholtz Memorial Lecture." *Journal of the Chemical Society. Transactions. 69:2* (1896):885–912; reprinted in Joseph Larmor, ed. *The Scientific Writings of the Late George Francis FitzGerald.* Dublin: Hodges, Figgis; London: Longmans, Green, 1902. Pp. 340–77.

———. "Thoughts on the Work of Helmholtz, Being an Abridgement of the Helmholtz Memorial Lecture." Joseph Larmor, ed. *The Scientific Writings of the Late George Francis FitzGerald.* Dublin: Hodges, Figgis; London: Longmans, Green, 1902. Pp. 378–86.

Fortschritte der Physik. Redaktion und Verlagshandlung. "Zur Vollendung des 50. Jahrganges der 'Fortschritte'." *Fortschritte der Physik des Aethers im Jahre 1894. Dargestellt von der Physikalischen Gesellschaft zu Berlin 50:2* (1894):i–xi.

Frankfurt, U.I. "Helmholtz's Electrodynamics and Its Evolution." [In Russian.] *Voprosy Istorii Estestvoznaniia i Tekhniki 14* (1963):49–55.

Franklin, Samuel S. "The Effect of Modified Correlations between Eye Movement and Head Movement on the Perception of Visual Direction: An Examination of Helmholtz's Eye Position Hypothesis." Ph.D. diss., University of Kansas, 1966.

Freudenthal, Hans. "The Main Trends in the Foundations of Geometry in the 19th Century." Ernest Nagel, Patrick Suppes, and Alfred Tarski, eds. *Logic, Methodology and Philosophy of Science: Proceedings of the 1960 International Congress.* Stanford, Calif.: Stanford University Press, 1962. Pp. 613–21.

Friedenwald, Harry. "The History of the Invention and of the Development of the Ophthalmoscope." *The Journal of the American Medical Association 38* (1902):549–52.

Fullinwider, S.P. "Hermann von Helmholtz: The Problem of Kantian Influence." *SHPS 21:1* (1990):41–55.

———. "Darwin Faces Kant: A Study in Nineteenth-Century Physiology." *BJHS* 24 (1991):21–44.
Galaty, David H. "The Emergence of Biological Reductionism." Ph.D. diss., The Johns Hopkins University, 1971.
———. "The Philosophical Basis of Mid-Nineteenth Century German Reductionism." *JHMAS* 29 (1974):295–316.
Garber, Elizabeth. "Thermodynamics and Meteorology (1850–1900)." *Annals of Science* 33 (1976):51–65.
Gariel, C.M. "Les travaux de H.-L.-F. Helmholtz." *Revue scientifique* 4th ser. 54 (1894):429–32.
Geissler, Hans-Georg. "Wahrnehmung als Rekonstruktion von Umgebungszuständen—Helmholtz im Ursprung der modernen Psychophysik." *WZHUB* 22:3 (1973):315–18.
Gerlach, Walther. "Hermann v. Helmholtz als Naturforscher." *Berichte der Deutschen Ophthalmologischen Gesellschaft* 56 (1950):3–12.
———. "Hermann von Helmholtz 1821–1894." Hermann Heimpel, Theodor Heuss, and Benno Reifenberg, eds. *Die Grossen Deutschen. Deutsche Biographie.* 4 vols. Berlin: Propyläen-Verlag, 1956. 3:456–65.
———. "Helmholtz, Hermann Ludwig Ferdinand v." *Neue Deutsche Biographie.* 16 vols. Berlin: Duncker & Humblot, 1953–90. 8:498–501.
Goetz, Dorothea, ed. *Hermann von Helmholtz über Sich Selbst. Rede zu seinem 70. Geburtstag eingeleitet und mit Anmerkungen versehen von Dorothea Goetz, Potsdam.* Leipzig: B.G. Teubner, 1966.
Goldschmidt, Ludwig. *Kant und Helmholtz. Populärwissenschaftliche Studie.* Hamburg and Leipzig: Leopold Voss, 1898.
Goldstein, Eugen. "Helmholtz. Erinnerungen eines Laboratoriumspraktikanten." *Die Naturwissenschaften* 9 (1921):708–11.
———. "Aus vergangenen Tagen der Berliner Physikalischen Gesellschaft." *Die Naturwissenschaften* 13 (1925):39–45.
Goodspeed, Arthur W. "Contributions of Helmholtz to Physical Science." *The Journal of the American Medical Association* 38 (1902):562–66.
Greeff, Richard. "Historisches zur Erfindung des Augenspiegels." *Berliner klinische Wochenschrift* 38:48 (1901):1201–2.
———. Dr. "Augenheilkunde und Optik. Helmholtz' bahnbrechende Arbeiten." "Helmholtz zum 100. Geburtstag. Erinnerungsblatt der Vossischen Zeitung." *Vossische Zeitung.* No. 404. Beilage 4. 28 August 1921.
Gregory, Frederick. *Scientific Materialism in Nineteenth Century Germany.* Dordrecht and Boston: D. Reidel, 1977.
Grell, Heinrich. "Über das Helmholtz-Liesche Raumproblem." In Scheel, ed., *Gedanken von Helmholtz über schöpferische Impulse* (1972):69–78.
Gross, Theodor. *Robert Mayer und Hermann v. Helmholtz. Eine kritische Studie.* Berlin: M. Krayn, 1898.
———. *Kritische Beiträge zur Energetik.* 2 vols. Vol. 1: *Die Verwandlungen der Kraft nach Robert Mayer.* vol. 2: *Hermann von Helmholtz und die Erhaltung der Energie.* Berlin: M. Krayn, 1901–2.
Gruber, Howard, and Valmai Gruber. "Hermann von Helmholtz: Nineteenth-Century Polymorph." *The Scientific Monthly* 83 (1956):92–9.

Gumprecht, F. *Leben und Gedankenwelt großer Naturforscher.* Leipzig: Quelle & Meyer, 1927.

Günther, Siegmund. *Geschichte der anorganischen Naturwissenschaften im Neunzehnten Jahrhundert.* Berlin: Georg Bondi, 1901.

Guralnick, Stanley M. "The Contexts of Faraday's Electrochemical Laws." *Isis* 70 (1979):59-75.

Haas, Arthur Erich. *Die Entwicklungsgeschichte des Satzes von der Erhaltung der Kraft.* Vienna: Alfred Hölder, 1909.

Haëne, Robert, d'. "La notion scientifique de l'énergie, son origine et ses limites." *Revue de métaphysique et de morale* 72 (1967):35-67.

[Hall], [A.] Wilford. *Evolution of Sound: A Part of the Problem of Human Life Here and Hereafter. Containing a Review of Tyndall, Helmholtz, and Mayer.* New York: Hall & Co., 1878.

———. *The Problem of Human Life: Embracing the 'Evolution of Sound' and 'Evolution Evolved,' with a Review of the Six Great Modern Scientists, Darwin, Huxley, Tyndall, Haeckel, Helmholtz, and Mayer.* New York: Hall and Company, 1880.

Hall, Granville Stanley. *Aspects of German Culture.* Boston: James R. Osgood, 1881.

———. *Founders of Modern Psychology.* New York and London: D. Appleton, 1912.

Hall, Thomas S. *Ideas of Life and Matter. Studies in the History of General Physiology 600 B.C.-1900 A.D..* 2 vols. Vol. 2: *From the Enlightenment to the End of the Nineteenth Century.* Chicago and London: The University of Chicago Press, 1969.

Hall, Vance Mark Dornford. "Some Contributions of Medical Theory to the Discovery of the Conservation of Energy Principle during the late 18th and early 19th Centuries." D. Phil. diss., University College, London, 1977.

Hall, Winfield S. "The Contributions of Helmholtz to Physiology and Psychology." *The Journal of the American Medical Association* 38 (1902):558-61.

Hamm, Josef. *Das philosophische Weltbild von Helmholtz.* (Inaugural diss., Bonn University.) Bielefeld: Beyer & Hausknecht, 1937.

Hansemann, David. "Ueber das Gehirn von Hermann v. Helmholtz." *Zeitschrift für Psychologie und Physiologie der Sinnesorgane* 20 (1899):1-12.

Hargreave, David. "Thomas Young's Theory of Color Vision: Its Roots, Development, and Acceptance by the British Scientific Community." Ph.D. diss., University of Wisconsin, 1973.

Harig, Gerhard. "Hermann von Helmholtz. 1821-1894." Gerhard Harig, ed. *Von Adam Ries bis Max Planck. 25 große deutsche Mathematiker und Naturwissenschaftler.* 2nd ed. Leipzig: VEB Verlag Enzyklopädie, 1962. Pp. 84-8.

Harman, P.M. *Energy, Force, and Matter: The Conceptual Development of Nineteenth-Century Physics.* Cambridge, London, New York: Cambridge University Press, 1982.

———. *Metaphysics and Natural Philosophy: The Problem of Substance in Classical Physics.* Brighton: The Harvester Press, 1982.

———. "The Foundations of Mechanical Physics in the Nineteenth Century: Helmholtz and Maxwell." Clemens Burrichter, Rüdiger Inhetveen, and Rudolf Kötter, eds. *Zum Wandel des Naturverständnisses*. Paderborn, Munich, Vienna: Ferdinand Schöningh, 1987. Pp. 35–46.

Harms, Robert T. "What Helmholtz Knew about Neutral Vowels." *In Honor of Ilse Lehiste/Ilse Lehiste Pühendusteos*. Robert Channon and Linda Shockey, eds. Dordrecht, Providence, Rhode Island: Foris, 1987. Pp. 381–99.

Hatfield, Gary. "Mind and Space from Kant to Helmholtz: The Development of the Empiristic Theory of Spatial Vision." Ph.D. diss., University of Wisconsin-Madison, 1979.

———. "Spatial Perception and Geometry in Kant and Helmholtz." *PSA 2* (1984):569–87.

———. *The Natural and the Normative: Theories of Spatial Perception from Kant to Helmholtz*. Cambridge, Mass., and London: The MIT Press, 1990.

Haubold, H.J. and R.W. John. "Albert A. Michelsons Ätherdrift-Experiment 1880/1881 in Berlin und Potsdam." *NTM 19* (1982):31–45.

Heimann, P.M. "Helmholtz and Kant: The Metaphysical Foundations of *Über die Erhaltung der Kraft*." *SHPS 5* (1974):205–38.

Heinig, Karl. "Einige Bemerkungen zu Helmholtz' wissenschaftlichem Wirken und der Chemie des 19. Jahrhunderts." *WZHUB 22:3* (1973):357–61.

Heinzmann, Gerhard. "Zwischen Objektkonstruktion und Strukturanalyse: Überlegungen zur Philosophie der Mathematik von Jules Henri Poincaré und Charles Sanders Peirce." Habilitationsschrift, University of Saarland, 1989.

Helm, Georg. *Die Energetik nach ihrer geschichtlichen Entwickelung*. Leipzig: Veit, 1898.

Henderson, Linda Dalrymple. *The Fourth Dimension and Non-Euclidean Geometry in Modern Art*. Princeton, N.J.: Princeton University Press, 1983.

Hendry, John. *James Clerk Maxwell and the Theory of the Electromagnetic Field*. Bristol and Boston: Adam Hilger, 1986.

Hermann, Armin. "Von Paul Erman zu Hermann von Helmholtz: Die Anfänge der Physik an der Universität Berlin." Wilhelm Treue and Gerhard Hildebrandt, eds. *Berlinische Lebensbilder. Naturwissenschaftler*. Berlin: Colloquium, 1987. Pp. 17–26.

Hermann, L., and P. Volkmann. "Hermann von Helmholtz. Rede gehalten bei der von der physikalisch-ökonomischen Gesellschaft zu Königsberg in Pr. veranstalteten Gedächtnissfeier am 7. December 1894." *Schriften der Königlichen Physikalisch-ökonomischen Gesellschaft zu Königsberg im Preussen 35* (1894):63–83.

Herneck, Friedrich. *Abenteur der Erkenntnis: Fünf Naturforscher aus drei Epochen*. Berlin: Buchverlag Der Morgen, 1973.

———. "Die Stellung von Hermann von Helmholtz in der Wissenschaftsgeschichte." *WZHUB 22:3* (1973):349–55; reprinted in idem, *Wissenschaftsgeschichte: Vorträge und Abhandlungen*. Berlin: Akademie-Verlag, 1984. Pp. 68–77.

Hertz, Heinrich. "Zum 31. August 1891." *Münchener Allgemeine Zeitung,* 1891. No. 241. Beilage-Nummer 202. Pp. 1–3; reprinted in *Gesammelte Werke von Heinrich Hertz.* 3 vols. Leipzig: J.A. Barth, 1894–95, vol. 1: *Schriften vermischten Inhalts.* Pp. 360–68.

———. *Erinnerungen, Briefe, Tagebücher.* Johanna Hertz, comp. Leipzig: Akademische Verlagsgesellschaft, n.d. [1927]; *Memoirs. Letters. Diaries.* 2nd enl. and rev. ed. Mathilde Hertz and Charles Susskind, eds. and trans. San Francisco: San Francisco Press, 1977.

Heyfelder, Victor. *Über den Begriff der Erfahrung bei Helmholtz.* (Inaugural diss., Berlin University, 1897.) Berlin: n.p., 1897.

Hicks, W.M. "Report on Recent Progress in Hydrodynamics. Part I. General Theory." *Report of the Fifty-First Meeting of the British Association for the Advancement of Science; Held at York in August and September 1881.* London: John Murray, 1882. Pp. 57–88.

High, Richard P. "Does James's Criticism of Helmholtz Really Involve a Contradiction?" *JHBS* 14 (1978):337–43.

Hirosige, Tetu. "Origins of Lorentz' Theory of Electrons and the Concept of the Electromagnetic Field." *HSPS 1* (1969):151–209.

Hirschberg, Julius. "Hermann v. Helmholtz." *Deutsche Medicinische Wochenschrift* No. 38 (20 September 1894):733–34.

[———]. "Hermann v. Helmholtz." *Zentralblatt für praktische Augenheilkunde 18* (1894):258–68.

———. "Die Reform der Augenheilkunde I." (= *Handbuch der Gesamten Augenheilkunde.*) A. Graefe, Th. Saemisch, C. Hess, et al., eds. 2nd rev. ed. 15 vols. Berlin: Springer, 1918. *15.*

———. "Hermann von Helmholtz. Ein Gedenkwort." *Berliner Klinische Wochenschrift 58* (1921):1116–18.

Höber, R. "Zu Helmholtz' 100. Geburtstage." *Deutsche medizinische Wochenschrift 47* (1921):1001–2.

Hochberg, Julian E. "Nativism and Empiricism in Perception." Leo Postman, ed. *Psychology in the Making: Histories of Selected Research Problems.* New York: Knopf, 1962. Pp. 255–330.

Hoff, H.E., and L.A. Geddes. "Ballistics and the Instrumentation of Physiology: The Velocity of the Projectile and of the Nerve Impulse." *JHMAS 15* (1960):133–46.

Hoffmann, Dieter. "Wissenschaft und Bürokratie: Hermann von Helmholtz und Friedrich Althoff im Spiegel ihres Briefwechsels." Bernhard vom Brocke, ed. *Wissenschaftsgeschichte und Wissenschaftspolitik im Industriezeitalter: Das 'System Althoff' in historischer Perspektive.* Hildesheim: August Lax, 1991. Pp. 245–50.

Hoffmann, Erich. "Hundertjährige Wiederkehr des Eintritts von H. v. Helmholtz und R. Virchow in die militärärztliche Akademie." *Münchener medizinische Wochenschrift 86* (1939):1560–62.

Holland, Gerhard. "Über das Helmholtzsche Phakoskop." *Pflugers Archiv für die gesamte Physiologie der Menschen 268* (1958/59):412–14.

Holländer, Eugen. "Hermann Ludwig Ferdinand von Helmholtz." In idem,

Anekdoten aus der Medizinischen Weltgeschichte. 2nd ed. Stuttgart: Ferdinand Enke, 1931. Pp. 186–87.
Höller, Hubert. *Der Gegenstand der Sinneswahrnehmung bei Aristoteles und Helmholtz. Zur Geschichte und Erläuterung des Wahrnehmungsproblems.* Ph.D. diss., Giessen University, 1925.
Homer, William Innes. *Seurat and the Science of Painting.* Cambridge, Mass.: The MIT Press, 1964.
Hörz, Herbert. "Über die Erkenntnistheorie von Helmholtz." *Aufbau 13* (1957):423–32.
———. "Hermann von Helmholtz." Otto Finger and Friedrich Herneck, eds. *Von Liebig zu Laue: Ethos und Weltbild grosser deutscher Naturforscher und Ärzte.* Berlin: Deutscher Verlag der Wissenschaften, 1963. Pp. 111–22.
———. "Die philosophischen Auffassungen von Hermann von Helmholtz." *WZHUB 22:3* (1973):279–86.
———. "Helmholtz und Boltzmann." Roman Sexl and John Blackmore, eds. *Ludwig Boltzmann. Internationale Tagung anlässlich des 75. Jahrestages seines Todes 5.–8. September 1981. Ausgewählte Abhandlungen.* (= *Ludwig Boltzmann Gesamtausgabe*, Band 8). Graz: Akademische Druck- und Verlagsanstalt; Braunschweig and Wiesbaden: Friedrich Vieweg & Sohn, 1982. Pp. 191–205.
———, and Andreas Laass. "Hermann von Helmholtz und die Physikalisch-Technische Reichsanstalt." *Perspektiven interkultureller Wechselwirkung für den Fortschritt. Beiträge von Wissenschaftshistorikern der DDR zum XVII. Internationalen Kongress für Geschichte der Wissenschaften in Berkeley (USA).* Akademie der Wissenschaften der DDR. Institut für Theorie, Geschichte und Organisation der Wissenschaft. Kolloquien. Heft 48. Berlin, 1985. Pp. 123–30.
———, and ———. *Ludwig Boltzmanns Wege nach Berlin. Ein Kapitel österreichisch-deutscher Wissenschaftsbeziehungen.* Berlin: Akademie-Verlag, 1989.
———, and Siegfried Wollgast. "Zu den philosophischen Auffassungen von Hermann von Helmholtz." *Deutsche Zeitschrift für Philosophie 19* (1971):40–65.
———, and ———. "Einleitung." Herbert Hörz and Siegfried Wollgast, eds. *Hermann von Helmholtz. Philosophische Vorträge und Aufsätze.* Berlin: Akademie-Verlag, 1971. Pp. v–lxxix.
———, and ———. "Hermann von Helmholtz und Emil du Bois-Reymond: Wissenschaftsgeschichtliche Einordnung in die naturwissenschaftlichen und philosophischen Bewegungen ihrer Zeit." In Kirsten, ed., *Dokumente einer Freundschaft* (1986):11–64.
Huber, P. "Hermann von Helmholtz." *Experientia 7* (1951):356–60.
Hurvich, Leo M., and Dorothea Jameson. "Helmholtz and the Three-Color Theory: An Historical Note." *American Journal of Psychology 62* (1949):111–14.
Hyslop, J.H. "Helmholtz's Theory of Space-Perception." *Mind. A Quarterly Review of Psychology and Philosophy 16* (1891):54–79.

Israel, Walter. *Substanzbegriff und Energieproblem in der modernen Physik. Eine erkenntnistheoretische Studie über Robert Mayer, Hermann von Helmholtz und Max Planck.* Berlin: Arthur Collignon, 1921.
Jaeger, Wolfgang. "Formen kreativen Denkens an Beispielen bahnbrechender Entdeckungen in der Augenheilkunde." *Klinische Monatsblätter für Augenheilkunde 162* (1973):268–77.
———, ed. *Die Erfindung der Ophthalmoskopie dargestellt in den Originalbeschreibungen der Augenspiegel von Helmholtz, Ruete und Giraud-Teulon.* Heidelberg: Brausdruck, n.d. [1977].
Jaki, Stanley L. "Goethe and the Physicists." *American Journal of Physics 37:2* (1969):195–203.
James, Frank A.J.L. "Thermodynamics and Sources of Solar Heat, 1846–1862." *BJHS 15* (1982):155–81.
James, William. *The Principles of Psychology.* 3 vols. (*The Works of William James.*) Cambridge, Mass. and London: Harvard University Press, 1981.
———. *Essays in Psychology.* (*The Works of William James.*) Cambridge, Mass. and London: Harvard University Press, 1983.
———. *Psychology: Briefer Course.* (*The Works of William James.*) Cambridge, Mass. and London: Harvard University Press, 1984.
Jammer, Max. "Energy." Paul Edwards, ed. 8 vols. *The Encyclopedia of Philosophy.* New York: Macmillan and The Free Press; London: Collier Macmillan, 1967. 2:511–17.
Jones, Ernest. *The Life and Work of Sigmund Freud.* 3 vols. New York: Basic Books, 1968.
Jungnickel, Christa, and Russell McCormmach. *Intellectual Mastery of Nature: Theoretical Physics from Ohm to Einstein.* 2 vols. Vol.1: *The Torch of Mathematics 1800–1870.* vol.2: *The Now Mighty Theoretical Physics 1870–1925.* Chicago and London: The University of Chicago Press, 1986.
Kahl, H. Russell. "The Philosophical Work of Hermann von Helmholtz. The Philosophy and Epistemology of the German Scientist." Ph.D. diss., Columbia University, 1951.
———. "Helmholtz, Hermann Ludwig von." Paul Edwards, ed. *Encyclopedia of Philosophy.* 8 vols. New York: Macmillan and The Free Press, 1967, 1972. 3:469–71.
———, ed. *Selected Writings of Hermann von Helmholtz.* Middletown, Conn.: Wesleyan University Press, 1971.
Kaiser, Walter. *Theorien der Elektrodynamik im 19. Jahrhundert.* Hildesheim: Gerstenberg, 1981.
Kangro, Hans. *Vorgeschichte des Planckschen Strahlungsgesetzes: Messungen und Theorien der spektralen Energieverteilung bis zur Begründung der Quantenhypothese.* Wiesbaden: Franz Steiner, 1970.
Karlson, Paul. "Hermann von Helmholtz 1821–1894." Willy Andreas and Wilhelm von Scholz, eds. *Die Grossen Deutschen. Neue Deutsche Biographie.* 4 vols. Berlin: Propyläen-Verlag, 1936. 3:524–41.
Karpinski, Louis C. "Hermann von Helmholtz." *The Scientific Monthly 13* (1921):24–32.

Katz, B. "Hermann von Helmholtz, der Überwinder der Romantik." *Deutsche Medizinische Wochenschrift 58* (1932):603-4.
Kayser, Heinrich. *Erinnerungen aus meinem Leben.* N.p. Manuscript. 1936. Copy at the American Institute of Physics, New York.
Kemp, Martin. *The Science of Art: Optical Themes in Western Art from Brunelleschi to Seurat.* New Haven, Conn., and London: Yale University Press, 1990.
Kidwell, Peggy Aldrich. "Solar Radiation and Heat from Kepler to Helmholtz (1600-1860)." Ph.D. diss., Yale University, 1979.
Kirsten, Christa, and Hans-Günther Körber, eds. *Physiker über Physiker: Wahlvorschläge zur Aufnahme von Physikern in die Berliner Akademie 1870 bis 1929 von Hermann v. Helmholtz bis Erwin Schrödinger.* Berlin: Akademie-Verlag, 1975.

———, and ———, eds. *Physiker über Physiker II: Antrittsreden, Erwiderungen bei der Aufnahme von Physikern in die Berliner Akademie. Gedächtnisreden 1870 bis 1929.* Berlin: Akademie-Verlag, 1979.

———, et al., eds. *Dokumente einer Freundschaft. Briefwechsel zwischen Hermann von Helmholtz und Emil du Bois-Reymond 1846-1894.* Berlin: Akademie-Verlag, 1986.
Klauss, Klaus. "Ein neuentdecktes frühes Dokument zur Geschichte der Erfindung des Augenspiegels durch Hermann v. Helmholtz." *NTM 18:1* (1981):58-61.
Klein, Felix. *Vorlesungen über die Entwicklung der Mathematik im 19. Jahrhundert.* R. Courant and O. Neugebauer, eds. Teil 1. Berlin: Julius Springer, 1926.
Klein, Martin J. "Mechanical Explanation at the End of the Nineteenth Century." *Centaurus 17* (1972):58-82.

———. "Boltzmann, Monocycles, and Mechanical Explanation." Raymond J. Seeger and Robert S. Cohen, eds. *Philosophical Foundations of Science* (= Boston Studies in the Philosophy of Science, vol. 11). Dordrecht, Boston: Reidel, D. 1974. Pp. 155-75.
Klix, Friedhart. "H. v. Helmholtz' Beitrag zur Theorie der Wahrnehmung." In Scheel, ed., *Gedanken von Helmholtz über schöpferische Impulse* (1972):25-36.

———. "Hermann von Helmholtz' Beitrag zur Theorie der Wahrnehmung—Bleibendes und Vergängliches in einem großen Lebenswerk." *WZHUB 22:3* (1973):307-14; reprinted in Georg Eckardt, ed. *Zur Geschichte der Psychologie.* Berlin: VEB Deutscher Verlag der Wissenschaften, 1979. Pp. 44-59.
Knapp, Hermann. "A Few Personal Recollections of Helmholtz." *The Journal of the American Medical Association 38* (1902):557-58.
Knott, Cargill Gilston. *Life and Scientific Work of Peter Guthrie Tait. Supplementing the Two Volumes of Scientific Papers Published in 1898 and 1900.* Cambridge: Cambridge University Press, 1911.
Koenigsberger, Leo. *Hermann von Helmholtz's Untersuchungen über die Grundlagen der Mathematik und Mechanik.* Leipzig: B.G. Teubner, 1896; reprinted Niederwalluf bei Wiesbaden: Martin Sändig, 1971.

———. "The Investigations of Hermann von Helmholtz on the Fundamental Principles of Mathematics and Mechanics." *Annual Report of the Board of Regents of the Smithsonian Institution . . . to July, 1896.* Washington, D.C.: Government Printing Office, 1898. Pp. 93–124.

[———]. "Das Elternhaus von Hermann Helmholtz." *Beilage zur Allgemeinen Zeitung*, Munich. Beilage No. 253. 1902. Pp. 228–30.

———. *Hermann von Helmholtz.* 3 vols. Braunschweig: Friedrich Vieweg und Sohn, 1902–3.

———. *Hermann von Helmholtz.* Trans. Frances A. Welby. Oxford: Clarendon Press, 1906; reprinted New York: Dover, 1965.

———. "Über Helmholtz's Bruchstück eines Entwurfes betitelt 'Naturforscher-Rede'." *Sitzungsberichte der Heidelberger Akademie der Wissenschaften, Mathematisch-naturwissenschaftliche Klasse* 14 (1910):3–8.

———. *Mein Leben.* Heidelberg: Carl Winters Universitätsbuchhandlung, 1919.

Köhnke, Klaus Christian. *Entstehung und Aufstieg des Neukantianismus. Die deutsche Universitätsphilosophie zwischen Idealismus und Positivismus.* Frankfurt am Main: Suhrkamp, 1986.

Kohut, Adolf. "Hermann Ludwig Ferdinand v. Helmholtz." *Westermanns Illustrierte Deutsche Monatshefte* 55 (1884):720–28.

Konen, Heinrich, and August Pütter. *Hermann von Helmholtz 1855–1858 Professor der Physiologie und Anatomie an der Universität Bonn. Reden, gehalten am 17. Dezember 1921 zum hundertsten Geburtstag.* Bonn: Bonner Universitäts-Buchdruckerei Gebr. Scheur, [1921?].

König, Arthur Peter, ed. *Beiträge zur Psychologie und Physiologie der Sinnesorgane. Hermann von Helmholtz als Festgruss zu seinem siebzigsten Geburtstag dargebracht von Th. W. Engelmann, E. Javal, A. König, J. von Kries, Th. Lipps, L. Matthiessen, W. Preyer, W. Uhthoff.* Hamburg and Lepzig: Leopold Voss, 1891.

König, Gert. "Der Wissenschaftsbegriff bei Helmholtz und Mach." Alwin Diemer, ed. *Beiträge zur Entwicklung der Wissenschaftstheorie im 19. Jahrhundert.* Meisenheim: Anton Hain, 1968. Pp. 90–114.

König, W. "Helmholtz als Physiker." *Bericht der Oberhessischen Gesellschaft für Natur- und Heilkunde zu Giessen, naturwissenschaftliche Abteilung* 8 (1922):1–8.

Kosing, Alfred. "Zu den philosophischen Anschauungen von Helmholtz." In Scheel, ed., *Gedanken von Helmholtz über schöpferische Impulse* (1972):49–55.

Krause, Albrecht. *Kant und Helmholtz. Über den Ursprung und die Bedeutung der Raumanschauung und der geometrischen Axiome.* Lahr: Moritz Schauenberg, 1878.

Kreidl, Alois. "Zur Geschichte der Hörtheorien." *Archives néerlandaises des sciences exactes et naturelles* sér. 3c 7 (1922):502–9.

Kremer, Richard L. "The Thermodynamics of Life and Experimental Physiology, 1770–1880." Ph.D. diss., Harvard University, 1984. Reprinted New York and London: Garland, 1990.

———. "From Psychophysics to Phenomenalism: Mach and Hering on Color Vision." Mary Jo Nye, Joan L. Richards, and Roger H. Stuewer, eds. *The Invention of Physical Science. Intersections of Mathematics, Theology, and Natural Philosophy since the Seventeenth Century. Essays in Honor of Erwin N. Hiebert.* (= Boston Studies in the Philosophy of Science, vol. 139). Boston and Dordrecht: Kluwer Academic, 1992. Pp. 147-73.

———., ed. *Letters of Hermann von Helmholtz to His Wife, 1847-1859.* Stuttgart: Franz Steiner Verlag, 1990.

Kries, Johannes von. "Hermann v. Helmholtz." *Deutsche medicinische Wochenschrift 17:35* (1891):1025-27.

———. "Helmholtz als Physiolog." *Die Naturwissenschaften 9:35* (1921):673-93.

K[rigar]-M[enzel], O[tto]. "Hermann von Helmholtz †." *Elektrotechnische Zeitschrift 15* (1894):613-16.

Kronecker, Hugo. "Hermann von Helmholtz." *The Electrician 27* (1891):437-39, 465-68.

———. "Hermann von Helmholtz. Akademischer Vortrag gehalten im Saale des Museums zu Bern." *Schweizerische Rundschau 4* (1894):510-38; reprinted Bern: Lack Scheim, 1894.

Krückmann, E. "Gedenkworte zum hundertsten Geburtstag von Helmholtz." *Medizinische Klinik 18* (1922):30-2.

Krüss, H. "Hermann von Helmholtz." *Zeitschrift für Instrumentenkunde 14* (1894):341-45.

Kuczynski, Jürgen. "Das Haus Helmholtz." In idem, *Gelehrtenbiographien* (= *Studien zu einer Geschichte der Gesellschaftswissenschaften*, Band 6). Berlin: Akademie-Verlag, 1977. Pp. 103-24.

Kühnert, Herbert. "Ein unbekannter Brief von Ernst Abbe an Hermann von Helmholtz: Ein Beitrag zur Vorgeschichte des JENAer Glaswerks Schott und Genossen." *Sprechsaal für Keramik, Glas, Email 94:9* (1961):213-16.

Kühtmann, Alfred. *Zur Geschichte des Terminismus. Wilhelm v. Occam, Etienne Bonnet de Condillac, Hermann v. Helmholtz, Fritz Mauthner.* (= *Abhandlungen zur Philosophie und Ihrer Geschichte*, Heft 20). Leipzig: Quelle & Meyer, 1911.

Kuhn, Thomas S. "Energy Conservation as an Example of Simultaneous Discovery." Marshall Clagett, ed. *Critical Problems in the History of Science.* Madison, Wis.: University of Wisconsin Press, 1959. Pp. 321-56. Reprinted in Kuhn's *The Essential Tension: Selected Studies in Scientific Tradition and Change.* Chicago and London: The University of Chicago Press, 1977. Pp. 66-104.

K[und]t, [August]. "Zum 70. Geburtstag von Hermann von Helmholtz." *Nationale Zeitung 44:501*, Sonntag, 30. August 1891.

Kusch, Ernst. "C.G.J. Jacobi und Helmholtz auf dem Gymnasium. Beitrag zur Geschichte des Victoria-Gymnasiums zu Potsdam." *Beilage des Schulprogramms des Victoria-Gymnasiums zu Potsdam.* Potsdam: Kramer'sche Buchdruckerei, 1896. Pp. 1-43.

Kuznetsov, B.G., and U.I. Frankfurt. "On the History of the Law of Conservation and Transmutation of Energy." [In Russian.] *Trudy Institut Istorii Estestvoznaniia i Tekhniki* [Akademiia Nauk SSSP] *28* (1959):339–76.

Lamb, Horace. "On Reciprocal Theorems in Dynamics." *Proceedings of the London Mathematical Society 19* (1887–88):144–51.

Landolt, E. "H. de Helmholtz. Esquisse biographique." *Archives d'ophtalmologie* (December 1894):721–42.

Lange, Frederick Albert. *History of Materialism and Criticism of Its Present Importance.* 3 vols. Trans. Ernest Chester Thomas. Boston: Houghton Mifflin, 1880–82.

Langen, Peter. "Hermann von Helmholtz. Ein Lebensbild nach den neuesten Quellen." *Westermanns illustrierte deutsche Monatshefte 94* (1903):782–97.

Lau, Ernst. "Helmholtz." *Sozialistische Monatshefte 58* (1922):48–9.

Laue, Max von. "Hermann von Helmholtz 1821–1894." Friedrich Herneck, ed. *Forschen und Wirken. Festschrift zur 150-Jahr-Feier der Humboldt-Universität zu Berlin, 1810–1960.* 3 vols. Berlin: VEB Deutscher Verlag der Wissenschaften, 1960. *1*:359–66.

Lauer, Hans H. "Wissenschaftshistorische Aspekte bei Helmholtz und Virchow." Heinrich Schipperges, ed. *Die Versammlung Deutscher Naturforscher und Ärtze im 19. Jahrhundert.* Stuttgart: A.W. Gentner, 1968. Pp. 87–101.

Law, Frank W. "The Origin of the Ophthalmoscope." *Ophthalmology 93:1* (1986):140–41.

Lawergren, B. "On the Motion of Bowed Violin Strings." *Acustica 44* (1980):194–206.

Lazarev, Petr Petrovich. *Helmholtz.* [In Russian.] Moscow: Akademia Nauk USSR, 1959.

Leake, Chauncey D. "Hermann von Helmholtz (1821–1894)." Webb Haymaker and Francis Schiller, eds. *The Founders of Neurology: One Hundred and Forty-Six Biographical Sketches by Eighty-Eight Authors.* 2nd ed. Springfield, Ill.: Charles C. Thomas, 1970. Pp. 225–29.

Lebedinskii, A.V., U.I. Frankfurt, and A.M. Frank. *Helmholtz (1821–1894).* [In Russian.] Moscow: Akademia Nauk USSR, 1966.

Lemmerich, J. "Die Hertzsche Entdeckung im Briefwechsel zwischen Hermann von Helmholtz, Emil du Bois-Reymond und Karl Runge." *Physikalische Blätter 44* (1988):218–20.

Lenard, Philipp. *Great Men of Science. A History of Scientific Progress.* New York: Macmillan, 1933.

Lenin, V.I. *Materialism and Empirio-Criticism.* (= V.I. Lenin, *Collected Works*, vol. 14). Moscow: Foreign Languages Publishing House, 1962.

Lenoir, Timothy. *The Strategy of Life: Teleology and Mechanics in Nineteenth-Century German Biology.* Dordrecht and Boston: D. Reidel, 1982; reprinted Chicago and London: The University of Chicago Press, 1989.

———. "Models and Instruments in the Development of Electrophysiology, 1845–1912." *HSPS 17* (1986):1–54.

———. "Social Interests and the Organic Physics of 1847." Edna Ullmann-Margalit, ed. *Science in Reflection*. (The Israel Colloquium: Studies in History, Philosophy, and Sociology of Science, vol. 3). Dordrecht, Boston, London: Kluwer Academic, 1988. Pp. 169–91.

Lenz, Max, ed. *Geschichte der Königlichen Friedrich-Wilhelms-Universität zu Berlin*. 4 vols. in 5. Halle a.d.S.: Buchhandlung des Waisenhauses, 1910–18.

Lenzen, Victor F. "Helmholtz's Theory of Knowledge." M.F. Ashley Montague, ed. *Studies and Essays in the History of Science and Learning Offered in Homage to George Sarton on the Occasion of His Sixtieth Birthday 31 August 1944*. New York: Henry Schuman, 1946. Pp. 301–19.

Levin, Julius. "Hermann von Helmholtz. Zu seinem siebzigsten Geburtstag." *Die Gegenwart. Wochenschrift für Literatur, Kunst und öffentliches Leben* 40 (1891):135–38.

Ley, Hermann. "Gesetz und Merkmal bei Hermann von Helmholtz." *WZHUB* 22:3 (1973):287–88.

Liedman, Sven-Eric. *Das Spiel der Gegensätze: Friedrich Engels' Philosophie und die Wissenschaften des 19. Jahrhunderts*. Frankfurt and New York: Campus, 1986.

Lloyd, J.T. "Background to the Joule-Mayer Controversy." *Notes and Records of the Royal Society of London* 25 (1970):211–25.

Loewy, Maurice. [Obituary remarks on Helmholtz.] *Comptes rendus hebdomadaires des séances de l'Académie des Sciences* 119 (1894):1044–46.

Lr. "H. von Helmholtz †." *Der Mechaniker. Zeitschrift zur Förderung der Mechanik, Optik, Elektrotechnik und verwandter Gebiete* No. 24 (20 September 1894):343–46.

Ludwig, C. "H. v. Helmholtz, der Arzt." *Aerztliches Vereinsblatt*, Nr. 289 (1894).

L[umme]r, [Otto]. "H. von Helmholtz †." *Der Mechaniker* No. 24 (20 September 1894):343–46.

———. "Der Physiker und die Technik." "Helmholtz zum 100. Geburtstag. Erinnerungsblatt der Vossischen Zeitung." *Vossische Zeitung*. No. 404. Beilage 4. 28 August 1921.

Lungo, Carlo del. "L'Opera Scientifica di Ermanno Helmholtz." *Nuova Antologia di Lettere, Scienze ed Arti* 4th ser. 87 (1900):710–27.

Mach, Ernst. *Einleitung in die Helmholtz'sche Musiktheorie. Populär für Musiker dargestellt*. Graz: Leuschner and Lubensky, 1866.

———. *History and Root of the Principle of the Conservation of Energy*. Trans. Philip E.B. Jourdain. Chicago: Open Court; and London: Kegan Paul, Trench, Trübner, 1911.

———. *Popular Scientific Lectures*. 5th ed. Trans. Thomas J. McCormack. La Salle, Ill.: Open Court, 1943.

———. *The Science of Mechanics. A Critical and Historical Account of Its Development*. 6th ed. Trans. Thomas J. McCormack. LaSalle, Ill.: Open Court, 1960.

McKendrick, John Gray. *Hermann Ludwig Ferdinand von Helmholtz*. New York: Longmans, Green & Co., 1899.

———. "Helmholtz, Hermann Ludwig Ferdinand von." In *Encyclopedia Britannica*. 11th ed. 28 vols. Cambridge: Cambridge University Press, 1910–11. *13*:248–49.

Magie, W.F. "Hermann von Helmholtz.—The Man and the Teacher." *The Electrical World 24* (1894):329–30.

Mainzer, Klaus. *Geschichte der Geometrie*. Mannheim, Vienna, Zurich: B.I.-Wissenschaftsverlag, 1980.

Mamlock, G. "Helmholtz als Klassiker der Naturwissenschaften. Zu seinem 100. Geburtstag am 31. VIII. 1921." *Deutsche medizinische Wochenschrift 47* (1921):1002–3.

Mandelbaum, Maurice. *History, Man, & Reason: A Study in Nineteenth-Century Thought*. Baltimore and London: The Johns Hopkins University Press, 1971.

Mann, C. Riborg. "Professor von Helmholtz." *Scribner's Magazine 18* (1895):568–70.

Margenau, Henry. "Introduction" to Helmholtz's *On the Sensations of Tone as a Physiological Basis for the Theory of Music*. Trans. Alexander J. Ellis. 2nd English ed. Reprinted New York: Dover, 1954.

Margolis, Joseph. "Goethe and Psychoanalysis." Frederick Amrine, Francis J. Zucker, and Harvey Wheeler, eds. *Goethe and the Sciences: A Reappraisal*. Dordrecht, Boston, Lancaster: D. Reidel, 1987. Pp. 83–100.

Matschoß, Conrad, ed. *Werner Siemens. Ein kurzgefaßtes Lebensbild nebst einer Auswahl seiner Briefe*. 2 vols. Berlin: 1916.

Maxwell, James Clerk. "Hermann Ludwig Ferdinand Helmholtz." *Nature 15* (1877):389–91; reprinted in W.D. Niven, ed. *The Scientific Papers of James Clerk Maxwell*. 2 vols. Cambridge: Cambridge University Press, 1890. 2:592–98.

Meijering, Theodoor. "Naturalistic Epistemology: Helmholtz and the Rise of a Cognitive Theory of Perception." Ph.D. diss., University of California-Berkeley, 1981.

Meinel, Christoph. *Karl Friedrich Zöllner und die Wissenschaftskultur der Gründerzeit. Eine Fallstudie zur Genese konservativer Zivilisationskritik*. Berlin: SIGMA, 1991.

Meinel, U. "Einige Bemerkungen zur visuellen Empfindung im Blickpunkt der marxistisch-leninistischen Abbildtheorie unter Berücksichtigung der Helmholtzschen Zeichentheorie." *Zeitschrift für ärztliche Fortbildung 66* (1972):587–89.

Mendelsohn, Everett. "Revolution and Reduction: The Sociology of Methodological and Philosophical Concerns in Nineteenth Century Biology." Y. Elkana, ed. *The Interaction Between Science and Philosophy*. Atlantic Highlands, N.J.: Humanities Press, 1974. Pp. 407–26.

Mendelssohn, Kurt. *The World of Walther Nernst: The Rise and Fall of German Science 1864–1941*. Pittsburgh: The University of Pittsburgh Press, 1973.

Mendenhall, T.C. "Helmholtz." *Science* n.s. *3:58* (1896):189–95; reprinted in *Annual Report of the Board of Regents of the Smithsonian Institution . . . to July 1895*. Washington, D.C.: Government Printing Office, 1895. Pp. 787–93.

Merz, John Theodore. *A History of European Thought in the Nineteenth Century.* 4 vols. 1904–12. Reprinted New York: Dover, 1965.
Meyer, Max F. "Tartini *Versus* Helmholtz Judged by Modern Sensory Observation." *The Journal of the Acoustical Society of America 26* (1954):761–64.
Meyering, Theo C. *Historical Roots of Cognitive Science: The Rise of a Cognitive Theory of Perception from Antiquity to the Nineteenth Century.* Dordrecht, Boston, London: Kluwer Academic Publishers, 1989.
Meyerson, Emile. *Du cheminement de la pensée.* 3 vols. Paris: Felix Alcan, 1931.
Mie, Gustav. "Die mechanische Erklärbarkeit der Naturerscheinungen. Maxwell.—Helmholtz.—Hertz." *Verhandlungen des naturwissenschaftlichen Vereins in Karlsruhe 13* (1895–1900):402–20.
Miller, John David. "Henry August Rowland and His Electromagnetic Researches." Ph.D. diss., Oregon State University, 1970.
———. "Rowland and the Nature of Electric Currents." *Isis 63* (1972):5–27.
Millington, E.C. "History of the Young-Helmholtz Theory of Colour Vision." *Annals of Science 5* (1941):167–76.
Moulines, Carlos-Ulises. "Hermann von Helmholtz: A Physiological Approach to the Theory of Knowledge." H.N. Jahnke and M. Otte, eds. *Epistemological and Social Problems of the Sciences in the Early Nineteenth Century.* Dordrecht, Boston, London: D. Reidel, 1981. Pp. 65–73.
Mulligan, Joseph F. "The Influence of Hermann von Helmholtz on Heinrich Hertz's Contributions to Physics." *American Journal of Physics 55:8* (1987):711–19.
———. "Hermann von Helmholtz and His Students." *American Journal of Physics 57:1* (1989):68–74.
Munk, Hermann. "Hermann v. Helmholtz †." *Berliner klinische Wochenschrift. Organ für practische Aerzte 31:38* (1894):859–60.
Münsterberg, Hugo. "The Helmholtz Memorial." *Science* n.s. *1* (1895):547–48.
Murphy, Gardner. *Historical Introduction to Modern Psychology.* Rev. ed. New York: Harcourt, Brace and Co., 1949.
Murray, Robert H. *Science and Scientists in the Nineteenth Century.* London: Sheldon; New York and Toronto: Macmillan, 1925.
Die Naturwissenschaften. Special issue: "Dem Andenken an Helmholtz. Zur Jahrhundertfeier seines Geburtstages." *Die Naturwissenschaften 9:35* (1921):673–711.
Nernst, Walther. "Die elektrochemischen Arbeiten von Helmholtz." *Die Naturwissenschaften 9:35* (1921):699–702.
Neumann, Carl. *Die elektrischen Kräfte. Darlegung und Genauere Betrachtung der von hervorragenden Physiker entwickelten mathematischen Theorien.* 2 vols. Leipzig: B.G. Teubner, 1873 and 1898.
Neumann, Josef. "Wahrnehmung und Kausalität in den Schriften der Sinnesphysiologen Hermann von Helmholtz, Johannes von Kries und Viktor von Weizsäcker: Ein Beitrag zum Verhältnis von Erkenntnistheorie und Wissenschaftstheorie in der Medizin." *Gesnerus 44* (1987):235–52.
N[ichols]., E[dward]. L. "Hermann von Helmholtz." *The Physical Review 2* (1894):222–27.

Nösselt, Volker. "Real-Phonetik oder Pseudo-Phonetik als Basis der Phonologie? Hermann contra Helmholtz t-Strukturalismus contra Spektralismus." *Communication and Cognition 12:1* (1979):107–39.

Ormerod, F.C. "Research in Otology—One Hundred Years after Helmholtz." *The Journal of Laryngology and Otology 79* (1965):845–69.

Ostwald, Wilhelm. *Elektrochemie: Ihre Geschichte und Lehre.* Leipzig: Veit, 1896.

———. *Erfinder und Entdecker.* Frankfurt am Main: Literarische Anstalt Rütten & Loening, 1908.

———. "Über den Ort der elektromotorischen Kraft in der voltaschen Kette." *Abhandlungen und Vorträge allgemeinen Inhaltes.* New ed. Leipzig: Akademische Verlagsgesellschaft, 1916. Pp. 160–82.

———. *Grosse Männer.* 6th ed. Leipzig: Akademische Verlagsgesellschaft, 1927.

Paalzow, A. "Helmholtz." *Allgemeine Deutsche Biographie.* 56 vols. Munich and Leipzig: Duncker & Humblot, 1872–1912. *51:*461–70.

Partenheimer, Maren. "Die Tragweite Goethes in der Naturwissenschaft: Hermann von Helmholtz, Ernst Haeckel, Werner Heisenberg, Carl Friedrich von Weizsäcker." Ph.D. diss., University of Utah, 1987.

———. *Goethes Tragweite in der Naturwissenschaft: Hermann von Helmholtz, Ernst Haeckel, Werner Heisenberg, Carl Friedrich von Weizsäcker.* Berlin: Duncker & Humblot, 1989.

Pastore, Nicholas. *Selective History of Theories of Visual Perception: 1650–1950.* New York, London, Toronto: Oxford University Press, 1971.

———. "Helmholtz's 'Popular Lectures on Vision.'" *JHBS 9* (1973):190–202.

———. "Reevaluation of Boring on Kantian Influence, Nineteenth Century Nativism, Gestalt Psychology and Helmholtz." *JHBS 10:4* (1974):375–90.

———. "William James: A Contradiction." *JHBS 13:2* (1977):126–30.

———. "Helmholtz on the Projection or Transfer of Sensation." Peter K. Machamer and Robert G. Turnbull, eds. *Studies in Perception: Interrelations in the History of Philosophy and Science.* Columbus, Ohio: Ohio State University Press, 1978. Pp. 355–76.

Paul, Robert. "German Academic Science and the Mandarin Ethos, 1850–1880." *BJHS 17* (1984):1–29.

Pernet, Johannes. "Hermann von Helmholtz. 31. August 1821 bis 8. September 1894. Ein Nachruf." *Neujahrsblatt der Naturforschenden Gesellschaft in Zürich 97* (1895):1–36.

Perry, Walter. "Professor Helmholtz' Rectoratsrede und die Englischen Universitäten." [With a reply by Helmholtz.] *Deutsche Rundschau 14* (1878):332–36.

Pielert, Walter. *Die erkenntnistheoretischen Ansichten von Helmholtz und Fick.* Inaugural diss., Friedrichs-Universität Halle-Wittenberg, 1924.

Planck, Max. *Das Princip der Erhaltung der Energie.* Leipzig: B.G. Teubner, 1887.

———. "Das Institut für theoretische Physik." In Lenz, ed. *Geschichte der Universität zu Berlin.* 3:276–78.

———. "Helmholtz's Leistungen auf dem Gebiete der theoretischen Physik." *Allgemeine Deutsche Biographie.* 56 vols. Munich and Leipzig: Duncker & Humblot, 1872–1912. *51:*470–72; reprinted in idem, *Physikalische Abhandlungen und Vorträge.* 3 vols. Braunschweig: F. Vieweg, 1958. *3:*321–23.

———. "Erinnerungen an Anna von Helmholtz." *Velhagen und Klasings Monatshefte 44:2* (1929/30):37–9

———. "Persönliche Erinnerungen." In "Festsitzung zur Feier des 90jährigen Bestehens der physikalischen Gesellschaft zu Berlin am 25. Januar 1935 im großen Hörsaal des Physikalischen Instituts der Universität." *Verhandlungen der Deutschen Physikalischen Gesellschaft 16* (1935):1–17, on 11–16.

———. *Wissenschaftliche Selbstbiographie.* Leipzig: J.A. Barth, 1948. Reprinted in idem, *Physikalische Abhandlungen und Vorträge,* 3:374–401.

———. "Persönliche Erinnerungen aus alten Zeiten." *Vorträge und Erinnerungen.* 5th ed. Stuttgart: S. Hirzel, 1949. Pp. 1–14.

Podolsky, Edward. "Physician Physicists." *Annals of Medical History* n.s. *3* (1931):300–7.

Polak, L.S. "Hidden Motion in Helmholtz's Theory of Heat." [In Russian]. *Voprosy Istorii Estestvoznaniia i Tekhniki 3* (1957):62–73.

———, and Iu. I. Solov'ev. "On the History of Physical Chemistry. The Studies of Helmholtz in the Field of Physical Chemistry." [In Russian)]. *Voprosy Istorii Estestvoznaniia i Tekhniki 8* (1959):48–56.

Pupin, M[ichael] I. "Hermann von Helmholtz." *The Electrical World 24* (1894):541–42.

———. I "Reminiscences of Hermann von Helmholtz." *Journal of the Optical Society of America 6* (1922):336–42.

———. *From Immigrant to Inventor.* New York: Charles Scribner's Sons, 1960.

Pyenson, Lewis. *Neohumanism and the Persistence of Pure Mathematics in Wilhelmian Germany.* Philadelphia: American Philosophical Society, 1983.

Querner, Hans. "Die Versammlung der Gesellschaft deutscher Naturforscher und Ärzte 1869 in Innsbruck." *Berichte des Naturwissenschaftlich-Medizinischen Vereins in Innsbruck 58* (1970):13–34.

Rabinbach, Anson. *The Human Motor: Energy, Fatigue, and the Origins of Modernity.* N.p.: Basic Books, 1990.

Radau, R[odolphe]. "Sur la base scientifique de la musique. Analyse des recherches de M. Helmholtz." *Moniteur scientifique–Quesneville.* No. 197 (1 March 1865):193–214.

Rakowski, Andrzej. "The Magic Number Two: Seven Examples of Binary Apposition in Pitch Theory." *The Humanities Association Review 30:1/2* (1979):24–45.

Ramsey, R.R. "The Helmholtz-Koenig Controversy." *Science 81:2110* (1935):561–62.

Randall, B. Alex. "The Debt of Otology to Helmholtz." *The Journal of the American Medical Association 38* (1902):561–62.

Rau, Albrecht. *Empfinden und Denken. Eine physiologische Untersuchung über die Natur des menschlichen Verstandes.* Giessen: Emil Roth, 1896.

Rauh, Hans-Christoph. "Zu den erkenntnistheoretischen Auffassungen von Helmholtz." *WZHUB 22:3* (1973):293–302.

Rechenberg, Helmut. "Persönlichkeiten aus der Frühgeschichte der Physikalisch-Technische Reichsanstalt: Werner von Siemens, Hermann von Helmholtz, Friedrich Kohlrausch und Emil Warburg." Wilhelm Treue and Gerhard Hildebrandt, eds. *Berlinische Lebensbilder. Naturwissenschaftler.* Berlin: Colloquium, 1987. Pp. 45-60.

Red[aktur]. "Helmholtz, Hermann Ludwig Ferdinand von." Hirsch, August, ed. *Biographisches Lexikon der hervorragenden Ärzte aller Zeiten und Völker.* 5 vols. Munich and Berlin: Urban & Schwarzenberg, 1962. 3:151-52.

Reinecke, Hans-Peter. "Naturwissenschaftliche Grundlagen der Musikwissenschaft." Carl Dahlhaus, ed. *Einfuhrung in die systematische Musikwissenschaft.* Cologne: Musikverlag Hans Gerig, 1971. Pp. 9-51.

Reinecke, Wilhelm. "Die Grundlagen der Geometrie." *Kant-Studien 8* (1903):345-95.

Reiner, Julius. *Hermann von Helmholtz.* Leipzig: Theod. Thomas, 1905.

Reiprich, Kurt. *Die philosophisch-naturwissenschaftlichen Arbeiten von Karl Marx und Friedrich Engels.* Berlin: Dietz, 1969.

Reissmann, August. "Helmholtz, Hermann (Ludwig Ferdinand)." *Musikalisches Konversations-Lexikon.* Ed. Hermann Mendel. 11 vols. Berlin: L. Heimann; New York: J. Schuberth, 1870-79. 5:191-92.

Rezneck, Samuel. "An American Physicist's Year in Europe: Henry A. Rowland, 1875-76." *American Journal of Physics 30* (1962):877-86.

Richards, Joan L. "The Evolution of Empiricism: Hermann von Helmholtz and the Foundations of Geometry." *The British Journal for the Philosophy of Science 28* (1977):235-53.

Riehl, Alois. "Hermann von Helmholtz in seinem Verhältnis zu Kant." *Kantstudien 9* (1904):261-85; reprinted Berlin: Reuther & Reichard, 1904.

———. "Helmholtz et Kant." *Revue de Métaphysique et de la Morale 12* (1904):579-603.

———. "Helmholtz als Erkenntnistheoretiker." *Die Naturwissenschaften 9:35* (1921):702-8.

———. *Führende Denker und Forscher.* Leipzig: Quelle & Meyer, 1922.

Riese, Walther, and George E. Arrington, Jr. "The History of Johannes Müller's Doctrine of the Specific Energies of the Senses: Original and Later Versions." *Bulletin of the History of Medicine 37* (1964):179-83.

Rintelen, F. "Hermann Helmholtz zum Gedächtnis." *Schweizerische Medizinische Wochenschrift 81:1* (1951):16-9.

R[odenberg], J[ulius]. "Hermann von Helmholtz." *Deutsche Rundschau 80* (1894):129-31.

Rohrschneider, W. "Die Erfindung des Augenspiegels durch Hermann Helmholtz vor 100 Jahren." *Deutsche medizinische Wochenschrift 76* (1951):1409-10.

Rompe, Robert. "Einige Worte über Helmholtz und die Physikalisch-Technische Reichsanstalt." In Scheel, ed., *Gedanken von Helmholtz über schöpferische Impulse* (1972):57-63.

Roscoe, Henry. *The Life and Experiences of Sir Henry Enfield Roscoe, D.C.L., L.L.D., F.R.S. Written by Himself.* London and New York: Macmillan, 1906.

Rosenfeld, B.A. *A History of Non-Euclidean Geometry. Evolution of the Concept of a Geometric Space.* Trans. Abe Shenitzer. New York, Berlin, Heidelberg: Springer-Verlag, 1988.
Rössler, F. "Vom Helmholtz'schen Kobaltversuch." *Ophthalmologica 18* (1949):149–60.
R[össler, Helmut]., and Günther Franz. "Helmholtz." *Biographisches Wörterbuch zur Deutschen Geschichte.* Munich: R. Oldenbourg, 1952. Pp.336–37.
Rothe, Rudolf. "Helmholtz als Mathematiker." "Helmholtz zum 100. Geburtstag. Erinnerungsblatt der Vossischen Zeitung." *Vossische Zeitung.* No. 404. Beilage 4. 28 August 1921.
Rothschuh, Karl E. "Hermann von Helmholtz (1821 bis 1894)." R. Dumesnil and H. Schadewaldt, eds. *Die Berühmten Ärzte.* 2nd German ed. Cologne: Aulis Verlag, 1966. Pp. 280–82.
———. *History of Physiology.* Trans. Guenter B. Risse. Huntington, NY: Robert E. Krieger, 1973.
Rowbottom, Margaret, and Charles Susskind. *Electricity and Medicine. History of Their Interaction.* San Francisco: San Francisco Press, 1984.
Roy, Louis. *L'électrodynamique des milieux isotropes en repos d'après Helmholtz et Duhem.* Paris: Gauthier-Villars, 1923.
Rubens, Heinrich. "Das physikalische Institut." In Lenz, ed. *Geschichte der Universität Berlin.* 3:278–96.
Rubner, M. "Helmholtz als Physiologe." In Warburg, et al., *Helmholtz als Physiker* (1922). Pp. 14–29.
Rucker, C. Wilbur, and Thomas E. Keys. *The Atlases of Ophthalmoscopy (1850–1950).* N.p., n.d. ["An exhibit in commemoration of the centennial of Helmholtz's invention of the ophthalmoscope, prepared for the Section on Ophthalmology, American Medical Association's meeting in San Francisco. June 26 to 30, 1950."]
Rüchardt, E. "Zum Erscheinen der Helmholtzschen Arbeit 'Über die Erhaltung der Kraft' vor hundert Jahren." *Die Naturwissenschaften 33* (1946):321–25.
Rücker, Arthur W. "Hermann von Helmholtz." *Fortnightly Review* n.s. *56* (1894):651–60; reprinted *Annual Report of the Board of Regents of the Smithsonian Institution . . . to July 1894.* Washington, D.C.: Government Printing Office, 1896. Pp. 709–18.
———. "Physical Work of Hermann von Helmholtz." *Nature 51*:1324 (1895):472–75; *51:1325* (1895):493–95.
———. [Obituary notice of Hermann von Helmholtz]. *Proceedings of the Royal Society of London 59* (1896):xv–xxx.
Ruff, Peter W. *Emil Du Bois-Reymond.* Leipzig: BSB B.G. Teubner Verlagsgesellschaft, 1981.
Rummenhöller, Peter. "Die philosophischen Grundlagen in der Musiktheorie des 19. Jahrhunderts." *Beiträge zur Theorie der Kunste im 19. Jahrhundert 1* (1971):44–57.
Ruska, Julius. "Helmholtz-Dokumente." *Süddeutsche Monatshefte 9:1* (1911–12):683–87.

Russell, Bertrand. *An Essay on the Foundations of Geometry*. Cambridge: Cambridge University Press, 1897.
S. A. ["Hermann Ludwig Ferdinand von Helmholtz."] *Memoirs and Proceedings of the Manchester Literary & Philosophical Society* 4th ser. 39 (1894):230–32.
Scalinci, Noè. "Il meccanismo accomodativo dell'Helmholtz (1855) è sostanzialmente quello esposto dal Morgagni (1719)." *Bollettino dell'Istituto Storico Italiano dell'Arte Sanitaria 10* (1930):43–51.
Schadewaldt, Hans. "Der Durchbruch der naturwissenschaftlichen Methode in der Medizin." *Medizinische Welt 26:9* (1975):407–12.
Schaefer, Hans. "Hundert Jahre Physiologie in Heidelberg." *Ruperto-Carola. Mitteilungen der Vereinigung der Freunde der Studentenschaft der Universität Hamburg e.V. 24* (1958):140–46.
Scharlau, Winfried, ed. *Rudolf Lipschitz: Briefwechsel mit Cantor, Dedekind, Helmholtz, Kronecker, Weierstrass und anderen.* Braunschweig and Wiesbaden: Friedr. Vieweg & Sohn, 1986.
Scheel, Heinrich, ed. *Gedanken von Helmholtz über schöpferische Impulse und über das Zusammenwirken verschiedener Wissenschaftszweige.* (= *Sitzungsberichte des Plenums und der Klassen der Akademie der Wissenschaften der DDR* Nr. 1, 1972). Berlin: Akademie-Verlag, 1972.
Scheel, Karl. In "Festsitzung zur Feier des 90jährigen Bestehens der physikalischen Gesellschaft zu Berlin am 25. Januar 1935 im großen Hörsaal des Physikalischen Instituts der Universität." *Verhandlungen der Deutschen Physikalischen Gesellschaft 16* (1935):1–17, on 2–11.
Schlick, Moritz. "Helmholtz als Erkenntnistheoretiker." In Warburg, et al., *Helmholtz als Physiker* (1922). Pp. 29–40.
———. *Philosophical Papers.* Henk L. Mulder and Barbara F.B. van de Velde-Schlick, eds. 2 vols. Vol. 1: *(1909–1922).* Trans. Peter Heath. vol. 2: *1925–1936.* Trans. Peter Heath, Wilfrid Sellars, Herbert Feigl, and May Brodbeck. Dordrecht, Boston, London: D. Reidel, 1979.
Schlick, M., and P. Hertz. "Erläuterungen" to Helmholtz's *Schriften zur Erkenntnistheorie.* Translated and reprinted in Cohen and Elkana, eds., *Hermann von Helmholtz. Epistemological Writings.* Pp. xxxiii–xxxvii.
Schlotte, Felix. "Beiträge zum Lebensbild Wilhelm Wundts aus seinem Briefwechsel." *Wissenschaftliche Zeitschrift der Karl- Marx-Universität Leipzig, Gesellschafts- und Sprachwissenschaftliche Reihe 5* Heft 4 (1955/56):333–49.
Schnädelbach, Herbert. *Philosophy in Germany 1831–1933.* Trans. Eric Matthews. Cambridge, London, New York: Cambridge University Press, 1984.
Schober, Herbert A.W. "Hermann von Helmholtz." *Journal of the American Optometric Association 40:5* (1969):518–21.
Schoen, Max. "Helmholtz. Zum hundertjährigen Geburtstag des großen Naturforschers." *Die Neue Zeit. Wochenschrift der Deutschen Sozialdemokratie 39* (1921):505–10.
Scholz, Erhard. *Geschichte des Mannigfaltigkeitsbegriffs von Riemann bis Poincaré.* Boston, Basel, Stuttgart: Birkhäuser, 1980.

Schreier, W. "Über einige Beziehungen zwischen den Arbeiten von Helmholtz und Maxwells Aufbau der elektromagnetischen Feldtheorie." *WZHUB 22:3* (1973):345–47.
Schröder, Hermann. "Helmholtz. (Zu seinem hundertsten Geburtstage am 31. August.)." *Münchener Medizinische Wochenschrift 68* (1921):1086–87.
Schüle, Prof. "Zwei Briefe aus dem Leben von Hermann v. Helmholtz." *Psychiatrisch-Neurologische Wochenschrift 35:51* (1933):618–19.
Schuster, Arthur. "Biographical Byways." *Nature 115* (1925):342–43.
Schwertschlager, Joseph. *Kant und Helmholtz erkenntniss-theoretisch verglichen.* Inaugural diss., University of Würzburg, 1883. Freiburg i.B.: Herder'sche Verlagshandlung, 1883.
Sepper, Dennis L. *Goethe contra Newton: Polemics and the Project for a New Science of Color.* Cambridge, New York, New Rochelle: Cambridge University Press, 1988.
Shakow, David, and David Rapaport. *The Influence of Freud on American Psychology.* New York: International Universities Press, 1964.
Shankland, R.S. "Michelson-Morley Experiment." *American Journal of Physics 32* (1964):16–35.
Shapiro, S.L. "The Universal Scientist." *The Eye, Ear, Nose and Throat Monthly 47* (1968):524–28.
Sherman, Paul D. *Colour Vision in the Nineteenth Century: The Young-Helmholtz-Maxwell Theory.* Bristol: Adam Hilger, 1981.
Siemens, Werner. *Aus einem reichen Leben: Werner von Siemens in Briefen an seine Familie und an Freunde.* Friedrich Heintzenberg, ed. 2nd ed. Stuttgart: Deutsche Verlags-Anstalt, 1953.
Siemens-Helmholtz, Ellen von. "Mein Elternhaus." "Helmholtz zum 100. Geburtstag. Erinnerungsblatt der Vossischen Zeitung." *Vossische Zeitung.* No. 404. Beilage 4. 28 August 1921.
———., ed. *Anna von Helmholtz. Ein Lebensbild in Briefen.* 2 vols. Berlin: Verlag für Kulturpolitik, 1929.
Sigerist, Henry E. "Hermann Ludwig Ferdinand von Helmholtz. 1821–1894." In idem, *The Great Doctors: A Biographical History of Medicine.* Trans. Eden and Cedar Paul. New York: Dover, 1971. Pp. 322–29.
Silliman, Robert H. "William Thomson: Smoke Rings and Nineteenth-century Atomism." *Isis 54* (1963):461–74.
Sodi-Pallares, Demetrio. "The Influence of the Investigations of Helmholtz on Electrocardiography." *American Heart Journal 51* (1956):647–53.
Spender, Natasha, and Rosamund Shuter-Dyson. "Psychology of Music." Stanley Sadie, ed. *The New Grove Dictionary of Music and Musicians.* 20 vols. London, Washington, D.C., and Hong Kong: Macmillan, 1980. *15:*388–427.
Stallo, J.B. *The Concepts and Theories of Modern Physics.* Percy W. Bridgman, ed. Cambridge, Mass.: Harvard University Press, 1960.
Stein, Howard. "Some Philosophical Prehistory of General Relativity." John Earman, Clark Glymour, and John Stachel, eds. *Foundations of Space-Time Theories.* (= Minnesota Studies in the Philosophy of Science, vol. 8). Minneapolis, Minn.: University of Minnesota Press, 1977. Pp. 3–49.

Sternberg, Maximilian. "Helmholtz und die Methodik der medicinischen Forschung." *Wiener Medizinische Presse 35* (1894):1729-32, 1765-67.
Stiller, Heinz. "Zur Bedeutung der Arbeiten von H. v. Helmholtz für die geophysikalische Hydrodynamik und für die Physik des Erdinnern." In Scheel, ed., *Gedanken von Helmholtz über schöpferische Impulse* (1972):45-8.
Strohl, E. Lee, Willis G. Diffenbaugh, and Robert W. Jamieson. "Herman von Helmholtz, 1821-1894: Physician, Musician, and Versatile Scientist." *Illinois Medical Journal 144* (1973):240-41.
Stromberg, Wayne H. "Helmholtz and Zoellner: Nineteenth-century Empiricism, Spiritism, and the Theory of Space Perception." *JHBS 25* (1989):371-83.
Strutt, John William [Lord Rayleigh]. *The Theory of Sound*. 2nd rev. ed. 2 vols. Reprinted New York: Dover, 1945.
Stumpf, Carl. "Hermann von Helmholtz and the New Psychology." *The Psychological Review 2:1* (1895):1-12; German trans.: "Hermann von Helmholtz und die neuere Psychologie." *Archiv für Philosophie. I. Abtheilung: Archiv für Geschichte der Philosophie 8:3* (1895):303-14.
———. "Sinnespsychologie und Musikwissenschaft. Helmholtz' Grundlegungen." "Helmholtz zum 100. Geburtstag. Erinnerungsblatt der Vossischen Zeitung." *Vossische Zeitung*. No. 404. Beilage 4. 28 August 1921.
Sully, James. "The Question of Visual Perception in Germany." *Mind. A Quarterly Review of Psychology and Philosophy 3* (1878):1-23, 167-95.
Sutrin, V.D. "Hermann Helmholtz and Ophthalmology." [In Russian.] *Oftal'mologicheskii Zhurnal [Odessa] 30:5* (1975):390-91.
Tait, P[eter]. G[uthrie]. *Sketch of Thermodynamics*. 2nd rev. and enl. ed. Edinburgh: David Douglas, 1877.
Taylor, Sedley. *Sound and Music: A Non-Mathematical Treatise on the Physical Constitution of Musical Sounds and Harmony, including the Chief Acoustical Discoveries of Professor Helmholtz*. London: Macmillan, 1873.
Terhardt, E. "Ein psychoakustisch begründetes Konzept der Musikalischen Konsonanz." *Acustica 36* (1976/77):121-37.
Thiele, Joachim. "Aus der Korrespondenz Ernst Machs: Briefe deutscher und englischer Naturwissenschaftler." *NTM 7:1* (1970):66-75.
———. *Wissenschaftliche Kommunikation: Die Korrespondenz Ernst Machs*. Kastellaun: A. Henn Verlag, 1978.
Thompson, Silvanus P. *The Life of William Thomson Baron Kelvin of Largs*. 2 vols. London: Macmillan, 1910.
Thomson, William. [Lord Kelvin]. ["President's Address."] *Proceedings of the Royal Society of London 57* (1895):37-41.
Thouret, Georg. "Nomen atque omen. Unser Name und seine Vorbedeutung." *Helmholtz-Realgymnasium in Schöneberg. Jahresbericht I. Ostern 1903*. Schöneberg: Emil Hartmann, 1903. Pp. 3-15.
Tingwaldt, C.P. "Über das Helmholtzsche Reziprozitätsgesetz in der Optik." *Optik 9* (1952):248-53.
Tobias, Wilhelm. *Grenzen der Philosophie, constatirt gegen Riemann und Helmholtz, vertheidigt gegen von Hartmann und Lasker*. Berlin: G.W.F. Müller, 1875.

Torretti, Roberto. *Philosophy of Geometry from Riemann to Poincaré.* Dordrecht, Boston, London: D. Reidel, 1978.

Treder, Hans-Jürgen. "Helmholtzsche und relativistische Elektrodynamik," *Beiträge zur Geophysik 79* (1970):401–20.

———. "Die Bedeutung von Helmholtz für die theoretische Physik." In Scheel, ed., *Gedanken von Helmholtz über schöpferische Impulse* (1972):37–43.

———. "Helmholtz' Elektrodynamik und die Beziehungen von Kinematik, Dynamik und Feldtheorie." *WZHUB 22:3* (1973):327–30.

———. "Die Einheit der Physik bei Helmholtz, Planck und Einstein." In idem, ed. *Einstein-Centenarium 1979: Ansprachen und Vorträge auf den Festveranstaltungen des Einstein-Komitees der DDR bei der Akademie der Wissenschaften der DDR vom 28.2. bis 2.3. 1979 in Berlin.* Berlin: Akademie-Verlag, 1979. Pp. 240–55.

———. "Descartes' Physik der Hypothesen, Newtons Physik der Prinzipien und Leibnizens Physik der Prinzipe." *Studia Leibnitiana 14:2* (1982):278–86.

Troland, Leonard Thompson. "Helmholtz's Contributions to Physiological Optics." *Journal of the Optical Society of America 6* (1922):327–35.

Truesdell, C. *The Tragicomical History of Thermodynamics, 1822–1854.* New York, Heidelberg, Berlin: Springer, 1980.

Tsuneishi, Kei-ichi. "On the Abbe Theory (1873)." *Japanese Studies in the History of Science 12* (1973):79–91.

Tuchman, Arleen. *Science, Medicine, and the State in Germany: The Case of Baden, 1815–1871* (Oxford and New York: Oxford University Press, 1993).

Turner, R. Steven. "Helmholtz, Hermann von." *DSB* 6:241–53.

———. "The Ohm-Seebeck Dispute, Hermann von Helmholtz, and the Origins of Physiological Acoustics." *BJHS 10* (1977):1–24.

———. "Hermann von Helmholtz and the Empiricist Vision." *JHBS 13* (1977):48–58.

———. "Helmholtz, Sensory Physiology, and the Disciplinary Development of German Psychology." William R. Woodward and Mitchell G. Ash, eds. *The Problematic Science: Psychology in Nineteenth-Century Thought.* New York: Praeger, 1982. Pp. 147–66.

———. "Paradigms and Productivity: The Case of Physiological Optics, 1840–94." *Social Studies of Science 17* (1987):35–68.

———. "Fechner, Helmholtz, and Hering on the Interpretation of Simultaneous Contrast." Josef Brožek and Horst Gundlach, eds. *G.T. Fechner and Psychology.* Passau: Passavia Universitätsverlag, 1988. Pp. 137–50.

———. "Vision Studies in Germany: Helmholtz vs. Hering." Gerald Geison and Frederic L. Holmes, eds. *Research Schools. Osiris 8.* Forthcoming.

Uhlig, Gustav. "Das Abiturientenzeugnis von H. Helmholtz." *Das humanistische Gymnasium 23* (1912):40–1.

Uthoff, Prof. "Bemerkungen zur Erfindung des Augenspiegels vor 50 Jahren." A. Wagenmann, et al., eds. *Bericht über die neunundzwanzigste Versammlung der ophthalmologischen Gesellschaft. Heidelberg 1901.* Wiesbaden: J.F. Bergmann, 1902. Pp. 3–8.

Ullmann, Dieter. "Helmholtz-Koenig-Waetzmann und die Natur der Kombinationstöne." *Centaurus 29* (1986):40–52.

———. "Ohm-Seebeck-Helmholtz und das Klangfarbenproblem." *NTM 25* (1988):65–8.

Unna, Paul Gerson. *Helmholtz und unsere heutige Weltanschauung. Zwei Vorträge gehalten am 27. März und am 10. April 1908 auf Veranlaßung des Deutschen Monistenbunds.* ... (= Flugschriften 3 und 4 der Ortsgruppe Hamburg e.V., Deutscher Monistenbund). Hamburg: Verlag Deutscher Monistenbund, 1908.

Van Rhyn, G.A.F. "Sketch of Professor Helmholtz." *The Popular Science Monthly 5* (1874):231–34.

Vogelsang, K. "100 Jahre Helmholtzsche Akkomodationstheorie." *Klinische Monatsblätter für Augenheilkunde 126* (1955):762–65.

Voit, C. von. "Nekrolog auf Hermann von Helmholtz." *Sitzungsberichte der mathematisch-physikalischen Classe der k. b. Akademie der Wissenschaften zu München 25* (1895):185–96.

Voločková, Jaroslava. "130 let Helmholtzova oftalmoskopu." [In Czech.] *Československá Oftalmolgie* [Prague] *38:1* (1982):51–3.

Wachsmuth, R[ichard]. "von Helmholtz, Anna, geb. von Mohl." Anton Bettelheim, ed. *Biographisches Jahrbuch und Deutscher Katalog.* 18 vols. Berlin: Georg Reimer, 1897–1917. *4*:14–20.

Wagner, Kurt. "Zur richtigen Deutung der Helmholtzschen Zeichentheorie." *Deutsche Zeitschrift für Philosophie 13* (1965):162–72.

Wahsner, R. "Das Helmholtz-Problem oder Zur physikalischen Bedeutung der Helmholtzschen Kant-Kritik." Akademie der Wissenschaften der DDR. Einstein-Laboratorium für Theoretische Physik. Potsdam. Caputh. PRE-EL 88-06. July 1988.

Walker, Gilbert T. "Helmholtz or Kelvin Cloud Waves." *Nature 129* (1932):205.

Warburg, Emil. "Helmholtz als Physiker." In Warburg, et al., *Helmholtz als Physiker* (1922). Pp. 3–14.

———. "Zur Geschichte der Physikalischen Gesellschaft." *Die Naturwissenschaften 13* (1925):35–9.

———, M. Rubner and M. Schlick. *Helmholtz als Physiker, Physiologe und Philosoph. Drei Vorträge gehalten zur Feier seines 100. Geburtstags im Auftrage der Physikalischen, der Physiologischen und der Philosophischen Gesellschaft zu Berlin.* Karlsruhe: C.F. Müllersche Hofbuchhandlung, 1922.

Warren, Richard M. "Helmholtz and His Continuing Influence." *Music Perception 1:3* (1984):253–75.

———, and Roslyn P. Warren, eds. *Helmholtz on Perception: Its Physiology and Development.* New York, London, Sydney: John Wiley & Sons, 1968.

Wassermann, Michael. *Otto Lilienthal.* Leipzig: BSB B.G. Teubner Verlagsgesellschaft, 1985.

Weale, R.A. "New Light on Old Eyes." *Nature 198* (1963):344–46.

Weech, Friedrich von. "Anna von Helmholtz." Friedrich von Weech and U. Krieger, eds. *Badische Biographien.* Heidelberg: Carl Winter, 1906. *5*:294–301.

Wenger, R. "Helmholtz als Meteorologe." *Die Naturwissenschaften 10* (1922):198–202.
Westheimer, G. "Helmholtz on Eye Movements." *Human Neurobiology 3* (1984):149–52.
Wetzels, Walter D. "Versuch einer Beschreibung populärwissenschaftlicher Prosa in den Naturwissenschaften." *Jahrbuch für Internationale Germanistik 3:1* (1971):76–95.
Wever, Ernest Glen, and Merle Lawrence. *Physiological Acoustics.* Princeton, N.J.: Princeton University Press, 1954.
Weyl, Hermann. *Philosophie der Mathematik und Naturwissenschaft.* Munich and Berlin: R. Oldenbourg, 1926. Rev. English ed.: *Philosophy of Mathematics and Natural Science.* Princeton: Princeton University Press, 1949.
Whittaker, Edmund. *A History of the Theories of Aether and Electricity.* Rev., enl. ed. 2 vols. Vol. 1: *The Classical Theories.* Vol. 2: *The Modern Theories, 1900–1926.* London, Edinburgh, Paris: Thomas Nelson, 1951–53.
Wiedemann, Eilhard. "Hermann von Helmholtz." *Sitzungsberichte der physikalisch-medicinischen Societät zu Erlangen 25* (1893):54–67.
Wiedemann, Gustav. "Hermann von Helmholtz. Wissenschaftliche Arbeiten." *AP 290* (1895):i–xxiv.; reprinted in *WA 3*:xi–xxxvi.
Wien, Wilhelm. "Hermann von Helmholtz zu seinem 25-jährigen Todestage." *Die Naturwissenschaften 7*:36 (1919):645–48; reprinted in idem, *Aus der Welt der Wissenschaft: Vorträge und Aufsätze.* Leipzig: J.A. Barth, 1921. Pp. 86–94.

———. "Helmholtz als Physiker." *Die Naturwissenschaften 9:35* (1921):694–99.

———. *Aus dem Leben und Wirken eines Physikers. . . .* Leipzig: Barth, 1930.
Wilkins, Robert H. "Neurosurgical Class—XXXVIII." *Journal of Neurosurgery 23* (1965):241–43. [Followed by an English translation of Helmholtz's *Beschreibung eines Augenspiegels* (1851) by Thomas Hall Shastid. Pp. 243–61.]
Williams, H.W. "Hermann Ludwig Ferdinand von Helmholtz." *Proceedings of the American Academy of Arts and Sciences* n.s. *22* (1895):592–96.
Winters, Stephen M. "Hermann von Helmholtz's Discovery of Force Conservation." Ph.D. diss., The Johns Hopkins University, 1985.
Wise, M. Norton. "German Concepts of Force, Energy, and the Electromagnetic Ether: 1845–1880." G.N. Cantor and M.J.S. Hodge, eds. *Conceptions of Ether: Studies in the History of Ether Theories 1740–1900.* Cambridge, London, New York: Cambridge University Press, 1981. Pp. 269–307.

———. "On the Relations of Physical Science to History in Late Nineteenth-Century Germany." Loren Graham, Wolf Lepenies, and Peter Weingart, eds. *Functions and Uses of Disciplinary Histories.* Dordrecht, Boston, and Lancaster: Reidel, 1983. Pp. 3–34.
Wittich, Dieter. "Das Gesetz der sogenannten spezifischen Sinnesenergie und die beiden philosophischen Grundrichtungen." *Deutsche Zeitschrift für Philosophie 12* (1964):682–91.
Wöhlisch, Edgar. "Adolf Fick und die heutige Physiologie." *Die Naturwissenschaften 26* (1938):585–91.
Wollgast, Siegfried. "H. v. Helmholtz in der Sicht M. Schlicks." *Rostocker Philosophische Manuskripte.* Heft 8, Teil 2 (1970):51–63.

———. "Naturwissenschaftlicher Materialismus und Pantheismus bei Hermann von Helmholtz." *WZHUB 22:3* (1973):289-92.
Wood, Casey A. "Hermann von Helmholtz—The Inventor of the Ophthalmoscope." *The Journal of the American Medical Association 38* (1902):552-57.
Woodruff, A.E. "Action at a Distance in Nineteenth Century Electrodynamics." *Isis 53* (1962):439-59.
———. "The Contributions of Hermann von Helmholtz to Electrodynamics." *Isis 59* (1968):300-11.
Woodward, William R. "From Association to Gestalt: The Fate of Hermann Lotze's Theory of Spatial Perception, 1846-1920." *Isis 69* (1978):572-82.
Wucnsch, Gerhard. "Hugo Riemann's Musical Theory." *Studies in Music 2* (1977):108-24.
Wundt, Wilhelm. [Review of H. Helmholtz, Professor der Physiologie zu Heidelberg, *Handbuch der physiologischen Optik.*] *Deutsche Klinik 19* (1867):326-28.
———. *Grundzüge der physiologischen Psychologie.* 5th rev. ed. Leipzig: Wilhelm Engelmann, 1902-3.
———. *Erlebtes und Erkanntes.* 2nd ed. Stuttgart: Alfred Kröner, 1921.
Wussing, Hans. "Der philosophische Kampf um den Energiesatz." *Naturwissenschaft, Tradition, Fortschritt,* Beiheft to *NTM.* Berlin: VEB Deutscher Verlag der Wissenschaften, 1963. Pp. 99-111.
Yamaguchi, Chûhei. "Helmholtz's Electrochemical Researches and the Concept of 'Free Energy'." [In Japanese.] *Kagaku shi Kenkyu 21* (1982):1-9.
———. "On the Formation of Helmholtz' View of Life Processes in His Studies of Fermentation and Muscle Action—in Relation to His Discovery of the Law of Conservation of Energy." *Historia Scientiarum 25* (1983):29-37.
———. "A Study of Hermann von Helmholtz's Earliest Physiological Research Leading to the Discovery of the Law of Conservation of Energy." [In Japanese.] *Kagaku shi Kenkyu 23* (1984):169-76.
Zenkl, Luděk. "Helmholtz's Interpretation of Consonance and Czech Music Theory." [In Czech.] *Hudební Věda 12:3* (1975):328-50.
Zenneck, J. "Zum 50. Todestag von Hermann von Helmholtz. 31. August 1821 - 8. September 1894." *Stahl und Eisen 64* (1944):581-84.
Ziegenfuss, Werner, ed. "Helmholtz, Hermann von." *Philosophen-Lexikon: Handwörterbuch der Philosophie nach Personen.* 2 vols. Berlin: Walter de Gruyter, 1949. *1*:498-502.
Zöllner, Johann Carl Friedrich. *Über die Natur der Cometen. Beiträge zur Geschichte und Theorie der Erkenntnis.* 2nd ed. Leipzig: Wilhelm Engelmann, 1872.
———. *Wissenschaftliche Abhandlungen.* 4 vols. Leipzig: L. Staackmann, 1878-81.

Index

Note: Unless otherwise indicated, all references are to Hermann von Helmholtz.

Academic freedom. *See* Helmholtz, Hermann von; Universities
Académie des sciences, Paris, 91
Academy of Sciences, Hungarian, 432
Accommodation, theory of. *See* Optics, physiological
Acoustics: and aesthetics, 522; experiments, 271; and musical science, 522; pre-Helmholtzian, 259, 261–66; as research field, 261–62; and sound transmission in telephone, 576. *See also* Acoustics, physiological
Acoustics, physiological, 5; aerial vibrations, 259, 281; consonance, 277–80, 582; consonance, beat theory of, 260; dissonance, theory of, 269; graphical analysis, use of, 278n; hearing, resonance theory of, 267–68, 269, 276, 278–81, 526–27; instruments in, 260; mathematical analysis in, 260, 271, 272, 273, 278, 281; and Müller's law of specific sense energies, 118; and music, 267, 581–83, 586; and perception theory, 260; pipe theory, 269, 281; reformation of, 260, 267–70, 278, 281, 284; research program in, 267–68, 281; resonance in, 281; resonators, and upper partials, 526, 529; and sensory processes (tripartite nature of), 282–85; siren and acoustical research, 262–63; siren and Helmholtz, 262–63, 267; siren and tone definition, 263; siren as sound generator, 262; siren experiments, 263; sound (defined), 286; sound perception, role of psychology in, 286, 287; spherical resonators, 260, 275–76, 277, 281; synthesis of, 9; theories of, 260; tuning fork, 260, 268–69; tuning fork as sound generator, 263; tuning-fork experiments, 274–76, 277, 281; vowels, theory of, 260, 274–77, 281. *See also* Acoustics; Epistemology; Illusions; *Lehre von den Tonempfindungen*; Music; Tone(s)
Adrian, Edgar, 206
Aesthetics: acquaintance with theories of, 523–24; aesthetic induction, defined, 545; artistic induction, and unconscious inference, 547; artistic induction, defined, 545–46; artistic styles, cognitive status of, 540; "Classical," 554n; classicist, 524, 553–58; cognitive status of, 523; comprehensibility, as classical aesthetic, 558; differences with Romantic, 554; as *Geisteswissenschaft*, 542–43; Hegelian, 524, 551n; as historically conditioned, 523, 535, 542, 543, 582; imagination, and the lawful,

637

Aesthetics *(continued)*
558; imagination, and understanding, 524, 558; inductive inference, and laws of science, 549; Kantian, 543n, 551; "laws" of, 535; limited theory of, 522, 533–34, 551, 553, 556, 558, 586; methodology of, 524–25, 535, 543; and natural science, 523; neo-Classicist vs. Romantic, 524; pleasure, artistic, 583; post-Kantian, 551; and reason in art, 586; Romantic, 553–54; scientific status of, 523; sources of ideas on, 579; thought vs. feeling, 524; truth in, 551; understanding and imagination, 547; understanding vs. imagination, 524. *See also* Aesthetics, musical; Art; Artists; Beauty; Ideal type; Inference, unconscious; Intuition; Intuition, aesthetic; Judgments, unconscious; Laws, scientific; Music; Painting; Science; Scientists; Tone(s); Truth; Truth, universal

Aesthetics, musical: as historically conditioned, 584; limited scientific explanation of, 584–85; pleasure, 583. *See also* Aesthetics; Art; Music; Tone(s)

Affinity. *See* Thermochemistry

Afterimages. *See* Ludwig; Müller; Optics, physiological; Scherffer; Volkmann

Airy, George: criticizes Brewster, 211, 214; and Purkyně effect, 214

Akademie der Künste, Berlin: and Brücke, 81; and Helmholtz, 17, 28, 29, 74, 81, 579

Akademie der Wissenschaften, Bayrische, Munich, 274–75

Akademie der Wissenschaften, Berlin, 219, 370; and Helmholtz, 89–91, 120, 267; and Planck, 429–30; prize problem of, 352

Ampère, André-Marie: influence on Helmholtz, 312, 330; law of, 345, 347, 349, 381, 382, 391

Anatomy of the ear, 267, 273, 279–80, 526, 582

Animal heat. *See* Becquerel; Breschet; Despretz; Dulong; Force, conservation of; Liebig; Physiology

Annalen der Physik und Chemie, 262; and Helmholtz, 267, 304

Anti-Semitism, rise of, 596. *See also* Zöllner

Anti-socialist laws, 596

Archiv für die Holländische Beiträge zur Natur- und Heilkunde, 132

Archiv für Ophthalmologie, 33, 35–6, 128, 132, 137, 149; and Helmholtz, 128, 139

Archiv für physiologische Heilkunde, 37

Archiv für rationelle Medicin, 132

Arithmetic: counting, 518–19; and empiricism, 500, 517–19; and psychological process of counting, 517–18, 519; as science of time, 517. *See also* Geometry; Mathematics

Armstrong, Henry: on Helmholtz as chemist, 424; hostility to theoretical chemistry, 426

Arrhenius, Svante, 427; and atomic nature of electricity, 429; dissociation theory of, 428–29; and Helmholtz, 406; on Helmholtz, 428; Helmholtz's influence on, 429; and theoretical chemistry, 405

Art: aims and purposes of, 539–40, 551, 552–53, 558; cognitive processes of, 523–24, 540, 545–47, 552, 557; cognitive status of, 540, 552; general formulae for, 537; and lawlike behavior, 578–79, 581; laws behind, 583–84; limited scientific explanations of, 523, 525, 533, 534–35, 536, 542, 578; and mental well-being, 581; as morality, 584, 586; non-representational nature of, 539, 556; and science, cognitive identity of, 558; and science, unity of, 540; as truth, 578. *See also* Aesthetics; Aesthetics, musical; Science; Truth; Truth, universal

Artists: sensibility and imagination in, 552; similarities to scientists, 522–23, 540, 585; thought processes in, 523, 525, 545–47, 551, 552, 553n. *See also* Aesthetics; Scientists

Associationism: British, 198; and Helmholtz, 197; and ideal type, 556–57; and local signs, 550. *See also* Associationism, laws of; Psychology

Associationism, laws of: and artistic imagination, 552; as laws of psychology, 550–51. *See also* Associationism; Psychology

Association of German Mathematicians, 361

Atkinson, E., 588n

Atomism, 335–36, 442, 465, 473; in chemistry, 415; critique of, 479; and dynamicism, 473; and German electrodynamics, 364–65; and German physicists, 335; and Helmholtz, 324–25, 442. *See also* Atoms

Atoms: forces of and heat, 324–25; and Helmholtz, 442. *See also* Atomism

Austrian War, 596

Avogadro, Amadeo: law of, 427

Bacon, Francis, 575-76
Barycentric curve (of the eye). *See* Maxwell; Optics, physiological
Barycentric diagram. *See* Newton
Beauty: basis of artistic, 539; expression of artistic, 556; and lawful truth, 553; nature of artistic, 583; perception of, 553; as reason, 583; theory of, 556-57. *See also* Aesthetics; Truth; Truth, universal
Becker, Carl, 600
Becquerel, Antoine-César: animal heat, 61; concentration cells, 412; influence on Helmholtz, 412; measurements of, 69, 71. *See also* Breschet
Beethoven, Ludwig van, 581n; preference for, 554
Beetz, Wilhelm, 297
Bell, Charles, 38
Ben-David, Joseph: on German university system, 162
"Bericht" (of 1846), 297-304, 311, 322; and Hess's law, 325; and mechanical theory of heat, 324. *See also* Force, conservation of
Berkeley, Bishop George, 157
Berlin: as center of medicine, 21
Berlin, University of: address of 1878, 484; appointment to, 410; and Helmholtz, 479; lectures at, 423; physics chair, 363; professor of physics at, 28, 47, 546-47; rectoral address at, 372; students at, 257; study at, 22. *See also* Friedrich-Wilhelms-Institut; Helmholtz, Hermann von
Bernard, Félix: and Brewster, 216; criticizes Brewster, 211, 214, 219
Bernoulli, Daniel: heat theory of, 434-35
Berthollet, Claude Louis: on heat, 323-24
Berthelot, Marcellin: and chemical theory, 405; thermal theory of affinity, 425; thermochemistry of, 404
Berzelius, Jöns: influence on Helmholtz, 55-6, 57; and Mitscherlich, 55; muscle analysis, 55-6, 57, 59
Besitzbürgertum, 597. *See also* Science, popular
Bessel, Friedrich Wilhelm: method of least squares, 97; personal equation, 94
Betti, Enrico: and relativity theory, 402
Bevilacqua, Fabio, 8
Bierhalter, Günter, 7, 11
Bildung: defined, 562; and social class, 597. *See also* *Geisteswissenschaften*; Science

Bildungsbürgertum, 596-97; vs. *Besitzbürgertum*, 597. *See also* Science; Science, popular
Binocular fusion. *See* Optics, physiological; Panum; Volkmann
Binocularity. *See* Optics, physiological; Stereoscope; Wheatstone
Binocular vision. *See* Dove; Hering; Optics, physiological; Perception, spatial; Wheatstone
Biot, Jean-Baptiste: law of, 380
Bismarck, Otto von: attacks social democracy, 589; and *Kulturkampf*, 589
Blindness. *See* Optics, physiological; Perception, spatial
Böhm, Ludwig: and du Bois-Reymond, 132; and Helmholtz, 132
Boltzmann, Ludwig, 3, 381, 400-1, 402; and atomism, 335-36; in Berlin, 360-61, 395; criticizes Helmholtz, 11, 433, 446, 447, 448, 450; on Darwin, 361; and entropy, 421; gas statistics of, 362; on goals of physics, 361-62; on the Hall effect, 343; and Helmholtz, 3, 374, 395; and Helmholtzianism, 343, 360-62; Helmholtz's influence on, 456; and Hertzian physics, 361; influence on Helmholtz, 433, 444-46, 447; kinetic energy and temperature, 444-45, 449-50; Maxwell's distribution law, 448; methods of theoretical physics, 360-62; models in physics, 449-50; tests electrodynamic theories, 375-76; tests Maxwell's theory, 395-96. *See also* Thermodynamics, second law of
Bonn, University of: anatomical institute, 40; career at, 40-1; and Helmholtz, 163, 247, 267, 268, 365, 593; support for anatomy and physiology at, 40-1. *See also* Helmholtz, Hermann von
Bonnet, Charles, 225; theory of nerve-excitability, 224
Boscovich, Roger, 224
Botanischer Garten, Berlin, 21
Bowman, William, 127
Breschet, Gilbert: animal heat, measurements of, 69, 71. *See also* Becquerel
Brewster, David: absorption of solar spectrum, 208-10; challenges Newton on color theory, 207, 209; color mixing, 227-28; criticizes spectrum theory, 211-14; Helmholtz praises, 216; Helmholtz refutes spectrum theory of, 215-20, 231-32, 237, 239, 284; and

Brewster, David *(continued)*
Herschel, 210; influence on Helmholtz, 208–9, 215, 228; Keith Prize, 211; as projection theorist, 158–59, 161; responds to critics, 214–15; response to stereoscope, 161; theory of white light, 225; triple-spectrum theory of, 210–12, 216, 220; and Young's theory, 211
British Association for the Advancement of Science, 220, 426
Brücke, Ernst, 24, 25, 28, 29, 30, 74, 77, 78; and empiricist theory of vision, 153; on eye movements, 161; and Helmholtz, 215, 219, 244–45, 253, 254, 297; influence on Helmholtz, 206–7, 219, 220, 231, 251; leaves Königsberg, 81; and Ludwig's *Lehrbuch*, 129–30; and Müller, 161; simultaneous contrast experiments, 219
Buchwald, Jed Z., 7, 8, 10, 11, 389, 390, 391
Budge, Julius, 40
Buffon, Georges-Louis Leclerc, Comte de: on subjective colors, 243
Busch, Alexander, 83

Cahan, David, 12–3
Cambridge University: admiration for, 588; criticizes, 588; honorary degree from, 415
Carnot, Lazare: and theoretical engineering, 293
Carnot, Sadi: correlation principle, 301; force-equivalent of heat, 323, influence on Helmholtz, 309, 323, 412
Cauchy, Augustin-Louis, 262
Causality. *See* Philosophy; Science, philosophy of
Causes. *See* Philosophy; Science, philosophy of
Challis, James: dispute with Stokes, 220
Charité (hospital), Berlin, 21; intern at, 26–7
Chemical Society, London, 414–15, 416; Helmholtz memorial meeting, 426–27
Chemical thermodynamics. *See* Thermodynamics, chemical
Chemistry: and atomic nature of electricity, 406, 410, 414–17; atoms and molecules in, 404–5; attitude toward, 407; British, and chemical thermodynamics, 426; British, hostility to theory, 426–27; education in, 406–7; electrochemistry, 403–4, 406, 408–17; electrolysis, 409–10, 410–17; French, influence on Helmholtz, 425, 426; galvanic polarization, 406, 409, 410–11; in Helmholtz's career, 430; influence on, 406, 424–28, 430–31; influence on American, 426; influence on German, 426; "New era in," 424–31; opinion of, 429; and physicists, 404, 405; positivistic nature of, 404–5; pre-1880 state of, 403; reaction rates, 403; reputation in, 424; rise of organic, 403; standing in, 405–6, 414, 424; as student of, 406–7; traditional, 430; transformation of, 404, 405–6; understanding of, 430. *See also* Chemistry, physical; Chemistry, physiological; Thermodynamics, chemical
Chemistry, physical: at Berlin University, 428; Boyle-Mariotte law, and water pressure, 413; chemical equilibrium, 403–4; dissociation theory and Helmholtz, 406, 429; energetics, 406; free energy, 406, 418–21, 423; free energy as analogue to potential energy, 420; free energy in France, 425; Gibbs-Helmholtz equation, 420–22; Helmholtz, 10; influence on, 413–14; limited role in, 427–28; osmotic pressure theory and Helmholtz, 431; reservations about, 427–29, 430–31; rise of, 427; solution theory, 427–28; solution theory of Helmholtz, 427, 431; support of, 406, 428. *See also* Chemistry; Thermodynamics, chemical
Chemistry, physiological, 406, 407–8; and animal heat, 407; interest in, 407. *See also* Chemistry; Physiology
Chevreul, Michel Eugène: color contrast, 536; and Helmholtz, 253; influence on color theory, 250; influence on Helmholtz, 536; and pigment mixing, 221–22; on subjective colors, 254
Chladni, Ernst Florens Friedrich: acoustics, 261–62; *Akustik*, 261; influence on Helmholtz, 285; influence on Weber brothers, 262; view of acoustics, 264
Civilizing power of science. *See* Science
Clapeyron, Benoit-Pierre-Emile: and Carnot's theory, 409; correlation principle, 301; force-equivalent of heat, 323; Helmholtz on, 409; influence on Helmholtz, 309, 321, 323, 325–26; law of, 321, 325–26; results compared with Holtzmann's, 326
Classen, August: as empiricist, 191

Clausius, Rudolf, 404; criticizes Helmholtz, 313, 326, 333; criticizes Weber, 382–83; dissociation hypothesis of, 415; entropy and chemical thermodynamics, 419–20; heat theory, 434–35; and Helmholtz, 327–28; vs. Helmholtz on work, 317–18; influence on Helmholtz, 412, 415, 418–19, 421, 444, 447; potential, 327–28; quantized electricity, 416; second law of thermodynamics, 435–36
Coccuis, Ernst Adolf, 33
Cohn, Emil: analogy in physics, 362; and Maxwell's theory, 375
Colding, Ludwig: and energy conservation, 292; and Helmholtz, 321
Collège de France, Paris, 587
Collegium medico-chirurgicum, Berlin, 21
Color. *See* Optics, physiological
Color blindness. *See* Dalton; Herschel; Maxwell; Optics, physiological; Young
Color contrast. *See* Chevreul
Color mixing. *See* Brewster; Da Vinci; Forbes; Foucault; Grailich; Grassmann; Mayer, Tobias; Newton; Optics, physiological; Plateau; Wünsch; Young
Color research. *See* Optics, physiological
Colors, subjective. *See* Buffon; Chevreul; Fechner; Optics, physiological; Plateau; Scherffer
Color theory. *See* Brewster; Chevreul; Goethe; Newton; Optics, physiological
Color vision. *See* Maxwell; Melloni; Newton; Palmer
Color wheel. *See* Optics, physiological
Columbia World Exhibition, 457; represents Germany at, 575
Combination tones. *See* Hällström; Koenig; Ohm; Tone(s)
Comptes rendus, 91, 92
Concentration cells. *See* Becquerel; Thermodynamics, chemical
Congruency. *See* Geometry
Consciousness. *See* Fichte
Consonance. *See* Acoustics; Acoustics, physiological; Music; Vischer
Corti, Alfonso, 279; influence on Helmholtz, 279–80
Coulomb, Charles-Augustin de: law of, 327, 382
Counting. *See* Arithmetic
Cramer, Antonie: and Donders, 149
Cranefield, Paul, 25
Crelle, A. L.: *Journal für die reine und angewandte Mathematik*, 380
Croonian Lecture, 182

Culotta, Charles A., 462n
Culture. *See* Helmholtz, Hermann von; Science
Cummings, William, 30
Czapski, Siegfried: confirms Gibbs-Helmholtz equation, 422

D'Alembert, Jean, 531–32; influence on Helmholtz, 456
Dalton, John: on color blindness, 226
Danish War, 596
Darwin, Charles, 2, 361; Helmholtz compared to, 2, 458; praise for, 570. *See also* Evolution, theory of
Data analysis. *See* Physiology
Da Vinci, Leonardo: color mixing, 227
Deductive method. *See* Science, philosophy of
Depth perception. *See* Hering; Optics, physiological; Painting
Descartes, René, 501n
Desmarres, Louis-Auguste: and Albrecht von Graefe, 127
Despretz, César-Mansuète, 408; animal heat, 62–3, 299; and Helmholtz, 300, 302; influence on Helmholtz, 332. *See also* Dulong
Deutsche Chemische Gesellschaft: elected honorary member of, 424
Deutsche Gesellschaft der Naturforscher und Ärzte: Innsbruck lecture, 592
Deville, Henri Etienne Sainte-Claire: thermochemical affinity, 417
DiSalle, Robert, 6–8, 11
Dissociation theory. *See* Arrhenius; Chemistry, physical; Clausius; Williamson
Dissonance. *See* Acoustics, physiological; Music
Donders, Franciscus Cornelius, 35n; and accommodation, 149; and Bessel, 149; and empiricist theory of vision, 153; eye movements, 128, 132; and Gauss, 149; and Albrect von Graefe, 127, 128; and Helmholtz, 11, 9n, 128, 172, 182, 189; and Helmholtz's dioptrics, 149–50; influence on Helmholtz, 140–41, 147, 149, 171; law of, 6, 128, 140–41, 147, 170, 172, 182, 199; mathematical limitations of, 149–50; on the ophthalmometer, 148–49; ophthalmometry, 36
Dove, Heinrich Wilhelm, 91; and colored top, 222; explains Purkyně shift, 248–49; eye movements, 161; and Helmholtz, 248–49; Helmholtz confirms his results, 236; improvement of

Dove, Heinrich Wilhelm *(continued)*
siren, 262–63; influence on Helmholtz, 206–7, 219, 220; and Purkyně effect, 219; *Repetorium*, 262; stereoscope and binocular vision, 223
Draper, John: criticizes Brewster, 211
Du Bois-Reymond, Emil, 24, 25, 77, 78, 593; criticizes Helmholtz, 91–2, 94, 480; electrophysiology, 64; as experimentalist, 92n; force as fiction, 480; and Heidelberg University, 42, 44; and Helmholtz, 46, 78–9, 83, 84, 85, 89n, 106, 183, 191, 267, 297, 304, 377; and Helmholtz's nerve-impulse work, 91–3; influence on Helmholtz, 89; and Ludwig's *Lehrbuch*, 129–30; and medicine, 39; molecular theory of stimulated effect, 89, 93; nerve and muscle physiology, 70–1; praises Fick, 133; reductionism of, 480; support for Helmholtz, 94; *Untersuchungen*, 78–9, 480; and Volkmann, 150
Duhem, Pierre, 408; chemical potentials, 425; and chemical theory, 405; and chemical thermodynamics, 423; Helmholtz's influence on, 425; opposition to affinity theory, 425
Dulong, Pierre: animal heat, 62–3, 299; and Helmholtz, 300, 302; influence on Helmholtz, 332; specific heats of gases, 326. *See also* Despretz
Dynamicism. *See* Physics

Ear. *See* Anatomy; Physiology
Easiest orientation, principle of. *See* Hering; Optics, physiological
Ebbingshaus, Hermann: and nativism, 201–2
Ebert, Hermann: praises Helmholtz, 416n
Ecole des Etudes Supérieures, 587
Edinburgh. *See* Royal Society of
Ehrenberg, Christian G. von: ganglionic cells, 24
Ehrenfest, Paul: on irreversibility, 455–56
Einstein, Albert, 2, 497; on Helmholtz, 2; Helmholtz compared to, 2
Electrical Congress, Paris: represents Germany at, 575
Electricity. *See* Chemistry; Mossotti; Physics; Riemann; Weber, Wilhelm
Electrochemistry. *See* Chemistry; Thermochemistry; Thermodynamics, chemical; Thermodynamics, second law of

Electrodynamics, 3, 9–10, 389–95; background to, 375; Biot-Savart law, 381; and cathode rays, 401; and charge, 392; compared to Maxwell's, 391–94; and differential volumes, 370; early theories of, 381; electrophysiology, as background to, 375, 376–77; electrostatic effects, in Fechner-Weber, 340, 342; electrostatic effects, in field theory, 340, 342; electrostatic effects, in Helmholtzianism, 340–41, 342; and the ether, 334, 336, 392, 394; experiments, 345; field theory, 334–35, 337, 338, 339, 342, 346–47, 358–59, 392, 393; as foreshadowing relativity, 401–2; framework of, 334; generality of, 390; general potential, 387–89, 390–91; *Gleitstelle* (slip-joints), 347, 348, 349; goal of Helmholtzian, 341; Helmholtzian, 334–37, 339, 340–41, 342, 344, 345, 346–48, 350–51, 358–59, 365, 370, 373; influence in, 374–75, 376, 389–90, 401; Maxwellian, 375, 388, 389, 391, 392–93; models in, 341, 369; obsolescence of, 401; open currents, 386; as outmoded, 374–75; particle theory of (= Fechner-Weber), 334–35, 337, 339, 342, 346–47, 363–64, 365; polarization, 390–91, 392, 393–94; potential function and Helmholtz, 344–45; rival to Maxwell's, 389–90; role in, 375; testability of, 393; time propagation in, 394; wave equations of, 393–94; Weberean, 382–83, 387; and X-rays, 401. *See also* Helmholtzianism; Physics; Physics, field; Weberean physics
Electrolysis. *See* Chemistry
Electromagnetic induction. *See* Faraday
Electrophysiology. *See* Du Bois-Reymond; Electrodynamics
Electrostatic effects. *See* Electrodynamics
Elektrotechnischer Verein, Berlin: report to, 575
Empiricism. *See* Arithmetic; Epistemology, empiricist; Geometry; Mathematics; Philosophy; Science, philosophy of
Energetics. *See* Chemistry, physical; Physics
Energy. *See* Energy, conservation of; Physics
Energy, conservation of, 432–33; and conversion processes, 292, 294; and dynamical theory of heat, 294; and engineering practice, 293; French engineering tradition of, 294–95, 296, 310, 311–12, 327; and heat engines, 292,

293; and Helmholtz, 368–69; Helmholtz as discoverer of, 292; histories of, 291–92; human implications of, 567–68, 570–71; as ideology, 600; Kuhn on origins of, 292–95; limitations of, 496; and natural processes, 433, 440; and *Naturphilosophie*, 292, 294; and perpetual motion, 294; and potential theory, 293–94; status of, 383; and thermochemistry, 404; and work, 293–94. *See also* Force, conservation of; Thermodynamics; *Ueber die Erhaltung der Kraft*
Engineering. *See* Carnot, Lazare; Energy, conservation of
Enlightenment, German, 599. *See also* Helmholtz, Hermann von
Entropy. *See* Boltzmann; Clausius; Thermodynamics, second law of
Epistemology, 118–19, 119–20, 155, 549; and action, 494–95; evolution of, 282–85; Kant's influence on, 461–62; metaphysical, 462; and physiological acoustics, 282–87; and physiological optics, 284–85; role of practice in, 119–20; and sensory physiology, 260; summarized, 487–88; "Thatsachen in Wahrnehmung," 484–88, 492. *See also* Acoustics, physiological; Epistemology, empiricist; Philosophy; Science, philosophy of
Epistemology, empiricist, 7, 203, 255, 260–61, 462–63; and geometrical knowledge, 499; and nativism, 499; as organizing principle, 260; and perception, 283–87. *See also* Acoustics, physiological; Epistemology; Philosophy; Science, philosophy of
Erdmann, Bruno, 2
Ergodic systems. *See* Physics
Error analysis. *See* Physiology
Ether. *See* Electrodynamics; Helmholtzianism
Ettingshausen, Albert von: and Boltzmann, 343; and the Hall effect, 343
Euler, Leonhard, 224; heat theory, 434–35; Helmholtz's criticism of, 531, 532; improvement on, 278; influence on Helmholtz, 224, 531, 532, 534; musical consonance, 531
Evolution, theory of, 372; at German universities, 589; and Helmholtz, 199; human implications of, 570–71; as ideology, 600; as progress, 567; and vitalistic forces, 570. *See also* Nativist-empiricist controversy
Experiment. *See* Science, philosophy of
Experimental interactionism. *See* Science, philosophy of
Explanation. *See* Philosophy
Extraterrestrial phenomena. *See* Helmholtz, Hermann von
Eye, optics of. *See* Optics, physical
Eye(s). *See* Optics, physiological
Eye as a measuring device. *See* Optics, physiological; Wundt
Eye measurement. *See* Optics, physiological
Eye movements. *See* Brücke; Donders; Dove; Fick; Graefe, Albrect von; Hering; Meissner; Optics, physiological; Volkmann; Wundt

Faraday, Michael, 11; electrodynamics, 337, 338; electromagnetic induction, 376, 378, 382, 390; on force and matter, 462; Helmholtz compared to, 458; influence on Helmholtz, 11, 302, 376, 410–11, 415, 462–63, 480–82, 495, 588; laws of electrolysis, 302, 411; praises, 480–81, 564. *See also* Electrodynamics, field theory; Faraday Lecture
Faraday effect. *See* Maxwellianism
Faraday Lecture, 414–15, 416, 429, 554n; influence of, 416. *See also* Faraday
Favre, Pierre, 408
Fechner, Gustav Theodor, 364; and atomism, 364; attacks Helmholtz on subjective colors, 255; current theory, metaphysical meaning of, 364; electrodynamics, 337–38; *Elemente*, 638; and Helmholtz, 253; Helmholtz compared to, 476; Helmholtz contrasted to, 477; influence on Helmholtz, 206–7, 237, 241, 242, 243–44, 245, 251–52, 254, 255, 257; influence on Hering, 153, 197; influence on Weber, 337, 364; and psychophysical law, 138–39; and subjective colors, 254–55; and Weber, 382; and Zöllner, 364. *See also* Electrodynamics, particle theory; Positivism
Feeling. *See* Aesthetics; Fichte
Fermentation. *See* Physiology, putrefaction and fermentation
Fichte, Immanuel, 125
Fichte, Johann Gottlieb, 11, 125; apparent differences with Helmholtz, 491–92; *Bestimmung*, 489; feeling vs. intuition, 489–90; idealism of, 462, 483; influence on Helmholtz, 11, 370n, 462, 463, 482–83, 485, 487, 488, 489, 490–91, 493, 494; philosophy of action,

Fichte, Johann Gottlieb *(continued)* 463, 490; praises, 485; theory of consciousness, 489–90; theory of perception, 489; theory of self-consciousness, 488–89, 490
Fick, Adolph, 37; and du Bois-Reymond, 133; and empiricist theory of vision, 153; eye movements, 130, 133–34; influence on Helmholtz, 139–40; and Ludwig, 130, 133; measurements corrected, 151; neglects Donders's law, 140; and "organic physics," 133; rejects Listing's law, 171
Field physics. *See* Physics, field
Fischer, Emil: on Helmholtz as chemist, 424
FitzGerald, George Francis: Helmholtz memorial lecture, 427
Fleck, Ludwig, 107
Föppl, August: *Einführung*, 375, 390
Forbes, James D.: color-mixing theories, 227–28; influence on Helmholtz, 227–28, 239; and Maxwell, 239; on Young's hypothesis, 227–28
Force: chemical, 403–4; concept of, 367–69, 468, 469, 470–71, 474–75, 476–77, 479, 495; as hidden cause, 464; *Kraft*, 313–18; as law, 495; living (*lebendige Kräfte*), 314–16, 319; and matter, 366–68, 464–65, 467–70, 471, 473, 474, 475, 477, 481, 495; meaning of, 463; measurement, Helmholtzian, 339; measurement, in field theory, 337, 338–39; measurement, in particle theory, 337–38; and mechanical theory of heat, 324; metaphysical status of, 465; as "objective power," 566–67; relational view of, 366–67. *See also* Force, conservation of; Matter; Science, philosophy of
Force, conservation of, 295, 304, 305–6, 313–19, 368–69; and Ampère's law, 330; and animal heat, 8, 27–8, 303, 304; applications of, 319–32, 567; applied to animal world, 332–33; applied to electrical processes, 378–79; applied to electrodynamics, 330–32; applied to galvanism, 328–29; applied to magnetism, 330; applied to muscular activity, 77; applied to physiology, 332–33; applied to plant world, 332; applied to static electricity, 327–28; applied to thermo-electric currents, 329; *Arbeitskraft*, 295; and batteries, 295, 296, 328–29; and causality, 368; chemical aspect of, 324–25, 406, 408–10; and conservation of *vis viva*, 294, 309–16, 317–18, 319, 368; contact theory, and batteries, 295; and electrical movements, 329; and force-equivalent of heat, 320–26; and friction, 319, 321; and galvanic currents, 328–29; goals for principle of, 332–33; and Grassmann's law, 330; and gravitational motion, 319–20; and heat, 320, 325; Helmholtz's abandonment of, 333; heuristic power of, 331–32; and Holtzmann's law, 325–26; impossibility of perpetual motion, 295, 297, 299, 300–1, 302–4, 305–6, 307, 309, 310, 311, 312, 319; and inelastic bodies, 319, 321; and Joule's law, 329; lack of experimental confirmation, 332–33; lectures on, 593; and Lenz's law, 330; and light, 320; meaning and power of, 567–68; and mechanical energy, 377–78; and mechanics, 319–20; as metaphysical theorem, 472; *Muskelaction*, 300, 303; and Neumann's law, 330–31; and Peltier effect, 329; and physiology, 297, 300, 302–3, 408; and polarization, 410; potential, 312, 317–18, 327–28; as progress, 567; relation to muscle physiology, 67n, 73–4; and standards, 568; and tension forces (*Spannkräfte*), 314–16, 317–18, 319, 320; and virtual velocities, principle of, 315; and *vis viva*, 296; and *vis viva* loss, 321; and *vis viva-vis mortua*, 316; and *vis viva*-work theorem, 313–14; and vitalism, 595; and Weber's law, 330–31, 383; and work, 294, 296, 309–15, 317–18, 327. *See also* "Bericht"; Energy, conservation of; Force; Forces; Heat; Heat, caloric theory of; *Ueber die Erhaltung der Kraft*
Forces: central, 295, 303, 305, 306, 307, 308–9, 312–13, 315, 316, 317–19, 464, 466; central, and Maxwell's theory, 496; central, in physics, 472; central, role in nature, 471; natural, unification of, 458; natural, unity of, 408; and sensations, 468–69; vital, 298, 299–300, 302, 303–4, 307. *See also* Force, conservation of
Fortschritte der Physik, 262, 297; and Helmholtz, 92, 241, 297, 300
Foucault, Léon: priority challenge in color mixing, 232
France: hostility to, 589
Franco-Prussian War, 596
Frankland, Edward: on Helmholtz as chemist, 424
Fraunhofer, Josef: lines of, 211, 214, 216

Frederick William I (of Prussia), 21
Frederick William II (of Prussia), 21
Free energy. *See* Chemistry, physical; Gibbs; Thermodynamics, chemical; Thomson, William
Free will. *See* Philosophy
Fresnel, Augustin: influence on Helmholtz, 320; laws of light, 320
Freud, Sigmund, 601
Friedrich-Wilhelms-Institut, Königliches medizin-chirurgisches (Pepinière), Berlin, 18, 19–23, 372; Helmholtz at, 406–7; professor of physics at, 48. *See also* Berlin, University of

Gadamer, Hans-Georg, 546n
Galileo: and time measurement, 577
Galvani, Luigi: and animal electricity, 577
Galvanometer. *See* Hertz; Instrumentation; Nerve impulse, velocity of
Gauss, Carl Friedrich: current interaction, 381; electrodynamics, 382; and exact measurement, 363; influence on Helmholtz, 327, 376; and method of least squares, 148; potential function, 327; principle of least constraint, 137–38
Geisteswissenschaften: attitude towards, 562–64, 571; and *Bildung*, 564, 571; characterized, 562, 563–64; and modernization, 576–77; vs. *Naturwissenschaften*, 523, 524–25, 543–47, 562, 564, 565, 571, 594; value of, 563–64, 571. *See also* Aesthetics; *Bildung*; *Bildungsbürgertum*; Moral sciences
Geometry: axiom of monodromy, 511–12; axioms of, popularized, 514; congruency, 511–12; congruency as foundation of geometry, 513; as deductive structure, 499–500; empirical basis of, 513; empirical character of, 402; empirical criticism of, 477; and empiricism, 499; Euclidean, 498, 514; factual basis of, 513; foundations of, 499, 515; general metric and rigid bodies, 498–99; and mathematics, 499; non-Euclidean, 498, 510, 514–15; and perception, 499, 509–10; and physical experience, 499, 513; and physical laws, 507, 514; and physical measurement, 498; physiological roots of, 512–13; possibility of, 513; Riemann's and Helmholtz's, 510. *See also* Arithmetic; Mathematics; Perception, spatial; Relativity; Space

Germany: cultural climate of, 202; illiberalism of, 599; and *Kultur*, 596–97; as *Kulturnation*, 596–97; social and political divisions in, 596. *See also* Helmholtz, Hermann von
Gesellschaft Deutscher Naturforscher und Ärzte: Innsbruck address, 474
Gibbs, Josiah Willard, 426; and chemical theory, 405, 425; and chemical thermodynamics, 418, 423; and free energy, 422; and Helmholtz, 418; on thermochemical affinity, 417
Gibbs-Helmholtz equation. *See* Chemistry, physical; Czapski; Horstmann; Van't Hoff
Goethe, Johann Wolfgang von: and artistic induction, 552; as color theorist, 481, 552; color theory of, 246; contrasted to Helmholtz, 465; criticism of, 481, 552, 576, 578, 582; *Farbenlehre*, 206, 209, 250; Helmholtz on, 481, 522–23, 525, 551–53; influence on color theory, 250; influence on Müller, 246; lectures on, 551, 552, 553, 558; and logical induction, 552; as natural historian, 551–52; ridicules Wünsch, 223
Goldstein, Eugen: electric discharge, 359–60; evacuated tubes, 344; and Helmholtz, 342; as Helmholtzian, 342, 359–60
Graefe, Albrecht von, 33, 35, 36, 161; and *Archiv für Ophthalmologie*, 128; at Berlin University, 127; and Bowman, 127; and Brücke, 127; clinical practice of, 127, 128; and Desmarres, 127; eye movements, 130, 132; and Alfred Graefe, 119n; and Helmholtz, 47, 119n, 127, 149, 189; influence on Helmholtz, 141; and Müller, 127; and Müller's identity theory, 128–29; ophthalmology, 34; ophthalmology clinic of, 127, 128–29; and ophthalmoscope, 128–29
Graefe, Alfred, 119n
Graefe, Carl Ferdinand von, 22
Graham, Thomas: influence on Helmholtz, 411; occlusion, 411
Grailich, Joseph: criticizes Helmholtz on color mixing, 232–33
Graphical method. *See* Muscle contraction; Nerve impulse, velocity of; Physiology
Grassmann, Hermann Günther: *Ausdehnungslehre*, 232; criticizes Helmholtz on color mixing, 232; Helmholtz's criticism of, 233–35, 236;

Grassmann, Hermann Günther *(continued)* influence on Helmholtz, 248, 330, 519; laws of, 381
Grassmann, Robert: influence on Helmholtz, 519
Gravitation: law of universal, 567n. *See also* Force, conservation of; Newton
Greeff, Richard, 35
Green, George: and Helmholtz, 327
Grube, Wilhelm, 20n

Hall, Edwin, 342–43. *See also* Hall effect
Hall effect: discovery of, 342–43; and field theory, 344; and Helmholtzianism, 343–44; and particle theory, 342–43
Hällström, Gustav: combination-tone theory of, 270–71; influence on Helmholtz, 271
Halske, Johann Georg, 68
Hamilton, William Rowan: influence on Helmholtz, 327; potential function, 327. *See also* Least action, Hamilton's principle of
Handbuch der physiologischen Optik, 3, 7, 9, 37, 119, 126–27, 146n, 153, 154, 155–56, 162–63, 168, 173, 177, 179, 181, 182–204, 206, 207, 208, 236, 237, 239–40, 241, 247, 248, 249, 251, 252, 254, 255, 256–57, 268, 269–70, 284, 287, 509, 513, 530n, 538, 547, 550–51
Hanslick, Eduard, 533n; influence on Helmholtz, 541n
Harcourt, A. G. Vernon: on Helmholtz as chemist, 424
Hargreave, David, 225, 239
Harmony, color. *See* Optics, physiological
Harmony, musical. *See* Music
Harmony, pre-established. *See* Nativism; Philosophy
Harris, Moses: and pigment mixing, 221–22
Hartley, David, 224; influence on Helmholtz, 224
Hassenfratz, Jean: solar spectrum, 233, 235
Hatfield, Gary, 11–2, 179
Hearing. *See* Acoustics, physiological
Heat, 300; electrochemical, 301–2; equivalents, 299, 301; formation in muscles, 52, 69–70, 71, 72–3, 73–4; mechanical equivalent of, 300, 303, 318, 320–23; mechanical theory of, 318, 324–26; and metabolism, 298, 300; models of, 298–99, 301; as molecular motion, 433, 434, 435, 438, 440, 441, 442, 454–55, 456; physical nature of, 61–2; and vital processes, 61–2; and work, 294, 296, 301. *See also* Force, conservation of; Heat, caloric theory of; Physiology
Heat, caloric theory of, 323–24; Helmholtz's criticism of, 325–26; rejection of, 323–24, 409. *See also* Heat
Heaviside, Oliver, 362; on Hertz, 374, 399; and Maxwell's theory, 375, 544
Hegel, Georg Wilhelm Friedrich: and *Geisteswissenschaften*, 544; hostility towards, 565; influence on Goethe, 465; opposition to, 463, 523, 599; rejection of his metaphysics, 544
Hegelianism, 125
Heidelberg, University of: address at, 522–23; appointment to, 18, 41, 44, 45, 47; chair of physiology at, 42, 44, 47; and Helmholtz, 163, 247, 268, 363, 365, 418, 593; institutionalization of science at, 41–2, 42–3, 45, 47; laboratory teaching at, 46; lectures at, 46; medical context of, 45–6, 48; physiology institute, 18, 41; prorectorial lecture at, 543–44, 576; salary at, 41; students at, 257. *See also* Helmholtz, Hermann von
Heidelberger, Michael, 8, 11
Heimann, Peter, 294
Heintz, Wilhelm, 297
Heliometer. *See* Optics, physiological
Helm, Georg: and chemical theory, 405; and Helmholtz on atomism, 442; on Helmholtz on heat, 324
Helmholtz, Caroline (mother), 554, 562–63, 579
Helmholtz, Ferdinand (father), 19–20, 30, 32, 554, 562–63, 579; as admirer of J. G. Fichte, 491; and I. Fichte, 125; and Hermann von Helmholtz, 100, 120; and nerve-impulse velocity, 93–4
Helmholtz, Hermann von: on academic freedom, 372; admiration for England, 588–89, 593; as ambassador of science, 560, 593; architectural taste of, 554; as army doctor, 27, 54, 162, 297–98, 574; artistic taste of, 554; on astrology, 596; as *Besitzbürger*, 574; career, chronicle of, xxi–xxix; career-building, 593; Charlottenburg home, 554; classical taste of in art and science, 556; criticizes Baden medical regulations, 46; criticizes English academics, 590–1; criticizes French academics, 587, 589, 590–91; and culture,

563; as cultural leader, 597; on culture and the nation-state, 576–77; death of, 401; and the electrical industry, 574–75; and the Enlightenment, 561–62; Enlightenment vision of, 599–600; fame, 574; on the German Empire, 589; on German scientists, 591–92, 593; on German students, 590; and God (mentioned), 541, 584; gymnasium studies, 563; habilitation speech, 244, 491; historical studies of, 4; on history of human life, 568–69, 570–71; as ideologue, 560, 599, 600–1; influence of, 206; intellectual authority of, 561; Kant memorial lecture, 109–10, 118–19, 121, 125, 126, 244–46, 247, 283, 287, 485, 504, 533; as *Kulturphilosoph*, 561; as *Kulturträger*, 561, 562; laboratory in Bonn home, 40–1; laboratory in Potsdam barracks, 27; and language study, 563, 582; lectures, 46, 442, 574, 577n, 582, 593; lectures in Berlin, 535; lectures in Cologne, 535; lectures in Düsseldorf, 535; as liberal, 561, 576, 598, 599, 600; life, chronicle of, xxi–xxix; love of literature, 563, 579; as mandarin, 561, 599; on the meaning of life, 569, 570–71; medical dissertation of, 24, 26–7, 53; medical education of, 18, 19–20, 22, 23–4, 32, 48, 162, 406–7, 573, 574; as medical practitioner, 17, 18, 19, 26, 27, 28; and the military, 27, 28; as modern, 401–2; on the modern nation-state, 576–77; and money, 593; musical taste of, 554; as musician, 522, 579; as non-Hegelian, 523; as non-Romantic, 523; opposition to mysticism, 595; opposition to spiritualism, 595–96; on origins of life, and extraterrestrial phenomena, 569; painters, contact with, 579; on painters and nature, 535–36; as *philosophe*, 600; as polemicist, 154, 155, 185, 560, 576, 591; as popularizer, 270; and Potsdam, 27, 297–98; priority concerns of, 89; Prussian values of, 599; reputation, 3–4; on research as basis of teaching, 586; as rhetorician, 155, 156, 190–91, 193, 195, 197, 204, 416; as science leader, 2–4, 374–75, 376, 401, 402; on science writing, 92–3; self-disappointment of, 400; and Sommering prize, 573; as spokesman of science, 561; status of, 44–5, 48, 395, 428, 593, 598; students of, 132, 257, 315, 335, 342, 374, 397, 413, 418, 429; as synthesizer (in general), 5, 8–13; taste in scientific theory, 554–55, 557; as teacher, 46, 203–4, 358, 374, 375, 376, 398–99, 522; teachers of, 20, 24, 26, 52–3, 297, 319, 406–7, 526; on teaching, 586; as traditionalist, 402; and travel, 563, 573; as uncritical of science, 562, 600–1; upbringing of, 562–63. *See also* Berlin, University of; Bonn, University of; Heidelberg, University of; Königsberg, University of; Science; Science, natural; Science, popular

Helmholtz, Katharina (daughter), 88, 105, 579

Helmholtz, Marie (sister), 579

Helmholtz, Olga (wife), 40; death, 163; and Hermann von Helmholtz, 79, 83; as research assistant, 67–8, 83, 88, 101, 105

Helmholtz, Robert von (son), 478n; advice to, 407

Helmholtzianism, 7, 10; and Ampère's law, 348, 349–50; anti-atomism of, 365; characteristics of, 344; definition of, 335–36, 339, 340–41, 342, 344, 365; implications of, 335, 340, 341, 346; influence of, 339, 360–63; laboratory practice of, 335, 337, 339, 341–42, 344, 345, 346, 350, 351, 352, 357, 389; and measurement, 351, 357; nature of, 346, 350–51, 357, 358–59, 360, 365; no models in, 341; and open-ended circuits, 345; and particle theory, 345; and polarizable ether, 345; and potential function, 346–48, 349–51; predictive range of, 344; as response to Zöllner, 372–73; roots of, 337, 363–73; and theory, 341, 344–45; threatened with failure, 345; use in discovery, 344–45; and variational principles, 346. *See also* Electrodynamics; Helmholtzians

Helmholtzians, 336–37, 357–58; weak community sense of, 358. *See also* Helmholtzianism

Henle, Jacob, 20, 21, 24–5, 26; and disease process, 43; and physiology at Heidelberg, 42; ties to medical community, 39

Henry, William: criticism of, 323–24

Herapath, John: heat theory, 434–35

Herbart, Johann Friedrich: aesthetics of, 523; empirical psychology of, 110; as empiricist, 191; influence on Helmholtz, 110, 125–26; and Kant, 125; at Königsberg University, 126; on space,

Herbart, Johann Friedrich *(continued)* 125–26; and unconscious inference, 126; and Zimmermann, 523
Hering, Ewald, 7, 168; agreement with Helmholtz, 177; attacks projection theory, 174, 177; attacks Volkmann, 180; attacks Wundt, 180; on binocular vision, 176, 180; career, 173; controversy with Helmholtz, 152–53, 155, 156, 168, 173, 176–78, 179–83, 185–87, 191–96, 197, 199, 203, 255, 257, 498; criticizes Helmholtz, 146n, 172, 494, 531; cyclorotation, 169; defends identity theory, 174; defends nativism, 203; depth-perception theory of, 174–75; and Descartes's *Optics*, 501n; easiest-orientation principle, 152–53; and evolutionary theory, 199; on eye movements, 199; and German psychology, 201; and Hankel, 153; and Helmholtz, 176, 182, 248; Helmholtz's victory over, 197–98; and horopter problem, 168, 176; hostility to Berlin physiology, 673; influence on Helmholtz, 177; intellectual roots, 197; and Kant, 503; law of identical visual direction, 174, 175, 181–82; *Lehre vom Sehen*, 199; and Listing's law, 172–73; and local signs, 175–76; and neo-Kantianism, 202–3; phenomenological approach of, 202–3; and physiological optics, 179; as polemicist, 173, 177; polemics with Volkmann, 183, 185; polemics with Wundt, 183, 185; praises Helmholtz, 172–73; and retinal depth values, 175–76, 192–97, 501; retinal geometry of, 501; retinal space values, 180–81; retinal theory of identity, 177; self-image of, 173; and sense perception, 173–74, 177; and spatial perception, 501; as teacher, 203, 204; and Wheatstone's results, 174. *See also* Nativism; Nativist-empiricist controversy; Perception, spatial
Hermann, Jakob: heat theory, 434–35
Hermann, Ludimar: criticizes Helmholtz's combination-tone theory, 272n
Hero of Alexandria: and work, 293
Herschel, John: on color blindness, 248
Hertz, Heinrich, 373, 496; and Akademie der Wissenschaften, Berlin, 352, 396, 397–98; analogy in physics, 362; analytical mechanics of, 457; electrodynamic theories, 375–76; electromagnetic waves, 336, 374, 399, 400; and electromagnetism, 352; and Faraday, 400; and galvanometer, 353–55; and Helmholtz, 335, 352, 374, 397; Helmholtz on, 397, 398, 399–400, 457; as Helmholtzian, 7, 336, 342, 352–57; on Helmholtz's electrodynamics, 394, 399, 400; Helmholtz's influence on, 339, 345, 352–53, 398–99, 456–57; and Helmholtz's laboratory, 352; inductive couplings experiment, 353–57; at Kiel University, 397; and Maxwell, 400; and Maxwell's theory, 375, 396–97, 397–400; and Newtonian force, 457; and parents, 352; and polarization, 345; *Prinzipien*, 397; and propagation velocity, 395; and Röntgen, 398; *Untersuchung*, 398; and Wheatstone bridge, 353–54. *See also* Electrodynamics; Helmholtzianism; Least action, Hamilton's principle of; Physics, relational
Herwig, Hermann, 371; as anti-Helmholtzian, 346–51; Helmholtz's response to, 349–50; as Weberean, 346–48, 350. *See also* Weberean physics
Hess, Germain Henri: and caloric theory, 324–25; influence on Helmholtz, 301–2, 409; law of, 299, 324–25, 409
Hillebrand, Franz: defends Hering, 202
Hittorf, Wilhelm: influence on Goldstein, 359
Hoek, Martinus, 150
Hofmann, August Wilhelm von: praises Helmholtz, 424
Holmes, Frederic L., 5, 6, 148, 208
Holtzmann, Karl: and Carnot's theory, 409; heat law of, 321, 325–26; and Helmholtz, 323; Helmholtz on, 409; influence on Helmholtz, 321, 325–26; mechanical equivalent of heat, 301; results compared with Joule's, 326
Horopter. *See* Meissner; Müller; Optics, physiological; Prevost; Vieth
Horstmann, August: attends Helmholtz's lectures, 418; and chemical thermodynamics, 418, 422–23; and Clausius's entropy theory, 418; and "Gibbs-Helmholtz equation," 422; and Helmholtz, 422–23; and theoretical chemistry, 405. *See also* Thermodynamics, chemical
Humboldt, Alexander von: and Helmholtz, 89n; Helmholtz compared to, 561; influence on Helmholtz, 599; and nerve-impulse report, 92; praises Helmholtz, 90–1; supports Helmholtz, 28

Index   649

Hume, David: on causation, 505; influence on Helmholtz, 505, 520
Huygens, Christian: conservation of *vis viva*, 315; influence on Helmholtz, 310
Hydrodynamics, 3; as background to electrodynamics, 375, 379–80; influence of, 402, 426–27; theory of, 379–80

Idealism: debt to German, 463; and Helmholtz, 11; Helmholtz's, 492–94; opposition to, 118–19, 125, 370, 565n. *See also* Philosophy; Science, philosophy of
Ideal type: in art, 524, 536, 539–40, 551, 552, 553, 554, 556–57; and lawfulness, 556–57; neo-Classical aesthetic of, 554n; in science, 555. *See also* Painting
Identity theory of vision. *See* Müller; Nagel; Ophthalmoscope; Wheatstone
Illusions: optical, 507–8; sensory, 185, 283, 284
Imagination. *See* Aesthetics; Artists
Induction. *See* Aesthetics; Induction, aesthetic; Induction, artistic; Induction, logical
Induction, aesthetic. *See* Aesthetics; Science, philosophy of
Induction, artistic. *See* Aesthetics; Goethe; Science, philosophy of
Induction, logical: defined, 545, 546; and Goethe, 552; in natural science, 565–66. *See also* Goethe; Science, philosophy of
Induction coils, use of. *See* Physiology
Inductive coupling. *See* Hertz
Inductive method. *See* Science, philosophy of
Industry, German, 596. *See also* Helmholtz, Hermann von; Science
Inference, unconscious, 120, 124–25, 184, 186n; and Herbart, 180; Hering's opposition to, 180; and Kant, 180; and lawful sensations, 555; and laws of association, 557; vs. logical inference, 184; and one-fiber/one-sensation doctrine, 548–49; and perception, 286, 287, 523, 550; psychology of, 524, 547–51; theory of, 525, 548–49; and Wundt, 180. *See also* Science, philosophy of
Instrumentation, 25–6, 79, 148–49; acoustical, 268–69, 582; and ballistic galvanometer, use of, 376; heliometer, and Helmholtz, 148; and Helmholtz, 18, 25, 28, 29–33; Helmholtz's use of, 376; measuring, 509–10; microscope, use of, 79, 88; multiplicator, use of, 65–7; in muscle physiology, 52, 65–7, 68; musical, 277; myograph, 103–4, 107; ophthalmometer, 111; ophthalmometer, invention of, 35–6; ophthalmometer and Helmholtz, 148–49; ophthalmotrope, 111; ophthalmotrope and Helmholtz, 148; physics, 7; physiological, 5–6, 111, 573; precision, 575; telescope, use of, 88, 219, 229; telestereoscope, invention of, 163. *See also* Acoustics, physiological; Hertz; Ludwig; Muscle contraction; Nerve-impulse, velocity of; Ophthalmoscope; Optics, physiological; Poiseuille; Stereoscope
Intelligibility. *See* Painting
Intuition. *See* Fichte, Johann Gottlieb; Intuition, aesthetic; Intuition, artistic
Intuition, aesthetic, 533; in art, 539–40; in science, 553; sensuous, 579; spatial, 507; as thought, 485, 491, 519–20, 550, 579. *See also* Intuition, artistic
Intuition, artistic: and experience, 579; and lawful regularity, 552–53, 556–57; and laws, 583–84; psychological theory of, 524. *See also* Intuition, aesthetic
Intuitionists. *See* Nativism
Irreversibility. *See* Ehrenfest; Poincaré; Thermodynamics

Jacobi, Carl Gustav: and the *Erhaltung*, 473; influence on Helmholtz, 327; and method of least squares, 97; potential function, 327
Jacobi, Joachim: and Helmholtz, 579
Jacobson, Julius, 35n
Jaeger, Eduard, 35n
Jardin des Plantes, Paris, 587
Joule, James Prescott: and energy conservation, 292; and heat equivalence, 329, 409; and heat experiments, 325; and Helmholtz, 301, 302, 303, 320, 324, 328; and Helmholtz's conversion errors, 322–23, 326; Helmholtz's criticism of, 321–23, 329; and Holtzmann compared, 326; influence on Helmholtz, 321–23, 325, 329; mechanical equivalent of heat, 321–22, 322–23; paddle-wheel experiment, 321, 322; work-equivalent of heat, 296
Judgments, unconscious, 254, 255; in aesthetics, 533–34; and Helmholtz, 207–8, 236, 249n; and perception, 547
Jurin, James, 242–43

Kaiser, Walter, 9–10
Kant, Immanuel, 109, 125, 462; aesthetic judgments, 524; aesthetics of, 523, 540; on art and genius, 540; concept of matter, 466–67; critique of nativism, 503; on Euclidean geometry, 502–3, 506; on the foundations of geometry, 502; and Helmholtz, 109, 125; vs. Helmholtz on arithmetic, 503; vs. Helmholtz on geometry, 503; Helmholtz's break with, 153; Helmholtz's criticism of, 462, 469, 477, 478, 485, 506–7, 551; Helmholtz's opposition to, 110, 111, 504; Helmholtz's reinterpretation of, 505–6, 520; influence on Helmholtz, 11, 294, 295, 296, 306, 307–9, 461–62, 463, 466, 467–68, 472, 473, 484–85, 504–6, 509, 513–14, 519–20, 543n, 599; influence on Land, 515–16; intentionality in art, 533n; *Kritik der reinen Vernunft*, 112–13, 125; *Kritik der Urteilskraft*, 125; metaphysics of nature, 462, 466–67, 468; *Metaphysische Anfangsgründe*, 308, 466, 469, 472, 473; rules vs. laws, 466; spatial knowledge, 506; spatial perception, 500; spatial theory of, 112–13, 157, 178, 179, 498, 502, 504. *See also* Neo-Kantianism
Kant memorial lecture. *See* Helmholtz, Hermann von
Karsten, Gustav, 216, 297; *Allgemeine Encyclopädie*, 162–63, 284; and Brewster's theory, 215; and Helmholtz, 162, 215, 284
Kelvin, Lord. *See* Thomson, William
Kerr effect. *See* Maxwellianism
Kinetic theory of gases. *See* Kirchhoff; Maxwell; Monocyclic systems; Physics
Kirchhoff, Gustav Robert, 481; goals of mechanics, 478; and Helmholtz, 495; Helmholtz compared to, 476, 478; influence on Helmholtz, 376; and kinetic theory, 385; laws of, 376
Klein, Felix: and Helmholtz, 457; on Helmholtz, 458
Knape, Christoph, 22
Knapp, Hermann Jakob, 35n, 47; and ophthalmometry, 36; validation of accommodation theory, 39
Knoblauch, Hermann, 297
Koenig, Rudolph: on combination tones, 273n; and tone quality, 276n
Koenigsberger, Leo, 87n, 304n, 471, 472–73, 491; biography of Helmholtz, 4; on Helmholtz's essay writing, 93n

Köhler, Wolfgang: criticizes Helmholtz, 530n
Kohlrausch, Friedrich: Helmholtz's influence on, 116; as Weberean, 351n
Kölliker, Albert, 238n
König, Arthur, 146n, 203n
Königsberg, University of: Helmholtz at, 28, 29, 40, 81–2, 83, 162, 215, 219, 247, 267, 365, 574; lectures at, 46–7, 227. *See also* Helmholtz, Hermann von
Königsberg Gesellschaft für wissenschaftliche Medizin, 36; Helmholtz as president, 36
Königsberg Physikalisch-ökonomische Gesellschaft: and Helmholtz, 106
Kopp, Hermann: on state of chemistry, 403
Kraft. *See* Force
Kragh, Helge, 7, 10
Kremer, Richard L., 8–9
Kries, Johannes von: and Helmholtz, 198; defends Helmholtz, 202–3
Krönig, August: heat theory, 434–35
Kuhn, Thomas S., 310n, 316, 328; on energy conservation as simultaneous discovery, 292–96
Külpe, Oswald: and nativism, 201–2
Kulturkampf, 596
Kulturnation, 596–97
Kulturträger, 562
Kymograph. *See* Ludwig

Lagrange, Joseph Louis: influence on Helmholtz, 46, 310
Lagrange's equations. *See* Monocyclic systems; Physics, field
Lamb, Horace: on Helmholtz and Listing's law, 146n
Laplace, Pierre-Simon de, 262; and positivism, 364n
Laplacian programme. *See* Physics
Land, J. P. N.: and Helmholtz, 515–16
Laws, scientific: and art, 540; and artistic insight, 585; as basis of science, 564–67; and causality, 475; conditions of, 495; and experience, 479; as force, 474, 475–77, 566–67; and German unity, 598; Helmholtz on, 465–66; and mental processes, 557–58; and natural processes, 557–58; and natural science, 571; nature of, 566; as objective power, 475; objectivity of, 520; and progress, 567; of psychology, 535; and reality, 477–78; and standards, 568; unifying function of, 554–55. *See also* Aesthetics; Ideal type; Intuition; Science, philosophy of

Least action, Hamilton's principle of, 333, 437, 451, 458; applied to electrodynamics, 400–1; attempt to generalize, 451–54; and energy conservation, 452, 496–97; as foundation of physics, 373n, 451–52; Hertz's view of, 457; and Lagrange's equations, 451; meaning of, 445–46, 451; and natural processes, 433, 440, 457n-58; and Newton's laws, 452; Planck's use of, 457–58; and relativity theory, 497; and statistical mechanics, 451; teleological tendency of, 457
Least constraint, principle of. *See* Gauss
Least squares, method of: and Helmholtz, 96–100, 101, 105, 107, 145–46, 147, 148. *See also* Wundt
Lebenskräfte. See Ueber die Erhaltung der Kraft
LeBlon, Jakob: and pigment mixing, 221–22
Le Chatelier, Henri: Helmholtz's influence on, 425, 426
Die Lehre von den Tonempfindungen als physiologische Grundlage für die Theorie der Musik, 3, 9, 11, 251, 269–70, 273, 277, 278, 281, 284–85, 287, 522, 525, 528–29, 531, 533, 540, 541, 542, 544, 547, 551, 572–73, 581–83
Lehrfreiheit, 590
Leibniz, Gottfried Wilhelm: causality of, 318; conservation of work, 317; Helmholtz's differences with, 316; influence on Helmholtz, 294, 316–17, 318, 319, 527n; and *petite perceptions*, 527; and work, 293
Lenoir, Timothy, 5, 6, 8, 9
Lenz, Heinrich Friedrich Emil: influence on Helmholtz, 302, 329, 330; law of and conservation of force, 329–30
Lernfreiheit, 590
Levi-Civita, Tullio: and relativity theory, 402
Liberalism, German, 561
Lie, Sophus: Helmholtz contrasted to, 499; Helmholtz's influence on, 511–12
Liebig, Justus von, 54; and animal heat, 298, 299; and correlation of force, 298–99, 300; Helmholtz on, 407, 408; influence on Helmholtz, 208, 298–99, 300, 302, 407, 408; and vital forces, 298, 299
Liebreich, Richard, 35n
Life, human. *See* Helmholtz, Hermann von; Science
Light, theory of. *See* Brewster; Force, conservation of; Newton; Physics; Wünsch; Young

Lipschitz, Rudolf: criticizes Helmholtz, 313
Listing, Johann Benedict: Helmholtz and Listing's law, 143–48, 170–73, 182; influence on Helmholtz, 140–47; law of, 6, 140, 151, 152, 199, 170–71; and Ruete, 132
Locke, John: influence on Helmholtz, 113, 468–69, 484; influence on Steinbuch, 113; qualities and sensations, 484
Logical empiricism. *See* Philosophy; Science, philosophy of
Lohff, Brigitte, 92n
Lorentz, Hendrik Antoon, 400–1; electron theory of, 401; and the Hall effect, 343–44
Lorenz, Ludvig Valentin: and relativity theory, 402
Loschmidt, Josef: influence on Helmholtz, 451, 453–54; on reversibility of mechanical processes, 451
Lotze, Rudolf Hermann: as empiricist, 191; Helmholtz contrasted to, 507; Helmholtz's influence on, 110; influence on Helmholtz, 110, 122, 187; influence on Wundt, 179; as Kantian, 503; local sign theory of, 110, 122–23, 178–79, 503, 507; as nativist, 179; reaction to non-Euclidean geometry, 514n; unconscious mental states, 527
Ludwig, Karl, 25; afterimages, 247; challenges Müller's vision theory, 130; and empiricist theory of vision, 129–30, 153; graphical method of, 52, 77–8; and Helmholtz, 297; and Henle, 39; influence on Helmholtz, 77–9, 80, 129–30; kymograph, invention of, 37; *Lehrbuch*, 129–31, 134; and physiology at Heidelberg, 42; praises Fick, 134; praises Helmholtz, 91; as teacher, 46; vision theory of, 130–31

McClelland, Charles E., 587n
McCormmach, Russell: *Night Thoughts*, 557
Mach, Ernst: anti-atomism, 335–36; contrasted to Helmholtz, 477; as critic of Helmholtz, 203; and Helmholtz, 203; Helmholtz compared to, 476; and Helmholtz on atomism, 442; and Hering, 202–3; opposition to Boltzmann, 335; sensationalist philosophy, 335. *See also* Positivism
Magnus, Gustav, 91, 297; as Baconian empiricist, 472; death of, 363; on the *Erhaltung*, 472–73; and Helmholtz,

Magnus, Gustav *(continued)*
479; Helmholtz as student of, 52–3; influence on Helmholtz, 93–4; memorial lecture on, 370; students of, 297
Mandelstamm, Emanuel, 47
Manometer. *See* Poiseuille
Marić, Mileva, 2
Massieu, François: and free energy, 422
Materialism: and German physicists, 335; and Helmholtz, 11, 463; hostility towards, 565n; opposition to, 370n, 493, 494, 595. *See also* Metaphysics; Philosophy; Science, philosophy of
Mathematics, 3; and empirical truths, 500, 517–19; empiricist philosophy of, 6–7, 7–8, 11; and experience, 499–500, 517; and Helmholtz, 458, 499; and laws of nature, 500, 519; and perceptual learning, 511; psychological process of, 500, 519. *See also* Arithmetic; Geometry; Least squares, method of; Measurement; Perception, spatial
Matter: concept of, 467–70, 471, 473; and force theory, 308; metaphysical status of, 465; as metaphysics, 479; vs. mind, 493. *See also* Force; Science, philosophy of
Maupertuis, Pierre Louis: influence on Helmholtz, 456
Maximilian, King of Bavaria: Helmholtz's debt to, 268–69, 573
Maxwell, James Clerk, 411; analogy in physics, 361–62; barycentric curve of the eye, 239; on color blindness, 248; color-vision research, 239–41; his demon, 454; electrodynamic theory of, 10, 337, 338, 374, 375, 381, 395–400, 416; equations of and least action, 400; followers of, 337, 357–58; gas theory of, 361; and Grassmann, 239; and Helmholtz, 435, 495–96; on Helmholtz, 389; and Helmholtz's hydrodynamics, 380; Helmholtz's influence on, 380; Helmholtz's praise for, 395; influence on Helmholtz, 206, 207, 237, 239, 240, 241, 246, 247, 248, 374, 389, 480, 588; and least action, 400; "molecules of electricity," 416; and Müller's law of specific sense energies, 240; and physical analogy, 380, 381; praises Helmholtz, 331n; and probabilistic methods, 435, 442; and Purkyně shift, 240, 249; status, 389; *Treatise*, 342, 358; velocity distribution law, 435, 448, 454; and Weber's law, 381; and Young, 226. *See also* Electrodynamics, field theory; Maxwellianism

Maxwellianism: and analogies in physics, 362; and the Faraday effect, 358; and the Hall effect, 358; and the Kerr effect, 358; and Poynting's theorem, 358. *See also* Maxwell
Mayer, Julius Robert: *Annalen der Chemie und Pharmacie*, 318; and conservation of force, 292, 318, 377; on heat, 320, 321; and Helmholtz, 301, 303, 318, 320, 321; Helmholtz on, 564; and Leibnizian causality, 318; mechanical equivalent of heat, 318; neglect of battery, 295
Mayer, Tobias: color mixing, 227; and pigment mixing, 221–22
Measurement: geometrical, 508, 509; and the human body, 508; the personal equation, and Helmholtz, 100, 120; in physics, 7; in physiology, 5, 6–7, 51–2, 65–7, 68; precision, 51–2; and rigid bodies, 513–14; units and standards of, 575. *See also* Force; Geometry; Helmholtzianism; Instrumentation; Mathematics; Muscle contraction; Perception, spatial; Space
Mechanical worldview. *See* Nature; Philosophy
Mechanics, 3; and action, 456–57; and foundation of physical science, 432–33; and foundation of physics, 457; and foundation of science, 474; fundamental reversibility in, 454; and Helmholtz, 456; Hertz's view of, 457; statistical, Helmholtz's influence on, 456. *See also* Least action, Hamilton's principle of; Science, philosophy of; Thermodynamics, second law of
Mechanism: defined (in biology), 408
Medicine: experiment and observation in, 38, 48; German community of, 5–6, 28, 44–5, 47–9; and Helmholtz, 5–6, 28, 44–5, 47–9; Helmholtz's impact on, 17–8, 19, 28, 33–8, 48–9; Helmholtz's status in, 44–5, 48; microscopy, 46; pathology, disinterest in teaching, 40; and physiology, 38–9, 43–4, 48; practice of, 34, 36–7, 39, 47, 48–9; reform of in Baden, 41–6, 48; scientific, 41–2, 43–4, 48. *See also* Ophthalmology; Ophthalmoscope; Physiology
Meissner, Georg: 161; and Donders's law, 140; eye-movement studies, 130, 132; and the horopter, 165; influence on Helmholtz, 144; and Listing's law, 140, 171; and retinal incongruity, 165
Melloni, Macedonio: and Brewster, 216; and color vision, 226; criticizes Brewster's theory, 211, 214

Menzel, Adolph von: and Helmholtz, 579
Metabolism. *See* Heat
Metaphysics: as enemy of, 544; Hegelian, 564; hostility towards, 372, 565n; materialist, and German universities, 589; materialist, rejection of, 535; opposition to, 493, 494, 564, 595, 599; skepticism towards, 366–67, 368. *See also* Philosophy; Science, philosophy of
Meteorology, 3
Method. *See* Science, natural; Science, philosophy of
Methodology. *See* Aesthetics; Physics; Physiology; Science, philosophy of
Meyer, August, 20n
Microscope. *See* Instrumentation; Physiology
Microscopy. *See* Medicine
Mie, Gustav, 401
Mill, John Stuart: Helmholtz's differences with, 543–44; influence on Helmholtz, 544, 550; *Logic*, 544; methods of science, 543
Minkowski, Hermann: spacetime physics, 512n
Mitscherlich, Eilhard, 91; Helmholtz as student of, 52–3, 406–7; influence on Helmholtz, 23
Mittheilungen des badenischen ärztlichen Vereins, 37
Models. *See* Boltzmann; Electrodynamics; Heat; Helmholtzianism; Physics, field
Modernization. *See* Helmholtz, Hermann von; Science
Moigno, François: criticizes Brewster's theory, 211, 213
Mommsen, Theodor: and Helmholtz, 563
Monocular vision. *See* Optics, physiological
Monocyclic systems: as analogies, 434, 439, 440, 441, 447, 449, 450; Boltzmann's approach to, 444–46; Boltzmann's extension of, 447–50; Boltzmann vs. Helmholtz on, 445, 446, 447–48, 449; coupling of, 446; cyclic velocity and temperature, 442–44; defined, 433–34; energy restrictions in, 440–41; extension to complex, 446–47; and first law of thermodynamics, 435–36; kinetic energy and temperature, 446; and kinetic theory of gases, 434–35; and Lagrange's equations, 437–39; and least-action principle, 437; liquid in tube as, 434, 443, 448, 449; particles in container as, 448; physical meaning of, 448; probabilistic approach to, 435; relation to heat theory, 433; rotating sphere as, 438, 443; rotating top as, 434, 440–41, 448; Saturn's rings as, 448; and second law of thermodynamics, 435–36; statics of, 439–40; stationary motion in, 438, 440; and thermodynamics, 440. *See also* Boltzmann; Physics; Thermodynamics, second law of
Monodromy, axiom of. *See* Geometry
Moral sciences: method of, 544–45; subject of, 544, 545–46. *See also Geisteswissenschaften*
Moser, James: and Helmholtz, 413
Moser, Ludwig: color research of, 219; influence on Helmholtz, 219, 220
Mossotti, Ottaviano: and quantized electricity, 416
Motion, perpetual: impossibility of. *See* Force, conservation of
Moutier, Jules: and Duhem, 425
Mozart, Wolfgang Amadeus: preference for, 554
Mühler, Heinrich von, 238n; praises Helmholtz, 561
Müller, Johannes, 78; afterimages, 246; challenges to his identity theory, 130, 132; as empiricist, 179; his *Handbuch*, 114, 120, 131; and Helmholtz, 89, 91, 110, 111, 112–13, 118, 119, 120, 121, 159, 162, 207, 297; Helmholtz as student of, 52–3, 526; Helmholtz's break with, 153; Helmholtz's regard for, 53; and the horopter, 165; identity theory of vision of, 127, 159–60; influence on Helmholtz, 23–4, 27, 206–7, 208, 237, 238, 243, 244, 245–47, 248, 255, 256, 282–83, 484, 485; influence on Hering, 197; as Kantian, 114, 116–17; and Kant's view of space, 112–13; his laboratory, 24, 27; law of specific sense energies, 114–16, 118, 207, 238, 240, 244, 245, 246–47, 256, 282–83, 500, 526–27; as nativist, 112–19; opposition to Steinbuch, 113, 114; opposition to Tourtual, 114; and the personal equation, 94n; praises Helmholtz, 90, 120; qualities of sensation, 484–85; sign theory of spatial perception, 110, 112–19, 121–22, 127; on state of physiology, 81; students of, 24–5, 297; supports du Bois-Reymond, 81–2; supports Helmholtz, 28, 81–2; supports

Müller, Johannes *(continued)*
Ludwig, 81–2; supports "organic physics," 81–2; and velocity of nerve impulse, 87; and Young, 246–47. *See also Müller's Archiv*
Müller's Archiv für Anatomie, Physiologie, und wissenschaftliche Medizin, 71, 77, 89, 104
Multiplicator. *See* Instrumentation
Mursinna, Christian Ludwig, 19–20
Muscle contraction, 74, 76–81, 83–7; apparatus for measuring, 79–81; and graphical method, 83–4; *Muskelaction*, 300, 303; and precision measurement, 83–7; and sources of heat, 66–7; and temperature changes in, 66–7; and velocity of nerve impulse, 85–7. *See also* Berzelius; Instrumentation; Measurement; Nerve impulse, velocity of; Physiology
Muscle physiology. *See* Du Bois-Reymond; Physiology
Music: aesthetic function of, 541; consonance, explanation of, 525, 531–32; dissonance, theory of, 269; harmony, 269; harmony, causes of, 277–78; harmony, lecture on, 541; harmony, theory of, 270; lawlike nature of, 583; melodic relations, aesthetics of, 533–34; non-representational character of, 535; perception and physiology of, 525–32, 534–35; perception and psychology of, 528–35; pitch, explanation of, 525, 531–32; pitch discrimination, 280; psychological aspects of, 585, 586; as pure sensation, 535; scientific analysis of, 584–85; theory of, 259; tones, explanation of, 525, 526–31, 534. *See also* Acoustics; Acoustics, physiological; Aesthetics; Tone(s)
Muskelaction. *See* Force, conservation of; Muscle contraction; Physiology
Myograph. *See* Instrumentation; Nerve impulse, velocity of; Physiology

Nagel, Albrecht, 161; criticizes Albrecht von Graefe, 132; criticizes Alfred Graefe, 132; criticizes Müller, 132; criticizes Volkmann, 132; as empiricist, 191; and empiricist theory of vision, 153; and Helmholtz, 119n, 149, 189; as Helmholtz's student, 132; and Müller's identity theory, 132; as projection theorist, 159; *Das Sehen*, 132; supports Helmholtz's empiricist theory of vision, 132; and Wheatstone's results, 132

Nasse, Hermann: and nerve velocity, 105n
Nativism: differences with Kantianism, 501, 502, 503; and Helmholtz, 499; Helmholtz's criticism of, 186, 187, 188, 191, 494, 503, 504, 506; Helmholtz's opposition to Hering's, 186; Helmholtz's opposition to Kant's, 186, 191; Helmholtz's opposition to Müller's, 120, 186, 191; Helmholtz's opposition to Panum's, 191; Helmholtz's partial acceptance of, 187–88; William James's, 198; and Kant, 500–1; Müller's, and nerve-velocity impulse, 120; Neo-Kantian elements of, 499; and pre-established harmony, 557n; resurgence of, 198–99, 203; and spatial learning, 501–2; and theory of spatial perception, 500–1. *See also* Ebbinghaus; Hering; Külpe; Müller; Nativist-empiricist controversy; Perception, spatial
Nativist-empiricist controversy, 78, 155, 177–78, 179–90, 191–204, 257, 283, 498; background to, 178–79; development of, 179–81; and Donders, 198–99; and du Bois-Reymond, 198–99; empirical causes of, 180; evolving views of, 179–80; Helmholtz as creator of, 156, 177, 183–84, 186, 197; Hering as creator of, 197; role of evolutionary theory in, 198–99. *See also* Hering; Nativism; Perception, spatial
Nature: "intelligibility" of as explanatory ideal, 555; as mechanical system, 437, 451, 457; as valued property, 574; world picture of, 458. *See also* Philosophy
Naturforscherversammlung, Karlsruhe, 269; and Helmholtz, 277
Naturphilosophie: and energy conservation, 292, 294; and French physics, 364; and Helmholtz, 296; Helmholtz's hostility to, 565; Helmholtz's rejection of, 366–67; influence on Goethe, 465; and Kant, 294, 295. *See also* Philosophy; *Ueber die Erhaltung der Kraft*
Naturwissenschaften. *See Geisteswissenschaften*
Neeff, Christian Ernst, 68
Neo-humanism, 576; Helmholtz's support for, 597n; and industrialization, 597; and the sciences, 597, 598–99; values challenged, 597
Neo-Kantianism, 202; and Helmholtz, 110, 202–3, 462, 565n; rise of, 125. *See also* Kant; Philosophy

Nernst, Walther, 427; and Ettingshausen, 343; and the Hall effect, 343; on Helmholtz as chemist, 424; and the Helmholtz-Gibbs equation, 422, 426; Helmholtz's influence on, 413–14, 426

Nerve impulse, velocity of, 83–95, 394; and ballistic galvanometer, use of, 376; discovery of 3, 105–9, 376–77; dissemination of results on, 101–5, 107; and graphical analysis, 94–5, 101–5, 107; influence on experimental physiology, 106–7; meaning of, 105–9; and the myograph, 103–4, 107; and non-precision techniques, 94–5, 100–5, 106; and precision measurement, 88–95, 98, 100, 101, 103, 104–5, 107–8; and propagation measurement, 52; propagation in time, 377; reaction to, 90–3; telescope, use in, 88. *See also* Instrumentation; Measurement; Muscle contraction; Physiology

Nerve physiology. *See* Du Bois-Reymond; Physiology

Neumann, Carl, 385; debates with Helmholtz, 385, 386; potential propagation, 394–95; on thermochemical affinity, 417

Neumann, Franz Ernst, 220; correlation principle, 301; electrodynamic theory of, 378; and Helmholtz, 83; influence on Helmholtz, 93–4, 97, 220, 229; method of least squares, 97; potential, 381, 387

Newton, Isaac: barycentric diagram of, 229; colored top of, 222; color-mixing claims rejected, 223; color-mixing experiments of, 222; color-mixing theory of, 229, 231; color theory of, 207, 216–17, 221, 222; color-vision theory of, 224; Helmholtz on, 567n; Helmholtz's confirmation of his color-mixing theory, 221; Helmholtz's correction of, 239; Helmholtz's criticism of, 235; Helmholtz's study of, 216; influence on color-mixing studies, 222–23; influence on Helmholtz, 206, 216–17, 220–21, 224, 231, 244–45, 308, 312, 316, 317, 456; law of universal gravitation, 596; *Optical Lectures*, 216, 221; *Opticks*, 205, 209, 216, 221, 224, 229, 232, 235; as originator of three-receptor hypothesis, 239n; theory of light and colors, 205, 209, 214; his worldview eroded, 402

Niederrheinische Gesellschaft: and Helmholtz, 267

Non-Euclidean geometry. *See* Geometry, non-Euclidean

Ohm, Georg Simon: analysis of sound waves, 526; on combination tones, 266; and conservation of force, 329; definition of tone, 262, 264, 265; influence on Helmholtz, 262, 277, 281, 285–86, 302, 329, 376, 526, 527; law of, 302, 526, 530n, 329; reinterprets Seebeck's experiments, 264; sensations of tone, 526. *See also* Acoustics, physiological; Ohm-Seebeck dispute; Tone(s)

Ohm-Seebeck dispute, 259, 261, 264–66, 528; and Helmholtz, 273, 285–86. *See also* Ohm; Seebeck; Tone(s)

Olesko, Kathryn M., 5, 6, 148, 208

One-fiber/one-sensation doctrine. *See* Inference, unconscious; Physiology; Sensations

Ontology. *See* Philosophy

Opelt, Friedrich: invention of siren, 262–63

Ophthalmology: clinical practice of and Helmholtz, 111, 149–53; and the ophthalmometer, 111; ophthalmometer, invention of, 35–6; ophthalmometer and Helmholtz, 148–49; research in, 119; revolution in, 33–4, 37, 547; use of evidence from, 189. *See also* Medicine; Ophthalmoscope

Ophthalmometer. *See* Donders; Instrumentation; Ophthalmology; Optics, physiological

Ophthalmometry. *See* Donders; Knapp

Ophthalmoscope, 111; diffusion of, 127; and Helmholtz, 215, 216, 217; impact on German medical community, 18, 23, 28, 33–5, 36; impact on Helmholtz's career, 39–40; improvement of, 37, 127; invention of, 3, 23, 28, 29–33, 127, 162, 574, 577; and Müller's identity theory, 128; and revolution in ophthalmology, 128; use in color mixing, 231. *See also* Medicine; Ophthalmology

Ophthalmotrope. *See* Instrumentation; Optics, physiological; Ruete

Optics, physical, 207; and the eye, 215, 216, 217, 219; telescope, use in, 229

Optics, physiological: accommodation, theory of, 35–6, 149; afterimages, 241–42, 243–44; afterimages, explanation of, 536–37; and art, 586; barycentric curve, 235, 236; barycentric curve of

Optics, physiological *(continued)*
eye, 239; binocularity, 163; binocularity, and depth perception, 158–59, 166; binocular fusion, 180, 189; binocular vision, 188–90; binocular vision, rise of interest in, 161–62; and blindness, explanation of, 187; color blindness, 247, 248; color harmony, 247, 250–51; color harmony and musical tones, 250; color mixing, 220–21, 226–37, 241–42; color mixing, painterly tradition of, 221–22; color-mixing experiments, 229–30, 231, 233–35, 245; color research, after Helmholtz, 256–58; color research, as synthesis, 8–9; color theory, 203, 207, 536–37; color theory, reaction to, 232; color theory, role of psychology in, 236; color theory, use of tops in, 232; color theory and Helmholtz, 256; color wheel, use of, 231; depth perception, 498; depth perception, early ideas on, 157; depth perception, role of psychology in, 181; easiest orientation, principle of, 111, 139–48, 152–53, 172; eye as a measuring device, 111, 126, 147–49; eye measurement, 148–52; eye movements, 127, 128, 130, 139–48, 149, 151–52, 169, 171–73; eye movements and local signs, 111; eyes as separate organs, 182; and geometry, 510; heliometer, use of, 148; Helmholtz as master of, 163; horopter, defined, 163–64, 166; horopter, experiments with, 166–69; horopter, mathematics of, 167, 177; horopter, meaning of, 177; horopter, problem of, 163–68; mathematical analysis in, 37; monocular vision, 187–88; and the ophthalmometer, 111; ophthalmometer, invention of, 35–6; ophthalmometer and Helmholtz, 148–49; ophthalmotrope, 111; ophthalmotrope, and Helmholtz, 148; the painterly tradition, and Helmholtz, 228, 235; and painting, 535, 536–39; and philosophical questions, 183; projection theory, 158–59; projection theory, Helmholtz's repudiation of, 177; purpose of, 110; retinal correspondence theory, 164–65, 177; retinal depth values, 192–94; retinal incongruity, 165–68, 190; rise of interest in, 161–62; stereoscopic lustre, 190; stereoscopic vision, and Helmholtz, 163; strabismus, 189, 200; subjectivity of color, 205–6, 207, 229, 247, 251–56, 282; telescope, use in, 219; telestereoscope, invention of, 163; vertical-needles experiments, 181; vision studies, debt to physiology and ophthalmology, 162; vision studies, disciplinary fragmentation of, 200–1; visual direction, Helmholtz on, 188; visual localization, compared to scientific discovery, 550; visual science, and explanation of depth representation, 538. *See also* Epistemology; *Handbuch*; Illusions; Least squares, method of; Ophthalmology; Painting; Stereoscope; Signs, theory of local; Vision, color theory of; Young-Helmholtz theory; Young's hypothesis

"Organic physics" group, 77, 78, 555n; aid from physicians, 39; differences with, 494; and Helmholtz, 27–8; program of, 25–6, 129–30. *See also* Brücke; Du Bois-Reymond; Helmholtz, Hermann von; Henle; Ludwig; Virchow

Ostwald, Wilhelm, 427; anti-atomism of, 335–36, 442; and chemical theory, 405; and Helmholtz, 406, 442; on Helmholtz as chemist, 424; Helmholtz's influence on, 426

Oxford University: admiration for, 588; criticism of, 588

Painterly tradition, the. *See* Optics, physiological

Painters. *See* Helmholtz, Hermann von

Painting: acquaintance with, 579; aerial perspective in, 538–39; aim of, 536; and analogy to retinal images, 538; brightness in, 536; color in, 535, 536–37; color contrast in, 536–37; depth perception in, 537–38, 538–39; distance representation in, 537–39; distinguished from music, 535; elements of, 535, 580; form in, 536; fundamental artistic effect of, 535; and the ideal type, 580–81; and intelligibility, 580–81; as interpretation of nature, 580; lack of motion perspective in, 538; lectures on, 541; purpose of, 536; representational character of, 535. *See also* Aesthetics; Optics, physiological

Palmer, George: color-vision theory of, 225

Panum, Peter Ludwig, 161; and binocular fusion, 180; Helmholtz vs., 185; Helmholtz's criticism of, 179; as projectionist, 179

Papal infallibility. *See* Universities

Particle theory. *See* Electrodynamics; Force; Hall effect
Pasteur, Louis, 54
Pathology. *See* Medicine
Perception, spatial, 3, 5, 155, 163, 177–78, 186–90; and artistic cognition, 547, 552, 556; binocular, 183; and the blind, 507–8; and bodily motion, 507–8; controversies in, 176–77; differences with Müller on, 111, 119; and distorting lenses, 514–15; early model of, 282, 284; empiricist theory of, 186–89, 198–201, 203–4; epistemological foundations of, 183–84; and foundations of geometry, 499, 513–14; Helmholtz's influence on theories of, 203–4; and laws of motion, 515; and laws of nature, 520; local sign theory of, 178–79; and mathematics, 499, 520; and measurement, 124, 126; and methodological tenets, 184–87; and practical utility, 123–24; projection theory, 158–59; projection theory, Helmholtz's repudiation of, 177; as psychological phenomenon, 119; and relations of sensation, 282; role of learning in, 185–86; role of psychology in, 120–21, 177, 184, 283, 284; and scientific cognition, 547, 549, 550, 552; seeing, as a psychological phenomenon, 110; seeing, empiricist basis of, 119–20; and sensory data, 124; and stereoscopic observations, 120–21; synthesis of, 9; and the telestereoscope, invention of, 163; theory of, 280, 283–84, 285, 498, 504–8; topogeneous factors, 517. *See also* Acoustics, physiological; Geometry; Helmholtz, Hermann von: Kant memorial lecture; Hering; Nativism; Nativist-empiricist controversy; Optics, physiological; Sensations; Signs, theory of local; Space; Vision, empiricist theory of
Perpetual motion. *See* Energy, conservation of; Force, conservation of
Personal equation, the. *See* Bessel; Measurement; Müller
Pfeufer, Karl, 39
Philosophy: abandonment of Kantian pure reason, 125; British empiricism, and Helmholtz, 197; and causality in science, 366–68, 566–67; causes, 308–9, 467; causes as basis of science, 565–66; causes as theoretical science, 466; despair over, 183; empirical causality, 316–17; empirical nature of causality, 505; empiricism, 7–8, 155, 183–85; empiricism in Anglo-American psychology, 198; free will, and causal law, 493; free will, and causal necessity, 504–5; free will, and science, 542; functional explanation, use of, 185; and Helmholtz, 565n; Helmholtz as leader in, 461–62; Helmholtz's empiricism vs. Berkeley's, 508; hidden causes, 466, 482–83; on hypotheses, 505; identity, opposition to, 493–94; law of causality, 307, 504–5, 555n; logical empiricism, Helmholtz's influence on, 520; mechanical, influence on Helmholtz, 224; mechanical worldview, 11, 308, 311, 318, 433, 493, 576; ontology, metaphysical, 477; opposition to metaphysics, 461; physical explanation, ideal of, 555; pragmatism (American), and Helmholtz, 198; pre-established harmony, rejection of, 557; psychological explanation, ideal of, 555; psychological explanation, use of, 8, 9; reality, as hidden from senses, 477–78; scientific explanation, "classicist" aesthetics of, 524; self-consciousness, and knowledge, 487; self-consciousness, and reality, 494; self-consciousness, and the will, 487; teleological explanation, 254; teleology, and Helmholtz, 167, 366; "Thatsachen in Wahrnehmung," 484–88, 492, 549; thinking, 570; will, and perception, 486; will, and self-knowledge, 484. *See also* Epistemology; Epistemology, empiricist; Idealism; Inference, unconscious; Judgments, unconscious; Materialism; Matter; Metaphysics; Nature; *Naturphilosophie*; Neo-Kantianism; Positivism; Realism; Reductionism; Science, natural; Science, philosophy of; Sensations; Signs, theory of local; Truth; Truth, universal; Vitalism
Physical chemistry. *See* Chemistry, physical
Physicists: and affinity theory, 417; British, influence on Helmholtz, 462–63, 480; German, 334. *See also* Physics
Physics: abstractions in, 482; action-at-a-distance, 378, 381–82; analogies in, 362; avoidance of probability theory in, 442; Berlin institute, 352; Berlin laboratory, 342, 358–59, 363, 398–99; British, 357–58; circuit science, state of, 353; "classical," shaping of, 376; as classicist in, 442; contributions to, 458; dispersion theory of light, 417; dynamicism, 473; electricity, atomic

Physics *(continued)*
  nature of, 406, 410, 414–17; emergence as theorist of, 295, 305–6; energeticism, and Helmholtz, 442; energy, Clausius's concept of, 369; energy, Thomson's concept of, 369; as energy based, 369–70; ergodic systems, 448–49, 450; and experience, 479; goals in, 10, 11; goals of, 341, 458; hypotheses in, 482; influence on, 336; and kinetic theory of gases, 434–35; knowledge of literature, 320; Laplacian programme and Helmholtz, 472; legacy in, 433, 456–58; mathematical, 316, 365, 380; and mathematics, 306; mechanical understanding of, 445; methodology in, 295, 296, 309, 311; microphysics, 402; microphysics, and Helmholtz, 24; motion of rigid bodies, 511; propagation in time, 377; relational, 359, 363, 365; research, meaning for civilization, 578; and sensory perception, 465; of sound, 525–26; as synthesizer of, 8–11, 309–13, 319, 321, 327–28, 333, 401; theoretical, 295, 305–6, 316, 328, 333, 402, 429; as theorist of, 8–11, 295, 305–6, 328, 333, 402, 429; thermoelectric circuits, 67, 69; vortex motion, theory of, 365. *See also* Acoustics; Acoustics, physiological; Chemistry, physical; Electrodynamics; Energy, conservation of; Force; Force, conservation of; Forces; Hall effect; Heat; Heat, caloric theory of; Helmholtzianism; Hydrodynamics; Instrumentation; Least action, Hamilton's principle of; Measurement; Mechanics; Monocyclic systems; Optics, physical; Physicists; Physics, field; Relativity; Science, physical; Space; Thermodynamics; Thermodynamics, chemical; Thermodynamics, second law of; *Ueber die Erhaltung der Kraft*; Weberean physics
Physics, field, 334, 378; attitude towards, 381; awareness of, 380–81; and Hamilton's principle, 358; and Lagrange's equations, 358; learning of in Britain, 358; and microphysical structures, 336; and models in, 341–42; problems with, 495–96; and variational principles, 358. *See also* Faraday; Maxwell; Physics
Physikalische Gesellschaft, Berlin, 132; founding, 297; and Helmholtz, 71, 89, 91, 92, 94, 106, 304
Physikalisch-Medicinische Gesellschaft, Würzburg, 401
Physikalisch-Technische Reichsanstalt, Charlottenburg, 554; as president of, 47, 575
Physiological optics. *See* Optics, physiological
Physiology: abandonment of, 203–4; and animal heat, 27–8, 299–300, 300–1, 408; as based on physical science, 377; data analysis in, 51–2, 90, 94, 95–100; disciplinary autonomy of, 18, 25–6, 39, 47; of the ear, 259; early research in, 5, 6; error analysis in, 51–2, 66, 72–3, 74, 90, 93–100, 101, 105, 107; experimental error in, 59; as experimentalist in, 50–3, 58, 59, 64, 66, 67–8, 73; graphs in, 66; growing disinterest in, 47; heat, 59–64, 70, 296, 298–301, 302; influence of physical science on, 50, 59, 61, 68–9, 78, 81, 83, 89, 93–4, 96–7, 148–49, 260; influence on psychology, 106; influence on physical science, 300; institutional growth of, 162; investigative enterprise in, 63–4; investigative strategy of, 101; limitations for understanding art, 542; mathematical analysis in, 5, 6, 32, 152–53, 163, 166, 269; and medicine, 38–9, 47, 48; methodology in, 300; microscope, use in, 79, 88; muscle, 50, 54–9, 65–6, 68–70, 72–3, 74; *Muskelaction*, 300, 303; and the myograph, 103–4, 107; nerve, 70–1, 72–3, 303; one-fiber/one-sensation doctrine, 530–31; precision measurement in, 56, 57–9, 65–6, 68–70, 73, 74; putrefaction and fermentation, 27, 50, 53–4, 408; rise of German discipline of, 162; role of psychology in, 247, 251, 252–56, 280; role of theory in, 73–4; and roots of music and painting, 577–78; as a science, 43–4, 48; sensory, and the arts, 524; *Stoffverbrauch*, 63; as synthesizer in, 8–9, 154–55, 163, 191, 204, 206–7, 237, 247, 254, 256, 258, 259, 260, 281, 285–86; as theorist in, 8–9; use of induction coils in, 68, 69; "vital force," and *Stoffwechsel*, 54–5, 59. *See also* Acoustics, physiological; Chemistry, physiological; *Handbuch*; Heat; Instrumentation; *Lehre von den Tonempfindungen*; Measurement; Medicine; Muscle contraction; Nerve impulse, velocity of; Ophthalmology; Optics, physiological; "Organic physics" group; Sensations
Pigment mixing. *See* Harris; LeBlon; Mayer, Tobias

Pipe theory. *See* Acoustics, physiological
Pitch. *See* Music
Planck, Max: and Helmholtz, 315, 429; and Helmholtz's chemical thermodynamics, 429–30; and Helmholtz's force conservation, 315–16; Helmholtz's influence on, 456–58, 497; Helmholtz's praise of, 429–30; Helmholtz's support for, 429–30; and least-action principle, 497; mechanical worldview, 457–58; nomination to Berlin Akademie, 429–30; praise for Helmholtz, 430; as spokesman of science, 561. *See also* Least action, Hamilton's principle of
Plateau, Joseph: and colored top, 222; Helmholtz's criticism of, 232, 248n, 255; on irradiation, 248n; priority in color-mixing results, 232; on subjective colors, 242, 243
Pleasure. *See* Aesthetics; Aesthetics, musical
Plücker, Julius: influence on Goldstein, 359
Poggendorff, Johann Christian, 91; as Baconian empiricist, 472; on Brewster's theory, 211; on the *Erhaltung*, 472–73; rejection of the *Erhaltung*, 564n
Poincaré, Henri: critique of Helmholtz on irreversible processes, 455
Poiseuille, Jean-Louis: manometer of, 78
Poisson, Siméon Denis, 262
Polarization. *See* Chemistry; Electrodynamics; Force, conservation of; Hertz; Röntgen
Populäre wissenschaftliche Vorträge, 363, 559n
Positivism, 476; and Helmholtz, 11, 442, 463, 476, 477, 478–79, 483, 564; in physics, 383. *See also* Mach; Philosophy; Science, philosophy of
Potential. *See* Clausius; Electrodynamics; Energy, conservation of; Force, conservation of; Gauss; Hamilton; Helmholtzianism; Jacobi; Neumann, Carl; Neumann, Franz Ernst; Weber, Wilhelm
Potsdam, 27, 297–98, 562, 579
Potsdam Kunstverein, 579
Pouillet, C. S. M.: influence on Helmholtz, 84–5, 87, 95n, 107; method of, 95n, 107
Poynting, John Henry: analogy in physics, 362
Pragmatism. *See* Philosophy

Prevost, Alexandre: and the horopter, 165
Preyer, William, 200
Privatdozenten, 590
Progress. *See* Science
Projection theory. *See* Brewster; Hering; Nagel; Optics, physiological; Panum; Perception, spatial
Psychology, 3; and aesthetic theory, 558; applied to music, 524; applied to painting, 524; associationist, 113; associationist, influence on Helmholtz, 224; Helmholtz and German psychology, 201; laws of and artists, 523; and physiology, 535; reformation of sensory, 183–84; self-consciousness, and knowledge, 487; self-consciousness, and reality, 494; self-consciousness, and the will, 487; theory of, 558. *See also* Associationism; Associationism, laws of; Inference, unconscious; Intuition, aesthetic; Intuition, artistic; Judgments, unconscious
Psychophysics: and Helmholtz, 139. *See also* Fechner; Weber, Ernst Heinrich; Wundt
Puhlmann, Wilhelm, 579
Purkyně, Jan: influence on Helmholtz, 206–7; his shift, 214, 219, 236, 247, 248–50
Purkyně shift. *See* Dove; Maxwell; Purkyně
Putrefaction. *See* Physiology

Rameau, Jean-Philippe, 531–32
Rankine, William John Macquorn, 404; "energetics," 474; Helmholtz's differences with, 474–75
Rathke, Bernhard: and Helmholtz, 105–6
Rayleigh, Lord. *See* Strutt, John William
Realism, 121, 124, 125, 566, 567; Helmholtz's, 464, 492; vs. idealism, 492–93; metaphysical, 462, 463–73, 473–82; scientific, 119–20. *See also* Philosophy; Science, philosophy of
Recklinghausen, Friedrich von, 161; rejection of Listing's law, 171; and retinal incongruity, 165
Reductionism: and idealism, 492–93; limits to, 493; opposition to in psychology, 555; in physics, 464, 465, 471, 473, 474, 476, 555; in physiology, 461, 493, 535. *See also* Philosophy; Science, philosophy of
Reich, German, 596
Reimer, Georg: publishes *Erhaltung*, 304

Relativity theory, 400–2; and Helmholtz, 400–1; Helmholtz and general, 520–21; Helmholtz and special, 512n. *See also* Geometry
Remak, Robert, 24
Resonators. *See* Acoustics, physiological
Retinal correspondence theory. *See* Optics, physiological
Retinal depth values. *See* Hering; Optics, physiological
Retinal incongruity. *See* Meissner; Optics, physiological; Recklinghausen; Volkmann
Retinal space values. *See* Hering
Retinal theory of identity. *See* Hering
Reversibility. *See* Loschmidt; Mechanics; Thermodynamics
Richards, Theodore W.: Helmholtz's influence on, 426
Richarz, Franz, 440–41; kinetic energy and temperature, 446; unitary theory of electricity, 417n
Riemann, Bernard, 328; and force-law derivation, 400; Helmholtz contrasted to, 499, 512–13; influence on Helmholtz, 512–13; metric of, 510, 512; priority in non-Euclidean geometry, 510; and quantized electricity, 416; and relativity theory, 402; space theory, 510–11
Riess, Peter, 91; influence on Helmholtz, 302
Rigid bodies. *See* Geometry; Measurement; Physics; Space
Roeber, August, 262
Rollet, Alexander, 161
Röntgen, Wilhelm Conrad: discovery of X-rays, 401; polarization experiments, 398
Root, Elihu: in Berlin physics laboratory, 411; and electrolytic convection currents, 411
Roscoe, Henry: and Helmholtz, 414; on Helmholtz, 414, 424
Rowland, Henry: analogy in physics, 362; and the Hall effect, 342, 343–44; in Helmholtz's institute, 397; tests electrodynamic theory, 397, 398
Royal Society of Edinburgh, 211
Rudolphi, Karl Asmund, 22
Ruete, Christian Georg Theodor: and Helmholtz, 33, 35n, 36; and Listing, 132; ophthalmotropic work, 132–33
Runge, Philip Otto, 221–22
Russell, Bertrand: criticizes Helmholtz, 516; criticizes Land, 516; *Foundations of Geometry*, 516; on space, 516

Sabatier, Paul: Helmholtz's influence on, 425
Sauver, Joseph, 531n
Savart, Félix: criticizes Chladni, 262; force law of, 380; invention of siren, 262–63
Schelling, Friedrich Wilhelm Joseph von: Helmholtz's opposition to, 463, 494; influence on Goethe, 465
Scherffer, Karl: on subjective colors, 243, 244; theory of colored afterimages, 224–25
Schering, Ernst, 510
Schiller, Nikolai: in Helmholtz's Berlin laboratory, 397; tests Maxwell's theory, 397
Schönlein, Johann Lukas, 20; influence on Helmholtz, 26
Schopenhauer, Arthur: Boltzmann's detestation of, 370n; Helmholtz's hostility to, 370n, 371, 372, 565
Schulze, Johannes, 40
Schuster, Arthur: in Helmholtz's Berlin laboratory, 415
Schwann, Theodor, 24; Helmholtz's influence on, 407; influence on E. Weber, 74n; and "organic physics" group, 78
Schwarzschild, Karl, 400–1
Science: aim of, 464; applied, 576–77; and art, relations of, 11–2; vs. art, 585; and the *Bildungsbürgertum*, 594–95, 597; civilizing power of, characterized, 560–61; civilizing power of, as enlightened values, 600; classical aesthetics of, 525, 556–57; and culture, 560; discipline formation of, 560; and the economy, 560; and education, 571; as foundations of aesthetic life, 560, 577–86, 599; and German culture, 596, 597; and German unity, 596, 597–99; goal of, 471, 474; and history, 570–71; and human progress, 572, 598; as ideology, 600; and industry, 573, 575–77; and intellectual understanding, 560, 562–71, 577, 599; and laws, 567; and meaning in life, 600; mission of, 572; and modernization, 598–99; and morality, 598; object of, 476, 490–91; opposition to romantic, 599; originality of views on, 561; political function of, 560–61, 586–99, 599–601; and political power, 576; and politics, 573; as power, 560, 571–77, 599; and practice, 572, 577; and progress, 592; progress in, 567; purpose of, 598, 599; and rationalism, 595–96,

598; as religion, 600; and research, 586–92; role of concepts in, 471–72; as the search for laws, 578; social function of, 560–61, 586–601; and society, 12–3; specialization of, 560; and the state, 573, 576–77, 598–99, 600–1; state support of, 573, 589, 591–92; and technology, 572, 576, 601; as truth, 578; vision of, 561. *See also* Aesthetics; Art; Helmholtz, Hermann von; Science, natural; Science, popular; Truth; Truth, universal

Science, natural: aim of, 566; and *Bildung*, 571; borders of, 2, 541; factual basis of, 564; goals of, 544; method of, 544–45, 546–47; rise of in Germany, 162; role in education, 594; and unification with philosophy, 109, 110, 120, 125, 126. *See also* Science; Science, popular

Science, philosophy of, 8, 11; aesthetic induction, defined, 545; artistic induction, and unconscious inference, 547; artistic induction, defined, 545–46; and causality in science, 366–68, 566–67; and causes, 308–9, 467; causes as basis of science, 565–66; causes as theoretical science, 466; deductive method, 544, 545, 546; deductive method, skepticism towards, 565; and empirical causality, 316–17; and empirical nature of causality, 505; experiment and causality, 483; experiment as basis of science, 564–65; experiment compared to observation, 483; experiment in geometry, 483; experiment in knowledge formation, 484; experiment in perception theory, 483; experiment vs. observation, 487, 494; experimental interactionism, defined, 463, 483; experimental interactionism, Fichtean roots of, 482–91; hidden causes, 466, 482–83; inductive inference, and laws of science, 549; inductive method, 544, 545, 546; inductive method, praise for, 565; influences on Helmholtz's, 462, 494; law of causality, 307, 504–5, 555n; logical empiricism, Helmholtz's influence on, 520; logical induction, defined, 545, 546; logical induction, in natural science, 565–66; methodology, 302; methodology, experimental, 195, 197; methodology, roots of, 463. *See also* Aesthetics; Arithmetic; Atomism; Epistemology; Epistemology, empiricist; Force; Idealism; Laws; Materialism; Matter; Mechanics; Metaphysics; Positivism; Realism; Reductionism; Sensations; *Ueber die Erhaltung der Kraft*

Science, physical: goal of, 306–7, 308. *See also* Physics

Science, popular, 155; addresses on, 559, 560; advancement of, 163; audiences for, 560; and the *Besitzbürgertum*, 560; and the *Bildungsbürgertum*, 560; in England, 594; function of, 593–99; lectures on, 593; promotion of, 593–95; purposes of, 560. *See also* Helmholtz, Hermann von; Science

Scientists: cognitive activity of, 523–24; cognitive processes of, 540, 545–47, 549, 550, 557; reason and understanding in, 552; thought processes in, 525, 545–47, 549, 500, 551, 552, 553n. *See also* Aesthetics; Artists

Seebeck, August, 262; and color vision, 226; definition of tone, 263, 264–65; Helmholtz's criticism of, 271–72; influence on Helmholtz, 248, 266, 271–72, 285–86; siren experiments of, 263; on tone quality, 266; view of acoustics, 264. *See also* Acoustics, physiological; Ohm-Seebeck dispute; Tone(s)

Seeing. *See* Perception, spatial

Self-consciousness. *See* Fichte; Philosophy; Psychology

Senckenbergische naturforschende Gesellschaft, Frankfurt am Main, 573

Sensations: and experience, 484, 492; external vs. internal, 485–86, 487; as mental events, 527; and neural activity, 530–31; one-fiber/one-sensation doctrine, 530–31; and perceptions, 489; vs. perceptions, 527–29; and spatial localization, 548–49; visual and motor, 507

Sherman, Paul, 215, 239

Signs, theory of local, 121, 122–25, 282, 507, 548–49; and Helmholtz, 110, 118, 147–48, 187–88; Helmholtz on, 508; Helmholtz's adoption of, 503. *See also* Epistemology; Lotze; Optics, physiology; Perception, spatial

Silbermann, Johann, 408

Siren. *See* Acoustics, physiological; Dove; Opelt; Savart; Seebeck; Tour

Solution theory. *See* Chemistry, physical

Sound. *See* Acoustics; Ohm

Space: characterization of, 509; and distance measurement, 510; geometrical, 510, 511; and laws of physics, 509, 516, 520; as a manifold, 510, 511; measurability of, and congruence, 512;

Space *(continued)*
  metaphysical questions about, 520; metrical structure of, 511–12; as physical fact, 511; pseudospherical, 515; relations of, and rigid motions, 508–9; and rigid body motion, 510, 511; three-dimensional, 512; topogeneous factors, 517; topological structure of, 511–12; uniformity of, and laws of motion, 516–17; visual, and learning, 506; visual, conception of, 504, 506; visual, problem of, 156; visualizability of, 515–16; visualizability of, and non-Euclidean geometry, 514–15. *See also* Geometry; Measurement; Perception, spatial; Relativity theory
Spatial localization. *See* Sensations
Specific sense energies, law of. *See* Acoustics, physiological; Maxwell; Müller; Vision, color theory of
Standards: electrical, 575; and German unity, 598. *See also* Force, conservation of
Stefan, Josef: and Boltzmann, 381
Steinbuch, Johann Georg: theory of space, 113–14
Steiner, Jakob: influence on Hering, 168
Stereoscope: and binocularity, 178; and challenges to Müller's vision theory, 160, 166; and depth perception, 178; and Helmholtz, 163; impact on vision theory, 158, 159, 166; purpose and function of, 157–58. *See also* Instrumentation; Optics, physiological; Perception, spatial; Vision, empiricist theory of; Wheatstone
Stereoscopic lustre. *See* Optics, physiological
Stoffverbrauch. *See* Physiology
Stoffwechsel. *See* Physiology
Stokes, George Gabriel: dispute with Challis, 220
Stoney, George Johnstone: electrolytic valency, 417; and the "electron," 416; priority dispute with Helmholtz, 416
Strabismus. *See* Optics, physiological
Strehlke, Friedrich: vibrating plates, 262
Strutt, John William: and chemical thermodynamics, 422; criticizes chemists, 418; and Helmholtz, 415; praises Helmholtz, 426; *Theory of Sound*, 358; and variational principles, 358
Stumpf, Karl: and nativism, 201–2
Szily, Coloman: on mechanical principles, 432

Tait, Peter Guthrie: influence on Helmholtz, 10, 369–70; *Treatise on Natural Philosophy*, 358; and Weber's force law, 312–13. *See also Treatise on Natural Philosophy*
Teleology. *See* Least action, Hamilton's principle of; Philosophy
Telescope. *See* Instrumentation; Nerve impulse, velocity of; Optics, physical; Optics, physiological
Telestereoscope. *See* Instrumentation; Optics, physiological; Perception, spatial
Tension forces. *See* Force, conservation of
"Thatsachen in Wahrnehmung." *See* Philosophy
Theatrum anatomicum, Berlin, 21
Thermochemistry: affinity, 403–4; chemical affinity, 417; death of classical, 423; and electrochemistry, 404; elimination of affinity, 406; measurement of affinity, 404; thermal theory of affinity, 404, 411–12, 414, 417, 420–21, 423, 425; and thermodynamics, 417; Thomsen-Berthelot principle of, 404, 417, 423. *See also* Physicists; Thermodynamics, chemical
Thermodynamics: as background to electrodynamics, 375, 377–79; and chemistry, 404; Helmholtz and Maxwell's demon, 454–55; and human limitations, 454, 456; and irreversible processes, 433, 450–56; laws of, 432; laws of, and human limitations, 441–42; mathematical analysis of, 7; mathematical foundation of, 3; mechanical understanding of, 435; mechanical view of, 11; reversible processes, mechanical interpretation of, 433–50. *See also* Energy, conservation of; Force, conservation of; Physics; Thermochemistry; Thermodynamics, chemical; Thermodynamics, second law of
Thermodynamics, chemical, 3, 405–6, 417–23; concentration cells, as background to, 406, 412–14; development of, 406; dissemination of, 423; and electrochemistry, 418; and energeticism, 430; free energy, 418–21; Helmholtz as founder of, 417–18; Helmholtz as theorist of, 10; Helmholtz's priority in, 421–23; mathematical analysis of, 7; and monocyclic systems, 423; and physical chemistry, 406, 424–26, 428; Planck's, 429–30; and principle of least action, 423; and thermal theory of affinity, 418; Thomson-Helmholtz rule, 409. *See also*

Chemistry; Chemistry, physical; Thermochemistry; Thermodynamics; Thermodynamics, second law of Thermodynamics, first law of. *See* Energy, conservation of; Force, conservation of

Thermodynamics, second law of, 404; applied to chemical reactions, 412; applied to chemistry, 417-18; Boltzmann's mechanical justification of, 445; Boltzmann's proof of, 447-48; and electrochemistry, 411-12; entropy, 432, 435-36, 455; entropy, for Boltzmann, 445-46; entropy, Helmholtz on, 421, 455-56; entropy, Planck's use of, 429; and Helmholtz, 455-56; human implications of, 568-69, 570-71; mechanical foundations of, 432-33; mechanical interpretation of, 436, 439, 440; mechanical proof of impossible, 447-48. *See also* Boltzmann; Energy, conservation of; Force, conservation of; Mechanics; Physics; Thermodynamics; Thermodynamics, chemical

Thomsen, Julius: thermochemistry of, 404

Thomson, J. J.: induction and energy conservation, 331n

Thomson, William, 404; analogy in physics, 361-62; applies conservation of energy, 379; and conservation of energy, 567; and free energy, 422; and heat equivalence, 409; and Helmholtz, 267-68, 415, 422; influence on Helmholtz, 10, 369-70, 480, 588; Thomson-Helmholtz rule, 409; *Treatise on Natural Philosophy*, 358; and Weber's force law, 312-13. *See also Treatise on Natural Philosophy*

Thought. *See* Aesthetics

Tone(s): characteristics of, 286; combination, 260, 267, 270-73, 278, 281, 285, 286, 287, 582; definition of, 259, 271-72, 285-86; musical, explanation of, 525, 526-31, 534; perception of and learning, 286-87; quality of, 267-68, 269, 273, 274, 275-76, 277, 278, 280, 281, 286, 582; theory of, 260; upper partial, 527-30, 582. *See also* Acoustics, physiological; Aesthetics; Aesthetics, musical; Ohm; Ohm-Seebeck dispute; Seebeck

Tour, Charles Caignard de la: invention of siren, 262-63

Tourtual, Caspar Theobald, 114

Treatise in Natural Philosophy: concepts of, 369-70; Helmholtz's translation of, 369; hostility to Weberean physics, 371; influence on Helmholtz, 369-70. *See also* Tait; Thomson, William

Treitschke, Heinrich von: and Helmholtz, 563; and *Preußische Jahrbücher*, 598

Truth: artistic, contrasted to scientific, 579; and beauty, 553; as goal of art and science, 552-53; objective, 471. *See also* Beauty; Truth, universal

Truth, universal: in art, 540; as classical aesthetic, 558; as goal of art, 551, 552-53; as goal of science, 551, 552-53; in science, 540. *See also* Beauty; Truth

Tuchman, Arleen, 5-6

Tuning fork. *See* Acoustics, physiological

Turner, R. Steven, 5, 7, 9, 239, 241, 255

Tyndall, John, 553n; *Faraday*, 593; *Fragments*, 593-95; *Heat*, 593; on Helmholtz on Joule, 322n; Helmholtz's forward to *Fragments*, 553n; Helmholtz's praise of, 594; *Sound*, 593; translated into German, 593-94

Ueber die Erhaltung der Kraft, 291-333, 377-78, 555; and chemistry, 408-10; debt to physiology, 377; and du Bois-Reymond, 464; Introduction, 304-9, 319, 463-64, 467-68, 471-72, 479; Kantian influence on, 294, 306, 307-9; lack of experimental results, 296, 303, 305, 320, 332, 333; and law, 495; and *Lebenskräfte*, 595; and Magnus, 304, 305; mentioned, 3, 216, 247; methodology, 304-6; and *Naturphilosophie*, 472-73; object of science in, 476; philosophical foundations of, 306-8, 311; and Poggendorff, 304, 305; publication of, 304; and realism, 463-64; reception of, 304; reprinting of, 474-75; structure of, 304-5; as synthesis, 8; as theoretical physics, 321. *See also* Energy, conservation of; Force, conservation of; Physics; Realism, metaphysical

Unconscious inference. *See* Inference, unconscious

Unconscious judgments. *See* Judgments, unconscious

Understanding. *See* Aesthetics; Science; Scientists

Universities: English, criticism of, 589, 591; English, failure of, 586-87; English, nature of, 587-89; French, criticism of, 586-87, 589, 591; French, failure of, 586-87; French, lack of research in, 587; French, nature of, 587;

Universities *(continued)*
German, academic freedom at, 589–93; German, function of, 586; German, isolation of, 592–93; German, as a model, 587; German, and papal infallibility, 589; German, political independence of, 589; German, problems of, 592–93; German, rapid growth of, 162; German, and reorganization of, 571; German, and scientific research, 589–91; German, superiority of, 586; German, threat to freedom at, 363; and scientific research, 586

Van der Waals, Johannes: and Gibbs's thermodynamics, 425
Van't Hoff, Jacobus Henricus, 427; and chemical thermodynamics, 423; and early chemical thermodynamics, 417; *Etudes*, 425; and Helmholtz, 406; and the Helmholtz-Gibbs equation, 422; Helmholtz's influence on, 425; osmosis theory, Helmholtz's dislike of, 427–28; praises Helmholtz, 416; and theoretical chemistry, 405
Velten, Olga von. *See* Helmholtz, Olga
Vierordt, Karl, 37
Vieth, Gerhard U. A.: and the horopter, 165
Vieweg, Friedrich, 269–70
Virchow, Rudolph, 24–5, 26; at the Charité, 26n; and disease processes, 43
Vischer, Friedrich Theodor: on consonance, 532; definition of aesthetics, 540; Hegelian aesthetics of, 543; influence on Helmholtz, 532
Vision, color theory of, 3, 163, 237–39, 244–45, 247; Helmholtz's abandonment of, 257–58; and law of specific sense energies, 118, 527; role of psychology in, 207–8. *See also* Optics, physiological; Vision, empiricist theory of; Young-Helmholtz theory; Young's hypothesis
Vision, empiricist theory of, 5, 6, 109, 110, 118–19, 119–20, 121, 123–26, 126–27, 147–48, 152–53; dissemination and acceptance of, 149–53; and painting, 580; relation to clinical ophthalmology, 111; role of psychology in, 111, 123, 127n, 147; as synthesis, 9; and Volkmann, 153. *See also* Helmholtz, Hermann von: Kant memorial lecture; Nativist-empiricist controversy; Ophthalmology; Optics, physiological; Perception, spatial; Signs, theory of local; Space; Stereoscope

Vision studies. *See* Optics, physiological
Visual direction. *See* Hering; Optics, physiological
Visualizability. *See* Space
Visual localization. *See* Optics, physiological
Visual science. *See* Optics, physiological
Visual space. *See* Space, visual
Vis viva. *See* Force, conservation of; Huygens
Vitalism, 493; challenge to, 377; defined, 407–8; vs. mechanism, 407–8. *See also* Heat
Vogel, Stephan, 5, 7, 9
Volkmann, Alfred W., 161; afterimages, 247; and binocular fusion, 180, 189; criticizes Müller, 246; criticizes Wundt, 151; on eye movements, 132, 150–51; and Helmholtz, 119n, 149, 150–53; influence on Helmholtz, 171; mathematical limitations of, 152; and method of least squares, 148; *Physiologische Untersuchungen*, 152; rejects Müller's identity theory, 131; and retinal incongruity, 165
Volta, Alessandro: and conservation of force, 328; contact law of, 328; and Helmholtz, 327; influence on Helmholtz, 328
Vorlesungen über Theoretische Physik, 374–75, 423n
Vorlesungen über Theorie der Wärme, 450
Vortex motion, theory of. *See* Physics
Vorträge und Reden, 559, 597, 598–99
Voss, Leopold: and Helmholtz, 270
Vowels, theory of. *See* Acoustics, physiological

Waetzmann, Erich: combination tones, 273n
Wagner, Richard: and Helmholtz, 579
Wagner, Rudolph: *Handwörterbuch*, 64, 74, 131; influence on Helmholtz, 64–5, 68
Weber, Eduard: influence on Helmholtz, 74, 76–7, 79; muscular motion, 74–6; "*Muskelbewegung*," 64; rotation apparatus, 52, 74–6
Weber, Ernst Heinrich: influence on Wundt, 135–36; and psychophysical law, 138–39; and pure sensations, 527; *Wellenlehre*, 262
Weber, Wilhelm: and conservation of energy, 383–84; electrodynamic theory of, 10, 308, 337–38; and exact measurement, 363; and Helmholtz, 312–

13, 368, 386; Helmholtz's criticism of, 375; Helmholtz's praise for, 98; influence on Helmholtz, 93-4, 330, 376; influence on Wundt, 135-36; law of, 312-13, 368, 383-87; and Maxwell, 312-13; potential, 387; quantized electricity, 416; reply to Helmholtz, 385; *Wellenlehre*, 262; and Zöllner, 386. *See also* Electrodynamics; Weberean physics

Weberean physics: atomism of, 365; as cowardly metaphysics, 363; and Fechner's hypothesis, 363-64; Helmholtz's opposition to, 363; and idealism, 364; laboratory practice of, 350, 351, 357; measurement in, 351, 357; and *Naturphilosophie*, 364; and positivism, 364. *See also* Electrodynamics

Wertheim, Emil: as Helmholtz's co-translator, 369

Westman, Robert, 156

Weyl, Hermann: criticizes Helmholtz, 517

Wheatstone, Charles: on binocularity, 160; his bridge, 376; challenges Müller's vision theory, 160-61; debate on spatiality, 157, 158; and Helmholtz, 120-21, 189, 376; invention of stereoscope, 157; refutation of identity theory, 176-77; stereogram of, 160-61; stereoscope and binocular vision, 223; stereoscope and depth relations, 538

Wheatstone bridge. *See* Hertz; Wheatstone

Whewell, William: criticizes Brewster's theory, 211

Wilde, Emil: rejects Brewster's theory, 211, 213

Will. *See* Philosophy

William I (of Prussia), 573

Williamson, Alexander, 592; dissociation hypothesis of, 415

Wissenschaft: defined, 562; as intellectual mastery, 571; and social class, 597

Wissenschaftliche Abhandlungen, 474-75, 559

Woinow, M., 47

Wollaston, William Hyde: influence on Helmholtz, 228; on solar spectrum, 210, 227

Work. *See* Energy, conservation of; Force, conservation of; Heat; Hero; Leibniz

World Exhibition, Chicago. *See* Columbia World Exhibition

Wüllner, Adolph: influence on Helmholtz, 413; and vapor pressure, 413

Wundt, Wilhelm, 161; *Beiträge*, 138; and du Bois-Reymond, 136; and empiricism, 198; as empiricist, 191; and empiricist theory of vision, 153; and eye as measuring device, 137, 139; eye-movement theory of, 134-39; Fechner's influence on, 138-39; and Fick, 134; Gauss's influence on, 137-38, 139; as Helmholtz's assistant, 46, 165; influence on German psychology, 201; influence on Helmholtz, 137, 139-40, 145-46, 146-47, 165, 187; and Listing's law, 140, 171; and method of least squares, 136-37, 139, 148; and Müller's theory of perception, 112-13; as nativist, 179; and the ophthalmoscope, 134-36; and psychophysics, 138-39; reaction against non-Euclidean geometry, 514n; role of psychology in vision theory, 138-39, 184; and unconscious inferences, 184, 185

Wünsch, Christian Ernst: color-mixing experiments, 222-23; ridiculed by Goethe, 223; tests Newton's theory of light, 222-23; theory of white light, 225

Young, Thomas: on color blindness, 248; color mixing, 227; and Helmholtz, 226; influence on Helmholtz, 206-7, 207-8, 231, 245-46, 247, 255; *Lectures*, 225-26, 227; on solar spectrum, 210; theory of light, 225-26. *See also* Young-Helmholtz theory; Young's hypothesis

Young-Helmholtz theory (of color vision), 237-38, 247-48, 536-37; origins of, 239-47. *See also* Optics, physiological; Young; Young's hypothesis

Young's hypothesis (of color vision), 225-26, 228, 237-38; Helmholtz's acceptance of, 207; Helmholtz's conversion to, 207; Helmholtz's extension of, 247-56; Helmholtz's rejection of, 207, 231-32, 240; Helmholtz's reversal on, 237, 238; Helmholtz's revision of, 241-42, 243-44, 245-46, 247; Helmholtz's skepticism towards, 236-37; Helmholtz's use of, 249-50; and Maxwell, 237. *See also* Optics, physiological; Young; Young-Helmholtz theory

Zeitschrift für physikalische Chemie: citations of Helmholtz in, 428

Zeitschrift für rationelle Medizin, 39

Zimmermann, Robert: aesthetics as a science, 540; aesthetics of, 523; appreciation of Helmholtz, 540

Zloczower, Awraham: on German university system, 162
Zöllner, Karl Friedrich, 10; as anti-Semite, 363, 371; attacks Helmholtz, 371, 386; attacks Helmholtzianism, 371; attacks Thomson and Tait, 370, 371, 386; career, 370; and Dove, 370; education, 370; hatred of du Bois-Reymond, 371; Helmholtz's counterattack on, 371–72; Helmholtz's critique of, 351; Helmholtz's hostility to, 10, 370; Helmholtz's polemics with, 493n; as idealist, 370, 371; and Magnus, 370; *Ueber Cometen*, 370–71, 372; *völkisch* ideology of, 371; and Weberean physics, 370, 371; as xenophobe, 363, 371
Zollverein, 596

Designer:	U.C. Press Staff
Compositor:	Impressions, A Division of Edwards Brothers, Inc.
Text:	10/12 Times Roman
Display:	Helvetica
Printer:	Edwards Brothers, Inc.
Binder:	Edwards Brothers, Inc.